Cutting metal with a laser.

Invitation to Physics

Jay M. Pasachoff & Marc L. Kutner

Williams College
Williamstown, Massachusetts

Rensselaer Polytechnic Institute
Troy, New York

W · W · NORTON & COMPANY · NEW YORK · LONDON

Published simultaneously in Canada by George J. McLeod Limited, Toronto.
Printed in the United States of America.
First Edition

Designed by Antonina Krass and set into Continental typeface by New England Typographic Services. The production editor was Frederick E. Bidgood. Donna L. Seldin and Cathryn Baskin served as permissions editors. Cover photo: Laser Beam, by Fritz Goro, Time-Life Picture Agency.

Library of Congress Cataloging in Publication Data
Pasachoff, Jay M.
Invitation to physics.
1. Physics. I. Kutner, Marc Leslie, joint author. II. Title.
QC21.2.P38 1981 530 80–17956
ISBN 0–393–95152–9

W. W. Norton & Company, Inc. 500 Fifth Avenue, New York, N.Y. 10110
W. W. Norton & Company Ltd. 25 New Street Square, London EC4A 3NT

1 2 3 4 5 6 7 8 9 0

Contents

PART IV *Wave Phenomena* 219

PART V *Electricity and Magnetism* 311

PART VI *Quanta and All That* 379

PART VII *Structure of the Universe* 471

Color Essays

Preface

Physics is exciting. Our aim in this book is to explain why. Many of the most interesting fields of physics are those that engage scientists today. By explaining these fields carefully and thoroughly, we try to bring across to students why these subjects in particular are exciting. For example, we devote full chapters to such topics of current interest as lasers, holography, quarks, general relativity, and astrophysics. And we deal with other contemporary topics in shorter sections and as examples throughout.

At the same time, we carefully and thoroughly explain why physics is of philosophical, fundamental, and practical importance. We show, for example, how the work and ideas of Aristotle, Copernicus, Galileo, Newton, Maxwell, Einstein, and a host of other pathbreaking scientists, including many from the twentieth century, have contributed to today's thinking about the world and universe around us. In the first part of the book we explain the fundamentals of mechanics in clear and thorough terms, so as to give students a good base. We give many examples of practical physics and "kitchen physics," involving such topics as motion, heat, electricity, and energy of all sorts. We discuss at some length the physics involved in the "energy crisis."

But why do physicists study physics? Most go into this field because they find it has irresistibly exciting problems to be solved. It is this spirit of excitement and adventure that we want most of all to carry across.

Too often physics books imply that basic physics ended in 1904, and that discoveries from 1905 or 1926 or even 1965 are new. We, instead, integrate the work of the twentieth century throughout the text. After all, it is over 75 years since Einstein's epochal 1905 papers. Today's students were all born after Sputnik. Indeed, they were young children when we not only started but also stopped sending people to the moon.

Students taking this course will not be professional physicists, and many will never take another science course. We hope that the background and overview that students get from our text will enable them to understand what is going on in modern physics. We hope that our text shows how interesting and important physics is not only for practical reasons but also for the intellectual challenge that invigorates our scientific society. We hope, too, through our phrasing and choice of examples and pictures, to show that physics is a science for students of both sexes.

We hope that the basic knowledge students get from this course will allow them to understand the developments in physics that we are certain will come in the years and decades ahead. Our newspapers and magazines are full of articles that deal with physics. Many issues of political and social importance depend on scientific data. Further, a knowledgeable person should have not only an understanding of the difference between science and technology, but also a feeling for why both are valuable.

Features of the Book

We approach physics in a descriptive way that does not require a background in mathematics. We draw a multitude of examples from the familiar world around us, while also using many contemporary examples.

Our extensive use of photographs should appeal to all students. The captions accompanying both black-and-white and color photographs function as ancillary text.

Invitation to Physics is unique among physics texts for including forty pages of color plates. We chose several topics for color essays to convey a sense of the beauty, elegance, and interesting nature of physics. Recent studies of the sun, demonstrations of optical phenomena and color, fluid-flow experiments, and some of the most recent technology in fusion research are among the subjects covered in these sections. Even with these pictures, professors and students will be glad to see that the price of the book is at the same level as other introductory texts.

Invitation to Physics incorporates several study aids that students and teachers will find valuable in using the text. New or technical words are printed in color where they are defined. These words reappear in the list of "Key Words" at the end of the chapter.

Also, the index is especially detailed to allow students to refer back to the pages or sections that would be most useful to them.

Each chapter is followed by a series of questions that highlight the important concepts and information that students should retain. We make some additional comments and provide additional questions in the *Teacher's Guide* that is available with this text.

Throughout the book, observations and facts tangential to the main discussion are highlighted as marginal notes. We have also, in some cases, used the margin to display the algebraic form of the word equations we discuss in the text. Some students may prefer the word equation and others the algebraic form; we leave it up to professors and to individual students. In addition to the marginal notes, we have included several essays. They vary in tone from the historical, such as biographies of Galileo and other scientists, to the more detailed pieces of analysis that some professors may wish to omit, such as the explanation of vectors.

We have numbered all sections to help professors make assignments and to help both professors and students to find references. We hope that professors will look into the teaching materials available in the *Teacher's Guide.*

Acknowledgments

One of the major features of our book is that we have placed a special premium on teaching physics accurately. We have gone to some lengths to check that the material in the text is accurate. We have also been careful to adjust the material to a level appropriate for college freshmen who have no scientific training. We have taken to heart the findings of Piaget about the necessity of concrete examples, and are keeping current in the literature in this important field of educational research. Still, we hold with Bruner that anything can be taught at some level to any group of people.

We wish to thank Joseph W. Connolly (Saint Bonaventure University), Bernard Kramer (Hunter College of the City University of New York), Irving Lazar (Queens College of the City University of New York), Thomas W. Listerman (Wright State University), Dennis Machnik (Rensselaer Polytechnic Institute), Robert B. Prigo (University of California at Santa Barbara), Martin K. Purvis (Barnard College), Lawrence C. Shepley (University of Texas), Arthur E. Walters (Rutgers University), Grace Marmor Spruch (Rutgers University), and Brian P. Watson (Skidmore College) for reading and commenting upon the entire manuscript. We thank Paul Stoler, Peter Kramer, John Lathrop, and Abraham Flexer for their comments upon particular sections of the book.

We also wish to thank Martin S. Weinhous and John Lathrop for their trial use of the book with their classes at Keene State College and Williams College, respectively. In addition, we thank Hugh Kirkpatrick (Williams College) for assistance with laboratory demonstrations during our own trial use of the manuscript. We are also grateful to the students in those classes for their comments.

We are grateful to the following for their hospitality: Christopher Quigg at Fermilab; Charles F. Keller, Jr., at Los Alamos; Jeffrey L. Linsky at the International Ultraviolet Explorer control center; and John Jefferies and Donald A. Landman at the Institute for Astronomy of the University of Hawaii.

Cathryn Baskin is due special thanks for her excellent work with us throughout the preparation of manuscript, pictures, and proof. The book has benefited immeasurably from her outstanding contributions.

At W. W. Norton & Company, we have many individuals to thank. We are very grateful to Joseph Janson II, and to Neil Patterson, for sharing our vision of the project and for getting us well under way. We are also very grateful to Edwin Barber for seeing the project through to completion, and to Donna Seldin and Fred Bidgood for excellent and expert editorial assistance on many stages of the manuscript and pictures. Others at Norton who have been of particular help are Donald S. Lamm, Christopher Lang, James Jordan, and Roy Tedoff.

Our homegrown editorial services boast of the expertise of Dr. Samuel S. Pasachoff, copy editor and proofreader extraordinaire, Anne T. Pasachoff, Nancy Kutner, and Naomi Pasachoff. Nancy Kutner is also responsible for the detailed index, which we view as a major student aid. We are grateful to Belle and Jack Nachwalter for their continued support. We thank Eric Kutner (age 7), Eloise Pasachoff (age 5), Deborah Pasachoff (age 3), and Jeffrey Kutner (age 1), for their inspiration and interest. We hope that they grow up finding physics as exciting as we do.

We look forward to hearing directly from readers, both professors and students. We will value your comments and use them to improve future versions of this work. We promise a personal reply to each writer.

Jay M. Pasachoff

Department of Physics and Astronomy
Williams College
Williamstown, Massachusetts 01267

Marc L. Kutner

Department of Physics
Rensselaer Polytechnic Institute
Troy, New York 12181

CHAPTER 1

The Physics of the Universe

"Would you like to learn physics?" asked the teacher.
"What does physics do?"
"Physics explains the principles of natural bodies and the properties of matter; it discourses on the nature of the elements, metals, minerals, rocks, plants and animals, and teaches us the causes of all the meteors, the rainbow, the aurora, comets, lightning, thunder, thunderbolts, rain, snow, hail, winds and whirlwinds."

This reply, from *The Would-Be Gentleman* by the seventeenth-century playwright Molière, is a fair reflection of the wide range of physics. Who would not want to be able to explain all those things? The pupil in the play *The Would-Be Gentleman* had much to learn, not only physics. When, in another lesson, he learned what poetry is, and that whatever is not poetry is "prose," he replied, "Do you mean that I have been talking prose all my life and not known it?"

In many ways, we are similar to Molière's Would-Be Gentleman, in that we have been using physics all our lives without necessarily knowing it. Whatever we do—climbing a flight of stairs, picking up our feet instead of shuffling when we walk, riding in a car, watching television, or even cooking dinner—involves physics in fundamental ways. We hope that by the time you finish this book, you will have come to understand how thoroughly physics affects your life.

But there is more to physics than practical explanations. Some of the most fundamental and fascinating questions about the uni-

Fig. 1–1.

verse and how it works are investigated by using the laws of physics. In fact, a definition of physics is the study of how and why things are the way they are, and how and why they work.

The study of physics leads us to consider what our universe is made of. We must work from the vanishingly small to the incredibly large to develop laws of physics that explain everything. In the course of these investigations, physicists have found a wide variety of strange and intriguing phenomena and continue to study them every day in their laboratories. Throughout this book, we'll be trying to give you the flavor of contemporary physics and show you its excitement. We shall do so not only by describing the physics itself, but also by introducing you to some of the people who do physics. (Physicists sometimes say that they "do physics.") In addition, we shall describe what it is like to be a physicist in these times of feverish activity.

Why study physics? The practical side by itself should be enough. But there is a more fundamental answer to the question, one that involves the nature of human beings, a species that has an inquisitive nature. There were always those—starting even before the discovery of fire—who would not be content until the workings of the world were understood. As the distinguished physicist Robert W. Wilson, the former director of the Fermi National Accelerator Laboratory, said when asked by a U.S. senator to state what the physicist's laboratory was contributing to national security, "It has nothing to do directly with defending our country except to help make it worth defending."

SCIENTIFIC NOTATION

The numbers we use in physics sometimes get incredibly large or small, and it would be tiresome to keep track of all those zeroes. For example, light travels 30,000,000,000 centimeters in each second, and if we ever wanted to do a calculation, we would have to get all the zeroes right. Scientists find it easier to use a form of notation in which we simply keep track of the number of zeroes.

To do this, we use powers of the number 10. After all, 10 is 10^1 and 100 is 10^2, so we see that the exponent (the raised number) is just the same as the number of zeroes. Similarly, 10^{37} would just be a 1 followed by 37 zeroes: 10,000,000,000,000,000,000,000,000,000,000,000,000. The number given in the first paragraph is 3×10^{10}; it is convenient to write numbers that aren't exact powers of 10 as one-digit numbers, with decimal points if necessary, times the powers of 10. Thus 4500 would be 4.5×10^3.

For numbers less than 1, we simply count the number of places the decimal point would have to be moved to the right to make 1. Thus $0.01 = 10^{-2}$ and $0.00024 = 2.4 \times 10^{-4}$.

If we ever want to multiply two numbers that are both powers of 10, we simply add the exponents, and make the sum the new power of 10. Thus $10^3 \times 10^4 = 10^7$ and $(3 \times 10^2) \times (4 \times 10^3) = 12 \times 10^5 = 1.2 \times 10^6$ (where we divided the 12 by 10 and multiplied the 10^5 by the same 10, a process that didn't change the final value, in order to leave the final form with only one digit before the decimal point).

1.1 *The World of the Small*

The next sections will provide a brief survey of topics that will be treated later in the book.

The ultimate structure of matter has long been a prime topic for thought. In ancient Greece, in the fifth century B.C., the philosopher Democritus suggested that matter is made up of tiny bodies that we call atoms. "The only existing things are atoms and empty space; all else is mere opinion," he wrote. The Roman philosopher Lucretius adopted this theory in his long poem on the nature of matter.

The fact that Lucretius could write a book-length poem in the first century B.C. on a topic that we would now consider to be physics, and that his discussion should lead up to a general philosophy of life, illustrates how intertwined science and the humanities have always been. Only comparatively recently, since the growth of knowledge has made it impossible for any one individual to be an expert in all fields, has science come to be considered as separate.

But even before Lucretius wrote, the Greek philosopher Aristotle had advanced his own theory of the universe and held that all was made of just four elements: earth, air, fire, and water. Aristotle's theories were generally believed for two thousand years; Democritus's theory of atoms was ignored and forgotten.

Aristotle's idea of astronomy, that the earth is in the center of the universe, was undermined in 1543 by the theory of Copernicus that the sun is in the center of the universe. In the following chapters, we shall see how in the centuries that followed, the work of Kepler, Galileo, and, triumphantly, Newton, banished Aristotle's astronomy to history.

But the atomic theory of matter was not revived until John Dal-

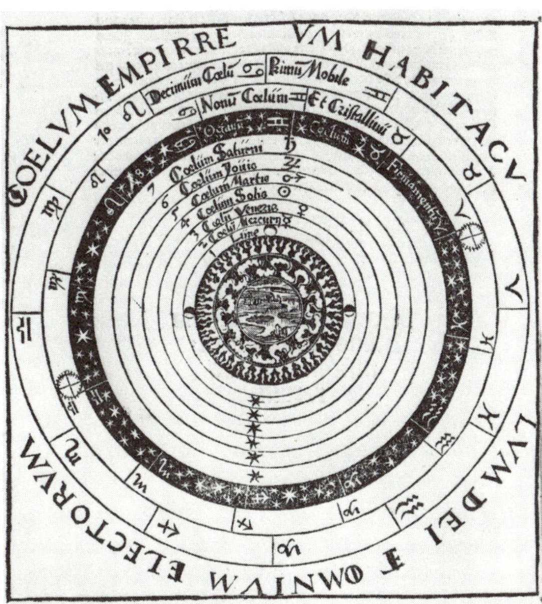

Fig. 1–2 Aristotle's view of the universe, from *The Cosmographicus Liber* by Petrus Apianus, published in 1524.

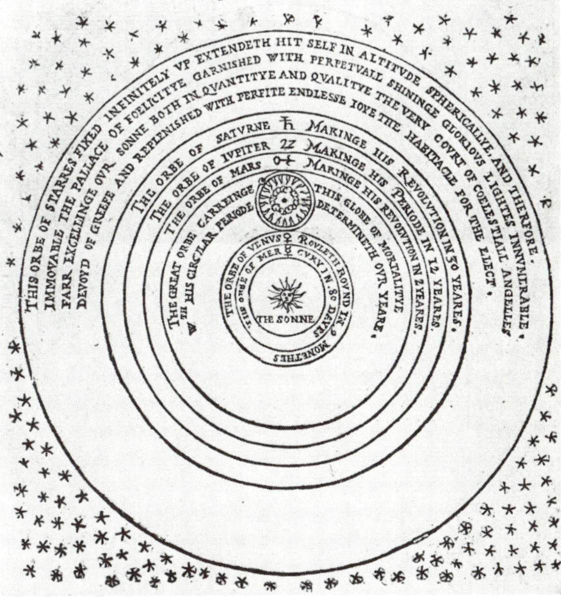

Fig. 1–3 The Copernican theory. The picture shows the first English diagram, due to Thomas Digges, which appeared in 1568.

Fig. 1–4 Nucleus and electrons orbiting it.

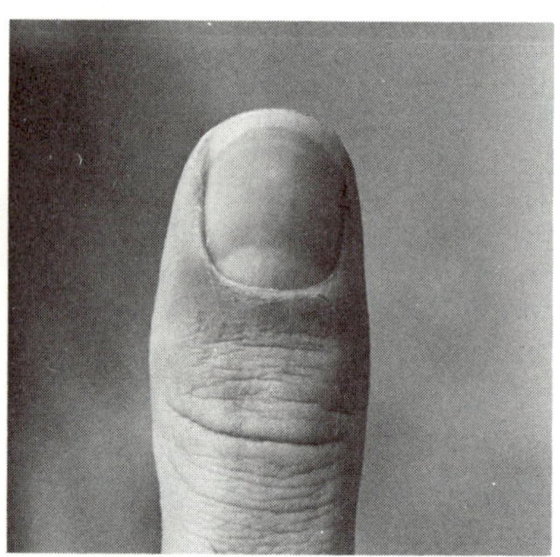

Fig. 1–5 A human thumb, shown about life-size. It is about 1 cm across. In the next pages we illustrate a variety of different scales of size both larger and smaller than this.

ton, in England in 1808, put forward a wide-ranging theory. Dalton suggested that:

- Matter consists of indivisible atoms.
- All the atoms of a given element are identical.
- Different elements have different kinds of atoms.
- Atoms are indestructible; chemical reactions merely rearrange them.
- A compound is made up of molecules, each of which contains a small and definite number of atoms of each element.

But the nature of atoms was not known. How do they attach themselves to each other? Do they have hooks on them? We shall see in Chapter 26 that in 1911, by shooting particles at atoms, Ernest Rutherford discovered that atoms were mostly empty space inside. Indeed, atoms seem to have a small, dense central portion called the nucleus, which is surrounded by outer layers filled with particles called electrons. The development of atomic and nuclear theory from this base occupied scientists from then until now, and has led to an understanding of the nucleus that has enabled us to create nuclear fission (the splitting of atoms) and fusion (the amalgamation of atoms) on earth. In Chapter 27 we shall discuss not only how fission and fusion might affect our energy supply for the future, but also what other means of energy we are using or might be able to use. A major physical and philosophical step along the way was the development of the quantum theory (discussed in Chapter 23), in which we find that the simplest particles often display an unusual, wavelike behavior.

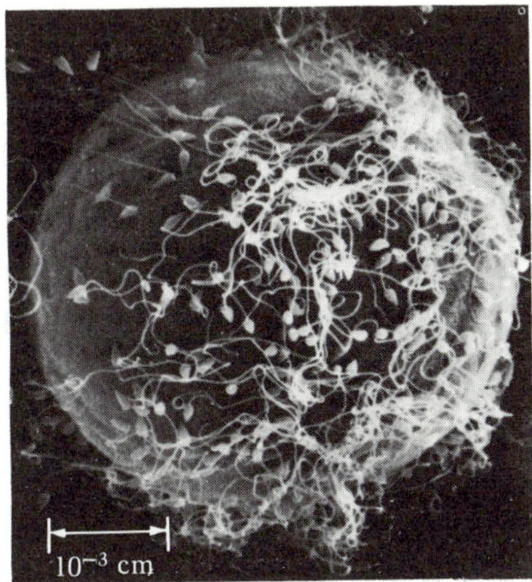

Fig. 1–6 A sea urchin egg surrounded by sperm, taken with a scanning electron microscope. The image is magnified about 850 times.

Atoms are so small that it is difficult for us to comprehend their size. In a cupful of water, there are as many atoms as there are grains of sand on all the beaches of the earth! Let us try to get a sense of scale by comparing sizes, beginning with more familiar scales.

Let us start with something life-size, say, the human thumb reproduced in Fig. 1–5. Figure 1–6 then shows a scale 1000 times smaller; it shows a sea urchin egg cell surrounded by sperm. (Pictures like this were made possible with an understanding of certain properties of the electron, which led to the development of the electron microscope, as we shall see in Chapter 23.)

We need still another factor of 1000 in order to see details within virus cells (Fig. 1–7). And we need a further factor of 10 before we start to see atoms.

But we are not yet all the way to the nucleus. The nucleus is more than 1000 times smaller than the overall size of an atom itself. Now we have no picture to illustrate this scale. The nucleus contains particles called protons and neutrons, which are roughly equal in mass, as was discovered in 1931. But the simple picture that we had for years of a nucleus consisting of a certain number of protons and a certain number of neutrons has long since proved to be grossly oversimplified. Other particles were found within the nucleus, and the number of different types we could observe increased as we built new and more powerful atomic accelerators ("atom smashers"). Still, we call these constituents elementary particles because of our lingering hope that the nucleus is composed of things that should be simple.

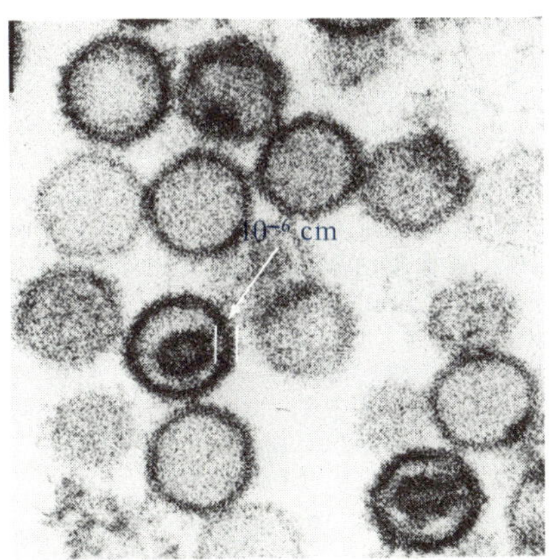

Fig. 1–7 The thickness of a membrane around a virus is about 1000 times smaller yet. This electron micrograph is of a herpes simplex virus in the nucleus of a human cancer cell.

Within the last decade, the number of types of elementary particles known has become so large—hundreds by now—that nuclear theory became more complicated than atomic theory. Could it be that all the different types of particles were in turn made up of a small number of truly simple particles?

In 1970 two scientists independently suggested that a new kind of particle, of which there are three varieties, make up all the previously known particles. These new particles were called quarks, from a line that one of the scientists, Murray Gell-Mann, liked in James Joyce's *Finnegan's Wake*, "Three quarks for Muster Mark." (Although literary detectives have tried, the origins of this phrase in Joyce are controversial. It is clear that "quark" is a kind of cottage cheese in Germany, and has been so even before Joyce's work.)

At first quarks came in three kinds—*up*, *down*, and *strange*. In Chapter 28 we shall see why the names are of descriptive value. Basically, all the known nuclear particles were combinations of these three quarks. The theory even predicted the existence of new particles, which were subsequently discovered. But although three quarks seemed an important simplification, in the last few years additional properties of matter have been found. They have been named, somewhat whimsically, *charm*, *truth*, and *beauty*. These properties, if real, would require the existence of three additional quarks. (At this stage it appears as though charm has definitely been found and truth probably so, with the sixth remaining a theoretical probability.) Are we now at our limit? Are there still more quarks to be found? Are there so many different types of quarks that we will wind up searching for a still more elementary (basic) kind of particle that will turn out to be the constituents of quarks? Will we, in fact, keep on finding smaller and smaller particles, as in the old verse,

And every dog has tiny fleas,
Upon its back to bite him.
The smaller fleas have smaller yet,
And on, ad infinitum.

The answers to these questions are so fundamental to our understanding of matter, and of the universe, that the field of elementary particle physics is a particularly active research area. In Chapter 28 we shall discuss elementary particle physics systematically.

PREFIXES FOR CERTAIN POWERS OF 10

To aid in comprehension, prefixes have been assigned to each power of 1000 (10^3). We attach the prefixes to the units; for example, milli- means 10^{-3}, and a millimeter is 10^{-3} meter while a milligram is 10^{-3} gram. The prefixes for powers smaller than one follow, along with examples that you may find not too useful but more memorable than metric units.

10^{-3} milli-	10^{-3} ink machine	=	1 millink machine
10^{-6} micro-	10^{-6} phone	=	1 microphone
10^{-9} nano-	10^{-9} goat	=	1 nanogoat
10^{-12} pico-	10^{-12} boo	=	1 picoboo
10^{-15} femto-	10^{-15} bismol	=	1 femtobismol

We shall also meet prefixes for powers greater than one:

10^3 kilo-	10^{12} tera-
10^6 mega-	10^{15} penta-
10^9 giga-	10^{18} exa-

The prefixes and their abbreviations are listed in Appendix 1.

1.2 *The Life-Size World*

The world may be made of atoms, and the atoms may contain quarks, but we usually experience the world on a larger scale. When we feel a brick, or a table, or a scrambled egg, we are interacting with the properties of matter in bulk. Similarly, when we talk about the temperature of the air, we are considering a large number of molecules together. Physics deals with these large-scale situations as well as with the small-scale ones discussed in the previous section.

Physics also deals with the properties of motion. It tells us how to throw a baseball, and how to make it curve. It tells us how to pole vault over a bar, how to fly, and how to streamline an airplane so that it will fly higher and faster. But physics is a universal science: the laws of physics hold not only on earth but also throughout the universe. The same laws that make an apple fall on earth apply (as Isaac Newton realized three hundred years ago) to the moon, and keep it in orbit around us. And now we have applied the same laws to the distant stars and galaxies. (We discuss the physics of astronomical objects—astrophysics—in Chapter 29.)

When we consider extremely rapidly moving objects—in particle accelerators, for example, or when we get to the largest scale—in cosmology (the study of the universe as a whole)—or when we consider weird states of matter (such as in a black hole, where there is so much mass concentrated in such a small volume that not even light escapes), we find that Newton's laws of physics are no longer

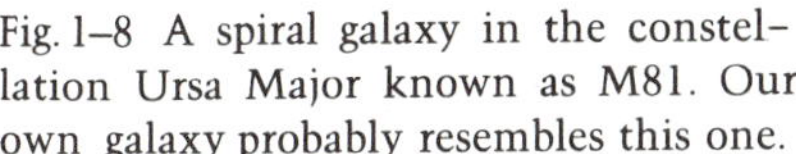

Fig. 1–8 A spiral galaxy in the constellation Ursa Major known as M81. Our own galaxy probably resembles this one.

Fusion Research

All modern technologies depend on the application of a variety of physical insights developed through research on basic physical principles. One of the most interesting technological studies now under way is the attempt to tame the fusion processes that fuel the stars, so that we can use fusion to yield energy for us here on earth. Let us look at some examples of fusion research in some detail, both to illustrate the types of physics involved and to give an idea of the grand scale of the operation and the role of physicists in it.

Fusion converts mass into energy. As we shall see in the chapter on the special theory of relativity, Albert Einstein proposed in 1906 that even a small amount of mass (m) may be converted into a large amount of energy (E) because $E = mc^2$, and the speed of light (c) is a very large number. (Obviously, our understanding of this important formula depends on the discussion of energy in its different forms, which takes up a large part of the first section of this book.)

The attempts to produce fusion involve some of the simplest types of atoms. (We will be discussing the makeup of atoms and their cores, called nuclei, in several places in this book.) The simplest atom, hydrogen, has a slightly more complex form called deuterium, and it is hydrogen and deuterium that can fuse at the lowest temperatures. This makes them obvious candidates for terrestrial fusion.

As we shall see in the chapter on electricity and magnetism, and in those on nuclear physics, hydrogen and deuterium nuclei are charged and thus tend to repel each other. We must find a way to overcome this natural repulsion.

We do not yet know which of several possible methods will be ultimately successful. One line of basic research into fusion uses lasers to provide energy. This research is being conducted at the Los Alamos Scientific Laboratory in New Mexico, at the Lawrence Livermore Laboratory in California, at the University of Rochester, and at KMS, Inc., in Ann Arbor, Michigan. At these research facilities, beams of energy in the form of laser light are shot at atomic nuclei from several directions simultaneously, as we see below. The energy from these beams is used to overcome the natural repulsion of the nuclei, forcing them close enough to allow them to fuse. Since the property of inertia, defined centuries ago by Galileo Galilei and Isaac Newton, keeps the nuclei centered in the oncoming beams, this method is a type of "inertial confinement." (We shall discuss inertia in the first chapter of this book.)

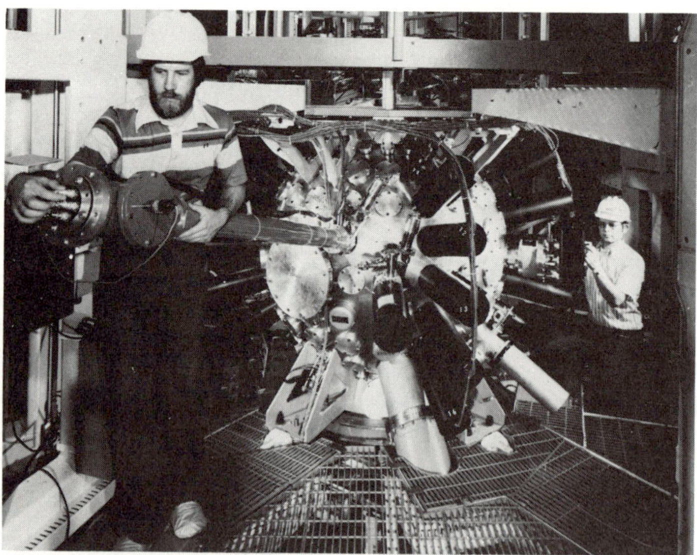

Fig. 1 Scientists at the Lawrence Livermore Laboratory prepare the Shiva target chamber to observe the fusion of a tiny amount of deuterium-tritium gas contained in a target smaller than a grain of sand. For less than one-billionth of a second, this vacuum chamber will hold fusion fuel heated and compressed by the world's most powerful laser to temperatures and densities like those found in the sun and stars.

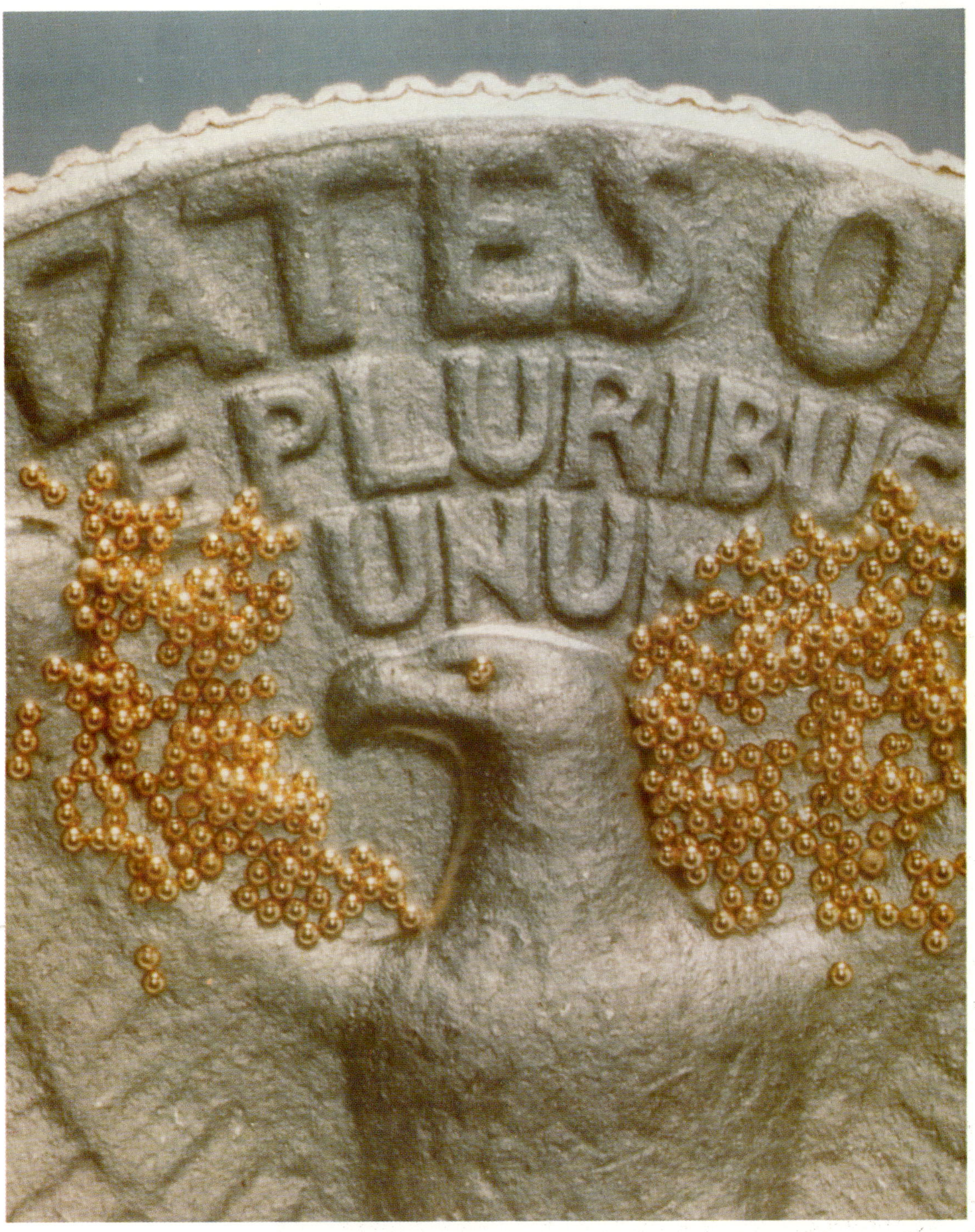

Plate 1 The hydrogen and deuterium nuclei to be fused are encased in a small glass balloon, called a "microballoon." Its tiny size is shown here in silhouette against a quarter. A liter jar holds a billion of them.

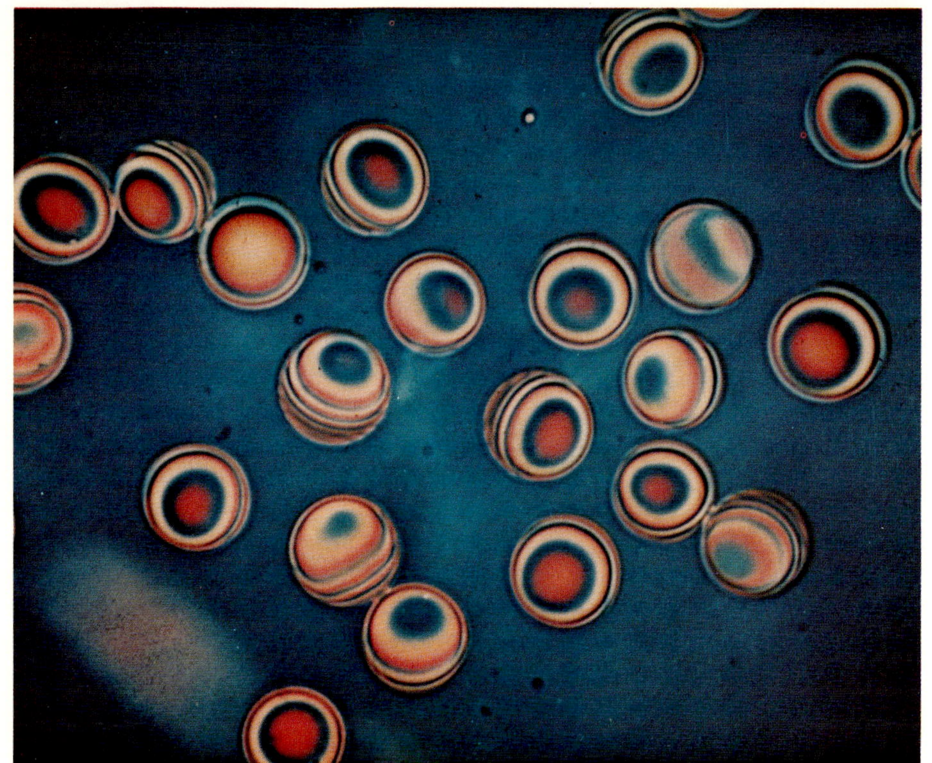

Plate 2 *(above)* The microballoon must be perfectly round, or else it will not allow the nuclei to be heated evenly enough to allow sufficient energy to be transferred to them to overcome their natural repulsion. Thus each batch of microballoons is inspected under a light beam. Because light acts as a wave, we see an "interference pattern" (which we shall discuss in the chapter on waves). The microballoons with a regular interference pattern are potentially usable. The others would break if we tried to use them because their surfaces are irregular. Only one of the microballoons here is sufficiently regular.

Plate 3 *(below)* To sort out the usable microballoons in the next stage, the best ones are placed in a vacuum tank. Microballoons that are not sufficiently regular explode, leaving only a few final candidates. These are then filled with a mixture of hydrogen and deuterium.

Giant lasers, which can send out a lot of energy concentrated in a small area, are a good way to direct energy at the microballoons. Lasers, to which we devote an entire chapter, are an invention of the 1960s. They work because light sometimes has the properties of a group of particles. The structure of the atom was worked out in a series of developments in the early part of this century, and led to the understanding that light and atoms did not work as simply as the classical picture and our intuitive understanding implied.

Investigations of atomic structure have taught us that atoms can be placed in a holding pattern of relatively high energy, and that the release of energy can be triggered from many atoms at the same time. This is the principle of laser action. It allows a lot of energy to be brought out in a pulse at one instant. It is these pulses of energy that we hope will allow us to fuse hydrogen and deuterium into helium.

Plate 4 *(above)* shows the current laboratory at Los Alamos, with several huge lasers all directed at a tiny microballoon. The microballoon is in a spherical chamber from which all air has been evacuated (Plate 5, *below*). Note the massive cables carrying electrical power and control signals. (Interference—like radio static—on cables would prevent the control signals from being reliable.) Fortunately, a new technology called "fiber optics" has been developed in recent years which virtually eliminates interference. The reflection properties of glass allow light signals from small lasers to be sent along wires of glass the thickness of a human hair. The signals can be made to vary, just as varying electrical signals carry radio waves or telephone conversations. Indeed, more and more telephone signals are now being transmitted by fiber optics.

The laser beams are generated by huge amplifiers (Plate 6, *at left*). In order to concentrate the laser beams on the microballoons, curved reflecting mirrors are used. Here we employ the laws of optics. These mirrors are unusual, compared to camera or telescope optics, because they are made of copper. We see some of these mirrors in Plate 7 *(below)*. Copper conducts heat especially efficiently, and so can carry away the heat transmitted to it by the laser beams without being excessively distorted. A knowledge of the properties of metals is thus also an important part of the arsenal of the laser-fusion scientist.

ANTARES

Laser-fusion tests are getting closer and closer to the point where as much energy is generated as is put in. Obviously, we need to develop the system to the point where much less energy is put in than is generated. Both Los Alamos and Berkeley are now developing much larger versions of their apparatus, which will come closer to the break-even point. The Los Alamos version, scheduled to open in 1983, is called Antares (Plate 8, *above*). We see one of the 12-beam amplifiers. Two will be used. Plate 9 *(below)* shows an earlier, 6-amplifier, design.

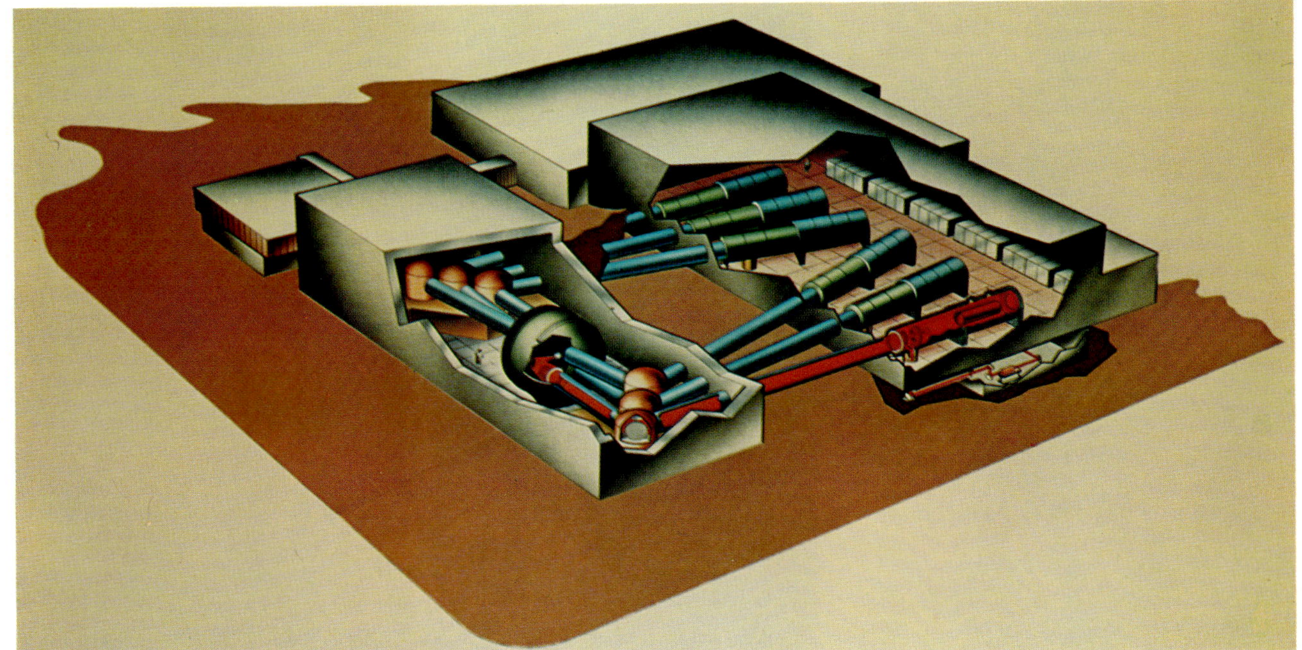

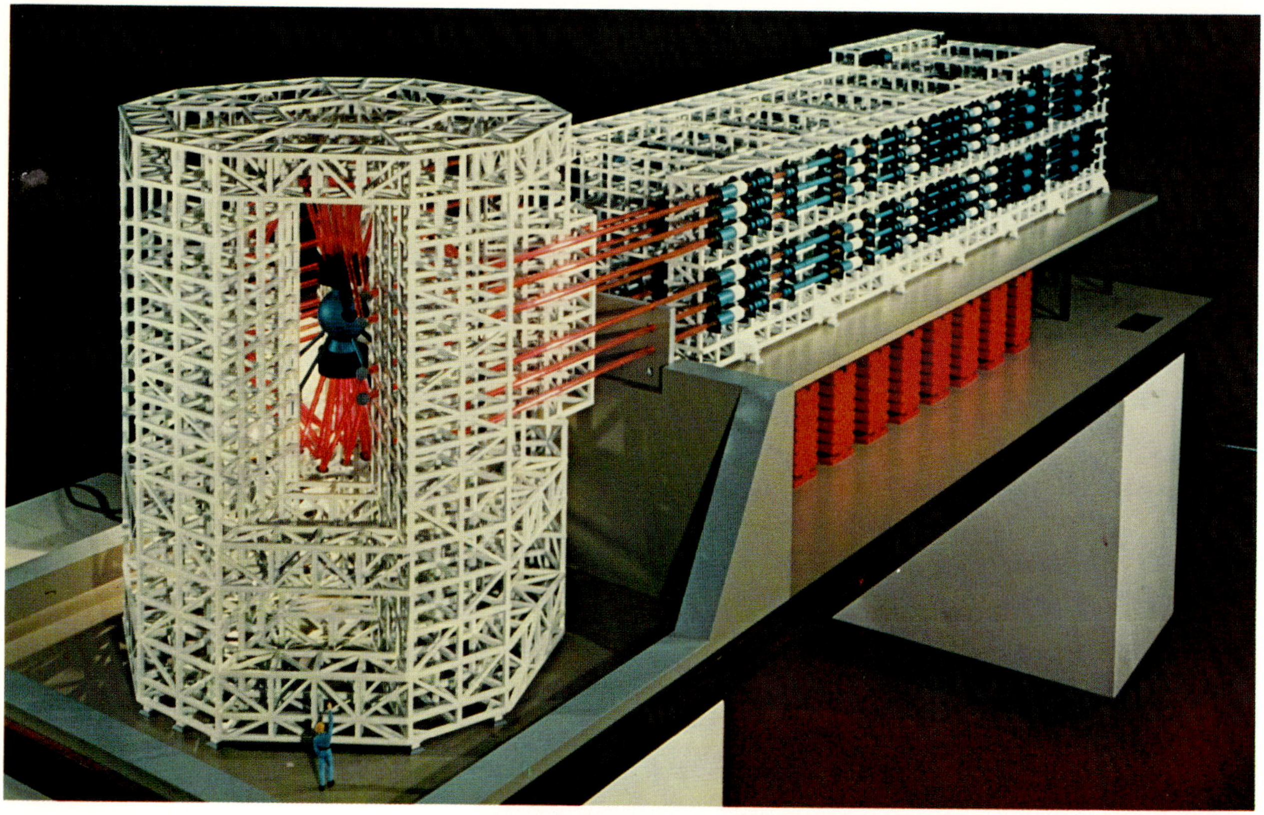

The Lawrence Livermore Laboratory's entry in laser-fusion research has twenty laser beams. It is called Shiva, after the Hindu god with many arms. Plate 10 *(above)* shows a model of Shiva's framework. The beams deliver more than 30 trillion (3×10^{10}) watts of optical power in less than one billionth of a second. A larger system, Nova, is under development.

Current nuclear power plants use fission instead of fusion. Fission means the breakdown of heavy elements, like uranium or plutonium, into lighter ones. The process generates much radioactivity. Fusion is much cleaner. Very light elements, like hydrogen, are built into other light elements, like helium. Plate 11 *(at right)* shows a copper plate used at the Lawrence Livermore Laboratory to test new materials for the fusion reactors of the future. Subatomic particles from the fusion will interact with the reactor walls in fusion reactors, and we must make certain that the reactor walls can withstand the bombardment. Still, very little radioactivity will be produced compared to fission reactors.

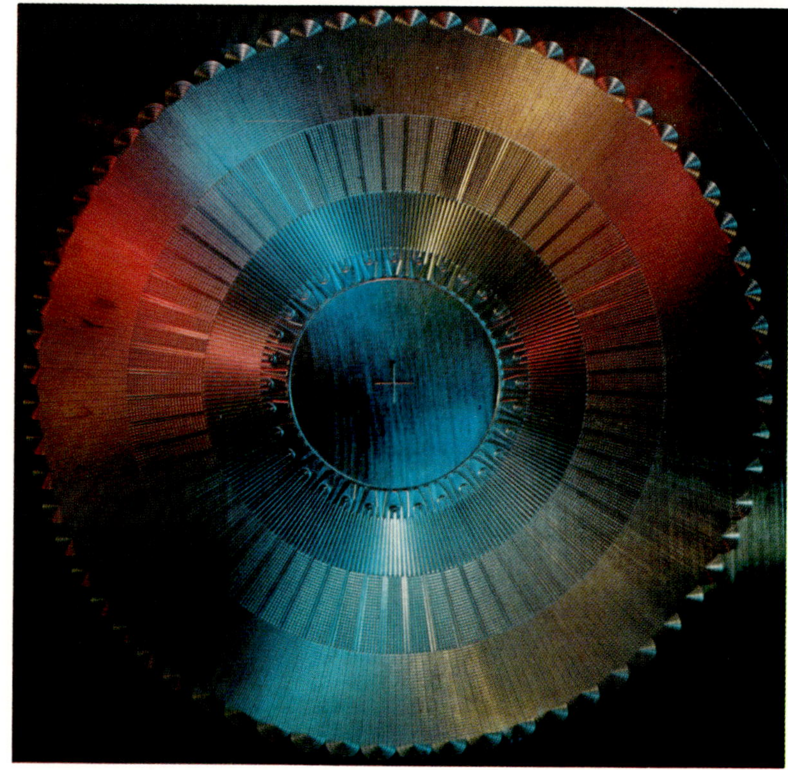

An alternative to lasers for an energy source is a set of beams of ions or electrons. Here we see the electron beam fusion apparatus at the Sandia Laboratory in New Mexico. Plate 12 *(at left)* shows the channels used to transport twelve high-energy electron beams to a central target. A thirty-six-beam apparatus is now undergoing testing. When one of the beams of electrons strikes an aluminum target, as in Plate 13 *(below)*, it produces about 250 billion watts of power in less than one ten-millionth of a second.

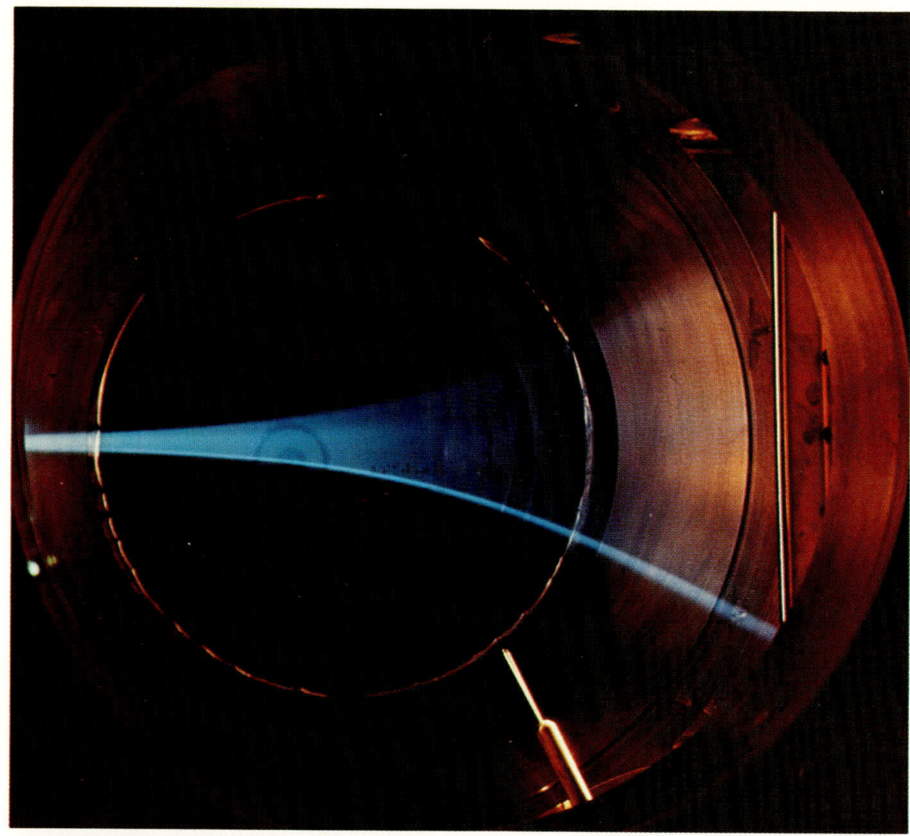

At some time, we must make a choice between various alternatives in inertial confinement, such as laser or particle beam, or type of laser. Eventually we must also evaluate whether inertial confinement fusion or another method now under study, the magnetic confinement method, is superior. This method uses magnetic fields to levitate hot gas without allowing it to touch the walls of its container. (After all, we know of no container able to withstand the extremely high temperatures that would be necessary to provide enough energy to overcome the natural repulsion of nuclei.) Plate 14 *(above)* shows a heated gas, called a plasma, produced in one test of magnetic confinement. Plate 15 *(below)* shows the Princeton Large Torus, a magnetic confinement apparatus.

Our energy supplies in the next century will come from various sources. Solar energy will be one of them, but even when it is developed to a high degree, solar will probably not supply more than about 20 percent of our energy needs. Coal, oil, and gas will run out, and in any case cause much harmful pollution. Whatever we decide about today's nuclear energy—fission—will still only be applicable in the near term. We have a limited supply of uranium, and fission power plants are only meant to be used for fifty years or so, unless breeder reactors come into general use. We will discuss these methods in the chapter on energy. There we shall also analyze safety, risk, conservation, efficiency, and other considerations for a variety of energy sources. Many scientists hope that the promise of fusion energy, a relatively pollution-free source, will be fulfilled and that fusion power plants will be built in the first decades of the twenty-first century.

Fig. 1–9 The Cray-1 computer at the National Center for Atmospheric Research in Boulder, Colorado. The thinking part of the computer, the Central Processing Unit, is in the foreground, with a person standing inside. The Central Processing Unit uses the tape and disk drives and magnetic tapes seen in the background.

Fig. 1–10 A computer chip. This chip holds more than 64,000 bits of information, and is used in computers. This chip alone holds as much information as used to be contained in the entire memory of an early room-sized computer.

A color spectrum appears as Plate 50.

adequate. Here is the realm of relativity, actually two theories—the special theory and the general theory of relativity—advanced by Albert Einstein in the early years of this century. We shall see time and time again in this book how the relativity theories and Einstein's other contributions affect our lives.

Sometimes we deal, on a life-size scale, with objects that are themselves subatomic. For example, the electricity with which we light our lights and run our machines is merely electrons flowing in a wire. These electrons help us at work and at play, but also enable entirely new kinds of activities. Electronic computers, for example, have opened new vistas by allowing us to make computations that would have taken hundreds of people centuries of time before computers were invented. And now computers continue to get faster and faster, more and more powerful, and physically smaller and smaller. They are so smart that they can do most anything, or so we begin to think until we realize that no computer program has yet been devised that can beat a grandmaster at chess. (There are, however, chess programs that can beat almost anyone else.)

The miniaturization of electronics is connected with the study of solid-state physics, which counts electronic equipment like TVs and hi-fis, large computers, pocket calculators, and digital watches among its contacts with the public. We will see that since Einstein's special theory of relativity says that nothing can travel faster than the speed of light (which can go seven times around the earth in a second), building computers that are physically smaller means that we can get signals through them more quickly, thus speeding up our calculations. We are almost at the point where computers are large enough and calculate rapidly enough to keep up with and ahead of the complexities of understanding our weather.

As we push matter to extreme conditions, sometimes strange and unexplained properties show up. For example, when we cool certain types of matter to hundreds of degrees below freezing, they become superconductive, that is, they lose all resistance to current flowing in them. The current can flow forever, without loss. Imagine how much energy we could save if currents kept flowing forever once we started them. And how much energy we would save if we could send power on superconducting transmission lines from our electric generators to our cities and then to our homes.

Optics, too, is an important field of physics, one in which we find out how to improve our view of the very large (with telescopes) and the very small (with microscopes), as well as sharpen and capture our view of our ordinary world (with regular cameras). Combine optics with spectroscopy, the breaking up of light into its rainbow of component colors, and we have a means to understand the atoms and the stars. We shall meet all these topics later in the book.

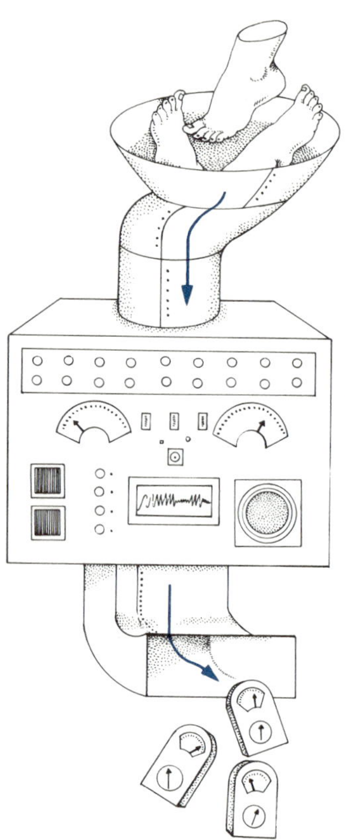

Fig. 1–11 Changing feet into meters.

SI UNITS

The units we use for measurement are discussed in Appendix 1. In this book we will use the current International System of units (called *SI* from the French *Système International*). These units are based on the *meter (m)* for length, the *kilogram (kg)* for mass, and the *second (s)* for time. In the so-called British system of units that has been most commonly used in the United States and Canada, 1 meter equals 39.37 inches (slightly over a yard) and 1 kilogram is approximately 2.2 pounds. The United States is gradually changing to SI units (which are the current version of the *metric system*) for many purposes; Canada has already done so to a greater extent (Fig. 1–11).

1.3 *The World of the Large*

Life on earth covers a wide range of sizes; an elephant, for example, is a thousand times larger than a butterfly. But the range of sizes and scales conceivable in the universe is so great that we get to a perspective from which we see elephants and butterflies as the cousins they truly are.

Let's consider, for example, your size, and try to realize the immensity of space by considering scales larger by factors of 1000.

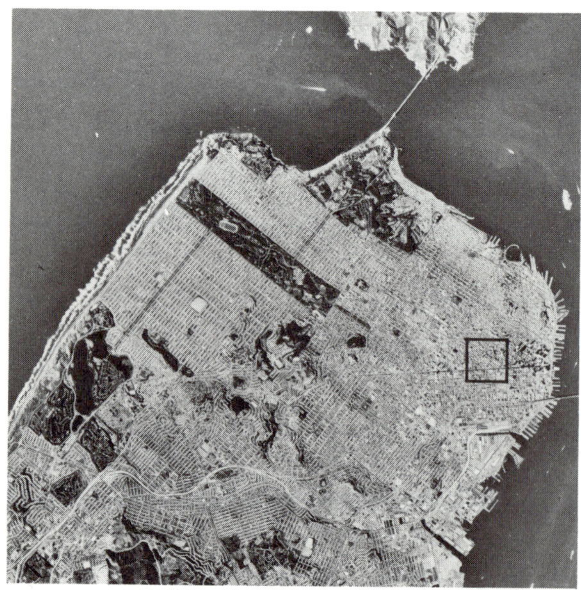

Fig. 1–12 The square marks 1 km in the center of San Francisco.

Fig. 1–13 The square marks a region on the earth 1000 km across.

Fig. 1–14 A million kilometers is about two and a half times the distance of the moon from the earth. Here we see the crescent moon and crescent earth photographed by the Voyager 1 spacecraft as it went to Jupiter.

Then 1000 times your size (humans are usually between 1.5 meters and 2 meters tall; let's round this off to 1 meter for convenience) is 1 kilometer (about $\frac{5}{8}$ mile), the size of a city center (Fig. 1–12).

So 1000 times bigger yet is 1000 kilometers (10^3 kilometers), which is the size of several states together (Fig. 1–13).

Another factor of 1000 shows an area 1,000,000 kilometers (10^6 kilometers) across, a distance that includes the orbit of the moon around the earth (Fig. 1–14).

When we go to a scale of 1,000,000,000 kilometers (10^9 kilometers), we see the orbits of the inner planets (Fig. 1–15).

The next jump, to 10^{12} kilometers (we get bored by now with writing out all the zeroes), shows us the entire solar system as a tiny spot, but we have not yet reached the other stars (Fig. 1–16).

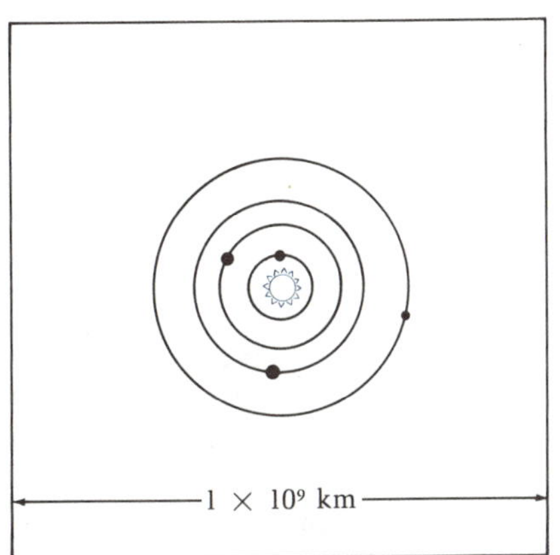

Fig. 1–15 The orbits of the four inner planets—Mercury, Venus, Earth, and Mars—can fit inside a box 10^9 km on a side. (The sizes of the orbits are drawn to scale; the sizes of the sun and planets are not.)

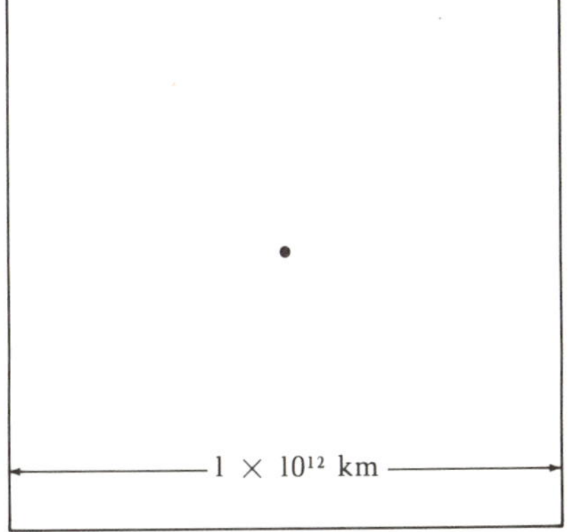

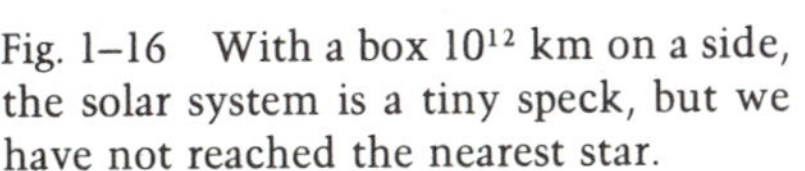

Fig. 1–16 With a box 10^{12} km on a side, the solar system is a tiny speck, but we have not reached the nearest star.

Finally, at a scale of 10^{15} kilometers, we reach the stars. By now, instead of using a kilometer scale, it becomes convenient to use a measure of distance related to how far light travels in a given time; 1 light-year is the distance that light travels in 1 year. Since light travels about 10^{13} kilometers in a year, we are now seeing 100

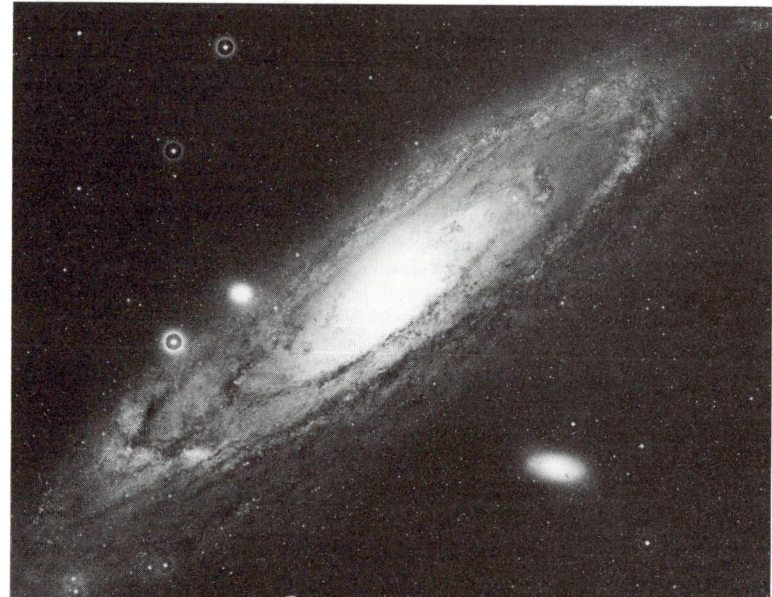

Fig. 1–17 The Great Galaxy in Andromeda, whose spiral arms unwind over a distance of about 50,000 light-years from its center.

Fig. 1–18 A cluster of galaxies in the constellation Hercules. Although a few foreground stars are visible (round images), most of the objects on this photograph are galaxies.

light-years. (Note that a light-year is a *distance,* not a time.) We see hundreds of stars, our sun only one of them, and an ordinary one at that.

When we reach a thousand times farther, we are seeing a region 10^5 light-years across, which shows an entire galaxy. Our sun is about two-thirds of the way out in our own galaxy. The galaxies are really fundamental building blocks of our universe.

We begin to reach our limit when we take our next view, of a region 10^8 light-years across. It includes giant clusters of galaxies; even galaxies turn out to be members of a crowd. Since we are looking 100 million light-years through space, the light from the farthest clusters of galaxies we see has taken 100 million years to reach us.

And now we reach our limit. We cannot get a view 1000 times larger, for it would be bigger than the universe itself, and we certainly cannot get outside of our own universe. We would have to look back 100 billion light-years, and long before we saw that far back we would reach our limit. Perhaps 13 billion years ago, we would see our universe being born in what we call the big bang.

The laws of physics that we determine on our puny earth must serve us in our study of the galaxies and beyond. Fortunately, there is every indication that the laws of physics are universal, which allows us to have hope that we can one day understand the universe.

1.4 Why Study Physics?

People study physics for different reasons. Some physicists are of a practical bent, and want to work out the applications of physics to the world of today. Other physicists want to work out the basic principles that will set the tone for the physics of tomorrow.

Most physicists are really involved in physics for its intellectual appeal and for its fundamental beauty. The idea that the laws of nature can be simplified sufficiently that we can understand them appeals to many of us in an esthetic sense, and it is perhaps this feeling more than anything else that draws us to this science.

Isaac Newton once said, "If I have seen farther, it is because I was standing on the shoulders of giants." Currently, the pace of physical discovery is so rapid that more physicists are alive today than the total number of physicists who worked in all previous times. Also, the number of discoveries and the rate of scientific revolutions are increasing at a similar pace. It has been said, "Nowadays we are privileged to sit alongside the giants on whose shoulders we stand."

We hope in this book to give you a view of contemporary physics that you will find as exciting as we do. The subject, the research, the ever-increasing rate at which exciting discoveries are made—we hope that all these will become part of you and of your mind. And now we set off on our tour.

Key Words

scientific notation
exponent
atoms
nucleus
electrons
protons
neutrons
elementary particles
quarks
universal science
relativity
solid-state physics
superconductive
optics
SI units
big bang

Questions

1. Estimate the size of the smallest objects that you deal with in everyday life. Estimate the size of the largest objects that you deal with in everyday life. How many powers of 10 do you cover from the largest to the smallest?
2. The speed of light is 3×10^{10} cm/s. What is the speed of light in km/s, given that there are 10^5 cm in a kilometer?
3. If we have 6.03×10^{23} protons, each of which has a mass of 1.67×10^{-24} gram, what is the total mass of all these protons?
4. Describe how quarks, protons and neutrons, nuclei, atoms, molecules, and chairs represent a list of increasing complexity. Specify how each is built out of the previous one on this list.
5. Why are *you* studying physics?

PART I

A World of Particles

Consider two very different scenes: At Fermilab, the Fermi National Accelerator Laboratory, a team of twenty physicists is preparing an experiment. They are placing a specially constructed target into position; everything must be perfectly aligned. Once everything is ready, they will shoot an invisible "beam" of protons at the target, and a myriad of exotic particles will emerge in all directions. Setting up this shot takes many weeks, and the results of the experiment may take years to analyze. Nevertheless these physicists hope that, when their analysis is completed, we will all know a little more about the fundamental particles that make up our world.

In a hotel ballroom in New York City, a pool player lines up a shot. The 7-ball should go in the corner pocket and the cue ball should rebound to a particular position on the table to set up a good position for the next shot. As in the case of the atom smasher, this shot must also be lined up with care. But the pool shot, however, may be lined up in a matter of seconds. The results of the shot will be immediately clear. If the 7-ball goes into the corner pocket, all is well for the shooter; if it doesn't, it's the opponent's turn.

What do these two situations have in common? In both cases, we are looking at collisions between particles; one kind of particle is as large as a billiard ball, and the other is even smaller than the nucleus of an atom. When particles collide, certain sets of basic physical laws allow us to predict what will happen. Because subatomic particles are so small, the laws that control them are more complicated in form than the laws that control pool balls, but there are many parallels.

On both the large and small scales, we will see that physicists can set up certain basic physical laws and apply them. (In particular, we shall study

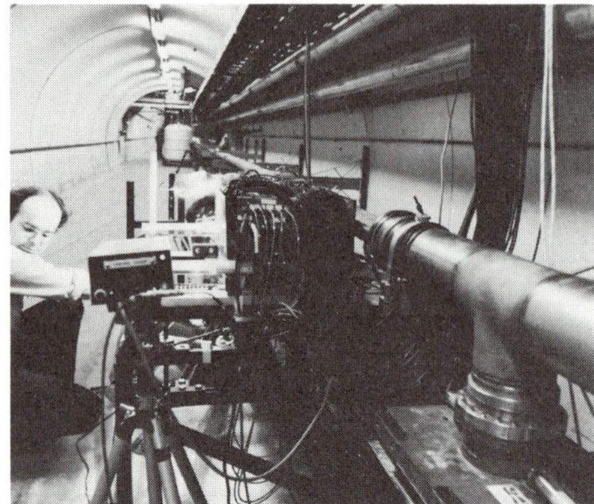

Fig. I–1 A scientist sets up an experiment in the main ring of the Fermilab accelerator.

Fig. I–2 Willie Mosconi lines up a shot in his match with Minnesota Fats televised on ABC's Wide World of Sports.

such laws as *the conservation of energy and momentum.)* Physicists apply these laws via detailed calculation. Billiard players appear to apply them intuitively, and they may never have learned about the conservation of energy and momentum formally. Nevertheless, based on watching many thousands of collisions between billiard balls, billiard players have acquired deep-rooted feelings for how these laws of physics work.

We like to talk in terms of particles because we think of particles as being basically simple. It is this simplicity that we are trying to exploit. But we find that the notion of a "particle" turns out to be fairly complex; finding "fundamental" particles and the ways in which they combine is one of the most important current research problems of physics.

What about the larger scale? A billiard ball may contain as many as 10^{24} atoms. Why do we talk about the ball as a particle? We can think of this terminology as a compromise between what we would like to know and what we can actually calculate. We can easily figure out what will happen to one particle in a given situation; keeping track of 10^{24} of them, even with a computer, is beyond our wildest dreams! However, as we will see in the following chapters, despite the complexities in the shapes of objects we encounter in everyday life, there is hope. We can learn a lot about how objects will react in various situations by studying the behavior of simpler objects—particles—in similar situations. Although considering the behavior of particles may seem like a drastic oversimplification, it is amazing how many phenomena it helps us understand.

In this section, we will consider the ways in which particles behave and interact. We begin, in Chapter 2, by seeing how we describe the motion of a particle. In Chapter 3, we consider the relationship between motion and forces. In Chapters 4 and 5, we show how it is useful to think in terms of two quantities—momentum and energy. As the final part of our classical treatment of particles, in Chapter 6 we consider objects that are rotating.

CHAPTER 2

Particles in Motion

Leaves stir in the wind. A stone rolls downhill. Stars move across the sky, and sun, moon, and planets move against a background of these stars. We need no special equipment to observe such events. One of the obvious facts of the physical world is that *things move.* Little wonder that mechanics, the study of motion, is one of the oldest topics in physics.

There are two parts to the study of motion: Kinematics deals with describing the motions of particles. (*Kinema* is Greek for

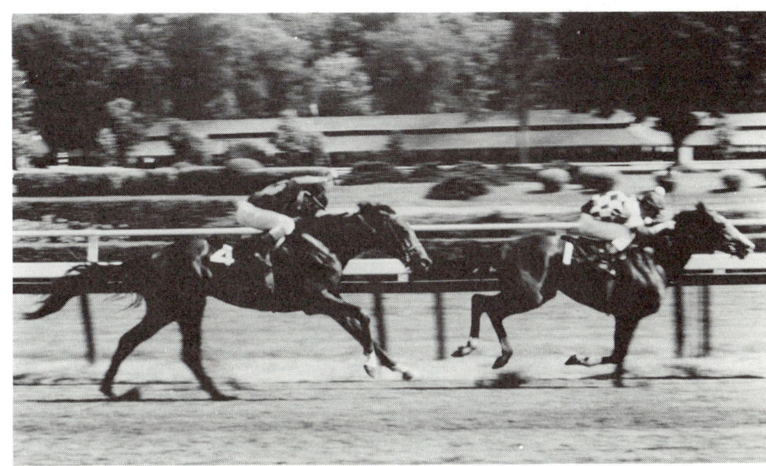

Fig. 2–1 Motion is sometimes very easy to sense.

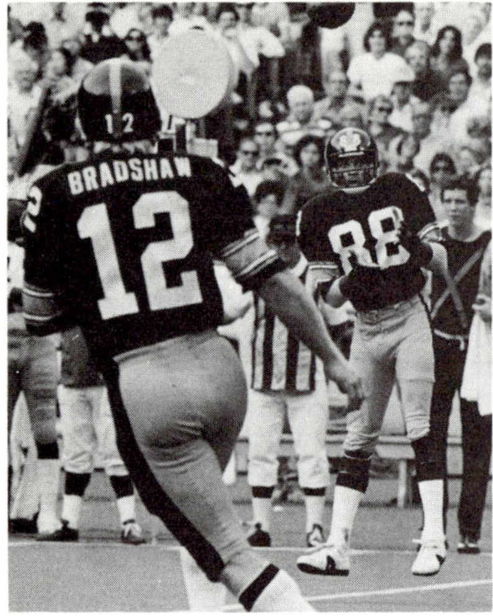

Fig. 2–2 A football pass play. Most of the important information deals with where the ball is at any time, and especially where it ends up. For these purposes, we can think of the ball as a point particle. However, in analyzing why the ball went where it did, it is then important to consider the ball as an extended object, which can, for example, rotate.

"motion"; "cinema"—moving pictures—comes from the same root.) Dynamics deals with the cause of motion, especially the relationship between motion and forces. (*Dynamikos* is Greek for "powerful.") We will discuss kinematics in this chapter and dynamics in the next.

Consider the motion of a football thrown by a quarterback. If we are fans watching a game, we care most about where the ball goes. A coach might also care about the behavior of the ball while it is in the air. Does it spin in a steady "spiral" or does it wobble or tumble? Motion from one place to another—in this case, from passer to receiver—is called translation; spiralling, wobbling, and tumbling are forms of rotation. In this chapter, we are going to discuss translation; rotation will be discussed in Chapter 6.

When we are interested in translation alone, we don't keep track of every particle of material in the ball. If we follow, or predict, the path of any one particle, we can be sure that the rest of the particles follow along. For this reason, when only translation is involved we analyze the motion of the ball (or of any other complex object) as if it were actually a single particle, with all its mass concentrated at a single point.

2.1 *Position*

By wanting to know where something is at a particular time and when it will arrive at a particular place, we must use the concepts of space and time. We have the advantage (and disadvantage) of our intuitive feel for these concepts, but we would be hard pressed to come up with precise definitions. Our intuitive notions are satisfactory for now. When we get to the theory of relativity, we will see how a more careful study of time and space and their relationship to each other will lead us to a revised understanding of these concepts and to some strange and exciting results.

How do we describe the motion of the Voyager spacecraft en route to Uranus? We must first note its position at several times. We will see in this chapter that we can then calculate how fast it is going at any time, and predict the path of its future motion in space.

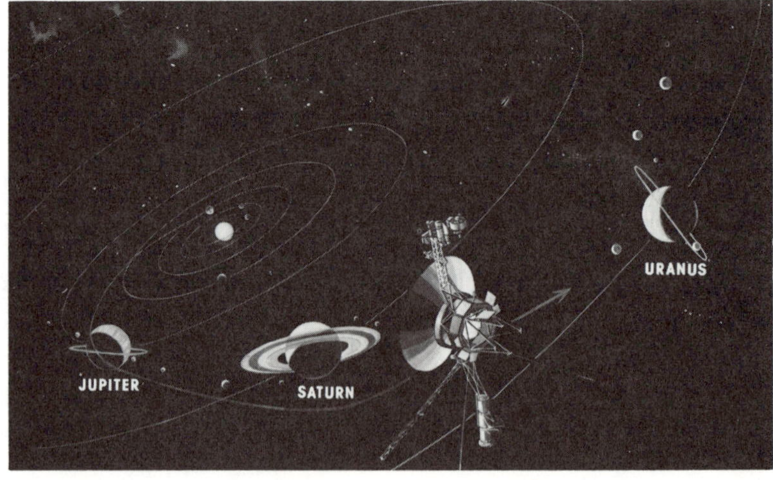

Fig. 2–3 A schematic drawing of the path of the Voyager spacecraft from the earth to Jupiter, Saturn, and beyond.

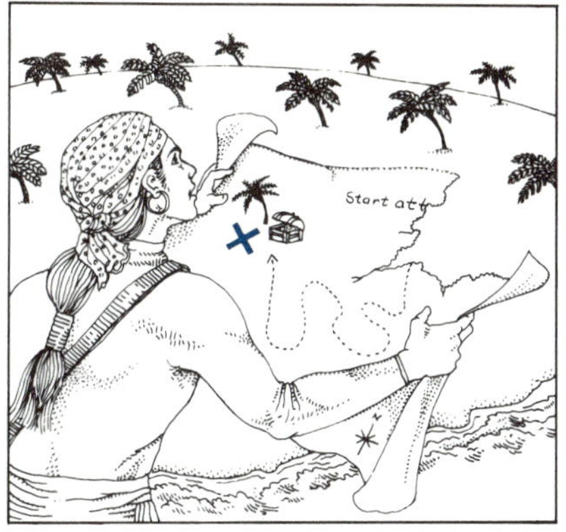

If you find a treasure map but the starting point has been torn away, that map is worthless (Fig. 2–4). In order to describe the position of an object, we must tell how far it is from something else. We call this "something else" the *reference point* or *origin.* The complete system for describing positions is called a coordinate system, because we can specify the position of an object by giving a set of numbers called *coordinates,* one coordinate for each dimension.

Fig. 2–4 A treasure map without a starting point is worthless.

Fig. 2–5 The reference line for longitude is the Prime Meridian, running through the Greenwich Observatory, England.

2.1a ONE DIMENSION

Even though our everyday world is three-dimensional, many motions are one-dimensional. Take, for example, a railroad track that runs north and south. We can pick one station to serve as the origin and give the position of a train by saying how far it is from that station. A small problem remains—the train can be north or south of the station. We can solve this by saying *arbitrarily* that distances south of the station are negative (less than zero). If the train is 2 km north of the station, its position is +2 km; if it is 1 km south, its position is −1 km. A single coordinate provides a complete description of the train's location (Fig. 2–6).

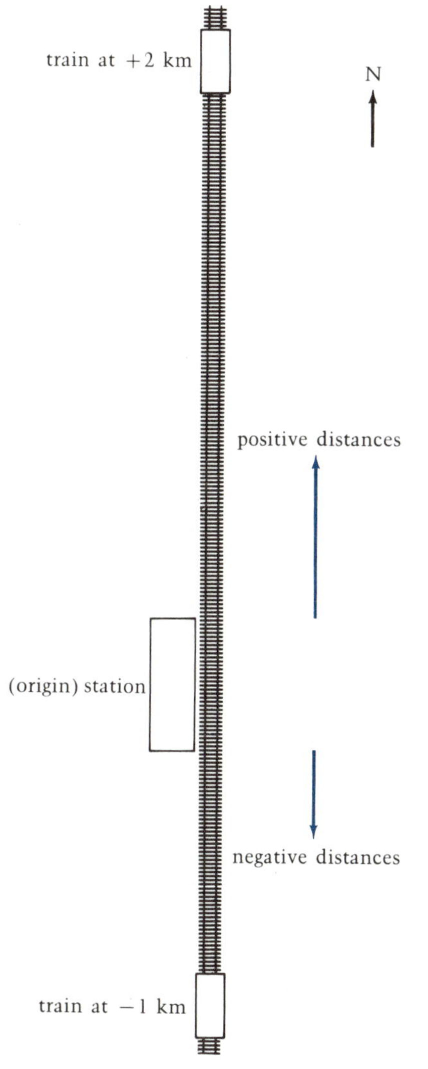

Fig. 2–6 One-dimensional position: train track and station.

Km	Mi			
		Train Number		**114**
		Train Name		**Metro-liner Service**
		Frequency of Operation		Daily
		Type of Service		® ✓ ☕
0	0	**Washington, DC** *(ET)*	Dp	1 00 P
15	9	Beltway Sta., MD *-Lanham*		
65	40	**Baltimore, MD** *-Penn. Sta.*		1 36 P
113	70	Aberdeen, MD ●		
175	109	**Wilmington, DE**		2 29 P
216	134	**Philadelphia, PA** *-30th St. Sta.*		3 03 P
223	139	North Philadelphia, PA ⊕		
268	166	**Trenton, NJ**		3 35 P
283	176	Princeton Jct., NJ ㊱ ⊕ *(Princeton)*		
309	192	New Brunswick, NJ ⊕		
322	200	Metropark, NJ *-Iselin*		
345	215	**Newark, NJ** *-Penn. Sta.*		D 4 14 P
361 361	225 225	**NEW YORK, NY** *-Penn. Sta.*	Ar Dp	4 30 P

Fig. 2–7 Metroliner schedule.

If we know when the train stops at each station, we can display this information as a timetable, a list of positions and associated times (Fig. 2–7). We can also plot the positions and times on a graph (Fig. 2–8). While the tabular form may be more useful for a rushing commuter, graphs offer advantages when we are studying motion, because we are often more interested in general trends than in exact numbers. If we wish to know whether a train is heading north or south, the direction of the line on a graph that represents the position of the train gives us a quick answer. The steepness of the line provides an idea of how fast the train is traveling.

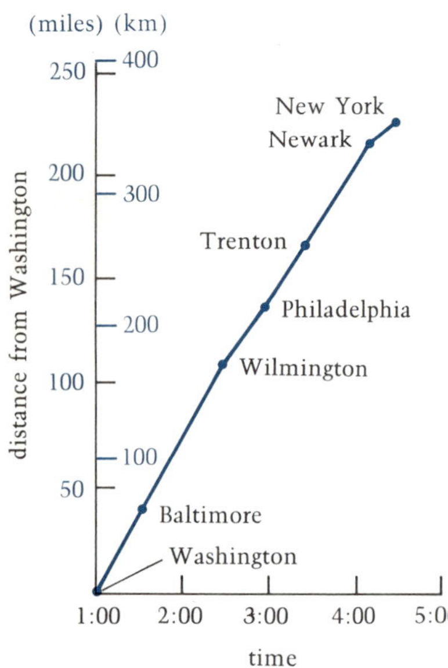

Fig. 2–8 Graph plotted from Metroliner schedule.

2.1b MORE THAN ONE DIMENSION

Knowing the distance of an object from a point of origin is not enough when dealing with situations involving more than one dimension. For example, if you know only that a treasure is buried 100 meters from a tree, you would have to dig up a circle of radius 100 meters around the tree. To locate the treasure by digging only at one spot, you must know in what *direction* to walk from the tree.

In physics we encounter a number of quantities that, like the position of an object, have a direction associated with them. Such a quantity is called a vector. To specify a vector, we must give two quantities: its size (called its magnitude) and its direction. A number that has magnitude but no direction associated with it is called a scalar. Some examples of each are given in Table 2–1.

TABLE 2–1 Examples of Vectors and Scalars

Vectors	*Scalars*
position	time
velocity	mass
acceleration	speed
force	amount of milk in a glass

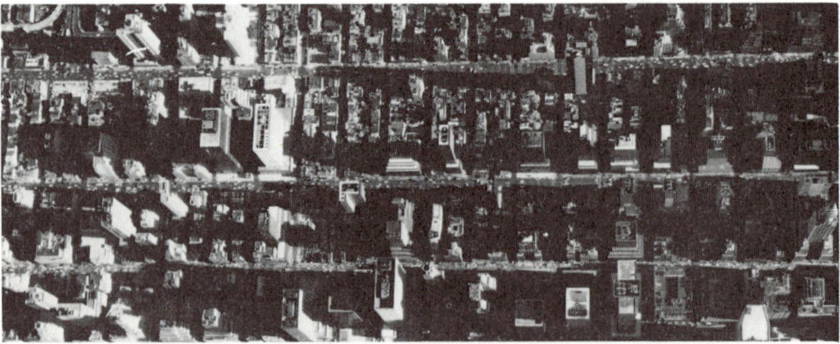

Fig. 2–9 Grid pattern of streets and avenues in midtown Manhattan.

A crucial feature of vector quantities, from a mathematical point of view, is that they all behave in the same manner. Once you know how to handle one vector quantity, you know how to handle any of them. For example, if you know how to add two positions, then you also know how to add two velocities. For this reason, mathematicians and theoretical physicists have spent a lot of time investigating the properties of vectors. Their studies go far beyond the scope of this book, but we mention them to underline the important role that vectors play in physics.

2.2 *Motion*

2.2a VELOCITY

We have seen how to describe the position of an object. Now we can discuss its motion, since motion is just a change in position. In fact, velocity, the quantity that reveals how things are moving, is defined in terms of changes in position. If we note the position of a moving ball at a particular time and then note its position some time later, its *average velocity* during this interval is defined as:

$$\textit{average velocity} = \frac{\textit{change in position}}{\textit{time interval}}.$$

This average velocity has a direction associated with it and this is a vector quantity. The direction can be shown as an arrow pointing from the initial to the final position of the ball in Fig. 2–10. If the ball travels in a straight line from initial to final position, the direction part of this vector will simply be identical to the direction in which the ball is traveling.

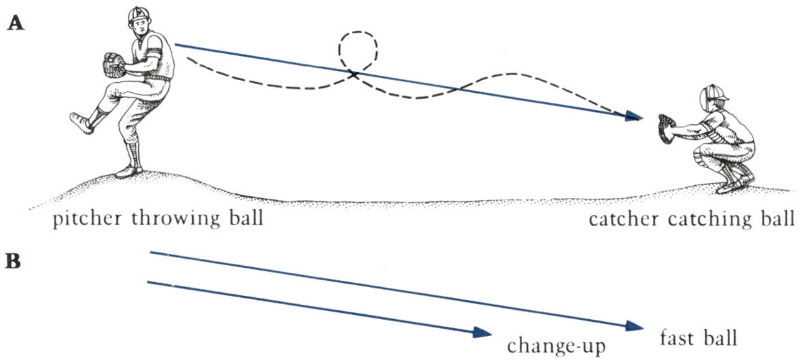

Fig. 2–10 A. Direction of the average velocity. The initial position of the ball is in the pitcher's hand; the final position is in the catcher's glove. The direction of the average velocity is that of a line drawn from the pitcher's hand to the catcher's glove (like a fast ball), even if the ball actually followed a very different path (like the dashed line or a change-up). B. The magnitude of the average velocity of the ball, represented by the length of the arrows, depends only on how quickly the ball covers the distance from the pitcher to the catcher.

ADDITION OF VECTORS

Suppose you start at some origin, drive 3 km east and then 4 km north. Where are you? To solve this, we first draw an arrow from the origin, 3 units long, pointed east. Then, from the tip of that arrow (since that is where you start your northward drive), you draw an arrow, pointed north, 4 units long. The tip of this arrow gives your final position. The vector representing this final position is simply an arrow drawn from the origin to your final position. Thus, *to find the result of adding two vectors (called the* resultant*) we place the tail of the second vector at the head of the first and then draw an arrow from the tail of the first vector to the head of the second.*

By comparing Figs. 2–11**A** and 2–11**B**, we can see that *when we take the sum of two vectors, it does not matter in what order we add them.* Ordinary numbers have this same property. When you add a column of numbers, it doesn't matter if you start at the top, bottom, or anywhere in between, as long as you add each number only once.

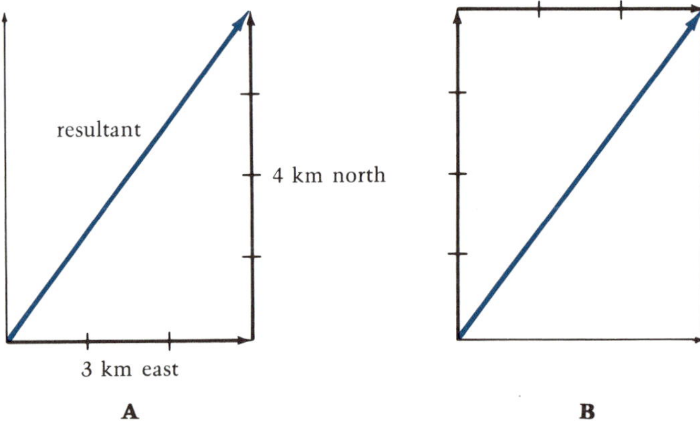

Fig. 2–11 Addition of vectors.

SUBTRACTION OF VECTORS

Actually, vector subtraction is a misnomer. We don't really subtract one vector from another; we just add the first vector to the *negative* of the second vector. What is the negative of a vector? *The negative of a vector is a vector of the same size, but pointing in the opposite direction.*

Suppose Eric walks 10 m north and Eloise walks 10 m east. What is Eloise's position as viewed by Eric? We are really asking for the difference between the vector representing Eloise's position and that representing Eric's position.

In Fig. 2–12**A** we see the vectors representing both positions, and we draw an arrow from Eric to Eloise. When we do the vector subtraction, we should get a vector that looks like this arrow.

In Fig. 2–12**B** we do the subtraction by first taking the negative of Eric's position and then adding it to Eloise's position.

In Fig. 2–12**C** we show that if we had done just the opposite, reversing Eloise's position and adding it to Eric's position, we get a result which is opposite to that in Fig. 2–12**B**. This makes sense, since the path from Eric to Eloise is just the opposite of that from Eloise to Eric. Thus, *if we reverse the order of subtraction of vectors, we get a vector pointing in the opposite direction, that is, we get the negative of the original result.* This is similar to the behavior of normal numbers. For example, $3 - 2 = 1$, while $2 - 3 = -1$.

Fig. 2–12 Subtraction of vectors.

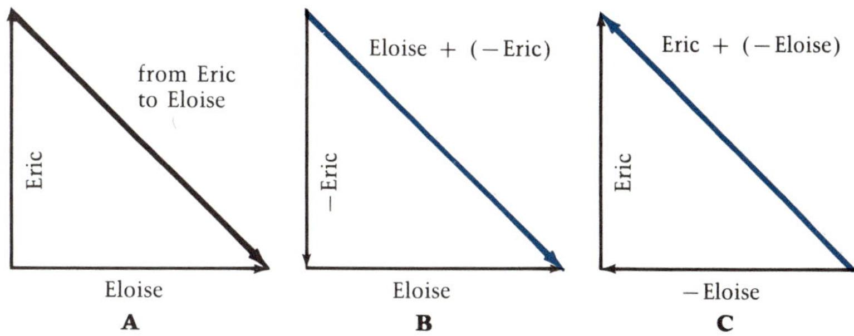

COMPONENTS OF VECTORS

Suppose we have a vector and we specify two directions. We can try to find two vectors, drawn along these directions, such that their sum is equal to the original vector. These two vectors are called the *components* of the vector along these directions. It is often convenient to represent a vector as the sum of its components along some directions. For example, in studying projectile motion we will see that the horizontal component of velocity remains constant while the upward component changes.

Figure 2–13A shows how to find the components of a vector. We just draw lines from the tip of the vector parallel to the specified directions (dotted lines) until they intersect the lines that give the specified directions. You can see in Fig. 2–13B that the two vectors we get in this way have a sum equal to the original vector.

Fig. 2–13 Components of vectors.

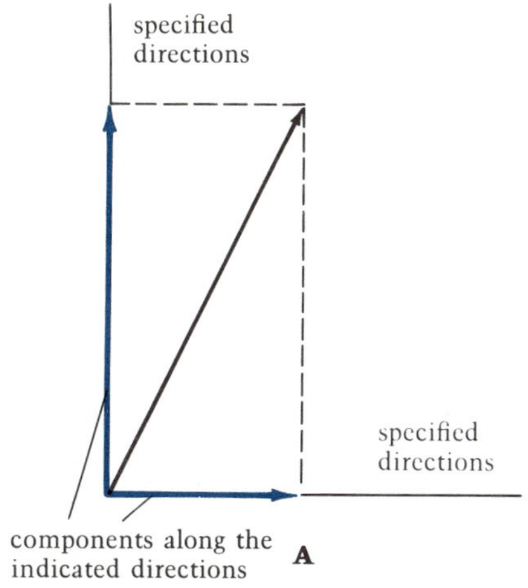

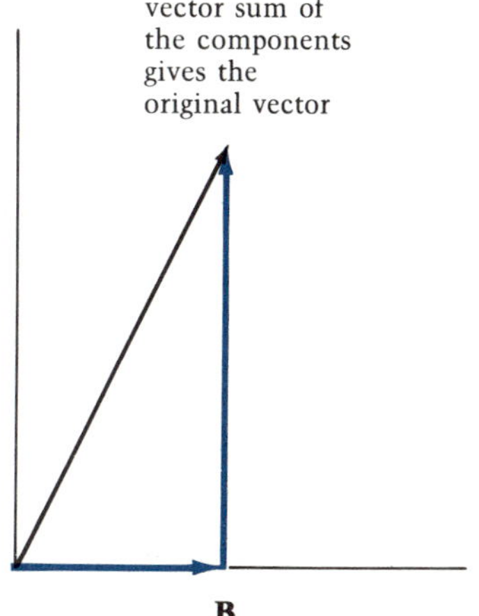

Notice that our definition of average velocity depends only on where the ball is at the beginning and end of the time interval. It tells us nothing about how the ball moved in between. It might have moved in a straight line or bounced off a wall. It might have started fast and then slowed down. If we want more information about this motion, we must look at the ball's average velocity over shorter time intervals. We can, for example, calculate its average velocity in the first and then the second half of the time interval. If

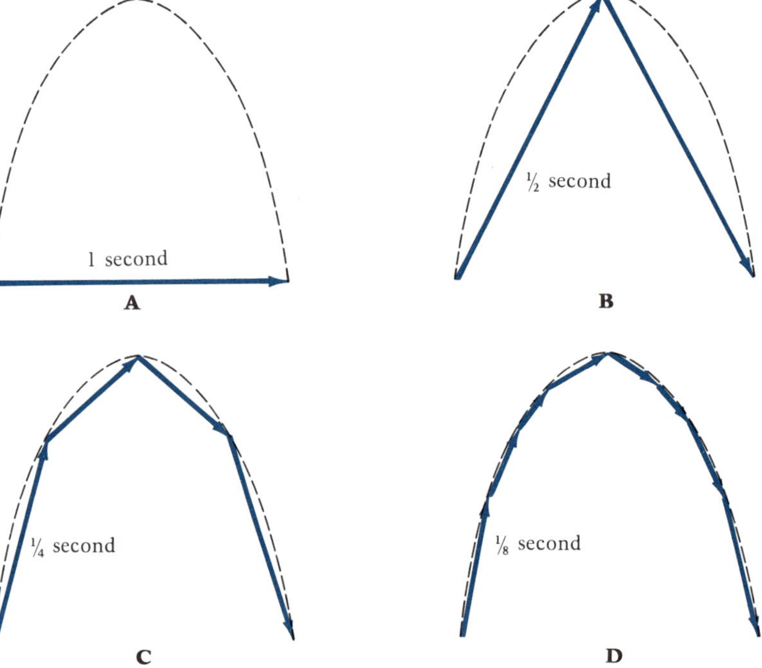

Fig. 2–14 The dashed line represents the path of a ball that spends 1 second in the air. In A, the arrow represents the average velocity in that 1 second. In B, we find the average velocity in two half-second intervals. In C, we find the average velocity in four quarter-second intervals, and so on. D. Notice that as we make the intervals shorter, our arrows come closer and closer to describing the actual motion of the ball.

we want even more detailed information, we can break the time interval into smaller steps (Fig. 2–14). When the time interval is so short that the ball's velocity cannot change by a significant amount, the average velocity over this very small time interval is called the *instantaneous velocity*. Strictly speaking, the word "instantaneous" is inaccurate, since we cannot define the velocity at one instant, but the time interval is so small that we can think of it as representing an instant. (Calculus deals with instantaneous quantities in a more precise way.)

The magnitude of velocity is called speed. Speed is a scalar; there is no direction associated with it. It tells us how fast something is moving but gives no information about direction. In colloquial usage, we often interchange "speed" and "velocity," but when discussing physics we must be careful to distinguish between them.

This is a common problem; there are a number of words for which the physicist needs to apply a precise definition but which we use in less precise ways in everyday speech; "momentum," that favorite word of sportscasters, is a prime example.

Speed is a *rate*, a quantity that tells us how fast something is happening. If we say that an object has a speed of 80 km/h, we do not mean that the object actually traveled 80 km or that we watched it for an hour. It means that *if* it traveled at that rate for 1 hour, it would go 80 km. It might travel 40 km in half an hour or 160 km in 2 hours or some other variation of this rate. Since speed is a rate, we can say that

$d = vt$ *distance = rate × time.*

Our common experience is confined to speeds in the meters/second or tens of meters/second range. The national speed limit of 55 miles per hour is equivalent to 80 km per 3600 s = 25 m/s. Let's look at a few simple examples of some not-so-common speeds:

An elementary-particles physicist may wish to find out how fast a proton is traveling after it collides with a nucleus. Devices to detect the passage of the proton can be set up along its path and the time of flight between the detectors measured. Suppose that two detectors are 10 meters apart and the proton covers the distance in 1 microsecond (10^{-6} seconds). The speed is then the distance divided by the time, which is (10 m)/(10^{-6} s), or 10^7 m/s. (Dividing by 10^{-6} is the same as multiplying by 10^6, and to multiply $10^1 \times 10^6$ we add the exponents, getting $10^{1+6} = 10^7$.) Things move fast in the subatomic world!

Light travels at a speed of 3×10^8 m/s. How far does light travel in one year? A year has about 3×10^7 seconds. (We find this by multiplying the number of seconds in an hour by the number of hours in a day and then by the number of days in a year.) Since the distance covered is (speed) × (time interval), distance is (3×10^8 m/s) × (3×10^7 s) = 9×10^{15} m. Since 9 is very close to 10, this is about 10^{16} m, or 10^{13} km!

The distance light travels in one year is called "one light-year." If you traveled one light-year, you still would be only one-fourth the distance to Proxima Centauri, the star closest to the sun.

It is often useful to graph the progress of a moving object. Let the horizontal axis represent time and the vertical axis represent distance. For any given time, we mark a point that gives the position of the object at that time. Let's see how different speeds show up on such a graph. Suppose two cars start out at the same time, a sedan at 50 km/h and a sports car at 100 km/h, both traveling in the same direction. After 1 hour, the sedan has gone 50 km and the sports car has gone 100 km. After 2 hours, they have gone 100 km and 200 km, respectively, and so on. We can see from the graph shown in

Fig. 2–15 that the line representing the faster car makes a greater angle with the time axis, that is, the line is steeper. Thus, the steepness of the line in such a graph—called the *slope* of the line—gives us an indication of how fast the object is moving.

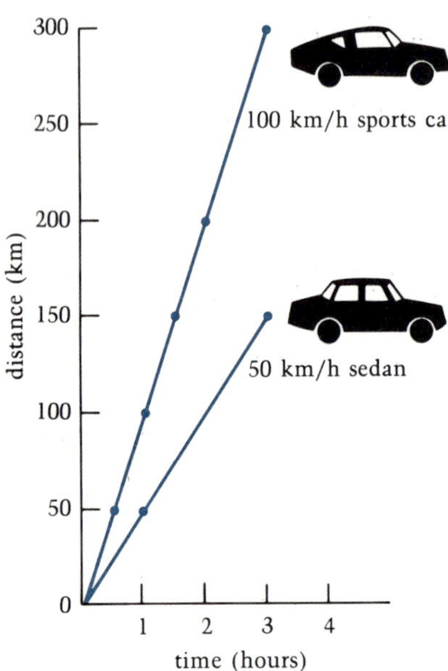

Fig. 2–15 When we plot a line showing the distance traveled at each time, we see that the line representing the faster car makes a greater angle with the time axis. Thus, the steepness of the two curves gives us a feel for the relative speeds of the cars. If the two cars start at the same place at the same time, the one traveling twice as fast will, in any time interval, have gone twice as far.

Let's make a graph of a simple trip. You leave your house at 8:00 A.M. and travel 40 km in the first hour and 80 km in the second hour. You stop 1 hour for a snack. You then reach your destination by traveling 80 km in the next hour. We can summarize your trip in Table 2–2 (and in the graph in Fig. 2–16).

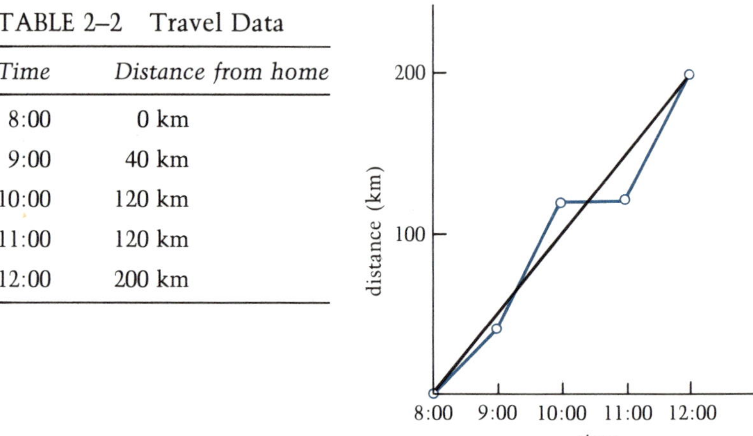

TABLE 2–2 Travel Data

Time	*Distance from home*
8:00	0 km
9:00	40 km
10:00	120 km
11:00	120 km
12:00	200 km

Fig. 2–16 You can get from one place (the origin) to another (upper right) either by traveling all the time at a constant average rate of speed (black line), or by traveling sometimes faster than average, sometimes slower than average, and sometimes not at all (blue line).

The total distance for the trip is 200 km and the total time is 4 hours, so the average speed for the whole trip is (200 km)/(4 h), or 50 km/h. Note that this is not equal to the average speed in any one of the hour intervals (40 km/h, 80 km/h, 0 km/h, and 80 km/h, respectively). The black line on the graph represents what we would get for a car traveling at a steady 50 km/h.

The question of average speed vs. instantaneous speed comes up in the case of a truck going up and down hills on an interstate highway. Under the current law in the U.S., the maximum speed is 55 miles per hour. (For this example, we are using miles per hour instead of km units in order to retain the familiar "55.") On downhill sections, a truck can easily do 55; uphill it might only be able to do 40. Its average speed is then well below 55 (see Fig. 2–17). Truck drivers claim that they should be allowed to average 55, which requires going over 85 on the downhill sections.

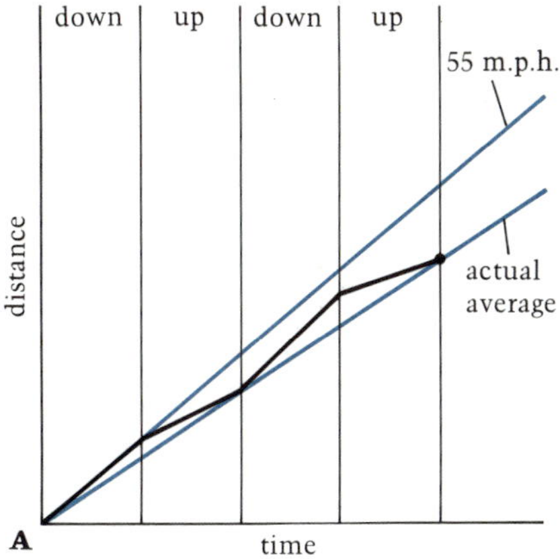

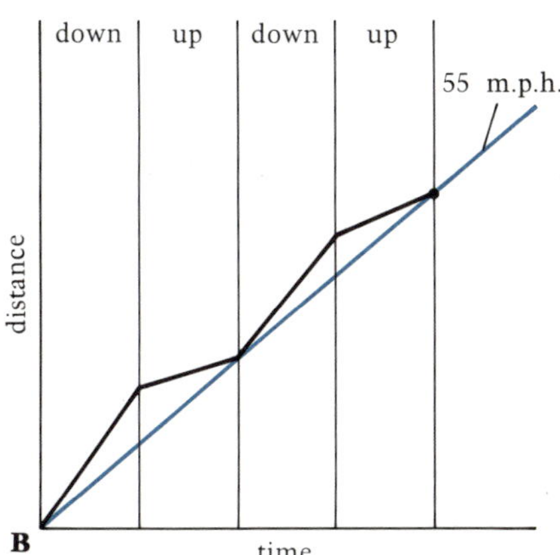

Fig. 2–17 Graphs of position as a function of time for trucks going up- and downhill. The solid line in each graph represents the actual position of the truck. Notice that the line is steeper, meaning that the truck is going faster, on the downhill sections. The blue lines represent average speeds. In A, the truck has a *maximum* speed (on the downhill sections) of 55 m.p.h. Its average speed is then somewhat less. In B, the truck *averages* 55 m.p.h. In order to do this, the speed on the downhill sections must be greater than 55.

Fig. 2–18 Trucks go slower than 55 m.p.h. going uphill, so they must go faster than 55 m.p.h. while going downhill if they are to average 55 m.p.h. for the trip.

2.2b ACCELERATION

There is a quantity called acceleration that tells how fast the velocity is *changing*. (We are now interested in changes in velocity, rather than changes in position.) The *average acceleration* over a time interval is found by noting the difference between the velocity at the beginning and the velocity at the end of the time interval and dividing this by the time interval:

$$\textit{average acceleration} = \frac{\textit{change in velocity}}{\textit{time interval}}.$$

As with average velocity, the average acceleration over an interval of time depends on the acceleration at the beginning and end of the interval. If you want more detailed information about the motion, you have to calculate for shorter and shorter time intervals. When the time interval is sufficiently short that the acceleration doesn't change by a significant amount, we say that we are then measuring the instantaneous acceleration.

If we know the acceleration, then we can compute the change in velocity in a given time. From the definition of acceleration we see that

$v = at$ $\quad$ *change in velocity* = *acceleration* × *time.*

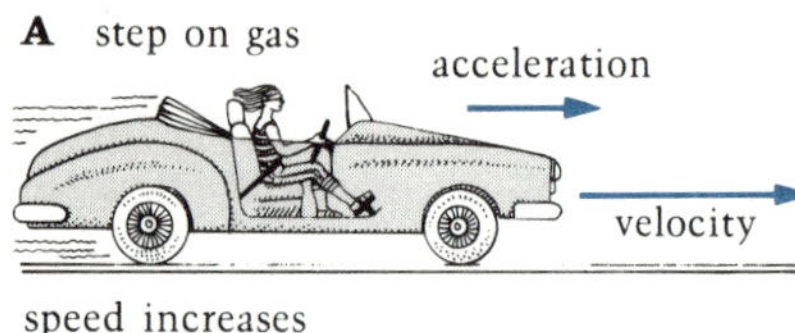

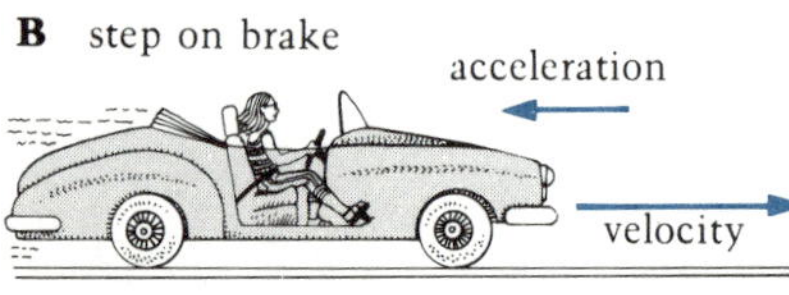

Fig. 2–19 A. When the driver steps on the gas, speeding up, acceleration is positive (forward). B. When the driver steps on the brake, slowing down, acceleration is negative (backward).

Acceleration is a vector, and its direction is the same as the direction in which change in velocity occurs. For example, for a car going in a straight line, if the speed is increasing, the acceleration will be in the same direction as the motion of the car. If the speed is decreasing (the car is slowing down), the acceleration is pointed opposite to the direction of the car's motion (see Fig. 2–19).

Even if a car's speed does not change, the car can still be accelerating. The velocity includes both the speed of the car and the direction of motion, so it is possible for the velocity to change just by changing the direction of motion—turning the car—even if the speed stays the same.

To see how we calculate acceleration, let's look at the simple case of a car accelerating from rest, so that after 10 seconds it is traveling at 30 m/s. What is the acceleration? The change in velocity is 30 m/s − 0 m/s = 30 m/s. Therefore the acceleration is (30 m/s)/(10 s) = 3 (m/s)/s. The units of acceleration may seem a little strange at first. You will usually see the above number written as 3 m/s/s or as 3 m/s². This comes from taking a velocity and dividing it by a time. However, the meaning is simply that for every second that the object travels, its speed increases by 3 m/s. For example, we know that after 10 seconds its speed is 30 m/s. What is its speed after another 10 seconds? In the second interval of 10 seconds, its speed will increase by another 30 m/s. The new speed is thus 30 m/s + 30 m/s = 60 m/s.

3 m/s/s is read "3 meters per second per second." 3 m/s² is read "3 meters per second squared." They mean the same thing.

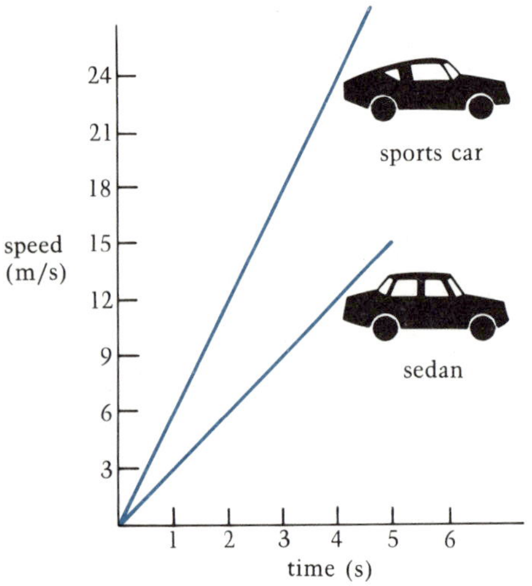

Fig. 2–20 Graph showing speed of sports car and sedan at different times. The steeper line (sports car) represents the greater acceleration.

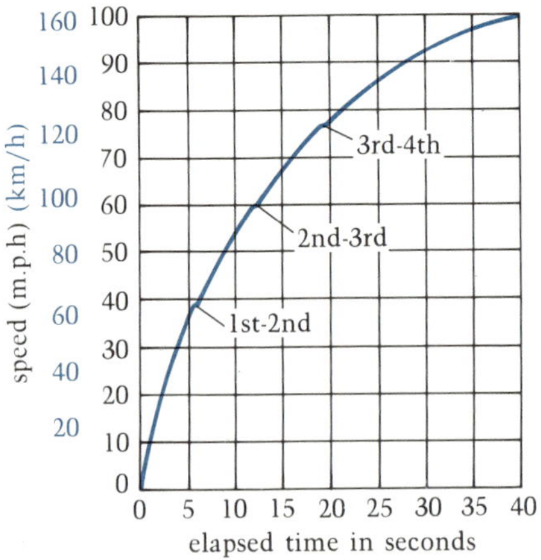

Fig. 2–21 An actual diagram for a Porsche—from *Road and Track* magazine. The graph shows how the car's speed increases over time (note the gear shifts).

Just as with the position of an object, we can mark the speed of an object on a graph. On the horizontal axis we plot the time at which we measure the speed, and on the vertical axis we plot the speed. Let's consider the example of the sedan and the sports car, starting at rest at the same time; the sedan has an acceleration of 3 m/s^2 and the sports car has an acceleration of 6 m/s^2. At any time, the speed of the sports car will be twice that of the sedan. We can see in Fig. 2–20 that the line giving the speed of the sports car is steeper than that giving the speed of the sedan. Thus, the steepness of the line gives us some feel for the amount of acceleration. (Remember, when we plotted distance of cars at different times, for cars of different speeds, the faster car had the steeper path.) Actually when real cars accelerate, the acceleration is not constant. Graphs of distance and speed vs. time for a real car are shown in Fig. 2–21.

2.3 *Falling Bodies*

We neglect air resistance.

Many of the features of motion with constant acceleration are illustrated by bodies falling near the surface of the earth. (In Chapter 7, we shall discuss the law of gravity.) Galileo first demonstrated that bodies of different masses, when dropped together from rest, will hit the ground at the same time. (The story has it that Galileo demonstrated this phenomenon by dropping such objects off the Leaning Tower of Pisa. There has been much debate about whether Galileo carried out this experiment; the current consensus is that he didn't.)

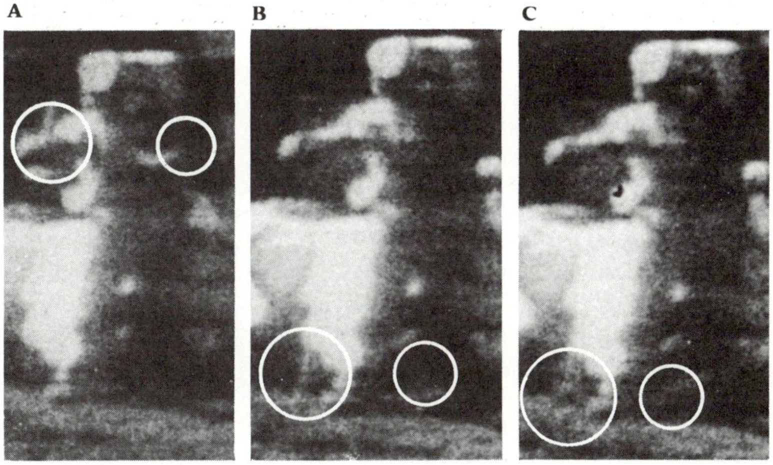

A B C

Fig. 2–22 An Apollo astronaut on the moon dropped a hammer and a feather simultaneously to demonstrate that in the absence of air resistance, both objects hit the ground at the same time. The hammer dropped by the astronaut's right hand and the feather dropped from his left hand are circled on these photographs of the television image that was relayed back to earth. The second and third pictures show adjacent frames.

TABLE 2–3 Time and Distance of a Ball Falling from Rest

Time	*Distance fallen*
0 s	0 m
1 s	5 m
2 s	20 m
3 s	45 m
4 s	80 m
5 s	125 m

Figure 2–23 shows a time-sequence photograph of a ball falling from rest. If we follow the motion for 5 seconds, we find that the distance fallen after each second is given by the data in Table 2–3. (These positions are also plotted in a graph in Fig. 2–24.) Notice that, as time goes on, the distance fallen in each additional second increases. This is not surprising since, if the ball is accelerating downward, the speed must be increasing as well.

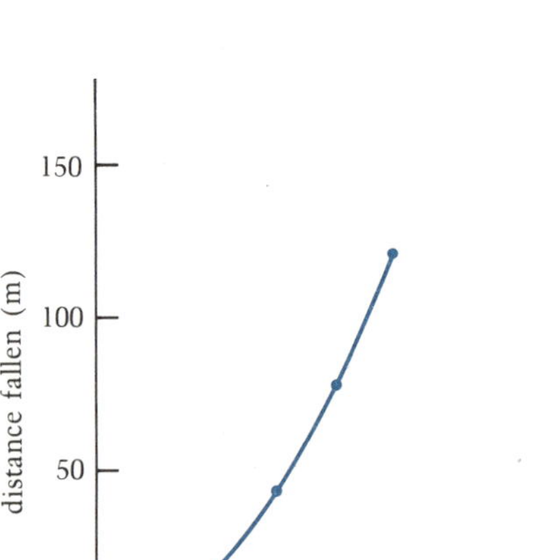

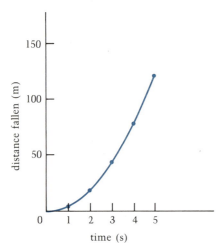

Fig. 2–24 Graph of a ball falling from rest (data in Table 2–3).

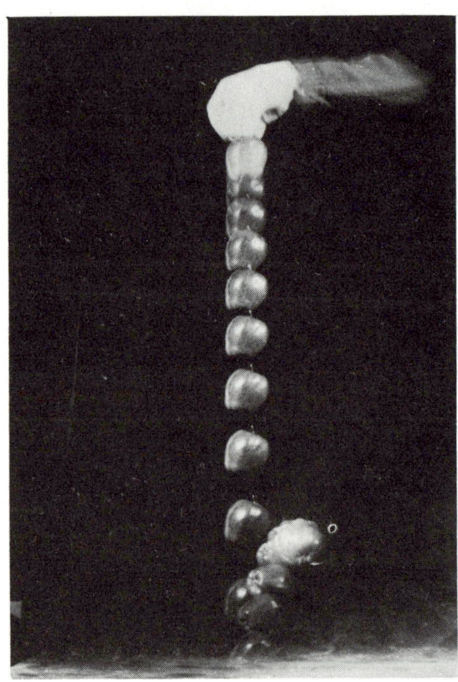

Fig. 2–23 A falling ball, here represented by an apple. The time intervals between exposures are equal, $\frac{1}{30}$ s.

GALILEO

Galileo Galilei was born at Pisa in 1564. When he was a seventeen-year-old freshman at the university, so the story goes, he noticed that a lamp swinging in the Pisa cathedral always took the same length of time to complete a swing, no matter how high it went. This fundamental discovery about pendulums marked the beginning of the scientific career of this Renaissance scientist.

Galileo later became a lecturer at the university, and began his research into the theory of motion. Most of his work was carried out when he was professor at Padua. During this time, he became a convert to the sun-centered theory of the universe that had been advanced by Copernicus, although he did not state this publicly.

When, in 1609, Galileo heard reports of the recent invention of the telescope, he quickly built several of improving quality, and was the first to turn the telescope to astronomical purposes. He discovered that Jupiter had moons around it, that the Milky Way was made up of individual stars, that the surface of the moon was pockmarked with craters, and that Venus went through a full series of phases. His series of observations also in-

Fig. 2–25 Galileo Galilei, 1564–1642.

cluded sunspots, which he studied for some time, although sunspots were also independently discovered by others at about the same time.

Galileo described his discoveries in popular works that he wrote in Italian, instead of Latin—the usual language of scholarly scientific works at that time—so as to be accessible to a wide readership. Because he held the relatively new and unorthodox view, first proposed by Copernicus, that the sun rather than the earth is at the center of the universe, he ran afoul of the Inquisition, the investigative arm of the Roman Catholic church, which was overwhelmingly concerned with enforcing religious orthodoxy. The details of his difficulties with the church are complex and have to do not so much with his teaching of the sun-centered theory as a method of calculating what goes on in the universe but rather with his holding that the theory was "true," in supposed contradiction to Scripture. Political and personal factors were also important.

In 1616, Galileo was warned by the Inquisition not to teach or hold to Copernicanism. He restrained himself for years, and after a while got permission to discuss the "world systems": sun-centered and earth-centered. When Galileo's work *Dialogue Concerning the Great World Systems* was published in 1632, however, it was clear that Galileo favored the theory of Copernicus and treated the opposing arguments—including an argument that had been personally advanced by the Pope himself—with scorn. This led to his prosecution and conviction by the Inquisition, although many feel that a forged document had been inserted into his file to provide false evidence against him.

Because of his age and fame, Galileo was put under house arrest instead of being sent to jail. He continued his scientific studies and published a major work on mechanics. With the telescope he discovered, among other things, that the moon wobbles a bit from side to side, allowing us to see part way around the back side. In 1637, Galileo—who had seen so far into the universe—went blind, but even so he continued writing. He died in 1642.

A

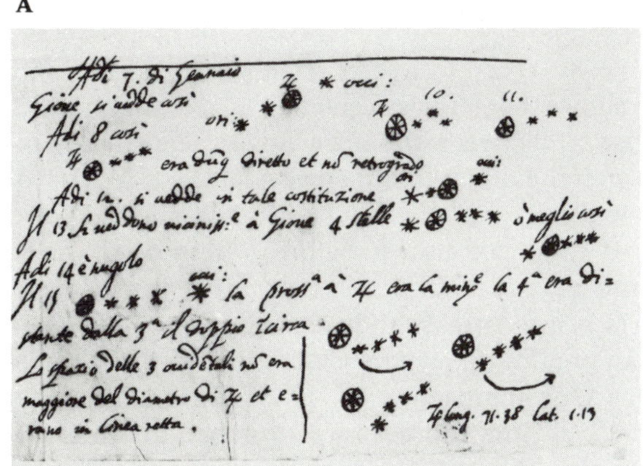

B

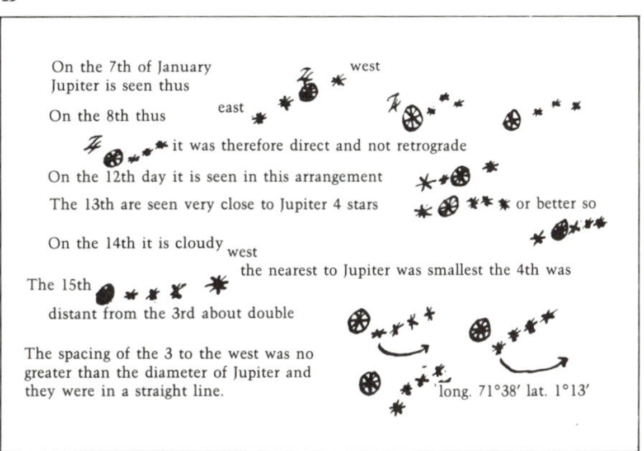

Fig. 2–26 A. Galileo's original drawing of the moons of Jupiter, from 1610. B. A translation.

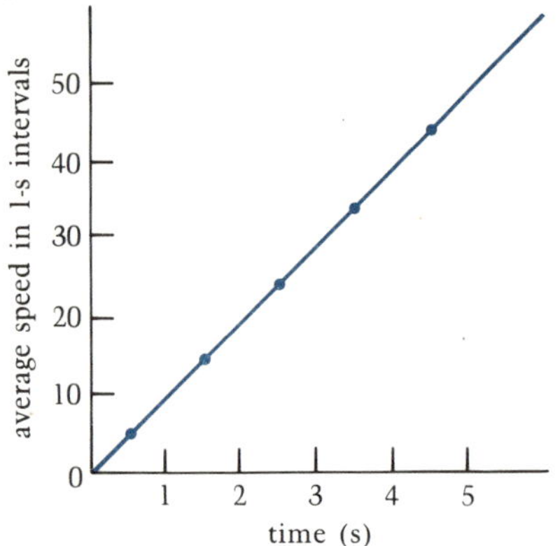

Fig. 2–27 In this graph, we plot the average speed in a 1-second time interval for a falling ball. The time intervals are centered at 0.5 s, 1.5 s, 2.5 s, and so on. Notice that from the time we drop the ball to the *middle* of the first interval only 0.5 s has elapsed, so the average speed increases by only half of its increase in the other intervals.

Let's calculate the average downward speed in each 1-second interval. It is simply the distance traveled divided by 1 second, the time interval. We summarize these results in Table 2–4. (These average speeds are also shown on the graph in Fig. 2–27.) The interesting thing to note is that in each full second, no matter which one we choose, the average speed has increased by 10 m/s. This means that the ball has an acceleration of 10 m/s/s. In the absence of air resistance all objects near the surface of the earth fall with the same acceleration. This quantity is called the acceleration of gravity. We usually designate this quantity by the symbol g. (The actual value of g on the earth's surface is 9.8 m/s/s, but for many calculations the approximate value of 10 m/s/s is sufficient.)

TABLE 2–4 Average Speed over Time of Ball Falling from Rest

Time interval	*Time at center of interval*	*Average downward speed during interval*
0 s to 1 s	0.5 s	5 m/s
1 s to 2 s	1.5 s	15 m/s
2 s to 3 s	2.5 s	25 m/s
3 s to 4 s	3.5 s	35 m/s
4 s to 5 s	4.5 s	45 m/s

Often the acceleration of an object is compared with g. We talk in terms of an acceleration of so many g's (pronounced "gees"). Since one g is an acceleration with which we are familiar, it provides a convenient reference.

How big is g? Is 9.8 m/s/s a large or small acceleration? The following example shows that bodies accelerate fairly quickly while falling. Hold a dollar bill between the open fingers of a friend, and tell your friend to grab the dollar bill once you let it go. (After you have tried this a few times, you may feel confident enough to offer friends the dollar if they can grab it before it gets below finger level.) You will find that the bill accelerates fast enough to fall beyond their fingers before your friends can react to its falling.

Let's see if we can predict the distance a ball will fall in a particular time, say, in 6 s. One approach to the problem is to look at the average speed of the ball from the time it is dropped to the end of any time interval. To do this, we divide the distance that the ball has fallen by the time it has been falling. The results are given in

Fig. 2–28 The engine of a car supplies a forward force greater than the force of friction. The acceleration of a racing car can be more than twice the acceleration of gravity.

TABLE 2–5 Average Speed over Distance of a Ball Falling from Rest

Time	*Distance fallen*	*Average speed from time of drop (distance/time)*
0 s	0 m	0 m/s
1 s	5 m	5 m/s
2 s	20 m	10 m/s
3 s	45 m	15 m/s
4 s	80 m	20 m/s
5 s	125 m	25 m/s
6 s	?	?

Table 2–5. This table shows that the average speed from the time of drop increases by 5 m/s every second.

We can use this table to predict the distance that a ball falls in 6 seconds. By looking at the table, we can say that the average speed for the first 6 seconds will be 30 m/s (25 m/s + 5 m/s). The distance fallen in 6 seconds is then

$$(\textit{average speed}) \times (\textit{time}), \textit{ or } (30\ \textit{m/s}) \times (6\ \textit{s}) = 180\ \textit{m}.$$

Notice that when we compute the average speed, it depends on the time during which the ball has been falling. If we double the time, we double the average speed, and so on. When we multiply *average speed* × *time*, we get a result that depends on the *square* of the time:

$$\textit{distance} = \textit{initial velocity} \times \textit{time} + \tfrac{1}{2}(\textit{acceleration} \times \textit{time}^2).$$

When we dropped the ball, it fell farther each second than in the previous second. Does our prediction agree with the results of the experiment? Let us make another table, in which we calculate the predicted distance (since the ball is falling from rest, initial velocity and time are both zero, so we can ignore this part of the equation). We can see from Table 2–6 that these numbers agree with the actual distances fallen.

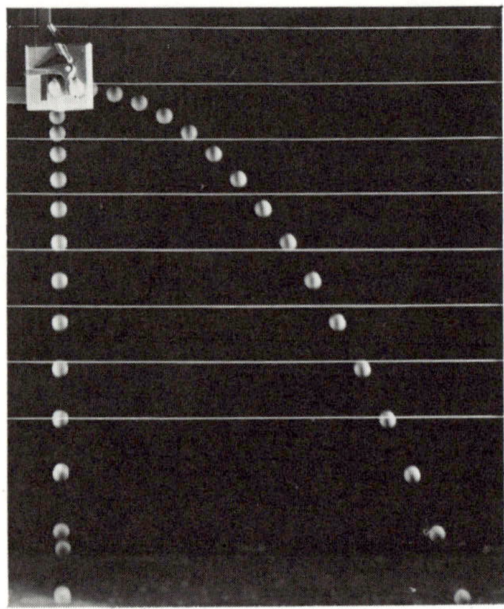

Fig. 2–29 Two golf balls were released simultaneously from the mechanism at the upper left and photographed at intervals of $\frac{1}{30}$ s. Notice that the balls drop at the same rate even though one of them was given a horizontal velocity.

TABLE 2–6 Distance over Time of a Ball Falling from Rest

Time	*Acceleration*	×	*Time*^{2}	÷ 2 =	*Distance*
0 s	(10 m/s/s)	×	$(0\ s)^2$	/2 =	0 m
1 s	(10 m/s/s)	×	$(1\ s)^2$	/2 =	5 m
2 s	(10 m/s/s)	×	$(2\ s)^2$	/2 =	20 m
3 s	(10 m/s/s)	×	$(3\ s)^2$	/2 =	45 m
4 s	(10 m/s/s)	×	$(4\ s)^2$	/2 =	80 m
5 s	(10 m/s/s)	×	$(5\ s)^2$	/2 =	125 m

2.4 *Projectile Motion*

We have already seen the behavior of a ball dropped from rest. What about a ball launched horizontally off a cliff? What about the path of any object, or *projectile*, launched near the surface of the earth? In Fig. 2–29 we see the results of simultaneously shooting a ball horizontally and dropping an identical one from rest. The positions of the two balls are shown after several equally spaced time intervals. We see two important features: (1) In each exposure, the first ball covers the same horizontal distance as in any other exposure—its horizontal motion is at a constant speed. (2) At any time, the distance fallen by each of the two balls is the same.

We assume that air resistance is negligible.

What can we conclude from this? The effect of gravity appears to increase the downward speed of each ball by 10 m/s every second, but gravity leaves the horizontal speed unchanged. *The horizontal and vertical motions are completely independent of each other!*

We are often told when driving to allow one car length of space between our car and the one in front for every 10 m.p.h. (10 m.p.h. = 4.5 m/s) of speed. This is supposed to allow for safe stops in emergency situations. If all cars were identical, and in a given situation *all brakes were applied at the same time,* the distances between the cars would remain fixed, since they would all travel the same distance while stopping. Thus, in this idealized situation it is not necessary to leave *any* space between cars. The problem is that all drivers do not apply the brakes at the same time. Each one waits to see, *and react to,* the tail-light signal from the car in front. Therefore, a small time (about 1 second) elapses between the time the car in front starts to slow down and the time the driver of the car behind reacts and steps on the brakes. During this time, a car going 4.5 m/s will travel a few meters. The gap between cars is to allow for this small distance, which is proportional to the speed of the cars. Automobile safety experts now tell us that the safest thing is to be 2 seconds behind the car in front of you. You can judge this 2-second interval by clocking the time it takes to pass some landmark.

What about the horizontal distance traveled before hitting the ground? We can first calculate how long it will take to reach the ground (assuming we know the height of the cliff). We then multiply this time of flight by the horizontal speed to get the horizontal distance. Being able to consider the vertical motion separately from the horizontal motion makes such a problem much easier to solve.

What about objects thrown up at some angle? Take, for example, a cannon on level ground. The cannon always shoots cannon balls out of the muzzle at the same speed. Suppose we want to try for maximum distance. We can try shooting horizontally; after all, this will give us the greatest horizontal speed. However, the ball drops to the ground very quickly and never gets a chance to go very far. We could keep the ball in the air a very long time by shooting it straight up. Unfortunately, it would then have *no* horizontal speed, so while it was in the air it would cover no horizontal distance. Clearly a compromise is necessary. We must shoot the ball with some vertical speed, so that it will stay in the air for a while, but we must also give it some horizontal speed, so it can cover some distance while it is in the air. It turns out that the compromise angle is 45°, giving equal amounts of vertical and horizontal speed. Figure 2–30 shows this best case and a few others that are not so good.

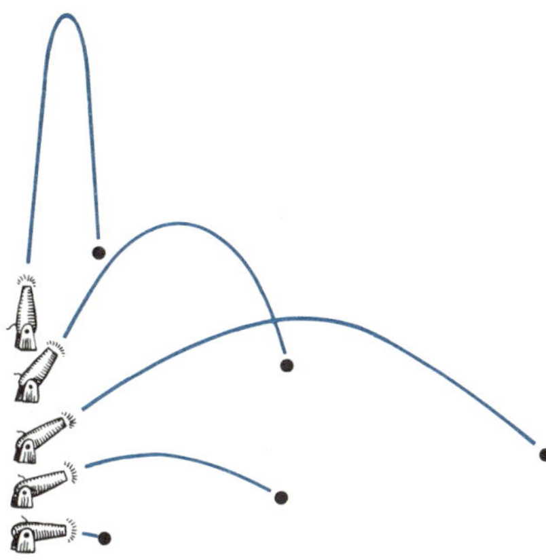

Fig. 2–30 Projectile motion. The range of a projectile depends on the firing angle. If you fire it close to the horizontal, it will have a high horizontal speed but won't stay in the air very long. If you fire it close to vertical, it will stay in the air a long time but won't have much horizontal speed. The compromise that gives the longest range is a 45° angle.

Fig. 2–31 In order to get maximum distance, this shot putter launches the shot at approximately a 45° angle from the ground.

Fig. 2–32.

Fig. 2–33 The rendezvous of Gemini VI and VII. The two spacecraft stayed very close to each other in orbit. For a long time, their relative speed was very close to zero, even though they were both moving around the earth at over 18,000 miles per hour!

2.5 *Relative Motion*

If you drop an object from the crow's nest of a moving ship, where will it land? Aristotle (who, as we mentioned in Chapter 1, was the Greek scientist from two thousand years ago whose ideas dominated science up to the sixteenth century) got this one wrong. He thought that the object would fall off the back of the boat when actually it will land at the base of the mast.

This is just one of many problems involving relative motion, the motion of one object as viewed from another. For example, you are driving along a highway at 80 km/h, overtaking a truck that is traveling at 50 km/h. At what rate are you overtaking the truck; that is, what is your speed *relative to the truck*? In this case, it is simply the difference between your speed and the truck's speed, or 30 km/h (Fig. 2–34).

What if you and the truck are traveling in opposite directions? Let us say you are traveling north and the truck is traveling south. To distinguish northward motion from southward motion, we can say arbitrarily that southward motion is in the minus direction. That is, the speed of the truck is -50 km/h while your speed is $+80$ km/h. To find the relative speed, we just take the difference between the two speeds and get $(80 \text{ km/h}) - (-50 \text{ km/h}) = (80 \text{ km/h}) + (50 \text{ km/h})$, since the negative of a negative number is a positive number. Thus, the relative speed is 130 km/h. It is not surprising that a head-on collision is more devastating than a head-to-tail one!

As another example, consider an airplane flying from San Fran-

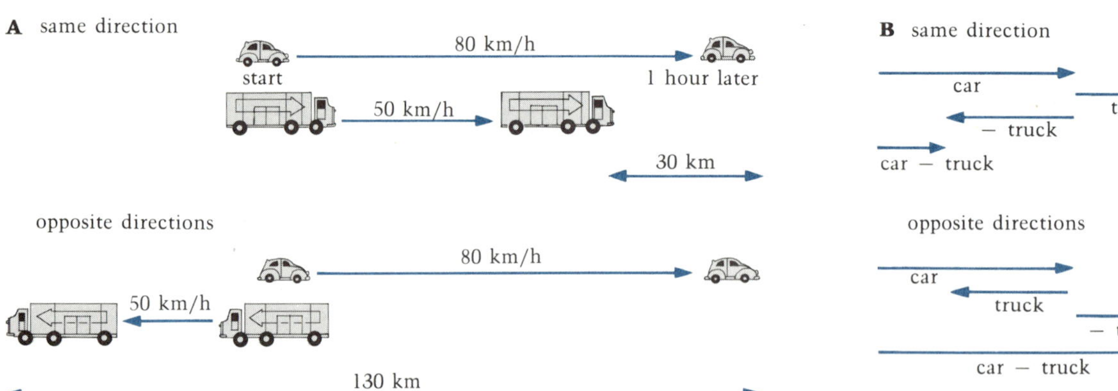

Fig. 2–34 The relative motion of a car and truck. In each case we are taking the difference between the velocity of the car and that of the truck. A. We can figure the relative velocity by seeing how far each would travel in an hour and by seeing how much the distance between them has changed. B. We see that we can get the same result by representing the velocity of each as a vector and taking (car velocity) − (truck velocity).

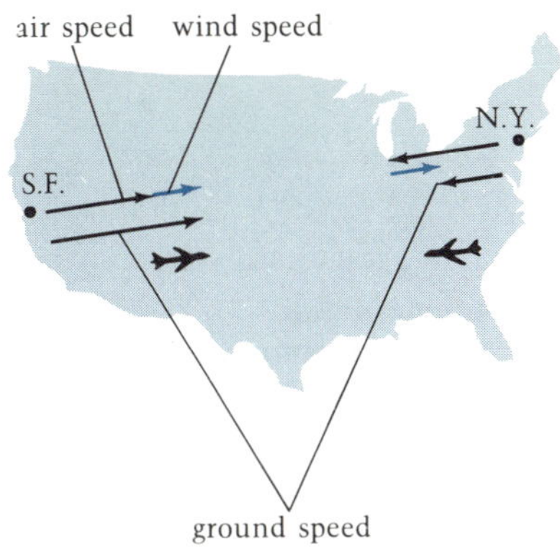

Fig. 2–35 Ground speed = air speed + wind speed (all vector quantities). The velocity of the airplane with respect to the ground is the vector sum of its velocity with respect to the air and the velocity of the air with respect to the ground. When the plane is flying with the wind, its speed with respect to the ground is increased over its air speed; when it is flying into the wind, the ground speed is less than the air speed.

cisco to New York. The design of the plane determines how fast it will fly relative to the air. This is called its *air speed.* Suppose the air speed is 800 km/h. Now assume that a steady wind is blowing from west to east, with a speed of 100 km/h. The time to get from one place to another on the ground depends on the speed of the plane relative to the ground, or its *ground speed.* The ground speed will be the air speed of the plane plus the wind speed. In such a circumstance, we say that the plane has a tailwind.

Suppose now we have an identical plane flying from New York to San Francisco. In this case the wind is in the direction opposite from the direction of flight of the plane. (This is called a headwind.) We must still add the two speeds together, but remember—they are in opposite directions. The ground speed is (800 km/h) − (100 km/h), or 700 km/h. Since the prevailing wind aloft is west to east, the time it takes to fly from New York to San Francisco is usually greater than the time to fly from San Francisco to New York (Fig. 2–35).

Fig. 2–36 A model of an aircraft in the test chamber of the 15-m × 30-m wind tunnel at NASA's Ames Research Center, California.

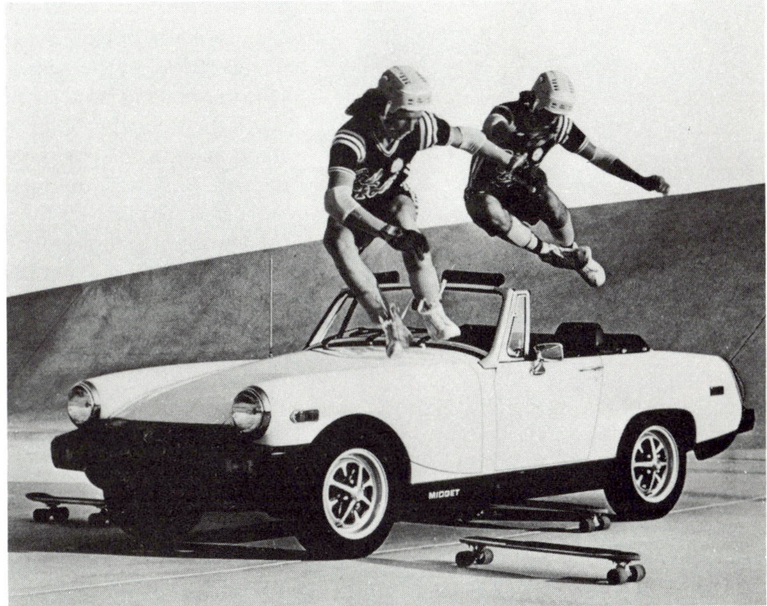

Fig. 2–37 The person on the right has jumped up from his skateboard. Since both he and the skateboard are traveling forward at the same speed, he will land on the skateboard. (The advertising agency that took this photograph to sell cars didn't like the spacing between the two people, so arranged for the second person to land on a new skateboard. To circumvent the laws of physics, they had to stop his original skateboard against the wheel of the car, as we see on the extreme left of the photo.)

Key Words

mechanics
kinematics
dynamics
translation
rotation
position
coordinate system
vector
magnitude
scalar
velocity
resultant
speed
acceleration
instantaneous acceleration
acceleration of gravity
g
relative motion

Questions

1. What is the difference between dynamics and kinematics?
2. Give two examples that differentiate between translation and rotation.
3. A train starts out on a trip on a straight track. In the first hour it travels 40 km. In the next half hour it travels 40 km. It then stops at a station for a half hour. It then travels 20 km in the next hour, and 20 km in the following half hour, finally reaching its destination. Draw a graph to show the position of the train at the end of each segment of the trip. (a) Use your graph to find the time when the train is traveling fastest. (b) Draw a line to indicate the average speed for the trip. (c) During which segments is it traveling slower than average? (d) During which segments is it traveling faster than average?
4. One car is traveling 40 km/h to the north. Another car is traveling 40 km/h to the south. (a) Are the velocities of the two cars equal? Explain. (b) Are the speeds of the two cars equal? Explain.
5. You throw a ball straight up in the air. It lands back in your hand 5 s later. What is the average velocity of the ball during its flight?
6. Draw an arrow 1 cm long, facing to the right. Let's assume that this represents the velocity of a car traveling 40 km/h to the east. Draw arrows to represent the following velocities (using the original arrow for scale): (a) 40 km/h to the north; (b) 20 km/h to the east; (c) a velocity whose components are 20 km/h to the west and 20 km/h to the south.
7. You are driving along a 2-km stretch of highway. For the first kilometer you travel at 30 km/h. You would like to travel fast enough during the second kilometer so that your average speed for the whole 2 km is 60 km/h. Can you do it? (*Hint:* Think in terms of the time it takes.)
8. If you are taking a trip in your car and want to know your instantaneous speed at any time, you can simply look at your speedometer. However, suppose you want to know your average speed for the whole trip. What information would you need?

9. The following table gives the position of a ball rolling down a ramp at various times:

Time (seconds)	1	2	3	4	5
Distance (meters)	0	1	4	9	16

What is the acceleration of the ball?

10. In Fig. 2–20, at which time does the car have the greatest acceleration?
11. Why was it interesting to watch the Apollo astronauts drop a feather and a hammer on the moon?
12. The acceleration of gravity on the moon is one-sixth that on the earth. Make a table to show how far an object would fall on the moon after 1, 2, 3, 4, 5, and 6 seconds.
13. We throw a ball off a cliff. At some time later, how does its horizontal speed depend on its vertical speed?
14. You are driving along in an open car. You throw a ball straight up in the air. While the ball is in the air, you step on the brakes. Where does the ball land relative to the car?
15. One train is traveling 40 km/h and another train is traveling 30 km/h. Is the relative speed of the two trains greatest when they are traveling in the same direction or in opposite directions?
16. Why does an airplane land flying against the wind?

CHAPTER 3

Laws of Motion

Fig. 3–1 Sir Isaac Newton.

If we were able to isolate a particle from all external influences, how would it behave? In other words, what is the "natural" state of motion of a particle? The study of dynamics—the relationship between force and motion—begins with the answer to this question. The question seems as much philosophical as physical in nature. Indeed, through the eighteenth century, the studies that we commonly call "physics" were referred to as "natural philosophy." The correct answer to the question of the "natural" state of motion did not appear until Sir Isaac Newton published his theory of motion in 1687.

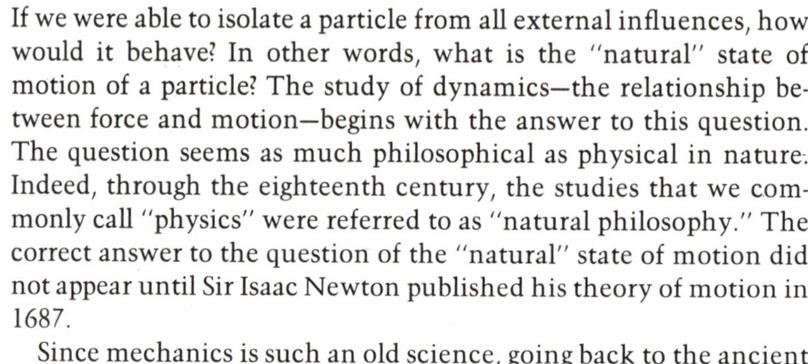

Since mechanics is such an old science, going back to the ancient Greeks, why did it take so long to come up with the right answer to such a fundamental question? The problem is that our everyday experience can be misleading. Think of what happens when you push a book along the surface of a table. The book slides at first, but shortly after you stop pushing, the book stops. If you want the book to keep going, you have to keep pushing. The Greeks interpreted similar observations as indicating that matter was naturally at rest. This is the view that prevailed until Newton's time.

What is wrong with this reasoning? After all, it seems to follow from careful observation of a variety of natural phenomena. The answer is that these phenomena do not give us the complete story. If you look back to the first sentence of this chapter, you will see that we want to know about the behavior of a particle *isolated from all external influences.* In the example we have discussed so far,

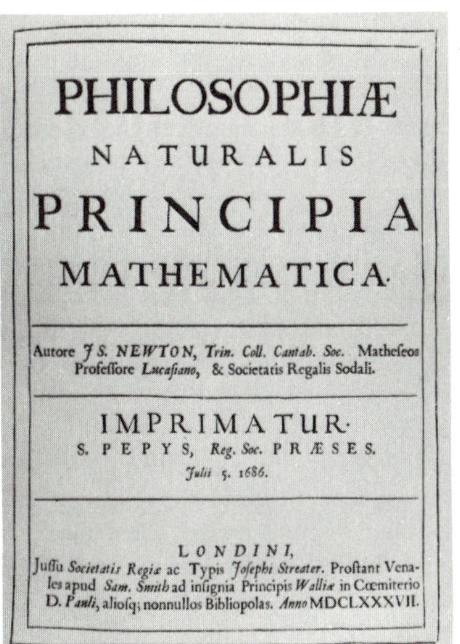

PHILOSOPHIÆ
NATURALIS
PRINCIPIA
MATHEMATICA.

Autore J S. NEWTON, *Trin. Coll. Cantab. Soc.* Matheſeos Profeſſore *Lucaſiano*, & Societatis Regalis Sodali.

IMPRIMATUR.
S. PEPYS, *Reg. Soc.* PRÆSES.
Julii 5. 1686.

LONDINI,
Juſſu *Societatis Regiæ* ac Typis *Joſephi Streater*. Proſtant Venales apud *Sam. Smith* ad inſignia Principis *Walliæ* in Cœmiterio D. *Pauli*, alioſq; nonnullos Bibliopolas. *Anno* MDCLXXXVII.

Fig. 3–2 Cover page of Newton's *Principia.*

the tendency to come to rest is being imposed by some influence that is less obvious than our pushing. In the case of the book sliding along the table, it is the force exerted on the book by the table, a force we call friction, that slows the book down and finally stops it.

Our common experience is not misleading in the sense of our seeing things that did not really take place—hallucinations, optical illusions, and so on. Rather, we have been misled by our failure to analyze properly what we have seen. We have not looked into all things that could possibly have influenced our experiment.

This mistake says something about experimental physics, whether ancient or modern. There is more to an experiment than putting some equipment together, throwing a switch, and reading a meter. The good experimentalist must be able to analyze an experiment to see if it is answering the question that is being asked. In a good experiment, outside influences must be eliminated as far as possible. If certain influences cannot be eliminated, their effect on the final result must be determined. An experimentalist may make the basic measurement for which an experiment was designed in only a few hours, but then have to spend months accounting for sources of error and uncertainty.

3.1 *Newton's Laws*

Newton's laws of motion were presented in 1687 in his book *Philosophiae Naturalis Principia Mathematica,* which is usually referred to as the *Principia* (prin-ki′-pē-a). These laws were actually a culmination not only of Newton's work but also that of some of his predecessors, most notably Galileo.

Fig. 3–3 Newton's home in Woolsthorpe, Lincolnshire, England.

ISAAC NEWTON

Isaac Newton was born in the rural town of Woolsthorpe, Lincolnshire, England, on Christmas Day, 1642. Galileo had died about a year earlier, and Newton was to pick up and extend Galileo's ideas about motion. Even as a boy, Newton was a scholarly type; the story has it that he once came home leading only a bridle. The horse had slipped out of it long before while Newton was lost in thought.

Newton attended the University of Cambridge, where he learned basic science and mathematics. A notebook he kept during some of his Cambridge years revealed that he was already starting to think about the fundamentals of physical phenomena.

When Newton was twenty-three years old, the university was closed because of the plague, and Newton went home to Woolsthorpe. During the next two years, in quiet and solitude, Newton developed many crucial ideas, including the basis of calculus, the foundations of optics,

Between about 1601 and 1619 Johannes Kepler had worked out three basic laws that described the orbits of the planets. His first law says that the paths of the planets around the sun are ellipses. The other two laws describe the speeds of the planets in their orbits. Kepler discovered his laws from extensive analysis of observations without any theoretical analysis. Newton later provided the theoretical basis from which Kepler's laws could be derived, and then derived the laws themselves.

and the inverse-square-law nature of gravity. Although Newton kept these ideas to himself, his reputation had grown, and soon after his return to Cambridge he was appointed professor.

In 1684, Edmond Halley—whose name we identify with the bright comet—visited Newton and learned that Newton had already derived basic parts of Kepler's laws on theoretical grounds. Halley encouraged Newton to publish his ideas, and Newton produced first a short publication and then, in 1687, his masterpiece. *Philosophiae Naturalis Principia Mathematica—The Mathematical Principles of Natural Philosophy*—is one of the great books of all time. It contains Newton's three laws of motion, and the law of universal gravitation. The importance of this work was immediately and widely recognized.

"I do not know what I may appear to the world; but to myself I seem to have been only like a boy, playing on the sea-shore, and diverting myself in now and then finding a smoother pebble or a prettier shell than ordinary, whilst the great ocean of truth lay all undiscovered before me."

Newton became involved in university politics and discovered the excitement of life in London. In 1696 he was appointed to the political position of Warden of the Mint and moved to the city. His research activity ended, although he became President of the Royal Society and ran it with a heavy hand. Newton spent much of his time arguing with the German mathematician Leibnitz over which of them had invented calculus; nowadays we credit them both with independent invention. Newton carried on many other battles of this type over the years, and it is sad to think that such a great man had so troubled a life. He died in 1727 with his reputation secure in history. In an uncharacteristically modest moment, he personalized an old saying to read: "If I have seen further it is by standing on ye sholders of Giants."

3.1a FIRST LAW—INERTIA

Newton's first law deals with motion when there are no external influences: If no force acts on a body, then it will remain at rest or continue to move with a constant velocity.

The phrase "constant velocity" means that neither the speed nor the direction of motion can change in the absence of a force. Remember that a particle at rest is just a special example of a constant velocity. The speed is zero (and its direction is not specified). If no forces are applied, its speed will stay zero.

We have already seen that if the velocity of a particle is constant, then the acceleration is zero. We can think of Newton's first law as saying that *in the absence of forces, there is zero acceleration.* Thus, the role of "force" has been changed from the earlier, incorrect notion. Instead of a force being necessary to *maintain* a certain velocity, a force is necessary only to *change* the velocity.

A "net force" is the force left over after as much as possible of all the forces present have balanced out.

Since an undisturbed object will move at a constant velocity, it is philosophically pleasing to regard the resistance to changes in velocity as an inherent property of matter. This property of matter is given the name inertia, and Newton's first law is often referred to as the *principle of inertia.*

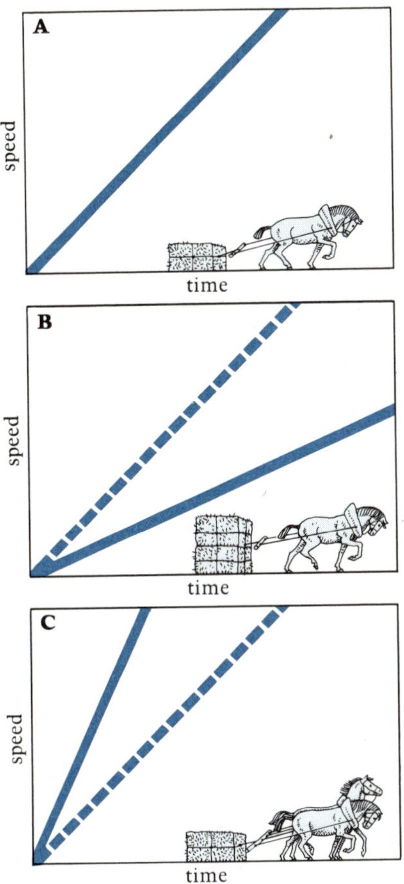

Fig. 3–5 A. Accelerating a block with one unit of force. The graph shows the speed at each time. The straight line means that there is a constant acceleration. B. Using the same force as in A, but pulling two identical blocks. The speed at each time is shown with the solid line in the graph. The blue line on the graph is the speed for the single block pulled by a single unit of force. Notice that at any time, the two blocks have acquired only half the speed of the single block. The two blocks have half the acceleration. C. Pulling one block with two units of force. The speed is shown by the solid line, and the single block with one unit of force is shown with the blue line. Now the acceleration is twice as great as for the single block with one unit of force.

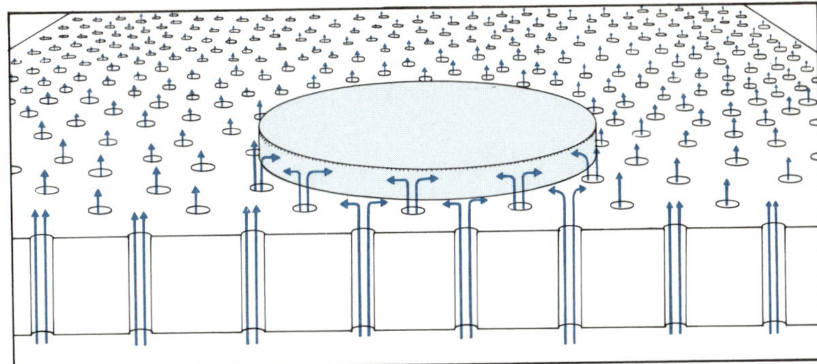

Fig. 3–4 In air tables such as the ones used in air-hockey games, jets of air are forced up through many small holes in the table. These jets exert an upward force on an object placed on the table, which reduces the contact force between the object and the table. A puck made of dry ice works in a similar manner. Since the dry ice sublimates (turns directly from a solid to a gas without becoming a liquid), gaseous carbon dioxide is forced out tiny holes in the bottom of the puck, which reduces the contact force between the puck and the table. Air tables and dry-ice pucks are often used in physics demonstrations when it is important to have very little friction (for example, when demonstrating that $F = ma$).

3.1b SECOND LAW—MASS

Newton's first law tells us that force and acceleration are related. His second law is concerned with how much acceleration is produced by a given force.

Let's look at a simple experiment to investigate this question. Suppose that we have a collection of identical blocks. We place one block on a frictionless table. (We can greatly reduce friction between the block and a real table, for example, by blowing jets of air through small holes in the table, thus allowing the block to sit on the air forced through the jets rather than on the table surface; see Fig. 3–4.) We pull the block horizontally with some fixed force, which we will call one unit of force. We can then graph the speed of the block at various times. The results are shown in Fig. 3–5A. The points lie along a straight line, which, as we have already seen, is what we get when we have motion with a constant acceleration.

Now we place a second block on top of the first and repeat the experiment, exerting the same force as before (Fig. 3–5B). This time we find that the speed after any time interval is only half of what it was in the case of a single block. In other words, the acceleration is only half as great.

What have we changed by adding the second block? We have changed the total amount of material present. The quantity that tells us how much material an object has is its mass. Doubling the mass has the effect of cutting the acceleration in half. We could repeat this experiment using any two objects of identical mass and we would find that as we increased the number of blocks (the mass), the acceleration would decrease. In fact, if we apply the same force to two bodies and measure the two accelerations, then

$$\frac{a_1}{a_2} = \frac{m_2}{m_1}$$

$$\frac{\text{acceleration of object 1}}{\text{acceleration of object 2}} = \frac{\text{mass of object 2}}{\text{mass of object 1}}.$$

If we exert the same force on an object of higher mass, that object will have a lower acceleration.

Since the relative amount of acceleration is proportional to the inverse of—"one over"—the relative mass, we say that the acceleration is "inversely proportional" to the mass.

Now let's go back to a single block. This time we pull it with two units of force (Fig. 3–5C). This means that we have now doubled the force on the block. In this case we find that, at any time, the speed is twice as great as in the case of one block and one unit of force. That is, doubling the force has doubled the acceleration. We can conclude that for any given object

$$\frac{a_1}{a_2} = \frac{f_1}{f_2}$$

$$\frac{\text{acceleration produced by force 1}}{\text{acceleration produced by force 2}} = \frac{\text{force 1}}{\text{force 2}}.$$

These two relationships are summarized in *Newton's second law,* as it is usually stated:

If F *stands for force,* m *for mass, and* a *for acceleration, then* $F = ma$.

Force = mass × acceleration.

The acceleration is proportional to the force. Sometimes, in order to be crystal-clear, we add a word of clarification and say that the acceleration is "directly proportional" to the force.

Both force and acceleration have directions associated with them (that is, they are vectors). Newton's second law not only tells us how the sizes of the force and acceleration are related but also tells us that the acceleration produced by some force is in the same direction as the force itself. If an object starts at rest and we exert some force on it, then the object will start moving in the same direction as the force. When you hit a golf ball with a driver, you

A

B

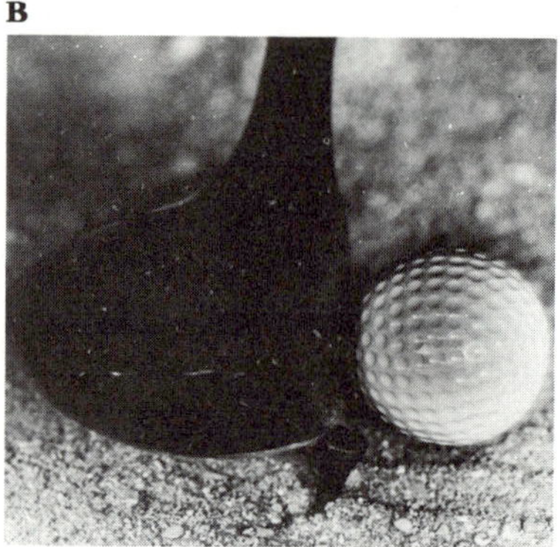

Fig. 3–6 The action of a golf club on a golf ball is equal to the reaction of the golf ball on the golf club, following Newton's third law.

Since we measure mass in kilograms (kg) and acceleration in meters per second per second (m/s^2), this tells us that we should measure forces in kg-m/s^2. Rather than write all that out, the name given to that quantity is, appropriately enough, one newton *(abbreviated* N*).*

would be surprised if the ball went backward instead of forward. If the object is already moving when the force is exerted, any change in velocity that results from an applied force must be added (as a vector) to the original velocity.

What happens if there is more than one force acting on an object? We add up all the (vector) forces to find the total force. The total is called the *net force* on the object. This method should not be too surprising. You know, for example, that to get a car out of a snowdrift it is useful to have as many people pushing as possible, and that you get the maximum force when everyone pushes in the same direction (Fig. 3–7).

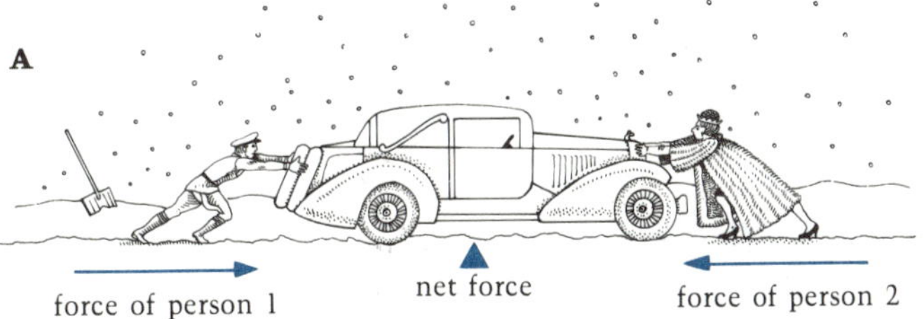

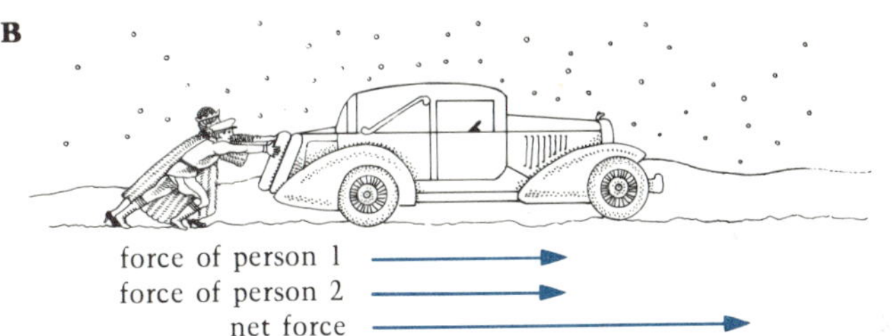

Fig. 3–7 The net force on a car in the snow is the vector sum of all the forces on it. A. If the two people push in opposite directions, then the net force is zero and the car will not move. B. If the two people push in the same direction, then the net force is larger than the force exerted by either one of them.

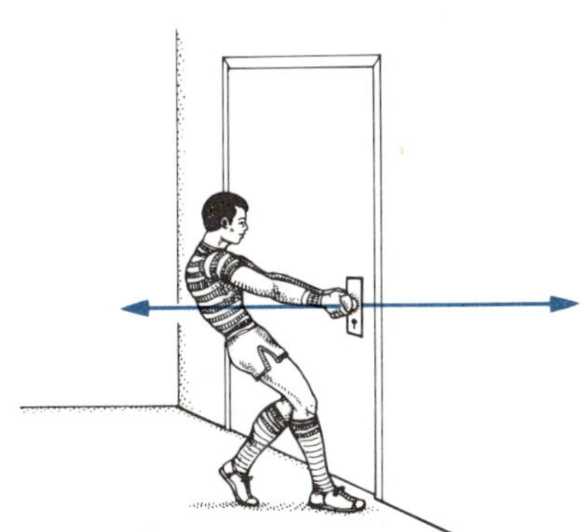

Fig. 3–8 As you lean back, exerting a force on the doorknob, it exerts an equal and opposite force in you, keeping you from falling.

3.1c THIRD LAW—ACTION/REACTION

We have been talking about how objects react to forces. In reality, any force exerted on an object must be exerted by another object. If a bat exerts a force on a ball, Newton's second law tells us what happens to the ball, but what happens to the bat?

You can get a "feel" for what happens by carrying out an experiment (Fig. 3–8). Close a door so that it is latched. Pull on the doorknob and, keeping a firm hold on it, lean back as far as you can. Clearly you are exerting a force on the knob. You can probably sense that if someone were to unlatch the door on the other side, it would swing toward you. On the other hand, *the doorknob is exerting a force on you.* After all, if you were to let go of the

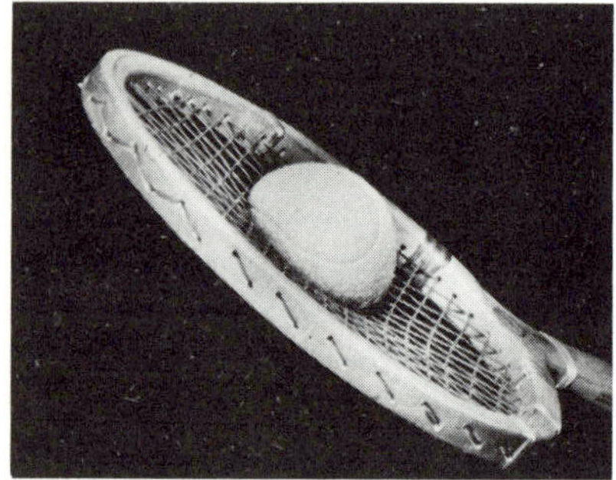

Fig. 3–9 The force that the ball exerts on the racquet is equal in strength, but opposite in direction, to the force that the racquet exerts on the ball.

doorknob, you would fall down. Therefore it must be the force that the doorknob is exerting on you that is keeping you up. This little experiment provides an example of Newton's third law: If object 1 exerts a force on object 2, then object 2 will exert a force of the *same strength*, but in the *opposite direction*, on object 1.

Newton referred to the force exerted by object 1 on object 2 as the *action*, and the force that object 2 exerts on object 1 the *reaction*. You may often hear Newton's third law stated as "For every action, there is an equal and opposite reaction." In any situation it is not necessary to know which force "came first." You just have to know that both are there. You must also remember that each force of the pair acts on a different object. For example, in the case of the bat hitting the ball, the force exerted by the bat acts on the ball, and the force exerted by the ball acts on the bat (see Fig. 3–11).

If you are on ice skates with both your feet facing straight ahead and you throw a snowball, you will tend to move backward because, as you exert a forward force on the ball, the ball exerts a backward force on you. A rocket works on the same principle. The engines of the rocket force hot gases out the rear of the rocket. In reaction to this, the gases exert a force on the rocket in the forward direction. Try blowing up a balloon and, without tying up the end, letting it go. You will see Newton's third law in operation.

There are examples of Newton's third law all around you. See how many you can recognize. (If you're sitting down while reading this book, you can start by explaining the force on the seat of your pants.)

Fig. 3–10 Newton's third law, as the Los Angeles Rams play the San Francisco 49ers.

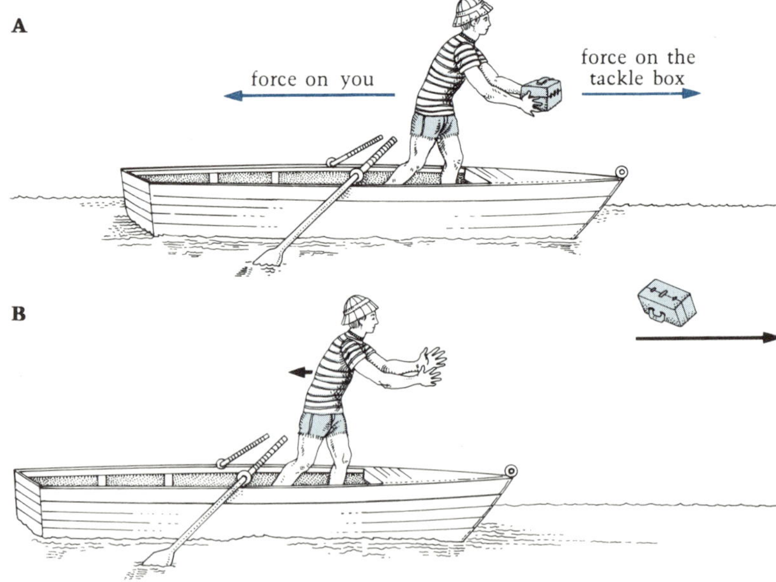

Fig. 3–11 A. What happens when you throw a tackle box from a rowboat. B. Because of Newton's third law, we see that the tackle box and the rowboat end up going in opposite directions.

Fig. 3–12 Launch of the Saturn V rocket that started the Apollo 17 mission on the way to the moon.

3.2 Touching

All around us we see things exerting forces on other things by "touching" them. We know forces are being exerted because we can see the accelerations that result. How do such "touching" or "contact" forces act? When objects seem to touch, there is no real "contact." The objects just get so close that an electric repulsion keeps them from coming any closer. By "repulsion," we mean a force that tends to push two objects apart. The details of this process need not concern us, but when you "feel" something, you are really feeling the pressure from the electric repulsion. (We will discuss electrical forces in greater detail later in the book.)

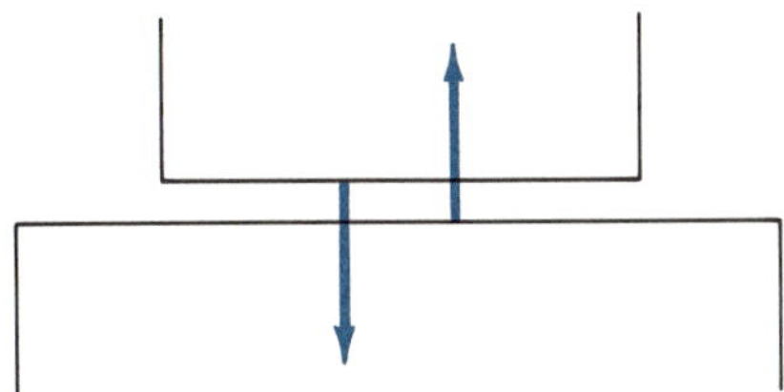

Fig. 3–13 Forces perpendicular to the contact surface. The arrows indicate the forces exerted *by* the objects. Notice that even though there are two forces shown, each object exerts only one force and each object has only one force exerted on it.

Even though we know that these forces are electrical, it is often convenient to ignore the details and just talk in terms of a contact force. When objects are brought close together, each exerts a force on the other (equal in strength and opposite in direction). The direction of this force is perpendicular to the surface of the object (Fig. 3–13). If you place a book on a table, it will exert a force perpendicular to the surface in which the book and table are in "contact." This will be true whether or not the table is flat or tilted at some angle. In turn, the table will exert a force on the book, of the same strength but in the opposite direction.

The harder you press sandpaper to a surface, the harder it is to move the paper along the surface.

As we have already seen, friction is another type of force that we encounter when two bodies are in contact. This is also just an electrical force between the atoms in the two bodies. Friction arises from the fact that the two objects are not perfectly smooth. On a microscopic scale, the surfaces can be rough, and jagged bumps on one object rub against those on the other (Fig. 3–14). The process is a very complicated one in which these little bumps can even stick together, which creates a force parallel to the boundary between the two surfaces. In most cases the friction force is related to the contact force that the objects exert on each other; as the contact force increases, the jagged edges are driven closer together, which increases the friction.

If we want to reduce the friction between two surfaces, we insert a lubricant, that is, a substance that has a much smoother surface, or possibly one that will fill in the rough spots in the contacting surfaces.

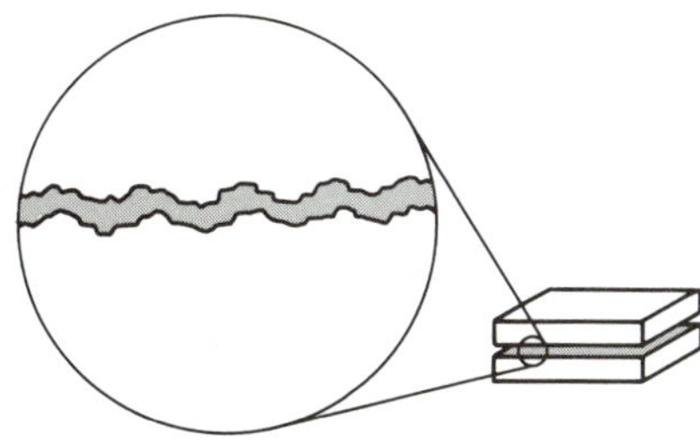

Fig. 3–14 Close-up of two surfaces in contact.

3.3 Statics

Many of the situations that we encounter every day involve motion with no acceleration. However, this does not mean that no forces are acting. It just means that all the forces exactly balance

Fig. 3–15 Vasily Alexeyev, the world champion Soviet weightlifter.

out. When this happens we say that the object is in *static equilibrium*. So the study of statics is based on Newton's first and third laws of motion.

To analyze static situations, we must identify all forces acting on each body, bearing in mind that *the (vector) sum of all the forces acting on each body is zero*. It is important that we consider each body separately and consider only the forces acting on that body.

Let's start with the simple case of a crate sitting on the floor (Fig. 3–16). What forces act on the crate? First there is the gravitational force exerted by the earth—the weight of the crate—which acts downward. The floor is supporting the crate and must therefore be exerting an upward force. There are only two forces acting on the crate—gravity acting downward and the support of the floor acting upward. For the crate not to accelerate, the upward force must exactly balance the downward force. The strength of the upward force is thus equal to the weight of the crate.

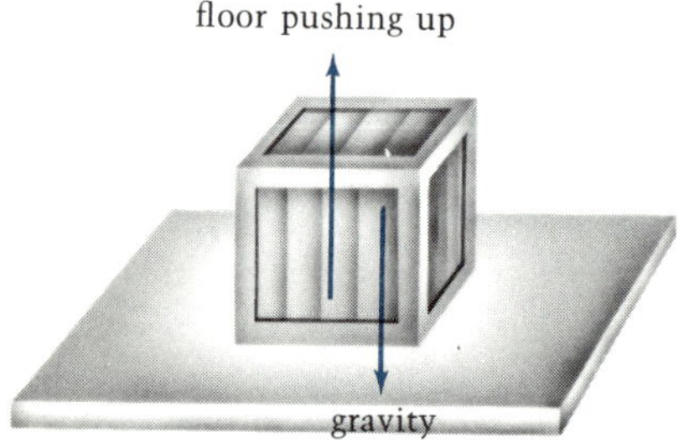

Fig. 3–16 The arrows indicate only the forces acting *on the crate*. The force of gravity shows up as the weight of the crate. The floor provides a reaction force up because the crate exerts a downward force on the floor. Newton's third law tells us that the floor must push up on the crate with a force of the same strength as the force with which the crate pushes down on the floor.

Now let's say that you are trying to push the crate across the floor. You push, exerting a horizontal force, but you can't get it to move. Why not? Your first reaction may be to say that it is too heavy. However, strictly speaking, this is not the case. It is really the static friction that keeps the crate from moving. Of course, the weight of the crate does come into play since the friction depends on the contact force between the floor and the crate, and the heavier the crate the stronger the contact force. But weight is not all that counts. The crate would move more easily if it were on wheels (since wheels eliminate the sliding friction).

The forces on the crate and the forces on you are shown in Fig. 3–17. Notice that every force on the crate has one that is equal and opposite so the crate does not accelerate. Also, every force on you has one that is equal and opposite, so you do not accelerate.

If the crate doesn't move, you may try to push harder and harder. Eventually, your feet may slip out from under you and the crate still may not move. At this point, you could ask someone to help. Together you could probably get the crate moving along the floor at a constant speed, this time without your feet slipping because neither of you, working together, has to push as hard as you did working alone.

Now let's apply our discussion to a car going at a constant speed on a highway. What forces are acting on the car? After our experi-

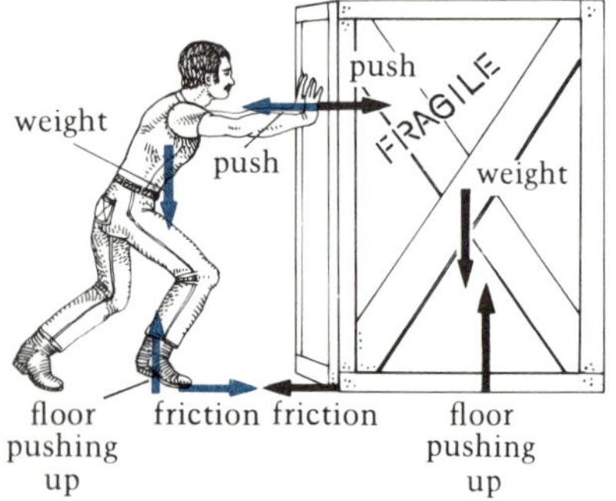

Fig. 3–17 The floor's pushing up on the crate is in reaction to the crate's pushing down on the floor. The floor's pushing up on the person is in reaction to the person's pushing down on the floor. The crate's pushing the person to the left is in reaction to the person's pushing the crate to the right. The force of friction that the floor exerts on the person's feet, to the right, is in reaction to the person's feet pushing back to the left. Blue arrows show forces acting on the man; black arrows show forces acting on the crate.

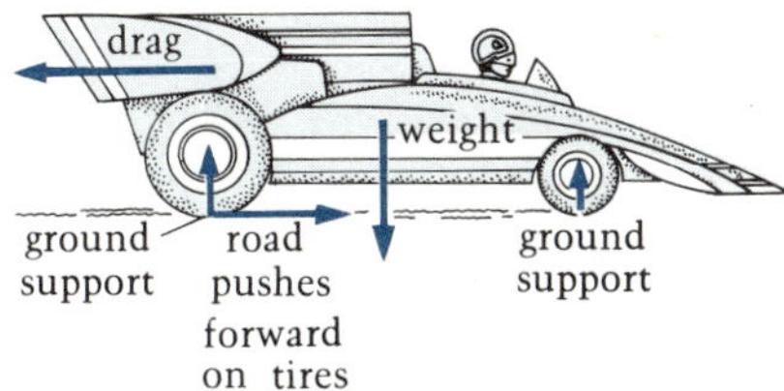

Fig. 3–18 The forces on a car, divided into their horizontal and vertical components.

ence with the crate, the up-down direction is now easy to recognize and understand (Fig. 3–18)—the weight of the car pushes downward and the road pushes upward. Also, the car is being pushed in the forward direction. How does this come about? The motor causes the wheels to turn and the wheels push backward on the road (a frictional force); the road therefore pushes forward on the tires. In this case we want as much friction as possible—we call it "good traction." What about forces in the backward direction? There is internal friction in the car among all the moving parts; this tends to slow the car down. There is also wind resistance ("drag"), which has the effect of friction.

If the car is moving at a constant speed, the push of the wheels must be balancing the internal friction and wind resistance. What happens if you step on the gas a little harder? The push will then exceed the wind resistance and there will be a net force in the forward direction—the car will accelerate. As it goes faster, the wind resistance (and internal friction) will increase. Eventually the speed of the car will have increased to the point where the wind resistance and friction again equal the forward push (although both will now be greater than before). The car will again travel at a constant speed. This crude analysis tells us why your car will not accelerate forever as you step on the gas. For any engine speed, there is a speed for the car at which the forces will be in balance and the acceleration will equal zero. It also explains why the engine must work harder (and so use more gas) to keep the car going at a constant high speed rather than at a constant low speed. (Internal engine friction also wastes gas when the car is going at a very low speed, so there is actually an optimal speed at which a car works most effectively.)

Fig. 3–19 Even when a rope is under tension, each piece of the rope "feels" no net force from the tension (and therefore does not accelerate). Let's look at one section of rope. Assume we are pulling it from the right. This section will then transmit the same force by pulling on the section on its left. However, by Newton's third law, the section on the left must exert an equal and opposite force on our section. Thus the section on the left pulls to the left. There is then a force pulling our section to the right and an equal force pulling it to the left. The result is no net force. Of course, even though there is no net force, the rope still "knows" it is under tension, just as you would know if people were pulling simultaneously on both your arms. Similar ropes under different amounts of tension behave differently. For example, if you pluck thin ropes, which we normally call strings, strings of different tension emit different notes.

3.3a ROPES AND STRINGS

Let's return to the problem of moving that crate across the floor. Suppose that instead of pushing it, you tie a rope to it and pull. Eventually, if you pull hard enough, the crate will start to move. How does the other end of the rope "know" that you are pulling at your end? Also, how does it know how hard you are pulling?

The answer is that each little section of the rope, on a microscopic scale, pulls on its neighboring section. Thus the force of the tug is transmitted from one end to the other (Fig. 3–19). Each piece of the rope knows that the rope is being pulled. This force that is transmitted down the rope is called the tension of the rope.

If you pull on a stationary rope, the tension in the rope equals the force with which you pull. This is true whether the other end of the rope is attached to a wall or to another person pulling in the opposite direction. Suppose the rope is attached to a wall and you

Fig. 3–20 In accordance with Newton's third law, if the weight on the right weighs the same as the person and swing on the left, the downward forces will balance each other. Bending the rope over a pulley does not change the forces being exerted.

Fig. 3–21 The Verrazano Narrows Bridge, gateway to New York City.

pull on the rope with a force of, say, 200 newtons (N). Thus, a rope with a 200-N tension must be pulled at each end with a force of 200 N. That pull can be supplied by the wall in reaction to the pull of the rope, or we can replace the wall with another person, or people, as in a tug-of-war. Their muscles would have to strain appropriately to exert that 200-N pull.

Ropes transmit their tension even if they are not in a straight line. For example, if you bend a rope over a pulley, you can pull down with a particular force on one side and have the rope pull up, with the same force on the other side, as in Figs. 3–20 and 3–22. (This assumes that there is no friction present.)

Fig. 3–22 Pulleys in use.

3.3b SPRINGS

A spring will transmit a force from one end to the other, but in doing so will change its visible appearance more than a rope will. The spring will either stretch or contract. The amount it stretches or contracts depends on the force exerted on it. As long as the force is not enough to distort the spring permanently, the amount by which the spring stretches or contracts is proportional to the force. We can write

$$f = k \times \Delta x$$

force = constant × change in length.

The Greek delta (Δ) stands for "the change in."

The constant is called the spring constant. A stiff spring will have a high spring constant, meaning that it takes a greater force to stretch or contract the spring a given amount. The spring constant depends on the material and construction of the spring.

Once you know the spring constant of a given spring, you can use that spring to measure forces. If you want to know how much something weighs, you can hang it from a spring and see how far the spring stretches. Most scales work this way.

These scales are "not legal for trade" because a spring can be easily stretched and rendered inaccurate, giving a reading larger than the correct weight.

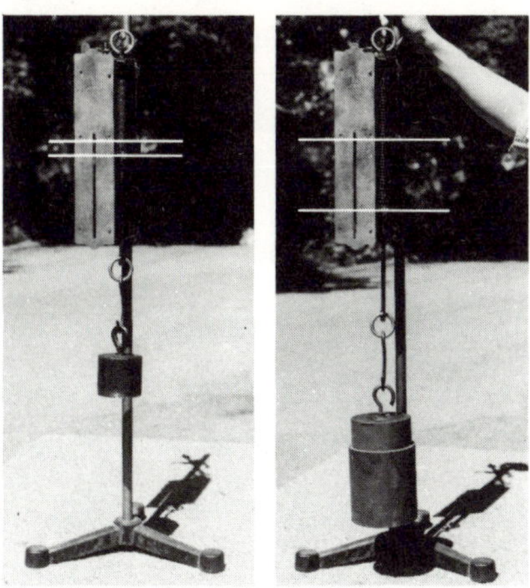

Fig. 3–23 The heavier the force on a spring, the more the spring stretches. Here one weight is five times the other, so the spring stretches five times as much. Note that it is the distance between coils that increases by this factor rather than the total length of the spring.

3.4 *Falling Bodies Explained*

In Chapter 2 we saw that, in the absence of air resistance, all objects near the surface of the earth fall with the same acceleration. (This acceleration, designated g, is 9.8 m/s^2.) The realization of this fact was an important step in the understanding of motion and gravity. Let's now briefly explain this result (we will discuss Newton's theory of gravitation more fully in Chapter 7).

Near the surface of the earth the gravitational force exerted on an object by the earth is simply the mass of the object times a constant; that is,

gravitational force = mass × constant.

Note that we have now encountered the mass of an object in two different roles. Earlier in the chapter we saw that mass is a measure of the ability of an object to resist being accelerated—it is a measure of inertia. We call it the inertial mass. Now we see that mass is also a measure of the degree to which the object can feel a gravitational force. The word "mass" in this context is a shorthand for gravitational mass. Some remarkably intricate experiments have shown that, to a very high degree of accuracy, the inertial and gravitational masses of an object are equal. It is important to remember that this is known as the result of an experiment and is not derived from any theory. There is no known reason why it *had* to be true.

Fig. 3–24 A spring scale.

This mass, m, cancels in the equation $ma = mg$, so $a = g$ for any mass.

What about the acceleration of an object? The fact that all bodies near the surface of the earth, no matter what their mass, have the same acceleration from gravity comes from the fact that the mass enters into the problem twice, once in computing the force of gravity and once in computing the acceleration that results from that force. A more massive object feels a greater gravitational force, but its greater mass means that it is more effective in resisting this force. So all falling bodies near the surface of the earth have the same acceleration (if we ignore the effects of air resistance) no matter what the mass of each body is.

$W = mg$, where W is the weight.

To make this consistent, the constant in the force equation must be equal to the acceleration of gravity, g. Thus, if you want to compute the gravitational force exerted by the earth on some body, just take the mass of the body and multiply it by g. We commonly call this force the weight of the object.

It is important to distinguish betweeen weight and mass. Mass is an inherent property of an object and does not change if you put the object somewhere else. Weight is just the gravitational force exerted on the object. When we colloquially speak of someone who "weighs" 70 kg on the earth, for example, we actually mean that the person has a mass of 70 kg, which corresponds to a weight on earth of 70 kg $\times$ 9.8 m/s^2 = 686 N. So if our 70-kg person steps on a metric bathroom scale, the scale feels a force of 686 N, but is calibrated to read 70 kg. This person would weigh only 255 N on Mars, where the acceleration of gravity is only 3.6 m/s^2, but would still have a mass of 70 kg.

In the United States, most people still use the "English system of measurement," in which we give the weight in pounds. The acceleration of gravity in the English system is 32 ft/s^2. Thus someone who weighs 154 pounds (which translates into 70 kg in the colloquial use of kilograms as a measure of weight) has a mass of (154 lb)/(32 ft/s^2) = 4.8 lbs^2/ft = 4.8 slugs, where the slug is a unit of mass (1 slug weighs 32 pounds). On Mars, the person would still have a mass of 4.8 slugs but would have a weight of 58 lb.

Actually, all bodies will fall with the same acceleration only when there is no air resistance. How would our observations be affected by air resistance? The important thing to note is that air resistance does not depend on the mass of the object; it only depends on the size and shape. If we have two objects of the same mass, but one large and the other small, such as an open sheet of paper and an identical sheet crumpled into a ball, the crumpled piece will fall faster since it suffers less air resistance.

3.5 *Accelerating Observers*

Suppose you jump out of a high-flying airplane while holding a ball in your hand. You attempt to describe the motion of the ball, relative to you (trying to keep your mind off your dangerous situation). As long as the ball is in your hand, it is at rest with respect to you. What if you let go? The ball will continue to accelerate downward at the same rate as you are (we are ignoring the effects of air resistance). If you released the ball at eye level, it would remain at eye level. You would conclude that the ball remained at rest.

Fig. 3–25 These skydivers are falling freely together, so they drop at the same speed.

So, you've released a ball and it doesn't appear to fall. Of course, if you look down, you will see the ground rapidly approaching, and you might say to yourself, "No wonder the ball doesn't appear to be falling. True, it is accelerating downward, but I am accelerating downward at the same rate." Although you may find your situation distressing, there is nothing strange in the physics.

Now suppose there were a windowless box placed around you

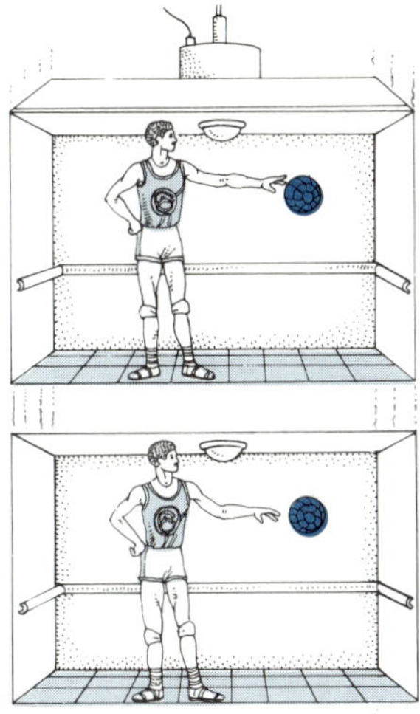

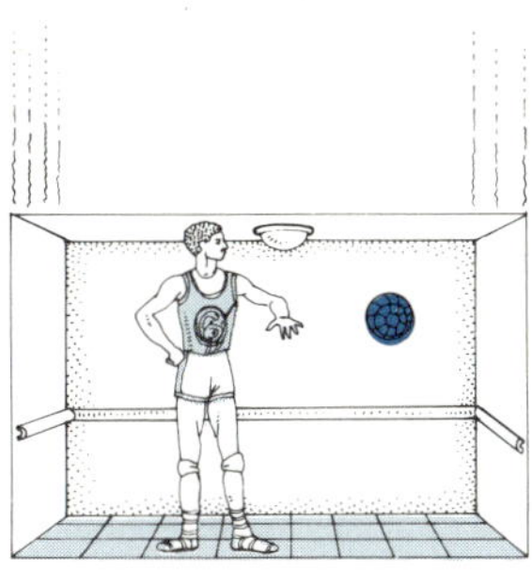

Fig. 3–26 What happens when you drop a ball in a box that is falling freely under the force of gravity.

Fig. 3–27 This NASA astronaut candidate is floating freely as the aircraft accelerates downward.

during all this. The ball, when released, would still remain at eye level. You would conclude that relative to you the ball had no acceleration. Unfortunately, now you would not be able to tell why it appeared to have no acceleration. There would be no visual reference to tell you that the ground was getting closer. If you were asked to explain why the ball doesn't appear to be falling, you might say that it's because you are falling at the same rate as the ball. However, you might also conclude that the ball isn't falling because there is no gravity. The frustrating thing about your situation is that apart from cutting a hole in the side of the box and peeking out to see the ground or some other external reference point—clouds, for example—or waiting for the box to hit the ground, *there is no experiment that you can do inside the box to tell you which of your explanations is correct* (Fig. 3–26).

In every way, your falling box would mimic a situation in which gravity had been turned off. If you suspend yourself 1 meter above the floor of this box, you will stay 1 meter above the floor! If you push yourself horizontally off one side wall, you will "float" straight across the box to the opposite wall. If you bring your feet to the floor, the floor is still accelerating away from your feet so that your feet will exert no force on the floor. If there is a scale on the floor, likewise, you will exert no force on the scale. Your weight will read zero. You will be experiencing weightlessness.

This is the same weightlessness experienced by astronauts orbiting the earth. The weightlessness does not arise from the fact that the spaceship is far from the earth's gravitational pull. It arises from the fact that the astronaut and the spaceship are in free fall together. They don't hit the earth because they are moving forward at the same time, and the earth curves away from them at the same rate at which they are falling.

This discussion tells us that we should be somewhat suspicious of observations made by observers who are accelerating. To see how suspicious we should be, let's look at a few more examples.

Take your ball onto a railroad car that is moving at a constant velocity. We have already seen that if you now drop the ball, it will land at your feet because it is moving along with you at the same horizontal speed as the train even though, because of gravity, it drops with increasing speed in the downward direction. As far as you are concerned, the ball has simply dropped at your feet, but an observer on the station platform would say that both you and the ball have kept pace with each other, moving along with the same horizontal speed.

Let's now say that the train is accelerating; its speed is increasing. Remember that an acceleration must be caused by a force. The car is accelerating because of the force exerted on it by the locomotive. You are accelerating along with the car because the floor of the car

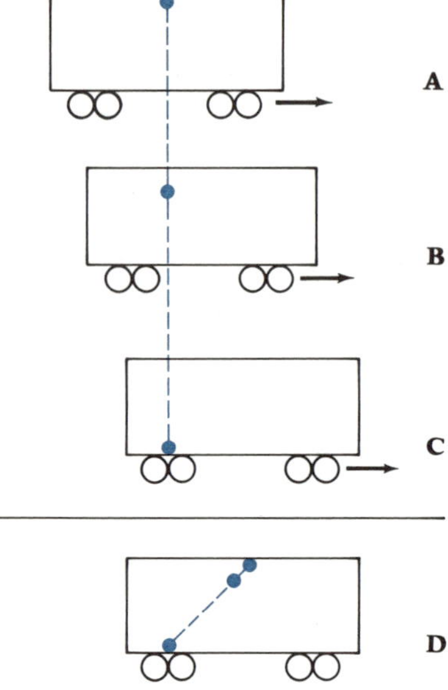

Fig. 3–28 A ball falling in an accelerating railroad car. A–C. As viewed from the railroad station. The figure shows the position of the ball and the car at three different times. The time interval between the A and B positions is the same as between the B and C positions. Notice that the ball falls farther in the second time interval than in the first, because it is accelerating. Notice, similarly, that the train travels farther in the second interval than the first, because it is also accelerating. As viewed from the station, the ball falls straight down, and the car accelerates forward. D. As viewed in the car. For the three pictures in A–C, we can see where the ball is in the car and draw these three positions into one picture of the car. Notice that the ball appears to slant backward. A person in the car might conclude that the car was going uphill.

is exerting a force on your feet. As long as the ball is in your hand, it will accelerate because your hand is exerting a forward force on the ball. As soon as you drop the ball, this force is removed. The car continues to accelerate forward and you continue to accelerate forward. But the ball no longer accelerates forward. Its horizontal speed remains whatever it was at the instant of release.

To an observer standing on the station platform, not accelerating with the car, the ball will appear to follow a trajectory, with a constant horizontal speed and increasing downward speed, just as before. However, the train, and you on it, accelerate away from the ball. The observer on the station platform sees the ball hit the floor of the car farther back than where you are standing (Fig. 3–28).

How does all this appear to you? You would just see the ball fall in a straight line. However, this line would slant backward, with the ball landing toward the rear of the car. If there were windows in the car, you could see the scenery rushing by faster and faster. This would tell you that the car was accelerating forward, allowing you to analyze the path of the ball properly. What if all the shades are down and you can't see out? You then have no way of discovering that the car is accelerating. You might just as well think that somehow a "horizontal gravity" had been turned on causing the backward motion of the ball.

You would even be able to "feel" this backward force. If you were sitting, you would feel pressed back in your seat. What you are really feeling is your seatback pushing against you, forcing you to accelerate forward.

Thus, we see that if we are accelerating, we may be fooled into thinking that there are forces present that are not in fact there. Such a force is called a pseudo-force. This is not really a force. It's just that, from our accelerated point of view, objects with no acceleration appear to have an acceleration; we are inclined to say that this acceleration must be caused by some force. *Pseudo-forces always appear to act in a direction opposite to our real acceleration. We are only seeing our acceleration mirrored in nonaccelerating objects.*

From this discussion you might guess that an accelerating railroad car might be a bad place to study the laws of motion. After all, we'd forever be confusing real and pseudo-forces. This is true, and we choose to study the laws of physics in laboratories that are not accelerating. We call such nonaccelerating systems *inertial systems.* (We shall say more about inertial systems when we discuss the theory of relativity in Chapter 8.)

One problem with pseudo-forces is that we sometimes talk about them as if they really exist, even when we know that they don't. This is usually done simply as a shorthand way of describing some situation.

A

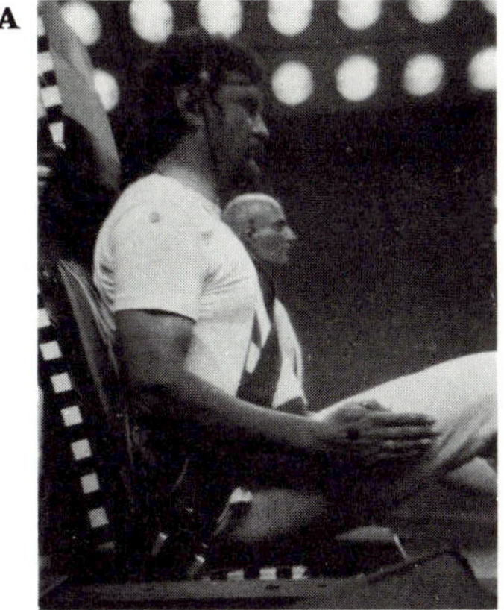

B

C

Fig. 3–29 When a car comes to a sudden stop, the seat stops but your body continues to move forward. If the seatbelt and shoulder harness didn't limit how far forward you could go, the windshield would.

Once we know about pseudo-forces, we can see how they crop up all around us. For example, suppose you are riding in a car at a constant speed and the driver suddenly steps on the brakes. You are "thrown" forward in the car. What force caused this? Actually, your motion has not changed. You are still moving forward with the speed that the car had before the driver hit the brakes. So you are actually sensing a pseudo-force. The force of the brakes has slowed the car down, but there is nothing to slow you down, so the dashboard is really accelerating toward you, rather than vice versa. (You can avoid being hit by the dashboard by wearing a seatbelt and a shoulder harness, which keep you tied to the car so that you decelerate along with the car.)

Similarly, the crushing "g-forces" felt by astronauts as they take off are really pseudo-forces. What the astronaut really feels is the seatback trying to accelerate right through the astronaut. Of course, the effect of this force is that the astronaut accelerates right along with the seat.

3.6 *The Role of Newton's Laws*

When we started our discussion of mechanics we said that one of our goals was to learn how to describe the position of a particle at any time. Have we reached that goal? Indeed we have. If we know the forces acting on a particle, Newton's second law tells us the acceleration at any time. Thus we know how the velocity is changing. And if we know the velocity at any time, we know how the position is changing. If we know the position at which an object starts and how that position changes, then we know the position at any time, which is what we were after in the first place.

Newton's contributions were of tremendous importance, and we can appreciate Alexander Pope's lines:

> Nature and Nature's laws lay hid in night.
> God said, Let Newton be! and all was light.

We shall see, however, in Chapter 9, when we discuss the theory of relativity, that Newton's laws are not valid in all circumstances, which led to John Collings Squire's rejoinder:

> It did not last: the Devil howling "Ho!
> Let Einstein be!" restored the status quo.

Key Words

friction
Newton's laws
Kepler's laws
force
inertia
mass
contact force
tension
spring constant
inertial mass
gravitational mass
weight
weightlessness
pseudo-force

Questions

1. Describe the motion of a ball with no forces acting on it.
2. A ball flies by, moving from left to right with a constant velocity. Is it possible that the gravitational force of the earth is the only force acting on it? Explain.
3. Why is Newton's first law so hard to observe in our everyday experience? Give two examples.
4. Draw a diagram that shows the forces acting on a baseball as it flies through the air. Explain how each of these forces affects the motion of the ball.
5. A certain force exerted on a 1-kg block produces an acceleration of 1 m/s^2. (a) What acceleration will the same force produce in a 2-kg block? (b) What is the strength of the force?
6. A person pushes a car with a horizontal force of 50 N, just overcoming the force of friction, so the car moves along with a constant speed. A second person now pushes alongside the first person, also exerting a force of 50 N. (a) What is the *net* force on the car? (b) If the car has a mass of 1000 kg, what is the acceleration of the car?
7. You jump straight up in the air. What force provides you with your upward motion?
8. You are hanging from a horizontal bar which is suspended between two walls. Draw diagrams to show all of the forces on (a) you, and (b) the bar.
9. You step on a scale that correctly reads 750 N (about 150 lb). What force does the scale exert on you?
10. What actually exerts the forward force on a rocket?
11. Suppose you are on ice skates and push off a wall on the side of the rink. Newton's law says that the force you exert on the wall is equal and opposite to the force that the wall exerts on you. Since these forces are equal and opposite, why don't they cancel out, leaving you motionless?

12. Why is it necessary to keep the engine of a car running if you only want the car to move along at a constant speed (that is, you don't want any acceleration)?
13. Two ropes are tied together, and each rope is pulled with a force of 100 N. What is the tension at the knot?
14. Suppose a spring is hanging from the ceiling. You weigh 750 N and hang from the bottom of the spring. (a) What force does the top of the spring exert on the ceiling? (b) What force does the ceiling exert on the top of the spring? (c) Draw a diagram showing the forces acting on the spring.
15. The acceleration of gravity on the moon is one-sixth that on the earth. (a) How does your weight on the moon compare with that on the earth? (b) How does your mass on the moon compare with that on the earth?
16. If you were riding in the railroad car depicted in Fig. 3–28, and measured the magnitude of the acceleration of the ball, how would this magnitude compare with the acceleration of gravity when the railroad car is not accelerating (that is, would it be greater, smaller, or the same)? Explain your answer.
17. In Jules Verne's *Journey to the Moon* a group of people is in a free-flying (that is, no rocket engine running) projectile from the earth to the moon. During the flight an object let out the hatch continues to stay just outside the window for the rest of the trip. Also, the people stand on the floor of the projectile. One of these phenomena cannot be right. Which is it and why?
18. If you find yourself suspended motionless in the midst of a space station's open space, can you get yourself to a wall? How?
19. You are in a railroad car going around a bend at constant speed, holding a ball in your hand. (a) Draw a diagram showing all of the real forces acting on the ball. (b) You now drop the ball. Draw a diagram showing all of the real forces acting on the ball.
20. List the things that you have to know in order to predict the position of an object at any time.

CHAPTER 4

Momentum

Newton's second law is usually stated as $F = ma$. However, this is not the way that Newton chose to phrase it. Instead, he talked about a quantity that he called the "motion" of an object. He said that if a force acts on an object for some interval of time, then the change in motion is equal to the product of the force and the time interval over which it acts.

Fig. 4–1 This train in *Silver Streak* may not have been moving as fast as a bullet, but it had a very large momentum because of its high mass. A simple wall was not enough to stop the train.

4.1 *Momentum*

Plural: momenta

What is the "motion" to which Newton referred, and how does his statement of the second law relate to the usual one? The quantity that Newton called motion we now call momentum. It is equal to the mass times the velocity of an object. That is,

If we let p be the momentum, v the speed, and m the mass, then p = mv.

momentum = mass × velocity.

Since velocity is a vector, momentum is too. The direction of the momentum is the same as that of the velocity.

The momentum of an object is a descriptive quantity, and we all have a good intuitive feel for the role that mass and speed play as they each contribute to momentum. For example, you know you can catch a baseball that is going 40 m/s (90 m.p.h.), but you can't stop a freight train going at the same speed (Fig. 4–1). And you know that a bullet can be devastating even though its mass is low.

Let's see how Newton's laws look when expressed in terms of momentum. The first law tells us that the velocity of an object will be constant if there are no forces acting on it. Under these circumstances, the momentum will also be constant. We can therefore restate the first law by saying that *in the absence of forces, the momentum of an object is constant.*

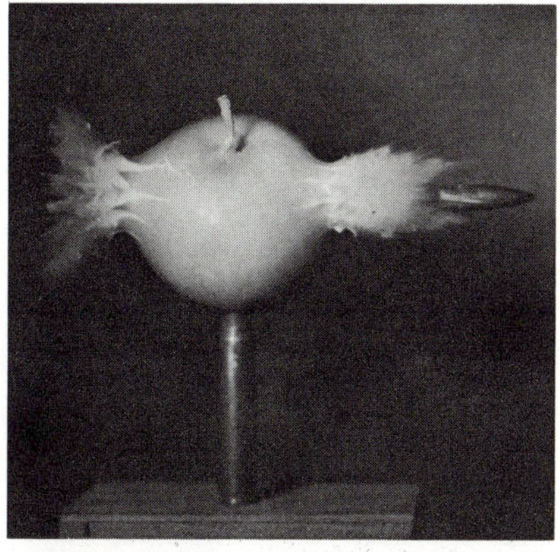

Fig. 4–2 A bullet has a very low mass, but carries a high momentum because of its high speed.

4.2 *Conservation of Momentum*

The formulation of Newton's third law (action/reaction) in terms of momentum has some important consequences. Assume that you and a friend are on a frictionless surface. You can push and pull each other, but you are isolated from all other influences. Suppose you push your friend. Newton's third law tells us that your friend exerts a force on you that is the negative of the force you exert on your friend. (By "negative" we mean equal in strength but opposite in direction.) The change in momentum for each person will be the force of your pushing multiplied by the length of time that you push. Since the forces are equal and opposite, your momentum change must be equal and opposite to that of your friend.

So for Newton's third law we can substitute an equivalent law: *if the only forces acting are those between two objects, whatever momentum is lost by one object is gained by the other.* The total momentum of the two objects remains the same.

This is an important result, which we call conservation of momentum. When we say that momentum is "conserved" we mean

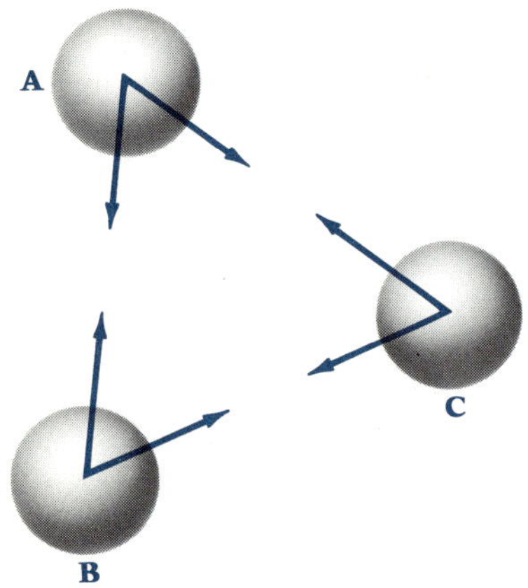

Fig. 4–3 The arrows indicate the forces that the balls exert on each other. In this case, the forces are attractive, as with gravity. For any pair, the forces are equal and opposite, so the change in momentum due to the forces between those two balls is equal and opposite. If we consider the pair A and B, the force that A exerts on B is equal and opposite to the force that B exerts on A. Therefore, the change in momentum of A due to the force exerted by B is equal and opposite to the change in momentum of B due to the force exerted by A. The same is true for any other pair. Therefore, these internal forces make no change in the total momentum of the three balls. In this example, there are no external forces, that is, no forces are executed by anything external to the objects we are considering.

that it remains the same (Figs. 4–3 and 4–4). In the case of you and your friend on the frictionless surface, in the absence of any external force, it is the total momentum of the two of you together that is conserved.

We can use conservation of momentum to explain air resistance. As a car moves through air, it collides with air molecules (as illustrated in Fig. 4–5). Say the car is moving from left to right. When it collides with an air molecule, the air molecule will move to the right. Since momentum must be conserved, the car will "recoil." That is, it will receive an impulse from right to left (opposite to the direction of motion of the car), and the car will slow down a little bit. (A more detailed analysis requires a study of how the air flows.)

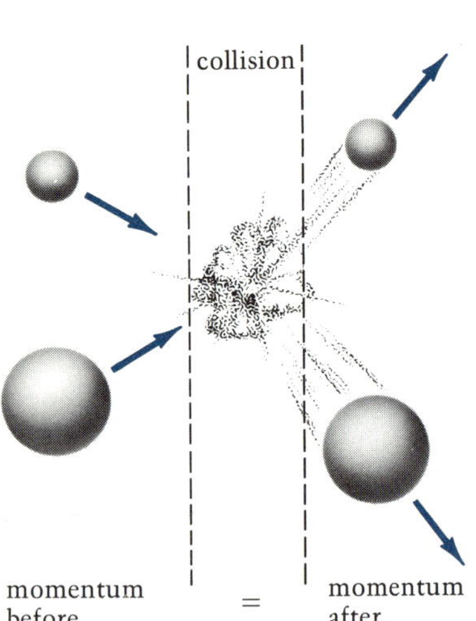

Fig. 4–4 If two balls collide and the only force acting is that of each ball on the other, then the total momentum before will equal the total momentum after the collision. We can say this without knowing any of the details of the collision. Using conservation of momentum allows us to solve many problems by looking at "before" and "after" without worrying about "during."

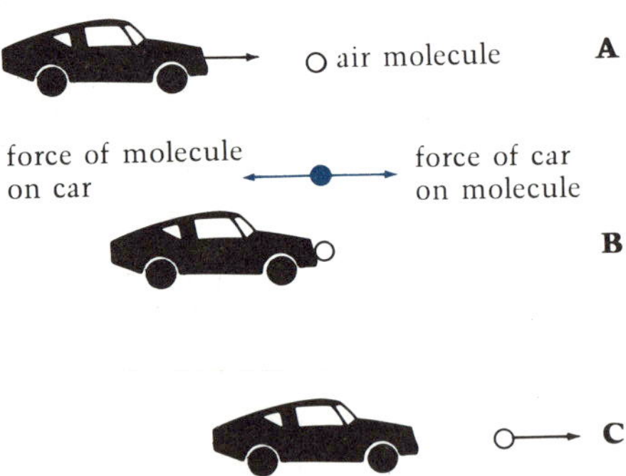

Fig. 4–5 This sequence of figures shows how air resistance arises. A. For simplicity, we'll assume that the air molecule is at rest. B. The car strikes the air molecule and exerts a force, from left to right, on the molecule. C. The molecule takes off to the right, but Newton's third law tells us that the car must recoil to the left. It will acquire a momentum to the left equal in size to the molecule's momentum to the right. The effect of one air molecule may not be too great, but in 1 second, a car may run into over 10^{23} such molecules, so the effects add up. When the car moves faster, the resistence is greater because it is hitting each molecule at a higher speed; it also runs into more molecules in a second because it is covering a greater distance.

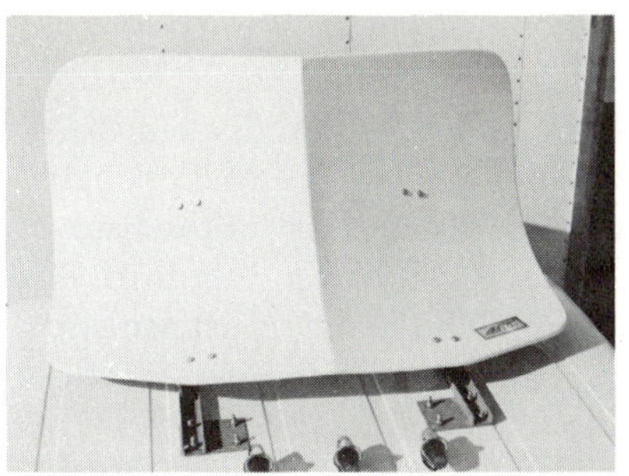

Fig. 4–6 The wind resistance of trucks can be decreased by the use of air deflectors such as the one shown in this photograph. With the deflector in place, the flat face of the trailer, which would give the greatest wind resistance, is protected by a wedge, which offers less resistance.

We can also see that the amount by which the car slows down depends on how fast it is going. The actual drag depends on the shape of the moving object, but for many objects the retarding force depends approximately on the cube of the speed. If this is the case, then the air resistance felt by a car at 40 m/s is about eight times that felt at 20 m/s (since $40/20 = 2$, and $2^3 = 8$), one reason why cars get poorer gas mileage at higher speeds.

Momentum conservation is only the first of a series of conservation laws that we will come across. One of the goals of physics is to present the simplest possible description of the physical world. Since they tell us about things that don't change, conservation laws play an important, simplifying role in physics. They also have predictive power. Later in this book we see how the application of conservation laws allowed physicists to predict the existence of certain elementary particles well before those particles were discovered.

4.3 Rockets

If two particles receive the same force for the same length of time, they acquire the same momentum. If the particles do not have the same mass, they will not have the same velocity. Since *mass* × *velocity* must be the same for both particles, the more massive particle will have the lower velocity. When two particles have the same momentum:

$$\frac{\textit{velocity of particle 1}}{\textit{velocity of particle 2}} = \frac{\textit{mass of particle 2}}{\textit{mass of particle 1}}.$$

In the last chapter we saw that if you stand in a rowboat and throw a tackle box off the back, the boat "recoils" in the opposite direction. We can understand this in terms of conservation of momentum. Whatever momentum the tackle box acquires, the boat must compensate by acquiring a momentum of equal size but in the opposite direction. Suppose the mass of the boat plus you plus the tackle boxes on board is 100 kg and the mass of one tackle box is 1 kg. If you are strong enough to throw the box at 10 m/s, what is the speed of the boat when it recoils? Since your mass is 100 times that of the box, your speed must be $\frac{1}{100}$ that of the box, or 0.1 m/s. This 0.1 m/s may not seem like much, but if you have a good supply of boxes and a strong arm, you can increase your speed with each box that you throw. In fact, for each box you throw off, you will pick up a little more speed than for the previous one because the recoiling boat contains one box fewer than the previous time and therefore has a lower mass. With a lower mass, it must pick up more speed to have the same recoil momentum.

This form of propulsion is the basic idea behind the operation of a rocket. The rocket carries all of its fuel with it. The rocket engine converts this fuel into hot gases, which are directed out the back of the rocket. These gases leave at a high speed, so even if their mass is not great, they have a large momentum. The rocket must acquire a momentum of the same magnitude, but in the opposite direction. As fuel is consumed, the mass of the rocket plus the remaining fuel decreases and the speed of the rocket increases. When a Saturn V rocket takes off, loaded with fuel, the thrust of its engines barely exceeds the weight of the rocket, and it starts to lift slowly; but within a few minutes it will be traveling at a few thousand meters per second! (Fig. 4–7).

A Saturn V rocket was the type used to send astronauts to the moon.

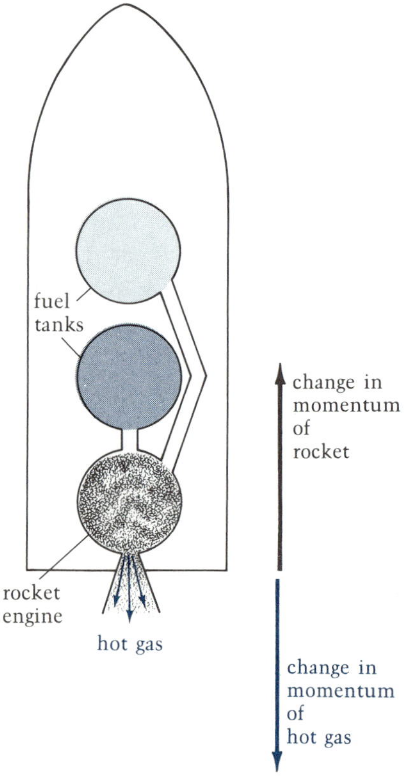

fuel tanks
rocket engine
hot gas
change in momentum of rocket
change in momentum of hot gas

Fig. 4–7 Operation of a rocket. Fuels are mixed together and ignited in the rocket engine, producing a stream of hot gas that is ejected from the rear of the rocket at very high speeds. In any given time interval, the rocket acquires a forward momentum equal to the backward momentum of the gas. The total momentum of the gas + rocket does not change (unless there is an external force, such as gravity).

Fig. 4–8 Skylab is launched into space. The forward momentum of the rocket equals the backward momentum of the hot gases.

4.4 *Center of Mass*

For any collection of particles, we can find a point that acts as the average location of all the mass. This point is called the center of mass. In studying translation only, we can treat our group of particles as though all the mass were concentrated at that one point. (Remember, we defined translation at the beginning of Chapter 2.) In fact, this is the reason why our treatment of motion for simple particles works so well for more complicated objects. The laws we derive for simple particles work equally well for things like cars and rockets (see Fig. 4–9).

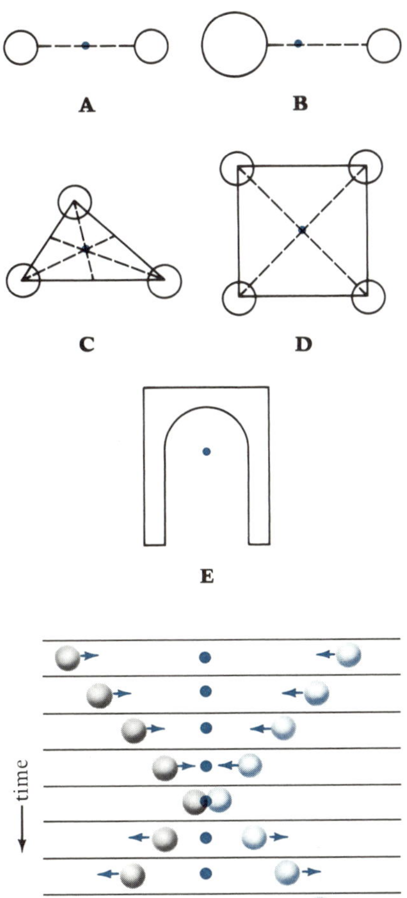

Fig. 4–9 The centers of mass for some collections of particles. A. For two particles of the same mass, the center of mass is just in the middle. B. If the two particles have different masses, the center of mass will be closer to the more massive particle. C. For a triangle with equal masses at the corners, draw a line from each corner to the middle of the opposite side, and the center of mass will be where these three lines intersect. D. For equal masses at the corners of a square, it is exactly at the center of the square. E. For the horseshoe shape, the center of mass will be midway between left and right, because the shape is the same to the left of center as it is to the right, but it will be closer to the top than to the bottom, since there is more mass on top.

To see the connection between center of mass and conservation of momentum, let's first look at two balls of identical mass. In this case the center of mass is located halfway between the two balls. (The center of mass is a fictitious point. It is not necessary that there actually be some mass at that point. In much the same way, the average of 1 and 3 can be 2, even though 2 is not one of the numbers averaged together.)

Suppose the two balls exert forces on each other and start to move apart. Since momentum is conserved, they will move apart with equal speeds. We can try to locate the center of mass when the balls have reached some other position. It will now be the point halfway between the new positions of the balls. However, since the balls have moved with equal speeds, they will have covered the same distance (although in opposite directions) and the center of mass will not have moved (Fig. 4–10).

So you see that there is an important connection between momentum conservation and the center of mass. *For any group of particles, internal forces can have no effect on the position of the center of mass.* We have already seen, from Newton's first law, that in the absence of any external forces a particle will continue with a

Fig. 4–10 Time-sequence drawing of two balls of equal mass initially at rest and then blown apart. In each frame, the • marks the position of the center of mass of the two balls. Since the balls have equal masses, the center of mass is always exactly at the midpoint between the two balls. At the start, the center of mass is fixed, and it remains fixed after the balls start to move. The two balls must move so that their center of mass stays at the midpoint between them, where it was before the collision.

Fig. 4–11 If two football players collide with equal and opposite momenta, their total momentum before the collision is zero, and their total momentum after the collision is zero. Here we see conservation of momentum as the Los Angeles Rams play the Kansas City Chiefs.

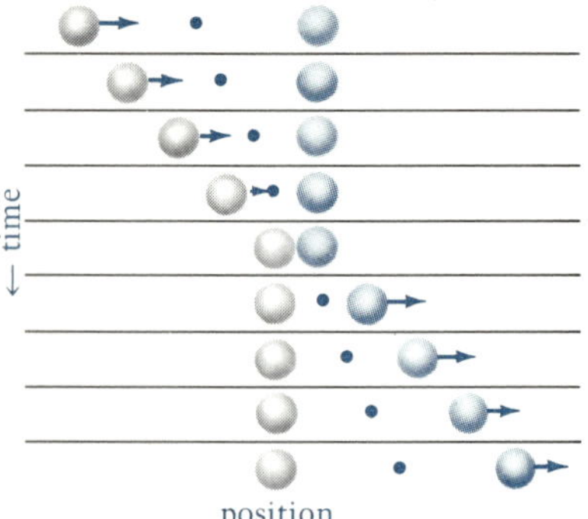

constant velocity. The same holds true for the center of mass. For example, suppose our two identical balls are coming toward each other head-on with equal speeds. In this case the velocity of the center of mass is zero again. After they collide, the velocity of the center of mass must still be zero.

We need not limit ourselves to situations in which the center of mass is at rest. For example, suppose one of the balls is at rest while the other one heads toward it (as in a theoretical billiard game in which the balls do not spin). As the moving ball gets closer to the one at rest, the center of mass must stay exactly halfway between the two balls, which means that it gets closer to the ball at rest.

Fig. 4–12 Time-sequence drawing of two equal balls colliding head-on, but with one ball initially at rest, as in a billiards shot. In each frame, the ● keeps track of the center of mass of the two balls. Since the balls are of equal mass, the center of mass is midway between the two balls. Notice that in each frame the center of mass moves half as much as the moving ball. In this way the center of mass always stays at the midpoint. After the collision, the center of mass must continue its motion. In the case of this collision, the ball that came in at the left ended up at rest. The struck ball must move in such a way that the center of mass moves as it did before the collision.

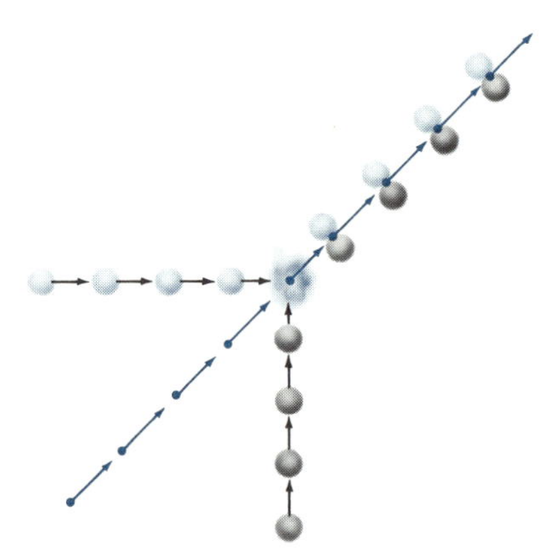

Fig. 4–13 What happens if there is a collision between two particles and they stick together? After collision (ignoring friction), the combined particle must move with the same velocity as the original center of mass. If it didn't, the center of mass after collision would not be moving with the same velocity as before collision. The center-of-mass time intervals are marked with a ●. Black arrows show the motion of each ball; blue arrows show the motion of the center of mass.

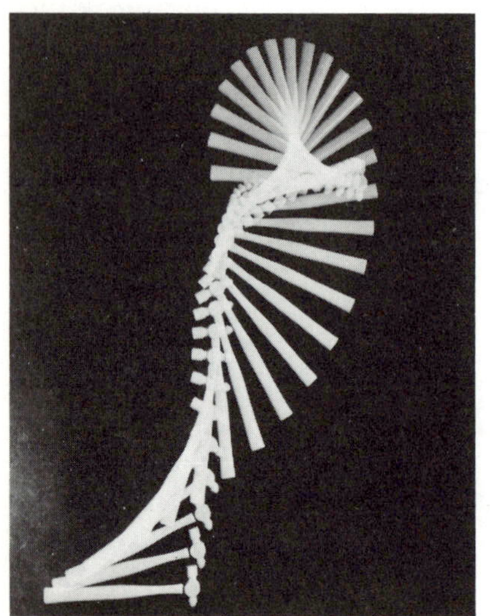

(left) Fig. 4–14 The center of mass of the hammer moves in a smooth path and is close to the head of the hammer, which is the most massive part.

(right) Fig. 4–15 Even though the diver may execute very complicated twists and tumbles, his center of mass follows a smooth path from the diving board to the water.

Fig. 4–16 As Dwight Stones jumps over the bar, some of his body is higher than the bar and some is lower than the bar. So at any time, the center of mass is lower than the highest point of the body, and the center of mass may not even pass over the bar.

Fig. 4–17 This supertanker doesn't travel at high speeds, but it is so massive that when it travels it has a very large momentum, and requires a considerable impulse to stop or turn. A supertanker like this must turn off its engines 25 km before it wishes to come to a stop! This property makes supertankers very hard to control, and makes it very difficult for their captains to respond to emergency situations.

After the two balls collide, the center of mass must continue to move at the same velocity as before. Various combinations of motions are possible after a collision, depending on the exact details of the collision. However, all of these possible motions must have the property that, in the absence of an external force, the center of mass continues to move with the same velocity (with the same speed and direction) as it did before the collision (Figs. 4–12 and 4–13).

When you have two particles of different mass, the center of mass is closer to the more massive particle. The exact placement of the center of mass depends on the ratio of the two masses. For example, the center of mass of the system composed of you and the earth is displaced from the center of the earth by only 10^{-23} of the radius of the earth. On the other hand, the moon's mass is 0.012

Fig. 4–18 A globular cluster in our galaxy. A globular cluster may contain up to about one million stars packed into a region a few hundred light-years across. (Although 100 light-years—about 10^{18} meters—may sound like a large distance, there are approximately a million stars in this cluster, so they are close together by normal standards.) Such a cluster is held together by the mutual gravitational attraction of all the stars within it. If we consider this cluster to be a system of particles, the forces among the stars are considered to be internal forces. They make no contribution to the net force on the cluster as a whole. Whatever force star 1 exerts on star 2, star 2 exerts an equal and opposite force on star 1. Although stars 1 and 2 may respond individually to these forces, this response has no overall contribution to the total momentum of the cluster. But there are also external forces acting on this system. All of the other matter in our galaxy is exerting a force on the cluster, keeping it from leaving the galaxy. In this system we can readily distinguish between internal and external forces. If we want to know about the motion of the cluster through the galaxy, we consider only the external forces. If we want to know how the stars move around within the cluster, we must consider just the internal forces.

times that of the earth. This places the center of mass of the earth-moon system 0.012 times the earth-moon distance, or about three-quarters of the way from the center of the earth to the earth's surface. We normally talk of the moon's going around the earth and the earth's going around the sun. It is really the center of mass of the earth-moon system, rather than the earth itself, that goes smoothly around the sun, while the earth and moon both move about that center of mass. Since the center of mass of the earth-moon system is so close to the center of the earth, the earth's motion is small compared to that of the moon.

In treating the motion of any complicated system, we can break the problem into two parts. We can first treat the motion of the center of mass, and then we can treat the motion of the individual particles with respect to the center of mass. Thus, if we follow the contorted motions of a high diver, we see that the diver's center of mass follows the same smooth path as a ball thrown with the same initial velocity. All of the tumbling can be treated separately, as motion about the center of mass.

Key Words

momentum
conservation of momentum
recoil
conservation laws
center of mass

Questions

1. Which has a greater momentum, an elephant standing still or a feather floating in the breeze? Explain.
2. An elephant and a flea, both initially at rest, are subjected to the same force for the same amount of time. (a) Which will have the greater momentum? (b) Which will have the greater speed? Explain both cases.
3. Compare the momentum of a 0.01-kg bullet traveling at 300 m/s, with a 1000-kg car traveling at 5 m/s.
4. A car traveling at 10 km/h has a stopping distance of 20 m (after the brakes are applied). What is the stopping distance for the same car traveling at 40 km/h?
5. An object is moving along with a constant momentum. What can we conclude about the forces acting on the object?
6. While sand is pouring through a funnel onto a scale, the scale reads a little higher than the weight of the sand on the scale. Why?

7. A pair of ice skaters performs a trick in which they skate toward each other, join hands, and spin around. If one skater has a mass of 50 kg and the other has a mass of 75 kg, how must they adjust their speeds before the collision to ensure that they will not drift across the ice while they are spinning? (Assume that the ice is frictionless.)

8. The earth and moon orbit around the sun. (a) Identify the forces that are internal to and those that are external to the earth-moon system. (b) Is the momentum of the earth-moon system constant?

9. Two skaters, one with a mass of 50 kg and the other with a mass of 100 kg, are standing still and holding hands. They push off each other so the 50-kg skater moves 2 m/s. (a) How fast is the other skater moving? (b) How fast is the center of mass of the two skaters moving? (Again, assume that the ice is frictionless.)

10. Two people are playing "catch" inside a closed railroad car. How would the car move as the ball is thrown back and forth? Why? (Assume that the car rolls with friction.)

11. You are on a railroad car with a pile of bricks on one end. You move the pile to the other end of the car. Assume the car is initially at rest, and is free to roll without friction. (a) What happens to the position of the car as you move each brick over? (b) Does the final position of the car depend on how fast you carry each brick? (c) Does it matter if you throw the bricks instead of carrying them? Explain.

12. The wedge shape of certain racing cars helps reduce wind resistance. It also helps keep the car pressed hard against the road, which provides good traction. Why does the wedge shape result in the car's being pressed against the road? (Draw a diagram to illustrate the answer.)

13. The longer a rocket engine burns, the greater the acceleration of the rocket. (a) Why? (b) As the acceleration increases, how does the momentum of the rocket change in any given second?

14. Give an example of an object with no mass at its center of mass.

15. As a rocket takes off from the earth, what happens to the center of mass of the rocket-plus-the-exhausted-fuel system?

16. (a) Can the following take place? One billiard ball is at rest and is struck by an identical ball. After the collision the two balls stay together, and stay at rest. (b) Explain.

17. (a) Can the following take place? Two identical billiard balls come at each other with the same speed. After the collision, one ball is at rest, while the other moves away. (b) Explain.

18. As we saw, the distance between the center of mass of the earth and you is about 10^{-23} of the radius of the earth from the center of the earth. (a) From this, what can you conclude about the mass of the earth relative to your mass? (b) What is the mass of the earth in kilograms?

19. We say that we can deduce the presence of unseen planets around stars if we can detect wobbling in the motion of the star. Why is this?

CHAPTER 5

Energy

Fig. 5–1 You can sense the energy released by the water as it goes over Niagara Falls.

If you stand at the base of a waterfall or under the hot glare of the noonday summer sun, you get a sense of energy. And we all have an intuitive sense of the energy that is stored in coal or in gasoline. Physicists define energy in technical terms, but the meaning is basically the same as our intuitive feeling.

We start by going back to Newton's laws. Knowing the forces on a particle at any time and using Newton's second law allows us to compute its acceleration. Once we know the acceleration, we can find the velocity at all times, and finally, once we know the velocity at all times and where the particle started its motion, we can find the particle's position at all times. However, although Newton's laws allow us to solve problems of this sort, they do not always provide us with the most convenient way of solving them.

For example, consider a ski jumper poised at the top of a ski jump chute (Fig. 5–2). The chute consists of a downward sloping ramp, a curved section at the bottom, and then a short upward sloping ramp. Say we want to know the speed with which the jumper will leave the end of the upward sloping section. Newton's laws give us all the tools that we need, but even in the ideal case of a frictionless chute, this problem is made hard by the fact that the acceleration of the skier is not constant throughout the run. As shown in Fig. 5–2, the skier's acceleration changes as the angle of the chute changes. We could solve this problem using Newton's second law,

Fig. 5–2 Acceleration of a skier on a jump chute. In this case we are only interested in that part of the acceleration that is directed along the slope. This is the part of the acceleration that will make the skier go faster or slower. The steeper the slope, the greater this acceleration will be. Notice that when the skier starts on a steep part of the chute (position 1), there is a large acceleration along the slope and the skier will gather speed quickly. As the slope levels out (position 2) the acceleration is less. As the skier starts to go back uphill (position 3) the acceleration is now directed backward, that is, the skier is slowing down. Just before the skier takes off (position 4) the uphill part of the chute is steepest, and the deceleration is greatest. Thus we see that the skier's acceleration depends on his position on the jump chute. If we wanted to find the speed of the skier at position 4 (or any other position), and just wanted to use $F = ma$, we would have to use calculus to solve the problem.

but we would have to use calculus to do it. This is but one example of a common situation in which the force you feel or the acceleration you experience depends on where you are. In this chapter, we'll see that such problems are more easily solved by introducing the concept of energy.

5.1 *Work*

We begin our discussion of energy by looking at the relationship between energy and work. Our everyday definition of work is related to the performance of some service. This service could involve physical exertion or it may only involve thought. The definition of work in physics is much more restrictive. It is entirely related to the action of forces.

Suppose you are pulling a crate across the floor. In order to do this you are exerting a force on the crate. If you are pulling the crate at a constant speed, then your pulling is supplying exactly the force necessary to overcome friction. If the box is accelerating, then part of your force overcomes the friction and part produces the acceleration. Suppose you drag the crate some distance across the floor. Then the work you do is related to the force you exert and to the distance you have pulled the crate.

Actually, we must be more specific about the *direction* in which the force is exerted relative to the direction in which the box moves. For example, what if a lion were sitting on top of the crate while you were pulling it (Fig. 5–3)? That lion's feet would be exerting a downward force on the crate while you were pulling it horizontally. Yet it doesn't seem quite fair to say that the lion on top has done work.

From the physicist's point of view, *the only forces that contribute to the work done are those forces acting along the direction of motion.* When we speak of a force acting along the direction of motion, we mean pointing in the direction of motion or in the exact opposite direction. These are forces that will affect the *speed*

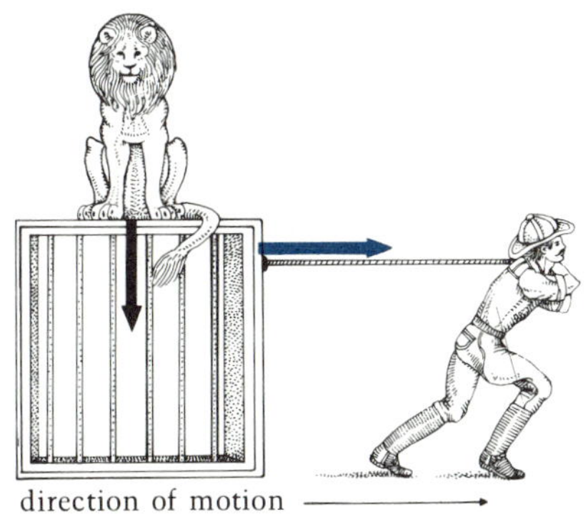

Fig. 5–3 The person pulling the box exerts a force along the direction of motion as indicated by the horizontal arrow. It is this force that is responsible for the speed of the box, and it is this force that is clearly doing the work. The lion sitting on top of the box is also exerting a force on the box, but this force is in the downward direction (as indicated by the downward arrow), or perpendicular to the direction of motion. The lion on top of the box is doing no work.

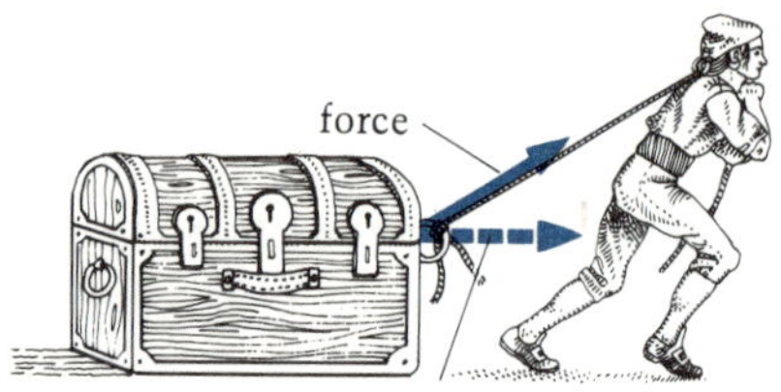

Fig. 5–4 If you are pulling at an angle, only the vector component of the force in the direction of motion does work.

of motion. If we exert a force in a direction that is perpendicular to the direction of motion, no work is done by that force.

What about a force that is neither directly along the direction of motion nor perpendicular to it? This might occur if you were pulling a low crate with a rope angled upward while you pulled. *The part of the force that points upward does no work, while the part of the force that points horizontally does work.* So when the force makes some angle with the direction of motion it still does some work, but less than if it were exactly aligned with the direction of motion (Fig. 5–4). We call the part of the force that is directed along the direction of motion the *vector component of force along the direction of motion.* (The components of vectors were described in Chapter 2.) We can now define the work done by any force as

$W = Fd$

work = component of force along direction of motion × distance.

If you pull the rope horizontally and exert a force of 50 newtons (50 N) while pulling the box 10 meters (10 m), then you will do an amount of work equal to (50 N) × (10 m) = 50 N-m. The combination N-m (newton-meter) occurs so often that we give it a special name—the joule (abbreviation J), after the nineteenth-century British physicist James Joule. How big is a joule? At the surface of the earth, 1 kg weighs approximately 10 N. So if you lift 1 kg 0.1 m, you have done 1 joule of work. Thus the joule is a convenient measure for describing work done in our everyday surroundings.

In the British system of units, the unit of work is the foot-pound.

It is often important to know how fast something or someone is doing work or can do work. The rate of doing work is called power.

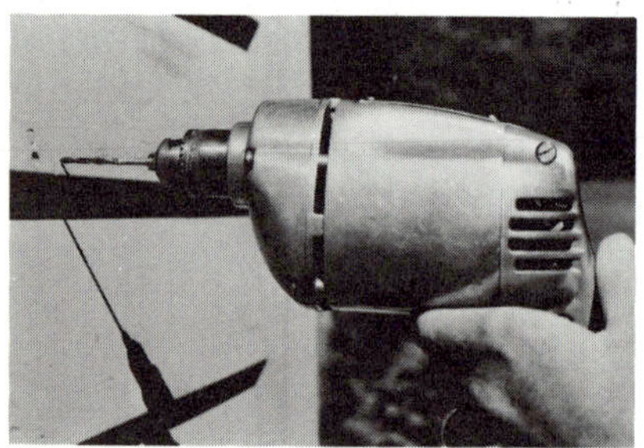

Fig. 5–5 This power drill can deliver one quarter of a horsepower.

Since power is the amount of work divided by the time interval, the units of power are joules per second. This is given a special name—the watt (abbreviated W), after James Watt, the eighteenth-century Scottish inventor who made the steam engine practicable. The watt is a familiar unit since it is the one used to indicate the rate at which our electrical appliances use up energy. At present, electric companies in the United States use the kilowatt-hour (kW-h) as their unit of energy, rather than the joule, although this will change sooner or later. One kilowatt is 10^3 watts, and an hour is 3600 seconds, so 1 kW-h is equal to 3.6×10^6 joules. Another traditional unit of power is the horsepower. This is equal to 746 watts, which is approximately as much power as a horse can put out per hour in the course of continuous work.

5.2 Work and Energy

Let's return to that crate. Assume that there is no friction and that you are pulling horizontally, thus doing the maximum work. The more work you do, the faster the box will move. If the box started at rest, then the speed you reach at any point is related to the work you've done by

$$\textit{work done} = \tfrac{1}{2} \times \textit{mass} \times \textit{speed}^2.$$

We can see, for example, that the work done to get an object to a given speed should be proportional to the mass of the object. To reach a given speed, a more massive object will require either a greater force or the force applied over a longer period of time. In either case, more work is done. Why should the work done depend on the square of the speed? Work is *force* × *distance.* Speed is a factor because the greater the force, the greater the speed we achieve, and speed is again a factor in the distance because the greater the distance over which the force acts, the greater the speed. Since speed is a factor twice, it enters the formula as speed^2. (The additional factor of ½ comes from the fact that the distance covered depends on the average of the speed at the beginning and the speed at the end, rather than simply the speed at the end.)

$K.E. = \frac{1}{2}mv^2$

We call the quantity (½ × mass × speed^2) the energy of motion, or kinetic energy of the object. So if you start pulling when the crate is at rest, then it will acquire a kinetic energy equal to the amount of work you do in pulling the crate (assuming there is no friction). What if the crate is not at rest when you start pulling? Then it will initially have some kinetic energy. As you pull the crate, you do some work and give the crate a new speed, and thus a new kinetic energy. You will find that the change in the kinetic energy is equal to the amount of work you have done. This applies equally well to increases and decreases in kinetic energy. If the force is in the same direction as the motion, then the speed, and with it the kinetic energy, increases. If the force is opposite to the direction of motion, the crate slows down and the kinetic energy decreases (see Fig. 5–6).

Work = change in kinetic energy.

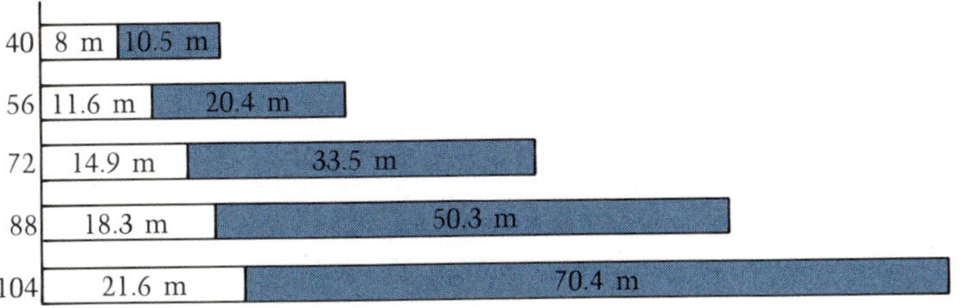

Fig. 5–6 Stopping distance. The open part of each bar is the distance traveled by the car from the time you decide to stop until your foot hits the brake. The colored part of each bar, which increases more rapidly than the left side, is the braking distance. It is proportional to the square of the velocity.

The fact that the kinetic energy depends on the square of the speed is very important. It means, for example, that it takes four times as much work to accelerate an object from rest to 100 m/s as it does to accelerate it to 50 m/s (since $100 = 2 \times 50$, and $100^2 = 2^2 \times 50^2$, and $2^2 = 4$). Likewise, it takes four times as much work to stop a car traveling at 100 m/s as it does when it is traveling at 50 m/s. Assuming the brakes always provide the same force, the only way that we can get the extra work to stop the faster car is to have the brakes exert that force over a longer distance. This means that the stopping distance will be longer for higher speeds, a distance that will increase as the square of the speed. The graph in Fig. 5–6 shows the stopping distances for a car going at various speeds (40 km/h is about 25 m.p.h.). Notice how quickly the "skid marks" get longer as the speed goes up.

In the SI system, the unit for energy is the joule.

Pulling the crate is an example of converting work into kinetic energy. We can also convert energy into work. A simple example is driving a nail into a piece of wood. Since the wood opposes the nail's entry, we have to do work to get the nail into the wood. The deeper we want to drive the nail, the more work we must do. We start with kinetic energy in the head of a hammer. The hammer head can have a lot of kinetic energy since it usually is fairly massive and we can get it to move at a high speed. When the hammer head strikes the nail, the kinetic energy disappears (the hammer stops), but, in exchange, we do work by driving the nail into the wood. Since work and energy are directly convertible into each other, we use the same units for energy as for work.

5.3 *Potential Energy*

Suppose you give a ball to a friend who then climbs a ladder and drops the ball back to you. The ball reaches your hand with some speed; it has acquired energy of motion—kinetic energy—in this process. Where did this kinetic energy come from? When the ball was dropped, it was pulled downward by the force of the earth's gravity. This gravitational force did work on the ball (equal to the gravitational force multiplied by the distance of fall). This work was converted into the kinetic energy that the ball had when it reached your hand. So it appears that we should give credit to the earth's gravity for doing the work that produced the kinetic energy.

This seems a little unfair to your friend. After all, the ball did not get to the top of the ladder by itself. And if the ball had not been at the top of the ladder, it would not have fallen and acquired kinetic energy. So if we give gravity the credit, we should at least give your friend an "assist." To lift the ball at a constant speed, your friend had to exert a force on it equal to its weight, therefore doing an amount of work equal to the weight of the ball multiplied by the height of the ladder. However, this work did not directly produce the increased kinetic energy, because the force exerted on the ball by your friend was opposed by the force of gravity, making the net force on it zero during the climb.

Even though the work did not directly produce the increased kinetic energy, it did put the ball in a position to be affected by gravity. It therefore seems fair to say that your friend's work increased the energy of the ball in some way. This new energy, associated with the position of the ball at the top of the ladder, can then be converted into kinetic energy by allowing the ball to fall. Since this energy has the potential to be converted into kinetic energy, we call it potential energy (see Fig. 5–7).

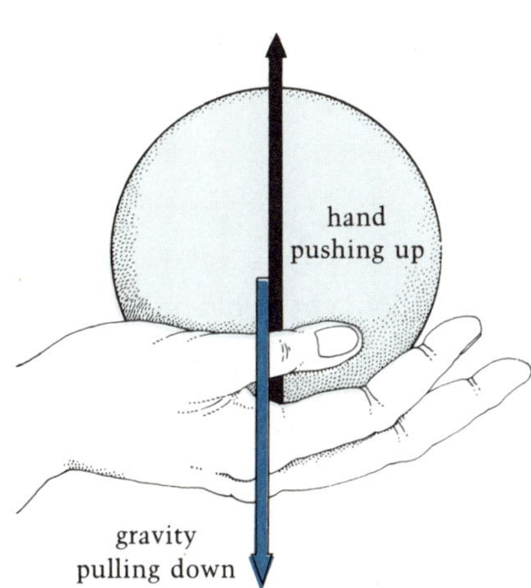

Fig. 5–7 Forces on the ball while your friend is carrying it up the ladder (with no acceleration). Gravity exerts a downward force on the ball, but your friend's hand exerts an upward force on the ball. The upward force is cancelled by gravity, so the ball does not accelerate, but it does move upward, so the hand is doing work (as it is exerting a force in the direction of motion).

To see how we can use the idea of potential energy, let's say you throw another ball up to your friend. The second ball reaches a height equal to the height from which your friend dropped the first ball, but your friend fails to catch it and it falls back to you. When it reaches you, it has the same speed as it had when you threw it upward. In fact, this speed is exactly equal to the speed that the first ball had when it reached your hand after being dropped by your friend. The speed with which it reaches your hand seems to depend only on the fact that at some point in its trip it was at rest at a height equal to that from which the first ball was dropped. It doesn't matter whether you threw it to that height or your friend carried it. Either way the ball acquired the same amount of potential energy.

When an object makes a round trip and returns to the starting point with the same kinetic energy as it had when it left, we say that the forces acting on it during the trip are *conservative.*

Potential energy that can be converted by gravity into kinetic energy is called gravitational potential energy. If you lift a ball a given distance, the force you exert is the weight of the ball (its mass times g), and the work you do is the force times the distance you raise it, so

change in gravitational potential energy
= mass × g × change in height.

ΔP.E. = mgh, where the Greek delta (Δ) stands for "the change in." We use m for mass and h for the change in height.

We can associate a potential energy with any conservative force. What is a *nonconservative force*? One example is friction. If you push a box across a floor at constant speed such that your push is exactly balancing the friction force, then even though you are doing work, the box is not accelerating. All of your work is being used up to overcome the frictional force, so when you finish pushing, the box has no additional kinetic energy. Therefore, in sliding the box from one side of the room to another, there has been no gain in potential energy. It appears that the work you have done against friction cannot be recovered.

This is actually an oversimplification. The work that you have done against friction does not completely disappear. If you feel the surface of the box that was in contact with the floor you will find that it feels hotter than before you started. Your work has been converted into heat. As we will see later, this heat is nothing more than an increase in the kinetic energy of the individual atoms that make up the box.

5.4 *Conservation of Energy*

The great advantage of introducing the concept of potential energy is that it provides a convenient way of keeping track of work done by a conservative force. This is work that can be recovered. Whenever the kinetic energy decreases, the potential energy increases. The change in kinetic energy is equal to the amount of work done, so this means that the increase in potential energy must be exactly equal to the decrease in kinetic energy.

In fact, we can associate an energy with any force. With gravity, we can associate gravitational potential energy. With friction, we cannot define a potential energy, but when something gets hotter from the action of friction forces, we can say that the internal energy of the object has increased. (By "internal energy" we mean the sum of the energies of the individual atoms that make up the object.)

Our results can now be summarized in a general law known as the law of conservation of energy. It says that the total energy of an isolated system will remain constant. By "total energy" we mean the sum of the kinetic energy, the potential energy, the internal (heat) energy, and any other forms of energy that may be present. Since the total energy remains constant, if any part changes, then the other parts must also change in such a way that the total energy remains fixed. For example, the kinetic energy could increase while the potential energy decreases by the same amount.

We came to the law of conservation of energy from the laws of motion, and one could solve problems without ever resorting to conservation of energy. However, in many cases, using conservation of energy simplifies the calculations that must be done. In addition, just as with conservation of momentum, there is often a

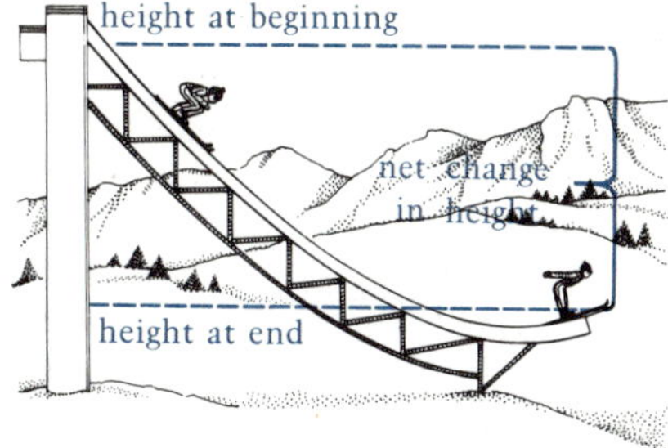

Fig. 5–8 Solving the problem of the ski jumper. When we use conservation of energy, the important thing to know is how much the potential energy has changed from the starting point to the finishing point. This only depends on the difference in height between the starting point and the finishing point and does not depend on how the chute curves in between.

great simplicity in expressing the laws of physics in terms of conservation laws.

We can now use the law of conservation of energy to answer the question about the ski jumper that we posed earlier in this chapter (Fig. 5–8). First, we must find the change in potential energy from the beginning position to the end position. This change depends only on the starting and ending heights, and does not depend on the nature of the slope in the middle. The potential energy is lower at the end than at the beginning because the end point is lower than the starting point. The loss in potential energy must be exactly offset by a gain in kinetic energy. From the gain in kinetic energy, we can calculate the skier's speed. (We have considered the ideal case, in which there is no friction. If there is friction, then the loss in potential energy is balanced by a gain in the kinetic energy plus the energy lost to friction. In such a case, to find the final kinetic energy of the skier, we have to know how much work was done by friction.)

This is just one example in which using energy conservation made a calculation simpler. There is really nothing new in the law of conservation of energy. It contains the same information as Newton's laws of motion—it just gives us another way of using the available information.

5.5 *Energy Conservation and Harmonic Motion*

The world around us is filled with things that oscillate. Oscillating systems start in some position, go through a sequence of motions, and return to the original position. After that cycle, the whole process is repeated again and again, exactly as before. The most familiar examples are timekeeping devices—pendulums, springs, oscillating quartz crystals, or even atoms. We call systems that repeat their motion in a regular fashion *periodic systems.* The time that it takes for the system to go through one full cycle is called the period. Such oscillating systems play a very important role on the atomic and subatomic scale. All oscillating systems have many features in common, so what we learn about a few oscillating systems can be applied to a variety of problems.

Oscillating systems provide good demonstrations of conservation of energy, for we have a constant flow of energy from kinetic to potential and back again. In this section we will look at the pendulum and then at the spring.

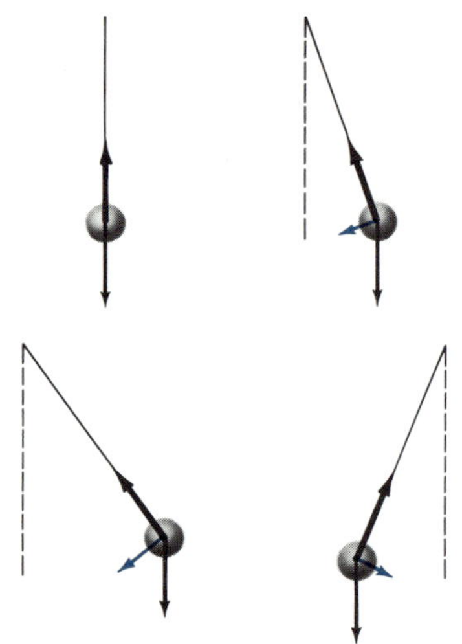

Fig. 5–9 The pendulum. A. When the ball is hanging straight down, there is a downward force (gravity) on it and an upward force (the tension in the string). These two forces balance, so there is no net force on the ball. B. When the pendulum is pulled to the side, gravity still pulls downward, but the tension in the rope now pulls upward and slightly to the side, so there is a net force that tends to pull the ball back toward the center. C. The farther we pull the ball back, the greater this force is. If we let go, the ball will start swinging back toward the center. When the ball reaches the bottom position, there is no force on it, but the ball has some momentum so it won't stop abruptly. Instead, it overshoots the bottom position. D. Now when it is on the other side, the force is in the opposite direction, so it will slow the pendulum down and eventually bring it back to the bottom, where it will again overshoot. It will repeat this motion over and over.

5.5a THE PENDULUM

A pendulum is simply a mass hung from a string. When we pull the mass to the side, it will swing back and forth in repeated motion, as illustrated in Fig. 5–9. Just as the acceleration of a falling object is independent of its mass, the period of the pendulum does not depend on the mass of the ball (for the same reason). It depends only on the length of the string—the longer the string, the longer the period. We credit Galileo with noting this feature of the pendulum; he timed a pendulum against his own pulse.

We can analyze the motion of a pendulum using conservation of energy (Fig. 5–10). When we pull the ball to the side, we are also pulling it up, so that its gravitational potential energy increases. The higher we pull it, the greater the potential energy. (We take the potential energy to be zero when the pendulum is at the bottom of the swing.) When we let go, the ball starts to fall, but the string makes it fall in an arc rather than straight down. As it falls, the potential energy decreases and the kinetic energy increases. When it reaches the bottom, the kinetic energy must equal the initial

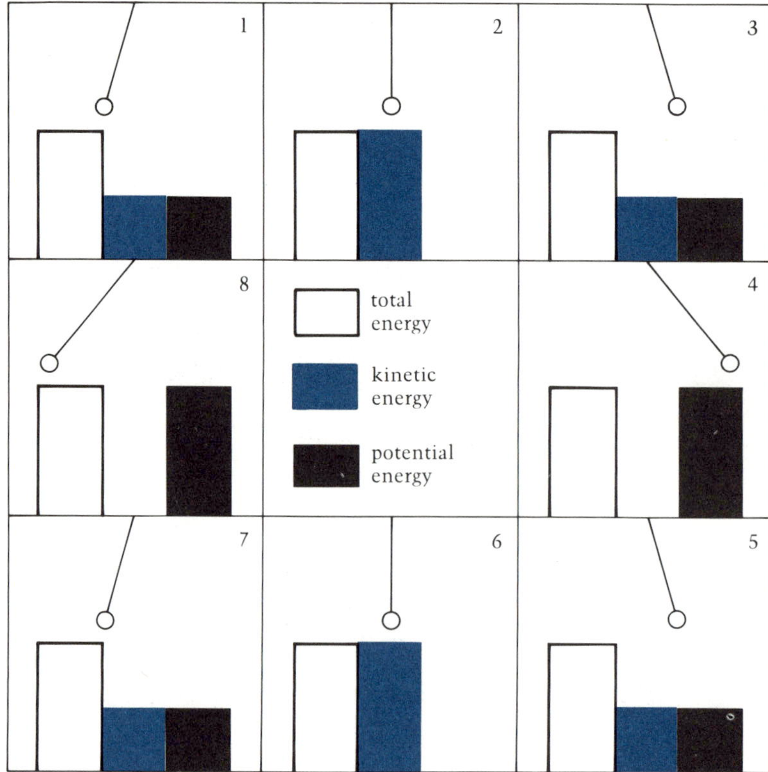

Fig. 5–10 The energy cycle of a pendulum. The clear bar is total energy, the blue bar kinetic energy, and the black bar potential energy.

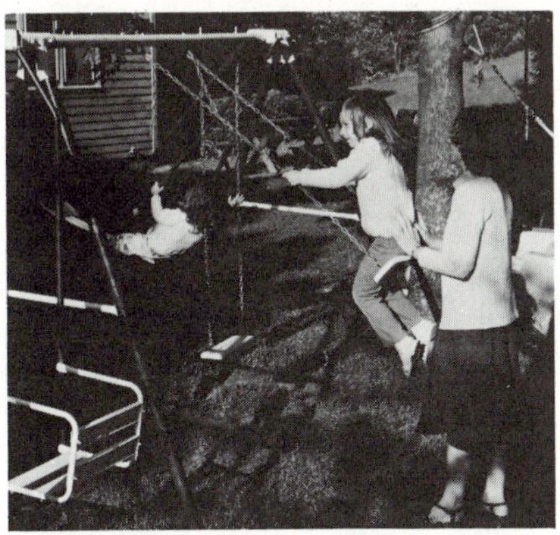

Fig. 5–11 When you push someone on a swing, push in resonance with the natural frequency of the swing. Notice, also, that at the highest point of the swinging motion, the chain is slightly bent. This is how you can raise yourself up, increasing your energy, even if no one is pushing.

potential energy. Thus, the higher you start the ball, the faster it will be traveling when it passes the bottom. As the ball starts to rise on the other side, the potential energy again increases, so the kinetic energy decreases. Eventually the ball rises to the same height at which it started, on the other side, and then starts back.

From this example, we see that we can think of an oscillating system as one in which the energy is converted back and forth from kinetic energy to potential energy. As long as there is no friction, the total energy is conserved and the swinging continues.

Suppose you have a swinging pendulum and you want it to swing higher. What is the most efficient way to get more energy into the pendulum? Think of what you do when you are pushing a child on a swing. You push the swing once each cycle, and you push it in the direction of the motion. That way you are doing work that will increase the energy of the swing. After each push, the swing will go a little higher than on the previous push. It turns out that one method of pushing the swing is really the most efficient. When your pushing keeps time with the swinging (that is, you push once per cycle, always in the direction of the motion), we say that you are pushing *in resonance* with the swing. The most efficient way to get energy into a system is to do it in resonance. Your pushing has the same period as the natural period of the swing.

5.5b SPRINGS

Many oscillating systems are based on springs. When you stretch or compress a spring, it exerts a force that is proportional to the distance the spring has been stretched or compressed. To stretch or compress a spring further you must exert a greater and greater force; in short, you must do some work. If you let the spring go, it will snap back, and the work you have done will be recovered as kinetic energy. Therefore, when a spring is stretched or compressed, it is reasonable to associate a potential energy with this new position, since when you let go the potential energy is converted into kinetic energy (Fig. 5–12).

Let's put a mass on the end of a horizontal spring, stretch it some distance, let it go, and follow it through a cycle. When you let the mass go, it has some potential energy, but the kinetic energy is zero. As the mass heads back toward its center position, its speed in-

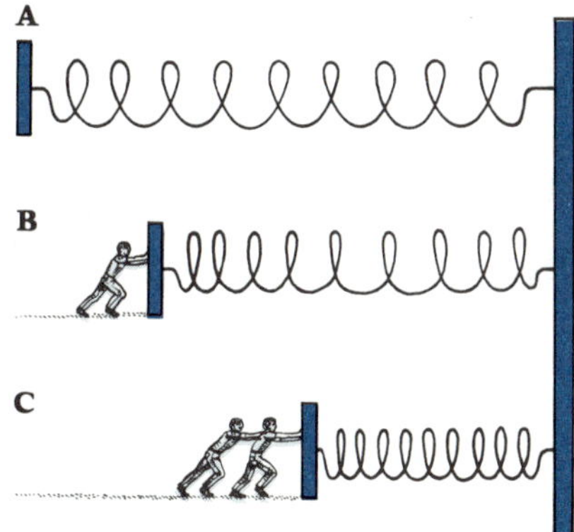

Fig. 5–12 Compressing a spring requires work. The distance that you can compress (or stretch) a given spring depends on how much force you exert. In A there is no force, and in B we apply a force and the spring is compressed a certain amount. In C the force is doubled, and the amount of compression is doubled.

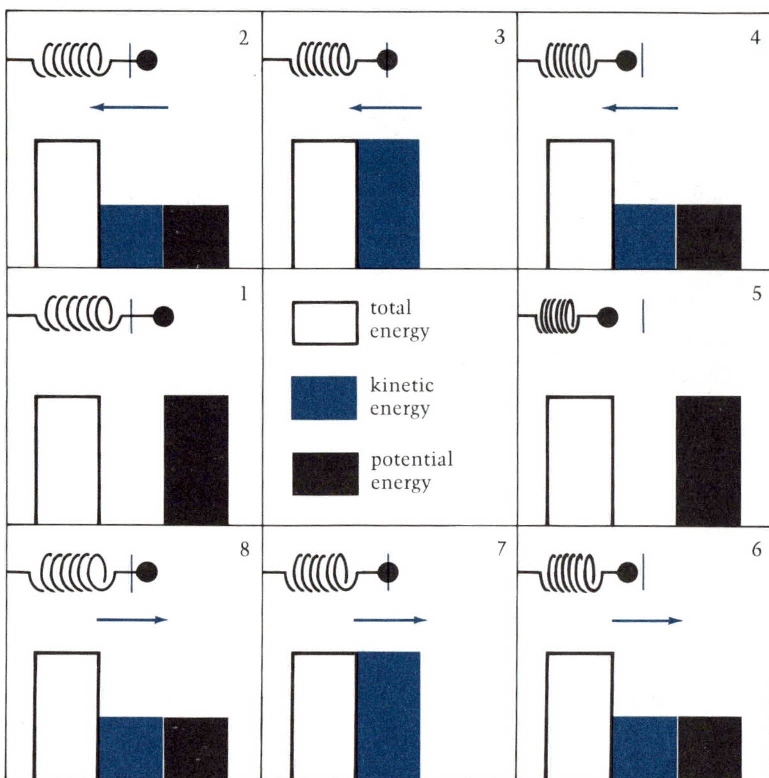

Fig. 5–13 Here we follow a spring through one full cycle of its oscillations. We use the bar graphs to keep track of the total (clear), kinetic (blue), and potential (black) energy. Notice that the total energy does not change, but the division of this total energy between kinetic and potential energy does change through the cycle. In each frame the blue vertical line represents the center position and the blue arrow represents the speed of the mass on the end of the spring.

creases. Thus the kinetic energy increases and the potential energy decreases. As the mass gets to the center position, the potential energy is zero, so the kinetic energy must equal the original potential energy. As the mass goes past the equilibrium position, the potential energy increases and the kinetic energy decreases. Eventually the mass stops, the energy is all in the form of potential energy, and the mass starts back. It will continue to go through this cycle, with the energy going back and forth between potential and kinetic, very much like the pendulum (Fig. 5–13).

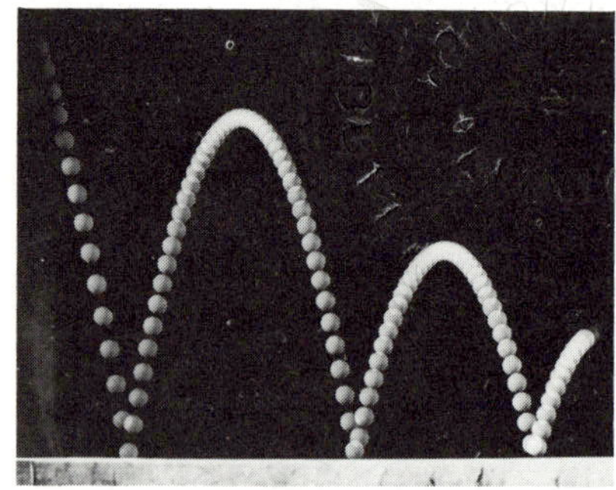

Fig. 5–14 Multiflash photo of a ball bouncing. Every time it bounces, it loses a little bit of energy (since the collision with the ground is not completely elastic). It also loses some energy to air resistance. We can tell that the energy is less for each bounce because the highest point of each successive bounce is lower than for the previous bounce.

5.6 *Energy Conservation and Collisions*

We have already seen (Chapter 4) that using conservation of momentum helps us simplify our analysis of collisions. If objects collide and there are no external forces acting, the total momentum of the objects before collision must equal their total momentum after collision. In much the same way, we can apply energy conservation to collisions and simplify calculations.

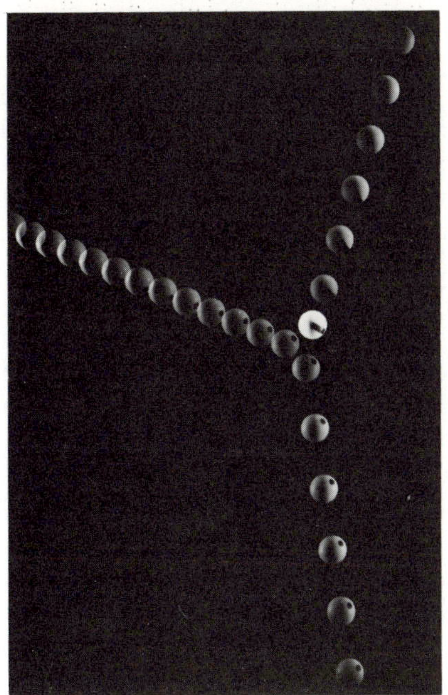

Fig. 5–15 Multiflash photo of a billiards-type shot. The striped ball was initially at rest. The dotted ball entered from the bottom of the picture and struck the other ball in an off-center hit. Notice that the images of the dotted ball are closer together after the collision than before the collision. This means that it is moving slower after the collision than before. Some of the energy of the dotted ball has been taken up by the striped ball. Also, the sum of the momentum of the striped ball and the dotted ball after the collision must equal the momentum of the dotted ball before the collision. (For this picture, the masses of the two balls are the same.)

In any collision between two objects, the total energy is conserved; that is, if we add up the energy before the collision and after the collision, we will find no change. However, it is possible that the energy will change forms during collision. For example, some kinetic energy may be converted into potential energy, or possibly some kinetic energy will go into the internal energy of one or more of the particles. The easiest cases to study are those in which the total *kinetic* energy does not change during collision. These are called elastic collisions. In an elastic collision, the sum of the kinetic energies of all the particles before the collision is equal to the sum of the kinetic energies of all the particles after the collision.

All is not lost if the collision is not elastic. We may often know why the collision is not elastic and can calculate the kinetic energy gained or lost during the collision. Often nuclear physicists learn about the structure of some nucleus by bouncing particles off the nucleus and seeing how much kinetic energy is lost in the process.

Inelastic collisions are when the particles involved are fundamentally altered during the collision. For example, if two particles stick together during collision (Fig. 5–16), some kinetic energy will be lost in the sticking process. Similarly, if one particle breaks into pieces during the collision, some energy may be released (Fig. 5–17). When elementary particles collide, the particles that emerge often bear no resemblance to the particles that came in, and the kinetic energy of the outgoing particles will not be equal to that of the incoming particles.

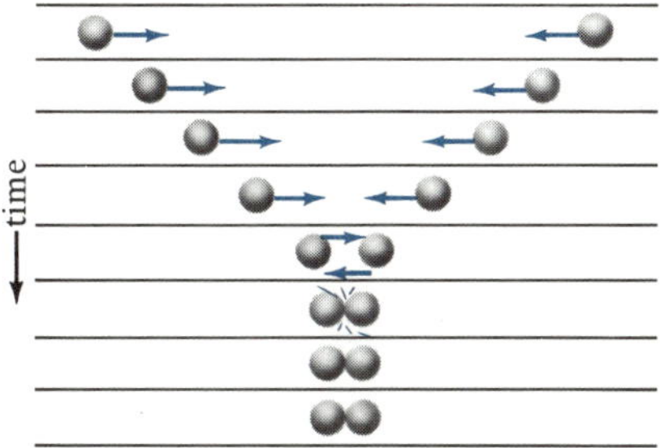

Fig. 5–16 A time-sequence drawing of an inelastic collision. Each vertical space shows the position of two balls at a slightly later time. The arrows represent the speeds of the balls. After they collide, they stick together, so some of the kinetic energy is lost in the sticking process. For the particular case shown here, the two balls have equal and opposite momenta, so the total momentum is zero. The two balls stay at rest after they have stuck together. If the initial momentum had not been zero, the two balls would have moved off together.

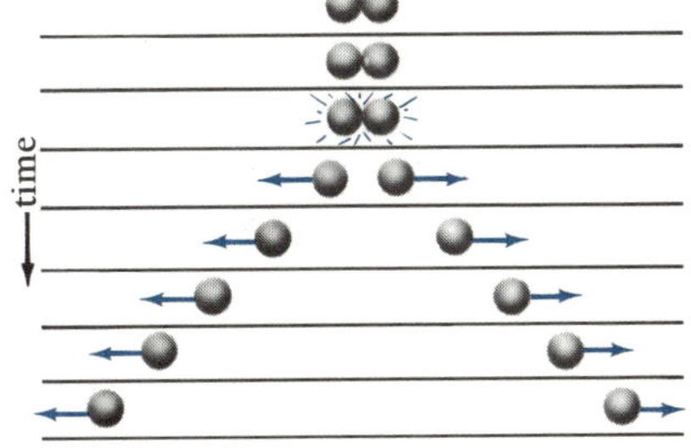

Fig. 5–17 This is the reverse situation from Fig. 5–16, with two balls breaking apart in an inelastic collision. In this case, some energy had to be added to break the two balls apart.

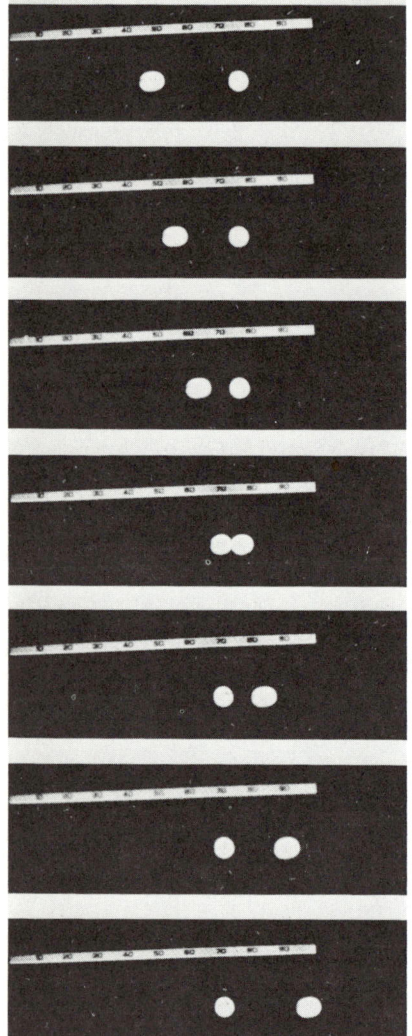

Fig. 5–18 An overhead view of a collision between two billiard balls of equal mass. The ball on the right is initially at rest. In the collision all of the momentum of the ball on the left is transferred to the ball on the right.

To see how conservation of energy really tells us something new about collisions, let's look at a very simple case from billiards. (Billiard balls are designed so that their collisions are very close to being elastic.) One ball is at rest on a table. The cue (white) ball heads toward it. The two balls have the same mass. Conservation of momentum tells us that the total momentum of the balls after collision must equal their total momentum before collision. The white ball strikes the blue ball head-on. We can think of many things that could happen in which the final total momentum is equal to the initial total momentum. For example, after collision the white ball may stand still while the blue ball moves away with the same speed that the white ball had coming in. On the other hand the two balls could stick together, each moving with half the velocity that the white ball had before the collision. Other combinations are possible. How do we know which combination is right?

In only one of the cases will the total kinetic energy after collision equal that before collision. In this case it happens to be the situation in which the white ball sits still and the blue ball moves away with the same velocity as the white ball initially had. If you figure out the kinetic energy in all of the other cases, you will find that it is not conserved.

If you have access to a pool or billiards table, try playing around with a few straight shots. See how the things we have talked about in conservation of energy and momentum seem to apply. (For the results of this chapter to apply, you must not put "english," or spin, on the ball, since there would also then be energy in the rotation, as we'll see in the next chapter.) Another system close at hand is a pair of pennies. Slide one across a table top so that it strikes the second squarely. The first will stop dead and start the other one sliding.

The application of conservation of energy and momentum to collisions is important in experimental nuclear and elementary particle physics. We know the energy of the incoming particles; once we see the directions of the outgoing particles, we can use the laws of conservation of energy and momentum to learn about the

A

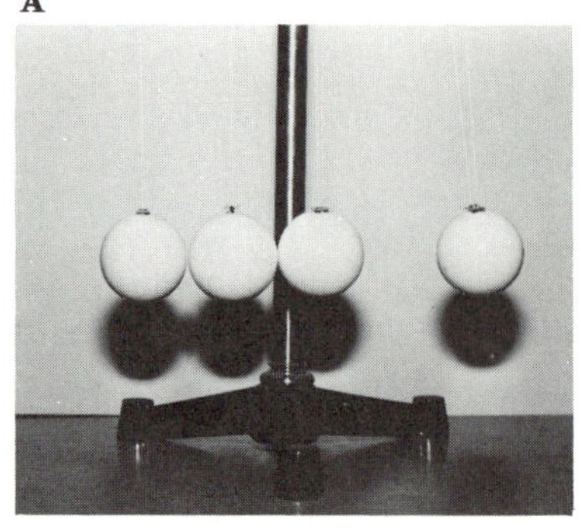

B

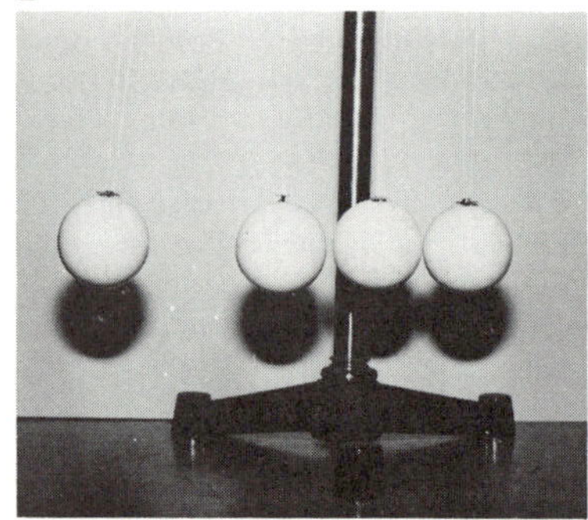

Fig. 5–19 A. When the ball on the right is allowed to swing and hit the other three balls, it transmits its momentum through the balls. B. The ball on the left is sent away with the same momentum that the incoming ball had.

A

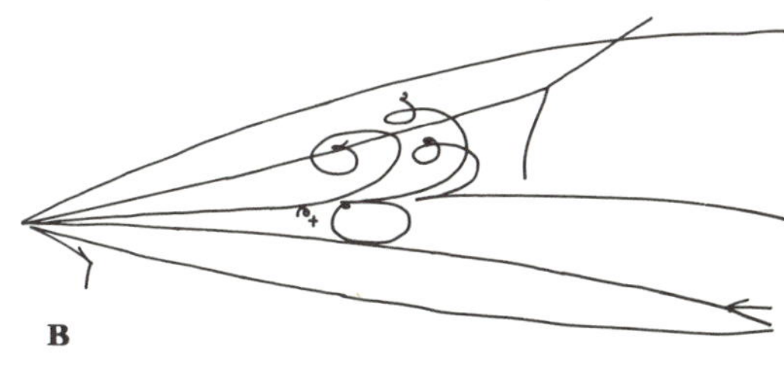

B

Fig. 5–20 A. Tracks in the 15-foot bubble chamber at Fermilab. An unseen neutrino hit a neon nucleus, producing several elementary particles. We can determine the properties of the particles coming out from the laws of conservation of energy and momentum. B. The drawing indicates which tracks come from the neutrino interaction.

particles coming out. Actually, conservation of energy is applied in a more sophisticated way to such problems. We must use the form that comes from the theory of relativity, including an accounting of the energy tied up in the mass of the particle according to Einstein's famous formula, $E = mc^2$. We will talk more about this in Chapter 8.

One interesting case occurs when a particle of low mass strikes a much more massive particle. After collision the more massive particle will recoil, but with a very small speed. This speed will be proportionately smaller than that of the recoil speed of the less massive particle by the ratio of the masses. Since the massive body has much less speed than the low-mass particle, the kinetic energy (which depends on the square of the speed) for the massive body will be much less than for the low-mass particle. Thus, if you throw a ball at a wall, which is very massive, the ball will come back with the same speed that it had when you threw it.

This analysis was based on the assumption that the wall was standing still and the ball was moving. What if the wall is moving too (Fig. 5–21)? Say we put the wall on the front of a train moving at 50 m/s. You throw the ball at 100 m/s. Now, relative to the wall,

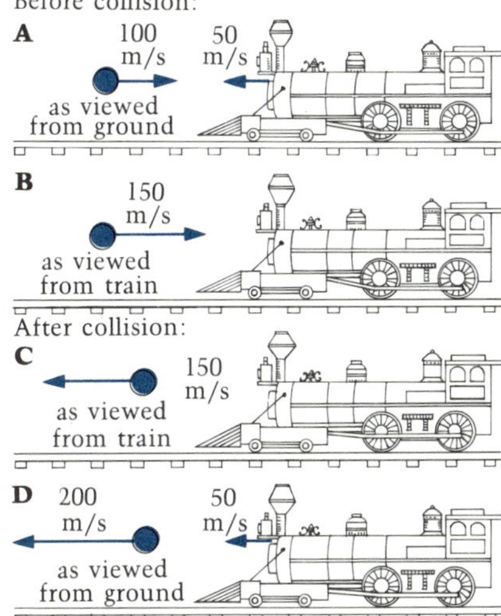

Fig. 5–21 A ball bouncing off a moving wall, with the wall attached to the front of the train. A. We see how the situation appears to an observer on the ground before the collision. The ball is moving toward the train at 100 m/s, while the train is moving toward the ball at 50 m/s. B. The speed of the ball relative to the train is 100 m/s + 50 m/s = 150 m/s; so as far as an observer on the train is concerned, the train is standing still and the ball is coming toward the train at 150 m/s. C. If the collision is elastic and the wall is much more massive than the ball, then an observer on the train will see the ball leave the train after the collision with the same speed as it came in—150 m/s. So now the ball is traveling away from the train at 150 m/s relative to the train. D. This means that the ball is now traveling 150 m/s faster than a train that is traveling at 50 m/s, meaning that relative to the ground the ball must be traveling at 200 m/s. Where did the extra kinetic energy for the ball come from? The train lost a little bit of its kinetic energy. Remember, the train started out with a much higher kinetic energy than the ball because of its much greater mass, so the train will hardly miss this little bit of kinetic energy.

the ball has a speed of 150 m/s. That means that it will bound off the wall with a speed of 150 m/s *relative to the wall.* But the wall is moving at 50 m/s, so the ball is now moving, relative to the ground, at 200 m/s. In other words, the new speed of the ball is the speed you gave it plus twice the speed of the wall.

Fig. 5–22 Reggie Jackson's strength keeps his bat from recoiling much and thus makes the ball travel relatively far.

We can apply this to hitting a baseball. If there were no recoil or distortion of the bat, the ball would leave the bat with a speed equal to the speed at which the pitcher pitched it plus twice the speed of the bat. Actually the bat recoils a little bit, so you don't do quite that well. This helps explain why a strong hitter has an easier time hitting a home run than a weak one. Your strength helps you keep the bat from recoiling. Ideally, you want to connect yourself rigidly to the earth so the recoil will be taken up by the earth, not just by you and the bat. Also, if you use a more massive bat, it will recoil less. However, you also want the bat to be moving quickly. The stronger you are, the faster you can accelerate the massive bat. It may be better to use a lighter bat and swing it faster. A lot of things work together to make a home run, but we can explain the basics by conservation of energy and momentum.

Part of NASA's exploration of the solar system employs this idea in the *gravity-assist* method. For example, if you send a properly aimed spacecraft to "bounce off" Jupiter, you can take advantage of the fact that Jupiter is so much more massive than the spacecraft that it will recoil very little. The spacecraft recoils with a speed equal to its original speed plus twice the speed of Jupiter in its orbit. Of course we can't just bounce something off Jupiter. Instead we let a gravitational encounter replace an actual collision. The spacecraft approaches Jupiter from behind on a path that brings it close, but not into direct collision. As the spacecraft is pulled in and then flung out, it is the same as though it were a rubber ball bouncing off another solid object. It really picks up a boost in speed at the expense of Jupiter (which will hardly miss the small amount of energy). This is the means by which NASA sent Pioneer 11 and the two Voyager spacecraft from Jupiter to Saturn (as we saw in Fig. 2–3).

Key Words

work
joule
power
watt
kilowatt-hour
kinetic energy
potential energy
gravitational potential energy
internal energy
law of conservation of energy
period
elastic collision
inelastic collision

Questions

1. In Chapter 3 we said that Newton's laws and a knowledge of the forces acting on a particle at any time are sufficient to deduce the motion of a particle. If this is true, what do we gain by introducing the concept of energy conservation into the study of particle motion?
2. In Fig. 5–2, suppose there were a straight chute, running from point 1 to point 4. In the absence of friction, how would the speed of the skier coming off this straight chute at point 4 compare with the speed of the skier at point 4 in the figure?
3. You push a box across a room. Your friend, exerting the same force on a similar box, pushes it across the room and then back to the starting point. How much work has your friend done, relative to the amount of work that you have done?
4. (a) Suppose you exert a force of 10 newtons in pushing a box across a floor at a speed of 1 m/s. How much work do you do in 10 seconds? (b) Suppose you now exert a force of 10 newtons in pushing a box across the floor at 2 m/s. How much work do you do in 10 seconds?
5. Two balls have different masses, but happen to have the same momentum. Which ball has the greater kinetic energy?
6. Two balls are dropped together from the same height. One ball is more massive than the other. If we ignore air resistance, which ball has the greater kinetic energy just before they hit the ground?
7. You have to carry a suitcase up a flight of stairs. Will you have to do more work if you run up the stairs than if you carry the suitcase up slowly? Explain your answer.
8. Two identical balls are at rest on top of a platform. One was carried to the top of the platform by a person climbing a ladder. The other was dropped from an airplane and caught by a person on top of the platform, who then placed it alongside the first ball. Which ball (if either) has the greater potential energy?
9. In falling through a distance of 10 meters, a particular ball gains 100 joules of kinetic energy. How much kinetic energy would the same ball gain if it fell through a distance of 20 meters?
10. A ball dropped from the third floor to the second floor of a house has a speed of 8 m/s just before it hits the second floor. The same ball is caught on the second floor, and then dropped, from rest, to the first floor. What is its speed just before it hits the first floor? (Assume that the distance between floors is always the same.)
11. When you drop a ball, the earth actually moves a little toward the ball, since momentum must be conserved. How does the kinetic energy acquired by the earth compare with that acquired by the ball? (*Hint:* The force that the earth exerts on the ball is the same in strength as the force that the ball exerts on the earth, and the distance that the earth moves is about 10^{-23} the distance that the ball moves.)

12. Two objects have the same mass, but one is traveling twice as fast as the other. How do their kinetic energies compare?
13. Two objects have the same speed, but one is twice the mass of the other. How do their kinetic energies compare?
14. One ball is moving north at 10 m/s. An identical ball is moving south at 10 m/s. How do the kinetic energies of the two balls compare?
15. Two identical cars are traveling along, one moving at 10 m/s and the other at 30 m/s. The brakes are applied and both cars are eventually brought to a stop. (a) How much work is done in stopping the faster car relative to the amount of work done in stopping the slower car? (b) How is this work related to the stopping distance for each car?
16. Explain the following statement: "If we view things on a small enough scale, there is no such thing as a nonconservative force."
17. During any part of the swing of a pendulum, which is doing more work on the ball, the string or gravity? (*Hint:* In Fig. 5–10, take note of the direction of motion at various times.)
18. In the pendulum discussed in Section 5.5a, why will a disaster result if the demonstrator gives the pendulum a little push, rather than just letting it go?
19. An outstretched spring is 1 meter long. You stretch it to a length of 1.01 m, then you stretch it to a length of 1.02 m. How does the work to go from 1.01 m to 1.02 m compare with the work to go from 1.00 to 1.01 m? (In other words, is the work more, less, or the same?)
20. Two billiard balls collide and one breaks apart. Is this collision elastic?
21. A baseball traveling at 40 m/s is struck by a bat traveling at 20 m/s. Assuming that the collision is elastic and the bat doesn't recoil (both of which are not really good assumptions), how fast does the ball leave the bat?
22. Would the gravity-assist method work if Jupiter were not orbiting the sun? Explain.

CHAPTER 6

Rotation

A merry-go-round, the spinning earth, a child's top, a hospital centrifuge—all are moving even though they stay in place. Rotating objects are common in our lives, and we must take an extra step in our analysis of motion to understand them.

Fig. 6–1 The hammer thrower. Before releasing the hammer (a ball at the end of a metal wire), the thrower gets it up to a high speed by whirling it around and around. The wire transmits the force to the ball to get the ball to move in the circular path, and the thrower's arms provide the force to the wire. When the thrower lets go, there is no longer a force to maintain the circular motion, and in the absence of forces the ball flies off in a straight line.

6.1 *Circular Motion*

Put an object on the end of a string and whirl it around. Feel the tug of the string on your hand. Whirl it around faster. How does the tug of the string change? Change the length of the string and try to whirl it around the same number of times per second as before. Does the tug feel stronger or weaker? As you whirl the string around, let it go. (Note: Try this part outdoors.)

In these experiments you are investigating the relationship between forces and circular motion. Why should there be a relationship between forces and circular motion? Because an object moving in a circle is accelerating. Its speed may not be changing, but the direction of motion *is* changing. Thus the velocity (a vector) is changing. A changing velocity means an acceleration, and to have an acceleration, you must have a force. Thus forces and circular motion are linked.

We can conclude that *whenever we see a particle moving in a circle (or any curved path), it must be under the influence of a force.* What about the object on the end of the string? What force

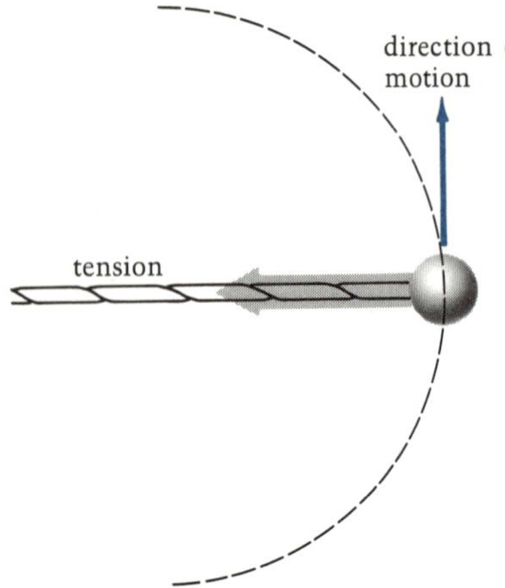

Fig. 6–2 The ball is attached to a rope and is moving around in a circular path. At the instant shown, it is curving to the left, so there must be a force to the left causing this change in the direction of motion. This force is provided by the tension in the rope. If we were to cut the rope, the force would be removed and the ball would fly off in its current direction of motion. Its path would be a straight line, since there would no longer be a force causing it to follow a curved path.

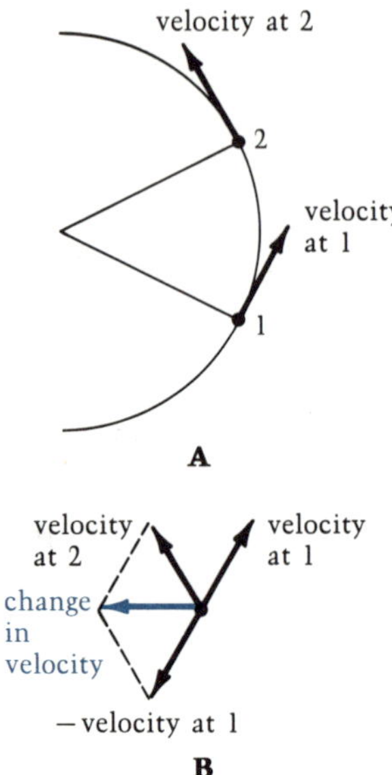

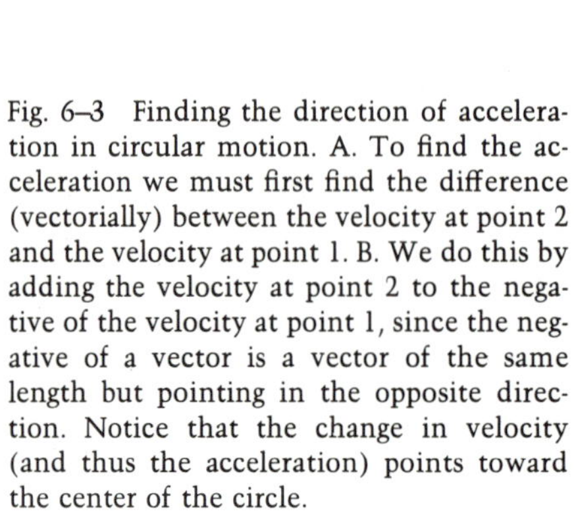

Fig. 6–3 Finding the direction of acceleration in circular motion. A. To find the acceleration we must first find the difference (vectorially) between the velocity at point 2 and the velocity at point 1. B. We do this by adding the velocity at point 2 to the negative of the velocity at point 1, since the negative of a vector is a vector of the same length but pointing in the opposite direction. Notice that the change in velocity (and thus the acceleration) points toward the center of the circle.

makes it go in a circle? It is the force of the string pulling on it, a force that is transmitted along the string as tension. The string pulls on your hand, and your hand pulls on the string, with equal and opposite force (by Newton's third law). Your hand can certainly feel any changes in the pull (Fig. 6–2).

To find the acceleration of an object moving in a circle, we take the velocity at some instant and subtract from it (as a vector) the velocity at an instant before, then divide this difference by the time interval that separates these two instants. When we do this we find that the magnitude of the acceleration is

$$a = \frac{v^2}{r}$$

$$\textit{acceleration} = \frac{\textit{speed}^2}{\textit{radius}}.$$

The speed, as we saw in Section 2.2, is the magnitude of the velocity. This acceleration is directed inward along the string (Fig. 6–3), perpendicular to the direction of motion. We know that the force required to produce an acceleration is equal to the *mass* multiplied by the *acceleration* ($F = ma$), and so

$$F = \frac{mv^2}{r}$$

$$\textit{force} = \frac{\textit{mass} \times \textit{speed}^2}{\textit{radius}}.$$

Thus, the greater the mass of the particle, the greater the force required to get it to move in a given circle at a given speed. When you try to whirl the object around faster, the force required increases quickly since it depends on the square of the speed. To keep something moving in a given circle at twice its current speed, you would have to exert four times the force.

The acceleration of an object moving in a circular path at a given speed decreases as the radius increases because the change in direction at each instant (the curve) becomes more gentle as the radius increases. Suppose we wish to double the radius and keep the speed the same. The force would be half of what it was before.

Example 1. When you drive your car at 20 m/s (about 45 m.p.h.) around a curve of radius 100 m, the acceleration is $(20\ m/s)^2/(100\ m) = 4\ m/s^2$.

Example 2. A NASA training device whirls astronauts around at a speed of 30 m/s in a circle of radius 30 m. The acceleration is then $(30\ m/s)^2/(30\ m) = 30\ m/s^2$, or about three times the acceleration of gravity.

Example 3. The blades of a kitchen blender spin at 30 times a second and the rotating part is 3 cm = 0.03 m in radius. We need the speed of the outside of this part, which must travel its circumference ($= 2\pi r = 0.2$ m) in $\frac{1}{30}$ s or at 6 m/s. The acceleration is thus $(6\ m/s)^2/(0.03\ m) = 1000\ m/s^2$, which is 100 times the acceleration of gravity.

A

B

Fig. 6–4 A. The 15-meter-long arm of the centrifuge at NASA's Johnson Space Center in Houston. B. These three astronauts in the ball at the end of the centrifuge's arm are undergoing an acceleration ten times that of gravity.

In all of these situations we are not talking about the force required to change the speed of the object. The force we are talking about is the one required to maintain the circular motion at a given speed.

What happens if the force responsible for the circular motion is removed? For example, what if you let go of the string? Now there are no forces acting on the object, so it goes off in a straight line. As Newton's first law tells us, if there are no forces on a moving object, it will move at a constant speed in a constant direction, i.e., in a straight line.

For example, let's look at a train going around a bend. What provides the inward force responsible for the acceleration? It is the rails pushing inward against the wheels. If they were not there, the train would continue to go straight (Fig. 6–5).

The train is accelerating when it goes around a bend, so some things will appear strange to the observer inside the train. Suppose you are on the train and you drop a ball. As long as the ball is in your hand, the hand is providing the force that keeps the ball

Fig. 6–5 When a railroad car goes around a bend, the rails supply the force that makes the car turn. The outer rail pushes in against the wheels of the train (arrow), causing the train to move in a circular path.

The Sun

The sun is a celestial laboratory we can use to study many topics of physics and astronomy. Like all stars, it is a globe of hot gas held together by its own gravitation, but is the only one of these celestial bodies close enough to earth for us to study it in detail.

The painting in Plate 16 *(at right, above)* shows the different parts of the sun. Deep inside, hidden from view is the solar interior, as we see in the right-hand segment of the painting. To understand the interior, we must understand how energy is transferred sometimes by radiation and sometimes by convection. At a temperature of 15 million kelvins and at pressures unattainable in laboratories on earth, hydrogen is fusing into helium. Energy is released as determined by Einstein's formula $E = mc^2$. Our only direct contact with the interior is through escaping neutrinos, subatomic particles that are released in the fusion reactions. We will see later in the text how recent work in atomic physics indicating that neutrinos may have mass applies directly to our understanding of the solar interior.

The level of the sun from which we receive light every day is called the photosphere. It is covered with small regions of convection called granulation. In the photosphere, we find cooler regions called sunspots. The bottom segment of the painting shows this ordinary view of the sun. Above the photosphere we find a region that glows colorfully pinkish at eclipses and is thus called the chromosphere. It is slightly hotter than the photosphere. We can study the chromosphere especially well in the red light from hydrogen atoms, a view we see in the left-hand segment of the disk.

At a similar temperature to that of the chromosphere, we find regions held together in beautiful shapes by the sun's magnetic field. When we see them projected onto the surface of the sun they are called filaments; when we see them off the edge of the sun projected against dark space they are called prominences. With the International Ultraviolet Explorer spacecraft, we are now able to study the chromospheres of many other stars besides the sun.

When the photosphere and chromosphere are covered by the moon at solar eclipses, we see a pearly white halo of light around the sun. It is called the corona, from the Latin word for crown. The corona contains gas at a temperature of two million kelvins held together by a magnetic field; this is a situation similar to the one that we want to control to harness nuclear fusion to generate energy on earth. The process of conduction is important in the corona for transferring radiation.

The top segment of the sun in the painting shows how the sun looks in x-rays; the radiation comes from the hottest regions of the corona. We can study x-rays from the corona with space satellites even without eclipses. Material escaping from relatively cool regions of the corona, called coronal holes, affects the upper atmosphere of the earth. With the Einstein Observatory spacecraft, we can study x-rays from many other stars, including some both hotter and cooler than the sun. We have recently found that many of the stars have stronger coronas than we expected, meaning that astrophysicists have a lot of work to do to explain this difference.

In the following pages, we will see a number of views of the sun from the ground and from space, each showing some different aspect of the sun and the physical processes in it.

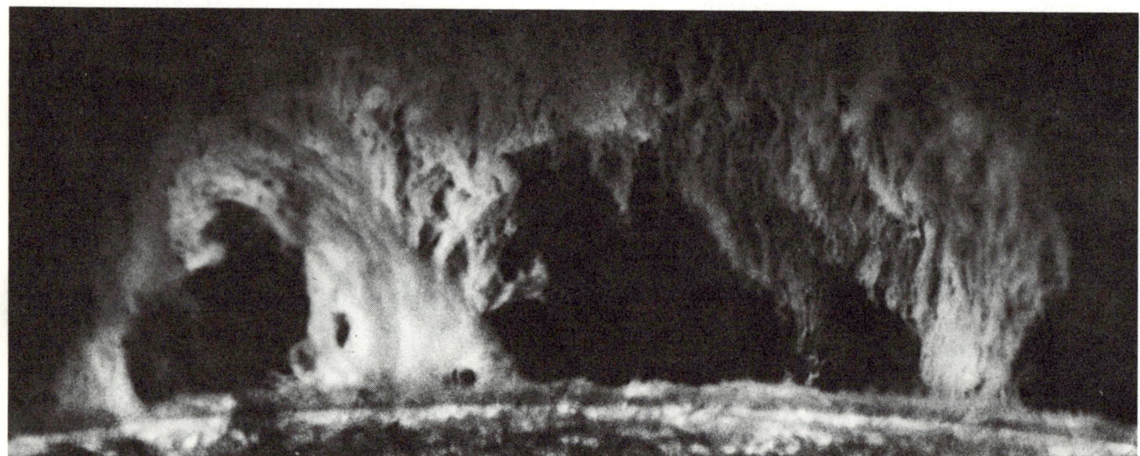

Fig. 2 A solar prominence, which takes its shape from the sun's magnetic field.

Plate 16 *(above)* is an artist's conception of the sun in x-rays (top), a cutaway of the interior (right), a view in white light (below), and in hydrogen light (left). Prominences as seen in the ultraviolet light of helium from a spacecraft surround the edge of the sun at right, an eclipse view appears outside the edge at left, and the x-ray view continues over the edge at top. (Painting by Anne Norcia)

Plate 17 *(below)* shows the drawing of the solar spectrum made by Joseph Fraunhofer in about 1814, on which he reported his discovery of the dark spectral lines crossing the band of color.

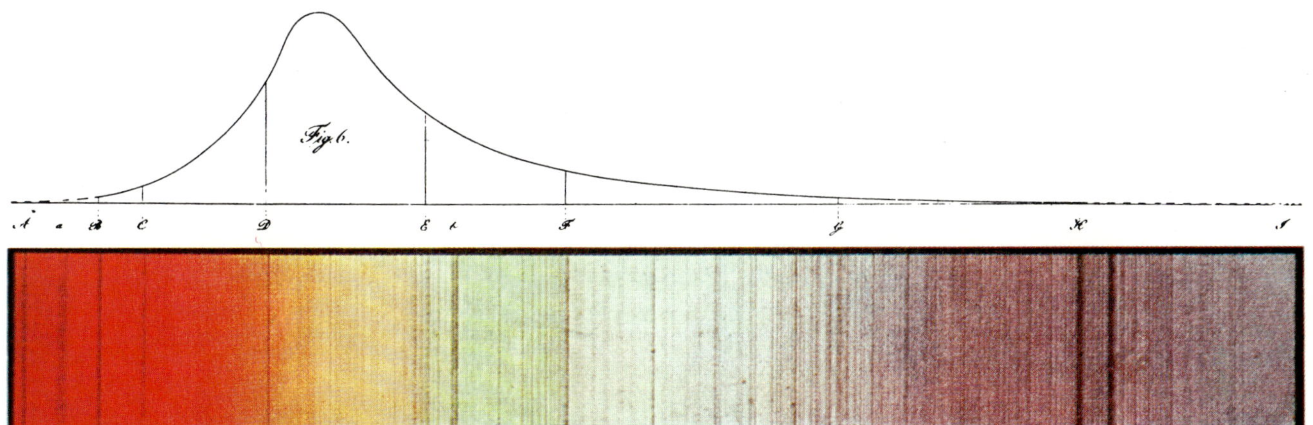

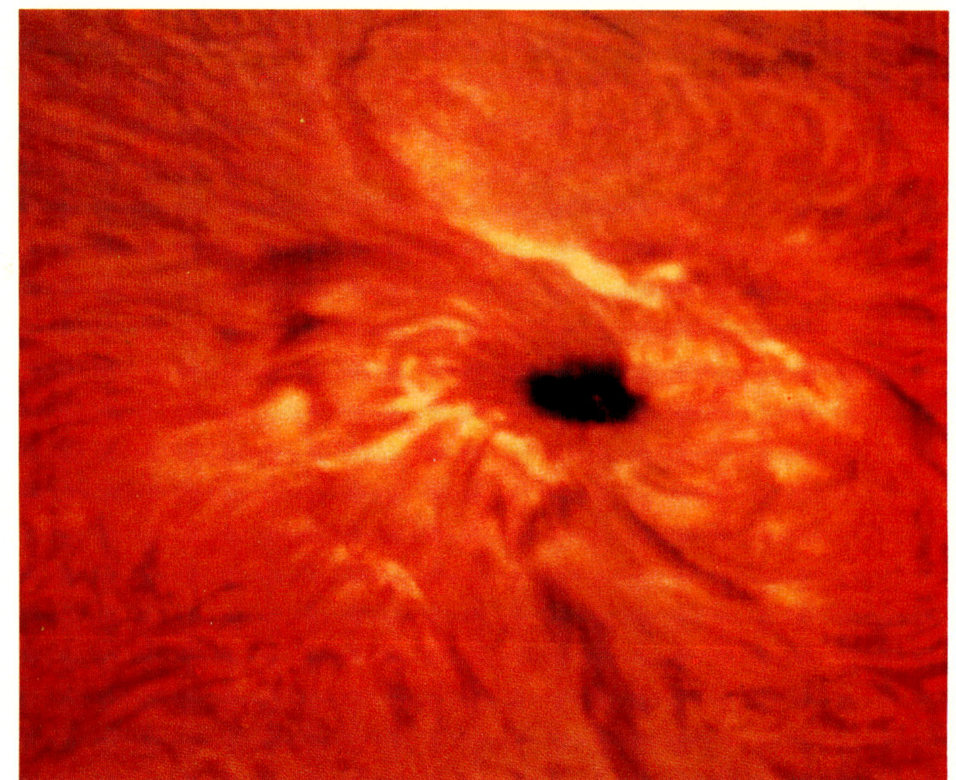

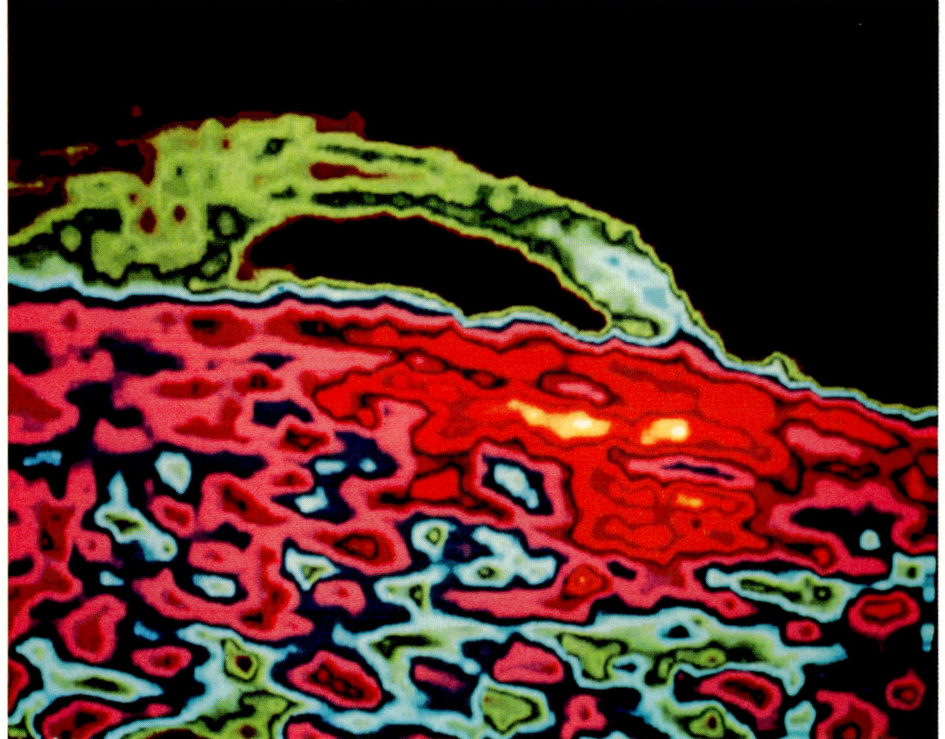

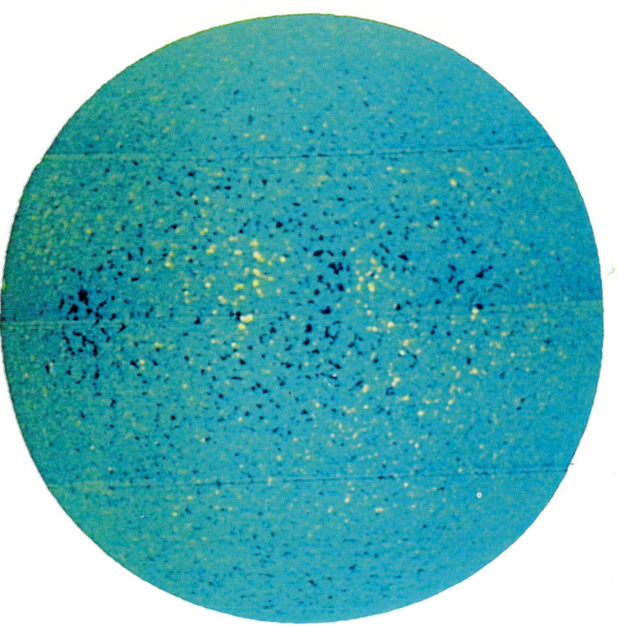

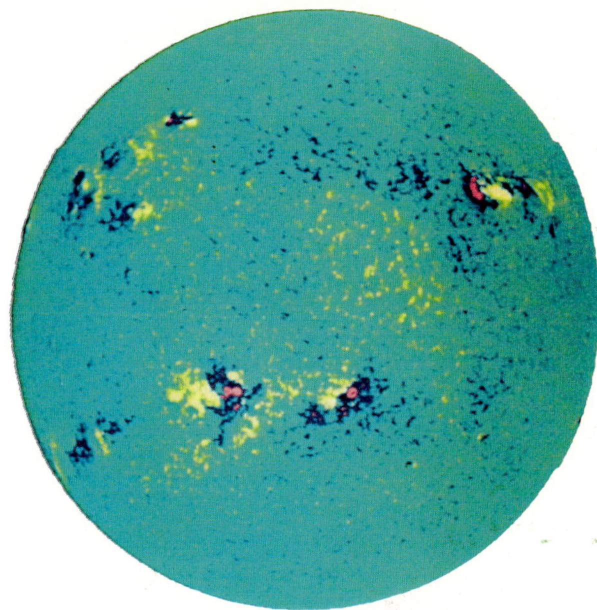

In Plate 18 *(facing page, above)* we see a view of a sunspot and a surrounding region in the red light from the first line of the Balmer series of hydrogen. A dark filament snaking downward from the top of the picture is following the magnetic field. You can also see other narrow dark structures following the magnetic field and pointing into the sunspot as iron filings point to the poles of a bar magnet.

In Plate 19 *(facing page, below)* we see a false-color view of a prominence assembled from spacecraft data. Each color shows gas at a different temperature, so the color composite shows how the temperature varies within the prominence. (Courtesy of Harvard College Observatory, from Skylab)

Plate 20 *(above)* shows the magnetic field of the sun at different parts of the solar activity cycle. One polarity is shown as purple and the other as yellow. At left, in 1976, the sun was relatively quiet and no regions of strong magnetic field appear. At right, in 1978, the number of sunspots was increasing with a vengeance. Note how each sunspot has regions of both polarities. Also, one polarity always precedes the other in the northern hemisphere while the other precedes in the southern hemisphere.

In Plate 21 *(below)* we see three maps of a sunspot region. The one on the left shows the magnetic field, the one in the center shows the white-light view, and the one on the right shows velocities measured with the Doppler effect. The yellow regions are receding from us. (Kitt Peak National Observatory photos, courtesy of William Livingston)

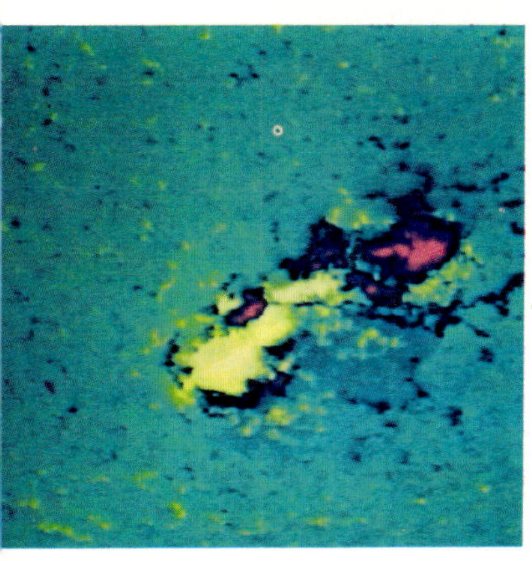

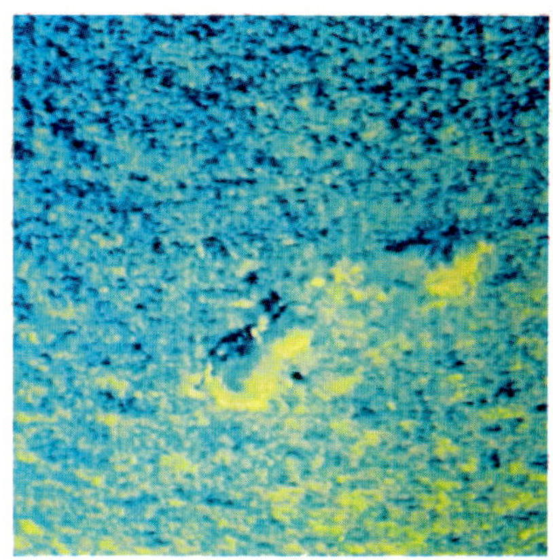

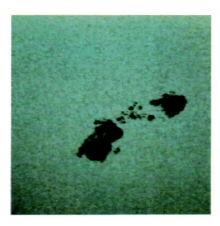

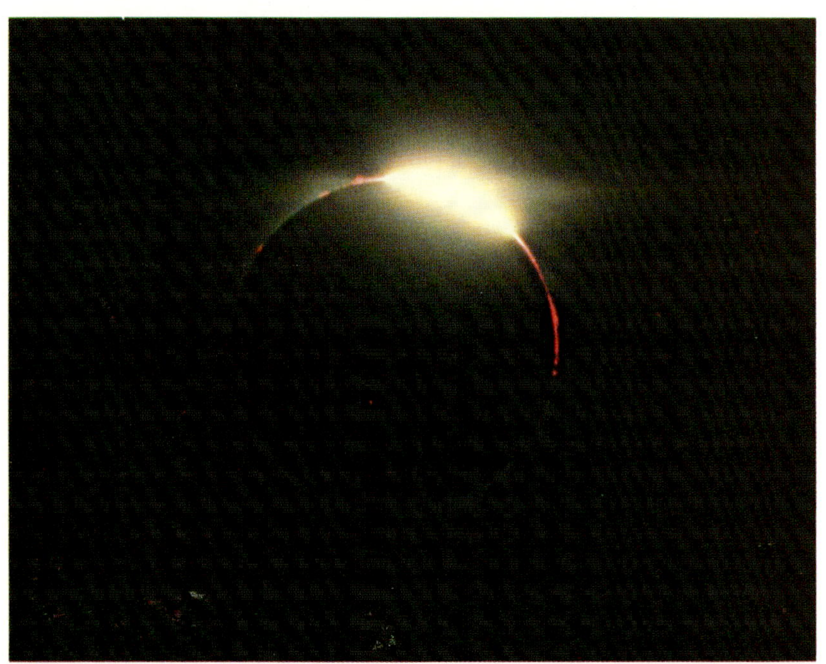

The diamond-ring effect is seen in Plate 22 *(at left)*, at the 16 February 1980 eclipse observed from India. The bright photosphere shining through a valley at the edge of the moon is so bright relative to the corona that it shines like a diamond. The pinkish chromosphere makes the diamond's setting.

The wide-angle view in Plate 23 *(below)* from the 1980 eclipse in India shows the dark daytime sky with the corona in it above. On the horizon, though, we see out of the zone of totality and observe a sunset-like effect. We can see the corona on a video screen from a television camera used to record the pointing of the telescopes, and data appearing in green on an oscilloscope screen.

Plate 24 The corona at the 1980 total solar eclipse. Coronal streamers extending in all directions are typical of this particularly active phase of the sunspot cycle. (Plates 22–24 by Jay M. Pasachoff and students, Williams College/National Science Foundation/National Geographic Society Expedition)

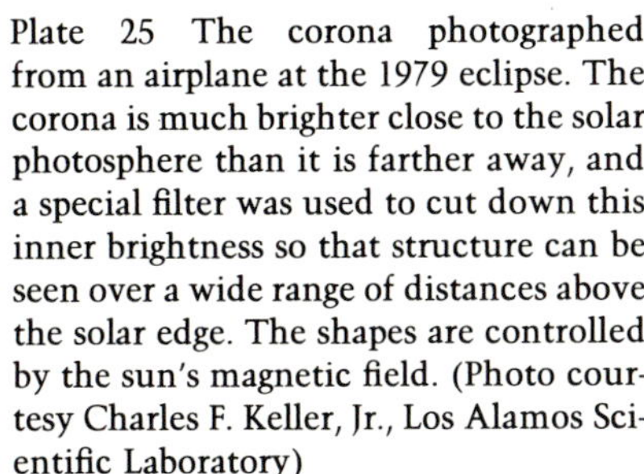

Plate 25 The corona photographed from an airplane at the 1979 eclipse. The corona is much brighter close to the solar photosphere than it is farther away, and a special filter was used to cut down this inner brightness so that structure can be seen over a wide range of distances above the solar edge. The shapes are controlled by the sun's magnetic field. (Photo courtesy Charles F. Keller, Jr., Los Alamos Scientific Laboratory)

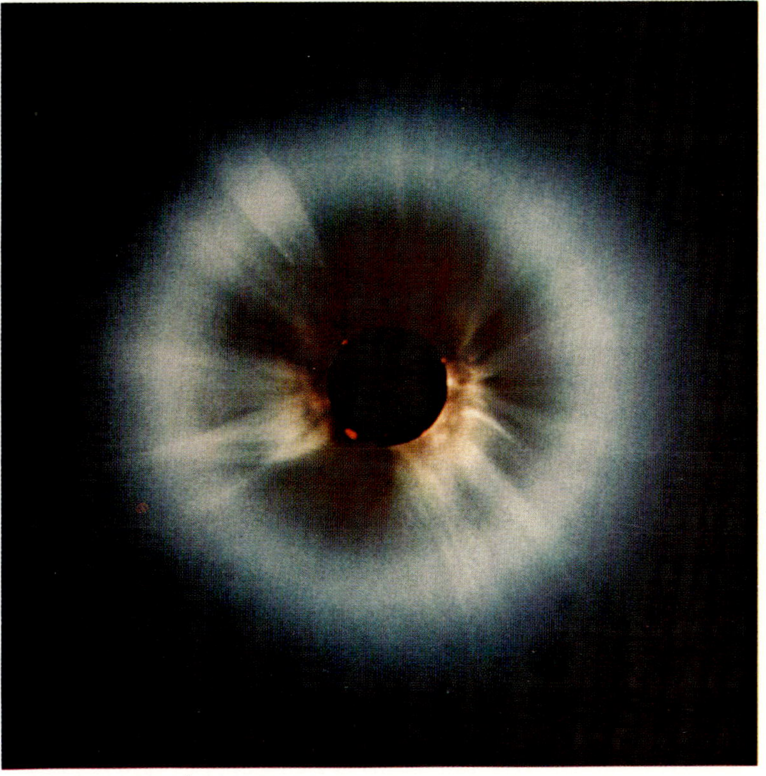

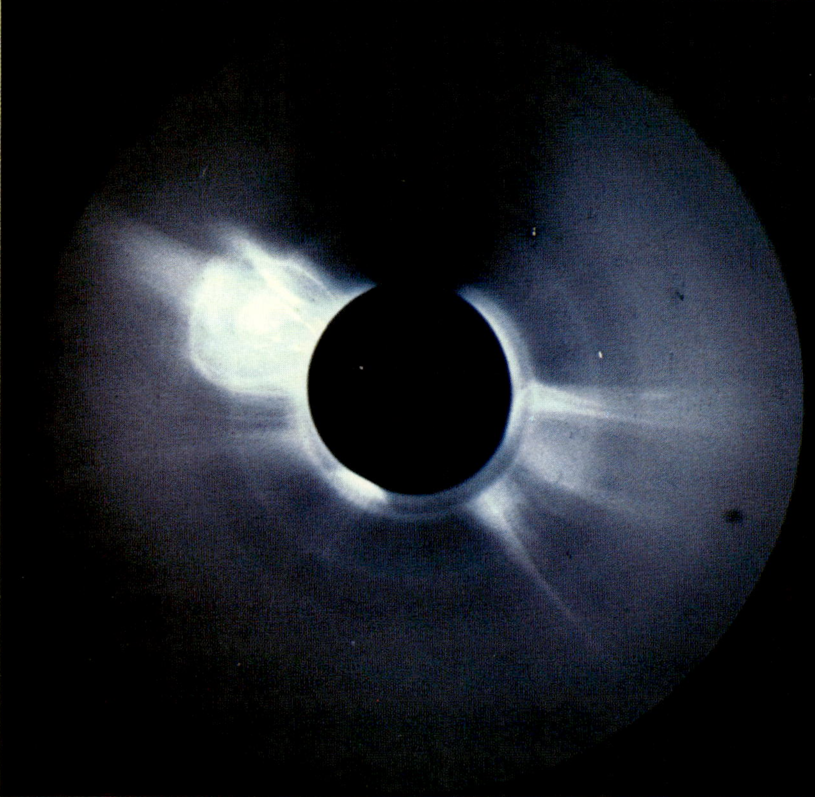

Astronauts made many solar observations from the Skylab spacecraft that was aloft in 1973 and 1974. Plate 26 *(above)* shows an eruption of a prominence viewed in the ultraviolet (inner reddish region) superimposed on a white-light picture of the outer corona made by blocking out the photosphere. The eruption of the prominence has obviously led to an outward-moving disturbance in the corona. (Ultraviolet observation by Naval Research Laboratory; white-light coronagraph observation by High Altitude Observatory; NASA spacecraft)

In Plate 27 *(at left)* we see a giant eruption moving outward, viewed with the Skylab white-light coronagraph. Particles and radiation from solar flares and other eruptions interact with the earth's magnetic field to make auroras and magnetic storms, to cause surges on power lines, and to disrupt radio communications. (High Altitude Observatory)

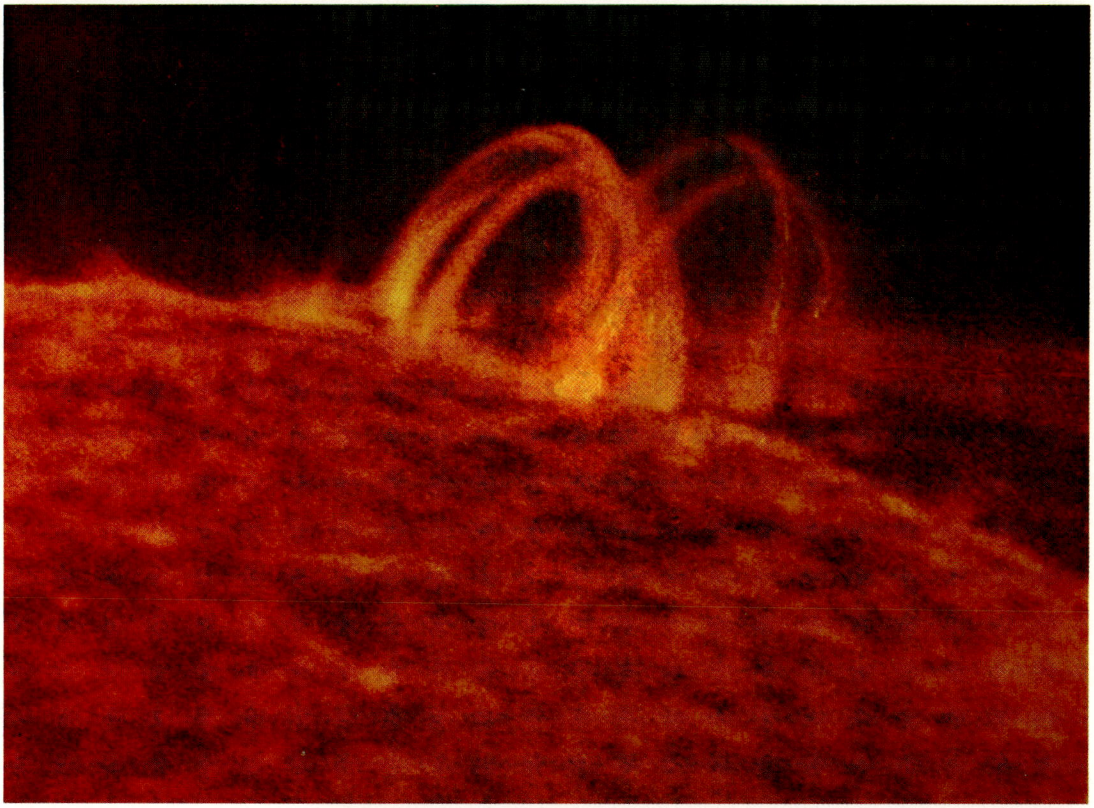

Plate 28 *(above)*, a close-up of prominences taken in the ultraviolet from Skylab, clearly shows how well material can be channeled into loops by the solar magnetic field. (Naval Research Laboratory)

The x-ray view of the sun in Plate 29 *(at right)* shows the hottest regions, over 4 million kelvins, which emit the x-rays. These regions are in the corona. A grid showing the solar surface has been added. Note at the edge of the disk how these hot regions of the corona are high above the photosphere. (R. M. Wilson, E. S. Reichmann and J. E. Smith, NASA/Marshall Space Flight Center)

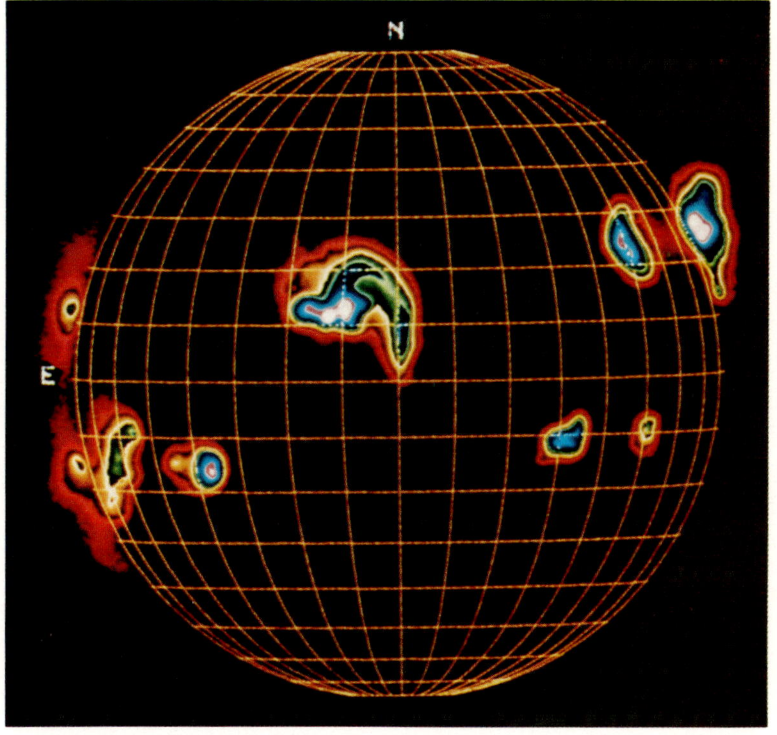

A new NASA spacecraft, Solar Maximum Mission, was launched in 1980 to study solar activity at this peak of the sunspot cycle. Plate 30 *(above)* shows a false-color view of what its coronagraph sees in visible light. Different colors show different intensity regions. The brighter regions, in general, have higher densities of gas. Blue shows the densest regions, and yellow shows the least dense regions.

The coronagraph contains a disk that blocks out the inner parts of the sun, which are too bright to study at the same time as these faint outer regions. The disk is 1.75 times the diameter of the sun, so the inner corona is covered. The series of narrow colored rings around the disk show the diffraction pattern of the disk. (Courtesy of Lewis House, Ernest Hildner, William Wagner, and Constance Sawyer, High Altitude Observatory/NCAR/NSF and NASA)

In Plate 31 *(below)* we see a series of images taken with the ultraviolet spectrometer aboard Solar Maximum Mission. The images in the top row show the intensity, and the images in the bottom row show the velocity measured from the Doppler shift. Red represents material moving away, and blue represents material approaching.

We see a jet of hot gas rising in a magnetic structure that has the shape of a loop. The velocity picture shows material coming toward us at the top of the loop, which indicates that material may be going over the top of the loop. (Courtesy of Bruce E. Woodgate, Einar Tandberg-Hanssen, and colleagues at NASA's Goddard and Marshall Centers)

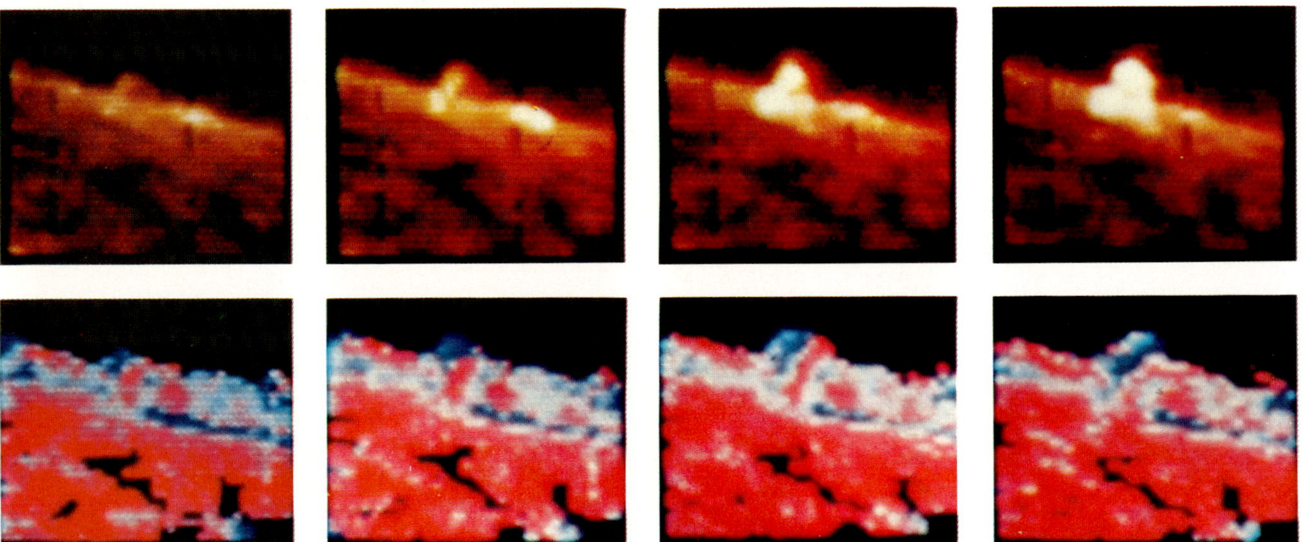

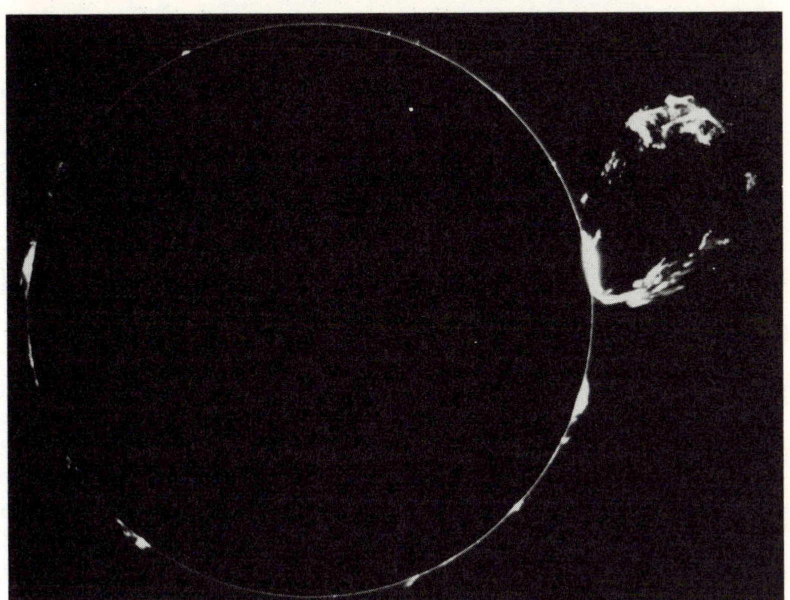

Fig. 3 Many of the events studied from Solar Maximum Mission can be observed in the visible part of the spectrum from the ground. This erupting prominence was photographed in 1979 with the coronagraph of the Institute for Astronomy of the University of Hawaii from a site high atop the volcano Haleakala. The air is so clear there that the dark disk we see at the center can be inserted to block the bulk of the solar radiation. The prominence spread a million kilometers in only about an hour.

(at right) Fig. 4 The solar magnetic field extends far into space and takes a spiral form. It guides the solar wind. Here we see an artist's conception of the warped spiral shape of the field as it extends to the orbit of Jupiter. The solar wind affects the earth's upper atmosphere and perhaps terrestrial weather. (Painting by Werner Heil under the direction of John M. Wilcox, Stanford University)

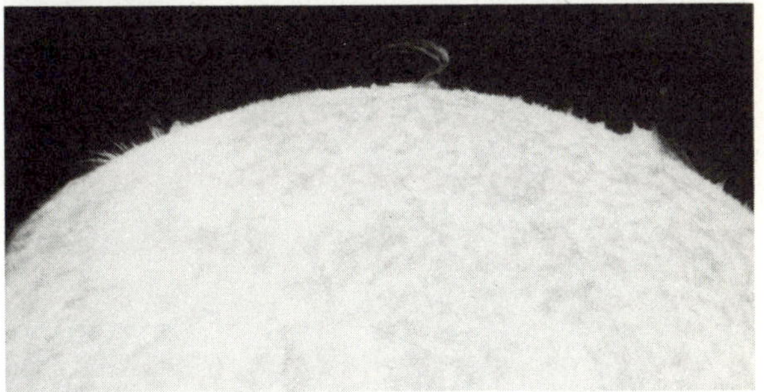

(below, right) Fig. 5 The sun is sometimes observed from rockets sent above the earth's atmosphere for brief periods of time. This observation shows the solar edge in the fundamental radiation of hydrogen known as Lyman alpha, which is so far in the ultraviolet that it has been little studied in the past. We see unprecedented Lyman-alpha detail in the filamentary structure of the solar prominences along the edge of the sun. We can also see very small eruptions, known as spicules, which are barely visible along the edge. These spicules probably entirely make up the solar chromosphere, and so are important in the transfer of mass and energy outward from the solar surface. (Courtesy of Roger M. Bonnet)

Solar research continues on a variety of fronts, including ground-based observations from optical and radio observatories, and at eclipses; observations from airplanes and rockets; and observations from space. In the mid-1980s NASA hopes to launch a pair of spacecraft known as Solar Polar Mission. These spacecraft will use Jupiter's gravity to propel them out of the plane of the orbits of the earth and the other planets in order to study the solar wind out of this plane and to view the solar poles. A Solar Optical Telescope, a large telescope devoted to high-resolution solar observations, is also on the drawing boards for launch by Space Shuttle.

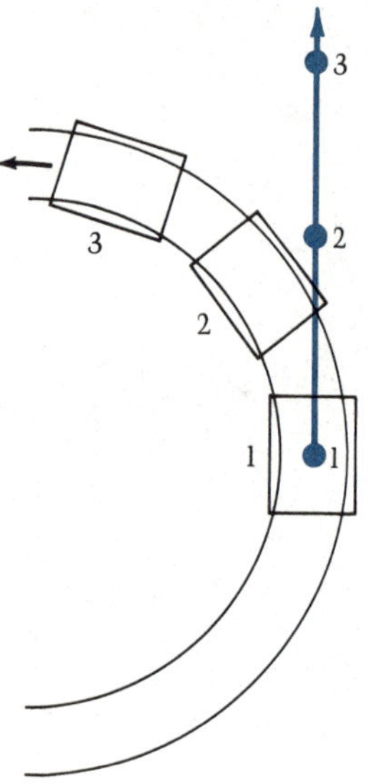

Fig. 6–6 A ball released on a flatcar going around a bend. As soon as the ball is released, there is no longer any force on it and it will move in a straight line as the train goes around the bend. The ball will move in the direction that the train was moving at the instant the ball was released and with the speed that the train was moving at the instant the ball was released. In this figure the ball was released at position 1. The positions of the ball and the train are shown at two later times (positions 2 and 3). Notice that the ball gets farther away from the train. An observer on the train would see the ball curving away from the train, but an observer off the train could tell that the train was actually curving away from the ball.

moving in a circle with the train. (Of course it is the friction of the floor that provides the force that keeps *you* moving in a circle with the train.) As soon as you let go of the ball, there is no force to keep it moving in a circle. As viewed from above by an observer outside the train, it will travel in a straight line. (This motion is just like that of an object that has been whirled around on a string that is then cut.) However, as the ball moves in a straight line, the train is still turning. To you, on the train, it will appear as if the path of the ball is curving in the opposite direction (Fig. 6–6).

If you didn't know better, you might think that there was a force on the ball, forcing it to curve that way. Of course, there is no force at all. It is *your* curving one way that makes it *appear* that the ball is curving the other. This is another example of the pseudo-forces that we talked about in Chapter 3. We even give this pseudo-force a special name—centrifugal force. Remember that although we speak of a "centrifugal force," sometimes for convenience and sometimes out of carelessness, *it is not a force.* It is an artifact of your acceleration.

To illustrate this, we can look at what happens to you inside a turning automobile. First of all, we might ask what provides the force that causes the car to turn. It is the tires. The actual force involved is the friction of the tires against the road. As you turn the wheels, the tires push outward on the road and the road pushes inward on the tires. As you make the turn tighter (decrease the radius of your circle) or try to go faster, a greater force is required since your acceleration is greater. Eventually you will reach the

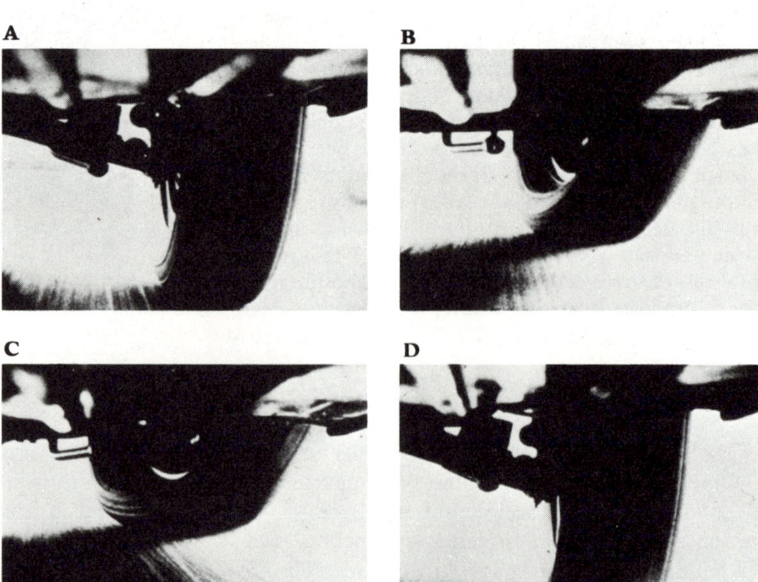

Fig. 6–7 If a car is to turn, a sideways force is necessary. This is provided by the friction between car and road. In these photos the force is so great that the tire is distorted.

Fig. 6–8 Banking the road helps you take a turn without skidding. Notice that the "upward" push of the road, which is directed perpendicular to the road surface, is now directed slightly inward on the turn. This provides a little extra inward force to help the car turn.

Fig. 6–9 A bobsled goes around a banked turn at high speed. Since the turn is banked, the ice provides an inward force on the sled allowing it to go in a curved path.

point for which the force required is greater than is provided by the static friction between the tire and the road. When this happens, the car cannot continue the turn and you skid. Obviously your car can make tighter and faster turns when the friction is greatest, as on a dry day rather than a wet day.

We can help traction along by tilting or "banking" the road inward. Part of the contact (nonfriction) force that the road exerts on the tire is now directed inward. This inward part of the contact force, added to the friction force, allows a greater acceleration. The turns for high-speed race tracks are often banked very steeply.

Now that we have the car going safely around the turn, let's look at you, the passenger (or driver). If you stay in your seat, then you will also be going around the turn and will be undergoing the same acceleration as the car. This acceleration must be produced by some force. This force is exerted on you by the seat, the seatbelts, and perhaps even the side wall of the car. If the car is making a left turn, you'll feel that force on the right side of your body. If you're not wearing a seatbelt and make a high-speed tight turn, the friction of the seat may not be enough to produce the required acceleration. As the car turns left, you will slide across the seat to the right until you hit the door, which will finally provide the force to pull you around. This sliding to the right is nothing more than your body's exerting its tendency to travel in a straight line when no force is applied. We sometimes say that your sliding results from "centrifugal force," but it really results from the *absence* rather than the presence of a force.

Fig. 6–10 Seventy meters above the ground these passengers with faith in physics are pressed into their seats.

Strange things can happen when you are going around in circles. Imagine that you are on a merry-go-round riding a horse in the outer circle of horses. A friend is riding a horse on an inner circle, between you and the center. You throw a ball directly at your friend. The ball travels in a straight line while you and your friend

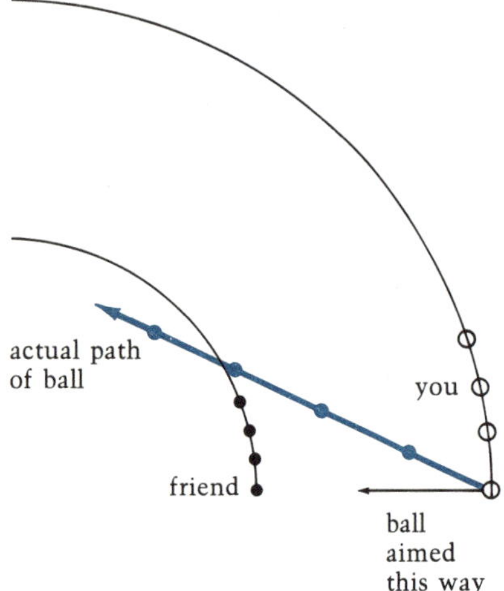

Fig. 6–11 This is what happens when you try to throw a ball to a friend riding on an inner ring of a merry-go-round. In this figure we show you the positions of you (open circles), your friend (closed circles) and the ball (blue circles) at equally spaced time intervals. Notice that you aim the ball directly for your friend, giving it some inward velocity. However, once the ball leaves your hand, it will also have a component of its velocity that is exactly equal to your velocity at the instant you released the ball. The ball will pass in front of your friend's horse and will appear to be curving away, even though we can see that the ball is really traveling in a straight line.

continue to circle. The velocity of the ball has two parts: (1) the sideways speed of your horse at the moment you threw the ball and (2) the inward speed you gave it with your throw. The ball will actually pass in front of your friend because you, on the outer circle, are traveling forward at a faster rate. If your friend throws a ball directly at you, it will also miss, passing behind you (Fig. 6–11).

If we are unaware of the strange effects of circular motion, we may think that some force makes the ball curve away from its target. We now know that nothing of the sort is happening. The ball is traveling in a straight line, but the target is curving away from it and moving at a slower speed than you. This is another pseudoforce. We call it the Coriolis force.

If you could hover high above the north pole of the earth and look down, you would see that our analysis of the merry-go-round applies equally well for things traveling from equator to pole or pole to equator. If you aim them in north-south lines, they drift either eastward or westward as a result of the earth's rotation. This drift is responsible for the circular pattern of winds associated with weather systems. The winds cycle one way on the northern hemisphere and the opposite way in the southern hemisphere. Venus rotates much more slowly than earth, so Coriolis forces have less effect on that planet's weather. Mars rotates at a similar speed and has a similar Coriolis force. Jupiter, on the other hand, rotates faster than earth and shows Coriolis effects in its Great Red Spot and in smaller storms.

Traditionally we say that water circles one way when draining from a bathtub in the northern hemisphere and the opposite way

Fig. 6–12 The Coriolis effect makes storms whirl in a counterclockwise direction in the earth's northern hemisphere and in the opposite direction in the southern hemisphere. Here 1980's huge Hurricane Allen in the Gulf of Mexico and Hurricane Isis in the Pacific west of Mexico are rotating counterclockwise. The storm near the southern end of Chile is rotating clockwise.

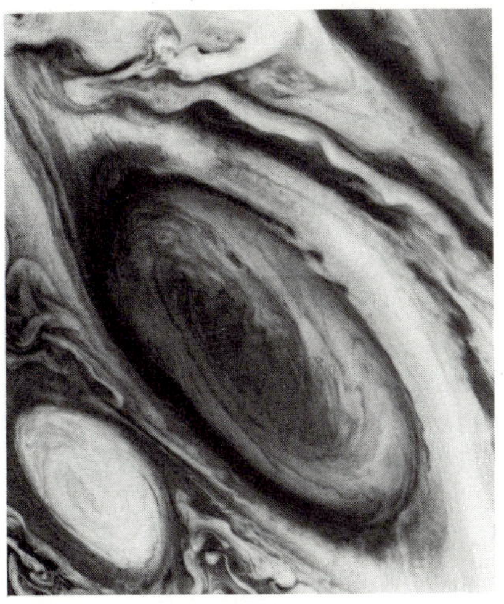

Fig. 6–13 The Great Red Spot on Jupiter was revealed by the Voyager 1 spacecraft to be a rotating storm many times larger than the earth. Notice how the Coriolis force makes the white oval to the lower left circulate in the same sense.

in the southern hemisphere; actually, on this scale other effects, such as friction between tub and water and turbulence in the water, usually dominate and we find only a random effect in our own baths. But in the name of science someone once did a careful test in a tub of water, using the smoothest tank and letting all eddies die out before opening the drain. Needless to say, he found the coriolis effect.

6.2 *Describing Circular Motion*

Even though circular motion requires a continuous change in direction, there is regularity to this change. For example, as long as the circling object goes at the same speed and on the same path, it will complete each cycle in the same time interval. The time required to make one cycle is called the period, just as it was for the pendulum.

6.2a ANGULAR VELOCITY

Sometimes it is more convenient to talk in terms of the number of cycles completed per second rather than the number of seconds per cycle. The number of cycles per second is called the frequency. Like speed, frequency is a rate. It tells us how fast something is going. These definitions reveal that:

$$\textit{frequency} = \frac{1}{\textit{period}}.$$

For example, if it takes ½ second for an object to complete one cycle, the frequency is 2 cycles per second.

So, even though the direction of motion is constantly changing, if the object's speed remains constant and it stays in the same circle, its frequency will remain constant. We use another quantity, related to the frequency, to tell us how fast the object is going around. This quantity is called the angular frequency or angular speed, which tells us how fast the object is sweeping through angles as it goes around. Since there are 360° in a circle, a frequency of 1 cycle per second corresponds to an angular speed of 360°/s and a frequency of 2 cycles/s corresponds to an angular speed of 720°/s.

When the angular frequency of circular motion changes, then we say that there is an angular acceleration. (This is not to be confused with the acceleration that is responsible for the circular motion in the first place.) Angular acceleration is related to the application of a force, but not all forces will produce an angular

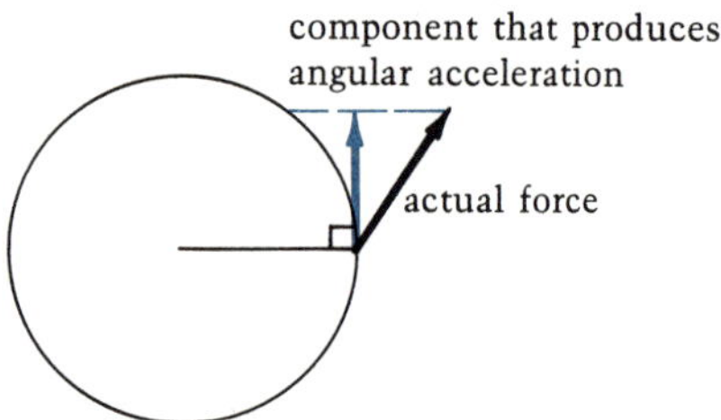

Fig. 6–14 An angular acceleration can only be produced by the component of the force that is perpendicular to the line joining the center of the circle and the object on which the force is being applied.

acceleration. In order to be most efficient at producing an angular acceleration, the force must be along the circular path, either forward or backward. This situation is shown in Fig. 6–14. Instead of saying that the force must be along the direction of motion, we often say that the force must be perpendicular to the line joining the moving object and the center of the circle.

6.2b TORQUE

We use still another quantity to gauge the effectiveness of a force in producing angular acceleration. This quantity is called torque. Torque is defined as follows:

$$\textit{torque} = \begin{bmatrix} \textit{component of force} \\ \textit{perpendicular to} \\ \textit{line joining center} \\ \textit{of circle and point} \\ \textit{where force is} \\ \textit{applied} \end{bmatrix} \times \begin{bmatrix} \textit{distance from center} \\ \textit{of circle to point} \\ \textit{where force is} \\ \textit{applied} \end{bmatrix}$$

We can think of the torque as a measure of the amount of "twisting" that we get from a given force. The torque of an automobile engine, for example, is a measure of its ability to cause the drive shaft of the car to turn around. A car with higher torque has greater pulling power than one with lower torque. Using a lower gear increases the torque.

Note that torque, although it is a force multiplied by a distance, is not the same as work. When we calculate the amount of work done, we have to measure the distance that an object moves while a force is applied. When we calculate torque, we need to know the distance from the point at which the force is applied to the center of the circle. The distance involved in calculating torque is sometimes called the lever arm. The same force can produce a greater torque if it acts through a greater lever arm.

There is an important relation between torque and angular acceleration, just as Newton's second law relates force and acceleration in a line. In fact, this relation looks very much like Newton's second law, and can even be derived from it. The force in Newton's second law is replaced by torque, and the acceleration is replaced by angular acceleration. In the place of mass, we find a quantity called the moment of inertia. That is,

$$\textit{torque} = \textit{moment of inertia} \times \textit{angular acceleration}.$$

For a simple orbiting particle, the moment of inertia is

mass of particle $\times$ *(radius of orbit)*^{2}.

The mass of a particle is a measure of its ability to resist acceleration when a force is applied. In the same way, the moment of inertia of a particle is a measure of that particle's ability to resist angular acceleration when a torque is applied. There is a difference between mass and moment of inertia, however: the mass of an object is an inherent quality; the moment of inertia must be specified with respect to some point or axis about which the particle is circling.

6.2c ANGULAR MOMENTUM

We have seen that we can come up with quantities to describe rotation that are analogous to the quantities we used to describe ordinary motion. These are summarized in Table 6–1. It would

TABLE 6–1 Analogous Quantities for Translation and Rotation

For translation	*For rotation*
speed	angular speed
acceleration	angular acceleration
mass	moment of inertia
force	torque
momentum	angular momentum

seem that we should also have a quantity that corresponds to momentum. Indeed there is such a quantity, and it is called angular momentum. It, too, is defined with respect to a reference point. Since ordinary momentum is *mass* $\times$ *velocity,* angular momentum must depend on the rotational quantity corresponding to mass (that is, moment of inertia) and the angular quantity corresponding to velocity (angular speed). So

angular momentum $=$ *moment of inertia* $\times$ *angular speed.*

An important quality of angular momentum is that when there is no change in torque, angular momentum does not change. We call this the law of conservation of angular momentum.

The conservation of angular momentum is one of the most important laws of physics. And like the other conservation laws, it can be expressed in simple terms—something doesn't change.

6.3 *Angular Momentum and Extended Objects*

In the previous section, we talked about individual particles, or objects that behaved like particles, traveling in circular paths about some point. This type of motion is called revolution. *For example, we say that the earth* revolves *around the sun once a year. When we deal with the spinning motion of an extended object—an object composed of many particles—we refer to it as* rotation. *For example, the earth* rotates *on its axis once a day. (See Fig. 6–15.)*

So far we have been discussing single entities—cars and particles and so on—moving in circular paths. In these simple cases, we could describe the motion adequately without recourse to angular momentum merely by using plain momentum and applying Newton's second law. The great value of the concept of angular momentum comes when we deal with objects that are more complicated.

Let's examine the rotation of an object made of many particles. We can begin by considering it to be the sum of all its constituent particles; but before going any further, we must make a distinction—some bodies rotate rigidly and others do not.

Let's stick with rigid bodies for now. We can calculate the angular momentum of such an object by adding up the moments of inertia due to each of the individual particles that make up the object. The answer we get depends on the shape of the object and how the material is distributed. Different objects of the same mass can have very different moments of inertia, even if the objects are approximately the same size. For example, if a bicycle wheel and a uniform disk have the same size and mass, the wheel will have the greater moment of inertia since more of its mass is concentrated far from its center.

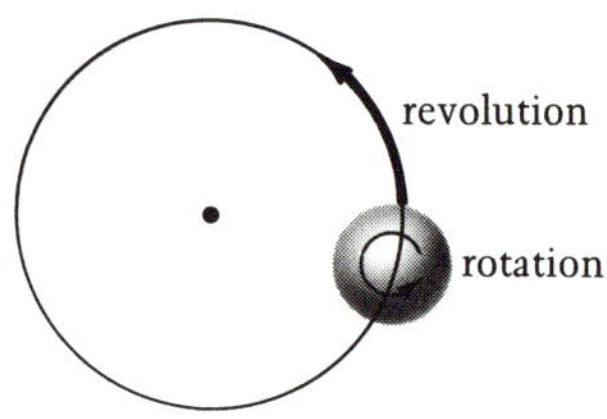

Fig. 6–15 Revolution and rotation.

6.3a APPLYING CONSERVATION OF ANGULAR MOMENTUM

If we know the moment of inertia of an object, we can calculate the angular momentum and apply the law of conservation of angular momentum. A familiar example is a spinning ice skater (Fig. 6–16). When she draws her hands close to her body, she reduces her moment of inertia (which depends on the square of the distance from the axis of the body). Since the product of the moment of

A

B

C

Fig. 6–16 U.S. silver medalist Lisa Marie Allen demonstrates how an ice skater applies conservation of angular momentum to a spin. (Notice that you can get an idea of how fast she is spinning in each frame by the fact that her ponytail extends out farther as she spins faster, so her speed is increasing in each frame.) A. She starts the spin turning very slowly, but with her arms and leg extended she has a large moment of inertia. B. She begins to bring her arm, and especially her leg, in, reducing her moment of inertia. To keep the same angular momentum, she will spin faster. C. With her leg close in and her arms over her head, her moment of inertia is greatly reduced, and she must spin even faster to have the same angular momentum.

Fig. 6–17 A circus tumbler thrown up at center. He is propelled upward when two colleagues at far right jump on the other end of a seesaw. He maneuvers his body to change his moment of inertia and thereby adjusts his rate of rotation.

inertia and the angular speed must remain constant, if the moment of inertia decreases the angular speed increases. The skater spins faster. She can control her rate of spin by changing the position of her arms.

A tumbler does a similar thing. When thrown up into the air, he gives himself a little angular momentum and starts rotating. If he wants to rotate faster, he curls his body into a compact form, a tuck position, thereby reducing his moment of inertia. To keep the angular momentum constant his rotation must speed up. Likewise, he can slow rotation by uncurling.

The law of conservation of angular momentum has many consequences in processes that are not part of our daily lives. For example, in any collision, angular momentum, as well as linear momentum and energy, must be conserved. In collisions among elementary particles in accelerators, the types of particles that can be produced are often limited by conservation of regular momentum. In fact, the existence of an elusive elementary particle— the neutrino—was deduced long before it was observed directly largely because its presence was needed to supply some energy and angular momentum.

6.3b PULSARS

Conservation of angular momentum may also explain, in part, a class of interesting and unusual astronomical objects—pulsars Pulsars were discovered accidentally in 1967 by Jocelyn Bell Burnell, then a graduate student at the University of Cambridge in England, and Anthony Hewish, her advisor. They found an object in the sky that emitted radio "pulses"—blips of static—at very regular intervals, one pulse every 1.3373011 seconds. Nothing was known to be capable of putting out such a regular signal whose position in the

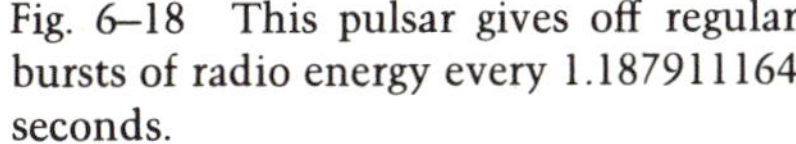

Fig. 6–18 This pulsar gives off regular bursts of radio energy every 1.187911164 seconds.

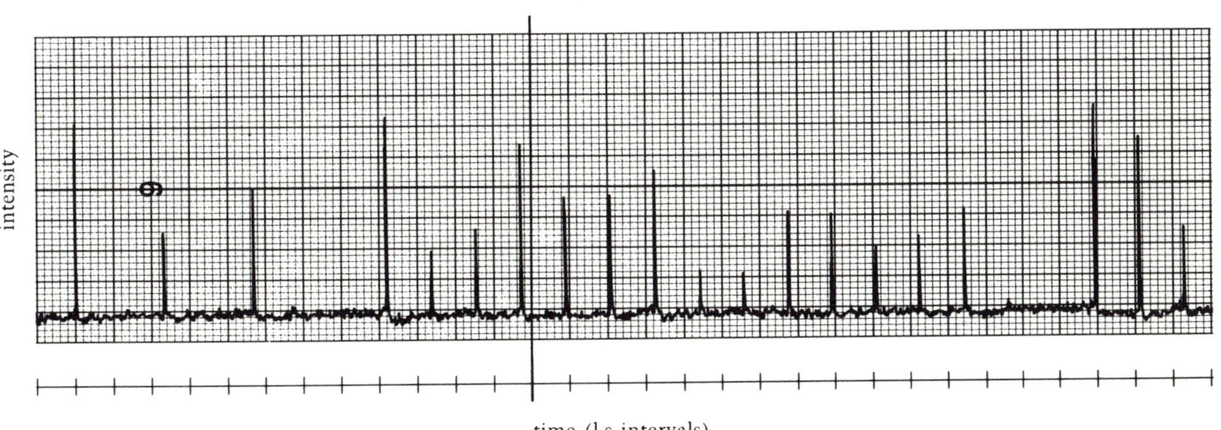

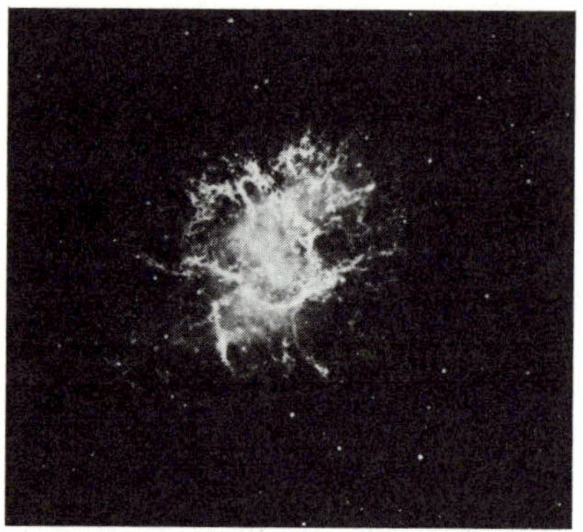

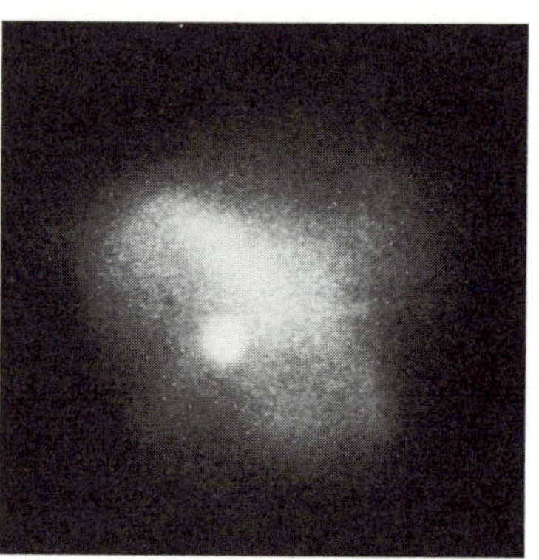

Fig. 6–20 The pulsar at the center of the Crab Nebula shows as the bright spot in the middle of this x-ray image made by S. S. Murray, R. Giacconi, and their colleagues with the orbiting Einstein Observatory, NASA's High Energy Astronomy Observatory 2.

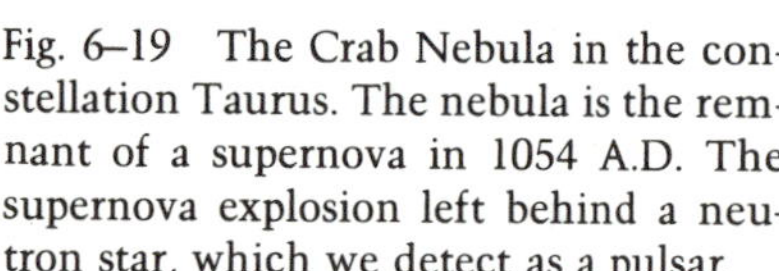

Fig. 6–19 The Crab Nebula in the constellation Taurus. The nebula is the remnant of a supernova in 1054 A.D. The supernova explosion left behind a neutron star, which we detect as a pulsar.

sky moved with the stars. It was even thought that the pulses might be a signal from some extraterrestrial civilization, and the source was referred to as LGM, for Little Green Men. This idea was quickly dispelled when three additional pulsing sources were found in other parts of the sky, each with a different pulsation period and each radiating over a wide range of frequencies. It seemed unlikely that four different civilizations were calling us at the same time and using so much energy to do so.

The question remained—what are pulsars, as these pulsating radio sources were called? Hundreds of them are now known. The question centered around what could produce a signal with such clock-like regularity. Speculation centered on a rotating object with a "hot spot" that passed by our view once per rotation, much like a lighthouse beacon. But what could be rotating so fast? One by one, various candidates were eliminated until the only remaining possibility was an object that had existed for many years only in the minds of theoreticians—a neutron star (Fig. 6–21).

A neutron star is believed to be the "leftover" bit when a massive star reaches the end of its life. The star begins to collapse so rapidly that it heats up and explodes in what we see as a supernova. For a brief time, the star may appear nearly as bright as a whole galaxy. Then the remnant of the core collapses and is compressed into neutrons, one of the basic particles of an atomic nucleus. This "neutron star" packs a mass equal to that of the sun into a ball only 20 km in diameter—a density of over 10^{17} kg/m^3, which amounts to a hundred billion tons in each cubic centimeter! While the star is being so compressed, its moment of inertia decreases, since it gets smaller and smaller and the mass is therefore increasingly concentrated at smaller distances from the center. If it has any initial angular speed at all, then, in order to conserve angular momentum, the star must rotate faster and faster. When it reaches its most compact size, this object with the mass of the sun may be rotating on its axis every quarter second!

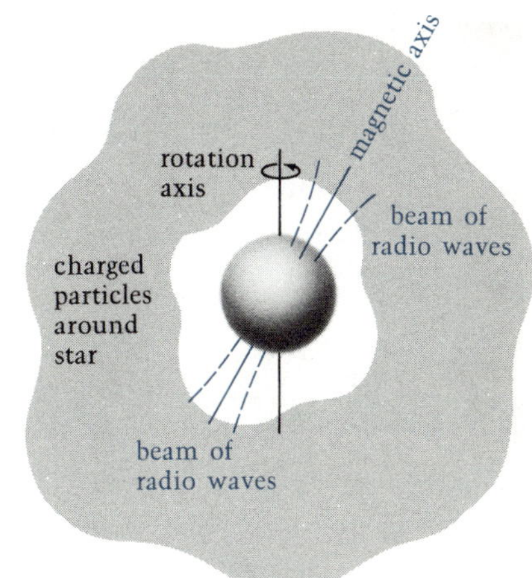

Fig. 6–21 A suggested picture for a neutron star. The neutron star rotates very quickly (more than once per second) around the rotation axis. The magnetic field axis of the star is not lined up with the rotation axis, and acts like a giant bar magnet sweeping around. The star is surrounded by a cloud of charged particles, and the magnetic field sweeping through these particles causes beams of radio waves to be emitted. To observe these waves, we must be lined up just right with the beam, so we only see a pulse once (and sometimes twice) for each rotation of the star.

After pulsars had been observed for some time, it was clear that they are gradually slowing down. This slowdown is related to conservation of energy. The pulsar is continuously giving up energy in the form of radio signals. Energy conservation tells us that this energy must come from somewhere. As the star radiates, its kinetic energy decreases. In the case of pulsars, the decrease in energy is taken up by a decrease in angular speed, which explains the observations.

Examining objects in terms of angular momentum and moment of inertia has allowed us to analyze a variety of exotic objects, from spinning neutron stars to spinning skaters.

6.4 *Torques and Extended Objects*

The amount of torque supplied by a force demands on how and where the force is applied. If you turn a bicycle over so that the front wheel is free to turn, you will find that it is much easier to start the wheel turning when you give it a push with your finger at the rim than when you apply the same force near the hub.

When calculating the acceleration of an object on which more than one force is acting, we need to add up all the forces acting on it. The same goes for torques. When more than one torque is applied, we add up the amount that each contributes to the angular acceleration. We must also keep track of directions. For example, if two torques are acting and each tends to cause a wheel to rotate clockwise, then we simply add the two torques together. However, if one torque causes clockwise rotation and the other causes counterclockwise rotation, we must subtract one from the other. The

Fig. 6–22 Two children of different weight can balance each other by sitting at different distances from the center of the swing. The heavier child sits closer so that distance × weight is equal for both.

net angular acceleration will be in the direction of the larger torque.

The action of a lever gives us a vivid example of how the resulting amount of torque depends on where force is applied; it also reveals the effects of several torques acting together. If you and a friend are on a seesaw (Fig. 6–22), which way will it rotate? Since you are on opposite sides of the point of support, the torques exerted by your weights will act in opposite directions. The seesaw will rotate in the direction that results from the action of the greater of the two torques. The lighter person can exert a greater torque by being farther from the point of support. If you weigh half as much as your friend on the other end, you can balance the board (no net torque) by being twice as far as she is from the center. If you move slightly farther away than that, the seesaw will tip in your favor. (The next chance you get, watch children on a seesaw. Notice how, without even knowing about torque, the child stuck on top moves closer to the end.)

6.5 *Balancing*

If you stand a pencil on its end and let go, the pencil falls over. On the other hand, if you give a block a slight tilt, it does not fall over; it returns to its original position. Since the pencil falls over with just the slightest push, while the block does not, we say that the pencil is unstable and the block is stable (Fig. 6–23).

Whether or not an object falls over depends on the torque on the object. For the purpose of calculating torques caused by gravitational forces, we can treat the force as though it acts at a single point, called the center of gravity. (For most practical situations, the center of gravity and the center of mass are in the same place.)

Once you start to tip a box, the torque on the box depends on the location of the center of gravity. If the box doesn't rotate too far, the torque will tend to rotate the box back. If you tip the box so far that the center of gravity passes over the point of support, then the torque changes direction and the box will continue to fall (Fig. 6–24).

If your body experiences a torque that tends to tip you over, you

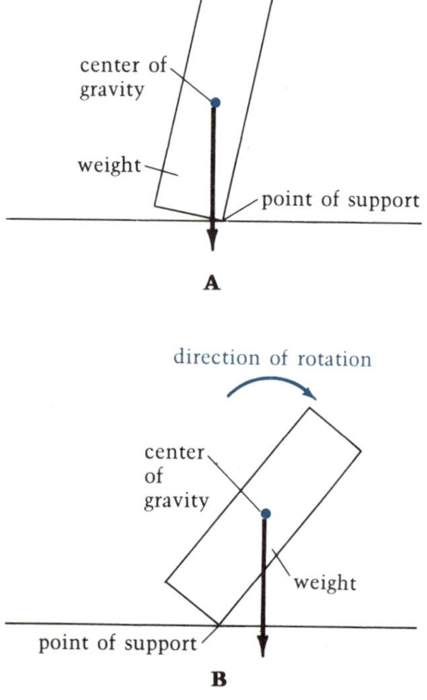

Fig. 6–23 A. If you tip a box over slightly, the center of gravity will remain on the same side of the point of support as when the box is flat on the table. This means that the torque will act in such a way that the box tips back to its original position. B. If you tip the box too far, then the center of gravity passes to the opposite side of the point of support. When this happens, the direction of the rotation caused by the torque changes, and the box tips over.

can still counteract its effects. You increase your resistance to a torque by increasing your moment of inertia. You can do this by holding your arms straight out to the sides. (This is the opposite of what you do if you are spinning on ice skates and want to decrease your moment of inertia to speed up the spin.) Notice how you instinctively throw your hands out to the sides when you start to lose your balance. The effect is even greater if you hold weights in your hands. The tightrope walker in Fig. 6–25 is enhancing his balance by artificially extending his arms with a balance pole, which increases the moment of inertia. People who have not tried this stunt cannot imagine how great the effect is on keeping your balance.

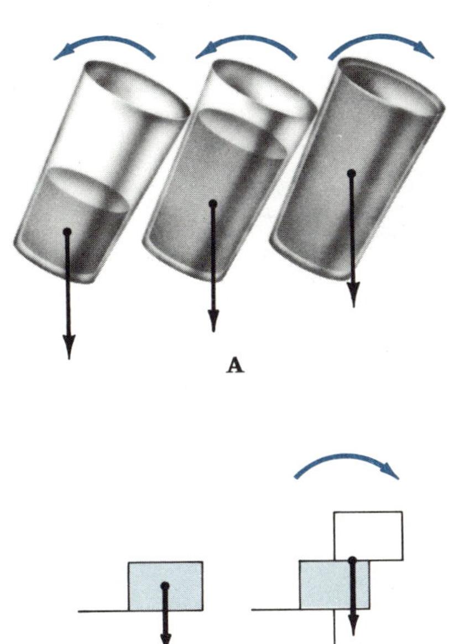

Fig. 6–24 Which of these items will tip over and which will not? In each case, the black dot represents the center of gravity and the straight arrow represents the force of gravity. The curved blue arrow represents the direction of rotation about the point of support. A. We are looking at a glass with three different levels of liquid. As we add more liquid, the center of gravity gets higher. In the first case the center of gravity is quite low. When the glass is tipped to the right, the center of gravity stays to the left of the point about which the glass is rotating. This means that the glass will rotate back to the left and will stay upright. The same is true for the second glass. However, for the third glass, the center of gravity is so high that a slight tip to the right pushes it to the right of the point of support. Now the torque tends to rotate the glass to the right and the glass tips over. B. We start with a brick whose center of gravity is right over the edge of the table. This means that there is no torque, so it does not tip over. (However, if someone were to give it a little push, it would keep on going.) When we add a second brick, the center of mass for the two bricks is now over the edge, and the bricks will rotate off the edge.

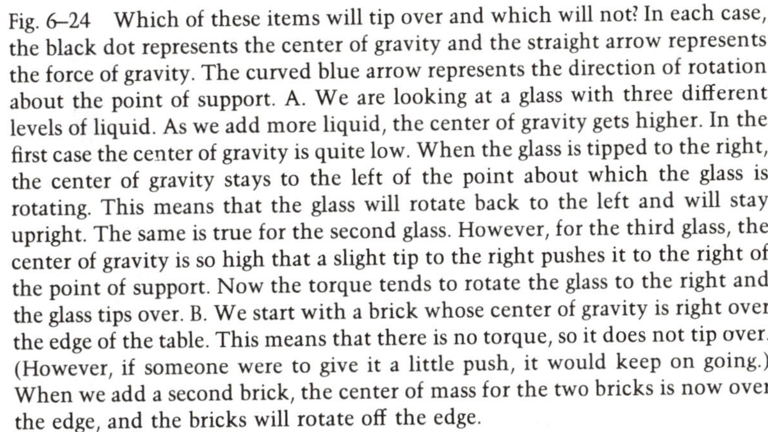

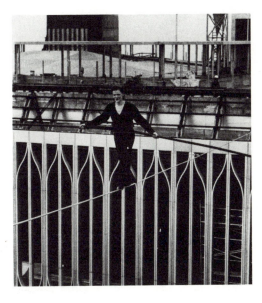

Fig. 6–25 French aerialist Phillipe Petit, using a balance pole as he crosses a cable stretched between the tops of the two World Trade Center towers in New York City, 110 stories above the ground.

Fig. 6–26

From *The New World* (Harper & Row) © 1962 Saul Steinberg. Originally in *The New Yorker.*

Key Words

centrifugal force
Coriolis force
period
frequency
angular frequency
angular speed
angular acceleration
torque
lever arm
moment of inertia
angular momentum
conservation of angular momentum
rotation
revolution
pulsar
supernova
lever
unstable
stable
center of gravity

Questions

1. How can a ball moving in a circle with a constant speed have an acceleration?
2. Suppose you are whirling a ball around on a string. If you want to keep that ball going around at twice the speed, but keep the length of the string fixed, by what factor will the tension in the string increase?
3. A car and a truck are going around the same curve at the same speed. The truck has twice the mass of the car. (a) How does the acceleration of the car compare with that of the truck? (b) How does the force on the car compare with that on the truck?
4. Why does a hammer thrower have to be strong?
5. For the NASA centrifuge discussed in Example 2, how fast would the astronauts have to move to feel a force equal to six times their weight?
6. A baby is in a carseat in a car. The car goes around a turn. What force causes the baby to move in a circular path along with the car?
7. Suppose you are inside a railroad car with a frictionless floor. The car goes around a left-hand bend. What happens to you?
8. Why do you have to slow down more to get around a given curve in your car on a rainy day?
9. A weather system starts north from the earth's equator. Describe the path of the storm as viewed by someone living in New York. In your description, use only real forces as explanations for what is taking place.
10. If we double the frequency of some motion, what happens to the period?
11. An object is moving around a circular track. If we increase the frequency of the motion, what happens to the force that the track must exert to maintain the circular motion?
12. You sit on one end of a seesaw. Where must someone twice your weight sit on the other side in order for the seesaw to balance?

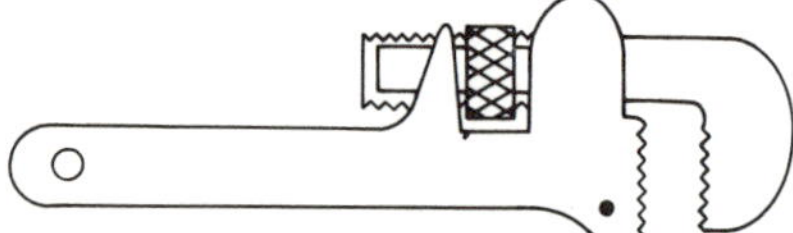

13. How does a monkey wrench allow you to twist a nut that you couldn't twist with your hands?
14. You wish to lift a 1000-N rock, but you can only exert a force of 100 N. Draw a diagram to show how you would use a lever to lift the rock. If you want to raise the rock 0.1 m in the air, how far down must you push on your side?
15. A turntable is turning freely, with a spider located at the center of the turntable. The spider walks from the center to the edge of the turntable. What happens to the rate at which the table is turning?
16. If you wish to use brakes to stop a turning wheel, is it better to apply the breaking force near the axle of the wheel or near the edge of the wheel? Explain your answer.
17. We have discussed the effect of a figure skater's pulling her hands in while spinning. What is the result if the figure skater is holding weights while bringing her hands in?
18. Under what conditions is angular momentum conserved?
19. What is the difference between rotation and revolution?
20. How might conservation of angular movement explain why all the planets orbit the sun in the same direction?
21. What effects of angular momentum conservation do we see in pulsars?
22. How is the slowing down of pulsars related to changes in rotational energy?
23. When the figure skater brings her hands in, angular momentum is conserved, so as the moment of inertia decreases, the angular speed increases. What happens to the energy of rotation?

PART II

Gravity and Relativity

What goes up must come down. We all have an intuitive feel for the concept of gravity. Gravity not only holds us to the surface of the earth; it also holds the earth in the solar system and the solar system in our galaxy. On the largest scale, gravity dominates the universe.

Some of the greatest scientists in history have attempted to explain gravity. Two of the giants in scientific and intellectual history—Isaac Newton in the seventeenth century and Albert Einstein in the twentieth—advanced theories of gravitation that form the base of our understanding. Where gravity is dominant—in accounting for the overall structure of the universe, for example, or for the formation of "black holes," where gravity is so strong that not even light escapes—Newton's theory of gravitation is not sufficiently accurate and Einstein's must be used. But Newton's theory remains useful for most of our current pursuits, so Einstein's work should be considered as a more generalized form of Newton's theory of gravitation rather than as a replacement for it.

We have already discussed Sir Isaac Newton and briefly described his life (Chapter 3). We should like now to say a few words about Albert Einstein, although we shall leave to the following chapters an examination of his scientific thought.

Einstein was born in Ulm, Germany, in 1879. He did not learn to speak until he was three, and some thought he might be retarded. As an older child, he seemed of average intelligence, but not a prodigy.

Einstein reported that his interest in science first arose when as a child of four or five he was shown a compass needle, ever seeking north. The other decisive event in the development of his scientific interests took place when he was twelve, at which time he discovered the logic involved in Euclidean geometry. By the time he was sixteen, Einstein was thinking about light, an interest that was to lead to his greatest works.

Einstein's family had moved from Germany to Switzerland, and he went to school there. Eventually he went to college in Zurich (although he failed the entrance exam when he first applied). He was an indifferent scholar, preferring his own thoughts to those he was being taught. (Little did his teachers realize that his own thoughts were deeper and more interesting that those that they were teaching.) Indeed, in Einstein's famous first paper on the theory of relativity, he had no bibliographic references.

Upon graduation from college in 1901, Einstein obtained a job in the patent office in Berne, Switzerland. Although this work was routine, it suited Einstein because it left his hours outside work free for him to think about science. Einstein had considerable mechanical ingenuity, and he was a good patent examiner.

Fig. II–1 Albert Einstein at the patent office in Berne, Switzerland, in 1905.

In a single year, 1905, Einstein published three epochal papers, each of which was sufficiently inventive that even alone it showed that he was a physicist of the first rank. We shall be discussing their content later on. The scientific papers made Einstein well known, and he was offered a university position. Over the next two decades, he held various university professorships in Zurich, Prague, and Berlin. By 1915 he developed his theory of gravitation, which is known as the general theory of relativity. When this theory was verified by certain astronomical observations (Chapter 9) in 1919, Einstein's name became a household word.

After Hitler rose to power in Germany, Einstein, who was born Jewish, came under attack. He didn't want to leave Germany, but he was eventually persuaded to accept an appointment to the newly established Institute for Advanced Study at Princeton, New Jersey. He arrived there in 1933.

Einstein's greatest work was then behind him, although he continued to work on major problems. In particular, he tried to work out the unification of all the forces of physics (gravity, electromagnetism, and the forces that hold atoms and subatomic particles together), although the solution eluded him and has not yet been found. He carried on because of his faith that a solution existed. "God is subtle," he said, "but He is not malicious."

Einstein was a modest person and remained modest even though he had gained the public eye as the wisest person in the world. Playing his violin was one of his greatest pleasures. His idiosyncrasies—he could not, for example, be bothered to wear socks or use different kinds of soap for washing and shaving—endeared him to the public. But Einstein sought to stay out of the public eye, and he lived quietly in Princeton. A first marriage, which produced two children, had failed in 1914; five years later he married his second wife, a cousin, who died only three years after they moved to Princeton.

Einstein's fame was such that he could not avoid involvement in political and scientific causes. When some scientists feared in the late 1930s that the Germans were developing an atomic bomb, they knew that only Einstein's signature on a letter to President Roosevelt would carry enough weight to bring the matter suitably to the president's attention. Einstein signed, Roosevelt listened, and the Manhattan Project was born, resulting in the atomic bomb.

It seems strange that Einstein, a pacifist who had refused even to serve in an army, should have wound up playing such an important role in war, although he deeply regretted the bomb. But Einstein had suffered in Germany, and he backed the American entrance into the war. When the original manuscript of Einstein's first paper on relativity could not be found, he showed his backing for the war by copying his paper out in longhand so that it could be sold at auction to raise funds for the war effort.

Einstein devoted himself to peaceful causes and was particularly interested in establishing a homeland for the Jews in what was then Palestine. He was even offered the presidency of the state of Israel, but declined. When Albert Einstein died in 1955, the whole world mourned.

CHAPTER 7

Gravity

When you look up at the pinpoints of light in the night sky you may be struck by how isolated each star appears. Yet every star feels the influence of billions of other stars pulling on it from all sides. That pull is as real as if giant hands reached across the void of space and grasped the star. This force, acting even over the great distances between stars and between galaxies, is the same force that holds each of the stars together. Indeed, it is the same force, gravity, that keeps you and this book tied to the surface of the earth.

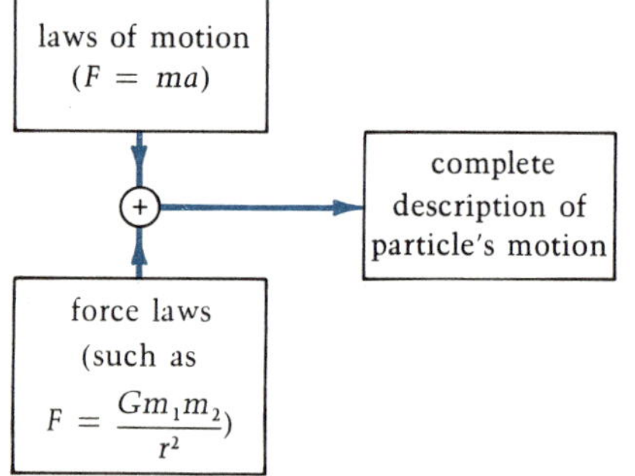

Fig. 7–1 If we know where a particle is and how it is moving (the so-called "initial conditions"), we can use the laws of motion and the force laws to work out a complete description of the particle's motion.

7.1 *The Apple and the Moon*

We are often told that Newton discovered the law of gravitation after being hit on the head by a falling apple. Although it is possible that Newton's thoughts were stimulated while watching falling apples, there is no evidence to suggest that any actually hit him. And anyway, the statement is somewhat misleading since it does not make clear the magnitude of Newton's scientific contributions.

Newton investigated gravity following his method of interpreting the physical world in terms of forces. Having made the connection between force and acceleration ($F = ma$) and noting that falling bodies accelerated, it was clear to him that some force was responsible for the acceleration of falling bodies. This brings up an important point. Newton's laws of motion, as discussed in Chapter

3, tell us only how a particle will behave in the presence or absence of a force. They tell us nothing about the specific forces that are actually acting. Therefore, if we are to describe the motion of a particle completely, we must arrive at descriptions that show what forces are acting on that particle at all times. These descriptions are called force laws. We must be careful to understand what we mean by the word "law." In this case, we mean something that describes how things are, not something that was legislated. (Remember the old joke, "What kept things on the earth before the law of gravity was passed?") In gravitation, we meet our first example of a fundamental force law (Fig. 7–1).

Newton did not discover that apples fall from trees. That fact was well established. But it was Newton who realized that the same force that causes them to fall might also reach out to more distant objects. He considered the possibility that whatever force causes an apple to fall is also responsible for keeping the moon in its path around the earth. Why is a force necessary for this? Since the path of the moon is curved, its direction of motion is continuously changing. Thus its velocity is changing (even if its speed isn't, since velocity includes both speed and direction). A changing velocity means an acceleration, and, having calculated the distance from the earth to the moon and the time the moon takes to go around the earth, Newton was able to calculate the acceleration of the moon and compare it with the acceleration of a falling apple, which is separated from the center of the earth by only the radius of the earth (Fig. 7–2).

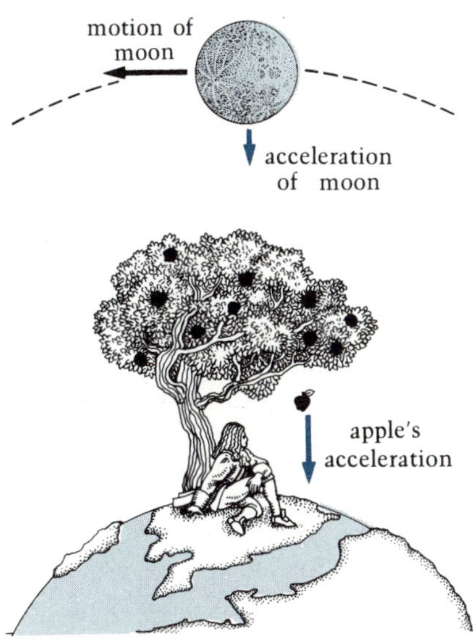

Fig. 7–2 Gravity provides the force that makes an apple fall out of a tree and keeps the moon in its orbit. The apple falls straight down because it has no sideways motion. The moon is moving to the side, and the gravitational attraction of the earth is just enough to provide the acceleration to keep it in its curved path.

7.1a THE MOTIONS OF THE PLANETS

Before we see how Newton proceeded from here, we must backtrack a little and consider how Newton's contemporaries explained the motions of the moon and planets. When we observe the motion of these objects, we do not have the luxury of observing from someplace outside the solar system, and so we find it difficult to get a clear picture of how things are laid out. Instead, we find ourselves right in the middle of all the action. All we can see is how the planets appear to move against the fixed background of distant stars. To appreciate the problem, think of a racing car driver in a race trying to sort out the motions of all the other cars, when all that can be seen is how they appear to move against the background of the grandstand. Spectators in the grandstand have a better view. The situation is even worse for the earthbound observer of the solar system. At least the driver knows about the motion of his own car, and, given a little time, can figure out the motions of the others by seeing how they move relative to it. The early astronomers did not even know how (or if) the earth was moving.

Around 350 B.C., the Greek philosopher Aristotle had proposed a picture of the system of planets that remained dominant for eighteen hundred years. Aristotle asserted, without proof, that the earth was at the center of this system. The sun, moon, and planets were on spheres, centered at the earth, and these spheres rotated around the earth, thus accounting for their motions across the sky.

There was a perplexing problem with this picture. If the planets were on spheres moving at constant speeds, the paths of the planets across the sky should be smooth and regular. This is not the case. In fact, the planets appear to spend part of their time moving backward! This backward motion, called retrograde motion, is illustrated for Mars in Fig. 7–3.

Fig. 7–3 The retrograde motion of Mars in the sky, reproduced in a planetarium.

In about 140 A.D., the Greek astronomer Claudius Ptolemy elaborated on Aristotle's basic ideas in an effort to make them consistent with observations. To account for the retrograde motion, he said that the planets moved on small circles, called epicycles. The centers of these circles were on larger circles, called deferents. These larger circles were centered near (although not quite on) the earth. Thus, a planet moved around its epicycle while the epicycle moved around the deferent. The speed of these two motions was arranged such that the planet spent most of its time moving in one direction, but part of its time moving backward (Fig. 7–4).

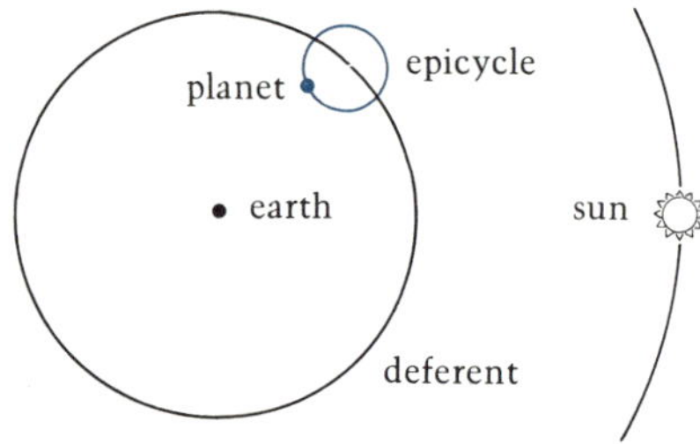

Fig. 7–4 The Ptolemaic system. As a planet travels around the earth, it is actually moving on a small circle called an *epicycle*, the center of which, in turn, travels on a large circle called a *deferent*. This accounts for the periodic retrograde motion.

Nicholas Copernicus, who worked in the early sixteenth century in what is now Poland, thought of a better explanation. He advanced a view of the solar system in which the sun was at the center and all the planets moved around it. This is called the sun-centered, or heliocentric, picture. (*Helios* is the Greek word for sun.) Copernicus felt that the heliocentric theory explained the observations in a simpler, more natural manner. In particular, it explained retrograde motion of the planets without the use of epicycles. Retrograde motion was seen rather as an effect of the *relative motion* of the earth and the other planets, as illustrated in Fig. 7–5.

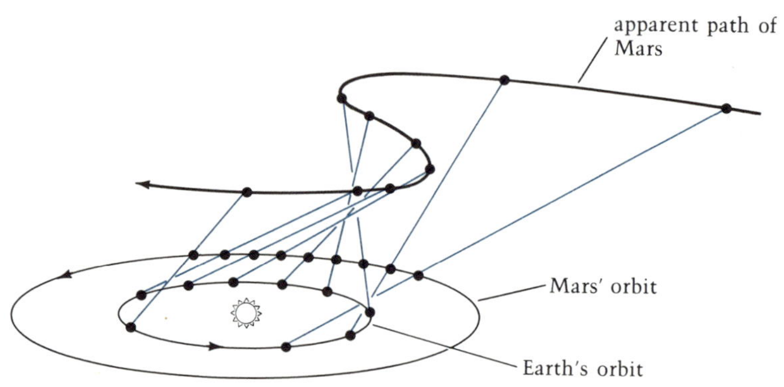

Fig. 7–5 Copernicus's explanation of the retrograde motion of Mars. The nine dots in the earth's orbit show the earth at nine different times, and the nine dots on Mars's orbit show the position of Mars at these same nine times. The colored lines represent the lines of sight from the earth to Mars at these times. Sometimes the line of sight points in the forward direction and sometimes it points in the backward direction, resulting in an apparent change in the direction of motion of Mars against the fixed background of stars.

Copernicus published his theory in 1543 in a book called *De Revolutionibus* (*On the Revolutions*). The general argument in favor of this new theory was its simplicity; it was able to explain the gross features of planetary motion without introducing large epicycles. (However, Copernicus still needed smaller epicycles to account for details of the motions.) On philosophical grounds, we generally say that if two theories explain existing data equally well, then we favor the simpler theory. This philosophical rule is called Occam's razor; it is a razor in the sense that it "cuts away" unnecessary complications. Of course, we must have some way of deciding which theory is simpler. For example, a proponent of the Ptolemaic system could say, "What can be simpler than having the earth at the center?"

William of Occam was a medieval philosopher.

By the "simpler" theory we usually mean the one with the smallest number of arbitrarily chosen assumptions. Each planetary theory involved one assumption—an answer to the question: which object is at the center? For Ptolemy it was the earth; for Copernicus it was the sun. However, in introducing epicycles, Ptolemy introduced an additional assumption in order to explain existing observations.

We also believe more strongly in a theory if it can predict the results of new experiments. Observational support for the heliocentric theory came from Galileo Galilei, who in 1610 became the first person to use a telescope for astronomical observations. He

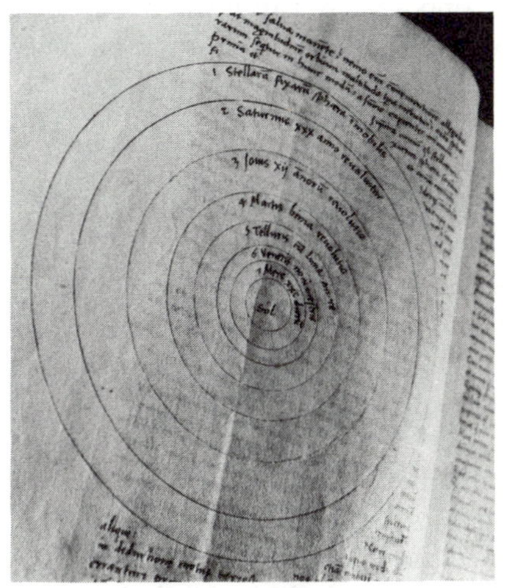

Fig. 7–6 Copernicus' drawing with the sun at the center, in his original manuscript of *De Revolutionibus*.

Fig. 7–7 Jupiter and its Galilean satellites observed from the earth.

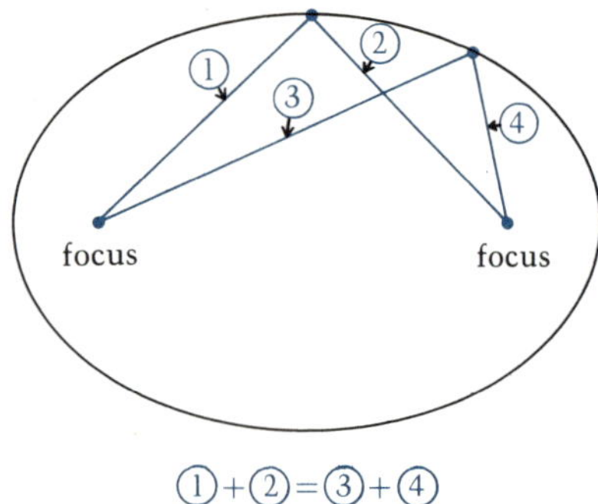

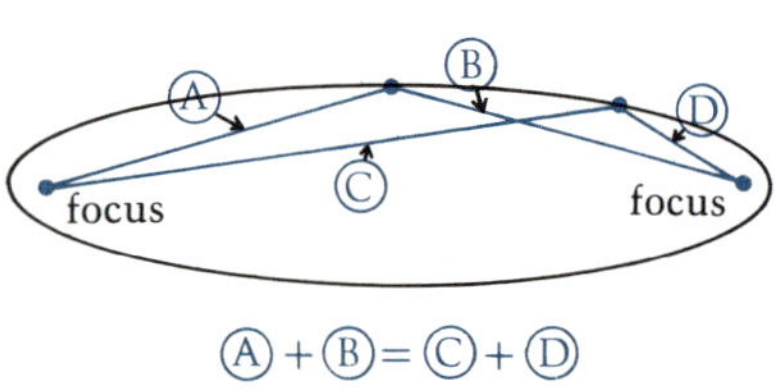

Fig. 7–8 Ellipses. The bottom ellipse is flatter than the top one. (We say it has a higher eccentricity.) For each ellipse, we can draw any pair of lines from one focus to a point on the ellipse and then to the other focus. It doesn't matter which point on the ellipse we choose. The sum of the lengths of the two lines is the same for each point.

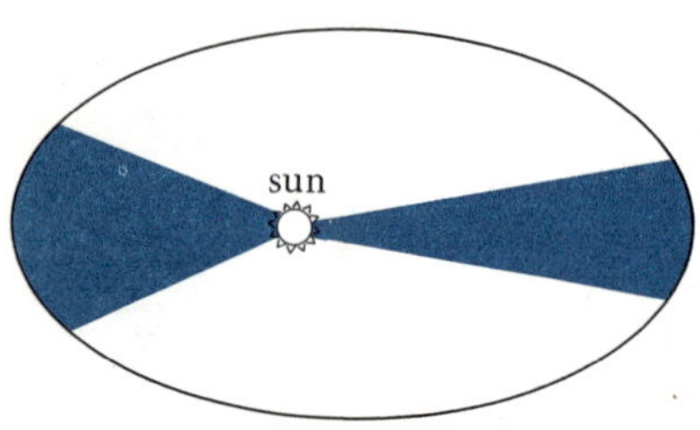

Fig. 7–9 Kepler's second law. When the earth is closer to the sun it moves faster than when it is far from the sun. The two shaded areas, swept out in equal time intervals, have the same area.

discovered four moons orbiting Jupiter, thus showing that heavenly bodies do not necessarily circle the earth. He also found that Venus went through an entire set of phases, just as the moon does. The Ptolemaic system could not reconcile the phases of Venus with the fact that Venus never appeared to move far from the sun. Unlike Venus, our moon could appear not only close to the sun in the sky at crescent phases but also on the opposite side of the sky from the sun when the moon has its full phase. Venus, in the Ptolemaic picture, would then always be a crescent because it is always close to the sun in the sky. The Copernican theory, however, could explain the full set of phases that Galileo observed.

Galileo proposed that his observational evidence strongly favored the Copernican theory. This assertion about the truth of the Copernican picture was strongly opposed by the Roman Catholic church, and Galileo was eventually forced to announce publicly his recanting of these "heretical" ideas.

7.1b KEPLER'S LAWS

A quantitative picture, that is, one based on numerical calculation, of the solar system required accurate observations. Such observations had been made near the end of the sixteenth century by Tycho Brahe, at his observatory, Uraniborg, on an island off the coast of Denmark. Tycho's observations marked the last gasp of naked-eye astronomy.

Tycho was plagued by a problem that still haunts scientists today—the need for financial support. When his support in Denmark ran out in 1597, Tycho moved to Prague. There he took on a young assistant, Johannes Kepler. After Tycho's death in 1601, Kepler gained access to Tycho's observations of the planets and went on to analyze them.

In the face of the better data provided by Tycho, Kepler found that neither Ptolemy's theory nor Copernicus's theory adequately predicted the observed positions of the planets. After years of analysis, doing laborious calculations without the benefit of modern computers or calculators, and often being sidetracked or delayed by computational errors, Kepler finally made some sense of the observations. Remember, all he had was the racing car driver's view of the apparent positions of the other planets while trying to put together the picture as it would appear from the grandstand. In 1609, Kepler presented two laws of planetary motion, and in 1618, a third law, all based *entirely on observation* (Figs. 7–8 to 7–10).

1. The planets orbit the sun in ellipses, with the sun at one focus.
2. The line from the planet to the sun sweeps out equal areas in equal times.

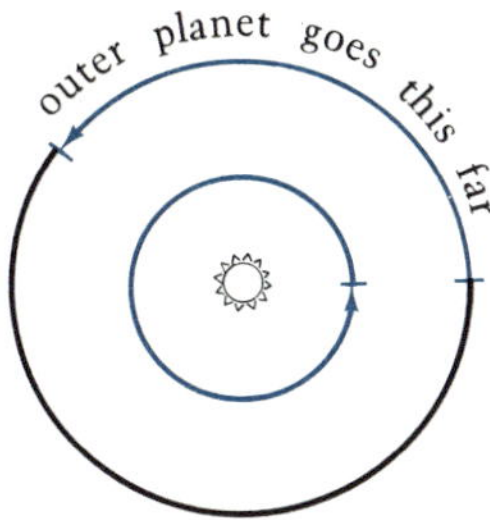

Fig. 7–10 Kepler's third law. In this figure, the outer planet is twice as far from the sun as the inner planet. In the time the inner planet makes one complete revolution, the outer planet goes through the distance indicated.

3. The square of the period of the orbit is equal to a constant times the cube of the semimajor axis (half the longest dimension) of the ellipse.

Kepler—in his first law—was the first to suggest that the orbits of the planets are not perfect circles. An ellipse is the curve drawn so that for any point on the curve, the sum of the point's distances from two fixed points, called the foci (singular, focus), is constant.

To draw an ellipse, place two tacks on a board and put a loop of string around the two tacks. (Obviously, the string must be longer than the distance between the tacks.) Then pull the string taut with a pencil and move the pencil along the string, keeping it taut (Fig. 7–11).

A

B

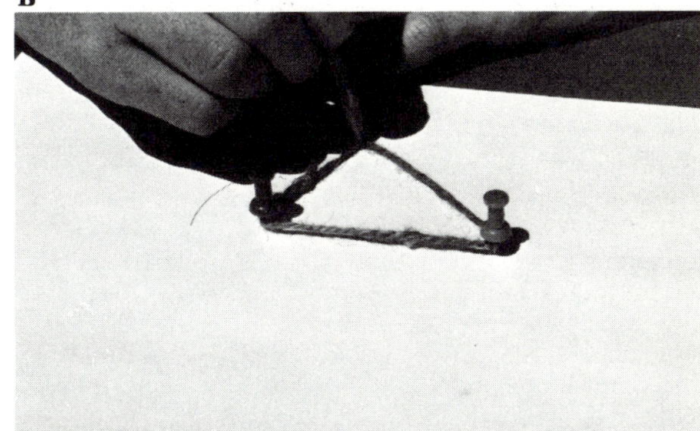

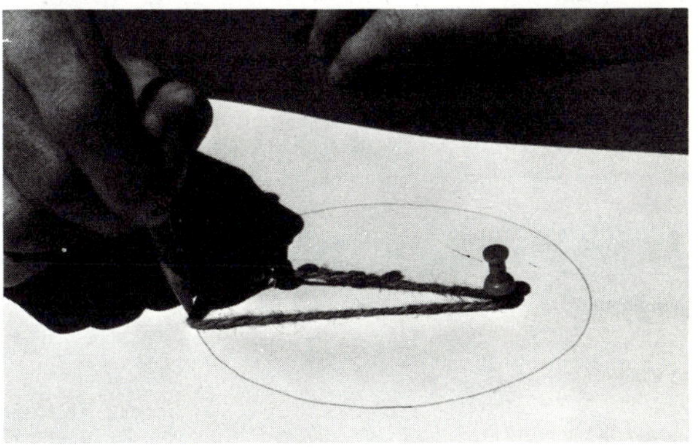

C

Fig. 7–11 Since the distance from any point on an ellipse to one focus plus the distance from that point to the other focus is constant, a piece of string of constant length allows you to draw an ellipse very easily. Simply move your pencil around, keeping the string taut.

Once Kepler introduced the idea that planetary orbits could be ellipses instead of circles, the accuracy of his predictions of planetary positions improved greatly.

Kepler's second law, called the law of equal areas, tells us that when a planet is in the part of its orbit that is close to the sun, it moves faster than when it is farther from the sun, and gives the relative speeds. This law, we now know, is equivalent to saying that angular momentum is conserved. We saw the same thing with the twirling skater. When her arms were close to her body she rotated faster than when they were far away. In the same way, when a planet moves closer to the sun, it must revolve faster to keep its angular momentum constant.

To illustrate Kepler's third law, we can look at two planets and compare the time it takes for them to go around the sun with their distances from the sun. For example, Mars is about 1.5 times as far from the sun as the earth. We call the distance of the earth from the sun one astronomical unit (abbreviated 1 A.U.), so we say that Mars is 1.5 A.U. from the sun. If we take the cube of 1.5 we get about 3.5. If we take the square root of 3.5 we get about 1.9. This means that Mars takes 1.9 times longer than the earth to orbit the sun, or 1.9 years.

Example: $p^2 = D^3$.
Since $D = 1.5$, $D^3 = 1.5 \times 1.5 \times 1.5 = 2.25 \times 1.5 = 3.5$.
Thus $p = \sqrt{3.5} = 1.9$.

7.1c THE LAW OF GRAVITATION

Sixty years after Kepler, we come to Newton, the apple, and the moon. We have already discussed Newton's insight that gravity, which causes the apple to fall, is also the force that keeps the moon in orbit around the earth. He was able to compare the acceleration of the apple with the acceleration of the moon, which told him the relative strength of the earth's gravity at the surface of the earth and at the moon. To see if his conjecture was correct, he needed a theory to predict how the earth's gravity would weaken as the object being pulled moved farther away from the center of the earth.

The answer was suggested by Kepler's third law, which was known to Newton. Given the third law's relationship between the period of the orbit and the radius of the orbit, Newton could calculate the force required to keep the moon in its orbit around the earth. He discovered that Kepler's third law is only satisfied if the force decreases as the square of the distance between the two objects decreases. For example, if planet A is twice as far from the sun as planet B, planet B feels four times the force felt by planet A.

In addition, Newton reasoned that the force of gravity between the sun and a planet must also be proportional to the mass of the planet and the sun. Otherwise, if two planets of different masses were at the same distance from the sun, they would have different accelerations (force divided by mass). But he knew from Kepler's third law, which doesn't depend on mass, that any objects in the same orbit around the sun have the same acceleration.

Finally, Newton reasoned that, by what we now call Newton's third law, the force that a planet exerts on the sun must be equal to the force that the sun exerts on the planet. Thus, if the force is proportional to one mass, it must also be proportional to the other.

With this reasoning, Newton was able to propose the law of universal gravitation. This law states that the gravitational force between two objects can be calculated from the following formula:

By writing the word "constant," we mean simply that the numerical value on the right side of the equation has to be multiplied by some number to put it in the proper units. Since the numerical factor does not change, even for different masses or distances, it is a constant. Later on we shall discuss the numerical value of the constant.

$$\textit{force of gravity} = \frac{\textit{constant} \times (\textit{mass of object 1}) \times (\textit{mass of object 2})}{(\textit{distance from object 1 to object 2})^2}.$$

This force is always directed along the imaginary line joining the two objects. It is directed such that each object attracts the other. Not only does this law give the strength of the force between two objects, but also states the universality of the law of gravitation. It is true for *any two* objects. It works equally well for the force between earth and apple and between earth and moon. In fact, Newton was able to show that the acceleration of the apple and the acceleration of the moon in its orbit are exactly in the ratio of the square of the distances of the center of the apple from the center of the earth and the center of the moon from the center of the earth. Newton was also able to use his laws of motion and the law of gravitation to derive all of Kepler's laws.

Newton's law of universal gravitation is

$$F = G\,\frac{m_1 m_2}{d^2},$$

where G is the *constant of universal gravitation*. The currently accepted value for G is 6.67×10^{-11} newtons-meters^2/kilograms^2 (6.67×10^{-11} N-m^2/kg^2). Note that when we work a problem with m in kilograms and d in meters, using G in the above units makes the force work out in newtons.

7.1d THE EARTH'S GRAVITY

Newton realized that one problem remained. The law of gravitation refers to the gravitational force between two point particles. Given the tremendous distances of the planets from the sun, it is reasonable to treat the planets and the sun as point particles. However, when you are standing in an apple orchard on the surface of the earth, the earth hardly appears as a point particle. Is it proper to treat the gravitational force exerted on you by the earth as if all of the earth's mass were concentrated at the center? To test this idea, Newton had to come up with a way of adding together the forces on you that result from all the individual particles in the earth. To do this, he devised the principles of calculus—a valuable side benefit of his research in gravitation (Fig. 7–12).

Using his new mathematical techniques, Newton was able to show that it is indeed proper to treat the earth as if all the mass were concentrated at the center. (In addition, he showed that if you were anywhere inside a hollow spherical shell of mass, the shell would exert no net gravitational force on you. You would be pulled equally in all directions.) The mathematical problems (as well as Newton's general shyness) caused a long delay in publication of these ideas, but they finally did appear in his *Principia* in 1687, two decades after his first thought of the common force experienced by the moon and the apple.

7.1e THE STRENGTH OF GRAVITY

When we discussed falling bodies in Chapters 2 and 3, we treated the force of gravity near the surface of the earth as though it does not change as you move higher above the surface of the earth. It is true that when we move to heights that are small compared to the radius of the earth, the changes in the force of gravity between ourselves and the earth are very small indeed. Under these conditions we ignore variations in distance and make the approximation that the force is constant. However, whenever we consider large changes in distance, such as in comparing the distant moon and the nearby apple, we must remember to take distance into account and use the law of gravitation in its complete form.

To calculate the actual gravitational attraction between two particles, we must first calculate the numerical value of the constant in the force law. Experimental physicists doing this experiment take two objects of known mass and place them a known distance apart; then they measure the force between the objects. The experiment is easier said than done, because of the extremely small gravita-

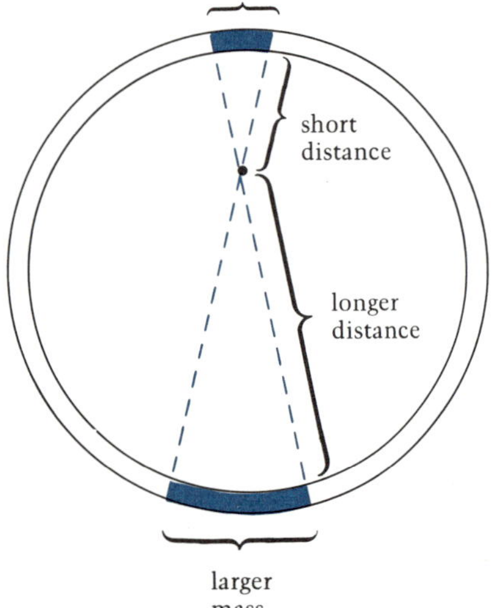

Fig. 7–12 The gravitational effect of a shell on an object inside the shell. We can consider the effects of sections on opposite sides of the shell, cut off by the pair of dashed lines. The smaller section has less mass but is nearer than the larger section, and the two sections actually exert the same force on an object at the position of the dot. This is true for any pair of shell sections. The forces cancel. Therefore, an object inside a hollow shell feels no gravitational force from the shell.

tional force between the two masses in the laboratory. For two masses of 1 kg each, the force between them will be about 10^{-11} times the force that the earth exerts on either mass. The first accurate measurement was made in England in 1798 by Lord Cavendish. He used a device that we now refer to as the Cavendish balance (see Fig. 7–13). The Cavendish gravitational measurement is equivalent to determining the value of the constant of universal gravitation, G.

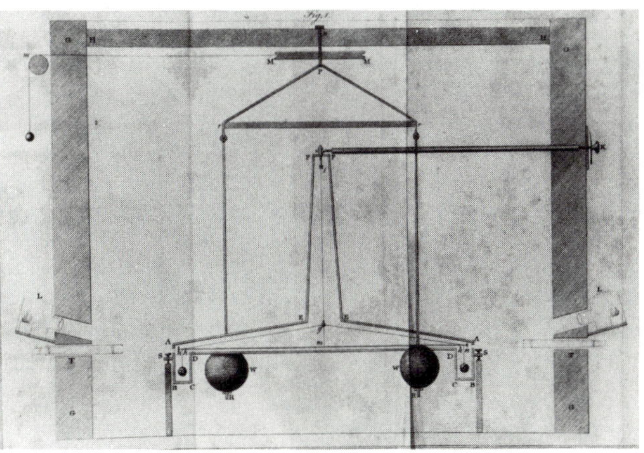

Fig. 7–13 Cavendish's engraving of his apparatus to measure the constant of gravitation. The large lead spheres, marked "W," were 30 cm in diameter and the smaller spheres, marked "X," were 5 cm in diameter. He was able to measure a small rotation of the framework that held the small spheres when the larger spheres were brought close to them.

Fig. 7–14 Superman and Lois Lane. Is it antigravity or merely "dense molecular structure," as is claimed?

The numerical value of G is rather small. As a result, the gravitational force between two protons, for example, is negligible compared to the electric force between them. Why, then, is gravitation so important in our lives? It is because the effects of many particles add up. As we pack more mass into two objects, their attraction for each other increases. Positive and negative electrical charges can cancel one another, but there is no negative mass to provide an antigravity force that would cancel the force of gravity.

Once we know the value of G, we can use Newton's more general form of Kepler's third law to calculate the mass of the sun. The term in Kepler's third law that we called a "constant" actually turns out to depend not only on G but also on the sum of the masses of the orbiting planet (in this case, the earth) and the mass of the sun. Since the mass of the sun is so much greater than that of the earth, we can ignore the mass of the planet in this calculation, which gives the form of the law that Kepler found by trial and error. Applying Kepler's third law, we find the mass of the sun to be 2×10^{30} kg—quite substantial, but, as we shall see, only modest for a star.

For Jupiter, however, a much more massive planet, the mass of the planet does have a measurable effect.

Kepler's laws apply to all orbiting bodies. An excellent example is provided by the moons of Jupiter. Currently 15 such moons are known, but the four largest—Io, Europa, Ganymede, and Callisto—were discovered by Galileo (and so are called the Galilean satellites). Their orbits also obey Kepler's third law: the square of the period of each orbit is proportional to the cube of the satellite's distance from Jupiter. The only difference between the system formed by Jupiter and its moons and the solar system as a whole is that the proportionality constant is different; it is now the mass of Jupiter, not that of the sun, that comes into account in addition to *G*. By observing the periods of the moons and their distance from Jupiter, we can calculate the mass of any planet that has a satellite we can observe. In 1978, a moon was discovered orbiting Pluto, giving us our first opportunity to get a reasonably accurate value for the mass of that distant planet.

From study of Jupiter's moons, we have found that Jupiter's mass is 318 times that of the earth. This result has been improved in accuracy only to a minor extent since the time of Newton. The most accurate value now comes from study of the effect of Jupiter's gravity on the Pioneer and Voyager spacecraft to Jupiter. The discovery of a moon of Pluto came as an absolute surprise. It was found with a ground-based telescope as a bulge on the otherwise round image of Pluto. Application of Kepler's third law to the Pluto system indicates that Pluto has only $\frac{1}{400}$ the mass of the earth.

7.2 A Closer Look at Orbits

Suppose we put a cannon on top of a tall tower and shoot projectiles horizontally. We start by firing them with slow speed and notice that they follow a curved path to the ground. If we fire them with slightly higher speeds, they travel farther before striking the ground. As we continue to increase the initial speed, we increase the distance they travel. When the initial speed of a projectile is high enough, it travels far enough that the curvature of the earth becomes an important factor. As the projectile curves toward the earth, the earth's surface curves away from the projectile, and the projectile travels farther than it would if the earth were flat.

There is a speed at which the curved path of the projectile exactly follows the curvature of the earth's surface. Thus the projectile never reaches the surface of the earth even though it is constantly falling. It always stays the same height above the surface. In this case, we say that the projectile is in orbit (see Fig. 7–16).

For any given height there is one speed at which an object must move to stay in circular orbit. For this speed, the gravitational force between the object and the body it is orbiting provides exactly the inward acceleration needed to sustain the circular motion. As mentioned earlier, this acceleration is equal to the *speed²/radius*. When we carry out these calculations, we find that the orbital speeds necessary to keep an object in orbit are those predicted by Kepler's third law.

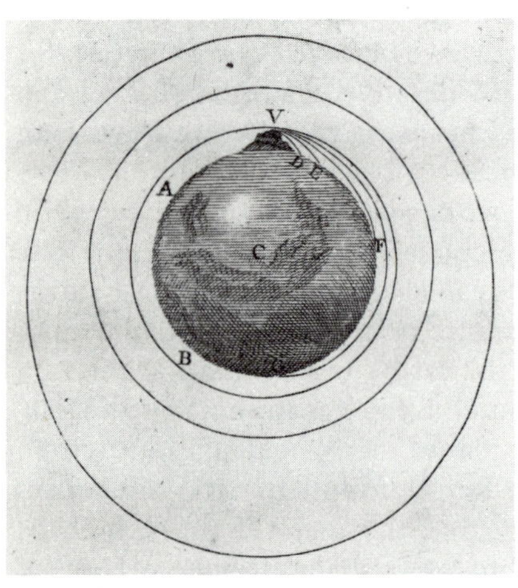

Fig. 7–15 Newton's diagram from 1731 of the launch of a projectile at greater and greater initial velocities.

A satellite in orbit just above the surface of the earth has a period of about 90 minutes. The speed of a satellite in such an orbit is high—about 8 km/s! The lowest practical orbits around the earth are those just above the atmosphere (about 150 km above the surface of the earth), where there is little friction to slow down the

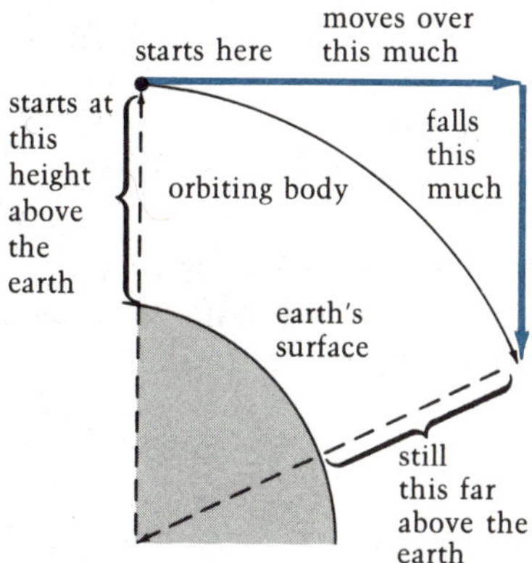

Fig. 7–16 An orbiting body is really falling around the earth, but the earth curves away at the same rate that the body falls.

Fig. 7–17 A close-up view of an Intelsat communications satellite.

satellite. The newspapers have reported the effect of such frictional drag on a Soviet satellite that came down in Canada and on NASA's giant Skylab.

When a satellite is launched, most of the energy expended by the rocket is not used to raise it 150 km above the earth but to get it moving at the speed required to keep it in orbit at that height. Satellites are usually launched toward the east to take advantage of the additional speed imparted to them by the earth's rotation to the east—about 0.5 km/s. Every little bit helps.

By Kepler's third law, period squared = distance cubed. This 24-hour period is about 16 times the period for orbits just above the surface of the earth. If we square 16 we get 256, which must then be equal to the cube of the distance in earth radii. Taking the cube root of 256 gives 6.4 earth radii.

What about higher orbits? Once we know the period for a satellite orbiting just above the surface of the earth, we can apply Kepler's third law to find the period for an orbit of any radius. For example, suppose we want a satellite always to stay above the same point on the surface of the earth, a useful position for communications and weather satellites. Its period must then be 24 hours, since it will revolve in its orbit once for every time the earth rotates on its axis.

The orbit that takes exactly one day must have a radius equal to 6.4 times the radius of the earth, which works out to be about 40,000 km. A satellite that stays over the same point on the surface of the earth is said to be in a synchronous orbit. (The orbit is sychronized with the earth's rotation.) From a height of 40,000 km, a satellite can see about one-third of the distance around the equator; with three satellites spaced properly in synchronous orbit you can cover almost all of the earth.

About eighty satellites are now in synchronous orbits.

Sometimes we are interested in sending a spacecraft beyond the gravitational influence of the earth. Suppose we try shooting it out

To minimize the effect of air resistance, real rockets are accelerated so that they reach escape velocity after they have risen above the earth's atmosphere.

In Chapter 29 we will see that there is what we might consider to be an escape velocity for the universe, which will determine whether the universe, now expanding from a "big bang," will continue to expand forever or will reverse itself and end up in a "big crunch."

of a cannon, as Jules Verne described in a turn-of-the-century novel (Fig. 7–18). We shoot it up in the air with some initial speed. As the projectile rises, it slows down because of gravity. If we have not shot it fast enough, it will eventually stop and then fall back to the ground. The faster we shoot the projectile, the higher it will go before it stops. If we shoot it fast enough, it will never stop. It will continue on, its speed getting slower and slower because of the earth's gravitational pull, but never reaching zero. It will never be pulled back under the influence of the earth's gravity. The minimum launch velocity that allows this to happen is called the escape velocity.

We find the escape velocity for an object by considering the conservation of energy. We want to give the spacecraft enough kinetic energy at launch so that, no matter how much energy is converted to gravitational potential energy as the spacecraft gets higher and higher, some kinetic energy will always be left.

Fig. 7–18 Zero gravity in the Jules Verne moon projectile at the "neutral point" where earth gravity balances moon gravity. Actual "floating" inside a space ship would occur when the motors were shut off.

When we do this calculation, we find that the escape velocity increases with the square root of the mass of the planet and decreases with the square root of the planet's radius. The earth's escape velocity is about 11 km/s; this is the speed that a spacecraft launched from earth must reach if it is to be able to turn off its engines and coast without ever returning to earth. Note that when we put a satellite in orbit not very far above the earth's surface, we give it a speed of 8 km/s—almost three quarters of the speed it needs to escape.

A

B

Fig. 7–19 The smoke allows us to follow the trajectory of a rocket that failed.

7.3 *Systems of Many Bodies*

Up to now, we have been considering the gravitational attraction between pairs of bodies. What if many objects are present? The gravitational force experienced by any one of them is the (vector) *(left)* Fig. 7–20 Uranus and its five satellites: (1) Ariel, (2) Umbriel, (3) Titania, (4) Oberon, and (5) Miranda. The spikes and faint large ring are artifacts caused by the overexposure of the planet.

(right) Fig. 7–21 Neptune with its larger satellite, Triton, close in and its small outer satellite, Nereid, far out.

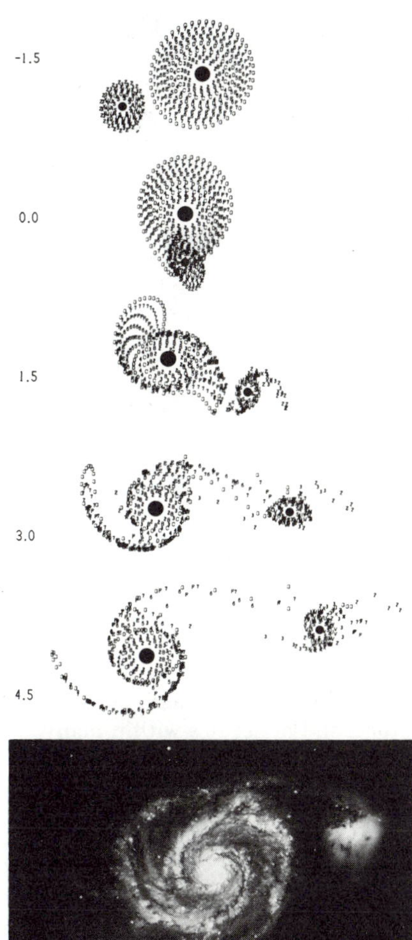

Fig. 7–22 A computer simulation of a collision between two galaxies, each consisting of many stars. The bottom photo of the Whirlpool Galaxy is for comparison with the final result.

sum of the forces exerted on it by all of the others. When one of the objects is much more massive than the others, the more massive object does not move very much under the influence of the others, so we can treat its position as being fixed. If we assume that it is fixed, we get an approximate answer. If we want a more precise answer, we modify the calculations to account for the slight motion of the more massive object.

One of the great triumphs of gravitational calculations involves the outer planets of our solar system. The planet Uranus was discovered in 1781 by William Herschel in England. After the discovery, astronomers realized that for a hundred years Uranus had been occasionally plotted on sky maps but had always been noted as a star, not a planet. Once Uranus was identified as a planet, it was determined that Uranus orbits the sun in about 84 years. Using this fact, Kepler's third law tells us that Uranus's distance from the sun is about 19 A.U. But Uranus's orbit was eventually calculated with great accuracy, including the small effects of the other planets, especially Jupiter and Saturn. Uranus did not quite follow the predicted orbit. Did these deviations from the predicted orbit result from weak gravitational effects of an undetected planet beyond the orbit of Uranus?

What happened next is a classic story of how personality can sometimes color judgment and interfere with scientific progress. John C. Adams in England calculated the predicted position of the new planet in 1845, but the Astronomer Royal did not take the results seriously, partly because Adams was only a young graduate student and partly because Adams did not respond to a question about the calculations with what the Astronomer Royal regarded as proper respect. The same calculations were being carried out independently by Urbain Leverrier in France. Leverrier's calculations were not greeted with much enthusiasm in France, but in 1846 he sent them to an acquaintance in Berlin, J. Galle. Galle immediately began observing, and discovered Neptune within hours. At the time there was an international feud over who deserved the credit, and we now credit both Leverrier and Adams with independently making the calculations that led to this discovery.

Think how complex it is to calculate the path of a spacecraft through the solar system. The Viking spacecraft to Mars was launched from a moving platform (the earth) and aimed at a moving target. The spacecraft's path was curved by the sun's gravity. Also, because current rockets can only carry fuel for a small fraction of the trip, the initial aim had to be accurate. As the two Viking spacecraft approached Mars, their positions were accurate to within 50 km after a total journey of 700 million km!

A

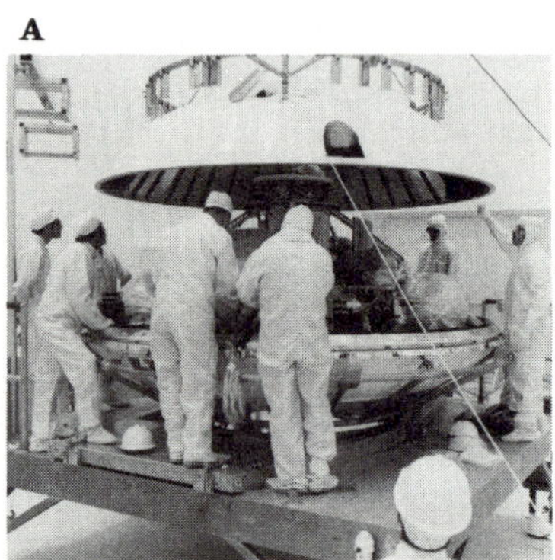

Fig. 7–23 A. The Viking 1 lander before launch. It was sterilized and treated under clean conditions to prevent contaminating the life-searching experiments, or even Mars itself, with microorganisms from earth. B. A view taken on the surface of Mars by this lander. We see sand dunes and large rocks. The lander's nuclear power generators, with American flags on them, appear in the foreground.

B

GRAVITY AND STARS

We realize that gravity plays an important role in our lives; after all, it keeps us on the surface of the earth. However, it plays another important role, one that is crucial to our very existence, since it is the force that molds tenuous clouds of gas and dust into stars. These clouds, scattered between the stars in our galaxy, are called interstellar clouds. They range from small clouds called globules, which have a mass only ten or so times that of the sun, to giant clouds of mass 100,000 times that of the sun. Sometimes the gas glows, heated by light from nearby stars. In some cases we can see these clouds because the dust within them reflects starlight. Many other interstellar clouds give off no light but we are able to see them as dark silhouettes against the bright background of stars. We can also detect radio emissions from the atoms and molecules within many of these clouds.

There are many questions about the exact manner in which such clouds evolve into stars, and different mechanisms may operate in different regions. Still, general pictures are emerging. In an interstellar cloud, the atoms in the gas are moving around at high speeds in all directions. At first, their speeds may be sufficient for them to move around freely. However, in some region there may be a slightly higher than average concentration of gas. This concentration may have come about quite randomly or may result from the action of

Fig. 7–24 The bright objects are 61 Cygni, a binary star in the constellation Cygnus.

other objects outside the cloud. (Perhaps a blast from a nearby exploding star, a supernova, compressed the matter.)

If these concentrations contain a lot of mass, they would have enough gravity to keep most of the gas particles in the immediate vicinity from escaping. Gradually, the particles start to fall into the center of the concentration. As the mass becomes more concentrated, gravity becomes stronger and the particles are accelerated faster into the center. We say that the cloud is in gravitational collapse. As the collapse continues, the cloud may break up into several sections, each of which may eventually become a star.

As each section collapses, it also heats up. Eventually the temperature in the center of each section becomes so high that nuclear reactions start and the energy generated in the nuclear reaction heats these new stars. This provides an outward force that fights gravity, which keeps the star from collapsing any further. The star is now in the prime of its life. A star like the sun spends about 10 billion years in this stage.

Because large clouds often break up into several parts, most stars are not formed individually. They come in pairs or even larger groupings. When we have two stars, we call the pair a binary star. In a binary star system, the stars are in orbit about each other, or, more properly, about their common center of mass (see Fig. 7–25).

Binary stars can tell us a lot about the masses of stars in general. We saw in Section 7.1 that we were able to find the mass of the sun from Kepler's third law if we knew the periods and sizes of the planetary orbits. The same is true with binary stars. We can learn about their masses by measuring the sizes and periods of their orbits around their center of mass and then applying Kepler's third law. From these studies, we find that the stars of the lowest mass have masses about one-tenth the mass of the sun and the most massive stars have about fifty times the mass of the sun.

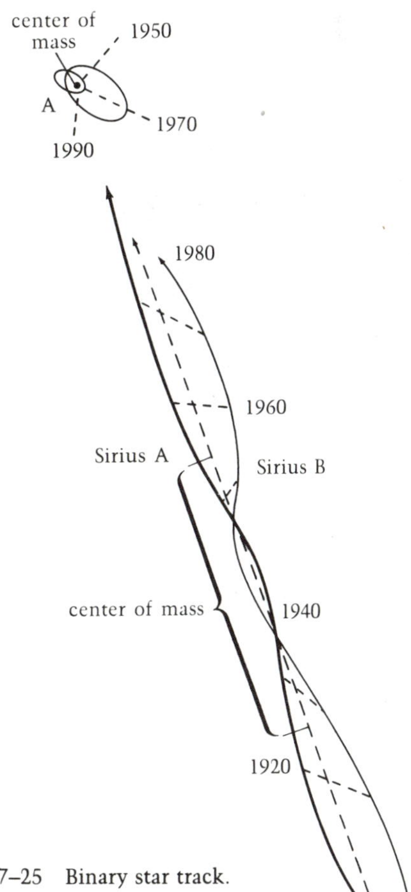

Fig. 7–25 Binary star track.

7.4 Tides

We have seen that the gravitational force exerted by one particle on another varies with changes in the distance between the two particles. Certain effects occur because of the difference in gravitational force between one point and some other point. One such effect is the ocean tides on the surface of the earth.

The main tidal effects on the earth are due to the moon, but the sun also plays an important role. Let us first consider the effect of the moon. For our simple example we will need to keep track only of the force that the moon exerts on three different parts of the earth: (1) the water on the side of the earth closest to the moon, (2) the center of the earth, and (3) the water on the side of the earth farthest from the moon. If these three positions were just three balls dropped from an earth-sized tower above the surface of the moon, they would, of course, all fall toward the moon's surface. Since the one closest to the moon would have the greatest acceleration, and the one farthest from the moon would have the least acceleration, as the three balls fell they would tend to spread out. An observer located on the middle ball would see each of the other two balls move farther away from the middle.

For the earth, the situation is similar, except that the three points would not spread out indefinitely. That is because the opposing gravitational pull on the water *by the earth* tends to pull the water back in, which limits how far apart the three points can be stretched. Also, in the real case, the three points will never fall into the moon because of the orbital motion of the moon and the earth.

The point on the earth that is closest to the moon will experience the greatest force, and the water on that side will experience the greatest attraction by the moon; the water on the opposite side of the earth will experience the least attraction. Since force causes acceleration, water on the side of the earth closest to the moon will experience the greatest acceleration, that on the opposite side of the earth will experience the least acceleration, and the center of the earth will experience an intermediate amount (Fig. 7–27).

Thus we find that the moon's exerting a different force on the near side, middle, and far side of the earth leads to the following results: (1) The water on the side closer to the moon is pulled away from the center of the earth, meaning the water level is higher there than average. We call this *high tide.* (2) The earth is pulled away from the water on the side opposite the moon. This also creates a high tide, which is why there are about two high tides every day. Actually, since the moon is moving in its orbit, it takes longer than 24 hours to return to the same position overhead, so the time between high tides is about $12\frac{1}{2}$ hours instead of 12 hours.

A

B

Fig. 7–26 Low and high tide in the Bay of Fundy in Nova Scotia, site of the world's highest tides.

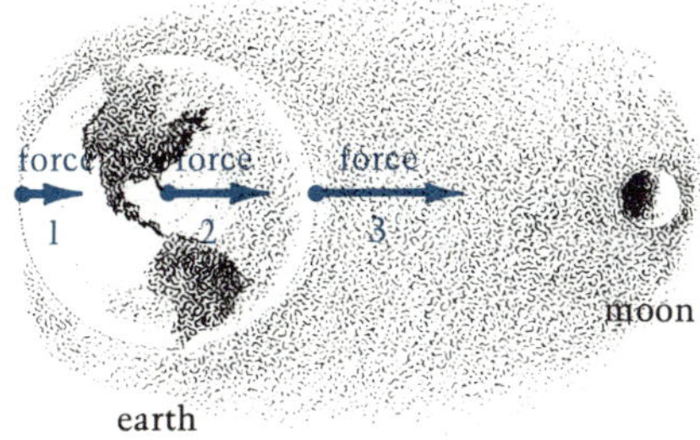

Fig. 7–27 Tides. The force that the moon exerts on point 1 is greater than the force it exerts on point 2, which in turn is greater than the force it exerts on point 3. The tendency of these different forces is to pull point 2 away from point 3 and point 1 away from point 2. (The differences in the forces have been exaggerated in this diagram.)

The sun also plays a role in creating the tides, but its effect is only about half that of the moon's because the sun is so far away that its force does not change very much from one side of the earth to the other. At the times of new moon and full moon, the tidal effects of sun and moon are in the same direction and high tides are particularly high. These high tides are called *spring tides*. At the time of the first and last quarter moon, the effects of sun and of moon are at right angles to each other, and the contrast between high and low tide is minimized. These are called *neap tides*.

We have discussed an idealized explanation of tides. The shape of the ocean bottom, the edges of continents and bays, and other factors have major effects on the tides at any given location.

When we speak generally of tidal effects, we mean an effect that varies from one side of a body to the other. Tidal effects are responsible for one of the most beautiful sights in our solar system—the rings of Saturn. These rings are 275,000 km across and probably only 10 km thick. The rings are not solid; they appear to be a collection of particles, each individually orbiting Saturn. We know that the rings are not solid because we can sometimes see stars through them. When we study the rotation of the rings, we find that the speed of rotation at each different distance from Saturn is the speed that would be expected according to Kepler's third law. This would be true only if each part of the ring were free to orbit on its own, because if the ring were solid, the speeds of different distances would vary in a different fashion.

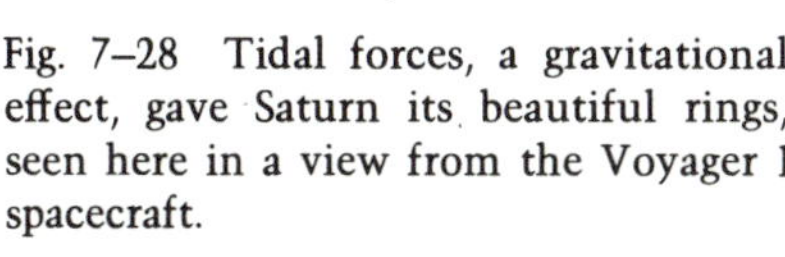

Fig. 7–28 Tidal forces, a gravitational effect, gave Saturn its beautiful rings, seen here in a view from the Voyager 1 spacecraft.

Fig. 7–29 When Voyager 1 passed close by Saturn in 1980, it discovered that Saturn actually has hundreds of rings.

We believe that the rings are chunks of rock and ice, ingredients of a moon that never formed. To understand this idea, think about two balls in orbit around Saturn, one a little closer to the planet than the other. Normally, we would expect the gravitational attraction that the two balls have for each other to pull them together. However, Saturn exerts a different pull on each of the balls, since they are at different distances from the planet. The one closer to the planet feels a greater acceleration. This tends to pull the particles apart.

Which tendency wins? Will the balls come together or will they be pulled apart? It depends on how close together they start out and how close to Saturn they are. If they are close to Saturn, the tendency to be pulled apart will be strengthened. On the other hand, if the balls are very close to each other, their attraction for each other will win. For any given material in orbit around a planet, there is a distance from the planet where the gravitational attraction of the orbiting particles for each other will not be sufficient to neutralize the tidal effects. This distance is called the Roche limit. If the material is outside the Roche limit, gravitational attraction between the two objects will win and a moon can form. When we calculate the Roche limit for Saturn, we find that the rings are inside the limit while the moons of Saturn are outside the limit. It was discovered in 1977 that Uranus also has rings and they, too, are within Uranus' Roche limit. A still bigger shock came in 1979, when Jupiter—which had been so carefully observed for so long—turned out to have a ring too. The ring, of course, is within Jupiter's Roche limit.

The Roche limit describes only the ability of a material to hold itself together against tidal effects by gravitational forces alone. Artificial satellites are within the Roche limit of the earth yet they don't fly apart. For that matter, *we* are within the Roche limit of the

Fig. 7–30 The newly discovered ring of Jupiter, as seen from the Voyager 2 spacecraft.

earth and we don't fly apart. That is because we (and artificial satellites) are not held together by gravitational forces. We are held together by the electrical forces between the atoms and molecules of our bodies.

Of course, there are places where even electrical forces would have trouble holding us together. For example, astronauts visiting a neutron star, which packs the mass of the sun into a sphere 20 km in diameter, would experience severe tidal effects. If they were standing on the surface, the gravitational attraction of their feet would be so much greater than the attraction of their heads that their bodies would be torn apart. If astronauts ever try to visit a neutron star, they would last a tiny bit longer if they arrived prone, stretched out horizontally.

Key Words

force laws
retrograde motion
epicycle
deferent
heliocentric
Occam's razor
Kepler's laws
focus of an ellipse
astronomical unit
law of universal gravitation
constant of universal gravitation
Cavendish balance
in orbit
synchronous orbit
escape velocity
interstellar clouds
globules
gravitational collapse
binary star
tides
tidal effect
Roche limit

Questions

1. What is the significance of the word "universal" in talking about gravitation?
2. Why did Newton reason that there has to be a force acting on the moon?
3. (a) In the debate over the earth-centered and sun-centered views of the planetary system, which do you think is the simpler theory and why? (b) Explain the significance of deciding which is simpler.
4. What are three ways in which Galileo's observations supported the heliocentric theory?
5. Which are more fundamental, Newton's laws or Kepler's laws? Explain your choice.
6. The earth is closest to the sun in January. Is the earth moving in its orbit at its fastest speed or its slowest then?

7. Jupiter takes 12 earth years to orbit the sun. What is the semimajor axis of its orbit?
8. Saturn's semimajor axis is 9.5 A.U. How long does Saturn take to orbit the sun in earth years?
9. The orbit of Jupiter's satellite Io has a semimajor axis that is 5.9 Jupiter radii; its period of revolution is 1.7 earth days, which is 4.2 Jupiter days. Ganymede's orbital semimajor axis is 15 Jupiter radii. (a) How long does Ganymede take to orbit Jupiter in earth days? (b) How long does it take in Jupiter days?
10. What do we mean when we say that the law of gravitation is consistent with Newton's third law?
11. Why was the development of calculus important to Newton's development of the law of gravitation?
12. Why can we treat falling bodies near the surface of the earth as though the acceleration of gravity is constant?
13. Why is the gravitational force between the earth and the sun greater than the electrical force between the earth and the sun, even though the electrical force between any two protons is much greater than the gravitational force between the two protons?
14. Why are satellites usually launched from west to east?
15. Is it possible for a synchronous satellite to stay directly above New York City? Explain your answer.
16. You go around the earth once per day as you are carried around by the earth's surface. Can we say that you are in a synchronous orbit?
17. Which is greater, the speed of a satellite in synchronous orbit around the earth, or the escape velocity of the earth?
18. As the earth goes around the sun, it does not travel in a simple ellipse. Instead, it wobbles in and out about that ellipse. Why?
19. What made astronomers think that there is another planet beyond Uranus?
20. As an interstellar cloud collapses to form a star, it heats up. Where does the energy come from?
21. How can we tell if a star has an unseen companion?
22. Explain why there are high tides on *both* the side of the earth nearest the moon and the side of the earth farthest from the moon.
23. Successive high tides are not exactly 12 hours apart. Explain why.
24. At what range of distance from Neptune, nearer or farther than its moons, would you concentrate your search for a ring?
25. Why is the study of tidal forces important for understanding the ring that has been discovered around Jupiter?
26. Draw a diagram to show what would happen to the earth if the sun's gravitational pull could be suddenly "turned off."

CHAPTER 8

Special Relativity

Fig. 8–1 Albert Einstein, on a visit to California in 1933.

The classical concepts of motion that we have discussed have been around since the time of Newton. It might seem that they have withstood the "test of time." The application of these basic concepts can explain many of our everyday experiences.

At this stage, we are like nineteenth-century physicists. They felt quite satisfied that the physics of that time was complete, and that it not only succeeded in explaining an amazing diversity of phenomena but also would continue to be able to explain the world in detail. This complacency was shattered by the events of the last part of the nineteenth century and the early twentieth century. We have already briefly mentioned the development of quantum mechanics, and will treat the "quantum revolution" of fifty years ago later in this book. In this chapter and the next, we will consider theories of relativity, which have particularly important consequences for mechanics as well as for gravitation.

Actually, what is often loosely called "relativity" has two quite different parts. The special theory of relativity was introduced in 1905 by Albert Einstein, and deals with the relationship between space and time. "Special relativity," as it is called for short, is limited in its application in that it does not take proper account of gravitational fields. Einstein generalized relativity in 1916 with the general theory of relativity, which deals with the effects of gravitation.

In this chapter, we discuss the special theory of relativity. We will see that it drastically alters some of our concepts of how to describe motion. In fact, some of special relativity's ideas seem to go against the grain of our everyday experience. These ideas forced Einstein to reevaluate some common assumptions that had seemed so obvious that nobody even bothered to state them.

Einstein showed us that our "everyday experience," despite its apparent variety, is really quite limited. For example, if you lived your whole life on a square kilometer of the earth without contact with the outside world, then it would be reasonable for you to believe that the earth is flat. Imagine your surprise and disbelief when you first saw a satellite weather picture on the TV news. As our experience increases, we are always liable to meet events that do not fit into our previous understanding.

These new events do not necessarily mean that our previous ideas are totally wrong. In the case of motion and gravitation, the old rules are still useful but only in a limited set of circumstances. In fact, any new, more general set of rules has to give the same answers as the old rules when applied to our old, limited experience. Thus, rules that help you navigate around a spherical earth should allow you to navigate around your own corner. We'll see, for example, that even though special relativity is especially important when velocities are very large, it also explains situations in which the velocities are small. For small velocities, we come up with (as we must) the traditional laws of mechanics.

The special theory of relativity is intimately related to light and how it travels. We will therefore start by looking at the theory of how light travels. Around the turn of the twentieth century, some aspects of the theory were proving troublesome.

8.1 *The Puzzle of Light*

The nature of light has been the subject of scientific discussion for over two thousand years. One of the perplexing problems is that in some ways light acts like a series of waves in the ocean, and in other ways it acts like a baseball headed for the bleachers. The explanation of this dual nature of light, sometimes wave and sometimes particle, was an important part of the birth of quantum mechanics and we will discuss it further in Chapter 23.

At the time Einstein presented the special theory of relativity, several experiments clearly indicated that light acts in certain situations as a wave rather than as a set of particles. And in addition, the British physicist James Clerk Maxwell in 1873 put forward a detailed theory that explained electricity and magnetism. We shall discuss Maxwell's theory when we discuss electricity and magnetism in Chapters 19 to 22, and need only comment here that this wide-ranging theory also explained light waves, so the notion that light acts as a wave was well established at the turn of the century.

8.1a THE SPEED OF LIGHT

A Danish astronomer, Olaus Roemer, was the first to demonstrate that light has a finite speed. In 1675, he showed that light does not get from one place to another instantaneously (although it does move rather fast). Roemer, who had been studying the moons of Jupiter, noticed that the eclipses of these moons by Jupiter did not always occur at the predicted times. Sometimes they were earlier than the prediction; sometimes they were later. He reasoned that the light from Jupiter's moons traveled different distances at different times of the year, depending on how far the earth was from Jupiter and its moons. If light travels at some finite speed, when the earth and Jupiter were relatively close, the light that left Jupiter as an eclipse occurred would arrive a little earlier than average. The eclipse would be visible ahead of its prediction. When the earth and Jupiter were relatively distant, the eclipses would occur behind schedule. Unfortunately, in Roemer's time the distance of the earth from Jupiter was not known accurately and he did not compute a value for the speed of light (see Fig. 8–2).

Jupiter
earth
4 3 2
sun 1

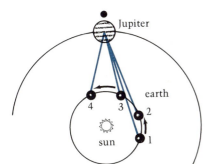

Fig. 8–2 When the earth is approaching Jupiter (position 1 to position 2), the eclipses of Jupiter's moons appear to occur ahead of the calculated schedule. When the earth is staying at about the same distance from Jupiter (position 3 to position 4), the eclipses appear on time. This phenomenon was observed by the Danish physicist Olaus Roemer in 1675. Roemer deduced from his observation that the speed of light is finite.

To measure the speed of light, we usually time the passage of a light beam over some known distance. Since light travels so fast, we must either be able to measure time intervals very accurately or else be able to make our measurements over a long distance. The earliest measurements of the speed of light were astronomical because of the long distances involved. In 1849, the French physicist Hippolyte Louis Fizeau made the first nonastronomical measurement. Ever since that time our knowledge of the speed of light (which is usually designated by the symbol c) has been steadily improved in a series of ingenious experiments. The currently accepted value for the speed of light is 2.99792458×10^8 m/s. We usually round this off to 3×10^8 m/s. (We did some calculations using this number in Chapter 2.) The speed of light corresponds to traveling seven times around the earth in a single second.

If light is a wave, what does it travel through? All the ordinary waves have to travel through *something*. For example, we cannot have waves in the ocean without the water. As we will see in Chapter 14, sound is a wave, but there is no sound if we pump all the air out of the room.

Fig. 8–3 You need water to have water waves.

To overcome this problem, nineteenth-century physicists postulated the existence of a material called the ether. This material had to be so thin that it had no effect on the planets in their motions, yet still supply enough material to carry light waves at such speed.

8.1b THE ABERRATION OF STARLIGHT

Not all the evidence pointed to light's being a wave traveling through the ether. Let us first consider the aberration of starlight, a delicate effect that we can describe with the following analogy.

Assume that you are standing in the rain, holding an umbrella over your head (Fig. 8–4). If the rain is coming straight down and you are standing still, you don't get wet. But if you get tired of standing in the rain and start to run for shelter, the front of your raincoat will begin to get wet. Your forward motion gives the rain a backward motion relative to you. When we add this backward motion to the downward motion of the rain, the rain seems to you to be slanting. The faster you run, the greater the slant will appear to be. You are running out from the protection of your umbrella.

You can think of this as an addition of vectors (Fig. 8–4E). Adding vectors was described in Chapter 2.

A similar thing happens with light. If you are moving, a light beam that crosses your path will appear slightly slanted. The slant will be very small since your speed is small relative to the speed of

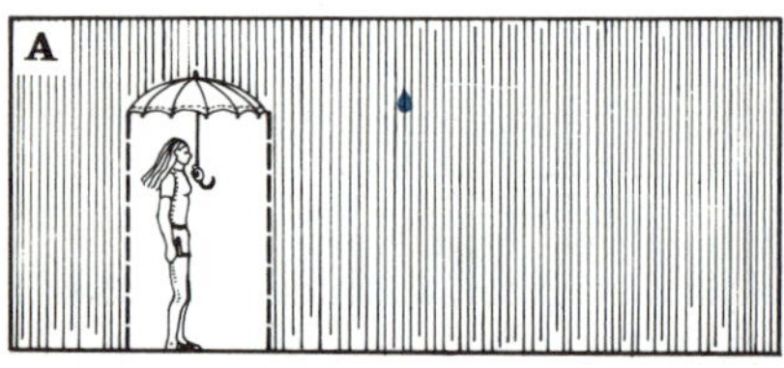

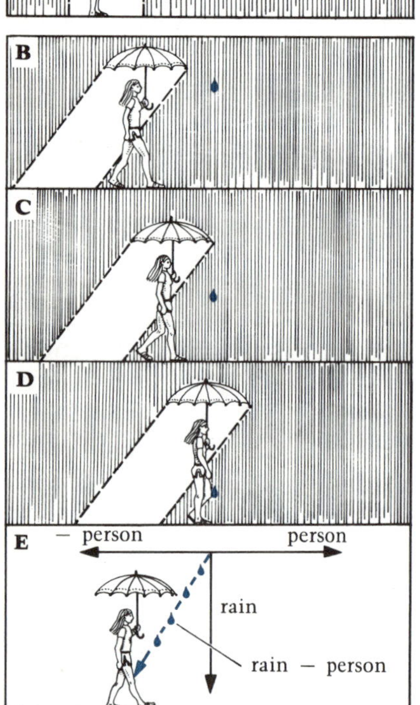

(left) Fig. 8–4 The aberration of starlight can be explained by the analogy of someone walking through the rain. A. Here the girl is stationary, so she is shielded from the rain by her umbrella. B–D. Once she begins to walk, she is no longer protected. Although the raindrops fall past her umbrella, her legs walk into them. E. The diagram shows the path of the raindrops from her point of view, and the vector explanation of this phenomenon. While she is walking, she would have to hold her umbrella forward to keep her entire body out of the rain. Because the earth is moving, telescopes have to be pointed in a slightly different direction than they would be if the earth were stationary.

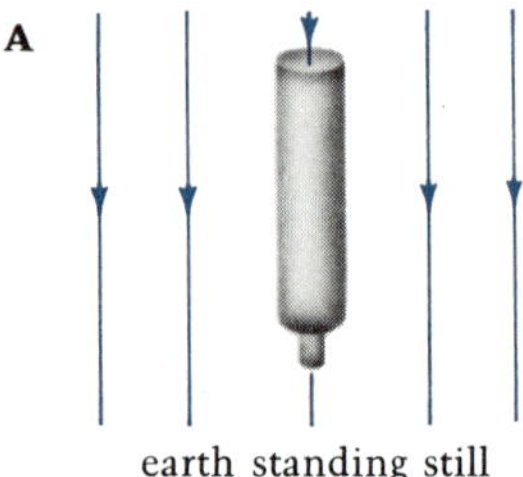

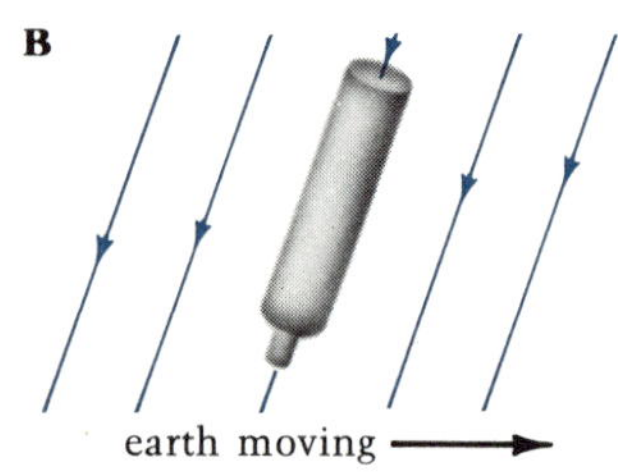

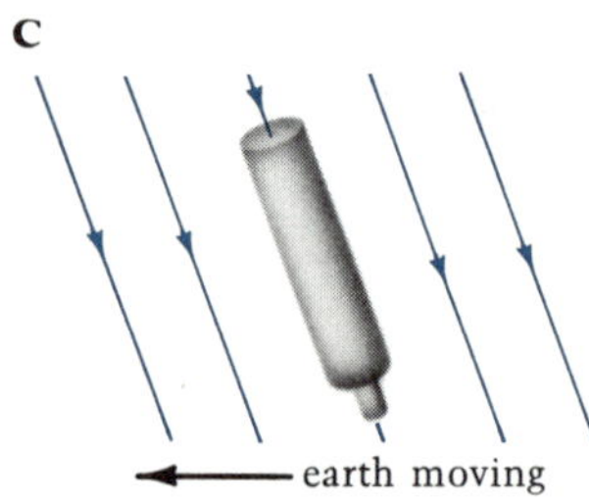

(right) Fig. 8–5 Effect of the aberration of starlight. A. This frame shows the direction in which we would have to point a telescope to view a star if the earth were not moving. B. If the earth is moving to the right, we must tilt our telescope slightly to the right to see the same star. C. If the earth is moving to the left (say, six months later), we tilt our telescope to the left to view the star.

light. However, it is possible to measure this tiny slant. If we look at a distant star with a telescope, the motion of the earth across the light beam from the star makes the star's position appear to change slightly. Six months later, when the earth is moving in the opposite direction, the star appears to shift in the opposite direction (Fig. 8–5). Over the course of half a year, the star's position will appear to change by a small amount, but an amount that is easily measurable by astronomers.

This effect is called the aberration of starlight and was first noticed in 1725 by James Bradley, a British astronomer. Bradley was able to use his discovery to measure the speed of light.

What does the aberration of starlight have to do with the ether? If light were made of particles, then the phenomenon can be easily understood by making an analogy between the light "particles" and the raindrops. However, if light were a wave that needed a medium through which to propagate, we can draw an analogy between the motion of light through the ether and the motion of a swimmer through water approaching a steamship. (The swimmer is analogous to the light and the steamship to the earth.) If the swimmer is swimming due south and the steamship is heading due east, then the passengers in the steamship will see the swimmer coming at them at an angle. However, if the steamship is so powerful that it drags all the water along with it, then the swimmer will also be dragged along eastward, will keep up with the steamship, and will come to the steamship perpendicularly. Similarly, if the earth dragged the ether along with it, then the light would also be dragged along and no aberration would be observed. Therefore, since we observe the aberration of starlight, the earth must be plowing through the ether like a ship through the water rather than dragging the ether along. The ether and the earth do not affect each other (Fig. 8–6).

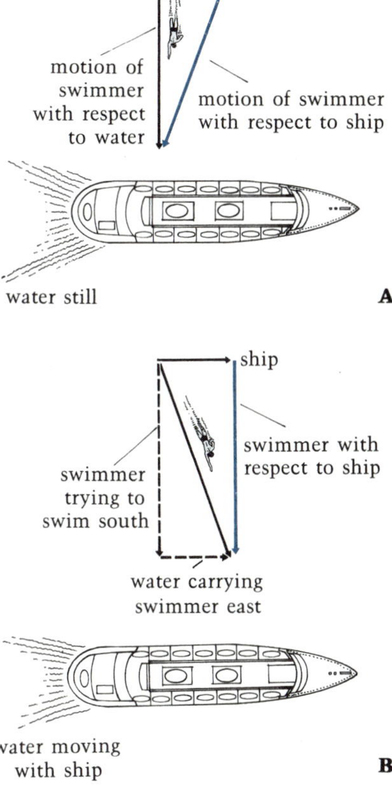

Fig. 8–6 An analogy showing the relationship between the aberration of starlight and the existence of the ether. In this analogy, we will let the boat represent the earth, the swimmer represent incoming starlight, and the water represent the ether. A. The water is still and the swimmer starts out southbound. However, the ship is moving to the east (right), so, to observers on the ship, the swimmer's path appears to be a diagonal. This apparent slanting of the swimmer's path due to the motion of the ship is like the aberration of starlight. B. Now, what happens if the water is dragged along with the ship? In this case, as the swimmer tries to swim south, the current will carry the swimmer to the east, and the swimmer will drift to the east at the same rate as the ship. Therefore, as far as viewers on the ship are concerned, the swimmer's path will not be slanted. There will be no aberration. This analogy tells us that if we observe aberration of starlight, the earth must be moving through the ether and not dragging it along. The Michelson-Morley experiment set out to measure the motion of the earth through the ether.

8.1c THE MICHELSON-MORLEY EXPERIMENT

Interferometers use a property of light called "interference," which we shall discuss when we discuss waves in Chapter 15. To study relativity, we need only know that Michelson's device worked and could measure differences in the speed of light to a high degree of accuracy.

Can we measure the motion of the earth through the ether? If light moves at a constant velocity with respect to the ether, then we on the earth should measure a different velocity of light depending on whether we are moving in the same direction as the light or in some other direction. The velocity of the *light* with respect to the *earth* should be equal to the difference between the velocity of the *light* with respect to the *ether* and the velocity of the *earth* with respect to the *ether*. Thus, the speed of light as measured on the earth would depend on the direction in which the light beam is traveling.

The speed of the earth is small compared to the speed of light, so we expect to measure only a very small change in the speed of light between light beams traveling in different directions. In 1881, A. A. Michelson, one of the first American experimental physicists, began to study the velocity of the earth with respect to the ether, which he proposed to do by analyzing the speed of light for light coming from different directions. He used a device he invented called an *interferometer*. This device could detect the difference in the speed of light traveling in two perpendicular directions. At the Case Institute of Technology in Cleveland in 1887, Michelson was joined by E. W. Morley in performing a more accurate version of the original experiment (see Fig. 8–7).

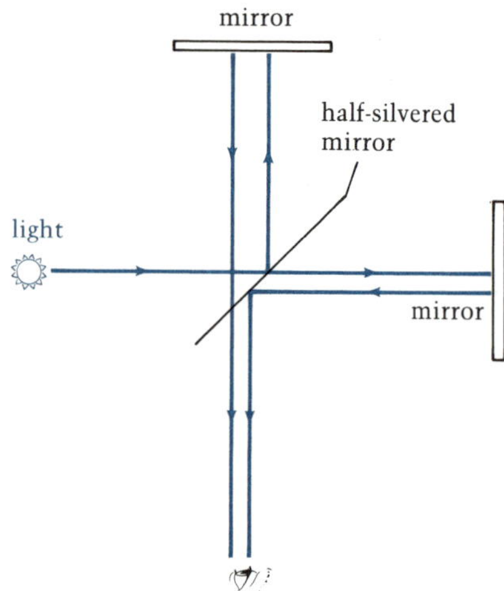

Fig. 8–7 Basic operation of the Michelson interferometer. Light enters from the left and strikes a half-silvered mirror, so some is passed through the mirror and some is reflected. One beam goes up, strikes a mirror and comes back down, with part again passing through the half-silvered mirror and reaching your eye. Similarly, the other part goes to the right and then is reflected back, with part of that being reflected to your eye. The two beams reaching your eye interfere with each other. If the earth is moving through the ether, then the speed of light should be different in different directions, and the interference pattern should change as the whole apparatus is rotated around.

The results of the Michelson-Morley experiment were a shock to most physicists. Michelson and Morley found that there was *no* detectible difference in the speed of light in the two perpendicular directions. Now, if light were propagating through the ether, the speed of light with respect to the earth would be different in the different directions. But Michelson and Morley found that the speed of light with respect to the earth is the same in all directions! Thus, the predictions of the ether theory were wrong.

Einstein, in his special theory of relativity, was the first to explain the Michelson-Morley experiment successfully. (Actually, Einstein was not trying to explain the Michelson-Morley experiment when he developed special relativity. He was primarily concerned with other problems. Einstein has even stated on some occasions that at the time he was not aware of the Michelson-Morley experiment.) Einstein pointed out that the ether was not necessary at all. It existed only by the assumption of late nineteenth-century physics that waves had to travel through something. Einstein held as a basic assumption that *the speed of light is independent of the velocity of the observer*. This is a strange idea since it seems to contradict our ordinary notion of relative motion, which we discussed in the first chapters of this book. However, this is just such a case in which our experience has been so limited that we have not been prepared for this new turn. Let us now take a further look at the special theory of relativity.

Note that we are ruling out acceleration but not velocities; it is perfectly all right for one inertial reference frame to travel at a constant velocity with respect to another. Ruling out acceleration is the same as ruling out gravity. We saw in Chapter 3, if someone locked you in a box with no windows, you could not tell whether the box was accelerating or whether it was in an external gravitational field. Actually, we can have an inertial frame when both a gravitational field and an acceleration are present, but only in the special case when the effects of the gravity cancel out the effects of the acceleration. An example of such a situation is the "weightlessness" felt by astronauts in orbit. For simplicity, we will ignore gravity in this chapter and just say that observers are in inertial frames if they are moving at constant velocities.

8.2 *The Principle of Relativity*

8.2a FRAMES OF REFERENCE

The special theory of relativity deals with the way that events appear to different observers, each of whom is moving at a different velocity. With each observer, we can associate a *frame of reference,* also called a reference frame. We can think of a frame of reference simply as a coordinate system that moves along with the observer so that this observer can record the times and locations of events.

The special theory of relativity shows that a certain kind of reference frame, called an inertial reference frame, has important properties. Inertial reference frames are reference frames that are (1) not accelerating and (2) not subject to external gravitational influences.

Einstein's fundamental postulate in formulating the special theory of relativity is called the principle of relativity. It is most commonly stated: the laws of physics are the same in every inertial frame. Thus, if two reference frames are in motion with respect to each other, there is no experiment that we can do to decide if one of them is "really" at rest while the other is moving. In other words, *there is no preferred inertial frame in which to do your physics.*

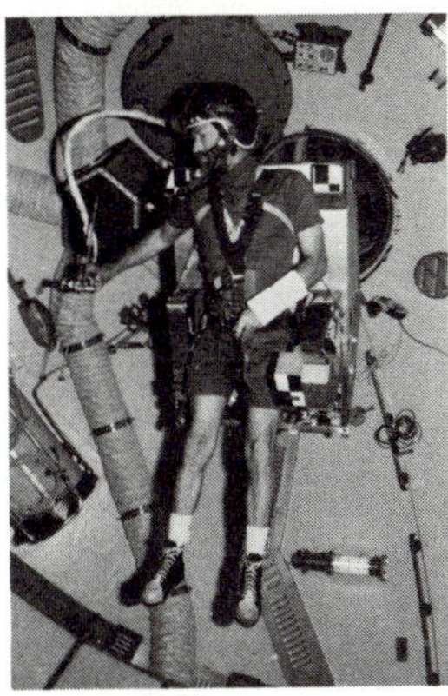

Fig. 8–8 Astronauts seem weightless because they are falling in a gravitational field at the same rate as the vehicle they are in. In this case we see an astronaut in Skylab.

8.2b THE SPEED OF LIGHT

To the principle of relativity, Einstein added the statement that the speed of light is the same for all observers. All the results of special relativity follow from these two postulates.

The notion that observers moving at different speeds would measure the same speed for a pulse of light does seem a little strange. Experimenters didn't leave the matter with Michelson and Morley. About a dozen years ago the speed of light was observed for the radiation emitted by a certain type of elementary particle produced by a particle accelerator. (The particles were pi-mesons, something that experimenters certainly knew but something that isn't part of the essence of this story. The pi-meson plays a central role in reactions among elementary particles involving the strong, or nuclear, force.) When pi-mesons decay, they produce a pulse of radiation. The speed of the pulse can be measured by detectors in the laboratory, which we say is "at rest."

In the particular experiment, pi-mesons were produced that were traveling at about 99.9 percent of the speed of light. The bursts of radiation are emitted in the direction in which the pi-mesons are traveling. Thus, if our old notion of relative motion applied, we might expect that the burst of radiation would be moving at nearly twice the speed of light when observed in the laboratory. (If a train is going 99.9 percent of 30 m/s, which is 29.97 m/s, and someone throws a ball off the front of the train at 30 m/s with respect to the train, then an observer in a station will see the ball traveling at nearly 60 m/s.) However, the experiment showed that the burst of radiation observed in the laboratory was still moving "only" at the speed of light, *not* at twice the speed of light. (Actually, by the time that this experiment was performed the special theory of relativity was so well established that no one really expected a different result. However, it is always reassuring to get as much confirmation as possible.)

Even if you are beginning to believe in your head that the speed of light is independent of the velocity of the observer, you may still not be satisfied in your heart. After all, it remains a strange idea. Our common intuition breaks down. Part of Einstein's great insight was not only that he realized that the fabric of classical mechanics was coming apart, but also that he was even able to identify where the rips were. He realized that in measuring speeds, we must time (with some sort of clock) how long it takes something to get from one place to another. He suspected that the real "problem" was with our common concept of *time*.

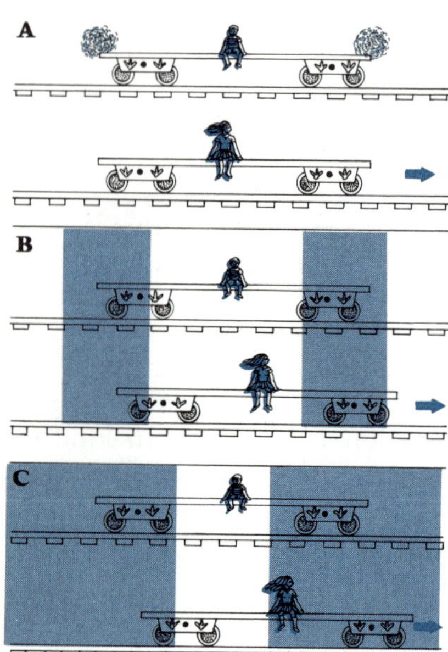

Fig. 8–9 The problem of simultaneity. Eric and Eloise are on identical railroad cars. With respect to Eric's car, Eloise's car is moving to the right. A. At the instant when Eric and Eloise are directly opposite each other, flashbulbs go off at opposite ends of Eric's car. The two flashes are simultaneous as seen by Eric. Are they simultaneous as seen by Eloise? B. We begin to watch how the light would spread out from the flashes, as indicated by the colored regions. (In this case we are only concerned with the spread of the light in the horizontal direction.) Eloise has moved over a little to the right, but still hasn't received the light from either flash. C. Light from the flash on the right reaches Eloise, but the light from the flash on the left has not reached her yet. She concludes that the flash on the right took place before the flash on the left. The two flashes were not simultaneous as viewed by Eloise.

8.3 *Problems With Time*

Einstein realized that physicists, along with everybody else, had always assumed that there was a universal time that all observers shared. This assumption was so ingrained that nobody even referred to it explicitly. Everybody always just spoke of "the time." Implicit in the common notion of time was the idea that if an event occurred, all observers would agree on the time of the event, that is, when the event occurred.

8.3a SIMULTANEITY

However, Einstein pointed out that time is not only a vague notion. It is an experimental quantity in that it is measured by clocks. (Conversely, let us define a "clock" as any device that measures time.) If we say that a train leaves at 7:00, we really mean that two

A

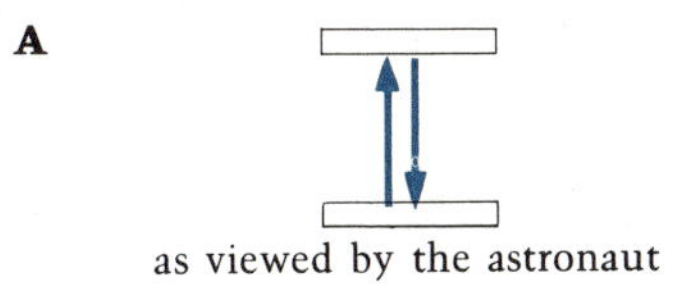

as viewed by the astronaut

B

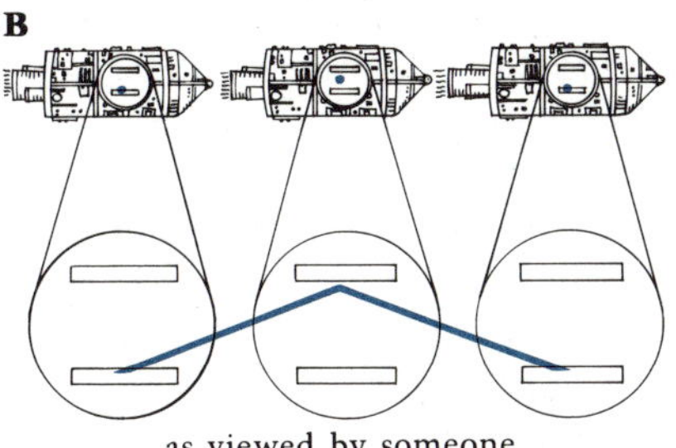

as viewed by someone, with respect to whom the rocket is moving

Fig. 8–10 Two views of a light clock inside a rocket. A. As seen by the astronaut inside the rocket. The light bounces back and forth between the two mirrors. B. As viewed by someone on the ground, who sees the rocket moving. Since the clock is moving to the side, as the light bounces back and forth it appears to be traveling in a diagonal track instead of moving straight up and down. To this viewer, the light travels a longer distance for a full cycle than it does for the astronaut. Since the speed of light is the same for both observers, each tick of the clock is longer as viewed by the observer on the ground than as viewed by the astronaut. (The horizontal motion has been exaggerated for clarity.)

Since distance = rate × time, $d = ct$, or $t = d/c$.

If t_a *is the proper time,* $t_a = d/c$.

events, (1) the leaving of the train and (2) the station clock reading "7:00," are *simultaneous,* that is, they happen at the same moment. If all observers are to agree that the train leaves at 7:00, then all observers must agree that these two events are simultaneous. Thus, the assumption that a universal time common to all observers in the universe exists is equivalent to the following assumption: if two events appear to be simultaneous to an observer in one inertial frame, then they must appear to be simultaneous to observers in all inertial frames.

This is a proposition that we can test fairly easily. Imagine the situation depicted in Fig. 8–9. There are two railroad cars on parallel tracks. For simplicity, we say that one is at rest. Eric sits in the center of this car. The other car, with Eloise in the center, moves past the stationary car at some high speed. At the instant that Eloise is directly opposite Eric, firecrackers go off at opposite ends of Eric's car. The two explosions are clearly simultaneous for Eric. However, the figure shows that they are not simultaneous for Eloise. Similarly, if the firecrackers had been set off in Eloise's car, they would have been simultaneous for her but not for Eric. It doesn't matter which car is moving. The important point is that events which appear simultaneous to one observer are not simultaneous to another observer moving with respect to the first.

This example shows that the concept of simultaneity is a relative one. If two observers cannot agree on the simultaneity of two events, then they can never agree on a single universal time. The passage of time will appear different to different observers.

8.3b TIME DILATION

Now let us see how different "time" can appear. We will consider a simple way of measuring time, called the *light clock* (Fig. 8–10A). This clock has two mirrors facing each other, and a pulse of light travels back and forth between them. We can think of each passage of the light pulse from one mirror to the other as one "tick" of the clock. How long is one tick?

The time for light to go from one mirror to the other is just the distance between the mirrors divided by the speed of light. (Remember, light travels at the same speed in all directions, so it doesn't matter if the light is going back or forth.

Now we put the clock in a rocketship moving at a constant speed (Fig. 8–10B). To an astronaut inside the rocket, the round trip of the light appears just as we described above, since the clock is at rest in the rocket. The time kept by the clock as viewed by an observer at rest with respect to the clock is called the proper time of the clock. Thus, in this example, the proper time is the distance between the mirrors, divided by the speed of light.

Suppose the clock is mounted in the rocket so that the light pulse travels back and forth in a direction perpendicular to the path of the rocket's motion. How would the round trip of the light appear to an observer on a planet, say Mars, that the rocket happened to be passing? From the point of view of this observer, in the time it takes the light beam to go from one mirror to the other, the mirror has moved over slightly. The faster the rocket is going, the more it has moved. Thus for each tick, the observer on the planet observes the light traveling farther than the astronaut observes it to travel. However, the speed of light is the same for both observers. Therefore, since the light travels a greater distance as seen by the Martian, the light pulse's trip between mirrors takes longer. The Martian sees time pass more slowly than the astronaut does.

From this example, we conclude that for the observer who sees the clock moving, each tick takes longer than for the observer who sees the clock at rest. Another way of saying this is that *moving clocks appear to run slow.* This phenomenon is called time dilation.

Dilation *is the proper word; "dilatation" is a common error.*

Before we go any further, let us anticipate one possible misconception. We have already seen that the principle of relativity tells us that if we have two observers moving with respect to each other, then no experiment exists that would tell us that one or the other is actually at rest. We might think that time dilation contradicts this, since the frame in which a clock keeps proper time seems to be special. However, it is not really a special frame, but only the frame in which a particular clock happens to be at rest. It is the frame that happens to be moving at the same speed as this particular clock. The fact that this clock may be at rest in the reference frame is not the same as saying that the reference frame itself is at rest. If you are traveling in an airplane, your watch is at rest with respect to the airplane and keeps proper time *for the airplane.* The ground appears to be moving, and clocks on the ground appear to you to run slow. If, on the other hand, you were on the ground, your watch would then keep proper time *for the ground,* and clocks on the moving airplane would appear to run slow.

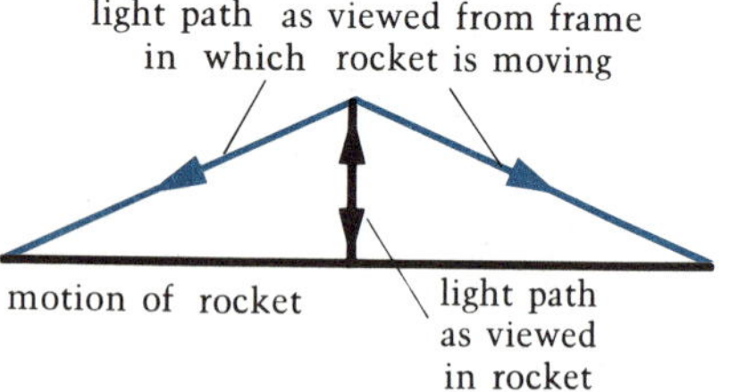

Fig. 8–11 Comparison of light paths in light clock.

To find out how the time kept by a moving clock compares with the proper time, we need only calculate the extra distance that the light travels. The light now travels on the hypotenuse of the triangle shown in Fig. 8–11 rather than one of the sides. When we work this out, we find that if v is the speed of the clock and c is the speed of light, then

$$\frac{\textit{measured time}}{\textit{proper time}} = \frac{1}{\sqrt{1 - v^2/c^2}}.$$

This special mathematical form, $1/\sqrt{1 - v^2/c^2}$, appears very often in relativity theory. Let us examine it carefully in order to get a feeling for how it behaves for different speeds.

Figure 8–12 shows the values of $1/\sqrt{1 - v^2/c^2}$ for various values of the ratio of the speed of the clock to the speed of light (v/c). We see that for low speeds, v/c is small and v^2/c^2 is smaller still. As a result, $1/\sqrt{1 - v^2/c^2}$ is so close to 1 that we would not be able to detect the difference (Table 8–1). This is important because it means that we cannot detect the effect of relativity until the speeds involved start getting close to the speed of light. It also means that at low speeds, the results of special relativity should agree with the old, classical results. We notice, however, that when the speed of the clock gets greater than 10 percent of the speed of light, the curve starts to shoot up quickly. For example, for a speed of 90 percent of the speed of light, the time measured in a laboratory (that is, the rest system with respect to which the clock is moving) is 2.3 times that recorded in the rocket (in which the clock is at rest).

As strange as this result seems, it is verified every day in particle accelerators around the world. Such accelerators provide good tests of the predictions of special relativity, since particles are accelerated to almost the speed of light. Also, as we shall see in Section 26.3b, several of these particles can serve as good clocks, since they

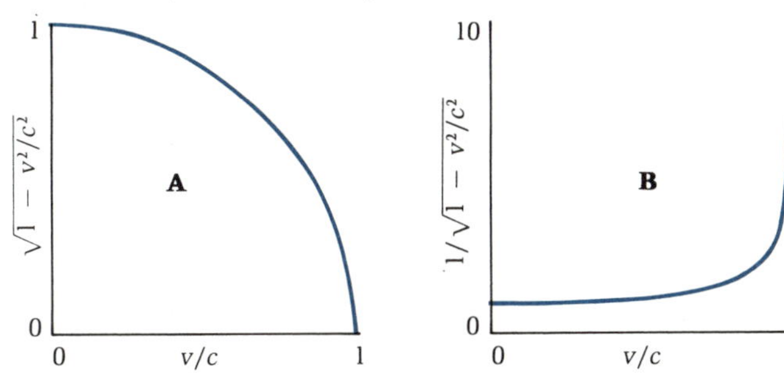

Fig. 8–12 In this figure, we plot the results given in Table 8–1. A. This figure shows how the quantity $\sqrt{1 - v^2/c^2}$ varies as v/c goes from zero to one (that is, the speed going from zero to the speed of light). For low speeds, the quantity is very close to one, and relativistic effects are very small. Once the speed becomes an appreciable fraction of the speed of light, the relativistic effects can be quite large. B. Similar behavior is seen in this graph in which we plot the quantity $1/\sqrt{1 - v^2/c^2}$ as v/c goes from zero to one.

TABLE 8–1 Dependence of Relativistic Effects on Speed

v/c	$\sqrt{1 - v^2/c^2}$	$1/\sqrt{1 - v^2/c^2}$
0.1	0.995	1.01
0.2	0.98	1.02
0.3	0.95	1.05
0.4	0.92	1.09
0.5	0.87	1.15
0.6	0.80	1.25
0.7	0.71	1.40
0.8	0.60	1.67
0.9	0.44	2.29
0.95	0.31	3.20
0.99	0.14	7.09
0.995	0.10	10.0
0.999	0.04	22.4

are unstable and decay into other particles. When the particles are at rest, we know how long these particles live, on the average, before such a decay. We can also measure their lifetimes when they are traveling at speeds close to the speed of light, and see how fast their internal clocks are running.

Fig. 8–13 Part of the accelerator at CERN where the muon decay experiment was performed.

TIME DILATION: AN EXAMPLE

One of the crews of Skylab astronauts orbited the earth for 80 days. When they landed, they had aged 0.002 second less than they would have had they stayed at home. They weren't traveling close enough to the speed of light (orbital velocity, as we saw in Section 7.2, is about 8 km/s, which is only 8/300,000 = 3/100,000 = 0.00003 = 0.003 percent of the speed of light) for time dilation to have much effect.

For example, a 1977 experiment at the CERN accelerator near Geneva studied the decay of a certain kind of unstable particle which, when at rest, decays in an average time of a little over 2

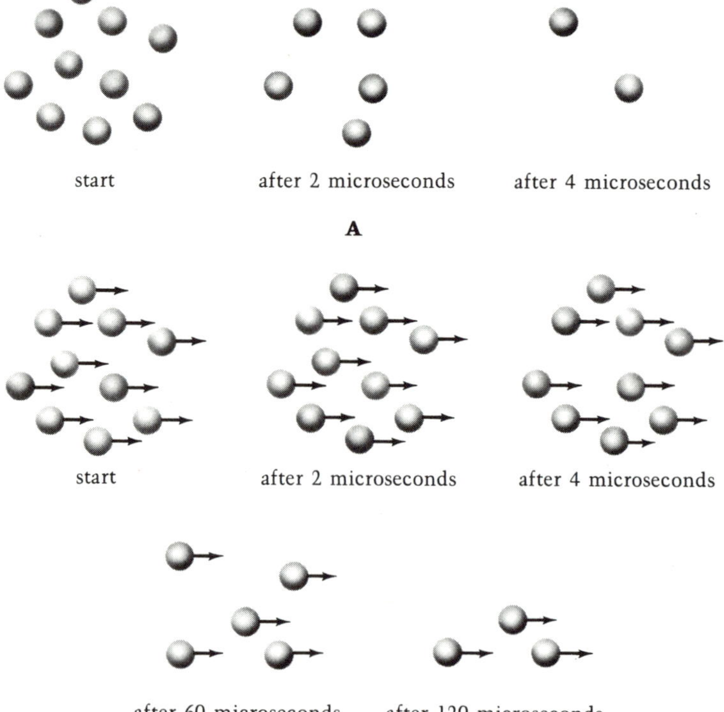

Fig. 8–14 The decay of muons. A. Muons at rest. After 2 microseconds (2×10^{-6} s), approximately half of the initial muons remain. After 4 microseconds, approximately one-quarter of the initial muons remain, and so on. B. For muons moving close to the speed of light (0.9995 times the speed of light), the decay appears to take longer, because of the effect of time dilation. Now it takes 60 microseconds, as measured in the laboratory, for half of the muons to disappear.

microseconds. (The particles happen to be mu-mesons, also called muons, but that is not important here.) In the CERN experiment the muons were traveling so fast that they lived for over 60 microseconds. Using special relativity, since they knew how fast the muons were traveling, the experimenters could calculate that the lifetime of a stationary muon must be 2.1948 microseconds. This number agrees to within 0.2 percent of the value measured directly for muons that are at rest. The experiment confirmed the predictions of special relativity to very high accuracy (Fig. 8–14).

Time dilation applies to all clocks, even biological clocks, the timekeeping involved with life processes. For example, astronauts age less during a spaceflight than their counterparts on earth! The difference is not very much for the astronauts who went to the moon—a fraction of a second—since they were not away for very long nor did they travel very fast compared to the speed of light. However, it has been suggested that the effect of time dilation could be used on long space trips, perhaps to other stars beyond our solar system. If a rocket travels fast enough, a trip lasting several hundred years in earth time would last only a few years for an astronaut, making it possible to carry out the mission in the astronaut's lifetime.

8.3c THE TWIN PARADOX

Further thought along these lines leads to a famous dilemma, the twin paradox (Fig. 8–15). One twin is an astronaut, and goes on a

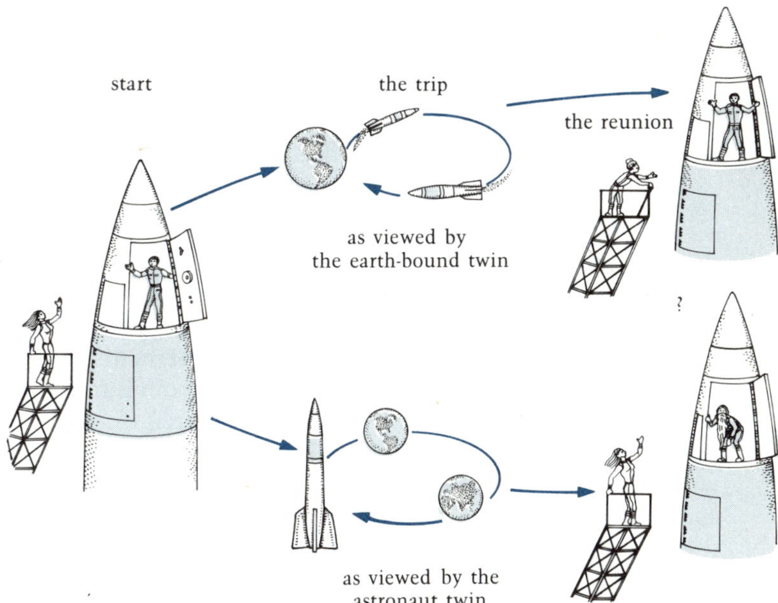

Fig. 8–15 The twin paradox.

space trip, traveling at high speeds. The other twin stays at home. Because of time dilation, the astronaut twin will age less than the twin on earth. When the astronaut returns, the twins will no longer be the same age. The astronaut will be younger than the twin who stayed home. But because motion is relative, we might just as well take the point of view of the astronaut. For the astronaut, the rocket stood still and the earth, carrying the twin, moved away and then eventually returned to the rocket. In this case, the twin who stayed on earth would be younger. Obviously they each can't be younger than the other. This is the paradox.

Further, if one were actually younger than the other, by comparing the twins and seeing which is the younger, we could decide which one was really at rest and which one was really moving. This might seem to violate the principle of relativity.

However, we are saved from the paradox by a closer look at the situation. Let us consider the astronaut. In order to return, the rocket had to slow down and turn around. Thus, for part of the trip it accelerated, which means that it was not an inertial frame. Therefore, it is proper for us to distinguish between the earth and the rocket, since one is an inertial frame and the other is not. The principle of relativity says only that we cannot distinguish between inertial frames. The astronaut twin really can come back much younger than the twin who stayed at home. Both have aged, but the twin who voyaged to the stars has aged less.

ASTRONAUT TWINS

An astronaut twin travels from the earth to the nearest star, Alpha Centauri, at six-tenths of the speed of light (180,000 km/s). The astronaut returns to earth. When the twins meet, the astronaut twin has aged 2 years and 8 months less than the stay-at-home twin.

8.4 *Spacetime*

The time coordinate is not exactly like the space coordinates. For example, as far as we know time changes in only one direction whereas we can move forward or back in space. The fact that time flows in only one direction is known as the arrow of time.

When we wish to discuss an event, we give its time and place. If we are talking about our normal three-dimensional world, then, as we saw in Section 2.1b, it takes three coordinates to describe the place. Now that we have seen that time is not so special, we can treat it as another coordinate. We say that it takes four coordinates to describe an event—three of space and one of time. This four-dimensional coordinate system is called spacetime.

It is not easy to draw four-dimensional pictures, so we can only draw certain aspects of spacetime. For example, we can plot any one

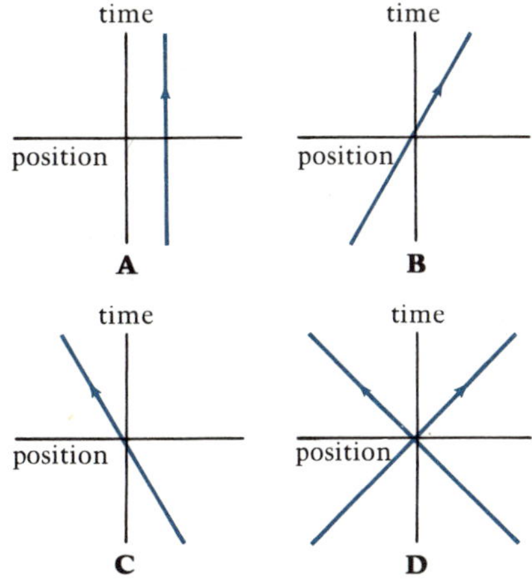

Fig. 8–16 Spacetime diagrams of world lines. A. Particle at rest. B. Particle moving at constant speed. C. Particle moving at constant speed in the other direction. D. Particles moving at the speed of light, but in opposite directions.

of the space coordinates and the time coordinate, as in Fig. 8–16. We can trace the path of any object in spacetime. Such a path is called a world line. The procedure is essentially the same as the one we used to plot the position of an object at various times (as discussed in Section 2.1).

Figure 8–16 shows a few simple cases. For example, the world line for an object standing still is a straight line at a constant position. Moving objects are represented by slanting lines, since their positions are changing with time. The faster the object is traveling, the greater the slant. We normally choose the scales on the axes of the graph so that a particle traveling at the speed of light makes the same angle with both axes, that is, its world line makes a 45° angle with the axes.

It is fun to draw the world lines of some more realistic activities. Figure 8–17 shows sample world lines drawn from life.

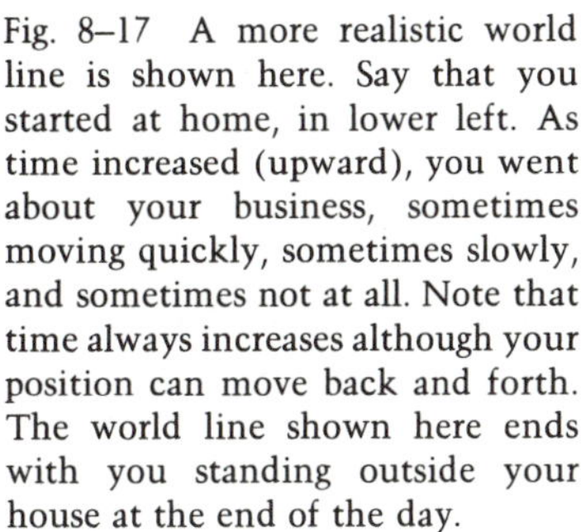

Fig. 8–17 A more realistic world line is shown here. Say that you started at home, in lower left. As time increased (upward), you went about your business, sometimes moving quickly, sometimes slowly, and sometimes not at all. Note that time always increases although your position can move back and forth. The world line shown here ends with you standing outside your house at the end of the day.

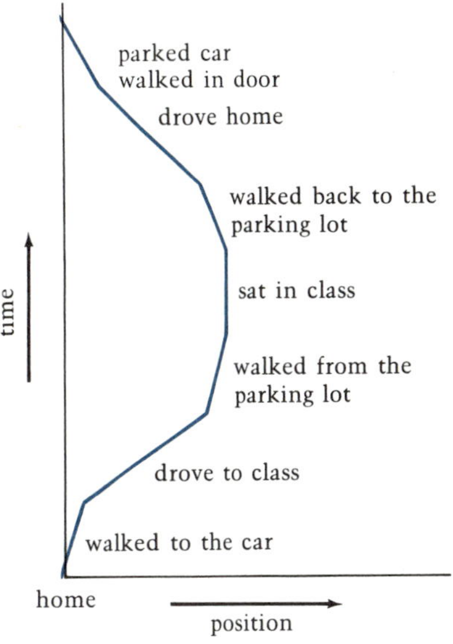

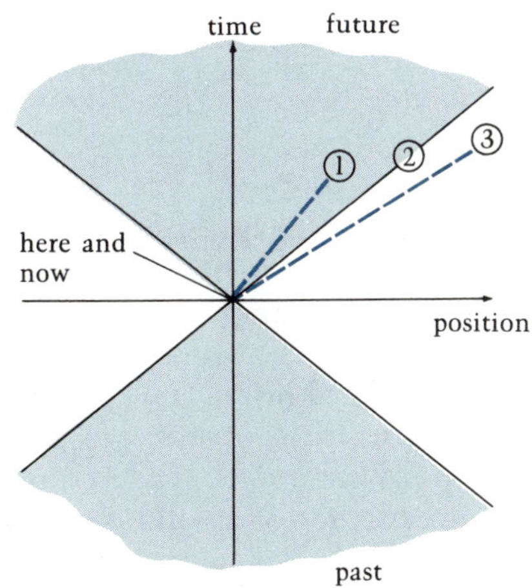

Fig. 8–18 Regions of two-dimensional spacetime. The line from the origin to point 1 is *timelike,* the line from the origin to point 2 is *lightlike,* and the line from the origin to point 3 is *spacelike.* Only events connected by timelike or lightlike intervals can be the causes of each other.

As we shall see later in this chapter, in Section 8.6, one of the conclusions of special relativity is that particles cannot travel faster than the speed of light. Under these conditions, we can think of spacetime as consisting of different regions. Let us consider some event at the origin of our spacetime diagram, and three events that occur at some later time (Fig. 8–18). We see that the first event can be connected to the origin by a line corresponding to a speed less than the speed of light. Thus, it is possible that a single observer could be present at both events, traveling from one to the other at a lower speed than the speed of light. It is even possible for one event to have caused the other. In that case, we say that the interval between the events is timelike.

Event 2 is connected to the origin only by a line corresponding to a particle traveling at the speed of light. In that case we say that the interval between the origin and event 2 is lightlike.

Finally, event 3 can only be connected to the origin by a line representing something moving faster than the speed of light, but nothing can travel so fast. Thus, a single observer who was present at the origin (a given position and time) could not possibly also be present at event 3. The time is somewhat different but the position is so different that nobody could travel fast enough to cover the distance in the time allotted. Also, because nothing can go faster than the speed of light, it is impossible for one of these events to have caused the other. We say that the interval between event 3 and the origin is spacelike.

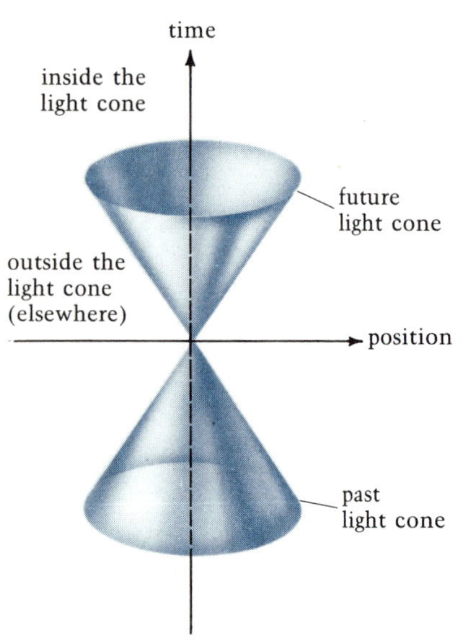

Fig. 8–19 Regions of spacetime.

If we consider two space dimensions instead of one, then the world lines of light, which appears as a V on Fig. 8–18, define a cone called the light cone (Fig. 8–19). We say that events connected to the origin by timelike intervals—which are possible to cover—are *inside the light cone.* We refer to events in the future as being inside the *future light cone,* and events in the past as being inside the *past light cone.* Events that are too far removed in space to have been in the observer's past or to be in the observer's future (spacelike intervals) are said to be *outside the light cone.* Any such world line could never occur.

8.5 *The Lorentz Transformations*

Any given event appears to have different sets of spacetime coordinates when viewed by observers in different inertial reference frames. Special relativity theory gives the relationship between these different sets of spacetime coordinates. The set of equations that expresses this relationship was discovered in the late 1800s by the nineteenth-century Dutch physicist H. A. Lorentz. These equations are known as the Lorentz transformations, because they transform one set of coordinates into another. We will not write out the Lorentz transformations in this book. Some of their properties can be visualized and we shall discuss them here.

We have already seen in Section 8.3b how time spans translate from one observer to another. In fact, the phenomenon of time dilation follows directly from one of the Lorentz transformations. We have not yet discussed how the motion of the observer affects the space coordinates. In this section, we will concentrate on determining the lengths of moving objects. This will give us three more Lorentz transformations, one for each spatial coordinate.

Every object has a proper length, its length measured in the frame in which the object is at rest. But the length can also be measured in frames in which the object is not at rest. Let us now see how these other lengths compare with the proper length.

8.5a THE LORENTZ CONTRACTION

We consider the length perpendicular to the direction of motion separately from the length along the direction of motion. The length perpendicular to the direction of motion is easy to treat because it does not change at all.

For the length parallel to the direction of motion, it turns out that moving objects shrink along the direction of motion. This is called the Lorentz contraction. The amount by which it shrinks can be given by

$$\textit{length} = \textit{proper length} \times \sqrt{1 - v^2/c^2}.$$

Note that we again encounter the same numerical factor, $\sqrt{1 - v^2/c^2}$, as we did for the time dilation. Consulting Fig. 8–12, we see that for speeds that are small compared to the speed of light, the length of the object doesn't change very much at all. However, as the speed gets close to the speed of light, the effect becomes quite large. For example, a meter stick (that is, a stick whose proper length is 1 meter) moving at 90 percent of the speed of light will appear to be only 44 cm long.

8.5b THE APPEARANCE OF MOVING OBJECTS

We have been talking about how one or another dimension of an object appears, but how does the *object* itself appear? We have just discussed how the length of a moving object along the direction of motion appears to be less than the proper length of the object. Does this mean that when you look at the object, you actually see it foreshortened? The answer to this question is not obvious and teaches us that here, as elsewhere in relativity theory, we must be very precise when we talk about making an observation.

When we "observe" the length of an object, as we did in the previous subsection, what are we actually doing? More precisely, we are measuring the locations of various points on the object (for example, the ends) *at the same time.* We can do this if we have enough observers so that each point at which we want to make a measurement has an observer. We can then locate various points on the object at a given time by seeing which observers were closest to those points at that time. From this information, we can reconstruct a picture of the object, point by point. When we do so, the

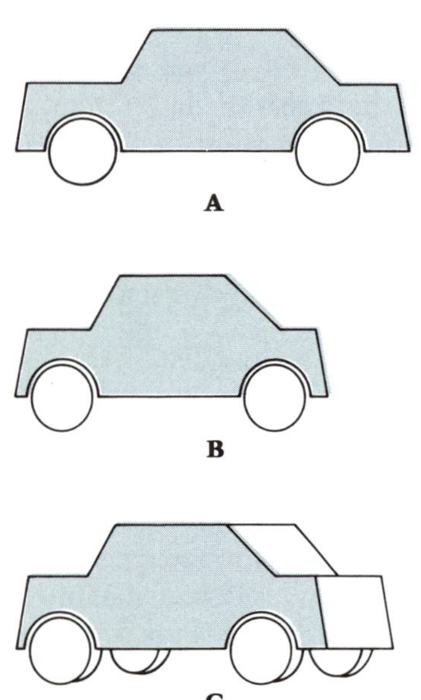

Fig. 8–20 A. Car at rest. B. Car Lorentz-contracted. C. Car rotated, but appearing on its near side to be contracted like B.

object appears to be shrunk in one dimension. The result is indicated schematically for a car in parts A and B of Fig. 8–20.

However, this observation procedure is not the same as "looking" at an object, or taking its picture. When we take a picture, we are recording light that *enters our camera* at one instant. However, our camera is not the same distance from all points on the object, so it takes light from different points different amounts of time to reach us.

To see how this applies to the car in Fig. 8–20, we'll just look at the two tail lights, one on the near side of the car and the other on the far side. It takes light a little longer to reach us from the far side of the car. We therefore see the far tail light as it was *slightly earlier than* the time for which we are seeing the near tail light. Thus, the far light will appear to be behind the near light, since the car continues to move ahead. As a result the back of the car appears as though it were rotated toward us. The faster the car is going, the greater the amount of apparent rotation. Now if we were looking at the same car while it was standing still and actually rotated, the near side from headlight to tail light would also appear foreshortened, because we would be viewing it at an angle. It turns out that the foreshortened length of the side of the car is exactly equal to the Lorentz-contracted length of that side!

Thus, if we simply relied on a snapshot of the car, we might think it was simply rotated and would not know that it was actually Lorentz-contracted. Nobody realized this point for over fifty years after the time Einstein presented the special theory of relativity in 1905. Until 1959, physicists believed that you could really "see" the Lorentz contraction of an object.

8.6 *Mass and Energy*

The position and time of an event are not the only quantities affected by the motion of the observer, according to the special theory of relativity. All the quantities we have dealt with—velocity, force, momentum, energy, acceleration—also vary as a result of the motion of the observer. Of course the variation must be consistent with the principle of relativity, that is, the quantities must vary in such a way that the basic laws of physics, such as Newton's laws, still hold when properly stated.

In our discussion of spacetime, we took the three coordinates describing the position of an event and the time of the event, and combined them into a single four-dimensional quantity to obtain the spacetime location of the event. It turns out that we also must do the same thing with the three components of the momentum of an object and with the energy of the object. The result is that the energy varies with the motion of the observer in a manner similar to the time. The energy of a moving particle is given by

$$Energy = \frac{m_0c^2}{\sqrt{1 - v^2/c^2}}.$$

where m_0 is the mass of the particle when it is at rest (called the rest mass; the subscript $_0$ can be taken to mean "at zero velocity"), v is the speed of the particle, and c is (as always) the speed of light.

Notice that when the particle is at rest, that is, when $v = 0$, Einstein was left with $E = m_0c^2$. As discussed in Section 5.6, this means that there is energy associated with the presence of mass. As the particle starts to move, its energy increases slowly. (It is just acquiring kinetic energy, as discussed in Chapter 5.) However, the energy rises quickly as the speed approaches the speed of light.

We can interpret the above equation by saying that the relation $E = mc^2$ is always true, but that as a particle moves faster, its mass increases, such that

$E = mc^2$

$$m = \frac{m_0}{\sqrt{1 - v^2/c^2}}$$

Einstein discussed these equations in a paper he published in 1906, a year after he presented the basic special theory.

$$mass = \frac{rest\ mass}{\sqrt{1 - v^2/c^2}}.$$

The relation among mass, rest mass, and velocity gives us an idea of how hard the job of a particle accelerator is. Of course, *force = mass × acceleration*. When the machine starts to accelerate a particle, the initial force required is only the rest mass times the desired acceleration. However, as the particle's speed approaches the speed of light, the mass increases and a greater force is required to produce the same acceleration. This is now routinely verified experimentally.

This relationship between the mass of a moving particle and its rest mass is the source of the common statement that no particle's speed can exceed the speed of light. Let us take a closer look at this statement.

The photon is an example of a particle that has no rest mass and travels at the speed of light. The neutrino, which we will discuss later on, had been thought to be another example. New evidence indicates that it may have some small amount of rest mass, however; if so, then it doesn't travel at quite the speed of light.

If we are accelerating a particle from rest, its mass increases as it goes farther. As its speed gets closer and closer to the speed of light, its mass increases without any limit. If its speed could really reach that of light, we would have $1 - v^2/c^2$ equal to zero, so the mass would be infinite. We say that this cannot happen since it would require an infinite amount of energy. Therefore, particles moving slower than the speed of light can never be accelerated to the speed of light.

However, we know that there are particles that travel at the speed of light. All of these particles have zero rest mass, so we are saved from the infinite-energy problem.

TACHYONS

The above argument really only tells us that the speed of light is a barrier that particles cannot cross. The possibility remains that there is a group of particles that are always traveling faster than the speed of light. For these particles, the "light barrier" means that they cannot *slow down* to the speed of light. Such particles would be called tachyons (from the Greek *tachys*, meaning "fast"). There are some subtle experimental techniques that should be able to detect tachyons if they exist, but so far such experiments have given indications against the existence of tachyons.

8.7 Some Concluding Remarks

In this chapter using the principle of relativity, and the constancy of the speed of light, we have come to a strange set of conclusions. Moving clocks run slow, moving rods appear to shrink along their direction of motion, and particles around us have an absolute speed limit. However, special relativity is not simply the story of flying clocks and sticks, fascinating as these things may be. The important point is that we must alter some of our preconceived notions about space and time. Events must be viewed in a four-dimensional spacetime, with time playing the role of just another coordinate (although one with some special properties).

We also saw that when we start to make measurements in spacetime, we get some strange answers—the distortion of times and distances. One way to think about this is to say that the rules of geometry in spacetime are not the same as they are in our normal three-dimensional world. This is actually a very fruitful approach to take since, as we will see in the next chapter, the presence of mass further distorts the geometry of spacetime.

Key Words

special theory of relativity
general theory of relativity
ether
aberration of starlight
Michelson-Morley experiment
reference frame

inertial reference frame
principle of relativity
proper time
time dilation
twin paradox
spacetime
world line
timelike
lightlike
spacelike
light cone
Lorentz transformation
proper length
Lorentz contraction
rest mass
$E = mc^2$
tachyon

Questions

1. Under what conditions do the results of Newtonian mechanics and the special theory of relativity agree?
2. How long does it take light to travel across your room?
3. A certain galaxy is one million (10^6) light-years away. How far is this in meters?
4. Some physicists have pointed out that since time is not so special, we should also measure time in meters, and not in a special unit (seconds). A time of 1 meter would correspond to the time for light to travel 1 meter. How many seconds is this?
5. Why did the early measurements of the speed of light involve astronomical observations?
6. Would the Michelson-Morley experiment have ruled out the existence of the ether if the aberration of starlight had not been observed? Explain.
7. What is "special" about the special theory of relativity?
8. In this chapter, we have seen that events that appear simultaneous to one observer may not appear simultaneous to another observer. How does this rule out the possibility of a "universal" time, which is the same for all observers?
9. You and a friend are heading toward each other at high speed. You carry identical clocks. While one second ticks off on your clock, you see half a second tick off on your friend's clock. As viewed by the friend, when one second ticks off on the friend's clock, how many seconds will appear to tick off on your clock?
10. You and a friend are heading toward each other at high speed. You carry identical meter sticks. Your friend's stick appears half a meter long to you. How long does your stick appear to your friend?
11. If there is no such thing as a preferred reference frame, what is the significance of proper time?
12. A clock flies by you at half the speed of light. If one second ticks off on a clock that you are holding, how much time ticks off on the moving clock? (Use Table 8–1 to help you.)

13. How fast (relative to the speed of light) must a meter stick be moving for it to appear half a meter long? (You may wish to use Table 8–1.)
14. A spaceship travels at 90 percent of the speed of light to a star 100 light-years away from the earth. (a) How long does this trip take as viewed from the earth? (b) How long does this trip take as viewed by the astronauts? (c) What might the astronauts conclude about the distance from the earth to the star?
15. Why do we say that if one event is outside the light cone of another event, then neither event could have caused the other?
16. How does the kinetic energy of Ron Guidry's fastball compare with the rest energy of the ball?

CHAPTER 9

General Relativity

Albert Einstein's triumphs in 1905 were substantial. In that single year, he published important scientific papers in three areas. One, the special theory of relativity, we discussed in the last chapter. A second, an explanation of what we now call the photoelectric effect (because it showed how light—*photos*, in Greek—could release electrons from certain metals), provided an important understanding of the nature of light, and was actually the work for which he was specifically cited when he was awarded the Nobel prize. (The Nobel committee thought, even in 1922, that relativity was too uncertain an idea to merit the prize.) Einstein's third 1905 paper described the seemingly random motions that are made by small particles suspended in a liquid; these ideas, too, improved our understanding of the atomic nature of matter.

But although Einstein's successes were important, Einstein did not rest on his laurels. He knew that his special theory of relativity was incomplete, because it did not contain any provision for the effects of gravitation. And gravitation certainly dominates the universe. Einstein set out to work out a more general theory of relativity, one that would include the effects of gravitation.

In the meantime, Swiss officials realized that Einstein's 1905 work was too important to leave him at the patent office, so arrangements were made for him to become professor at the University of Zurich. Since his formal duties were increased, although his salary was not, and since it was more expensive to live in the big city of Zurich than it had been in Berne, Einstein was not necessarily better off as a result of his new recognition.

In 1911, Einstein accepted a prestigious chair of physics at Prague, although the next year he returned to Zurich, where he became professor at the Polytechnic.

9.1 *The Principle of Equivalence*

Just as he had done with notions of simultaneity, Einstein started to think about the basic assumptions we make when talking about gravity. We have an intuitive notion of mass, but just what is mass? Is it truly an inherent property of matter, or is it affected by other objects in the universe?

Some years before, the German scientist Ernst Mach had suggested that the mass of an object was related to a framework provided by all the other matter in the universe. And since most of the matter in the universe is far away from us rather than nearby, it is the distant parts of the universe (we would now say the distant galaxies or clusters of galaxies) that determine the properties of local matter, that is, the matter that is in our neighborhood.

The only way to test this idea directly would be to remove all the distant matter from the universe and see if anything changed about our local matter. This is obviously an impossible thing to do. So Mach's principle, that the local matter is affected by the totality of all the distant matter, remained a theoretical principle rather than one that was observationally or theoretically verified.

Einstein realized that when we spoke of "mass," we used the word in two different contexts. "Mass" appears in Newton's laws, for example, in the equation that force is equal to mass times acceleration. This mass is what determines inertia, and Einstein called it inertial mass.

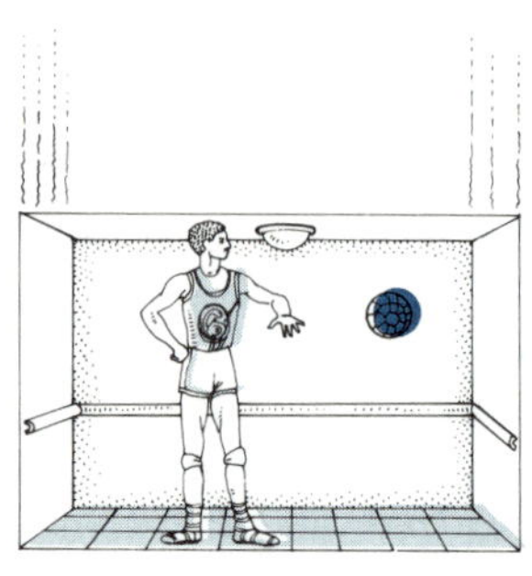

Fig. 9–1 Einstein's elevator provides a demonstration of the principle of equivalence.

Also, Newton's law of universal gravitation says that mass is affected by a gravitational field, and also creates a gravitational field. Einstein called the mass determined from its relation to gravitation by the name gravitational mass. If Mach's principle were true, then if all the distant mass were removed from the universe, we could not predict offhand that inertial mass and gravitational mass of nearby objects would change in the same way, for these two types of mass were defined differently.

Einstein then advanced his own idea: he wrote in a 1911 paper that, in fact, *gravitational mass and inertial mass were equivalent in that no experiment could ever tell one from the other.* This is called the principle of equivalence. Although this idea did not of itself provide a theory of gravitation, it is at the base of the theory that Einstein later derived.

9.1a EINSTEIN'S ELEVATOR

The principle of equivalence is best illustrated by "Einstein's elevator," which is a situation that we have already discussed in Section 3.5 and will review below. This is an example of a "thought experiment"—an experiment that is done in the mind rather than in a laboratory (see Fig. 9–1).

What if we were in a closed elevator, so that we could not see out? Our feet are pressed against the floor. Normally, we think that gravity is pulling us down.

But what if the elevator were magically transported out into a region of space where there is no gravity? If the elevator were not moving, we would float around inside. Indeed, if both the elevator and we were moving steadily, we saw in the previous chapter that we would not be moving with respect to the elevator; we would still float around inside.

Einstein realized, however, that if something were to accelerate this elevator (picture a rocketship tugging upward on the elevator), we would again be pressed against the floor. In fact, Einstein's principle of equivalence holds that there is no way we could tell whether we were in a gravitational field or were undergoing an acceleration. (We assume that we have no contact with the world outside the elevator; for example, it wouldn't be fair to be able to see a TV picture of the rocketship towing us along!)

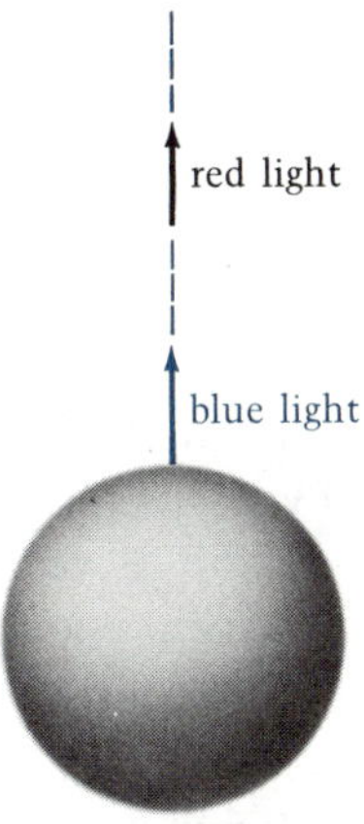

Fig. 9–2 Gravitational redshift. As light gets farther from a source of gravity, its color becomes slightly redder. The actual shift is quite small, and our eyes cannot detect the shift in color.

9.1b TESTS OF THE PRINCIPLE OF EQUIVALENCE

It is important to see that the principle of equivalence has survived some very close checking. The principle of equivalence was later incorporated into the general theory of relativity, and many of the observational tests that supposedly check the general theory really only test the principle of equivalence.

One type of experiment that has been carried out in the last two or three decades has provided direct tests of the principle of equivalence in a manner that is quite similar to the elevator thought experiment. We shall see in Section 15.7 that when something that emits light is moving with respect to us, the color (see note) of the light is changed, usually very slightly. Since the effect of a gravitational field is the same as the effect of an acceleration, when light escapes from a gravitational field it must be shifted in color, just as an accelerating object changes in velocity, also resulting in such a shift in color (Fig. 9–2). The presence of a shift in color caused by a gravitational field has been confirmed in light from the sun, although the effect is minuscule and other effects in the sun mostly mask it, and in light from a type of collapsed star known as a white dwarf (see Chapter 29), for which the effect is stronger.

Note: *For those of you who want to look ahead, we are loosely using the word "color" here to mean "wavelength of light," because each wavelength corresponds to a precise color, as we shall see later on. Changes in wavelength are also equivalent to changes in energy, so we are saying that the light changes in energy when it "falls" in a gravitational field, which is quite reasonable. The speed of light (in a vacuum) remains constant no matter what the gravitational field.*

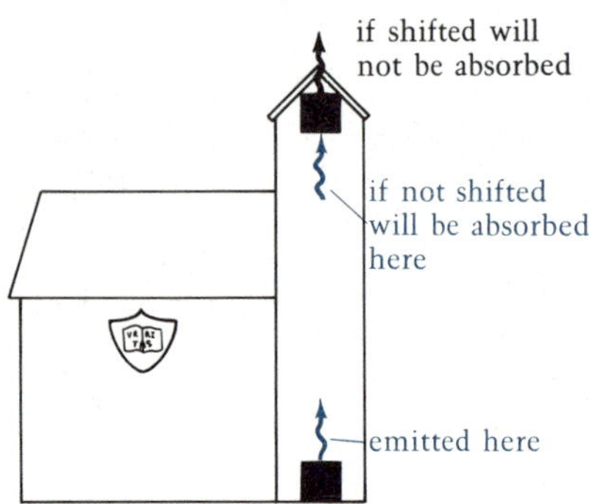

Fig. 9–3 Experiment to measure the gravitational redshift. A bit of radiation is emitted by a radioactive material in the basement of a building. A similar material is near the roof of the building. If the radiation arrives at the roof with the same color it had in the basement, it will be absorbed by the material near the roof. However, if its color changes, even a little bit, it will not be absorbed at the roof.

The shift in color that would be caused in a vertical distance of tens of meters, which is all that can be hoped for in an experiment on earth, is very small. So performing this direct experiment had to await the discovery of a process—the Mössbauer effect—that was sensitive to very small variations of the color of light. Rudolph Mössbauer, a German scientist, discovered in the 1950s that, under a particular set of conditions, a certain type of atom emits radiation at an exceedingly precise color and is also able to absorb radiation hitting it only if that radiation has the very same color, to a high degree of accuracy.

In a tower in the Harvard physics building, some of the material was placed and was made to radiate its precise color of light. At the other end of the tower, more of the same material was placed, and it was monitored to see if it was affected by the radiation. When the emitter was placed on the top of the tower and the receiver was placed at the bottom, the effect of the gravitational field would shift the color in the same sense as a downward acceleration. When, on the contrary, the emitter was placed on the bottom of the tower and the receiver was placed at the top, a shift of color in the opposite sense was expected. The shift expected was very small (Fig. 9–3).

The experiment was carried out over a period of many years, centered around 1960, and all the sources of inaccuracy that could be found were removed. The tower was even evacuated—made into a vacuum—to avoid the effects of air. The results confirmed Einstein's principle of equivalence to a high degree of accuracy.

9.2 Curved Space

Einstein would have liked to have had his new ideas about the principle of equivalence tested as soon as possible, but in 1914, soon after he moved to Berlin, World War I broke out. Einstein was a pacifist, and remained aloof from the war effort. He continued his scientific work, and in the next two years he was extraordinarily productive, finishing thirty scientific papers.

He continued to ponder the implications of Mach's principle, and how to account for gravitation. He found the answer in mathematical ideas that had been under development for years. They involve non-Euclidean geometry.

We are most familiar with *Euclidean geometry*, the basic geometrical ideas worked out in Greece about 500 B.C. by Euclid and his followers. Although space really has three dimensions—a cube, for example, has a height, a width, and a depth—let us carry out our discussion in two dimensions to make it easier to visualize.

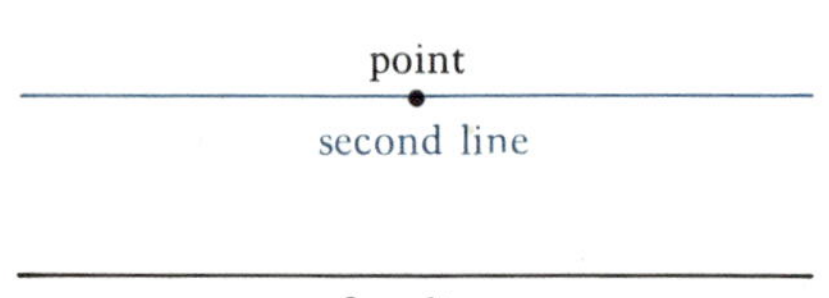

Fig. 9–4 The parallel postulate. On a flat surface, given a line and a point, there is only one new line through that point that is parallel to the first line.

A Euclidean surface in two-dimensional space is a plane, that is,

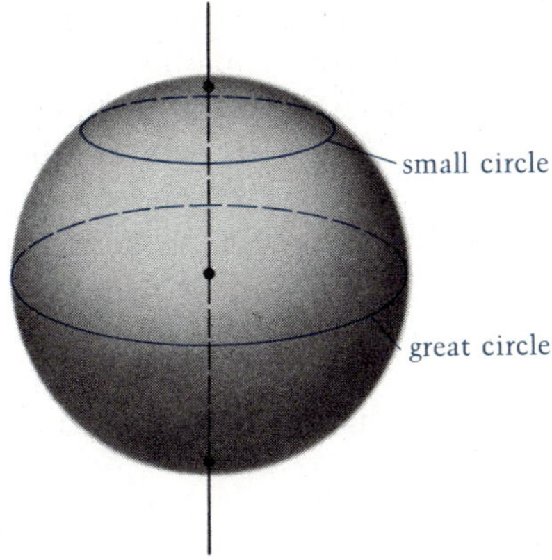

Fig. 9–5 Circles on the surface of a sphere. A great circle takes the place of a line on a flat surface. A great circle passing through two points is the shortest distance between those points.

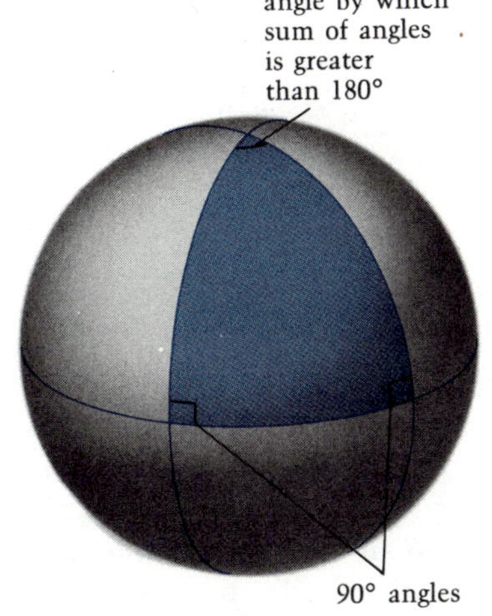

Fig. 9–6 A triangle on the surface of a sphere. The triangle must be made up of great circles, since they are the equivalent of straight lines. The horizontal line is like the equator of the earth and the vertical lines are like the meridians. The meridians cross the equator at right angles, so the two 90° angles are indicated. However, there is still another angle in the triangle, near the top. The sum of the angles of the triangle is greater than 180°.

what we would call a flat surface. If we draw a line on the plane, through any given point not on that first line, *we are able to draw one and only one line that is parallel to the first line.* (This is Euclid's "parallel postulate"; Fig. 9–4.) Any other line we draw through that point will get closer or farther away from the first line, and so cannot be parallel to it. Another property of a plane is that *the sum of the three angles of any triangle we draw on it is 180°.*

However, we are also somewhat familiar with one type of non-Euclidean surface. Consider, for example, the surface of a sphere. If a straight line is defined as the shortest distance between two points, we find that all straight lines are part of what we call *great circles* (Fig. 9–5). Great circles are the largest circles we can draw on a sphere. If we were to cut the sphere along any great circle, we would have two equal hemispheres. (The equator is an example of a great circle.) On the surface of a sphere, all great circles intersect, which means that *no two parallel straight lines can exist.*

Now let us also figure out the sum of the angles of a triangle we draw on the surface of a sphere, so that we can compare the answer with the 180° that the angles of a triangle on a flat surface add up to. Consider, for example, a quarter section of an orange (Fig. 9–6). Each of the angles is 90°, so the sum of the three angles is 270°. In fact, *the sum of the angles of any triangle we draw on the surface of a sphere is greater than 180°.* If we consider a small enough triangle the sum of the angles is not much greater than 180°, but it is always at least a little greater. Such a surface, where the sum of the angles of a triangle is greater than 180°, is called a surface of positive curvature.

Another type of non-Euclidean surface exists, and a saddle is an example. On it, *we can draw an infinite number of lines through a given point with each line just as parallel as any other to a given line.* Further, *the sum of the angles of any triangle is less than 180°.* This is called a surface of negative curvature.

These examples are all for two-dimensional surfaces, but the same definitions hold true for three-dimensional volumes as well. Einstein discovered, between about 1911 and 1916, that by carefully considering the properties of space as though it were curved,

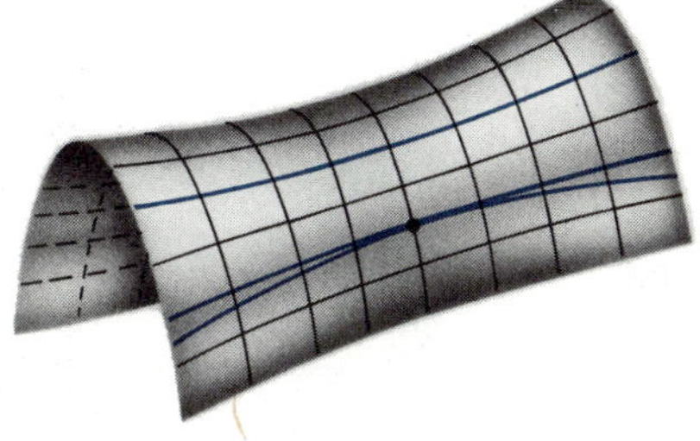

Fig. 9–7 A surface of negative curvature allows more than one line through a point to be parallel to a given line.

he could account for the effects of gravity. He considered three dimensions of space and one dimension of time, making a four-dimensional *spacetime* (Section 8.4). Einstein needed only the curvature of spacetime to explain the effects of gravity; none of the "tugging" that we intuitively associate with gravity is necessary.

Einstein actually had to make two different types of calculations. One dealt with the way in which the presence of mass warps the geometry of spacetime and the other to tell how particles (and light) travel through this warped spacetime.

Einstein actually worked with a set of mathematical equations, but we shall discuss their physical effects conceptually. Consider that the presence of a mass warps the space around it (Fig. 9–8). Then there is a tendency for nearby masses, as shown in the figure, to "fall" into the depression representing the warp. This has the same effect that gravity would have.

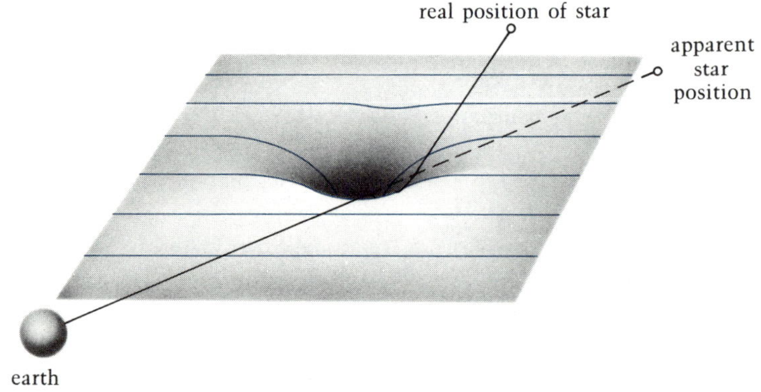

Fig. 9–8 In the general theory of relativity, we can think of mass as though it warps the space near it. In this two-dimensional visualization, the presence of the large mass of the sun warps the surface in the middle, so that when light from the star (solid line from top) passes near it, the light is deflected. It thus arrives at the earth coming in a slightly different direction than it would have had if the sun had not been there. Since we on earth know only the direction in which the light is coming, we project the light we receive backward; it appears that the star has been deflected slightly in the sky to the position marked by the upper end of the dotted line. Note that the light travels in a straight line as best it can on a warped surface.

Let us consider, for example, the effect that a warp in space would have on light. The principle of equivalence indicates that light should be affected by such a warp, because the effect of gravity must not be distinguishable from the effect of an acceleration, which would lead to a shift in color. As light travels through the warped area, its path appears curved to an outsider (if there could be an outsider with such a view). This is similar to an effect that takes place in a roulette wheel, as the ball rolls inward, or on a warped pool table. It appears that the light enters the region of the large mass heading in one direction and departs heading in a different direction. The light takes the shortest distance between two points, which is *by definition* a straight line. But to us, the light appears to bend.

Although the idea that light would bend when it travels near a large mass comes from the principle of equivalence, which dates from 1911, Einstein did not work out the correct way of calculating the amount of the bending until 1915. His 1915 answer was twice as great as he had calculated in 1911 using the incomplete theory.

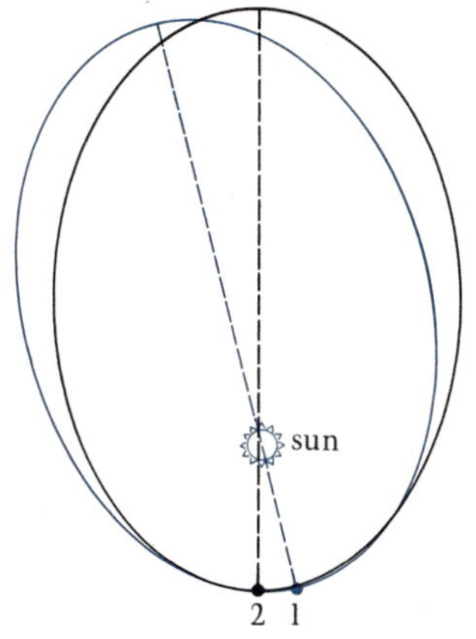

Fig. 9–9 Advancement of the perihelion of Mercury. The perihelion is the point of closest approach to the sun. In orbit 1, drawn in blue, the perihelion is at point 1. In the other orbit the perihelion is at point 2.

The warp of space would also affect any bodies that came near the sun, the nearer the better. Mercury is the closest planet to the sun, and orbits the sun once every 88 days. Mercury's orbit is slightly elliptical, so in some parts of its orbit Mercury would be a little further in on the warp than in other parts. As a result, on the basis of Einstein's theory, we would expect the orientation of the orbit to shift around slightly with respect to the distant stars: if we watched any particular point on the shape of the orbit, it would tend to move around the sun from year to year. The point that is easiest to watch is the nearest point of the elliptical orbit to the sun; this point is called the perihelion (from the Greek words *peri-*, "around," and *helios*, "sun"). Einstein's theory showed that the perihelion of Mercury should advance (move around) a little bit each year. The amount predicted was very slight—only 43 seconds of arc per century (see Fig. 9–9).

Nevertheless, a discrepancy had been known since the turn of the century in the orbit of Mercury. The perihelion of its orbit did advance, but by about 5000 seconds of arc (about 1.5°) per century. However, much of this advance could be explained by the gravitational effects of the other planets on Mercury, which are known. When these other effects were subtracted, only a small amount of unexplained advance was left—43 seconds of arc per century. This agreement was too exact to be a coincidence; Einstein's theory gave the same result for the advance of the perihelion of Mercury, and so (it might seem) Einstein's theory must be right.

When Einstein's general theory of relativity was published in 1916, the explanation of the perihelion advance of Mercury was a strong point in its favor. Still, the observational result had been known before the theory was advanced, and one can always think up some kind of theory to explain a result that is already known. It is more convincing that a theory is correct if the theory makes a prediction of some unknown measurement, and you can make new observations that turn out to agree with the prediction. (That the new observation agrees doesn't actually prove that the theory is correct, although it certainly makes one feel better about it. If the new observation were in *disagreement*, however, the theory would have to be discarded.)

Some philosophers of science say that this is how a theory becomes accepted—after testing its predictions against new observations. Other philosophers of science have found that only sometimes does science work so neatly.

Although Einstein's results still needed experimental verification, many scientists accepted them immediately. The results were thoroughly mathematical in form, but Einstein had been led to them more from his intuitive feelings about the structure of space than from formal mathematical derivations. In fact, the great mathematician David Hilbert said, "Every boy in the streets of our mathematical Gottingen [the university] understands more about four-dimensional geometry than Einstein. Yet, despite that, Einstein did the work and not the mathematicians."

9.3 *Experimental Tests of General Relativity*

We have already seen how the advance of the perihelion of the planet Mercury provides some observational verification of Einstein's general theory of relativity. But this was an old result; a new test was needed.

Einstein's prediction that light should be bent by a large mass provided that new test. The largest mass nearby is our sun, and Einstein proposed that astronomers should look at stars that the sun appears to pass closely in the sky, and see if the stars appear displaced.

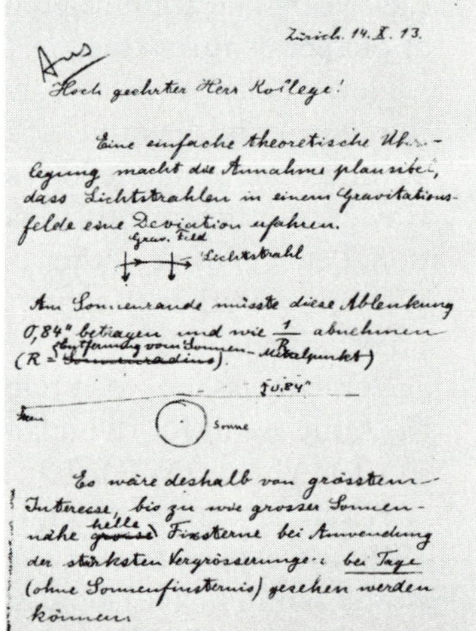

Zürich. 14. X. 13.

Hoch geehrter Herr Kollege!

Eine einfache theoretische Überlegung macht die Annahme plausibel, dass Lichtstrahlen in einem Gravitationsfelde eine Deviation erfahren.

Am Sonnenrande müsste diese Ablenkung 0,84" betragen und wie 1/R abnehmen (R = Entfernung vom Sonnen-Mittelpunkt).

Es wäre deshalb von grösstem Interesse, bis zu wie grosser Sonnennähe helle Fixsterne bei Anwendung der stärksten Vergrösserungen bei Tage (ohne Sonnenfinsternis) gesehen werden können.

A

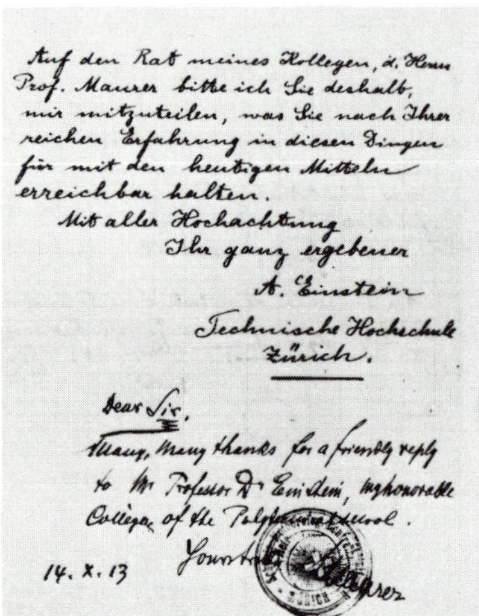

Auf den Rat meines Kollegen, d. Herrn Prof. Maurer bitte ich Sie deshalb, mir mitzuteilen, was Sie nach Ihrer reichen Erfahrung in diesen Dingen für mit den heutigen Mitteln erreichbar halten.

Mit aller Hochachtung
Ihr ganz ergebener
A. Einstein
Technische Hochschule
Zürich.

Dear Sir,

Many, many thanks for a friendly reply to Mr Professor Dr Einstein, my honorable College of the Polytechnical School.

Yours ...

14. X. 13

B

Fig. 9–10 Einstein's original calculation of the amount of bending of light around a large mass.

Einstein wrote to the great astronomer George Ellery Hale, at the Mt. Wilson Observatory in California, to ask if the stars could be observed near the sun, something that would necessarily have to be done in the daytime, when the sun is up. Hale replied that this would be impossible under normal circumstances. Both scientists realized that this meant that the experiment had to be done under the tricky circumstances of a total solar eclipse.

Once every year or two, somewhere on the face of the earth, the moon blocks the face of the sun so exactly that the entire visible surface of the sun is covered and the sky becomes as dark as night. This is called a total solar eclipse. (At other locations on earth, perhaps only a few hundred kilometers to one side or the other, the sun is only partly covered.) Fortuitously, the moon does not appear more than a few percent greater than the size of the sun, so we can see the stars near the sun. We also see the solar corona (Fig. 9–11) at that time.

Following Einstein's 1911 prediction that light from the stars would appear to be bent, a team of German scientists organized an expedition to the place in Russia where a total eclipse would occur in 1914. But the scientists were still en route to their site when war broke out, and they were interned as prisoners of war. As a result, they could not make their observations. This ultimately turned out to be fortunate (for Einstein, although not for the scientists on the expedition), because you may recall that Einstein's 1911 paper was an incomplete treatment of gravity. Although it predicted that starlight would be bent, it did not give the correct predictions for

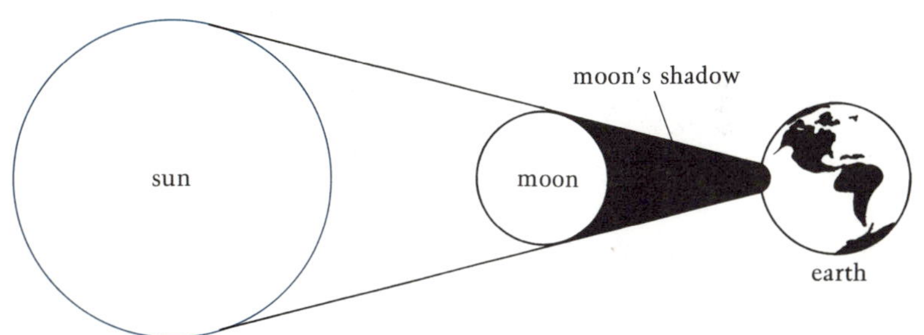

Fig. 9–11 In a total eclipse of the sun, the moon passes between the sun and the earth, so that the moon's shadow falls on the surface of the earth.

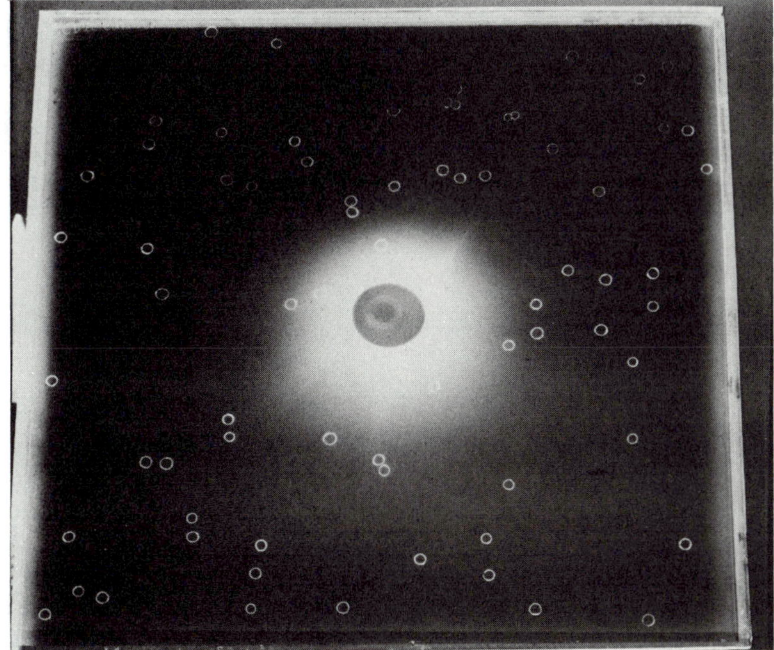

Fig. 9–12 The photographic plate for the experiment to verify the deflection of light by the sun. The solar corona surrounds the back side of the moon (center). The small circles indicate the location of the star images.

the amount of the bending. Had the scientists succeeded in their 1914 expedition, their results would have differed from those advanced by Einstein, and the theory would have been discredited.

But the 1914 expedition failed, and before a further expedition could be made (the German scientists were released soon), Einstein worked out the rest of his theory. Now his prediction for the amount of the bending of light from a star just at the edge of the sun was 1.74 seconds of arc, twice the earlier prediction. This is a very small amount, close to the size that the image of a star appears on a photographic plate because of the shimmering of the earth's atmosphere. And this value is the maximum possible for stars just at the edge of the sun; it decreases with distance out from the sun.

In 1919, two British expeditions were sent to observe that year's total eclipse. The observing sites were remote; one was on an island off West Africa and the other was in northern Brazil. Although the African site was partly cloudy, both sites provided successful observations, and within a few months the results from study of the photographs were known. Einstein's theory was confirmed (Fig. 9–13).

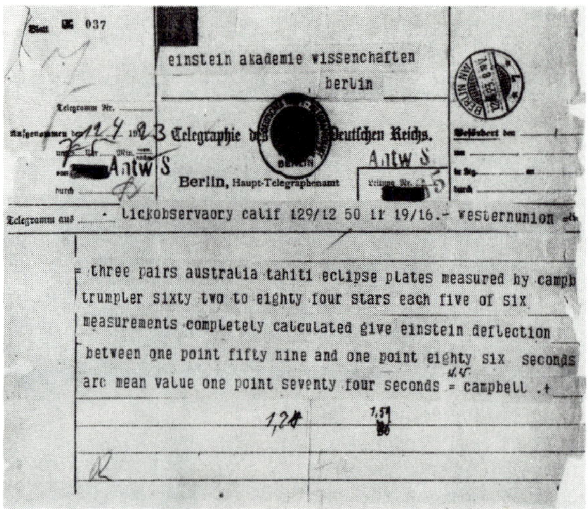

Blatt 037

einstein akademie wissenchaften
berlin

Telegramm Nr.
Aufgenommen den 12 4 1923
Telegraphie des Deutschen Reichs.
Antw S
Berlin, Haupt-Telegraphenamt
Antw S

Telegramm aus · lickobservaory calif 129/12 50 1r 19/16.- westernunion

= three pairs australia tahiti eclipse plates measured by campb
trumpler sixty two to eighty four stars each five of six
measurements completely calculated give einstein deflection
between one point fifty nine and one point eighty six seconds
arc mean value one point seventy four seconds = campbell .+

1,24

Fig. 9–13 The telegram to Einstein confirming his prediction and his theory.

We are particularly fond of the postcard that Einstein wrote his mother when he had received by telegram preliminary word that his theory had been confirmed (Fig. 9–14). "Joyful news today," he wrote. "H. A. Lorentz has telegraphed me that the English expedition has really proved the deflection of light by the sun."

But another anecdote reveals that Einstein knew in his heart all along what the result must be. One of his students reported that when she asked what if his prediction had not been confirmed,

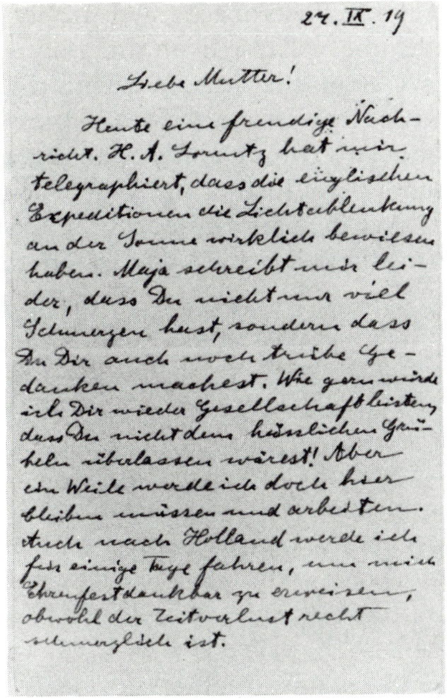

27. IX. 19

Liebe Mutter!

Heute eine freudige Nachricht. H. A. Lorentz hat mir telegraphiert, dass die englischen Expeditionen die Lichtablenkung an der Sonne wirklich bewiesen haben. Maja schreibt mir leider, dass Du nicht nur viel Schmerzen hast, sondern dass Du Dir auch noch trübe Gedanken machst. Wie gern würde ich Dir wieder Gesellschaft leisten, dass Du nicht dem hässlichen Grübeln überlassen wärest! Aber eine Weile werde ich doch hier bleiben müssen und arbeiten. Auch nach Holland werde ich für einige Tage fahren, um mich Ehrenfest dankbar zu erweisen, obwohl der Zeitverlust recht schmerzlich ist.

Fig. 9–14 Einstein's postcard to his mother reporting that he had heard of the verification of his general theory of relativity.

Einstein replied, "Then I would have been sorry for the dear Lord—the theory is correct."

Einstein's general theory of relativity has been subjected to test after test over the years, and all confirm its validity. Although the deflection-of-light test has been carried out many times over the years, most recently at the 1980 eclipse, this type of measurement is now most accurately carried out in the radio part of the spectrum. The positions of quasars in the sky can be observed very accurately with radio telescopes, and the bending of the radio waves as a quasar passes near the sun has been measured and found to agree with Einstein's prediction to better than 1 percent. The accuracy of the agreement is not only confirmation of Einstein's theory but also tends to rule out other alternative theories of gravitation that have been advanced from time to time since Einstein's work. (These different versions predict bending by different amounts.)

Recently signals were sent from earth to the Viking spacecraft in orbit around Mars and returned to earth, just as Mars was about to go behind the sun as viewed from earth (Fig. 9–15). The time it took the signals to travel to and from Mars was delayed 250 microseconds, in agreement with Einstein's prediction, just as anything would be slightly delayed as it traveled a longer distance because of the warp in space caused by the sun.

Another prediction of Einstein's theory, one that could not be tested directly in Einstein's time, is that clocks run slower when placed in a stronger gravitational field than they do in a weaker one. The effect is very small for differences in the gravitational field that we can get near the earth, and so very accurate clocks are needed to measure it. The most accurate timekeepers are now atomic clocks, so named because they use atoms to keep time. (Since the frequencies of radiation from atoms are given in cycles per second, if the length of a second changes as measured by the atomic clock, then it will show up a change in frequency, which we can measure relatively easily.) Scientists from the University of Maryland carried an atomic clock aloft in an airplane, in order to keep it at an average altitude above the earth's surface of 10 km, where the gravitation field is slightly lower than it is at sea level. The clocks that traveled ran the 47 nanoseconds (47 billionths of a second) faster each second that was expected on the basis of Einstein's general theory. (At the speed at which the planes flew, the change caused by special relativity's time dilation was only 10 percent as strong as the effect from general relativity.)

At last we are moving from the situation where delicate experi-

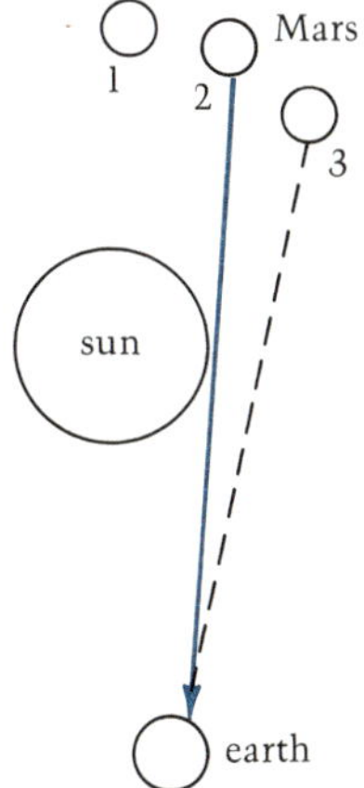

Fig. 9–15 The Viking test of relativity. When Mars is in position 1, there is no signal. At position 2, the effect of general relativity can be detected. No effect from general relativity is found when Mars is in position 3.

The most recent test of the contribution of general relativity to an advance of perihelion has been made not with a planet in our solar system but with a distant pulsar. Of the hundreds of pulsars now known, only three are in binary systems, orbiting around other objects. Since its discovery in 1973, one pulsar's perihelion has been advancing by about 4° per year, in agreement with the amount predicted by relativity.

ments had to be performed to detect the effects of general relativity. We think that some stars, when they die, collapse in on themselves so thoroughly that the mass of several suns is contained within a region a few dozen kilometers across. The warping of space in these regions is so extreme that even light can't get out, and we call these regions black holes. We shall discuss them further in Chapter 29, but here we can at least point out that the very existence of black holes as we are currently considering them depends on the validity of the general theory of relativity.

9.4 *Gravity Waves*

Einstein's general theory of relativity indicates that gravity must have many features in common with light. In particular, gravity should have some properties that make it seem like a particle and other properties that make it seem like a wave.

The particles that carry gravity would be called gravitons. None has yet been discovered. We have more hope of observing gravity waves in the near future.

As gravity waves pass by, they should distort the space they are passing through. Most of the current experiments designed to detect gravity waves involve a piece of some material that is kept as isolated as possible from all outside forces (except gravity, which we can't stop). If the material should start "ringing," that would indicate that a gravity wave has passed through.

The effect involved is very small, so particularly delicate experimental apparatus is needed. The original try, at the University of Maryland, used aluminum cylinders suspended in a vacuum. Now some of the apparatus holds the cylinder of matter up in a magnetic field. Other groups are using bars of sapphire crystal or silicon that should "ring" for an especially long time.

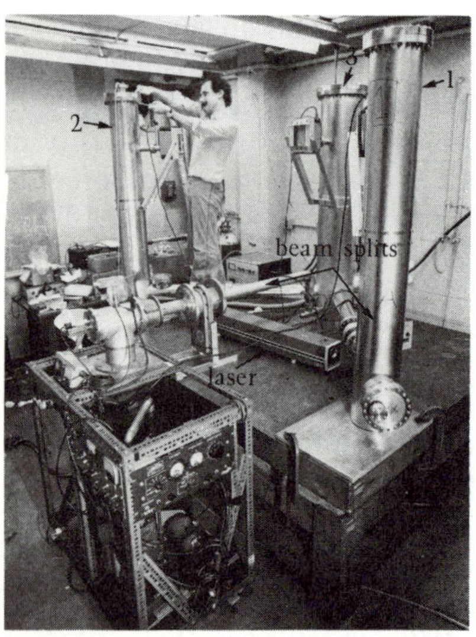

Fig. 9–16 The MIT gravity-wave detector. Light from a laser is bounced around by mirrors so that it enters vertical column 3 (background). There the laser beam is split. Part bounces back and forth between columns 1 and 3, while the other part bounces back and forth between columns 2 and 3. These multiple reflections make the system effectively larger than its actual size. When the two beams are allowed to combine, an interference pattern is set up (as we shall discuss in Chapter 15). This tells us if either of the paths has changed in length, even by a minute amount. One of the paths would be expected to change in length briefly if a gravity wave passed. The researchers hope that future larger versions of this laser interferometer would be sensitive enough to detect gravity waves from stellar explosions (supernovae, as will be described in Chapter 29) in nearby galaxies.

The strongest gravity waves would be generated when large masses are being accelerated very quickly. A dying star that is collapsing, for example, should give off a strong burst of gravity waves.

The first gravity-wave experiment reported events that could be gravity waves, but they occurred much more frequently than had been expected. Even setting up two cylinders separated by thousands of kilometers, on the ground that a gravity wave would start them both going but something more local (like a passing truck) would start off only one, did not settle why so many "events" were being observed. More recently, many experimenters all around the world have set up equipment that should be more sensitive, but have not detected gravity waves. (It is true, though, that they are looking at different wavelengths.)

Although the current experiments have not found gravity waves, most scientists feel that the next round of improvements in the apparatus should give a sensitivity that will detect gravity waves. So this is a problem that may be solved in the next few years.

In the meantime, an indirect indication that gravity waves exist has been found in the binary pulsar. Since the pulsar is being whipped around its partner so rapidly, it should give off strong gravity waves. These should rob the system of energy, making the pulsar and its neighbor spiral closer together. As a result, they would start orbiting each other slightly more rapidly. The effect is small—only one-ten-thousandth of a second per year—but it has been detected. This is good evidence that the system is giving off gravity waves.

Key Words

inertial mass
gravitational mass
principle of equivalence
non-Euclidean geometry
positive curvature
negative curvature
advance of perihelion
total solar eclipse
atomic clock
black hole
graviton
gravity wave

Questions

1. What is "general" about the general theory of relativity?
2. What is the difference between gravitational mass and inertial mass?
3. If gravitational mass were different from inertial mass, would objects of different masses still fall at the same rate?

4. We know that Skylab felt the gravitational force of the earth, but the astronauts in Skylab were still "weightless." How can that be?
5. List three experimental tests of general relativity.
6. (a) Why is an eclipse of the sun necessary to measure the bending of starlight? (b) How can the effect be measured for radio waves? Is an eclipse necessary?
7. What fraction of the total perihelion advance of Mercury results from effects of general relativity?
8. What is the status of direct evidence for the existence of gravity waves?
9. What indirect evidence indicates that gravity waves exist? Explain.
10. What astronomical evidence is there for a gravitational redshift?
11. What are two tests of general relativity that have been carried out from earth?

PART III

Particles Together

In the first section of this book, we studied the mechanics of individual particles. In our everyday experiences many objects behave as if they were simple particles. For example, if we want to know about the motion of a car, we can simply follow the motion of the center of mass of the car. After all, the doors and windows must keep pace with the center of mass. We have also seen how we can study the motion of a few particles when they get together. The motion of the planets around the sun was one of our examples.

Despite the far-reaching applicability of some of our simple concepts, sometimes we cannot escape the fact that the things around us are actually composed of many particles—atoms or molecules. It is fine to say that the engine of the car must move along at the same speed as the center of mass, but if we want to study how that engine works, we have to look at what goes on with the gases within the engine. These gases may contain something like 10^{23} particles! We might want to determine what happens to the gas by following each of the 10^{23} particles, applying Newton's laws of motion to each. Unfortunately, this is a hopeless task, even for the largest and fastest computers that we can envision. We need a different approach.

Instead of trying to follow the individual particles, we can usually satisfy ourselves by concentrating on the bulk properties of the materials, which depend on the average motions of all the particles. These are properties that we can measure in the laboratory. In this section, we'll see how we can describe, predict, and apply some of these bulk properties. In so doing, we will follow the average motions of large collections of particles, and see how the laws we have seen applied to individual particles—laws of motion, conservation of energy, conservation of momentum, and conservation of angular momentum—apply to these large collections of particles.

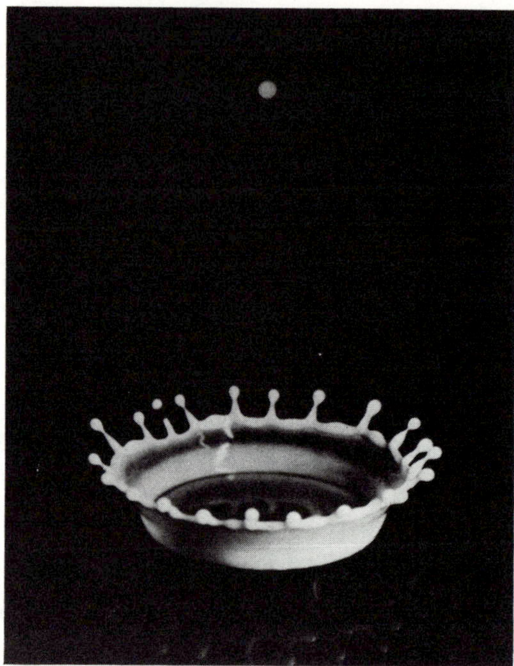

Fig. III–1 The splash from a drop of milk, photographed with an ultrashort-duration strobe light.

We'll start by looking at fluids—gases and liquids. We will talk about the structure of solids later on, in Chapter 25, after we have discussed some relevant atomic physics. We will talk about some of the bulk properties of solids in Chapters 11 and 12, when we talk about temperature and heat as bulk properties of materials.

In this section of the book, we'll see that we can do quite well in sticking to the large-scale properties of large numbers of particles. However, this may seem philosophically unsatisfying, since we know that these bulk properties should somehow be derived from the sum of all the motions of the individual particles. Even if we can't look at all 10^{23} particles individually, it would seem that there must be some way to make the jump from the individual particle motions to the bulk properties. In fact, there is a way; it is called statistical mechanics. We will not be discussing statistical mechanics in any detail in this book, but let us mention that, as the name implies, statistical mechanics is the application of statistical methods (methods for looking at large numbers of objects) to mechanics. One of the great triumphs of statistical mechanics is that we can start with the motions of individual particles, apply statistical techniques, and come up with the expected bulk properties. Most of the studies of the bulk properties came well before the development of statistical mechanics. Nevertheless, it's nice to know that there is some independent way of deriving the results that we find with the bulk-property approach.

CHAPTER 10

Fluids

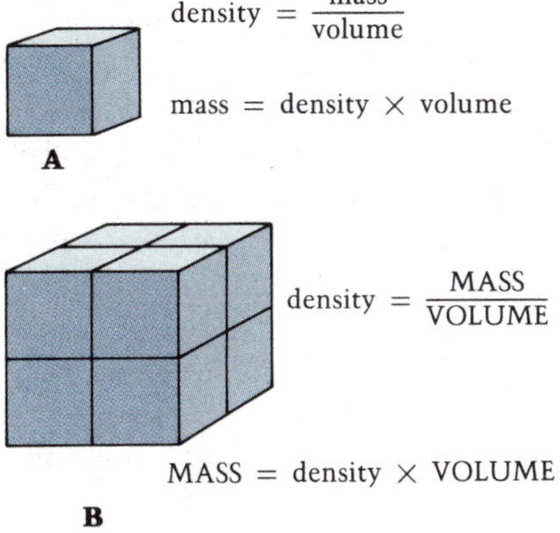

Fig. 10–1 Density is an inherent property of the material out of which the object is made. If all the cubes are made out of the same substance, whatever the volume and mass of A, B will have eight times the volume (since there are eight cubes) and also eight times the mass. The density is the same in the two cases.

We generally classify things as solids or fluids by saying that a solid is something that can hold its shape, and that everything that isn't a solid is a fluid. There are some hazy spots in these definitions, but, in most cases, we can easily tell whether something is a fluid or a solid. We generally divide fluids into gases and liquids. In liquids, the individual atoms or molecules are not completely free to move about. Each molecule is held loosely in place by its neighbors. In gases, the individual particles are very free to move about, and have their motions disturbed only when they bump into other particles or into the walls of the container holding the gas.

10.1 *Describing a Fluid*

The *density* and the *pressure* are the two most important quantities for describing the condition of a fluid. The density of any object is simply the mass of the object divided by the volume of that object. We can express it in kg/m^3, and it is, indeed, the number of kilograms that we would have if we had exactly 1 cubic meter of the material (Fig. 10–1).

Fluids are capable of exerting forces on their surroundings. Let's see how this happens in a gas. The molecules that make up the gas are in a state of continuous motion in all directions. Sometimes they collide with each other, at which time the two colliding parti-

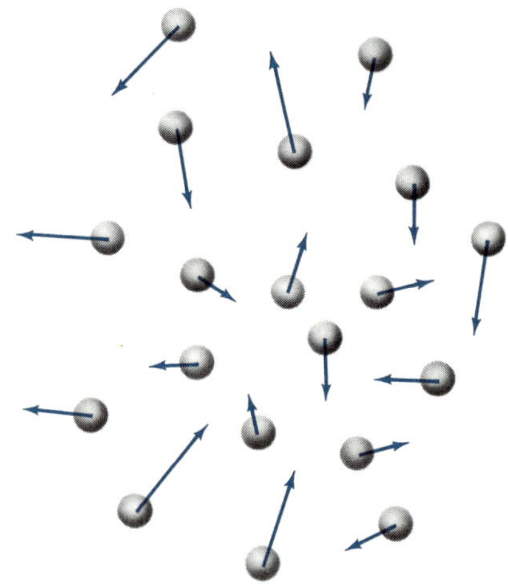

Fig. 10–2 In a gas, particles move with random speeds in random directions.

One cubic centimeter (1 cm^3) of water weighs almost exactly 1 gram, which was the basis for the original definition of "gram." Thus water has a density of 1 gm/cm^3, certainly an easy number to remember. This corresponds to 10^3 kg/m^3 in the now-mandatory SI units. (In the English system, 1 cubic foot of water weighs about 65 pounds.)

cles change direction. Sometimes a particle collides with the wall and changes direction. If you could watch this motion for a while, it would not seem to make any sense; the motion of any particle seems to be aimless. We call this random motion (Fig. 10–2).

We can't see this motion in the particles directly, since the particles are much too small to be observed. However, there is an effect that demonstrates that the random motion is present in fluids. It is called Brownian motion, since it was discovered by Robert Brown, a British botanist, in 1827. Brown noticed that when he examined pollen in water under a microscope, the pollen seemed to bounce around in a random fashion. The first complete theory of this motion was put forth in 1905 by Einstein. (It was a busy year for Einstein; the publication of his special theory of relativity followed a few months later.) The motion of the pollen results from its being struck in random sequence from all sides. It is being struck by particles too small to be seen, but which possess enough momentum to change the motion of the pollen.

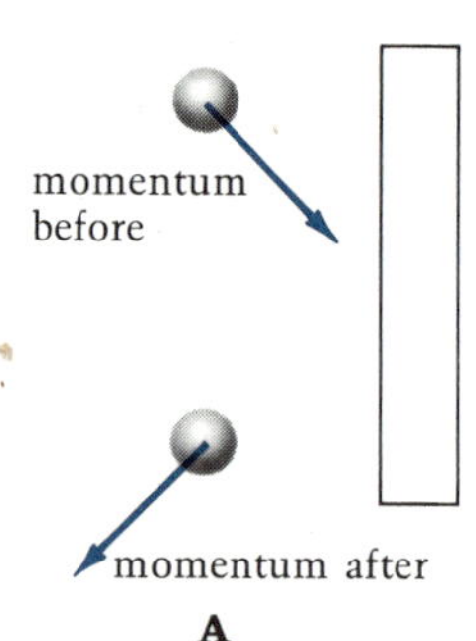

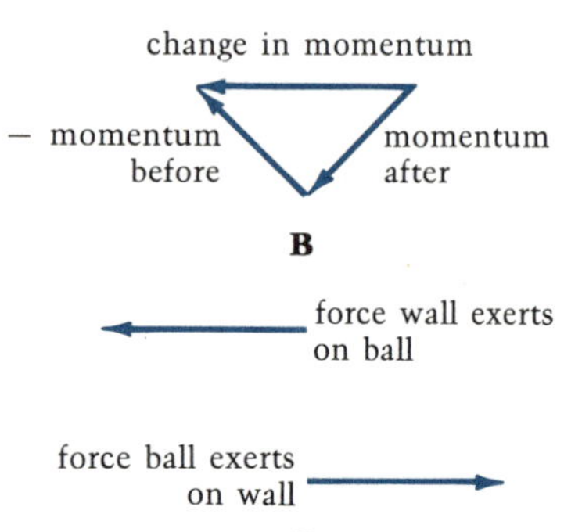

Fig. 10–3 The force that a molecule exerts on a wall while bouncing off. A. The momenta before and after the collision with the wall are shown. Notice that the momentum parallel to the wall has not changed, while the momentum perpendicular to the wall has reversed direction. B. To find the change in momentum, we subtract the momentum before from the momentum after. Notice, as we suspected above, that the change in momentum is perpendicular to the wall. C. The force that the wall exerts on the molecule will be in the same direction as the change in momentum of the molecule. After all, it is this force that caused the change in momentum of the molecule. Newton's third law tells us that the force that the molecule exerts on the wall is equal in strength to the force that the wall exerts on the molecule, but points in the opposite direction. So the forces to the side cancel out and the force that the ball exerts on the wall is perpendicular to the wall.

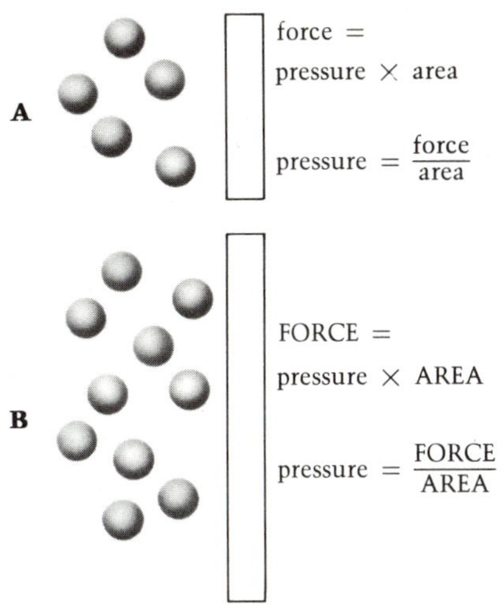

Fig. 10–4 The pressure in a gas is an inherent property of the gas. The force on a wall depends on how big the wall is. In B the wall is twice as big as in A, so the force on the wall will be twice as great, since twice the number of molecules can hit the wall in a given time. In each case the pressure is the same.

Fig. 10–5 The highest pressures ever reached were achieved in this diamond cell by Ho-Kwang Mao and Peter M. Bell. The diamonds are 0.3 carats each. The samples, less than 0.3 mm across, are squeezed between the visible diamond on top and a lower diamond (hidden behind a shield) that is driven upward by a piston. Since pressure is force divided by area, the fact that the area of contact is very small leads to very high pressure.

When the molecules in a gas bounce off the wall of their container, they exert a force on that wall. Newton's third law tells us that this force must be equal in strength but opposite in direction to the force that the wall exerts on the molecule in forcing it to change direction (Fig. 10–3).

The total force on the wall at any given time is the sum of the various forces exerted on it by all of the molecules hitting the wall at that time. The bigger the wall, the more molecules there will be hitting it at any time. We would like a quantity that gives us a measure of the ability of a fluid to exert a force on a wall. This should not depend on the size of the wall; it should depend only on the properties of the fluid. This specific property is called pressure (Fig. 10–4). We define pressure by saying that the force exerted on a given wall is given by

$$\textit{force} = \textit{pressure} \times \textit{area of wall.}$$

This means that if we want to measure the pressure in a fluid, we can find the force that it exerts on a given wall, and divide that force by the area of that wall. That is,

$$\textit{pressure} = \frac{\textit{force}}{\textit{area}}.$$

Just as a fluid exerts a force on a wall, each part of the fluid can exert, via the pressure, a force on the neighboring parts of the fluid. If the fluid is not moving, the force must be perpendicular to the wall of the container or to the imaginary boundary surface between any two segments of the fluid. If any force is exerted parallel to the surface, then the fluid will flow. After all, if it didn't flow, it wouldn't be a fluid.

When pressure is applied to a fluid, that pressure is transmitted in all directions and to all parts of the fluid (Fig. 10–6). This means that if you fall into a swimming pool, an alligator on the bottom of the pool would feel the slight increase in pressure. The idea that

Fig. 10–6 At each wall of the box, the pressure force points perpendicular to the wall and outward. The net force on this box is zero, since there are equal and opposite forces on opposite walls.

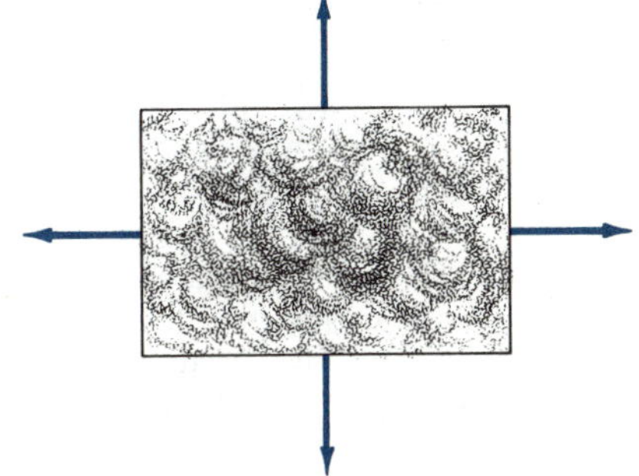

pressure is transmitted equally to all parts of the fluid follows directly from Newton's laws as applied to fluids, but we generally refer to it as Pascal's principle, after the seventeenth-century French scientist, Blaise Pascal, who first proposed it.

The SI unit for pressure, N/m^2, is also given the name the pascal.

Pascal's principle has an important practical application. Suppose we have two pipes that are connected at the bottom and then filled with water (Fig. 10–7), and assume that one pipe has a larger diameter than the other. We place a piston at the top of the water in each pipe. Suppose we push down on the piston in the smaller pipe. The pressure exerted on the water will be equal to the force we exert divided by the area of the smaller piston. By Pascal's principle, this additional pressure will be transmitted everywhere in the water. That means that the pressure everywhere will be higher by this amount.

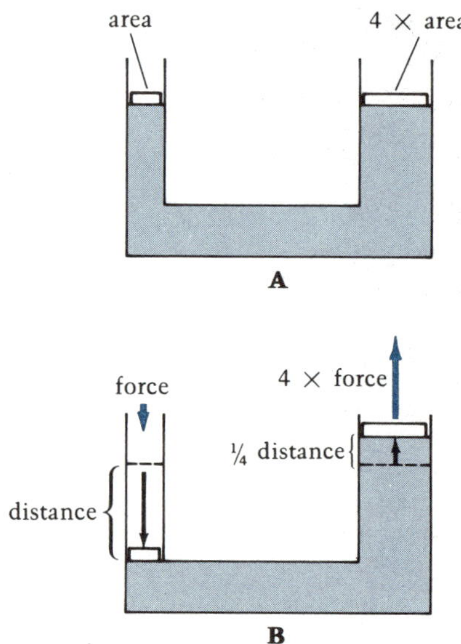

Fig. 10–7 Hydraulic lift. A. In this tank, the piston on the right has four times the area of the piston on the left. If we press down on the left piston, the pressure in the fluid will increase by the same amount all over. However, since that pressure is acting on four times the area on the piston on the right, the upward force on the piston on the right will equal four times the downward force on the piston on the left. This means that you can lift heavy objects. B. When we push down on the left, the piston will travel a certain distance. When this happens, some volume of fluid will be pushed over to the right side, raising the right piston. However, since the tube on the right has four times the area, the transferred volume of water will fill only a quarter the distance you pushed the left piston, since the volume of a cylinder is the *area of the base* × *height.* This means that the right piston moves up a distance equal to one-quarter the distance that you moved the left piston. So, the right piston feels four times the force but moves only one-quarter the distance. This means that the *work* done on both pistons (*force* × *distance*) is the same, and energy is conserved in this process.

Fig. 10–8 Hydraulic lifts are used in garages to raise cars so that they can be repaired conveniently. Because of the hydraulic assist, even a heavy car can be lifted by a small applied force.

Let's look at the water that is in contact with the large piston. Its pressure has increased by this amount. The upward force that it exerts on the large piston is the added pressure times the surface area of the large piston. Since this piston is larger, the water exerts a greater force on this piston than we exerted on the smaller piston. (If the large piston is 10 times the area of the small piston, and if we exert a force of 100 newtons on the small piston, then the water will exert a force of 1000 newtons on the large piston.) This system, called a hydraulic lift, acts just like a lever. *The force you exert gets magnified.* However, just as is the case with a lever, energy must be conserved. To lift something a given height, you must still do the same amount of work. You will exert one-tenth the force at your

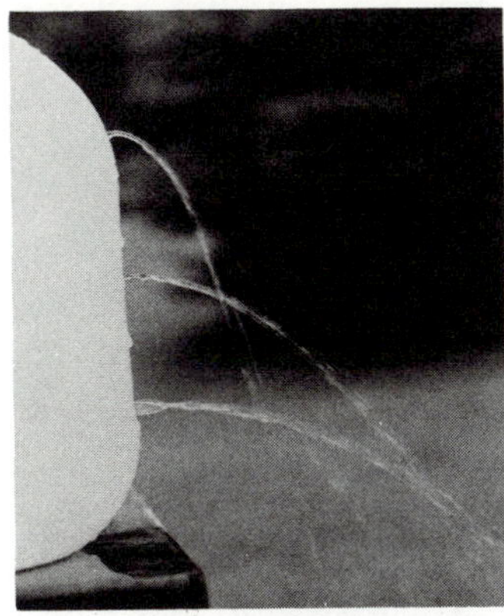

(above) Fig. 10–9 A side view of a milk container filled with water and punctured with three holes. Since pressure increases downward, the stream of water from the bottom hole spurts out farther. Try it.

(below) fig. 10–10 Notice that the horizontal supporting ribs of this silo are closer together at the bottom than at the top. This is because the pressure is greater at the bottom than at the top.

end, but for every 1 meter you lift the other side, you will move your piston 10 meters. We may normally think of fluids as sloshing around with no particular strength, but they actually play an important role in much heavy machinery.

10.2 *Fluids at Rest*

Suppose we have a tank of water. Is the pressure the same at various depths in the water? Is the force that the top of the water exerts on the air equal to the force that the bottom of the water exerts on the bottom of the container? Figure 10–9 shows a little experiment to demonstrate that the pressure is actually different at different depths.

We can see that the pressure at the bottom of the container should be greater than that at the top (Figs. 10–11 and 10–12). The

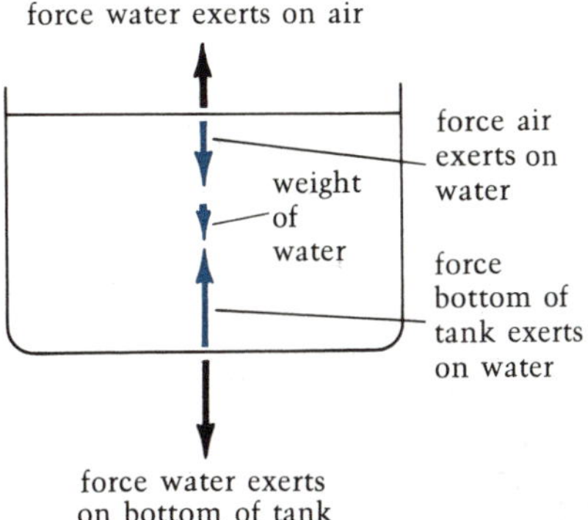

(above) Fig. 10–11 The up-down forces on water in a tank. The air exerts a downward force on the top of the water (equal in magnitude and opposite in direction to the force exerted on the air by the water). The water exerts a downward force on the bottom of the tank (equal to the force the air exerts on the water plus the weight of the water itself). So, Newton's third law tells us that the bottom of the tank must exert an equal but upward force on the bottom of the water.

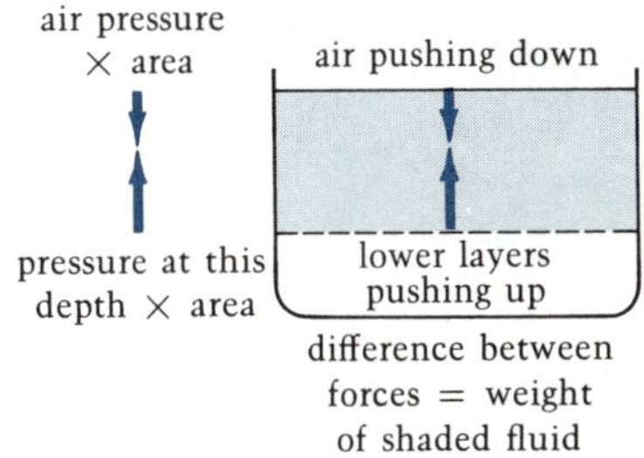

(at right) Fig. 10–12 The pressure at any depth in a tank of fluid must be greater than that at the surface, and it must be greater by enough to support the weight of the fluid above that depth. The pressure at any depth depends on the density of the fluid. The denser the fluid, the greater the weight that must be supported, so the greater the pressure supporting the weight.

Fig. 10–13 Deep-sea divers work under great pressure.

Fig. 10–14 The *Alvin*, specially designed for deep dives, has been used for many search-and-rescue missions.

A

B

Fig. 10–15 A. A full tanker. B. An empty tanker, riding high in the water.

pressure at any depth must be sufficient to support the weight of the water (plus the air) that is above that depth. As you go deeper into the water, there is more weight above to support, so the pressure must be greater.

The increase in pressure as you go deeper in the ocean has some very important effects. For example, divers cannot go too deep before the pressure is too great for them. Even at depths where divers can tolerate the pressure, there are other adverse effects. Under the high-pressure conditions, an increased amount of nitrogen dissolves in the blood. When divers resurface, the pressure on their bodies greatly decreases and the nitrogen leaves their blood. If it leaves too quickly, it forms little bubbles, which is very dangerous, and produces the condition known as the "bends." In order to prevent the bends, divers must rise very slowly, to allow the dissolved nitrogen to be released gradually from their blood. The great pressure at depths under the ocean surface also limits how far a submarine can dive under the surface.

10.3 *Buoyancy*

The change in pressure as you go deeper in a fluid is responsible for a very important phenomenon—buoyancy. Try the following experiment: Place some sand in the bottom of a paper cup and stand the cup in a basin of water. The cup will float. (If you don't add a little sand, the cup will float so high in the water that it is hard to keep it from tipping over.) Add a little more sand. The cup sinks lower in the water. If you keep adding more sand, the cup will sink lower until it no longer floats (Fig. 10–16).

What kept the cup afloat originally? Let's look at the forces acting on the cup. There is the weight of the cup, acting downward. The air pressure also exerts a downward force. (The air pressure is equal

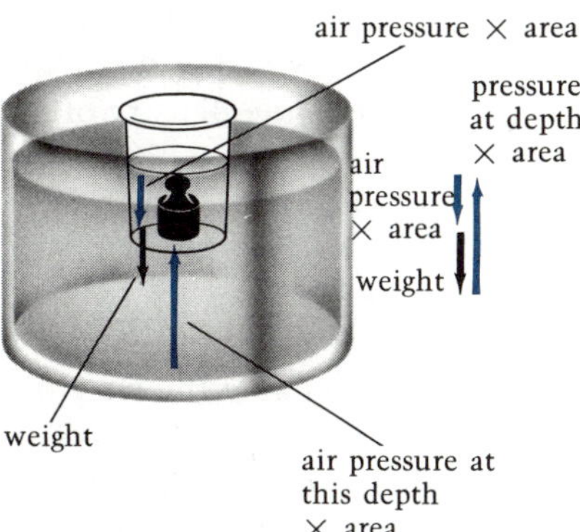

Fig. 10–16 Forces on a partially submerged cup. The upward force is due to the pressure at the depth of the bottom of the cup. The downward forces are due to the weight of the cup and its contents and the pressure of the air on the bottom of the cup. If the cup is to float, then the upward force must balance the downward forces.

to the pressure at the very top of the water.) There is also an upward force on the bottom of the cup. (For simplicity we assume that the walls of the cup are vertical so there are no upward or downward forces on them.) The pressure is greater the lower you go in the water, so the upward force on the bottom is greater than the downward force on the top. We call the difference between the upward force on the bottom and the downward force on the top the buoyant force. *If the buoyant force is equal to the weight of the object, then the object will float.* The weight is actually being supported by the difference in water pressure.

10.3a FLOATING

How can we improve the chances that something will float? We might increase the density of the fluid in which we are trying to float it. For example, salt water has a higher density than fresh water (which is why it is easier to float in the ocean, or, even better, in the Great Salt Lake or the Dead Sea). The higher density means

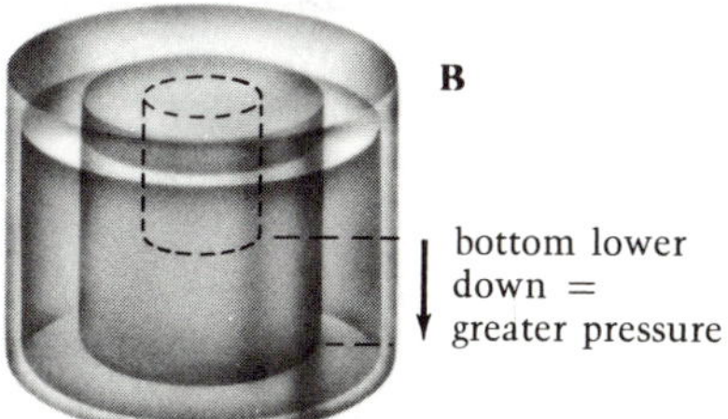

Fig. 10–18 When you increase the size of a floating object, you get two advantages: (1) The increase in the area of the bottom gives you an increased upward force, since the force is pressure × *area*. (2) By moving the bottom deeper, you are moving it to a region of greater pressure, so there will be a greater upward force.

Fig. 10–17 It is easy to float in the Dead Sea.

that the pressure increases faster with depth, so the pressure difference between the top and bottom of the object is now greater than when the density was lower.

Another way is by decreasing the density of the object that you wish to float. For example, you can lower the density by keeping the size the same and reducing the weight. Thus, a styrofoam cube floats better than a steel one. In this case the buoyant force is the same for the same amount of submerged volume, but the amount of weight that must be supported is reduced. Another way to reduce the density is to keep the weight the same but increase the size of the object. This has two effects. As you increase the surface area of the bottom, the upward force is increased (since the force is the pressure multiplied by the surface area). Also, as you increase the height of the object, the top and the bottom are further apart. This means that the pressure difference between the top and the bottom is now greater, with the result that the buoyant force is greater (Fig. 10–18).

We can tie these two ideas together to realize that whether or not something will float depends on the relative densities of the floating object and the fluid in which it is to float. If the density of an object is greater than that of the fluid, then the pressure difference between the top and the bottom will not be enough to support the weight of the object. If the density of the object is less than that of the fluid, then the buoyant force will be greater than the weight and the object will float to the surface (keeping just enough of the object in the water to make the buoyant force equal to the weight).

The story goes that Archimedes, in about 250 B.C., discovered the principle while he was trying to determine the density of a crown to see if it was made of pure gold. He thought of the method while he was in the municipal bath and ran naked through the streets in excitement shouting, "Eureka"—"I have found it."

You will often hear that "something will float if its weight is less than that of the water it displaces." What does that mean? By *displaced water,* we mean a volume of water equal to the volume of the submerged part of the object. Saying that an amount of water with the same volume as the object has a weight greater than the object itself is equivalent to saying that the water has a greater density than the object. Therefore this rule about the weight of the displaced water is the same as our statement above about the densities of the fluid and the floating object. You will often see the "displacement" of a ship noted in its description. This is simply the weight of water that could fill the volume of ship that is generally underwater. The description of buoyancy in terms of the weight of displaced water is called Archimedes' principle.

Fig. 10–19 The tip of an iceberg.

From the fact that ice floats in a glass of water, we can conclude that the density of the ice is lower than the density of water. (This is very unusual; usually the solid form of a substance has a much higher density than the liquid.) Actually the density of ice is about 89 percent of that of water at refrigerator temperature. As a result, even though the ice floats, only 11 percent stays above the surface of the water, which is why icebergs in the ocean can be so dangerous. We only see the "tip" of the iceberg—the rest lurks below the surface of the water. However, the fact that ice floats at all is very important. If ice sank to the bottom of the ocean, the freezing of the oceans would take place much more efficiently than it does, with dangerous results. And lakes and rivers would freeze solid, killing all the fish. We shall further discuss this important property of water in the next chapter.

10.3b AIR PRESSURE

Water is not the only fluid in which the pressure varies with depth. Our atmosphere has the same property. The pressure at the bottom of the atmosphere must be sufficient to support the weight of the atmosphere. The pressure at some point higher up need only be great enough to support the part of the atmosphere that is above that level. There is one additional complication in the atmosphere—the density is not constant. Instead, as the pressure goes

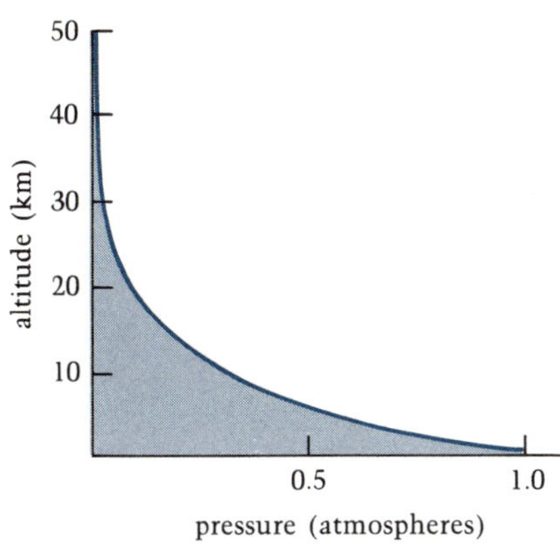

Fig. 10–20 Pressure in the earth's atmosphere as you get higher in altitude. The pressure at the ground is defined to be 1 atmosphere. Notice how quickly the pressure gets below half an atmosphere.

down, the density also goes down. This leads to the type of pressure distribution shown in Fig. 10–20 for the earth's atmosphere. The reduction in density at higher altitudes explains why we cannot breathe as easily in the mountains. For someone who has lived at sea level, breathing becomes uncomfortable at about 3 km, although a lot depends on the physical condition of the individual.

Although we live in the pressure of the earth's atmosphere, we normally overlook it. How can that be? When air pushes on both sides of a wall, the net force is zero. The time when we get a true feel for the force that air can exert is when there is air on one side of a wall and nothing on the other side. We call the absence of all substances a vacuum. In practice we can never achieve a true vacuum, but we can greatly reduce the pressure. When we suck the air out of a can, the air pressure from the outside is sufficient to crush the can.

Often in our discussions of gases, it is convenient to talk about ideal gases. The concept of an ideal gas is an approximation in which the individual particles have no attraction for each other, and the particles take up a very, very small amount of the available space (that is, there is a lot of empty space between particles). The ideal gas is an approximation that makes calculations easier, but that in any case happens to be a fairly good description of most real gases under normal conditions. In gases that behave like ideal gases, it has long been experimentally observed that at a given temperature, pressure is inversely proportional to volume. We shall discuss this further in Chapter 11.

Fig. 10–21 Sir Edmund Hillary (left) and Tenzing Norkay at the summit of Mount Everest on May 29, 1953, the first time that Mount Everest had been conquered. Note the oxygen masks.

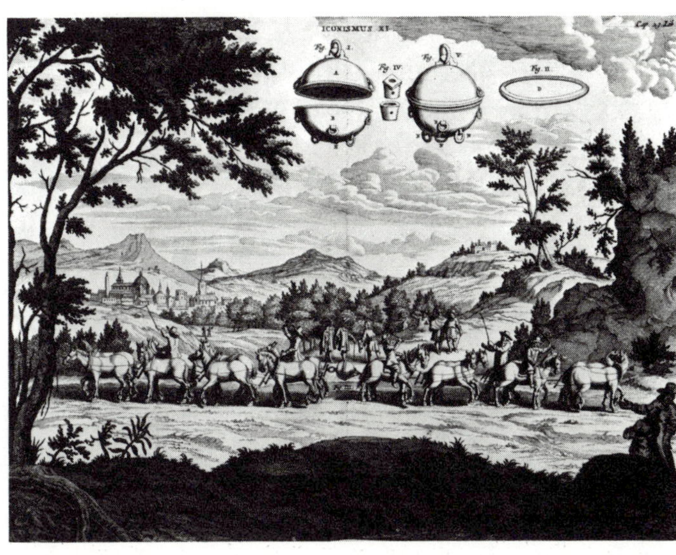

Fig. 10–22 When Otto von Guericke invented the vacuum pump in 1654, he showed that air pressure would hold two hemispheres together even against the pull of teams of horses. In this engraving, he has used his pump to evacuate the space between the two spheres.

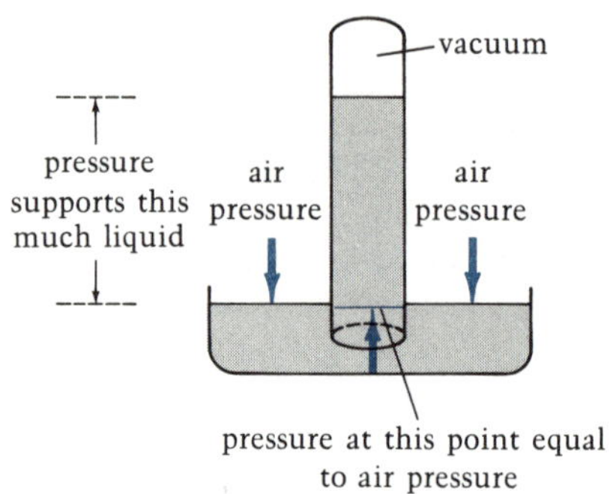

Fig. 10–23 The barometer. This simple barometer consists of a dish of liquid and a tube, partially filled with the same liquid. The tube full of liquid is placed upside-down in the dish, so the liquid is free to flow back and forth between the dish and the tube. Ideally, there is a vacuum above the liquid in the tube, so there is no force pushing down on the top of the liquid in the tube. Air pressure pushes down on the top of the liquid in the dish. This means that all the liquid that is on a level with the top of the liquid in the dish will have a pressure equal to atmospheric pressure. Therefore, the column of liquid in the tube, above the top of the liquid in the dish, is supported by a pressure equal to the atmospheric pressure. The amount of liquid in the tube is exactly that which can be supported by the atmospheric pressure. So, by measuring the height of the liquid in the tube, we can calculate the atmospheric pressure. The height of the column also depends on the density of the liquid. For example, average atmospheric pressure will support a column of mercury that is 76 cm high, but will support a column of water that is about 1000 cm high. This is because the density of mercury is about 13.6 times that of water. If we want a barometer that fits in one room, we use mercury instead of water. Other styles of barometers are also made. You just have to find something that moves when the atmospheric pressure changes.

How do we measure air pressure? We must have some way of measuring the force that the air exerts on some surface. One device to do this is a barometer, invented by Evangelista Torricelli in 1643. A barometer is illustrated in Fig. 10–23.

If you know the air pressure at sea level on a given day at some place on the earth, you can measure your height above sea level by using a barometer. All you need to know is how the atmospheric pressure varies with height. You use the barometer to measure pressure and then convert that pressure into height. Such an instrument is used, for example, by aircraft and is called a *pressure altimeter.* If a pilot wants to know the height of the airplane above a certain airport, the pilot must find out the pressure on the ground at the airport and then take the difference between that pressure and the pressure measured by the plane's altimeter. It is important to remember that the ground pressure is not always the same but changes with the weather; pilots get the latest ground pressure by radioing the tower. Keep track of the weather reports in your area for a while to see how much the pressure can vary.

Since the pressure is different at different heights in the air, air can supply a buoyant force. The density of an object floating in air

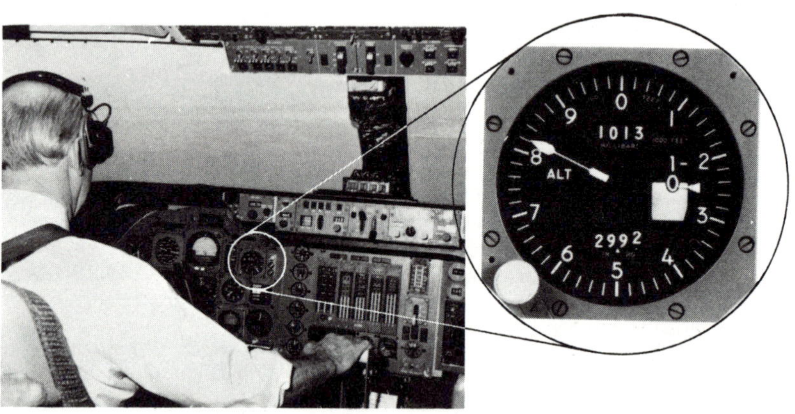

Fig. 10–24 Altimeters in airplanes measure barometric pressure.

Fig. 10–25 The ascent of the first hot-air balloon by the Montgolfier brothers in France in 1783.

There is a famous story told about a physics teacher who gave a test on this material and asked the students how they would measure the height of an apartment building using a barometer. The teacher was looking for an answer involving the measurement of the air pressure difference between the ground and the roof, but one student said that you could just drop the barometer off the roof and time the fall. You could then use your knowledge of falling bodies to calculate the height. The teacher was reluctant to fail the student since the answer did show some knowledge of physics; the student was given another chance.

On the second chance the student came up with some other methods. One was to tie the barometer to a string and lower it over the edge of the roof until it reached the ground. One then just measures the length of the string to get the height of the building. At last the student was told that the method could not employ any additional watches or meter sticks, so the final answer was to offer a fine barometer to the owner of the building in return for the information on how tall the building is.

---

must be less than the density of air. Such things are called *lighter-than-air objects.* One such device is the helium-filled or hydrogen-filled balloon. (We normally use helium because hydrogen is explosive, as the explosion of the dirigible *Hindenburg* showed in the 1930s.)

Helium and hydrogen are much lighter than the average constituents of the atmosphere, so their density is much less than that of the air (which is mostly nitrogen, with about 20 percent oxygen). A balloon will float when you have added enough helium, for example, such that the weight of helium plus balloon is less than the buoyant force of the air. Another way to accomplish the same effect is to heat the balloon (being careful not to burn it). As the gas

Fig. 10–26 This modern balloon, filled with helium, is carrying scientific instruments aloft.

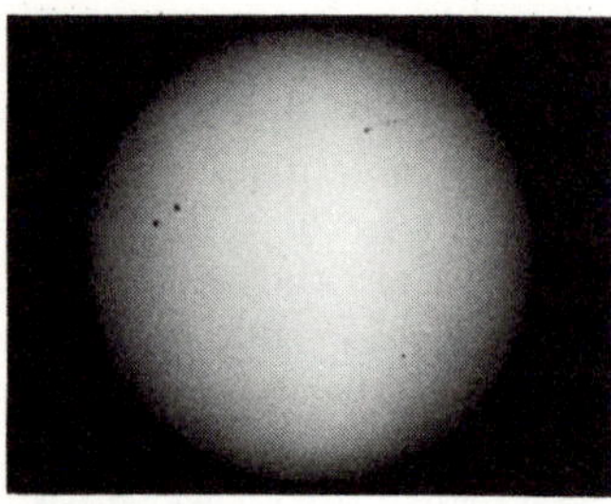

Fig. 10–27 The sun, observed in white light.

heats up inside the balloon, its pressure increases. As a result, it expands until its pressure is again equal to the atmospheric pressure. Then you have the same weight, but the balloon is bigger, which makes the buoyant force bigger.

The atmosphere is not the only gas that is supported by its own pressure. The sun is an even bigger example. Each layer of a star is supported by the gas below, through the pressure of the gas. It is this pressure that keeps the sun's properties so stable: it won't change appreciably in size for another five billion years.

10.4 *Fluids in Motion*

In the previous section we saw that more goes on in a fluid than you might have expected. (The next time you see a glass of water, you may treat it with a little more respect.) When fluids start to flow—move from one place to another—the results are even more amazing.

Let's look at water flowing through a simple hose with one opening at each end. There are no leaks and no places where water can come in along the way. We can then say rather confidently that whatever water flows in one end must end up flowing out the other. Similarly, whatever water comes out one end must have come in the other end.

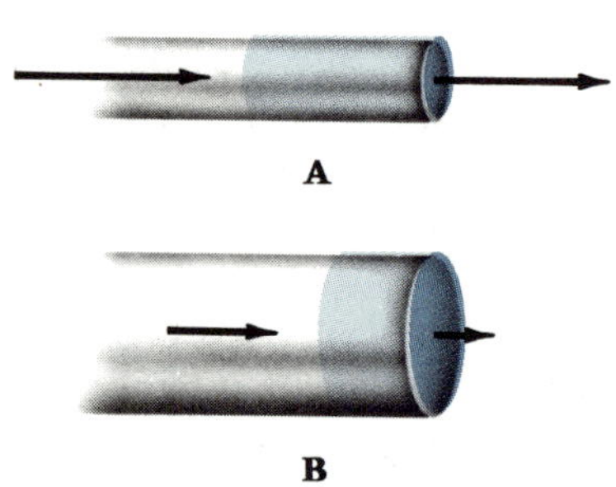

Fig. 10–28 Here we see two hoses, one of which (B) has an opening with twice the area of the other (A). We arrange the flow so that the same amount of water comes out of each hose in 1 second. The shaded region shows the amount of water that will come out in 1 second. Notice that the shaded regions have the same volume. A has half the area at the end, but has twice the length. In A, since the column that must get out in 1 second is twice as long, the fluid must flow twice as fast through the smaller opening for the same flow rates. Therefore, the larger the opening, the slower the fluid has to move to get a given amount of fluid out per second.

The amount of water that we can get through any section of the hose depends on how large the opening is. The greater the area of the opening, the more water we can force through. Let's assume that the opening into which the water flows is twice the size of the opening out of which the water flows. We know that we must still get the same amount of water out of the small hole as we put into the large hole. How can it be done? The answer is that by the time the water reaches the small hole, it must be flowing twice as fast as when it went into the larger hole. That way, in any given second, the same amount of water flows out of the small hole as goes into the large hole (Fig. 10–28).

A similar thing would happen if we reversed the situation. If we put the water in through the small hole, it would come out the large hole with half the speed that it went in. We can state this as a general rule. If the density of the fluid is constant, then

$$\textit{flow speed} \times \textit{area of opening} = \textit{constant}.$$

Note that as we decrease the size of the opening, we increase the flow speed, which should not be too surprising. If you have water flowing out of a garden hose (or a faucet) and you cover part of the

Fig. 10–29 Effect of the size of the tube opening on the speed of flow. A. The front of the hose is uncovered, and the water doesn't come out at a very high speed. B. Part of the front is covered, and the water comes out at a higher speed, as we can tell by the fact that it travels farther.

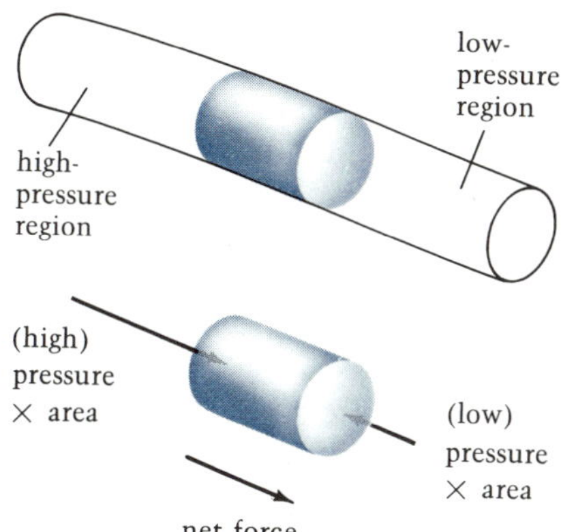

Fig. 10–30 A segment of fluid traveling between a high-pressure (left) and a low-pressure (right) region. Below, we show the forces on this segment of fluid. We have the high pressure × area pushing to the right, and the low pressure × area pushing to the left. There will therefore be a net force from left to right. If the fluid is flowing from left to right, this force will make it go faster, so it arrives at the right with a higher speed than it had on the left. If the flow is from right to left, the force will slow the fluid down. In either case, the fluid will have a higher speed in the low-pressure region than in the high-pressure region.

exit hole, the water will come out at a higher speed. This effect also occurs in canyons, where a gentle breeze that starts at one end can develop into a strong wind as the canyon narrows down. The same effect also explains why the streets in large cities, with a canyonlike effect of skyscrapers on both sides of the street, often have very high winds.

If the speed of flow of a fluid changes, then the fluid is accelerating. We know that an acceleration must be caused by some force. In a fluid, the forces result from the pressure that each part of the fluid exerts on its surroundings. Suppose we follow one section of fluid as it flows from a high-pressure to a low-pressure region (Fig. 10–30). The pressure on the back side will be higher than the pressure on the front side, so there will be a net force from back to front. The fluid accelerates as it moves this way, and shows up at the low-pressure area with a higher speed than it had in the high-pressure region. The opposite occurs if the fluid is flowing from a low-pressure region to a high-pressure region. In this case, the force will be from front to back and the fluid will slow down.

An equation that describes the speed of flow is called Bernoulli's principle, after Daniel Bernoulli, who lived in France in the eighteenth century. Although it has a special name, it really states nothing more than conservation of energy. The basic point to remember is that the higher the speed, the lower the pressure.

DEMONSTRATING BERNOULLI'S PRINCIPLE

You can easily demonstrate the reduction in pressure with increasing air speed. Hold two pieces of paper a few centimeters apart and let them hang down parallel to each other. Blow between the two papers. You will see them come together. They do this because your blowing has increased the speed of the air between the papers, thereby decreasing the pressure. Then the air pressure on the outside, which has not changed, is able to push the papers together.

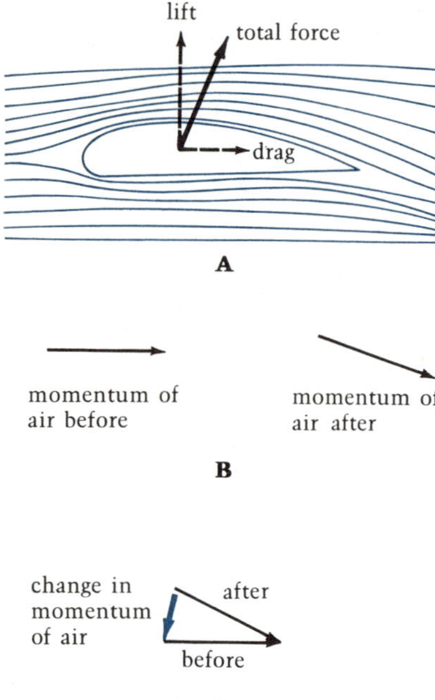

Fig. 10–31 The flow of air past a wing. A. Notice that the air starts out moving horizontally, but ends up with some downward motion. This can be understood in terms of conservation of momentum. B. The momentum of the air before and after striking the wing. There has been a small downward component added to the momentum in passing by the wing. C. This is confirmed here, where we take the difference between the momentum after and before. The change in momentum of the air is downward and slightly forward. This means that the wing has exerted a force on the air that is downward and slightly forward. Then, Newton's third law tells us that the force that the air exerts on the wing is upward and slightly back. The upward part of the force is called the lift and the backward part of the force is called the drag.

10.4a FLYING

The most familiar application of Bernoulli's principle is to the flow of air past the wings of an airplane. Actually, we must be careful in applying Bernoulli's principle because the interaction between the air and the wing is important, and the energy of the air is not really constant (although the energy of air plus airplane is). Figure 10–31 shows what happens as the air flows past the wing. The shape of the wing and the tilt of the wing cause the air to be deflected downward. Since the wing exerts a downward force on the air, the air must exert an upward force on the wing. This upward force is called the lift, and it is what allows the airplane to become and stay airborne.

In terms of Bernoulli's principle, we describe the situation in the following way. The air coming toward the wing separates, with some going over the wing and some passing under the wing, and then comes back together. The air flow is such that the air passing over the wing has to travel a longer distance than the air passing under the wing. This means that, relative to the surface of the wing, the air passing over the wing is going faster than the air passing under the wing. The higher speed corresponds to a lower air pressure, so the air pressure over the wing is less than that under the wing. The greater pressure under the wing then results in the lift.

Notice in Fig. 10–31 that, in addition to the lift, there is also a slight backward force. This is called the drag. This force tends to slow the plane down. When an airplane is flying at constant speed and at a constant height, the engines must do just enough work to overcome the drag and keep the plane from slowing down. The lift and drag depend on a number of factors including the speed of the plane and the shape of the wing. When an airplane is taking off, a lot of lift is important and the wing is held in a position that gives this lift (Fig. 10–32), even though it also gives a lot of drag. When the plane is cruising, less lift is needed and it is more easily provided, with the plane moving much faster than at takeoff. Then, to

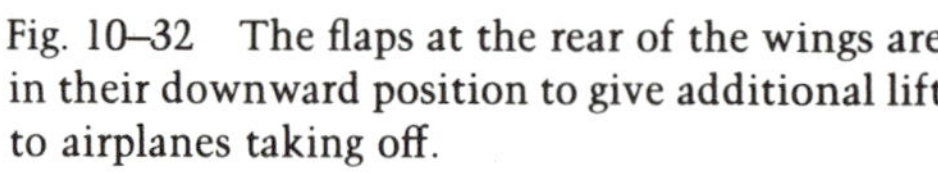
Fig. 10–32 The flaps at the rear of the wings are in their downward position to give additional lift to airplanes taking off.

conserve fuel, the wing is held in a position to minimize drag. In studying the properties of wings, it doesn't matter if the wing is moving through the air or the air is flowing past the wing. It is only the relative motion of the wing and air that counts. For that reason, wings are tested in wind tunnels.

Although the ideas of fluid flow have been well known for some time, fantastic improvements are still being made in the design of such things as airplane wings. Some examples are the new wings shown in Fig. 10–33 which will allow for more efficient airplanes. Other improvements are in the area called streamlining, in which things are designed to minimize the drag, whether it is on an airplane or a truck. Even the ski jumper in Fig. 10–35 is trying for maximum lift and minimum drag.

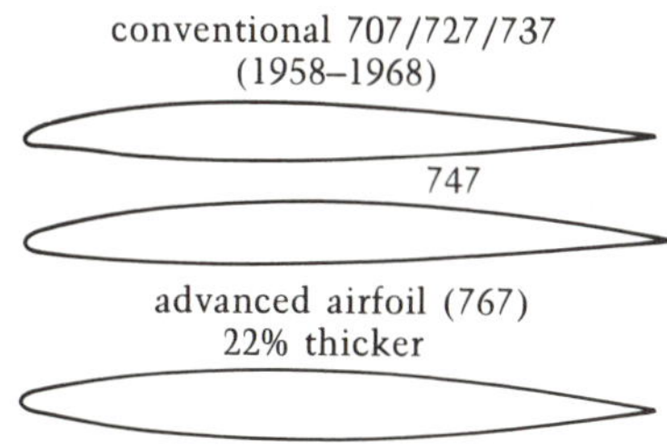

Fig. 10–33 New airplanes for the early 1980s will have a new wing design that will provide more lift with much less drag through the air. These new airplanes will thus be more efficient in their use of fuel. They will also be able to land at lower speeds, thereby permitting use of shorter runways and thus allowing service to more airports.

A

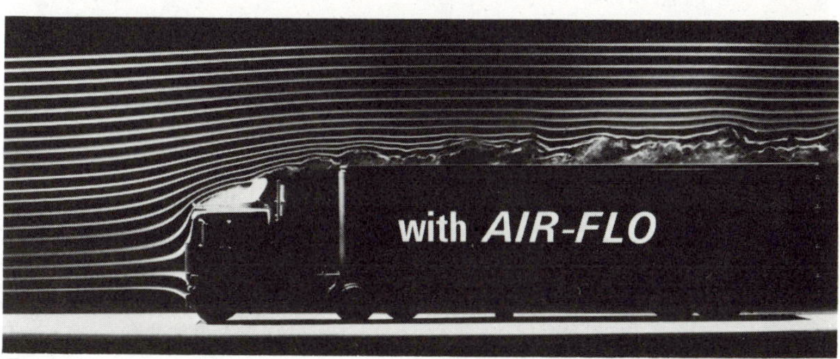

B

(right) Fig. 10–34 Air flow past a truck without and with an air deflector. In Chapter 4 we analyzed air deflectors from the point of view of conservation of momentum. Here we see that the air deflector also provides for a smoother flow of air past the truck. We say that the truck with the air deflector is more streamlined than the one without.

Fig. 10–35 This ski jumper at the Lake Placid Olympics tries to place his body and skis in the form of an airplane wing. If he does this properly, the air flow past his body gives him lift and helps him to jump farther.

10.4b AIR FLOW AND SPORTS

Our study of fluid flow allows us to describe more than just the flight of an airplane. We can also analyze Tom Seaver's curve ball, Billie Jean King's tennis shots that drop just in the court when it looks like they will surely be out, and the way that Jack Nicklaus's drives rise high in the air after going almost straight out from the tee. These situations are similar to the airplane wing in that we can use Bernoulli's principle to get an idea of what is happening, but we must also look carefully at the flow of air past the ball and the interaction between the ball and the air.

To see how these sports examples work, let's consider a ball traveling through the air with topspin. That is, the ball is spinning so that the top is being carried in the direction of motion of the ball and the bottom is being carried in the opposite direction by the spin. As the ball spins, some of the air will be dragged around in a circular pattern. This will happen as long as the surface of the ball is rough (which is why the fuzz on a tennis ball is important). To see how this affects the air flow past the ball, we can look at the situation from the point of view of an observer moving along with the ball, as shown in Figs. 10–36 and 10–37. At any point, the velocity of the air flow is the sum of the velocity of the air that is flowing past the ball and the velocity of the air that is rotating past the ball. This, along with some other effects, results in deflecting the path of the air upward. Thus, if the ball exerts an upward force on the air, the air must exert a downward force on the ball and the ball will be pushed downward, the opposite of lift. The analysis

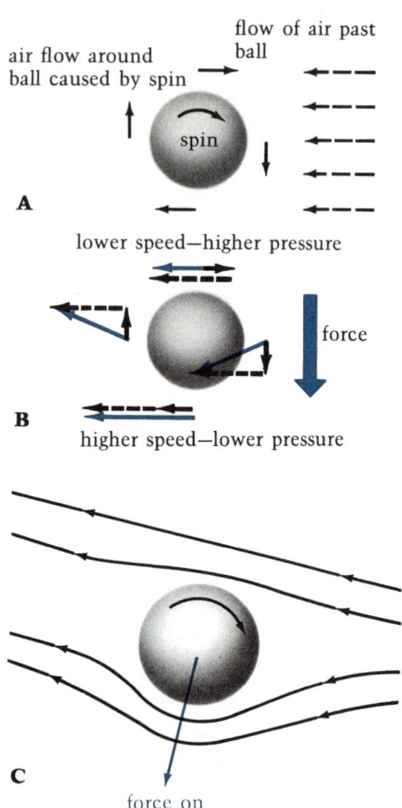

Fig. 10–36 A. We are looking at the air flow around a ball with "topspin." The ball is moving from left to right, and the ball is rotating so that the top of the ball is also moving from left to right. We are looking at the air flow from the point of view of someone riding along with the ball. The spin of the ball causes some of the air to rotate around with the ball, as indicated by the solid arrows around the outside of the ball. Also, there is air flowing past the ball, indicated by the dashed arrows. This results from the motion of the ball through the air, and the flow speed is equal to the speed of the ball. B. At any point around the ball, the velocity of the air is equal to the sum of the velocities of the air flowing past the ball and the air rotating with the ball. Here we add these two velocities together at various locations. The most interesting locations to notice are those directly above and below the ball. Above the ball, the two velocities are in opposite directions, so the net speed is lower than the speed of the straight air flowing past the ball. Below the ball, the two velocities are in the same direction, so the net speed is higher than the speed of the straight air flowing past the ball. C. The overall air flow past the ball. Notice that the air is deflected and that the force on the ball is opposite to the force on the air.

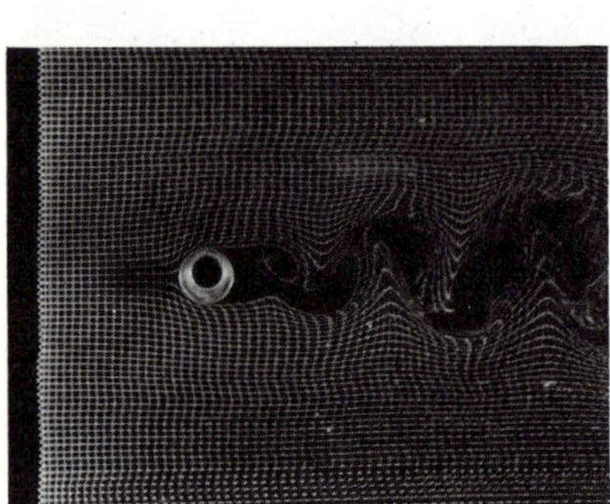

Fig. 10–37 A photograph showing the flow of water from left to right past a cylinder. Note how turbulence results. A pitcher's knuckleball moves erratically for similar reasons. If the ball were spinning, it would carry a little air around with it, causing the pressure to be higher on one side. This would cause the ball to curve. (Courtesy Yaksuki Nakayama, Tokai University)

(left) Fig. 10–38 Topspin makes Billie Jean King's drives curve downward and land within the court.

(right) Fig. 10–39 Backspin makes Jack Nicklaus' shots gain more distance than you would project from their original projectory.

from the point of view of Bernoulli's principle tells us that the air flow over the top of the ball is slower than under the bottom, and that the pressure on the top must therefore be greater than the pressure on the bottom. Thus, a good tennis player can hit the ball very hard, and know that as long as the racquet puts a lot of topspin on the ball, the ball will come down before it goes too far.

The opposite occurs if the ball has backspin. Then the air rushing under the ball is going slower than the air going over the ball. Thus, the pressure will be greater under the ball, causing the ball to rise more than normal. When a golfer hits a ball off the tee, it normally goes out without rising much at first. However, the ball will have a lot of backspin (with the proper air flow aided by the dimples on the ball), so that it will eventually rise, and travel much farther than you might have expected from looking at the original trajectory.

Fig. 10–40 Ron Guidry's pitch curves because of sidespin.

By now you can realize that if a pitcher wishes to make a ball curve, sidespin must be applied to the ball. The direction of the curve will depend on the direction of the sidespin—to the left or right. (In this case, the seams of the baseball are important in getting the air flow around with the ball. With a tennis ball, the fuzz on the ball achieves much the same effect. It is harder to throw a curveball with a very smooth ball.)

You can test these things for yourself. To use the effect of spin on the flight of the ball, you might first try a Ping-Pong ball, which is easy to spin. Also, it is so light that the forces of the air pressure produce large motions. You can see the relationship between the direction of spin and the direction of curve with a striped beachball. Just follow the rotation by watching the stripes.

A

B

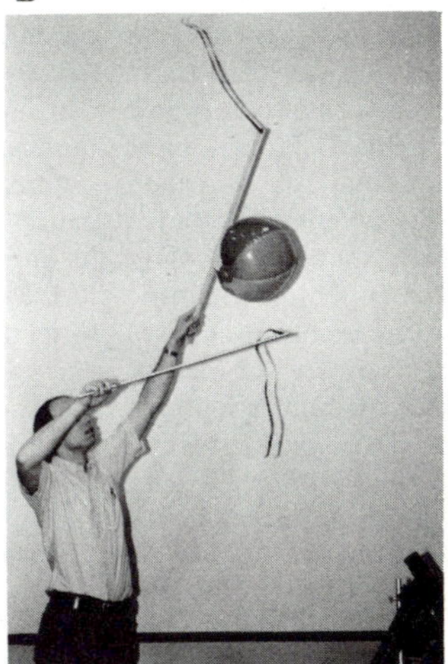

Fig. 10–41 Dynamic lift. A. The beachball is supported in the flow of air from the hose at the lower right. B. The streamer held under the ball falls straight down, but the streamer held over the ball is blown up, showing that the air is flowing mostly over the ball and not under it.

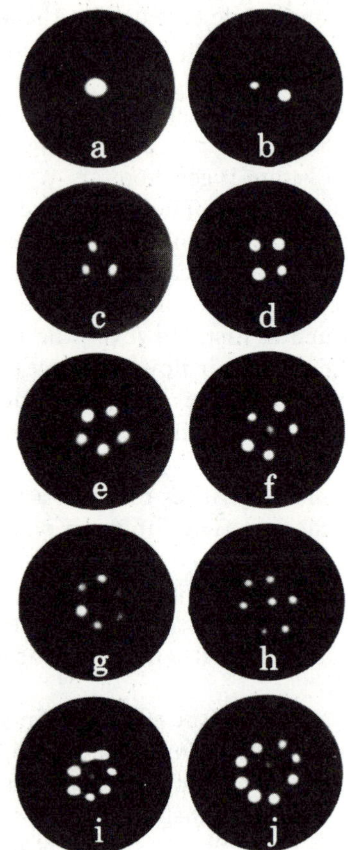

Fig. 10–42 One of the unusual things about a superfluid is that the properties of even large quantities resemble certain properties of individual atoms. For example, some properties of atoms are "quantized," in that they increase by jumps instead of continuously. In superfluid helium spun at increasing speed, the number of whirlpools increases abruptly whenever certain velocities are reached. These photos show the appearance of superfluid helium at 0.1 K between these abrupt changes. Water in a container rotated at increasing speed, in contrast, would simply rotate faster. (Photo by Edward J. Yarmchuk, M. J. V. Gordon, and Richard E. Packard, University of California at Berkeley)

SUPERFLUIDS

If you stir a cup of coffee or tea, you will notice that shortly after you have stopped stirring, the liquid stops swirling around. What causes it to stop? Where did the energy of rotation go? To friction. Fluids experience frictional forces, just as solids do. The details of the mechanism by which the friction operates in fluids is different from that in solids, but the result is the same—lost kinetic energy. The forces that act like friction in fluids are called *viscous forces*: the viscosity of a fluid is a measure of its ability to feel such forces.

Virtually all fluids have some viscosity. However, an unusual thing happens to liquid helium when it is cooled below 2.17 degrees above absolute zero, the theoretical limit of cooling. (We'll discuss absolute zero in the next chapter.) Part of the helium appears to lose all viscosity and flows in ways that normal fluids cannot. It becomes a superfluid.

What can a superfluid do? Under the right conditions it can flow right up and over the side of a bowl, all by itself. It can pass through the narrowest of tubes. In fact, the narrower you make the tube, the easier the helium will pass through. This unusual behavior comes from a quantum mechanical effect (which we will not go into here) that is reasonably well understood by physicists.

Key Words

solid
fluid
liquid
gas
density
random motion
Brownian motion
pressure
Pascal's principle
hydraulic lift
buoyancy
buoyant force
Archimedes' principle
vacuum
ideal gas
barometer
Bernoulli's principle
lift
drag
streamlining
viscosity
superfluid

Questions

1. Water has a density of 1000 kg/m³. What is the mass (in kg) of 3 cubic meters of water?
2. A 2-m³ block of a certain material has a density of 5000 kg/m³. What is the density of a 4-m³ block of the same material?

3. A 2-m^3 quantity of a certain material has a mass of 4000 kg. What is the density of this material?
4. Air exerts a downward force of 10^5 N on a certain tabletop. If the tabletop were twice as large, what would the downward force be?
5. Air exerts a pressure of 10^5 N/m^2 on a certain tabletop. If the tabletop were twice as large, what would the pressure be?
6. What factors influence the pressure of a gas in a container?
7. Why doesn't air pressure cause your house to collapse?
8. In Fig. 10–7, a certain amount of work is done in lifting the weight on the right. Relative to this amount of work, how much work is done in pushing the piston down on the left?
9. List three places in which you have seen or heard of hydraulic systems being used.
10. When a certain block is held just under the surface of a lake, the pressure difference between the top of the block and the bottom of the block is 10^4 N/m^2. Now suppose the block is pulled down so the top of the block is 10 m under the surface. What is the difference in pressure between the top and the bottom?
11. When a certain block is held just under the surface of a lake, the buoyant force is 10^4 N. Now suppose the block is pulled down so the top of the block is 10 m under the surface. What is the buoyant force?
12. Two freighters have the same weight, but because of their different shapes, one sits lower in the water than the other. Which freighter experiences the greater buoyant force?
13. Two freighters have the same shape, but one carries a heavier cargo, and therefore sits lower in the water. Which freighter experiences the greater buoyant force?
14. A freighter starts a journey on the Great Lakes (fresh water) and ends up in the Atlantic Ocean (salt water). During which part of the trip will the freighter ride higher in the water? Explain your answer.
15. If you heat the air in a balloon and it expands, what happens to the average density of the balloon? (That is, does it increase, decrease, or remain the same?)
16. Will an ice cube float on mercury?
17. You put some water and ice into a glass, and mark the level of the top of the water on the side of the glass. When the ice melts, will the level of the top of the water be higher, lower or the same as before? Explain.
18. The force that each tire of your car exerts on the ground is equal to the pressure in the tires multiplied by the area of tire in contact with the ground. The sum of the forces on all the tires must equal the weight of the car. If you add extra cargo to your car, how do the tires change to accommodate the extra weight?
19. In a commercial jetliner flying at high altitude, the cabin pressure is usually kept at about the atmospheric pressure at an elevation of about 2 km. What fraction of the sea level pressure is this? (Use the graph in Fig. 10–20.)

20. If you put a paper cup over your mouth and suck some air from the cup into your lungs, the cup will stay on your mouth without falling. Why?
21. Why do we usually use mercury in barometers instead of water?
22. An airplane is approaching an airport, and the pilot reads the pressure measured by the altimeter. What additional information does the pilot need to determine the exact height of the airplane above the airport?
23. In a large city, is the wind likely to be stronger on a narrow side street or on a main boulevard?
24. You adjust the nozzle of a garden hose to make the water come out faster. What have you done to the opening in the nozzle?
25. A wind flow is arranged so that wind flowing from north to south, passing on the east side of a ball, flows at 20 m/s, and wind flowing from north to south, passing on the west side of the ball, flows at 25 m/s. Will the ball be forced to the east or west?
26. A given wing is designed so that air flows 20 percent faster over the top than under the bottom. Why does the lift increase when the plane goes faster?
27. Why does the lift on a given wing increase as we make the wing larger?

CHAPTER 11

Temperature and Heat

When you open a window on a cold winter day, the temperature indoors falls sharply. If you want to bring the temperature back up to where it was originally, you have to expend some energy (which you usually get by burning some fuel). So it appears that by opening the window, the temperature difference between the inside of your house and the outside allowed your house to lose some energy.

The study of the flow of energy, when related to temperature differences or changes, is called thermodynamics. In Chapter 5 we talked about the energy of particles and saw how energy was conserved. We also mentioned the quantity "internal energy" for objects consisting of many particles. In this chapter and the next, we will further explore the energy of objects of many particles, and how that energy moves around. We must now add temperature to density and pressure as an important large-scale quantity for describing the state of an object.

11.1 *Temperature*

Most of us have a very good qualitative feel for temperature. We can touch something and immediately describe it as "hot" or "cold." You probably even have a fairly good quantitative sense of temperature. You can feel someone's forehead and tell 98.6° from 100° Fahrenheit (37.0° from 37.7°C) with ease. When you listen to the weather report in the morning, you decide how to dress based on the numbers the announcer gives you for the current and predicted temperatures. A difference of a few degrees might cause you to choose a different outfit. So you see, your sense of temperature is well developed.

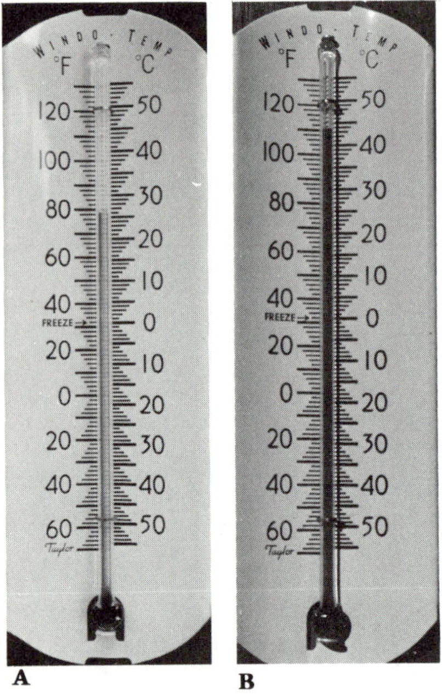

A **B**

Fig. 11–1 The height of mercury in these two pictures shows that the thermometer was hotter in B than in A.

Certain properties of objects, such as their size, may depend on their temperature. In fact, when we "measure" the temperature of an object, we are really measuring some property that depends on temperature and then converting that measurement to a temperature. For example, when you take your temperature with a mercury thermometer, you are really measuring the length of the mercury in a tube. Someone in a laboratory has already *calibrated* that thermometer, which means that the length of the mercury for various temperatures has already been established. So instead of painting numbers on the thermometer that give the actual length of the mercury, the numbers on the tube are the temperatures at which the mercury will have those particular lengths.

The temperature scales that we normally encounter are the Celsius (formerly called the centigrade) scale and the Fahrenheit scale. On the Celsius scale, the point at which water freezes at sea level is arbitrarily set at 0 degrees, and the point at which it boils is set at 100 degrees. (We generally write these as 0°C and 100°C.) The Fahrenheit-scale is passing out of use. On it, water freezes at 32°F and boils at 212°F. From this we can see that a change of 100°C corresponds to a change of 180°F, so 1°F is equivalent to 5/9°C.

$$°F = \frac{9}{5}°C + 32°.$$

$$°C = \frac{5}{9}(°F - 32°).$$

Although the Celsius scale is used for many scientific applications, there is another scale that is very important in studying physical systems. It is the Kelvin scale (abbreviation K). A change of 1°C is equal to a change of 1 K, but 0 K corresponds to −273.15°C. As we'll see later in this chapter, there is a real physical significance to zero on the Kelvin scale; we call it *absolute zero*. It corresponds to the lowest temperature that can possibly be reached.

$$K = °C - 273.15°$$

1 K is 1 kelvin.

11.2 *Thermal Expansion*

One of the most common properties that depends on temperature is the length of an object. As we increase the temperature of most objects, they expand. This expansion can be understood in terms of the atomic structure of the material. For example, we can think of a solid as a regular pattern of atoms. The atoms do not stand still, but are constantly vibrating around, held roughly in place. As we raise the temperature, the vibrations become more violent, driving the atoms farther apart. The average distance between any two atoms increases. This makes the whole object get larger. The amount of

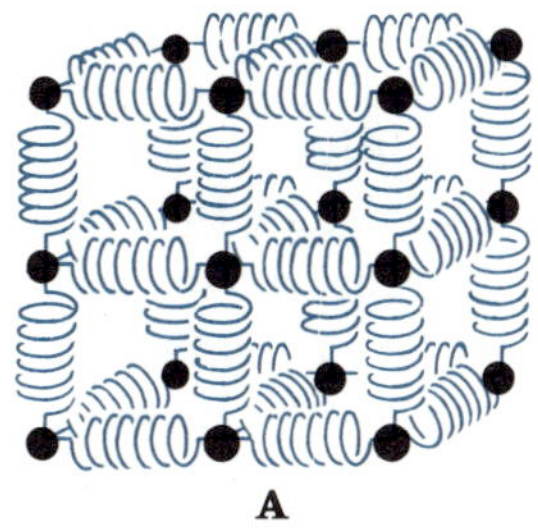

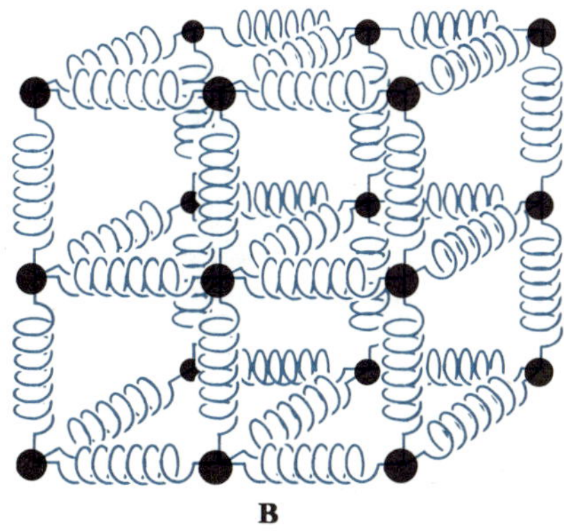

Fig. 11–2 A microscopic view of thermal expansion. We can imagine the atoms as being held together in a solid by a series of springs, with each atom attached to its nearest neighbors. The atoms don't sit still in these positions, but vibrate back and forth. A gives the arrangement at low temperature. When we raise the temperature, as shown in B, the vibrations of the atoms get more violent (as depicted by the larger black spots representing the atoms). When this happens, each atom pushes its neighbors farther away, getting more "elbow room." The effect is to have each "spring" stretch. Every atom ends up farther from every other atom than before, and the whole object expands.

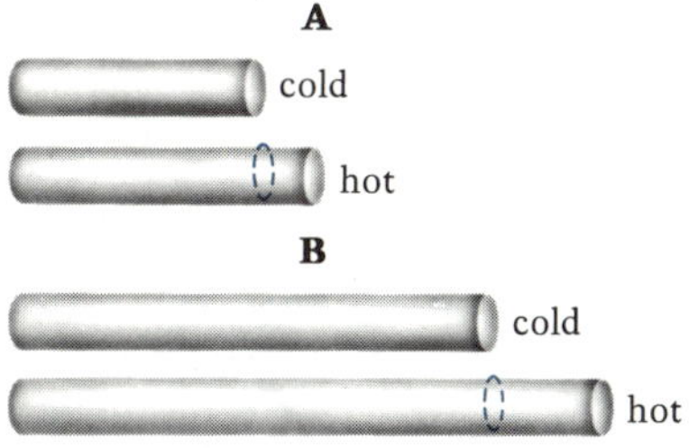

Fig. 11–3 The actual amount of expansion depends on the length of the rod. The rods in A and B are made of the same material, but the cold rod in A starts out half the length of the cold rod in B. They are both heated by the same amount and expand. They expand so that the hot rod in A is still half the length of the hot rod in B. In order for this to happen, the rod in B had to expand by twice as much as the rod in A. However, each rod expanded by the same fraction of its total length. The fractional change in length for a 1° temperature change is called the *coefficient of expansion*, and depends on the material out of which the rod is made.

expansion depends on the material involved and the amount of the temperature change (Fig. 11–2).

Suppose we have two rods made of the same material, with one rod 1 meter long and the other rod 2 meters long. We start them at some temperature and then increase that temperature by the same amount for each rod. Do both rods increase in length by the same amount or does one stretch more than the other? Let us think back to our atomic picture. If the space between every two atoms increases, then the rod with twice as many atoms will stretch twice as much. If the 1-meter rod stretches by 0.001 meter, then the 2-meter rod stretches by 0.002 meter. Notice that each rod stretches the same percentage—0.1 percent—of total length. For any rod, the percentage or fractional change in length depends on the material out of which the rod is made (Fig. 11–3).

We see examples of thermal expansion all around us. When it is

Fig. 11–4 The zig-zag pattern of the trans-Alaska pipeline allows for expansion and contraction of the pipe during temperature changes.

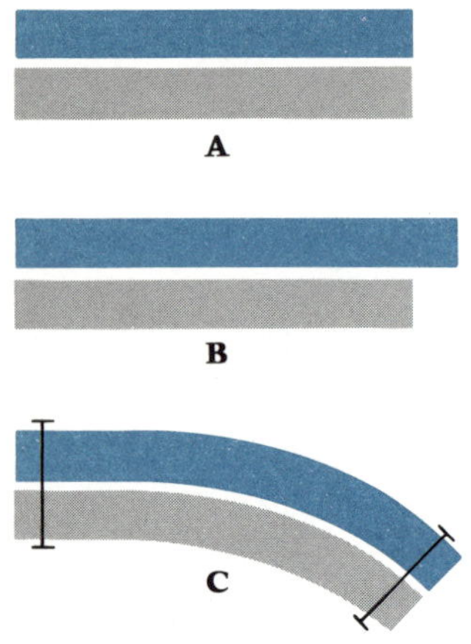

Fig. 11–5 Making a thermostat. In A, we start with identical lengths of two different metals. The metals have different coefficients of thermal expansion. (We call this pair of metals a bimetalic strip.) As a result, when they are heated, as in B, the upper one expands more than the lower one. However, if we clamp the ends together, as in C, then the upper one must still expand more, but the ends must stay together. This can only happen if the strip bends, with the longer metal going around the outside. If we hold one end fixed, the position of the other end will depend on the temperature. We can arrange it as part of a switch that will turn the heat on when it gets too cold, or off when it gets too hot.

very hot, sections of roadways may expand by so much that they no longer fit in the roadbed. When this happens, the road buckles. It is important in building a road or a bridge to leave spaces that allow for this expansion. We also put our knowledge of thermal expansion to practical use in making a thermostat, as shown in Fig. 11–5.

Not all materials expand when heated. At most temperatures, water will expand when the temperature is raised. However, there is one small temperature range—from 0°C to 4°C—for which water contracts as it gets warmer. This has a very important consequence. Suppose we are cooling a lake. At any time the densest water will sink to the bottom. Once all the water is cooler than 4°C, the warmest water will be the densest and will sink. The 0°C water will stay on the top, so the ice will first form in a layer at the top (Fig. 11–6).

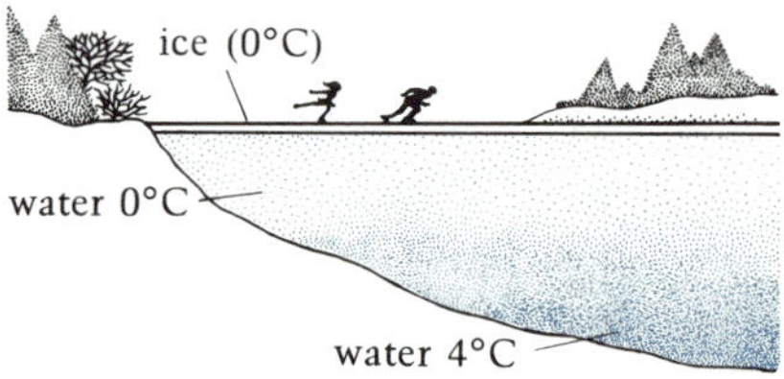

Fig. 11–6 The density of water depends on the temperature of the water. In any lake, the densest water will sink to the bottom and the least dense water will rise to the top. For most temperatures, the density of water decreases (the water expands) when the water is heated. The exception is between 0°C and 4°C. Once all the water in a lake is 4°C or less, the 4° water is the densest and sinks to the bottom. The cooler water will rise to the top. So, the 0°C water will be at the top, and this is where the ice will have to form. Once the ice forms, we will have 0°C water just under the ice, and the temperature will increase as we get lower in the lake, until we reach the 4°C water at the bottom.

11.3 *Heat*

When you place a hot brick in contact with a cold brick, the hot brick gets cooler and the cold brick gets warmer. Up through the seventeenth century it was assumed that when this happened, some warm fluid actually flowed from the hot body to the cold one. This flowing material was called *caloric.* However, it became clear from experiments that the caloric or total amount of heat was not always conserved. It could be manufactured and destroyed. As we have seen, you can "make heat" by rubbing your hands together. We have come to associate heat with work done (as in rubbing your hands together), or with changes in energy.

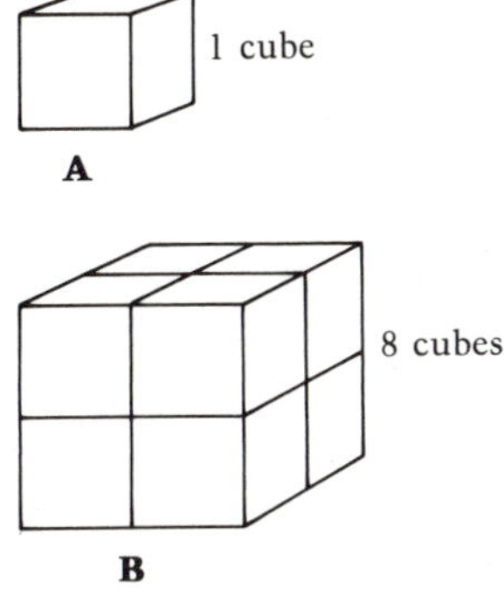

Fig. 11–7 The *specific heat* is a property of a given material. The two objects in A and B are made of the same material, and therefore have the same specific heats. The *heat capacity* depends on the material out of which an object is made (its specific heat), and how much of that material is present. Therefore, the eight-cube figure (B) will have eight times the heat capacity of the one-cube figure (A).

Another unit that is still used for heat is the British thermal unit (BTU), *which is the heat required to raise one pound of water from 63 to 64°F. So, 1 kcal is about 4 BTU.*

Fig. 11–8 To cool off the hot piece of iron that she is working on, this blacksmith dips the iron in a large bucket of water.

In fact, heat is only a form of energy. The concept of heat is useful in keeping track of energy transfers that are associated with temperature differences. (The details of the ways heat can be transferred from one object to another will be discussed in the next chapter.)

We even define our units of heat in terms of the temperature changes that are produced in certain substances. The kilocalorie is defined as the amount of heat required to raise 1 kilogram of water from 14.5 to 15.5°C. The "calorie" that we talk about in diets is actually a kilocalorie (kcal). A typical diet that includes 3000 kilocalories (a dieter would say "3000 calories") in one day contains approximately enough "food energy" to heat 30 kg of water from the freezing point to the boiling point!

We often wish to know how much heat is required to produce a given change in temperature in an object. For any object, the amount of heat required to raise its temperature by some specified amount (usually 1°C) is called the heat capacity of the object. The heat capacity will depend on the amount of material present. The more material you have, the more heat is required to raise its temperature by some given amount. The heat capacity will also depend on the type of material you are using. Any particular material has a property called specific heat. The heat capacity is then the specific heat multiplied by the amount of material present. Thus we can talk about the specific heat of water (which is 1 kcal/kg), but we talk about the heat capacity of Lake Superior (Fig. 11–7).

What happens when we bring a hot object and a cold object into contact? For example, suppose you drop a hot horseshoe into a bucket of cold water. The horseshoe will get colder and the water will get hotter. This will continue until the temperature of the water and that of the horseshoe are the same. What will the final temperature be? We can figure it out by noting that *the heat lost by the horseshoe must equal the heat gained by the water.* As the horseshoe loses a certain amount of heat, its temperature will fall by some amount. The water then gains this same amount of heat. If we start with a large trough of water, the heat capacity of the water will be much larger than that of the horseshoe, and the temperature of the water will change by less than that of the horseshoe. The final temperature of both water and horseshoe will be close to the initial temperature of the water.

This process also explains why your body begins to feel uncomfortable in bath water much hotter than 45°C although you can pull a piece of aluminum foil off a baked potato that has been in a 250°C oven. Because of its small size and low specific heat, the aluminum foil has not been able to store much heat to transfer to your hand.

11.4 *Heat and Work*

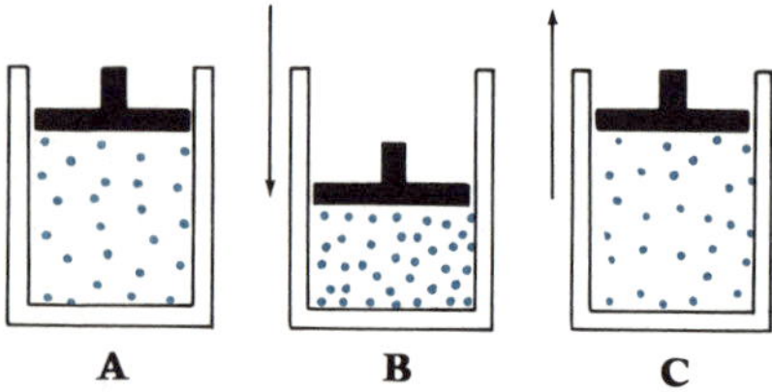

Fig. 11–9 In going from A to B the piston is pushed down, doing work on the gas. We can say that the work done *on* the gas is positive, or the work done *by* the gas is negative. In going from B to C, the gas expands, which pushes the piston up. This means that we can get work out. We can say that the work done *by* the gas is positive, or the work done *on* the gas is negative. Sometimes you will see the second law of thermodynamics expressed in terms of work done on the gas and sometimes in terms of work done by the gas. Rather than worrying about what this does to the signs, just try to reason whether the work done will increase the energy of the gas or decrease the energy of the gas. For example, when we push down on the gas, we are increasing the energy of the gas. When the gas expands, it is doing work, so its energy decreases.

Since heat is related to energy, it must also be related to work. In our study of mechanics (Chapter 5) we associated work with energy. We saw that if you did work on something, its energy increased. This increase in energy might show up as an increase in the kinetic energy, potential energy, or internal energy. (Remember, by internal energy we really mean the total kinetic energy of all the microscopic particles that make up the object.) In this chapter, we have seen that work, heat, and energy all seem to be related. This relationship is summarized in the first law of thermodynamics, which says that the change in the internal energy of a system is equal to the heat added to the system plus the work done on the system. Here, by "system" we mean any object or collection of objects. Actually, the first law of thermodynamics is really only a statement of conservation of energy.

The change in the energy of a system can either be positive or negative. That is, the energy can increase or decrease. First, let's look at the "heat added" to the system. If heat is actually added to the system, then this will be a positive number. If the heat is actually removed, then the "heat added" is a negative number.

The work done on a system can be plus or minus also. Let's look at some gas in a cylinder with a piston. The gas exerts a force on the piston, caused by the pressure of the gas. If we want to compress the gas by pushing down on the piston, we must exert a force, and we will do work on the gas while we are compressing it. This will increase the internal energy of the gas. What if we remove our force? Then the pressure of the gas on the piston will cause the piston to move out as the gas expands. Now it is the gas that is doing the work on the piston, and the internal energy of the gas decreases. To summarize—when we compress the gas by pushing on the piston, the work done on the gas is positive; when the gas pushes back and expands, the work done on the gas is negative. When we say that the work is negative, it means simply that we are getting work out of the gas rather than putting it in. (See Fig. 11–9.)

11.5 *Temperature and Energy in an Ideal Gas*

Notice that the first law of thermodynamics deals with changes in the internal energy of the gas, not changes in the temperature. However, internal energy and temperature are closely connected. We can illustrate this for an ideal gas. As we saw in the last chapter, an ideal gas is one in which the particles occupy a very small part of the available space, and in which the particles exert no forces on each other except when they bounce off each other like hard balls.

For an ideal gas, the equation of state is particularly simple. If P is the pressure, T the temperature on the Kelvin scale, V the volume occupied by the gas, and N the number of particles in the gas, then the equation of state for an ideal gas is $PV = NkT$. k is a constant, the Boltzmann constant (whose value is given in Appendix 1).

For any gas, the temperature, density, and pressure are all related. Since these three quantities describe the state of the gas, an equation relating them to each other is called an *equation of state.*

In Chapter 10 we saw that the pressure resulted from the random motions of the particles making up the gas. However, the internal energy also depends on these random motions, because the internal energy is the sum of the kinetic energies of all the particles. The greater the speeds of the random motions, the greater the kinetic energies of the particles. The important concept is that *the total kinetic energy of the particles in the gas is proportional to the (Kelvin) temperature of the gas.* (See Fig. 11–10.)

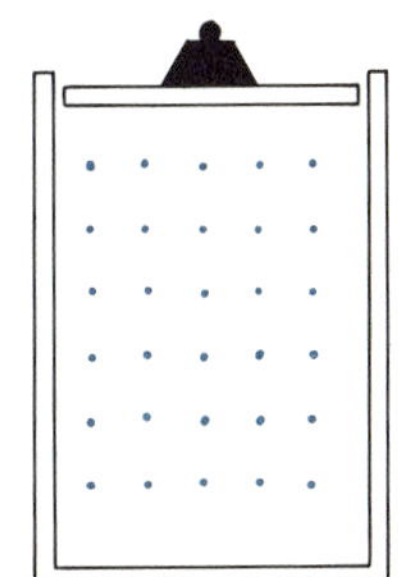

Fig. 11–10 The relationship between pressure and volume for an ideal gas at constant temperature. We can adjust the pressure on the gas by the amount of weight that we put on the piston. If we put on a weight, the piston will fall until the gas pressure is great enough to support the new weight. When we double the weight, we double the pressure, and the piston drops to half its original height. This means that the volume of the gas has been cut in half.

For example, in a typical room full of gas, the internal energy is about 10 million joules! The molecules in the room are moving pretty fast—a few hundred meters per second.

The internal energy in a gas can be substantial. Notice that if we divide the total energy by the total number of particles, we are left with the average energy per particle. This doesn't mean that each particle has this energy. Since the energies of the particles are random, there will be particles with energies higher or lower than the average. If we raise the temperature, the average energy increases, although there will always be some slowpokes and some speeders.

Fig. 11–11 The pressure from steam pushing a piston can move a train.

11.6 *The Second Law of Thermodynamics—Are Perfect Engines Possible?*

We have seen that it is a straightforward matter to convert work into heat. What about going in the opposite direction? We know that it can be done. One example is the steam engine, in which heat is used to boil water. The pressure of the resulting steam pushes a piston and does work, such as moving a train (Fig. 11–11). Of

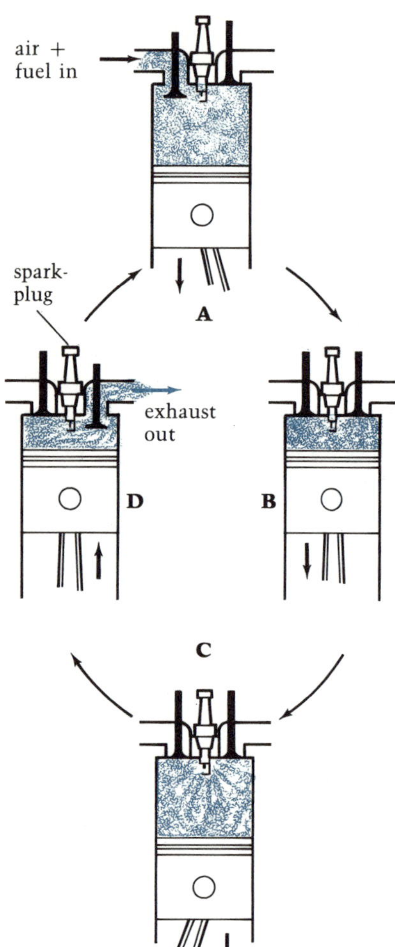

Fig. 11–12 Steps in the cycle of an automobile engine. A. An air-fuel mixture is taken in as the piston goes down. B. The piston comes up, compressing the mixture, making it hot. C. The spark plug fires, igniting the mixture, causing the gas to expand. This pushes the piston down and is when the work gets done. D. The hot exhaust is pushed out.

course, we would like to get out as much work as we can for any given amount of heat that we put in. We can define the efficiency of an engine as the ratio of the amount of work we get out to the amount of heat we put in. What are the limits on such an efficiency?

In most practical situations, we are limited to engines that go in cycles. By "cycle" we mean that the engine repeats a series of steps (taking in heat and doing work along the way). The internal combustion engine in an automobile is an example of a cyclic engine (Fig. 11–12). In this engine it takes two complete in-out cycles of the piston to make one full cycle in the engine, so this engine is known as a four-cycle or four-stroke engine. On the first outward move of the piston, a mixture of fuel and air is sucked into the cylinder. On the first inward move of the piston, the mixture is compressed. The spark plugs then ignite the mixture, causing the mixture to expand, which forces the piston outward. This is the step in which the work is actually done, when the hot gases push against the piston. Finally, on the second inward move, the hot gases are ejected from the cylinder and the process is ready to start again.

In any cycle, the engine ends the cycle in the same condition as when it started the cycle. Therefore the internal energy of the engine after the cycle is the same as that before the cycle. Since there is no change in energy, the first law of thermodynamics tells us that the work done by the engine must equal the net amount of heat taken in during the cycle. This tells us that the engine can be no more than 100 percent efficient.

The next question is, can the engine ever be 100 percent efficient? That is, can it convert all of the heat absorbed into work? Your first response might be to ask why an engine shouldn't be able to do that. In the automobile engine, we put in a certain amount of heat (in the form of the burning of the air-fuel mixture). We also got some work out, but is the work the only thing we got out? No. We also got some exhaust out. And that exhaust is hot—so hot that you cannot touch any part of the exhaust system (such as the muffler) of a car that has been running. Part of this exhaust gas went into the engine at the outside air temperature and was heated up during the engine cycle. Therefore, some of the heat that we put into the engine was expelled in the exhaust. (Still more heat is also carried away by the cooling system of the car.) Thus, the amount of heat that could actually be converted into work is the amount of heat we put in, less the amount of heat that was expelled with the exhaust (or carried away by the cooling system). The engine is less than 100 percent efficient (Fig. 11–13).

For any given engine the details may be a little different, but the net result is the same. Of all of the heat taken in, some is given back

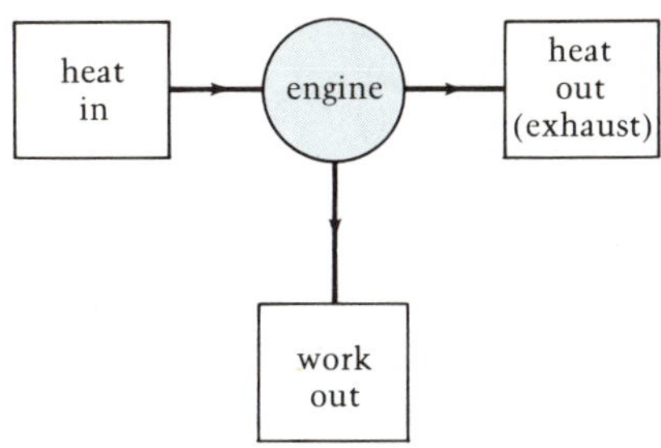

Fig. 11–13 Energy flow in a real engine. Some heat is taken in from a high-temperature source. (In an automobile engine, it is the ignition of the air-fuel mixture.) Some of this heat is converted into work, and the rest is given out as exhaust heat to a cold-temperature region (such as the outside air). Notice that in accordance with the first law of thermodynamics, the heat out + work out = heat in. That is, energy is conserved.

out again as exhaust. Therefore all the heat that is taken in is not available to do work. Is this just a drawback of the design of the automobile engine, or is there some fundamental physical problem here?

This question was studied extensively in the early part of the nineteenth century in France by Sadi Carnot. Carnot first considered the difference between reversible and irreversible processes. A *reversible* process is one for which the reverse process can also occur. For example, gas expanding against a piston and doing work is reversible, since we can do work from the outside and push the piston back in, so that the gas returns to its original state. Mixing milk in your coffee is an example of an *irreversible* process. There is nothing we can do with our spoon that will unmix the milk and coffee.

To study engines that used only reversible processes, Carnot introduced an engine based on a cycle (which he considered in theory only), now called the Carnot cycle. (This cycle is illustrated in Fig. 11–14.) The important feature is that the cycle consists of reversible processes and operates between two sources of heat, one at a high temperature and the other at a low temperature. Over the course of the cycle, heat is taken in from the high-temperature source and exhaust heat is given out to the low-temperature source. Carnot showed that all reversible engines operating between the same two temperatures had the same efficiency. Any irreversible engine operating between the same two temperatures has a lower efficiency than a reversible one.

Carnot's analysis also allowed him to define an absolute temperature scale, based on the heat taken in or given out in various parts of the cycle. Zero on this scale is called absolute zero (and is zero on

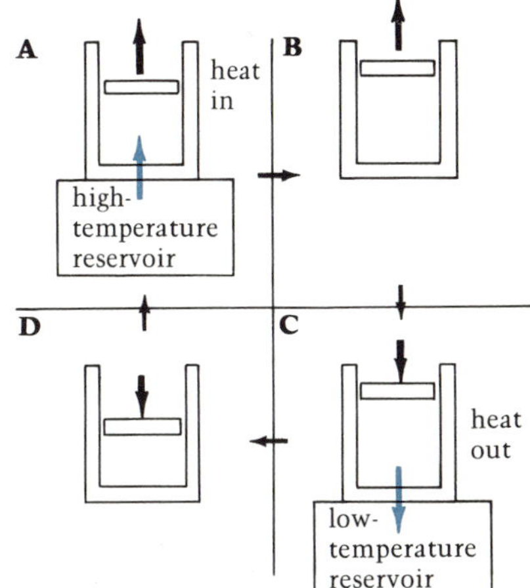

Fig. 11–14 The Carnot cycle. This was the cycle analyzed theoretically by Sadi Carnot in his efforts to understand the efficiency of engines. A. In the first part of the cycle, the gas is brought into contact with a high-temperature reservoir, and is allowed to expand. In the process of expansion, the gas does some work, and also takes some heat in from the reservoir. B. In the next step, the gas is isolated so that no heat can flow in or out, and the gas continues to expand. As it expands, it does more work, but since it cannot take any heat in, the energy of the gas decreases. C. Now the gas is brought into contact with a low-temperature reservoir. The gas is compressed, so we do work on the gas, and, in the process, the gas gives up some heat to the reservoir. D. In the final step, the gas is again isolated so that no heat can flow in or out, and we continue to compress the gas. We are doing work on the gas, and no heat can get out, so the internal energy of the gas must be increasing. When step D is finished, the gas is in the same state as it was at the beginning of the cycle, and is ready for step A again. In the process, heat was taken in at step A. Some of this heat left as exhaust in step C, and the rest showed up as work.

the Kelvin scale), and is the temperature at which the low-temperature source must be if the engine is to be 100 percent efficient. Note that the internal energy of an object is lowest at absolute zero. (However, the energy does not fall to zero, because of quantum mechanical effects.)

All of this consideration of the efficiency of engines leads to the second law of thermodynamics. The second law tells us that there are many processes that are allowed by the first law, but don't actually take place in the real world. There are many different statements of the second law of thermodynamics, but they all turn out to be equivalent. By "equivalent," we mean that we can show that if any one of the statements is true, then all of the others must be true.

One way of stating this law is to say that it is impossible to have a process whose net result is to convert totally into work the heat taken from a source at constant temperature. This is essentially

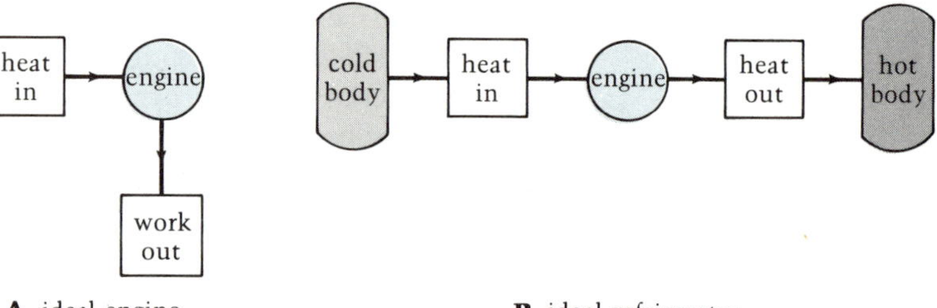

Fig. 11–15 These are processes that are allowed by the first law but *forbidden* by the second law of thermodynamics. A. An ideal engine takes heat in from some source and converts it all into work, with no heat exhausted. B. An ideal refrigerator takes some heat out of a cold body and transfers exactly that amount of heat (so no work is done) to a hot body.

what we found out in our look at engines. Some of the heat we take in must be rejected as exhaust to a lower temperature region. This statement of the second law was given by William Thomson (later to be Lord Kelvin) in England in about 1850. An alternative way of stating the same physical idea, given by Rudolf Clausius in Germany about fifteen years later, is that it is impossible to have a cyclic process whose only result is to transfer heat from a cold body to a hot body.

Kelvin's statement essentially says that *there is no such thing as a perfect engine.* Clausius's statement says that *there is no such thing as a perfect refrigerator.* The two are equivalent, since if we had a perfect engine we could run it backward and get a perfect refrigerator, and if we had a perfect refrigerator we could run it backward to get a perfect engine (Fig. 11–15).

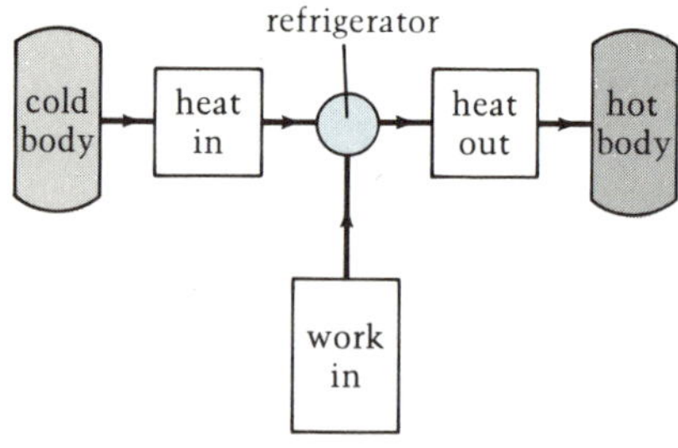

Fig. 11–16 Energy flow in a real refrigerator. We do some work, and remove some heat from a cold object. Then heat is exhausted to a hot object. The heat out must equal the heat in + work in. In a real refrigerator, we usually get this work from a motor, which gets its energy from the electric company.

A real refrigerator requires that we do some work to transfer the heat from the cold region (inside the refrigerator) to the hot region (outside the refrigerator; Fig. 11–16). That is why we keep our refrigerators plugged in; we get the work from the electric com-

Fig. 11–17 Electric generating plants are often placed beside rivers so that the water from the river can be used to absorb excess heat. This infrared aerial image shows warm regions as whiter, and we can see the thermal plume extending downstream.

pany. We can think of a refrigerator as an engine run in reverse. The result is that we have taken in heat at the lower temperature and given up a larger amount of heat at the higher temperature, but to do it, we had to do work along the way. Then the amount of heat given up at the higher temperature is equal to the amount of heat taken in at the lower temperature plus the net amount of work done. Instead of talking about the efficiency of a refrigerator, we use the coefficient of performance, which is the ratio of the heat removed at the cold source divided by the work required to run the refrigerator. Most real refrigerators have coefficients of performance of about 5.

The second law of thermodynamics tells us about the limitations on how much work we can get out of an engine. Knowing this has obvious practical consequences, as discussed above. Another consequence, which you might not think of right away, is thermal pollution. If we have an engine operating between a low-temperature source and a high-temperature source, in each cycle heat will be taken in at the high-temperature source, and, since it cannot all be converted to work, some heat will be given off at the low-temperature source. This heat will have the effect of raising the temperature of the low-temperature source. This temperature rise can have adverse effects.

Let's look at the case of an electric generating plant (Fig. 11–18). The hot source is a boiler in which steam is produced. (The heat for the boiler may come from burning coal or fuel oil, or from nuclear reactions.) The steam is then used to turn a turbine, which is, in turn, attached to a generator. A turbine is a wheel with blades attached in such a way that steam passing over the blades causes the wheel to turn. The turning effect generates the electricity (as we'll see in Chapter 23), and the turning comes about because the steam does work on the turbine. However, we still have some hot steam left, since it cannot give up all its energy into the work of the

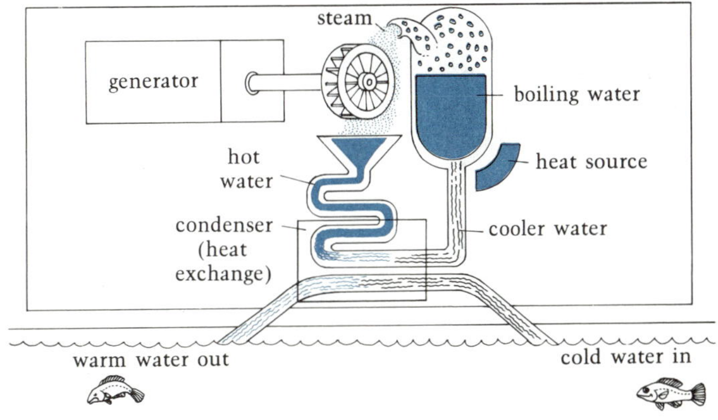

Fig. 11–18 Thermal pollution. The excess heat in the generation of electricity is carried out in water diverted from a river. This makes the river warmer than normal.

Fig. 11–19 These tall towers dump the excess heat from the power plant into the atmosphere through the process of evaporation.

turbine. We can exhaust this steam, or, for more efficient operation, we can use a condenser to cool it down and return the water to the boiler. In this condenser, we usually take cold water from a river and pass it in pipes through the steam, so heat is transferred from the steam to the water. We use much more water than there is steam, so the net result is that the steam all gets liquefied but the water heats up only a little bit. Nevertheless, when we return the water to the river, it is still hotter than when it left the river. Thus the river temperature will be higher than if the power plant were not there.

The temperature changes can cause significant ecological changes. Certain bacteria and algae will grow more quickly. There will also be adverse effects on the fish that naturally inhabit the river. Less oxygen can be dissolved in water that is heated, which reduces the oxygen supply for the aquatic life. For these reasons, alternative ways are being sought to dispose of the extra heat created by power plants. The heat must be dumped somewhere, and the only likely alternative to the waters seems to be the atmosphere. This is now done in many places, using evaporative cooling towers (Fig. 11–19).

As we start to become more aware of our environment, we can see that some of the not-so-subtle consequences of thermodynamics can be very important to us today.

11.7 *Phase Changes*

Fig. 11–20 The three phases of water: steam, water, and ice.

You can tell by comparing an ice cube, a glass of water, and the steam coming out of a kettle that these three substances seem quite different. However, they are all made up of the same constituent—water molecules. The difference is in how the molecules are put together. In ice, the molecules are tightly bound into a crystal structure. In water, the molecules are more loosely bound in an irregular pattern. In steam, the molecules are not bound to each

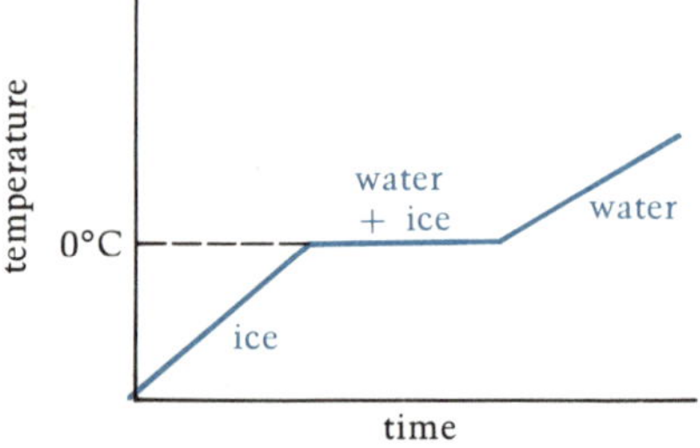

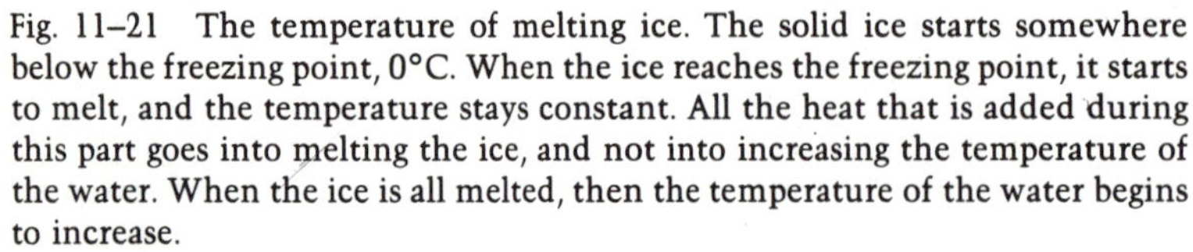
Fig. 11–21 The temperature of melting ice. The solid ice starts somewhere below the freezing point, 0°C. When the ice reaches the freezing point, it starts to melt, and the temperature stays constant. All the heat that is added during this part goes into melting the ice, and not into increasing the temperature of the water. When the ice is all melted, then the temperature of the water begins to increase.

other at all. It seems amazing that three such different substances can be made out of the same basic building block.

These three different forms—solid, liquid, and gas—are called the phases of a substance. They exist under different conditions, and the point at which one converts to the other can depend on a number of conditions. One important factor is the temperature, but such things as pressure also have an effect (see Fig. 11–21).

Fig. 11–22 When the steam strikes the cold plate, it condenses, and we get water droplets on the plate.

Let's look at what happens when you let ice cubes melt in a beaker. Once some of the ice has melted, put a thermometer in and monitor the temperature of the water continuously. As the melting continues, the temperature stays at 0°C. We can speed up the process of melting by placing the beaker in a dish of warm water. As long as we keep the ice and water well mixed, and as long as there is still some ice, there is no change in the temperature. It is still 0°C. Only when all of the ice is melted does the temperature of the water start to rise.

This shows that even though there is heat being added to the ice-water mixture, this heat does not go into increasing the temperature of the water until all the ice is melted. Therefore, we can deduce that this heat must be needed to convert the ice into water. This heat is not associated with any changes in temperature of the water and ice. It is associated only with the phase change. We call this heat required to produce a phase change the latent heat. In melting ice, it provides the energy necessary to break the bonds between neighboring water molecules in the crystal. In going from water to steam, the latent heat goes into overcoming the attraction between the water molecules and also to the work done as the newly formed steam expands until it reaches atmospheric pressure.

If we make a phase change in the reverse direction, we get the latent heat back out. This is why steam at 100°C will give you a much worse burn than water at 100°C. When the steam hits your skin, it begins to condense and gives up its extra heat.

The temperature at which a phase change occurs also depends on the pressure surrounding the material (Fig. 11–23). As we increase the pressure, the temperature at which the water will boil is in-

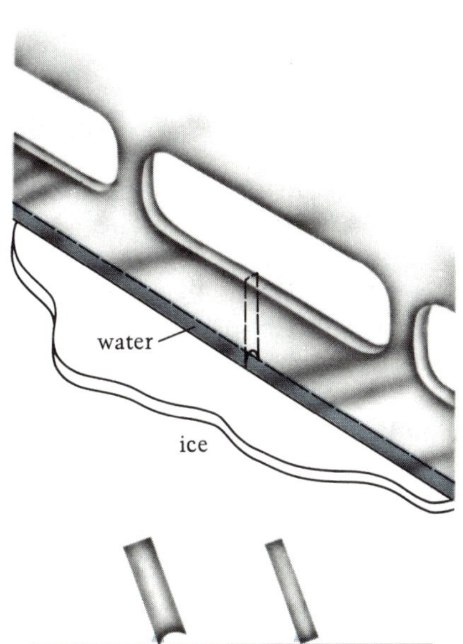

Fig. 11–23 There is an interesting effect when you apply pressure to water or ice near 0°C. Under pressure, the freezing point of water is lowered, so if you exert a lot of pressure on ice, it will melt. (Most solids work the other way; pressure raises the freezing point.) The lowering of freezing point with pressure is what allows you to ice skate. Your ice skate blades exert pressure on the ice, which melts the ice under the blades. Your blade then slides along on a very thin layer of water. The more pressure you can exert, the easier it is to skate. Ice skate blades are constructed so that only a small part of each blade is touching the ice at a given time. The pressure you exert is your weight divided by the surface area over which your weight is supported. If you are supported on a very small area, you will exert a very large pressure on that area.

Fig. 11–24 The pressure that builds up in a car radiator when it overheats can lead to the release of steam. The steam can be seen rising under the hood and also spreading out above the top of the hood.

creased. This explains why the radiator in an automobile engine is kept under pressure. It allows the water to stay at a temperature above 100°C without boiling away. Of course, when the pressure is suddenly removed, the boiling process can be quite rapid, since the water is well above 100°C. Because of this effect, you should never remove the radiator cap from an engine that has just been running. Give it some time to cool down.

We see the reverse effect when you try to use boiling water for cooking at high altitudes. Since the air pressure is lower than at sea level, the water boils at a lower temperature and never gets as hot as the water that you boil at sea level. If you try to boil an egg in Colorado, it will not cook in the same amount of time, since the water is cooler.

There are other ways in which phase changes occur. For example, if you leave a glass of water out, it will evaporate even though you never brought its temperature close to the boiling point. How can this happen? To find the answer, we must go back to our picture of atoms and molecules moving about in a random fashion. Although there is some average energy for this random motion, and this average energy is related to the temperature, many particles are moving with a little more than the average speed and many with a little less than the average speed. The most energetic of the particles may have just enough energy to break free of their bonds and fly away. In this way, little by little, the water can evaporate.

A

B

Fig. 11–25 Evaporation progressively changes water from a liquid state to a gaseous state.

When some water evaporates, the water that is left behind is cooler. The evaporation has removed the highest energy particles in the water, so the average energy of the remaining particles is less than it was with the higher energy particles present. In a glass of water in a room, the glass will absorb some heat from the surroundings to bring the temperature back up to where it was. However, if this absorption of energy is prevented, the water will continue to cool. This phenomenon is called cooling by evaporation.

Cooling by evaporation is very important for our bodies. One way the body gets rid of excess heat is by perspiring. The water droplets sitting on the surface of the skin evaporate, cooling the skin. The process is magnified if we use something that evaporates more rapidly than water, such as alcohol. Try rubbing some alcohol on your skin and see how cool it feels. We can also speed up the process of evaporation by blowing over the surface. Blowing removes some of the recently evaporated molecules from the vicinity of the surface and lets more get away. Notice that your skin feels cooler as you blow across the alcohol. With water, the rate of evaporation depends on how much water vapor is already in the air. On days with high humidity, the evaporation does not take place very quickly, your body cannot cool at a normal rate, and you feel uncomfortably warm.

11.8 *Cryogenics*

"Cryo-" from the Greek, means "icy cold."

Cooling by evaporation also plays an important role in a field of physics research known as cryogenics, the study of physics at very low temperatures. Cryogenic physicists will often cool various substances well below even 1 K. They can then study the properties of materials at these low temperatures or take advantage of properties we already know about. For example, many substances become superconductors at very low temperatures—all resistance to electricity disappears. A current in a superconductive wire will last forever. (Superconductors will be discussed in more detail in Chapter 25.)

To get to cryogenic temperatures, we use the thermodynamic properties we have been discussing. You can cool something down by bringing it into contact with something colder. For example, you can put it into liquid nitrogen, which boils at 77 K. If you want it even colder you can put it into liquid helium, which boils at about 4 K. Suppose you want it colder still. Then you have to make the liquid helium colder. If you let some of the helium evaporate, the remaining liquid will be colder. If you want the liquid to get even colder, you have to let the evaporation continue. You do this by attaching a *vacuum pump* to the vessel holding the helium. As the helium evaporates, the pump pulls out the helium gas, thereby allowing more to evaporate and the cooling to continue. Many scientific and technological advances have come about as a result of our ability to use such procedures to get to colder and colder temperatures.

Fig. 11–26 Four superconducting magnets at Isabelle, the new elementary particle accelerator at the Brookhaven National Laboratory.

How low can you actually cool something? We have already seen that you can't cool anything below absolute zero, but can you ever reach absolute zero? The third law of thermodynamics says that we cannot lower the temperature of a system to absolute zero in a

---

THE LAWS OF THERMODYNAMICS

First law: The change in the internal energy of a system is equal to the heat added to the system plus the work done on the system.

Second law (Lord Kelvin's version): It is impossible to have a process whose net result is to convert totally into work the heat taken from a source at constant temperature.

Second law (Clausius's version): It is impossible to have a cyclic process whose only result is to transfer heat from a cold body to a hot body.

Third law: We cannot lower the temperature of a system to absolute zero in a finite number of steps.

---

finite (that is, not infinite) number of steps. You can get closer and closer to absolute zero, but you can never quite get there.

THE LAWS OF THERMODYNAMICS: An Informal View

First law: You can't win, you can only break even.
Second law: You can break even only at absolute zero.
Third law: You can never get to absolute zero.

Key Words

thermodynamics
temperature
Celsius
Fahrenheit
Kelvin
coefficient of expansion
bimetallic strip
heat
kilocalorie
British thermal unit (BTU)
heat capacity
specific heat
first law of thermodynamics
equation of state
Boltzmann constant
efficiency
cycle
Carnot cycle
absolute zero
second law of thermodynamics
coefficient of performance
thermal pollution
phase of a substance
latent heat
evaporation
cooling by evaporation
cryogenics
third law of thermodynamics

Questions

1. Why are bridges built with small gaps between adjacent sections of the roadway?
2. When heated a given amount, a 1-meter bar of a certain material expands by 2 mm. How much will a 2-meter bar of the same material expand when heated by the same amount?
3. When a fluid expands, the volume increases by some amount, but each dimension may change differently, since the fluid will still conform to the shape of its container. How does this help us make a thermometer in which the length of mercury in the tube changes by a lot, even for a small temperature change? (*Hint:* Think about the use of the bulb full of mercury at the bottom of the thermometer.)
4. A metal sheet has a length of 1 meter and a width of 0.5 meter. It is heated until its length is 1.02 meter. What is its width?

5. Fig. 11–5 shows what happens to a particular bimetallic strip when it is heated. Draw a diagram to show what happens to the same strip when it is cooled.

6. A certain amount of heat is required to heat 1 kg of a material by 10°C. (a) Relative to this amount of heat, how much heat is required to heat 1 kg of the material by 20°C? (b) Relative to the answer to part (a), how much heat is required to raise the temperature of 2 kg of the material by 10°C?

7. Two blocks are made out of the same material. One block is twice the volume of the other. (a) How do the specific heats of the two blocks compare? (b) How do the heat capacities of the two blocks compare?

8. Two blocks are made, each out of a different material. One material has a high heat capacity and the other has a low heat capacity. The two blocks have the same mass. How do the specific heats of the two blocks compare?

9. If you drop a hot horseshoe into the Atlantic Ocean, the temperature of the horseshoe and the temperature of the ocean will each change until the two temperatures are equal. Will the final temperature be closer to the initial temperature of the ocean or the horseshoe? Explain your answer.

10. If no work is done on or by an object, how does the change in internal energy in a process compare with the heat taken in during the process?

11. We start out with two identical containers of gas, each at 20°C. The first container is heated to 99°C and then cooled to 60°C. The second container is heated to 70°C and then cooled to 60°C. How does the internal energy of the gas in the two containers compare when both are at 60°C?

12. A gas is placed in a thermally insulated container. That means that no heat can flow in or out. We compress the gas by pushing down on a piston. (a) What happens to the internal energy of the gas? (b) What happens to the temperature of the gas?

13. A gas is placed in contact with a large object at some temperature. This means that the temperature of the gas will be constant. We push down on the gas with a piston. How much of the work that we have done gets expelled as heat?

14. If we compress a sample of ideal gas, changing its volume to half the initial volume, how does the pressure change if the temperature remains constant?

15. If we compress a sample of an ideal gas, changing its volume to half the initial volume, how does the temperature change if the pressure remains constant?

16. We have three large objects at different temperatures, 250 K, 300 K, and 400 K. Between which two objects would you run a Carnot engine to get the most efficient performance?

17. In a particular Carnot engine, during one complete cycle, 10 joules of heat are taken in at the high-temperature reservoir and 6 joules of

heat are given off at the low-temperature reservoir. How much work is done during one cycle?

18. Are there processes that are allowed by the first law of thermodynamics but not allowed by the second law? If so, name one.
19. What is thermal pollution?
20. Is it possible to have a mixture of ice and water in which the amount of ice stays the same, that is, there is no melting or freezing?
21. The air around us contains water vapor, even though the temperature is well below 100°C. How is this possible?
22. Drops of water on a hot skillet will appear to dance around. What do you think causes this? (*Hint:* Think of what happens on the underside of the water drops, the part in contact with the hot surface.)
23. Bobsled racers will sometimes (illegally) heat the runners on their sleds to make them go faster. Why does this work?
24. Do you think that it would be an advantage to have heavy or light people in a bobsled race? Why?
25. On a hot humid day, why does turning a fan on make you feel comfortable, even if it is the hot air that is being blown around?
26. When weather forecasters give the humidity, they could give the absolute humidity, the total amount of water vapor in the air. Instead they give the relative humidity, which is how much water vapor is in the air relative to the amount of water vapor that could be held in the air at the current temperature. Why is the relative humidity more relevant to our comfort?

CHAPTER 12

How Heat Gets Around

How heat moves from one place to another is a vital concern in our lives. For example, energy is manufactured in the core of the sun. It must then get from the core to the sun's surface, which takes millions of years. This energy must then get from the sun to the earth, a process that takes just eight minutes. In the summer we try to keep this heat out of our houses. In the winter we try to keep it in.

There are basically three different ways in which heat gets from one place to another:

Conduction—energy transport via a chain reaction. A particle (for example, an atom or an electron) strikes the one next to it, giving up some of its energy. The second particle then strikes a third one, and so on. The individual particles stay pretty close to their starting places. Each one, though, passes on a little bit of energy.

Convection—the most direct way, since the energy is transported by the bulk motion of material. Hot material moves from one place to another, bringing its heat with it.

Radiation—energy is carried in the form of light, radio waves, infrared, ultraviolet, or any other type of electromagnetic radiation. (Electromagnetic radiation will be discussed in more detail in Section 22.5.)

Now let's look at how each of these ways of transporting heat operates.

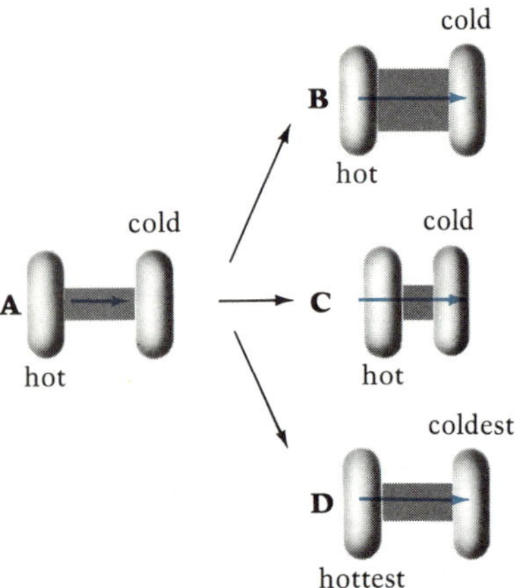

Fig. 12–1 Three ways to improve the rate of heat conduction. A. We start with a hot object and a cold object, and connect them with some substance through which the heat will flow (shaded region). The heat flow is indicated by the arrow. We now wish to increase the rate of heat flow. B. We can make the area through which the heat can flow larger. C. We can bring the hot and cold objects closer together, so the heat does not have to flow as far. D. We can make the hot object hotter and the cold object colder, increasing the temperature difference between them. (We could also connect the two objects with a substance of higher thermal conductivity; this is not shown here.)

Fig. 12–2 A thermogram of a house. The brighter regions are the locations where the most heat is escaping, such as around the windows and doors.

12.1 *Conduction*

Conduction is a flow of heat through a body without any parts of the body moving along with the heat. The heat is transmitted in collisions between the particles that make up the object. In a collision, each particle passes along some extra energy to the next particle (see Fig. 12–1).

We are most interested in the rate of heat flow, which is the quantity that tells you how fast energy is coming or going. For example, in deciding how much home heating fuel you will use, you want to know how fast the heat is leaving your house. Your furnace must then generate heat at the same rate if you are to maintain the temperature. Once you have the house at the temperature you want, you need the furnace only to replace the heat that is lost. If your house did not lose any heat (through windows, cracks in walls, under doors, and so on) then you would not have to use your furnace to maintain any given temperature.

In order for conduction to occur, there must be a temperature difference between the two ends of the object through which the heat will flow. The heat will then flow from the hotter end to the coldest end. The greater the temperature difference, the faster the heat will flow. The flow rate also depends on the area through which the heat has to flow. Just as more water can flow through a wider pipe, more heat will flow through a wider object. However, the heat flow rate will go down as we make the distance it must flow longer. These ideas can be summarized as follows:

$$\textit{rate of heat flow} = \frac{\textit{constant} \times \textit{area} \times \textit{temperature difference}}{\textit{length of flow}}.$$

Let's look at the various quantities on which the heat flow depends and see how they help keep your house from losing heat. First, the larger the area for the heat to flow through, the greater the flow. Often there will be a larger heat flow through your window glass than through the walls. If the windows are larger, then the heat loss will be larger. Next, the heat flow is proportional to the temperature difference. (This is one reason why it costs more to keep your house at a higher temperature.) With the higher temperature inside, the difference in temperature between the inside and outside is greater. As a result, the heat-flow rate from the inside to the outside is greater. Moreover, the heat-flow rate decreases as the flow has to go farther. Thus, if you are worried about heat flow between the inner and outer walls of a new house, building these walls as far apart as is practical will help reduce the heat flow.

We get many examples of heat flow in the kitchen. For example, each color of cooked meat represents meat that has reached a certain temperature. When you slice into the center of a cooked beef roast, you see the progression of color, from brown (well done) at the edge to pink (rare) at the center, which signifies that the temperature at the center never got as high as that at the edge. (Most problems with cooking times in standard ovens are purely questions of heat flow.)

Notice that the rate of heat flow depends on a constant. This constant is different for different materials and is called the thermal conductivity of the material. A high thermal conductivity means that the material conducts heat very well, while a low thermal conductivity means that it conducts poorly. A substance that conducts poorly is called an insulator. To reduce the heat flow between two points, you fill the space between them with a good insulator.

Insulation that you buy for your home now is labeled with an "R value"; higher R values signify better insulating qualities. If you add two layers of insulation, one on top of the other, you simply add their R values to get the R value of the total.

You may notice that when you step barefoot on a cold floor, the sensation is not always the same. A marble floor feels quite cold, but a wood floor may not feel that cold. The reason is a difference in thermal conductivities. Wood has a relatively low conductivity, and therefore does not conduct very much heat away from your foot. Marble has a higher conductivity and conducts heat away from your foot at a higher rate, which produces the colder sensation.

When we wish to chill a bottle of wine, it is best not to simply stick it into a bucket of ice. If we do, since the ice touches the bottle at only a few points, we are relying on the air to conduct the heat from the bottle to the ice. Instead, it is better to put the bottle in a bucket of ice water. The water is also at 0°C, and conducts heat away from the bottle very efficiently.

Fig. 12–3 Although ice alone packed around a bottle of wine or champagne will chill it, filling the bucket with ice water will allow better contact between the cool material and the bottle. This increases conduction so the bottle cools faster.

Air's thermal conductivity is lower than most of the metals by a factor of about 10^4. In fact, air is a fairly poor heat conductor. You can cook a chicken much faster in water that is at only 100°C than you can cook it in an oven at a somewhat higher temperature.

We can take advantage of the low thermal conductivity of air and get very good insulation by using layers of material with gaps for layers of air. This is why, when going out in the cold, it is better to wear many light layers of clothing than one heavy coat. Air helps keep you warm. Your skin gives off a little heat, warming the air that is close to it. This warm air then separates your skin from the cold air, reducing the flow of heat from the skin to the colder air. As a result, you can take your gloves off on a freezing winter day. The warming effect works best if the air is very dry because wet air is a better heat conductor. It also works better if there is no wind blowing. Wind blows the precious layer of warm air away from your skin, and replaces it with a layer of cold air. Your skin would then lose some heat to this layer, which would be blown away too.

The wind-chill factor combines the effects of temperature and wind, considered for the rate at which an average nude body in the shade would cool. Numerically, it is expressed as the loss of body heat in kilogram-kilocalories per hour per square meter of skin surface (kg-kcal/h/m²). For a wind speed of 40 km/h, for example, an air temperature of 0°C (32°F) would feel like −16°C (3°F).

So you feel colder on a windy day. In determining the rate at which your body will lose heat, you must take into account both the temperature and the wind. The number that does this is called the wind-chill factor (see Fig. 12–4).

A vacuum is an even better insulator than air, because a vacuum contains nothing to transmit heat. Although we cannot attain a perfect vacuum, we can reduce the conductivity of a given sample of air considerably by pumping most of the air out of its container. A Thermos bottle works this way. The bottle has an inner layer and an outer layer with a gap in between. Most of the air is removed from the gap, so the gap becomes a very good insulator. Very little heat flows across the gap. (In addition, a Thermos bottle uses an inner surface designed to reflect back any radiation, so there is no loss by radiation. See Fig. 12–5.)

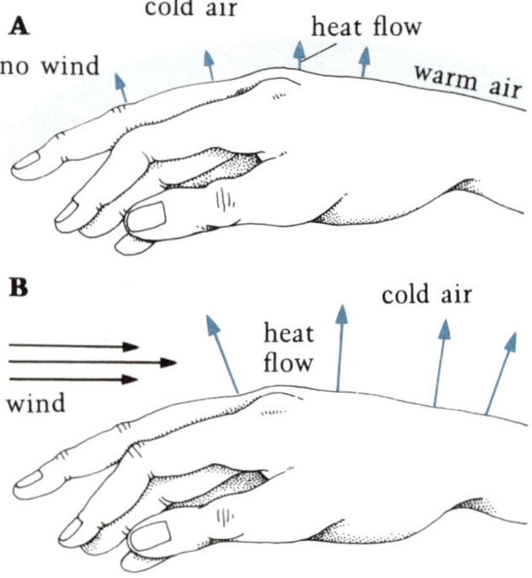

Fig. 12–4 When there is no wind, your skin will heat a little layer of air, and that warm air will stay next to your skin. If your skin is to lose more heat, that heat must get from the skin, through the warm air, to the colder air. The result is a low heat-flow rate. B. If a breeze removes the warm air layer, then the cold air is right next to the skin. Since heat leaving the skin does not have as far to travel, there is a large heat-flow rate.

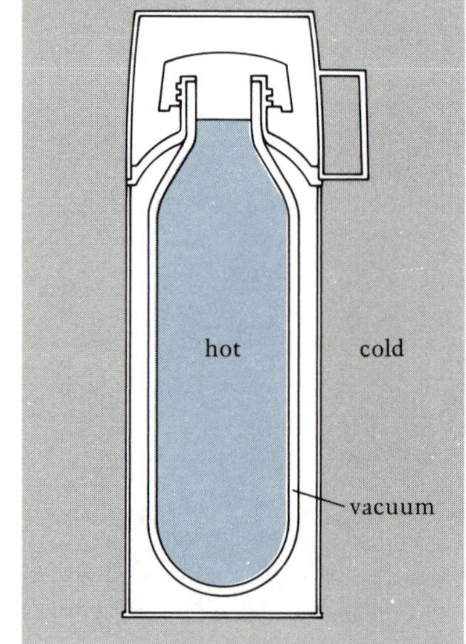

Fig. 12–5 The Thermos bottle uses a vacuum to provide insulation.

Only 0.06 joules/second flow through 1 square meter of surface on the average. A joule/second is a watt, so you would need the heat flowing through 1600 square meters of surface to light a common 100-watt light bulb.

Heat conduction is also responsible for many geological phenomena on or near the surface of the earth. The earth actually consists of a series of shells. The innermost region is called the core, and is composed primarily of iron and nickel. The core is a very dense liquid, although the innermost part may be solid. The core has been very hot ever since the earth was formed because of heat released in the process of formation. Most of the material outside the core composes the mantle, and on top of the mantle is a thin layer, the crust. Additional heat is now contributed by the presence

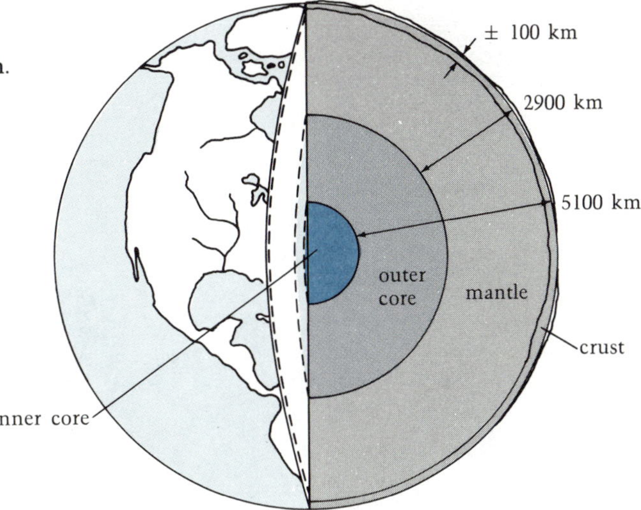

Fig. 12–6 The inside of the earth.

A

B

Fig. 12–7 Geysers (A) and volcanoes (B) are caused by heat generated inside the earth. In B, we see Mount St. Helens blowing its top.

of radioactive materials. These radioactive elements give off energetic particles that collide with the other rocks in the crust and mantle. In these collisions, the particles give up energy that heats the rocks. (See Fig. 12–6.)

Most of the earth's radioactive material is in its outer layers. This material acts as a heat source not too far below the ground. Rock is not a very good conductor of heat, but some conduction does take place, and there is a general heat flow up to the surface of the earth. This heat flow is very small. But there are areas on the earth, called *geologically active areas*, in which the heat flow (mainly due to

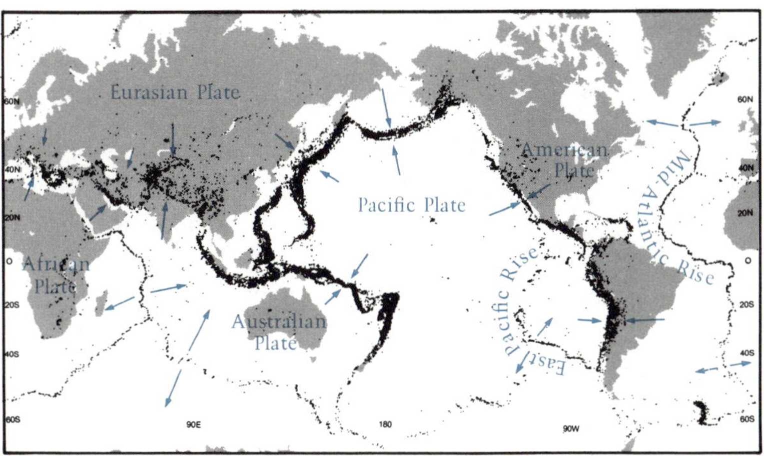

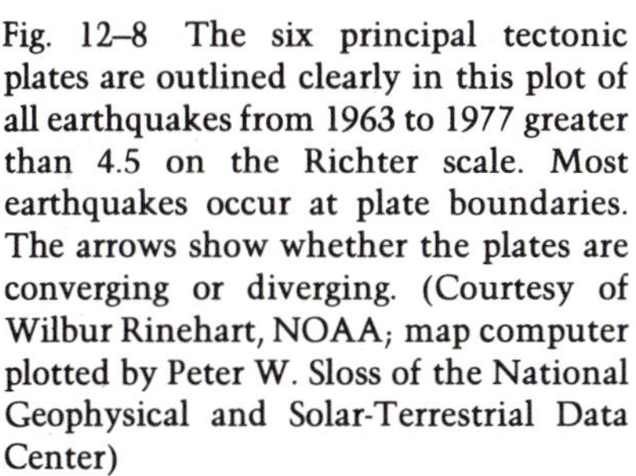

Fig. 12–8 The six principal tectonic plates are outlined clearly in this plot of all earthquakes from 1963 to 1977 greater than 4.5 on the Richter scale. Most earthquakes occur at plate boundaries. The arrows show whether the plates are converging or diverging. (Courtesy of Wilbur Rinehart, NOAA; map computer plotted by Peter W. Sloss of the National Geophysical and Solar-Terrestrial Data Center)

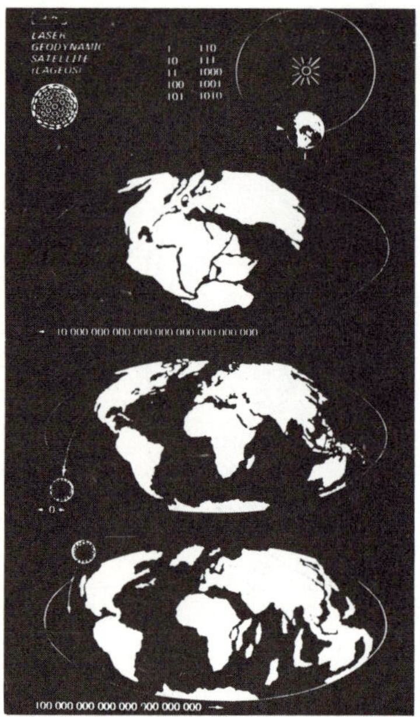

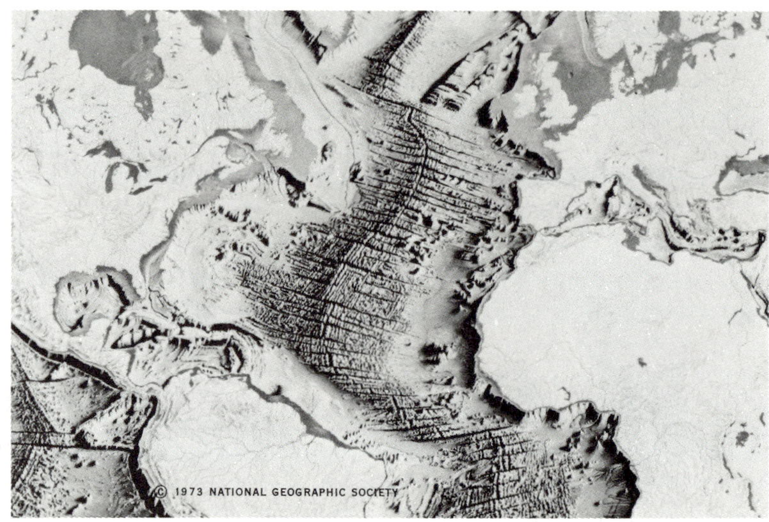

Fig. 12–10 The Atlantic Ocean floor marks the division between two continental plates. Upwelling molten matter in the mid-Atlantic ridge is being deposited as new sea floor as Europe and America drift apart.

Fig. 12–9 Continental drift, 200 million years ago (top), today (middle), and 10 million years in the future (bottom). We see a message that has been sent aloft in NASA's Lageos satellite, which carries reflectors for laser beams from earth. The satellite allows extremely accurate measurements to be made of movements of the earth's surface. The message (at top), prepared by Carl Sagan of Cornell, also includes the binary numbers and a drawing of the earth around the sun (labeled "1" for "1 year"). We see the satellite being launched from California in the middle panel and the satellite returning to earth in the lowest panel.

Fig. 12–11 Great mountain ranges, like the Himalayas, result from the shifting of the earth's plates. Mt. Everest is the tallest peak (with snow blowing off its top).

convection) is much greater than the average. Some scientists feel that these areas might eventually be tapped as sources of geothermal energy.

Another important consequence of the heat flow through the earth is that the layer immediately below the crust is heated to the point that the rock is no longer very hard. It does not quite melt, but becomes more like a plastic. The top layers of the earth actually float on top of this plastic layer. The rigid outer layer is segmented into plates (wide, shallow regions) only 50 km thick and thousands of kilometers in extent. The continents sit on top of these plates. Since the plates are floating, they actually move around the surface of the plasticized layer, carrying the continents with them. The phenomenon is called continental drift. In recent years there has been much evidence collected in favor of this theory.

We can tell where the plate boundaries are since they are very active areas in a geological sense. They are marked by earthquakes. When two plates move apart, the space between them is taken up by hot material rising from the interior. One famous example of this is the *mid-Atlantic ridge.* When two plates come together, one is pushed up and the other may be pushed down. According to the theory of continental drift, this type of action is responsible for the creation of the great mountain ranges.

12.2 *Convection*

Convection is the transfer of energy from one place to another by the motion of a mass of material between the two points. When you exhale on a cold day, the warm air from your lungs moves out of your mouth, transferring some heat from your body. You can even see the water vapor (gas) in the warm air condensing into small water droplets when the warm air exhaled mixes with the cold air.

You can see convection whenever you heat a pot of water on the stove. Look into the pot as the water gets hotter and as it approaches the boiling point. (Be careful not to get too close to the boiling water or the resulting steam.) The water is hottest at the bottom of the pot, where it is closest to the source of heat. As the water at the bottom of the pot gets hotter, it expands slightly and its density therefore decreases. It becomes buoyant and starts to rise gently. This rising water carries its heat to the top. You can see the effects of this churning motion in the water by seeing how the surface is disturbed. Once the water starts to boil, the motions will become even more rapid. When the water turns into water vapor, it forms bubbles. These bubbles form at the bottom of the pot, where the boiling takes place first. The bubbles are very buoyant and rise very quickly. This accounts for the frenetic motion that we see in a pot of boiling water (see Fig. 12–12).

A similar thing happens in the earth's atmosphere. The air in the lower atmosphere is not heated directly by the sun's rays (as we shall see in the next section). Instead, it is heated by the ground. Thus, the air is warmest closest to the ground—closest to its heat source. As the air is heated, it expands and its density decreases. This parcel of air becomes more buoyant than the surrounding air

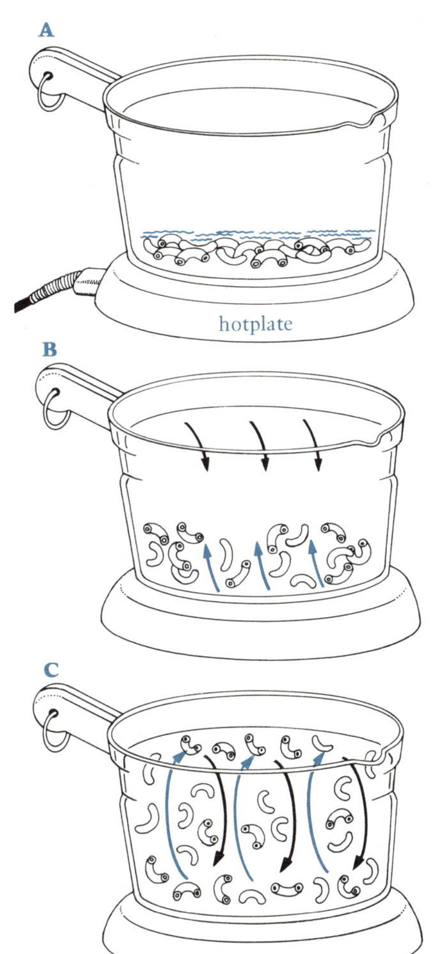

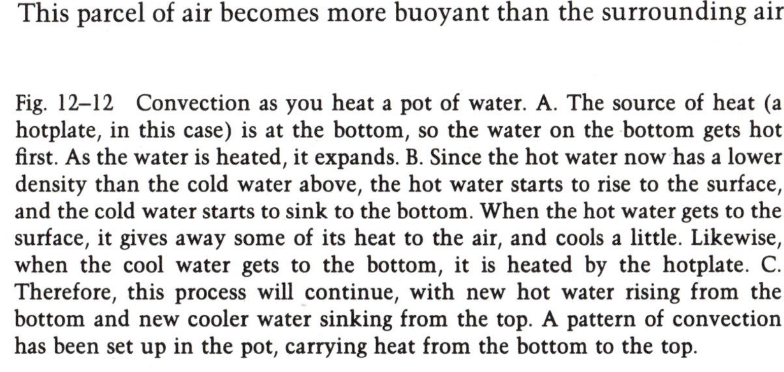

Fig. 12–12 Convection as you heat a pot of water. A. The source of heat (a hotplate, in this case) is at the bottom, so the water on the bottom gets hot first. As the water is heated, it expands. B. Since the hot water now has a lower density than the cold water above, the hot water starts to rise to the surface, and the cold water starts to sink to the bottom. When the hot water gets to the surface, it gives away some of its heat to the air, and cools a little. Likewise, when the cool water gets to the bottom, it is heated by the hotplate. C. Therefore, this process will continue, with new hot water rising from the bottom and new cooler water sinking from the top. A pattern of convection has been set up in the pot, carrying heat from the bottom to the top.

and begins to rise. (It is replaced at the ground level by cooler, denser air that is no longer buoyant and is falling down.) As the air expands, it cools. (In expanding, it does work and therefore loses energy.) Eventually it reaches a height at which it is at the same density as the surrounding air, and the upward motion stops. If the rising air contains a lot of water vapor, the water will condense as the air cools, forming clouds (see Fig. 12–13).

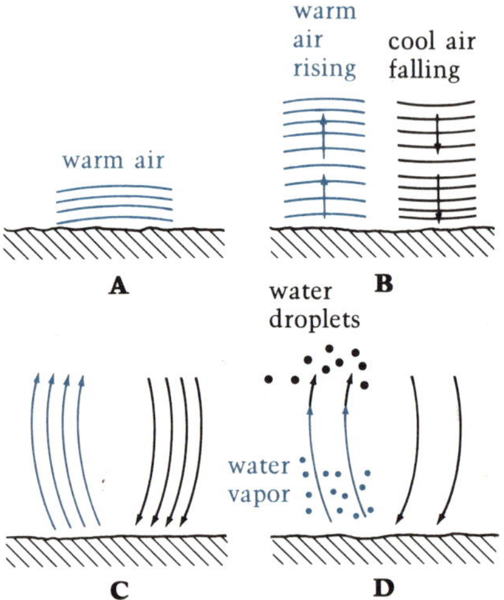

Fig. 12–13 The development of convection currents in the earth's atmosphere. A. The warm ground heats the air above it. B. The warm air expands, and since its density is lower than the surrounding air, it rises. At the same time, the denser cooler air falls down from above. C. Once the flow is established, it can go quite rapidly, with hot air rising and cool air falling to take its place. D. As the hot air rises, it cools. If there is a little water vapor in the air, when the air cools the vapor will form water droplets, which make up the clouds.

Fig. 12–14 This condor in Chile rides the rising warm air currents, or thermals.

Often the convection currents in the atmosphere can get quite strong. For example, when the conditions are right, a bird or the pilot of a glider can find rising warm air columns, called thermals, and ride these thermals to a higher altitude. Sometimes the convection can be more violent, as in the case of a thunderstorm. Within the thundercloud there are vicious convection currents that carry the air up to an altitude of as much as 20 km! Airplanes caught in these up or down currents will suddenly rise or fall several hundred meters. Keep your seatbelts fastened!

It is convection that governs our weather on a variety of levels. We can look at land and sea breezes, for example. On the seashore,

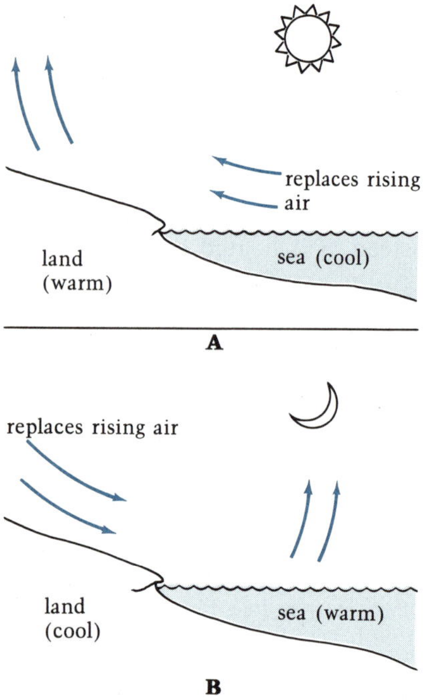

Fig. 12–15 Land and sea breezes. A. During the day, the land heats faster than the water, so the land is warmer than the water. The air just above the land is warmed, and begins to expand. As it expands, its density decreases and it rises up in the air. This leaves behind a region where the pressure is lower than the pressure over the water, so the cooler air from over the water blows in over the land. This is a *sea breeze*. B. At night the situation is reversed. The land cools quickly, so the sea is now warmer than the land. The warmer air rises above the water, and the cooler air from the land blows out to sea. This is called a *land breeze*.

during the day, the land heats up faster than the water. Thus the hot air rises faster over the land than over the water. The hot air rising over the land leaves behind a lower pressure region, and some of the air over the water rushes in, pushed by the higher pressure air over the water. We call this a sea breeze. At night, the water cools more slowly than the land, so the water is warmer and the process is reversed, resulting in a land breeze, one that blows from land to sea. The fact that water retains heat better than the land is also responsible for the more moderate climates that are present near the ocean. The winters are warmer and the summers are cooler than at nearby inland locations. (See Fig. 12–15.)

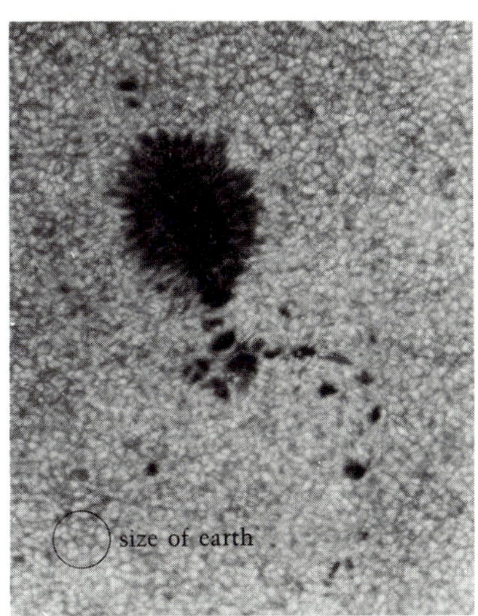

Fig. 12–16 Around the sunspot, the rest of the sun is covered with small regions of convection known as granules. Individual granules are only about 700 km across. (Courtesy of W. Livingston, Kitt Peak National Observatory)

The driving force for most of our weather is the temperature differences between various locations. These temperature differences eventually lead to pressure differences and the winds blow from the high-pressure to the low-pressure regions.

Convection plays an important role in the appearance of the sun. When we look at a photograph of the sun, we can see that its surface does not appear smooth, but has a mottled appearance. It looks as if the surface is divided into granules, like the grains of sand on a beach. This phenomenon is called *granulation*, and results from convection in the sun. The bright granules are places where hot gas (that is, hotter than average at the surface) is rising up at the top of a convection stream. The dark spaces between granules are the cooler gas that is starting to fall back down. It also appears that the sun has larger zones of convection 20,000 km across, *supergranules*, containing many granules. The existence of supergranulation suggests that there are even larger convection zones farther below the surface than those that cause the normal granulation.

12.3 Radiation

Light is but one example of radiation, or, more properly, *electromagnetic radiation*. (We shall leave the discussion of its relationship to electricity and magnetism until Chapter 22.) We call the range of various types of electromagnetic radiation the electromagnetic spectrum, which we divide into the radio, infrared, visible, ultraviolet, x-ray, and gamma-ray sections.

All objects can emit and absorb radiation, and radiation carries energy. When an object emits radiation, it is giving off energy, and when it absorbs radiation, it is taking in energy. Sometimes the emission or absorption will take place only in certain parts of the spectrum and sometimes emission and absorption are distributed all across the spectrum. When an object gives off some radiation,

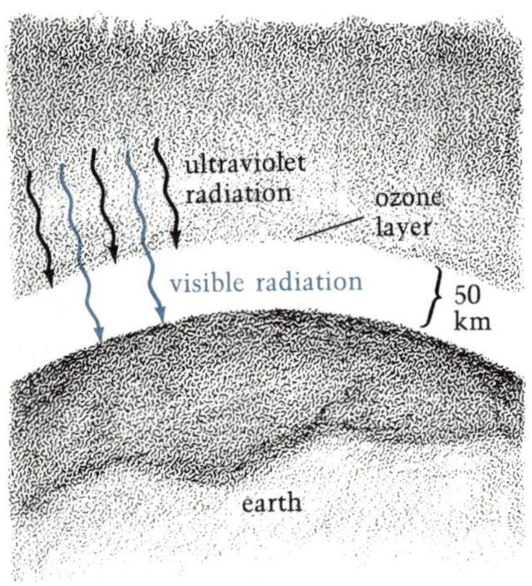

Fig. 12–17 The ozone layer blocks out the harmful ultraviolet radiation from the sun, but lets the visible sunlight through.

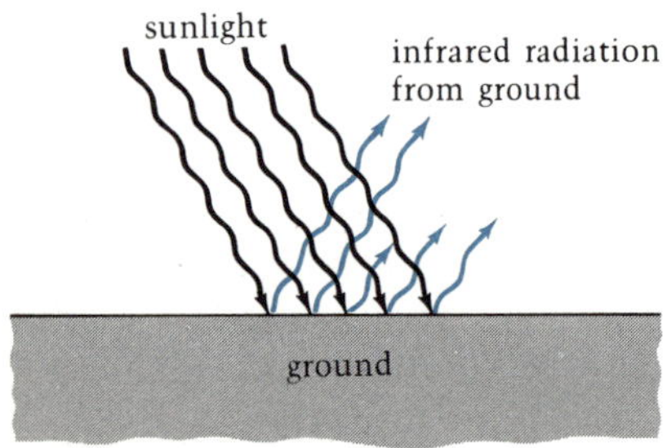

Fig. 12–18 Radiative heating of the lower atmosphere. Light from the sun (black arrows) comes through the atmosphere and heats the ground. The hot ground then gives off infrared radiation (blue arrows). However, the earth's atmosphere absorbs a lot of this radiation, and most of the radiation is absorbed close to the ground. (In this figure, the end of a ray represents the place where it is absorbed.) Since most of the radiation is absorbed near the ground, the air closer to the ground is heated more than the air higher up. Thus it will be warmer near the ground than higher up. (In real situations, when the air near the ground gets too much warmer than the air higher up, some heat will be brought higher up by convection.)

then the energy stored in the object must decrease by the amount of energy given off in the radiation. For example, the object might give off some radiation and get cooler in the process. The opposite occurs with absorption. When an object absorbs some radiation, the energy of the object increases by the amount of energy absorbed from the radiation. This increase in energy may show up as an increase in temperature. Actually, in real situations, emission and absorption are often going on simultaneously.

12.3a ATMOSPHERES

Energy gets from the sun to the earth by radiation. Most of this radiation is in the form of visible light, although there is some infrared and ultraviolet radiation as well. What happens to this radiation as it reaches the earth? The infrared and ultraviolet radiation are absorbed by the atmosphere, and never reach the surface of the earth. Especially in the case of the ultraviolet this is a good thing, since the ultraviolet radiation is potentially harmful, causing, among other things, skin cancer. The ultraviolet is absorbed by the *ozone* in the atmosphere. (Ozone is a form of oxygen in which three oxygen atoms are bound together into one molecule—O_3—rather than the more common form in which two atoms are bound together—O_2.) Most of the ozone is found in a thin layer, called the ozone layer, about 50 km above the ground. Thus, the presence of the ozone layer is extremely important to our health. (Many scientists are concerned about the possibility that the exhaust gases of supersonic transports, the propellant gases in some aerosol cans, or the refrigerants in air conditioners and refrigerators, when released, will destroy large amounts of ozone.) (See Fig. 12–17.)

Since some of the energy in the radiation from the sun is absorbed in the ozone layer, that layer is warmer than the layers below. The layers below receive only the visible radiation, but they cannot absorb it so it passes to the ground. The ground is a very good absorber of this radiation, and almost all of it is absorbed. As a result, the ground is heated. As the ground heats up, it also gives off radiation. However, the earth radiates primarily in the infrared rather than in visible light, and this infrared radiation is mostly absorbed by the water vapor in the atmosphere. This heats the atmosphere. The heat is then carried upward, either by convection or by more infrared radiation from the ground. Thus, the ground serves as the immediate heat source for the lower atmosphere. As you go higher, the air gets cooler, even though you are getting closer to the sun. (The temperature outside a jet at 7 km altitude is far below 0°C.)

In the atmosphere, infrared radiation is absorbed primarily by carbon dioxide (CO_2) and water vapor. When the amount of water

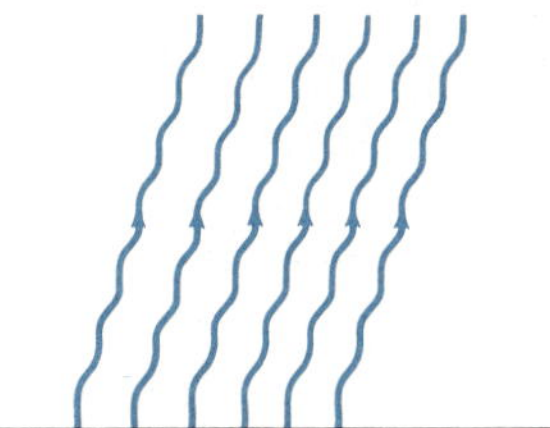

Fig. 12–19 Radiative cooling at night. Now there is no source of energy for the ground, but it still has some excess heat left over from the daytime. If the atmosphere could absorb the infrared radiation given off by the ground, then this radiation could heat the lower atmosphere, as in Fig. 12–18. However, when the conditions are very dry, a lot of the infrared radiation gets out. This is because much of the absorption of infrared radiation is by water vapor. When there is no water vapor, the radiation does not heat the atmosphere, and it gets very cold. This is why the nights can be very cold on the desert, even though the days are very hot.

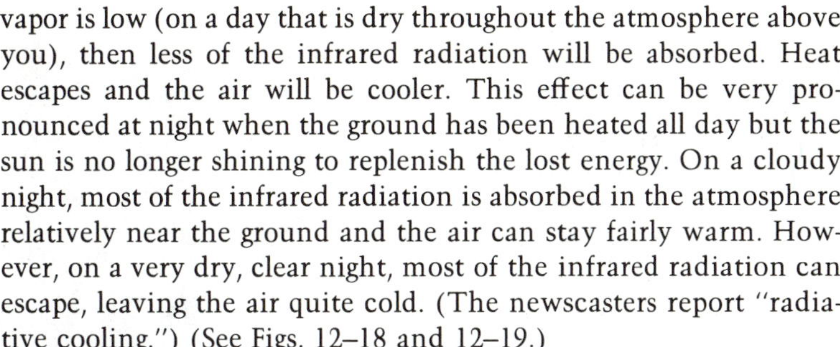

vapor is low (on a day that is dry throughout the atmosphere above you), then less of the infrared radiation will be absorbed. Heat escapes and the air will be cooler. This effect can be very pronounced at night when the ground has been heated all day but the sun is no longer shining to replenish the lost energy. On a cloudy night, most of the infrared radiation is absorbed in the atmosphere relatively near the ground and the air can stay fairly warm. However, on a very dry, clear night, most of the infrared radiation can escape, leaving the air quite cold. (The newscasters report "radiative cooling.") (See Figs. 12–18 and 12–19.)

This general sequence of events, in which visible radiation is absorbed by the ground and the energy is then reradiated as infrared radiation that is then blocked from escape, is called the greenhouse effect. It got its name because it used to be thought that greenhouses were heated in this way (Fig. 12–20). The glass in a greenhouse allows the visible light that makes up most of the incoming solar radiation to pass through. Glass blocks infrared, though, and does not pass the infrared radiation emitted by the heated ground. Scientists now realize, however, that this process is not the most important one for heating the air inside actual greenhouses. In real greenhouses, the fact that the glass stops the heated air from mixing with outside air is more important. So real greenhouses are not heated by the "greenhouse effect."

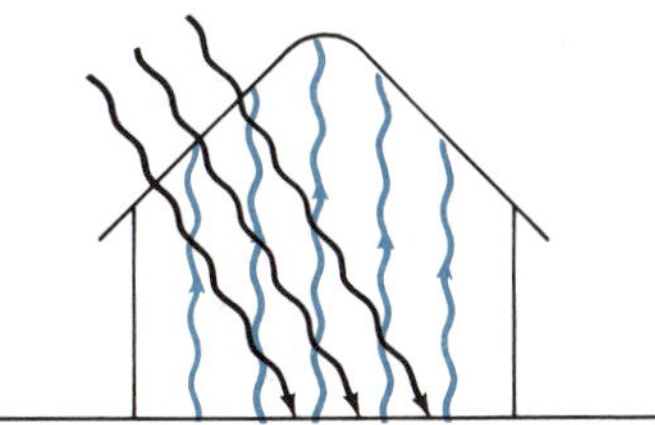

Fig. 12–20 The greenhouse effect. Sunlight can pass through a material and heat the ground (black arrows). The infrared radiation given off by the ground (blue arrows), however, cannot pass through the material. The energy is thus trapped inside the enclosure and the gas inside heats up. This process is important in the atmospheres of the Earth and other plants and moons, but not in actual greenhouses.

Fig. 12–21 The clouds of Venus observed from the Pioneer Venus spacecraft. Below the clouds the temperature is about 750 K.

We have an even more extreme example of the greenhouse effect on Venus (Fig. 12–21), where carbon dioxide is primarily responsible for absorbing the infrared radiation. The temperature on the surface of Venus is about 750 K (hot enough to melt lead), compared with the earth's 300-K surface. If there were no atmosphere, we calculate that Venus would be less than 375 K. Our knowledge of the surface temperature on Venus comes from studies of the radio waves emitted by the surface, as well as more direct observations from spacecraft, especially the Soviet spacecraft Venera 9 and

Fig. 12–22 An artist's conception of the U.S. spacecraft that arrived at Venus in 1978. One of the spacecraft separated into individual probes that descended through the clouds and sent back evidence that endorsed the greenhouse theory for Venus's high temperature.

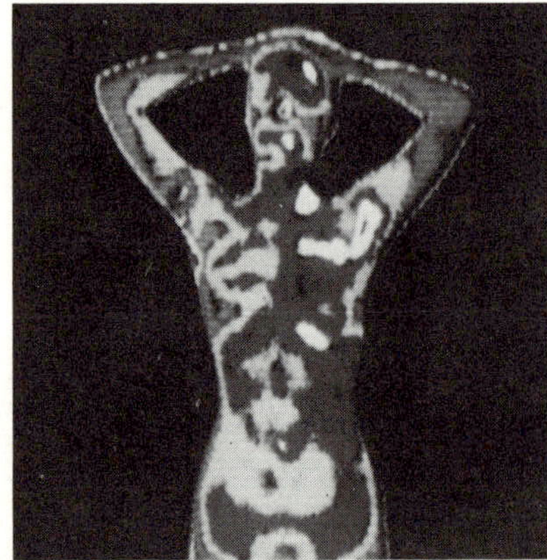

Fig. 12–23 Your body gives off infrared radiation. The hotter areas give off more radiation. Infrared detectors can make a map of your body, called a thermogram, to see where the hot and cold spots are. These are very helpful in diagnosing certain types of illness, especially certain cancers.

Fig. 12–24 A house heated with solar panels. Fluid heated within the panels circulates throughout the house to distribute the energy.

10, which landed on Venus in 1975, and the Soviet and American Venus probes of 1978 (Fig. 12–22).

It has long been a mystery why two such similar planets, Venus and the earth, should have evolved so differently. One explanation is that since Venus is closer to the sun, it started a little hotter. This extra heat released carbon dioxide from the rocks into the atmosphere, which started the greenhouse effect working. As the planet got warmer, more carbon dioxide was released, enhancing the greenhouse effect. This *runaway greenhouse effect* might explain why such small differences in their original conditions have resulted in two such different planets. The idea that small changes in an atmosphere can run away with disastrous effects has many scientists trying to keep a close watch on our own atmosphere.

12.3b HEATING AND COOKING

If we are careful, we can modify the effects of radiative heating. For example, if you don't want your house to get too hot in the summer, you can make sure that the visible radiation does not get absorbed. One way to do this is to put on a roof that reflects the incoming radiation. It would be even better if you had a roof that not only reflected the incoming visible radiation but also retained its ability to radiate away heat in the infrared. Certain white paints, especially those containing titanium oxides, have this property. On the other hand, in the winter, you want to increase the amount of radiation that your house absorbs. In the past few years many architects have placed an emphasis on houses that make efficient use of the solar energy that lands on them. Such houses require less electricity, oil, or gas for heating and cooling.

One of the best examples of radiative heating is provided by a microwave oven. In a normal oven, air is heated by conduction and heat is transferred to the food by conduction. In a microwave oven, microwaves (radio waves of a certain frequency range) carry the

Fluids, Heat, and Electricity

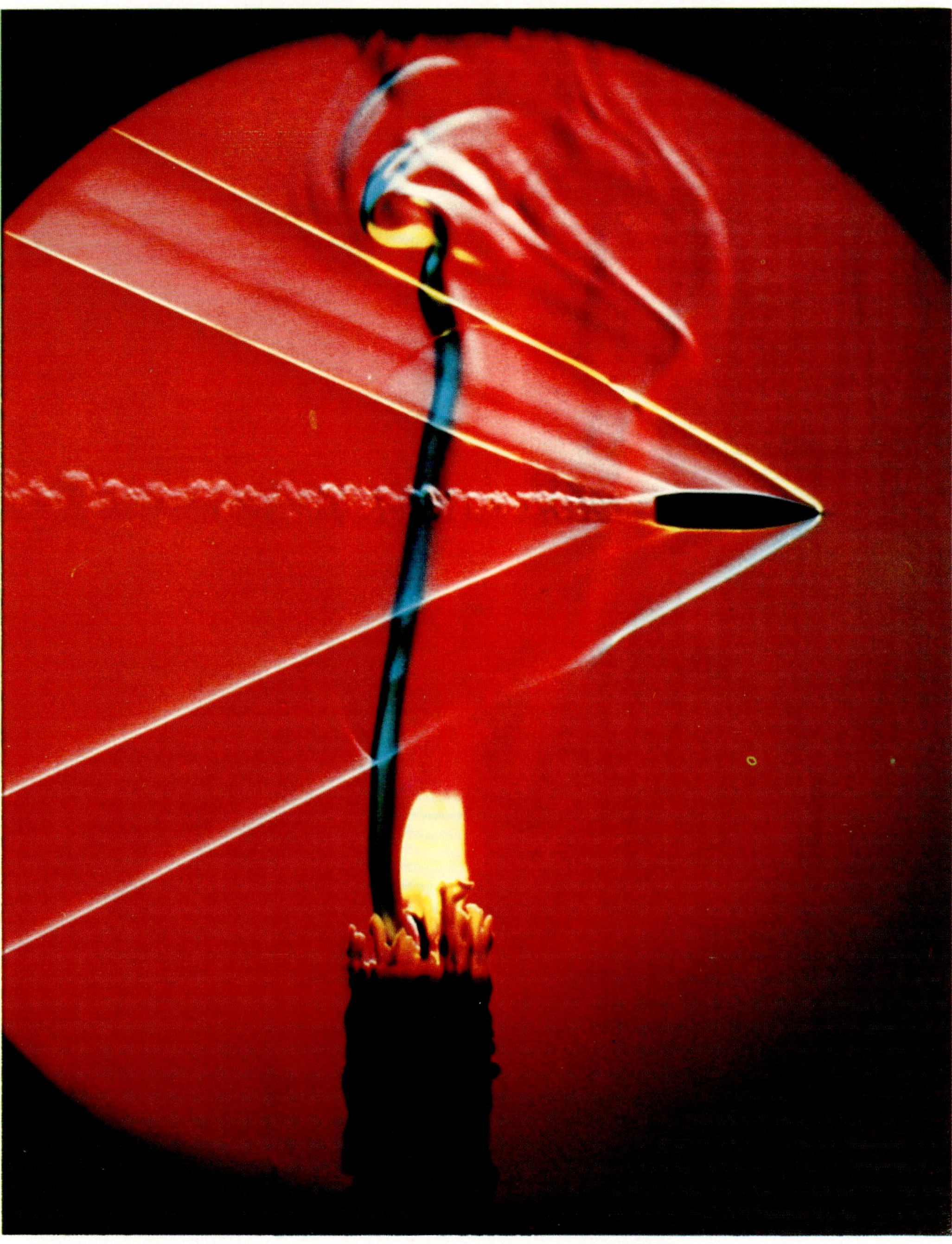

Plate 32 We see the shadow of the hot air rising from a candle. The shock wave caused when the bullet passes through is also visible. The exposure was one-third microsecond in duration. (Strobe photo by Kim Vandiver and Harold E. Edgerton, MIT)

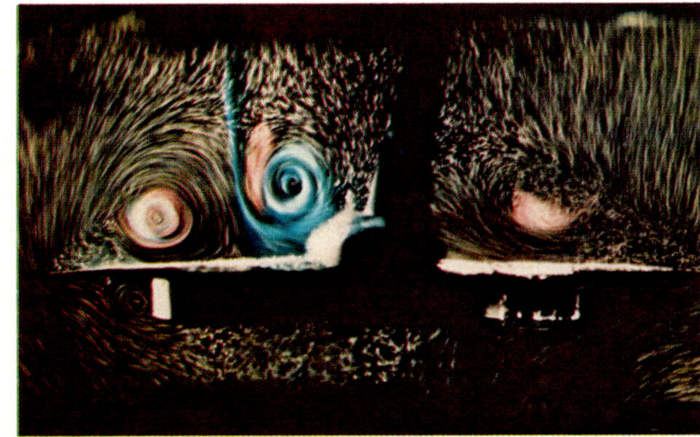

In Plate 33 *(top left)* we see regular eddies formed over the wings of a mockup of the supersonic Concorde airplane in a wind tunnel. The flow over the wings is smooth.

Plate 34 *(top right)* shows how the smooth flow can be disturbed by breaking up the regular flow of air as it passes over the airplane's surface. (Photos by ONERA, France)

Plate 35 *(at right)* shows the flow of air around a model of an airplane wing oriented at a slight angle. We see how a spark is displaced by the flow, and can follow changes in velocity. Note the air flow is separated from the wing at the rear; no spark crosses this region. (Courtesy of Y. Nakayama, Tokai University)

In Plate 36 *(below)* we see the flow of a liquid from left to right over a step. The interference fringes broaden as the velocity increases. (Photo by M. Horsmann, Ruhr University)

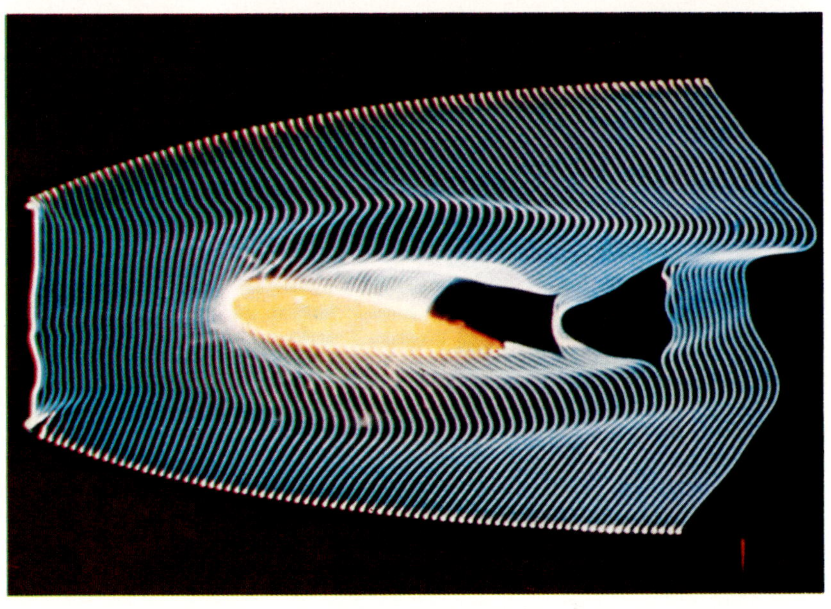

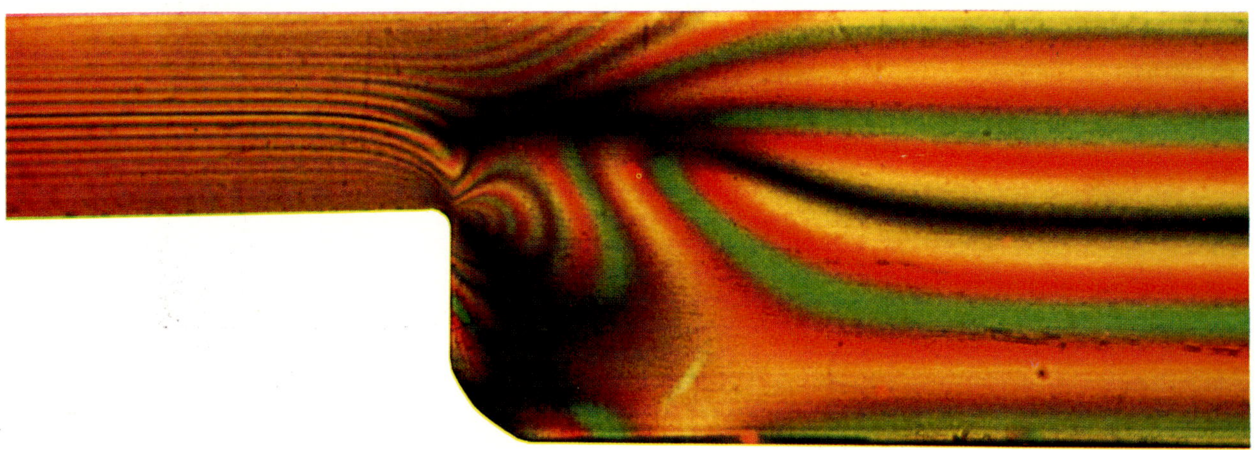

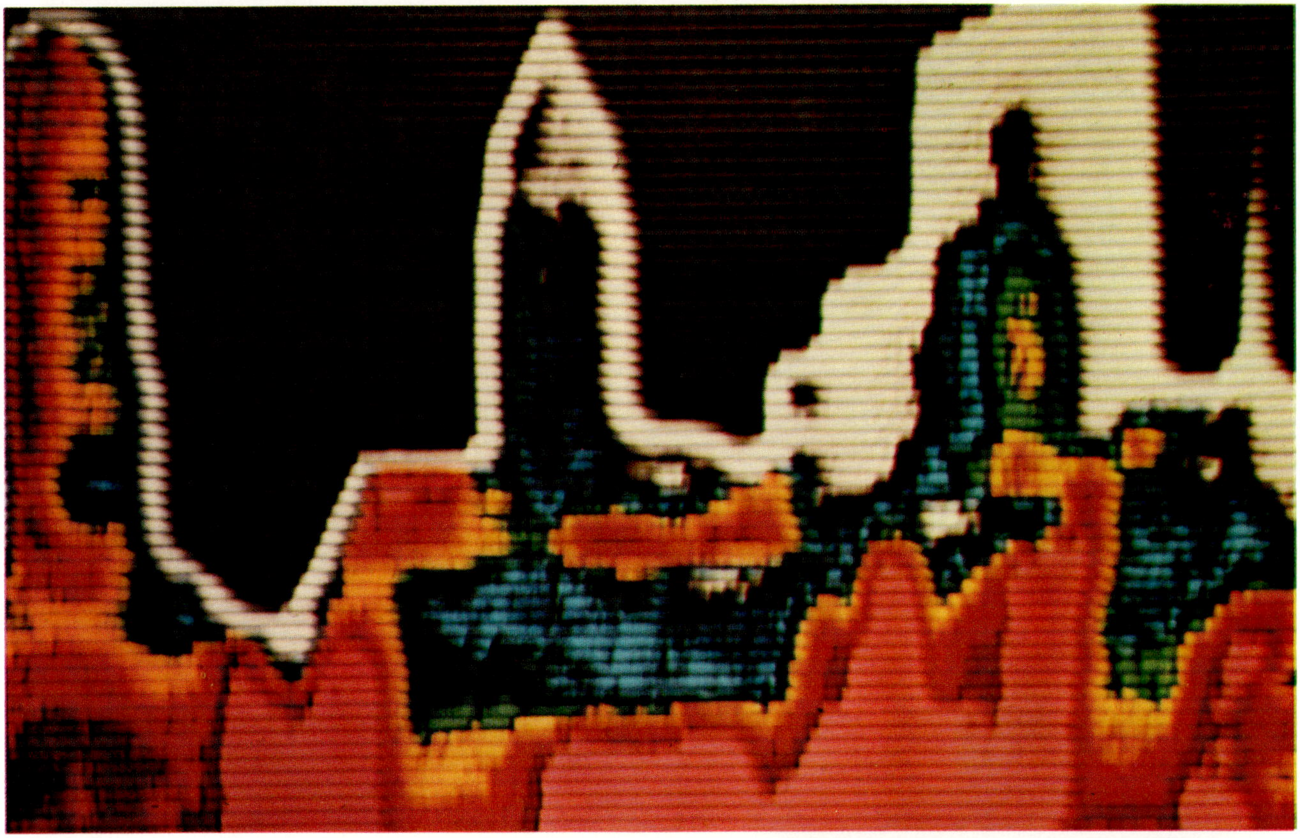

The thermogram (heat picture) in Plate 37 *(above)* shows the New York City skyline. The bright areas show the hottest regions.

The thermogram in Plate 38 *(at right)* shows a human body. Since cancers or other diseased sections of the body are often hotter than average, thermograms can be used for diagnostic purposes.

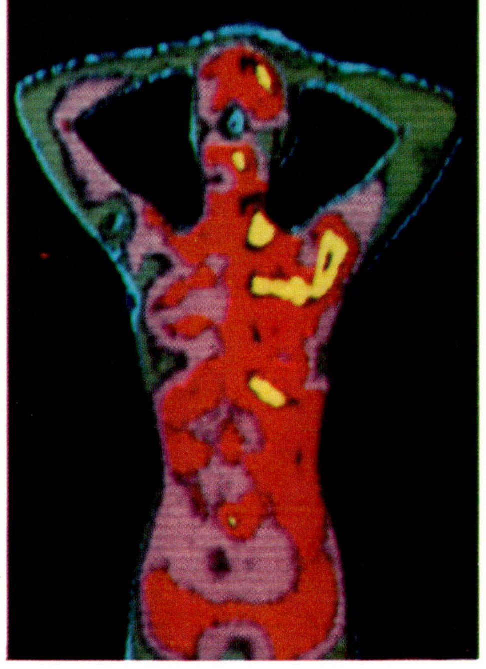

Plate 43 The largest accelerator for elementary particles (colloquially called an "atom smasher") in the United States is at the Fermi National Accelerator Laboratory, Fermilab, near Chicago. The aerial view above shows the 2-km-diameter ring in which protons are circulated around and around while they reach tremendously high speeds and energies up to 500 billion volts. They are then released down one of the straight tracks that extend diagonally across this photograph. Exploring the inside of a nucleus by bashing it with elementary particles is like finding out what is inside a watch by throwing it against a wall and watching what comes out (mainspring, etc.). But for nuclei and elementary particles, we don't have any alternative.

Plate 44 Protons are stripped from their electrons in hydrogen gas to start on their journey to high energies. The beam of protons is accelerated slightly and passed into later stages of acceleration. The equipment shown here provides the radio frequency "kick" of about one million watts millions of times each second that accelerates the proton beam in the main ring. The protons are guided around the ring by magnetic fields. They circle 200,000 times, traveling farther than the distance to the moon and back. Finally, they reach speeds over 99.999 percent that of light.

Nuclear Physics

The following four pages illustrate aspects of nuclear physics that will be discussed in Part VI.

At the Brookhaven National Laboratory on Long Island, New York, many experiments surround an experimental nuclear reactor (Plate 41, *above*) to make use of the particles it gives off.

In Plate 42 *(below)* we see a view inside the reactor. The blue light, known as Cerenkov radiation, is formed whenever particles travel faster than the speed that light is traveling. Although no particles can go faster than the speed of light in a vacuum, light goes more slowly in water. These particles are giving off energy as they slow down while going faster than the speed of light in water.

Plate 43 The largest accelerator for elementary particles (colloquially called an "atom smasher") in the United States is at the Fermi National Accelerator Laboratory, Fermilab, near Chicago. The aerial view above shows the 2-km-diameter ring in which protons are circulated around and around while they reach tremendously high speeds and energies up to 500 billion volts. They are then released down one of the straight tracks that extend diagonally across this photograph. Exploring the inside of a nucleus by bashing it with elementary particles is like finding out what is inside a watch by throwing it against a wall and watching what comes out (mainspring, etc.). But for nuclei and elementary particles, we don't have any alternative.

Plate 44 Protons are stripped from their electrons in hydrogen gas to start on their journey to high energies. The beam of protons is accelerated slightly and passed into later stages of acceleration. The equipment shown here provides the radio frequency "kick" of about one million watts millions of times each second that accelerates the proton beam in the main ring. The protons are guided around the ring by magnetic fields. They circle 200,000 times, traveling farther than the distance to the moon and back. Finally, they reach speeds over 99.999 percent that of light.

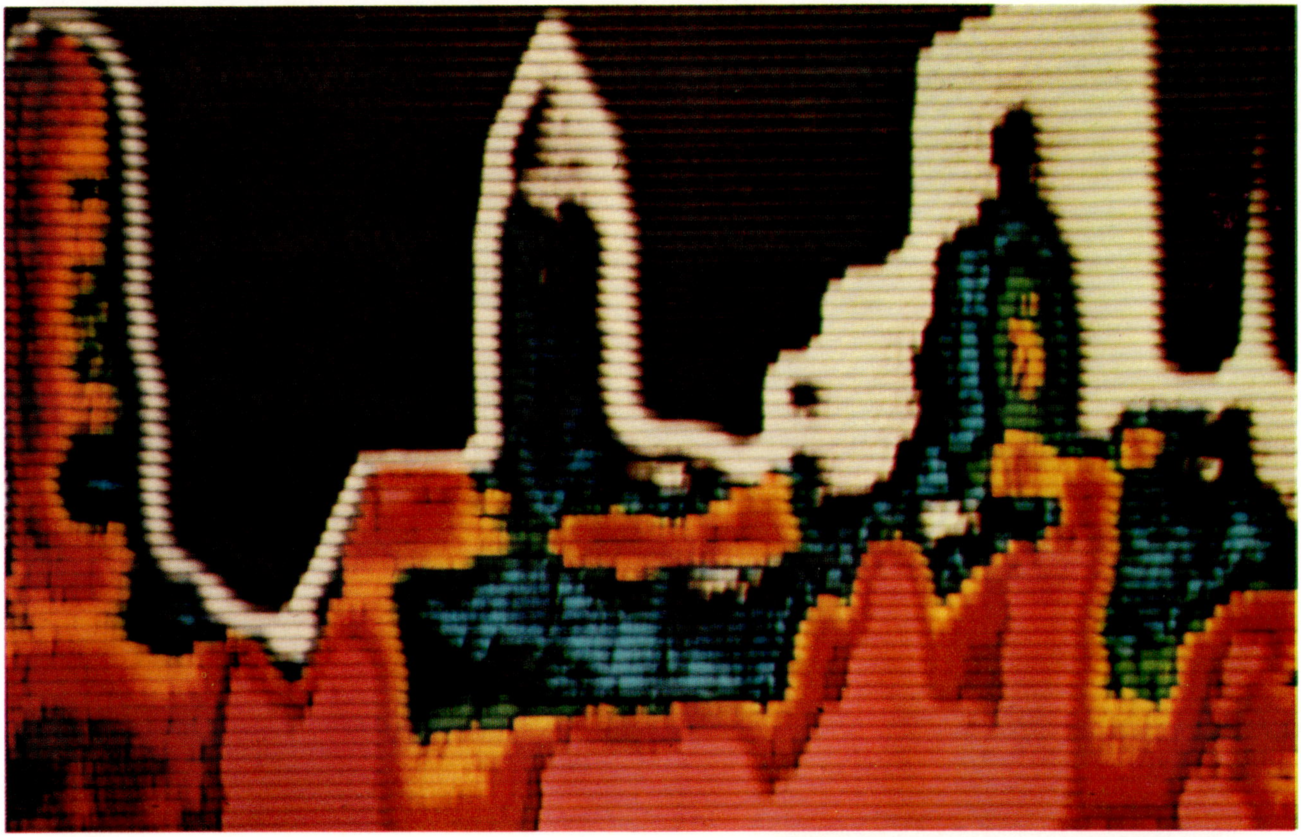

The thermogram (heat picture) in Plate 37 *(above)* shows the New York City skyline. The bright areas show the hottest regions.

The thermogram in Plate 38 *(at right)* shows a human body. Since cancers or other diseased sections of the body are often hotter than average, thermograms can be used for diagnostic purposes.

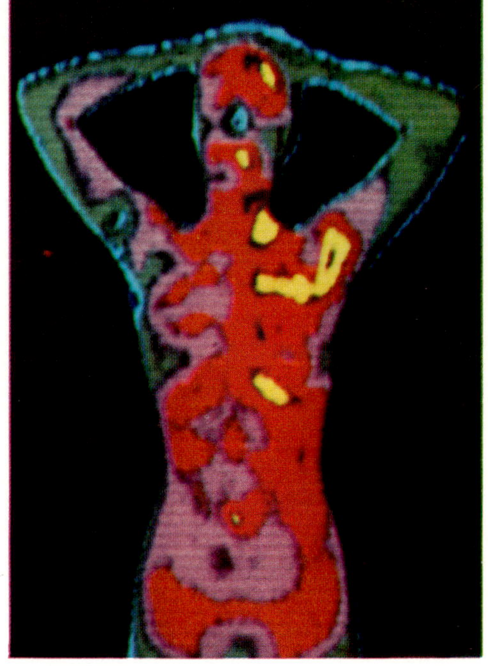

Plate 39 *(above)* shows sparks caused by a difference in voltage between two electrodes.

In Plate 40 *(below)* we see sparks caused by differences in voltage between various regions of the air—lightning—over the Kitt Peak National Observatory near Kitt Peak, Arizona. (© 1972 Gary Ladd)

Plate 45 In a tunnel underground, the main ring contains about 1000 magnets that guide the beam around in a circle 6.3 km in circumference. The beam is in a vacuum in the lower red containers. The upper ring of magnets and vacuum uses superconducting magnets to get higher magnetic fields and greater accelerations while using less electricity. Fermilab has become the largest user of liquid helium in the world in order to cool its superconducting magnets.

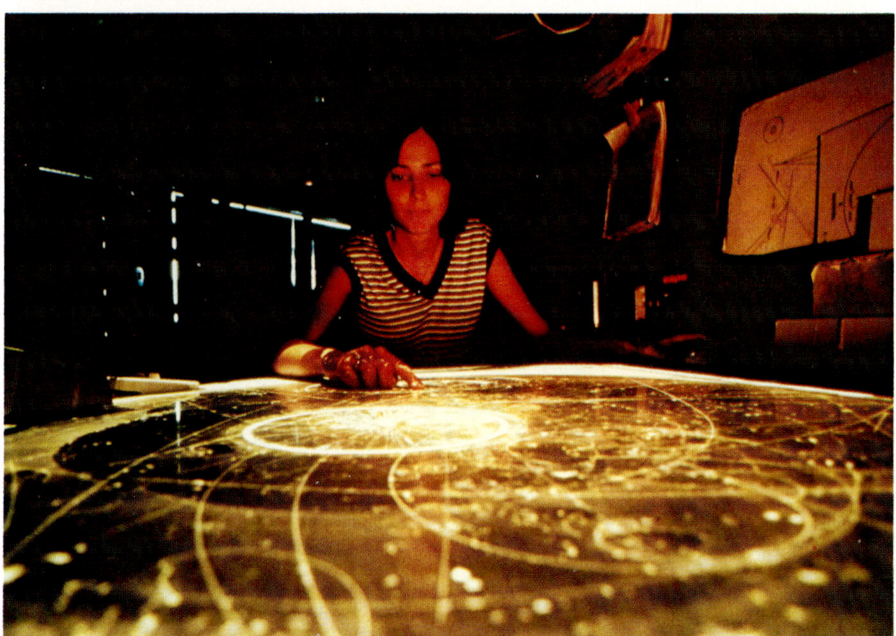

Plate 46 One way of studying particles is to direct the beam into a bubble chamber, a 4-m-diameter sphere of hydrogen in which particles leave trails of bubbles. Hundreds of thousands of photographs of these trails must be examined, as we see above, to find the telltale traces of new types of interactions. Elementary particles and quarks are studied in these ways.

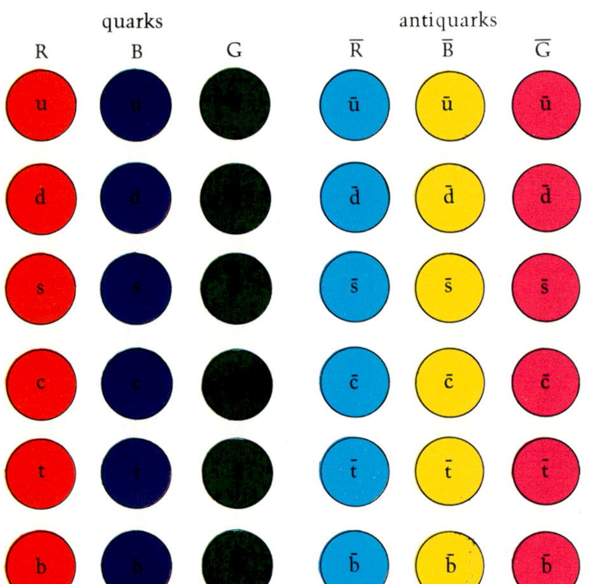

Plate 47 Quark colors and flavors. According to the currently accepted theory, the properties of matter follow from the flavors and colors of quarks. We now think that there are six different quark *flavors*—up (u), down (d), strange (s), charm (c), top, or truth (t), and bottom, or beauty (b). Each flavor comes in each of three colors—red (R), blue (B), and green (G). For each flavor of quark there is an antiquark, indicated by the bar over the flavor (for example, $\bar{u}$). Each antiflavor comes in an anticolor. As with actual colors, we have used cyan for the anticolor of red, yellow for the anticolor of blue, and magenta for the anticolor of green.

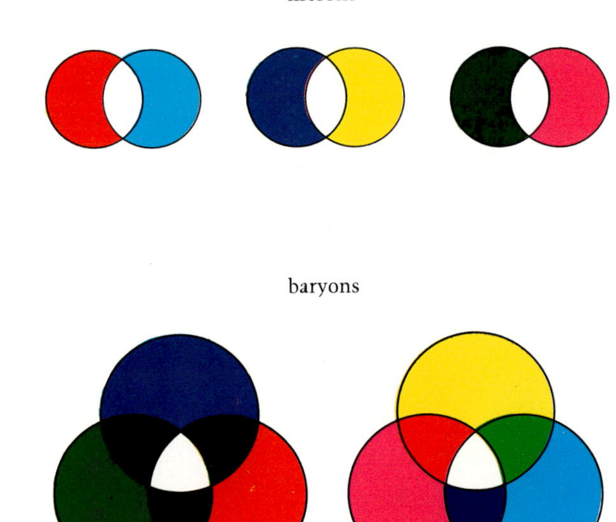

Plate 48 Combinations of quarks. The allowed combinations of quarks are those that have no net color (those whose colors add up to white, or colorlessness). Colored quarks can add to have no color in two ways: (1) *Mesons.* A quark of one color and an antiquark of the corresponding anticolor add up to white. The quark can have any of the six flavors and the antiquark can have any of the six antiflavors. (2) *Baryons.* Three quarks, one of each color (each quark can be any one of the six flavors), add up to white. So do three antiquarks, one of each anticolor (and each antiquark can be any one of the six antiflavors).

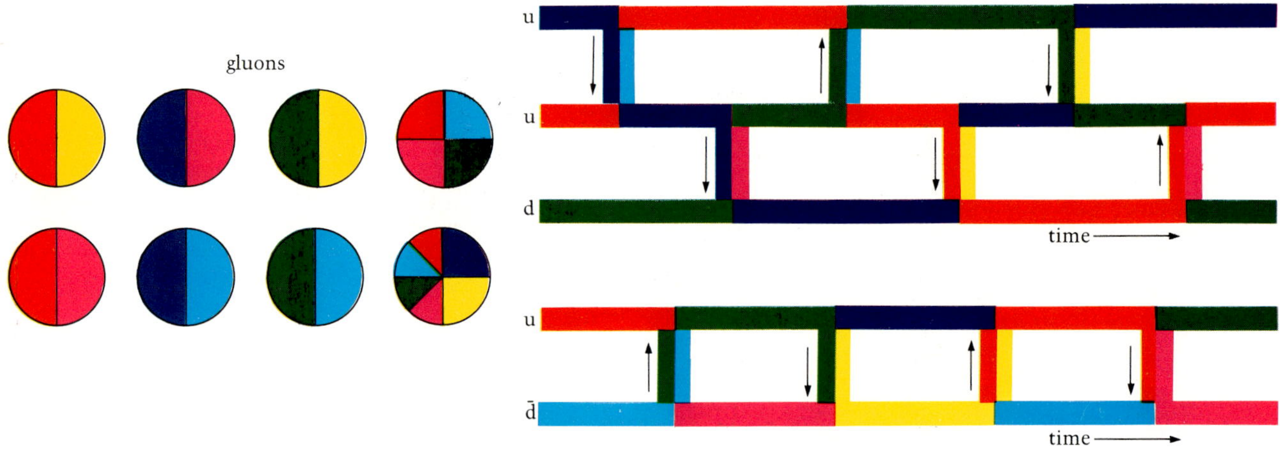

Plate 49 Gluons. The quarks are held together by gluons. There are eight different gluons *(at left).* Six have one color and a different anticolor. Two are more complicated combinations of colors and corresponding anticolors. Quarks exert forces on each other by exchanging gluons.

We follow three quarks in a baryon as time goes on *(at right, above).* Gluons move in the directions indicated by the arrows. When the gluon moves from one quark to the other the quark colors change, but the flavors remain the same. Also, at any time, each color must be represented in the baryon.

Below, we follow two quarks in a meson as time goes on. Again, the exchange of gluons leaves the flavors unchanged, but changes the colors. At any given time, we have a color and its corresponding anticolor.

energy and pass directly into the food. The microwaves are absorbed in the food (mostly by the water in the food) and thus give up their energy directly to the food. They penetrate to all parts of the food, so it all cooks at the same rate, and the required heat is deposited in the food rapidly (Fig. 12–25).

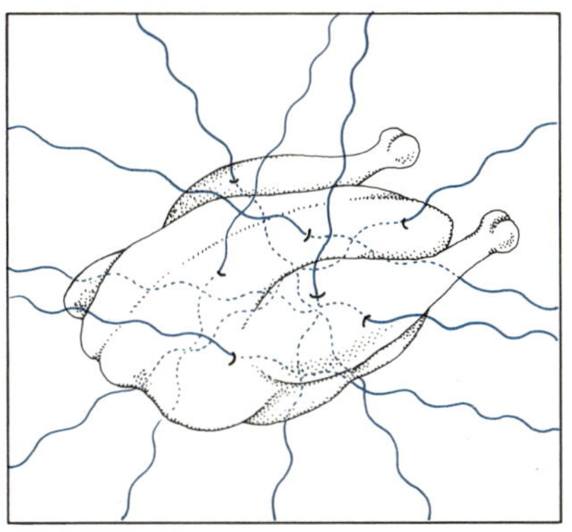

Fig. 12–25 Cooking in a microwave oven. The microwaves come from all directions, after bouncing around inside the oven. Some of the microwaves are absorbed near the surface of the food, and some pass through into the inside of the food. In addition, some pass right through the food, bounce off the walls, and get another shot at the food. In this way, all parts of the food are cooked at the same rate, and you don't have to wait for heat conduction from the air outside the food to the center of the food. The microwaves get there directly, so the food cooks very quickly.

Key Words

conduction
convection
radiation
thermal conductivity
insulator
wind-chill factor
core
mantle
crust
geothermal energy
plates
continental drift
thermals
electromagnetic spectrum
ozone layer
greenhouse effect

Questions

1. Why do you have to burn more fuel to keep your house at a constant 72°F in the winter than to keep it at a constant 66°F?
2. Why do you have to burn more fuel to keep your house at a constant 68°F when the outside temperature is 0°F than when it is 35°F?
3. On a cold winter day, why do you feel colder if your clothes are wet than if they are dry?
4. Why do storm windows, with two panes of glass and an air gap in between, keep your house warmer than windows with the two panes of glass touching each other?
5. Give one example that shows that water is a better conductor of heat than air is.
6. Can you have conduction of heat through a perfect vacuum? Explain your answer.
7. A Thermos bottle is designed to stop two types of heat flow. Which are they?

8. When you heat a pot of water so that it boils, what makes the hotter water rise to the top in the process of convection? That is, explain the physical processes behind boiling and convection.
9. Why does the temperature in our lower atmosphere decrease as you get higher in altitude?
10. We often see clouds arranged above the ground in a very neat layer. Why should this be? (*Hint:* Think of what happens to water vapor carried aloft by convection, as it gets to cooler and cooler parts of the atmosphere.)
11. The ozone layer is warmer than the layers directly above and below it. Where does the extra heating for the ozone layer come from?
12. In this chapter, we described how you would design a roof to minimize radiative heating and promote radiative cooling. How would you design a roof to do just the opposite, to maximize radiative heating and minimize radiative cooling? (This is what you would want to do in the winter.)
13. On a clear day, where does most of the sun's visible radiation get absorbed?
14. The sun's radiation is most intense when the sun is highest in the sky, around noon. However, the "heat of the day" is in the midafternoon. Why should this be the case?
15. For the following processes, state whether the primary energy transport is by radiation, convection, or conduction: (a) you cook a chicken in a pot of boiling water; (b) you bake a chicken in an oven; (c) you cook a chicken in a microwave oven; (d) you get a sunburn; (e) you open a window to cool off your house; (f) you lose heat through the walls of your house.
16. When there is a layer of ice on top of a lake, ice is made just at the place where the ice layer and water are in contact. As the ice layer gets thicker and thicker, it gets harder and harder to make more ice. Why?
17. If you want to use two pieces of glass, parallel to each other but with a gap in between, as an insulator in the wall of your house, would you fill the gap with water or air? (Don't worry about the water's running out.) Explain your answer.
18. If you want a large piece of ice to melt quickly, you should break it up into smaller pieces. Why is this? (*Hint:* Think of what happens to the total surface area of the ice as we break it into smaller and smaller pieces.)

PART IV

Wave Phenomena

Have you ever wondered why you don't feel a rush of air when a sound reaches you? Air carries sound, but none of the air actually moves across the room. Yet somehow the message gets across. This transmission of information or energy, without any bulk transfer of material, is a property that belongs to waves.

In this section we will see a variety of wave phenomena, starting with the simplest pulses traveling down a rope and ending with the behavior of light waves. We'll see how such seemingly different phenomena have a lot in common. We'll see how waves behave in very unusual ways. For example, we can put two waves together in such a way that no wave results. We'll discuss a technique called holography, by which we can capture three-dimensional images without using lenses. We'll also see how, when a wave strikes a wall, some energy gets through the wall and some bounces back.

As you read this section and the examples in it, you should become aware of how often we meet wave phenomena. Some waves are so familiar—those that bring us TV programs, for example—that we are barely aware of all the wave motions that are constantly going on around us. Yet sometimes we use our understanding of waves to help us study very remote phenomena, such as when we observe light waves, radio waves, or x-rays coming to us from the most distant objects in the universe.

Studies of the wave properties of light help us understand how we see. And even telephone calls are now carried more and more often by light beams instead of by electrical signals, using the techniques of fiber optics that we shall discuss. We shall also analyze how lenses and mirrors work, and how they are put together to make cameras, microscopes, and telescopes.

CHAPTER 13

Waves

Fig. 13–1 Regular ocean waves. When they pass by a boat or other object, the object bobs up and down but is not carried along with the wave. Neither does the ocean water itself get carried along with the wave. When the water gets too shallow in toward shore, the wave "breaks," ending the similarity between ocean waves and the pure waves we discuss in this chapter.

We are surrounded by wave phenomena. Waves in the ocean come in, one after the other, and eventually break on the beach. A violin concerto you hear on the radio starts with the sound waves generated by a violin string. The sound waves travel across the stage to a microphone, where they are converted to radio waves. Your radio converts them back into the sound waves that finally reach your ear. If the electronic equipment is very good, the sounds you hear still contain much of the richness present in the vibrations within the violin. (See Fig. 13–2.)

The properties of waves are of paramount importance to the physicist. Beyond the everyday wave phenomena like those mentioned above, twentieth-century physicists have learned that even those things around us that we consider most sturdy and solid can be thought of as nothing more than a complex of wave motions. This method of thinking of the world will become very important when we start to talk about quantum mechanics. For the present, we will start to look at some of the properties of the more common waves.

13.1 *Wave Pulses*

In a wave, energy or information is sent from one place to another without the actual transfer of matter. How can this happen? Think, for example, of a line of dominoes, standing next to one another. If

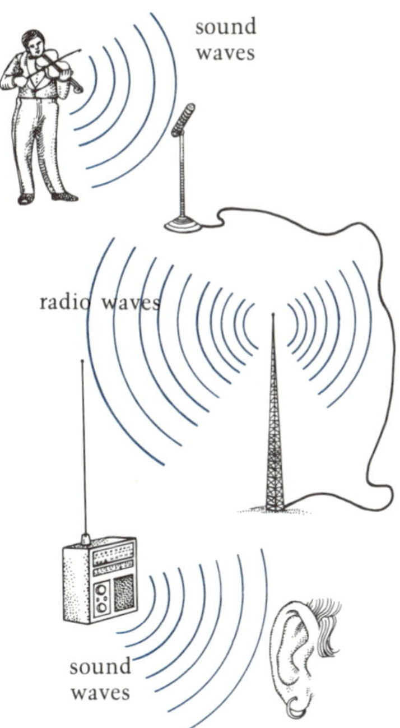

Fig. 13–2 When you listen to a violin concerto on the radio, different types of waves are involved. First there are the sound waves traveling from the violin to the microphone. Then the radio waves travel from the transmitter to your antenna. Finally, your radio converts these radio waves back into sound waves for you to hear.

Fig. 13–3 The record for dominoes felled is almost 250,000.

the first domino is knocked into the second, a chain reaction is started. (The current record is almost 250,000 dominoes.) Eventually the last one falls over. We have clearly transmitted a signal along the chain of dominoes. (For example, the first domino could be knocked over as a signal to people in another room that dinner is ready.) However, each domino has not moved very far; it has merely fallen over.

Such a single passage of a wave is called a wave pulse. Dominoes have to be set up again if you want to send another pulse. To send a series of wave pulses and study their properties, you need a material that springs back to its original state after the wave pulse has passed. Such a material is called elastic. Waves in elastic materials are called mechanical waves. The material through which the wave travels is the medium for the wave. Three common examples of elastic media are strings, springs, and the air. We'll look at each in turn.

Fig. 13–4 A wave traveling down a rope from left to right. Individual parts of the rope move up and down, but do not move forward with the wave.

13.1a STRINGS

Figure 13–4 shows a long rope tied to a post at one end. When we jerk one end of the rope, a wave pulse travels down the string. Although the pulse gets from one end to the other, the string has not gone anywhere. This surely fits our definition of a wave.

To visualize the pulse, we can think of the rope as a series of

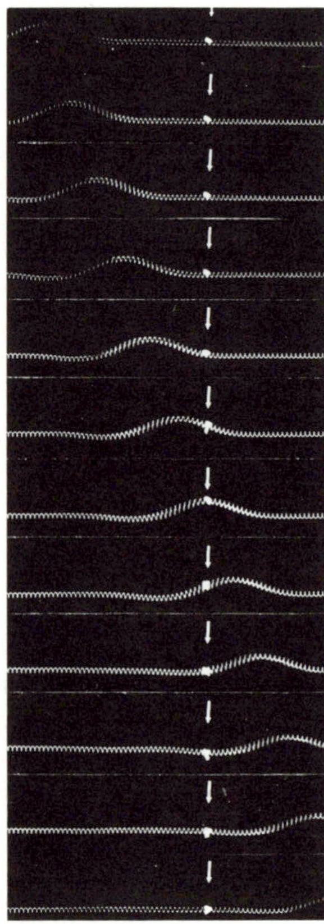

Fig. 13–5 Transverse wave on a spring. Each frame shows the wave pulse at a slightly later time than the previous frame. Notice that as the wave moves from right to left, the point on the spring marked by the ribbon only moves up and down.

The waves we have been discussing are called mechanical waves because the energy is transmitted when one part of a system pushes or pulls on another.

balls, each connected by springs to its nearest neighbors. The motion of each ball is governed by that of its neighbors. Each one starts its upward motion when it is pulled up by the ball on its left. Each ball copies the motion of the ball on its left, but with a slight time delay.

This example gives us a good idea of what a wave pulse actually is. Each part of the string moves briefly away from its normal position. In the case above, notice that each part of the string only moved up and down, but never moved along the string. This is an example of a transverse wave. In a transverse wave, the individual particles that are being disturbed move in directions perpendicular to the direction in which the wave is moving. The wave moves from left to right but the particles in the string move briefly up and down.

The velocity of the wave depends on both the tension of the rope and the density of the material in the rope. *The wave pulse will move fastest when the tension is greatest and the mass per length of rope is least.*

13.1b SPRINGS

If you have a spring or Slinky, you can demonstrate two different types of waves. One is the transverse wave, just as on a string (Fig. 13–5). Just stretch the spring a bit and strike it sideways and watch the pulse travel, with the individual particles on the spring moving perpendicularly to the direction in which the wave is moving.

There is another type of wave that you can send down the spring. Instead of striking the spring sideways, you can gather together a few of the coils at the end. When you release them, a wave will move down the spring (Fig. 13–6). This time the wave will be one in which the coils are pressed together at the point where the wave pulse is present. Each piece of the spring moves back and forth, along the length of the spring. Of course, each coil never strays very far from its initial location. Such a wave is called a longitudinal wave. At various points along the spring, we can have compressions, places where coils are closer together than normal, and rarefactions, places where the coils are farther apart than normal.

A

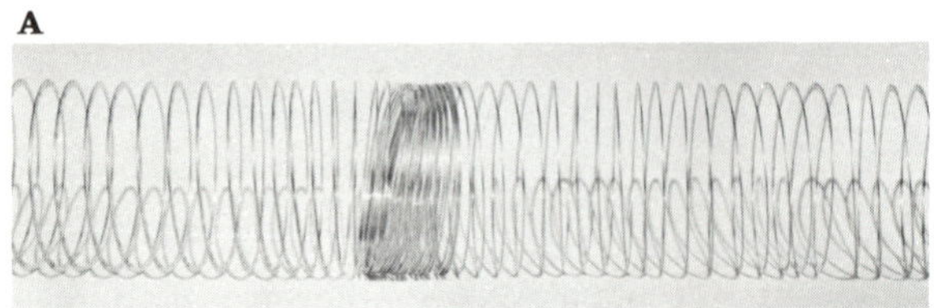

B

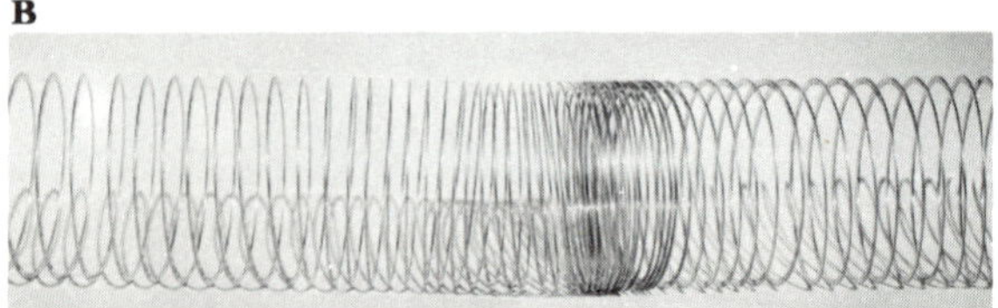

Fig. 13–6 A wave pulse traveling down a spring (namely, a Slinky) from left to right. When the pulse has completely passed, all the parts of the spring return to their normal positions.

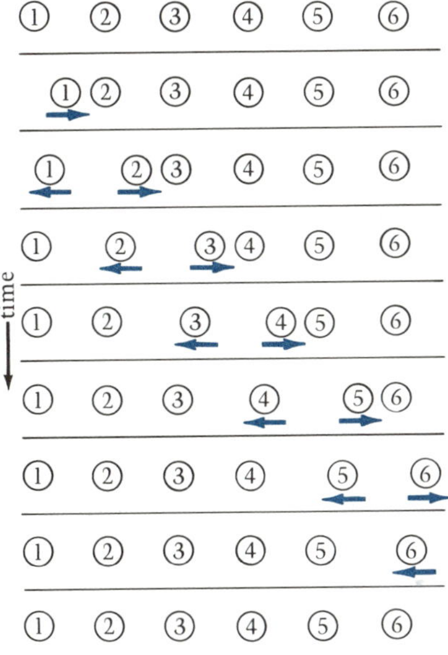

Fig. 13–7 Passage of a longitudinal wave pulse. Each row shows the positions of six particles at a successively later time. In the first row, all of the particles are at rest before the wave enters from the left. In the following rows, each particle is knocked to the right by the particle on its immediate left. After striking its neighbor, each particle returns to its original place. The pulse is transmitted without any of the particles traveling very far.

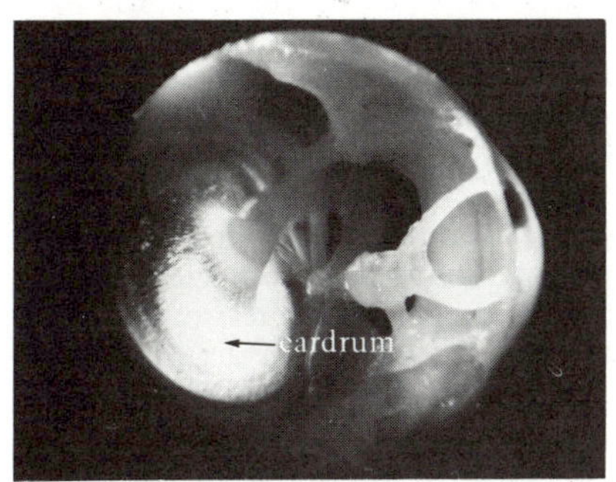

Fig. 13–8 This photograph of the middle ear was made possible by the use of fiber optics. The eardrum, curving inward slightly, can be seen at the left.

13.1c AIR

We can also send mechanical waves through the air. Under the right conditions, these are sound waves. Before we look at the connection with sound, let's see how we can send waves through the air. Remember, in the springs and strings, each particle could pull or push its neighbor. In the air, we have collisions in which each particle pushes the other, but there is no pulling. We can transmit waves only by knocking air molecules into each other. We push the first particle; it pushes the next, and so on. Each particle moves in the direction that the wave is traveling, so these waves will be longitudinal waves, similar to those in the spring (Fig. 13–7).

Just as with the longitudinal wave in the spring, in the air we will have regions where the air molecules are closer together than normal (compressions) and regions where the air molecules are farther apart than normal (rarefactions). In the compressions, the pressure is higher than the normal atmospheric pressure. In the rarefactions, the pressure is lower than the normal pressure.

With these longitudinal waves, the air pressure changes as the wave passes by. These changes in pressure can affect our eardrums. When the air pressure is higher, the thin membrane in the ear is pushed inward. When the pressure is lower, the higher pressure inside your ear pushes the membrane outward. Physicists have an answer to the famous question, "If a tree falls in the forest and there is no one to hear it, is there still a sound?" By "sound," a physicist means a variation in pressure. Newton's laws tell us that when a tree falls, it will disturb the air and create a certain pressure variation. The pressure variation is present even if nobody is present to receive the sound. If you encounter a fallen tree, you can safely conclude that there was a pressure wave at the time the tree fell.

How fast does sound travel? The exact speed depends on the properties of the gas involved because the transmission of the pulse depends on having one molecule knock into another. Remember that this process happens all the time because the molecules move about randomly, with an average speed that depends on the temperature. We would therefore expect the sound waves to travel through the gas at approximately the same speed as the average of the random speeds of the molecules.

For sea level air at 0°C, the speed of sound is about 331 m/s, much slower than the speed of light. It is easy to sense the time it takes for sound to travel any appreciable distance. At a football game, if the cheerleaders for one team set off a cannon and you are at the opposite end of the field, you will clearly see the flash (which travels at the speed of light, and takes only about one-millionth of a second to reach you) before you hear the bang. During a thunderstorm, you can tell how far away a lightning strike is by counting the number of seconds between the time you see the lightning (the light reaches you almost instantaneously) and the time you hear the thunder. At 331 m/s, the sound will travel approximately 1 kilometer for every 3 seconds (1 mile for every 5 seconds), so if the thunder occurs 6 seconds after the lightning, the stroke was about 2 kilometers away.

The speed of sound in the air tells us how fast the air molecules can respond to an applied pressure. What happens if something moves through the air at a speed greater than that of sound? (We call such motion supersonic.) Say that we are talking about an airplane. As it flies through the air, it pushes the molecules together faster than they can respond and move out of the way. Pressure builds up at the front of the plane. This very strong build-up of pressure is called a shock wave. This shock wave spreads outward from the nose of the plane. A second shock wave starts at the tail of the plane, corresponding to a sharp decrease of pressure as the plane moves out of a space faster than air can rush in and fill the space. The shock waves move along parallel to the plane at the speed of the plane, but spread outward from the path of the plane at the speed of sound. They therefore trail behind the plane in a pattern like the bow waves of a ship (see Fig. 13–9).

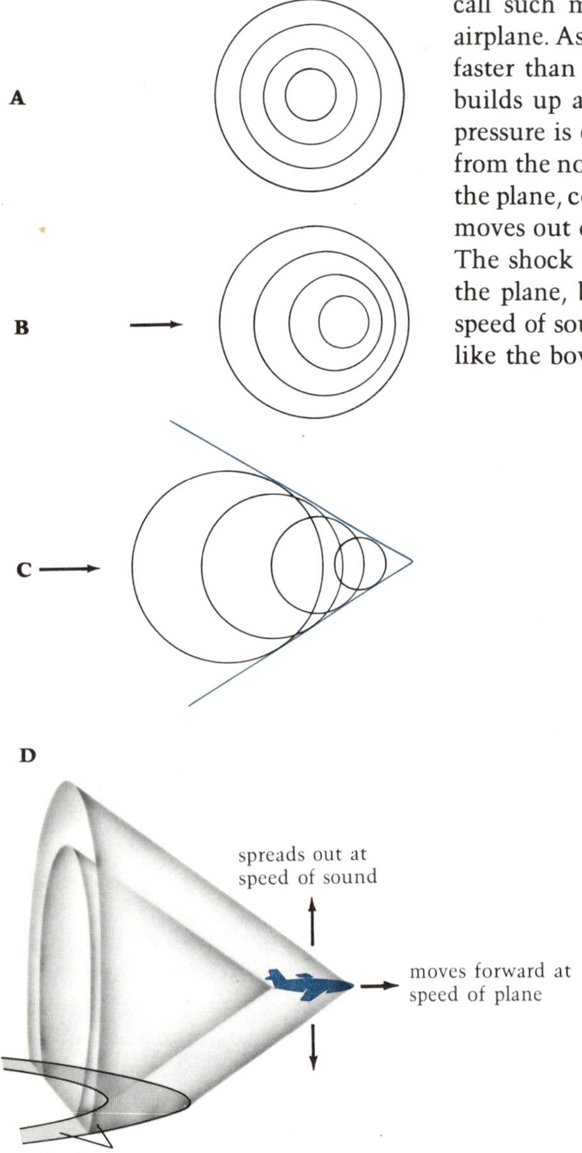

Fig. 13–9 Formation of a shock wave. A. If an object emitting sound waves is standing still, the pulses will spread out in a circular pattern. We can think of each dark circle as representing the high-pressure region of a wave. The distance between successive circles is the wavelength. B. We let the source of the waves move to the left with a speed less than the speed of sound. Now each wave spreads out in a circle from the point at which that wave was emitted. However, each wave was emitted at a different point, so the circles no longer have the same center. C. Now we let the source of the wave move faster than the speed of sound. The circles start to overlap. In particular, we get two lines along which the pressure build-up from each of the pulses adds to the others to give us a line of high pressure. D. When an airplane flies at supersonic speed, such a shock wave is formed. Actually, there are two shock waves, one building up at the nose of the plane, where the air is being compressed faster than it can move out of the way. This is a high-pressure region. At the rear of the plane, the plane is moving out of space faster than the air can flow back into the vacated space, so we have a low-pressure region. Just as in C, where the circles spread out with the speed of sound, but move along to the left at the speed of the object, the shock waves from the airplane spread out in a cone, moving forward at the speed of the airplane, and outward at the speed of sound. Wherever a listener passes through these cones, a double "crack" will be heard—a sonic boom.

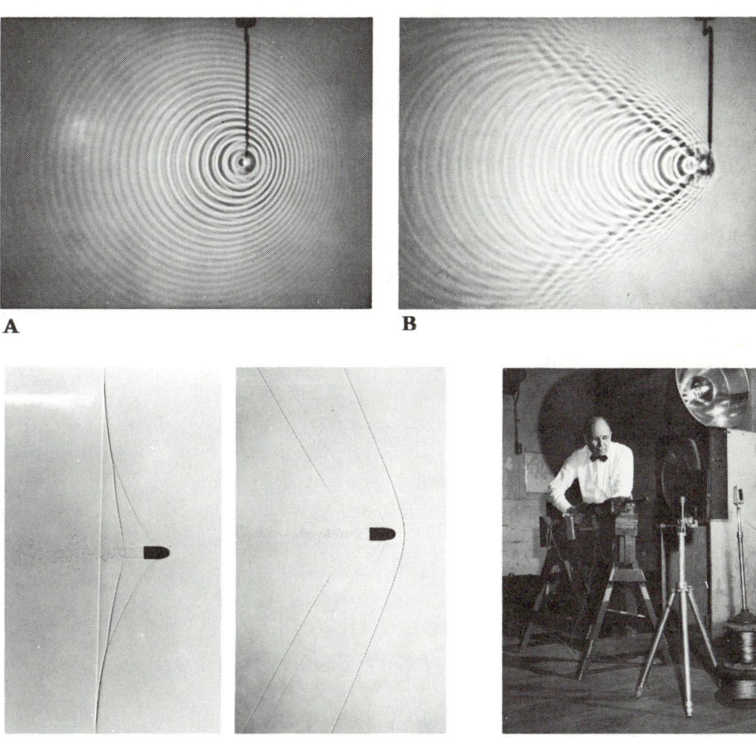

Fig. 13–10 Water waves in a ripple tank. The waves are produced by a small object at the end of the stick that is periodically dipped into the water. A. The source is moving slowly to the left, as in Fig. 13–9B. B. The source is moving to the left faster than the speed of waves in the water. A shock wave is formed, as in Fig. 13–9C.

(left and center) Fig. 13–11 Photographs of bullets traveling through the air, showing the sound waves. An extremely bright electronic flash of less than a millionth of a second in duration was used to take these pictures. The device is called a stroboscopic flash, or a "strobe." A. A .22-caliber "short" bullet traveling at less than the speed of sound. B. A .22-caliber "long" bullet traveling faster than the speed of sound. Compare the shock wave in this picture with those in Figs. 13–9C, 13–9D, and 13–10B.

(right) Fig. 13–12 Dr. Harold Edgerton of MIT, the inventor of the strobe, setting up to take pictures like those in Fig. 13–11.

Wherever the shock waves cross the ground, there is a sudden increase and then a sudden decrease in pressure, which makes a double "crack" sound. We call this sound a sonic boom. A sonic boom can be heard as long as the plane is moving faster than the speed of sound, and not just when the plane breaks through the "sound barrier." Besides being annoying and potentially harmful to hearing, the pressure variations associated with sonic booms can shake windows loose, rattle dishes, or even destroy buildings. Much of the question over whether to build a fleet of supersonic passenger planes centered on the question of the effects of the sonic booms as these planes flew over populated areas. The British/French Concorde flies at supersonic speeds only over the ocean, and goes at subsonic speeds as it approaches land and over land. The Soviet supersonic plane is used over land, but flies at supersonic speeds mostly when over the extensive unpopulated areas of the Soviet Union.

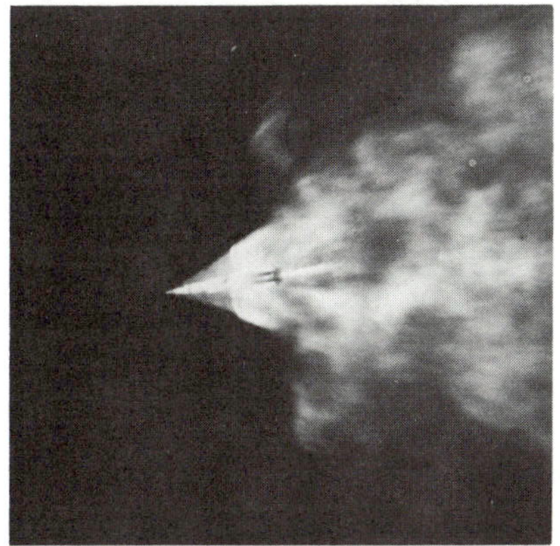

Fig. 13–13 The shock wave is visible as the second stage of Apollo 11 takes over en route to the moon.

Sound waves travel not only through gases, but also through any material substance, solids and liquids included. In most solids, sound travels very quickly since the atoms are very close together and quickly communicate any pressure changes. For example, the speed of sound in aluminum is 15 times the speed of sound in air.

DECIBELS

There is a large range in the intensity of sound waves traveling through the air. It is convenient to use a scale in which constant changes on the scale correspond to increasing the sound level by some multiplying factor. This scale is the decibel (abbreviated dB) scale. On this scale, the faintest sound that a human can hear is arbitrarily set equal to 0 dB. An increase in the sound intensity by a factor of 10 corresponds to adding 10 dB ("deci-" means one-tenth, so this is 1 bel, but decibels are more commonly used). The scale is based on logarithms, although we shall not use them explicitly. Every time the intensity doubles, the intensity measured in decibels increases by about 3. For example, 13 dB is twice as strong as 10 dB (since we have added 3 dB).

Although this scale has certain advantages, it often misleads people. For example, during the debate over whether to build an American SST (supersonic transport), one congressman remarked that the SST at some distance away would generate a 110-dB sound while a normal jet would generate 100 dB at the same distance. He couldn't see what all the fuss was about over a 10 percent increase. Of course, it is not a 10 percent increase in intensity; it is an increase in intensity by a factor of 10. Table 13–1 gives the decibel level of some common sounds.

TABLE 13–1 Decibel Level of Common Sounds

Sound	*Decibel level (dB)*
Faintest sound that can be heard	0
Soft whisper	30
Normal talking	60
Busy street	70
Subway train (harmful level)	100
Rock concert (painful level)	120
Jet takeoff nearby	150

13.2 *Harmonic Waves*

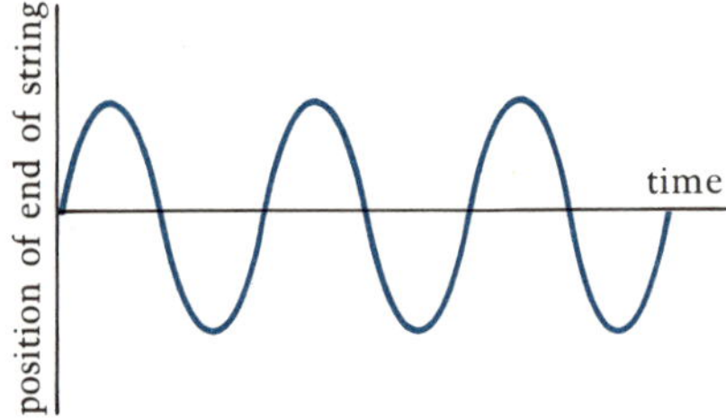

Fig. 13–14 A periodic wave. This drawing shows the position of the end of the string at various times when we drive the end of the string with simple harmonic motion.

To this point we have been talking about single pulses traveling through some medium—along a string or spring, or through the air. These are elastic media and return to their initial configurations quickly when a pulse passes through, so we can transmit a repeated string of pulses. For example, with a rope you can continue to move one end up and down. You will see pulses alternating up and down and traveling away from you. Compressions and rarefactions move down the Slinky. The same thing holds for sound waves. We can put a drumhead at the end of an air tube and move the drumhead back and forth, which sends compressions and rarefactions down the tube.

Let's concentrate on the transverse waves on the string. We keep the waves coming by moving the end up and down. If we repeat the same pattern over and over, we call the wave a periodic wave (Fig. 13–14). After a while we might get tired of moving the end up and down. We could attach the end of the string to a weight that bounces up and down on the end of a spring. We have already described the motion of a weight on the end of the spring (Chapter 5). We saw there that it undergoes simple harmonic motion. We therefore call the waves that it generates on the string harmonic waves.

If we graph the position of the end of the string at different times, it will look like the curve in Fig. 13–14. This is the characteristic curve that we associate with harmonic waves and harmonic motion. The time required to go through one full cycle of this motion is called the period of the motion. As we have discussed, it is sometimes more convenient to talk about the number of full cycles that we go through in 1 second. This number of cycles in 1 second is called the frequency.

$f = \frac{1}{p}$, or *frequency* = 1/*period*.

Suppose we look at a point farther down the string and ask about its position (up-down) at various times. The motion of any point on the string will mimic that of the end point. The only difference is that a point down the string will start its motion later than the end point does because it takes time for a wave to travel from the end to the point in question (Fig. 13–15). The motion of every point looks like the motion of the end point with the exception of the delay. *For a harmonic wave, each point on the string undergoes simple harmonic motion.*

If we take a picture of the string at some time, what would it look like? The position of any given point above or below the average position of the string will be the same as the position of the end point a certain time ago. That time is the delay time for the wave to reach that point from the end. For example, if the wave takes 2 seconds to reach that point, the height of the point will be the same as the height of the end 2 seconds ago. Each point on the string is at a different height from its neighbors.

Suppose we take our picture at a time when the end point is at maximum height (Fig. 13–15). As we go down the string we will find the neighboring points lower than the end point, and we are essentially looking at the position of the end point going back in

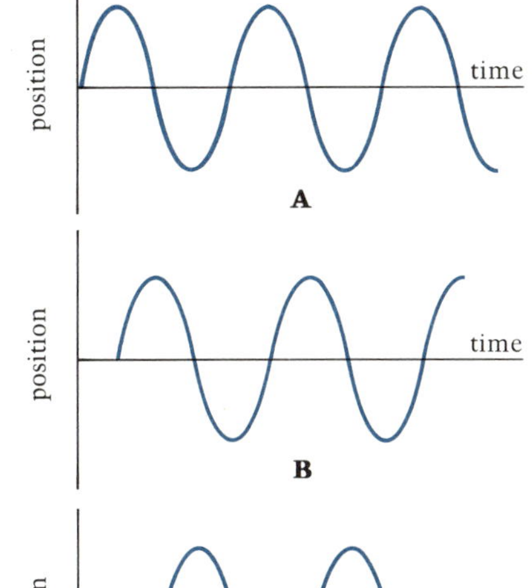

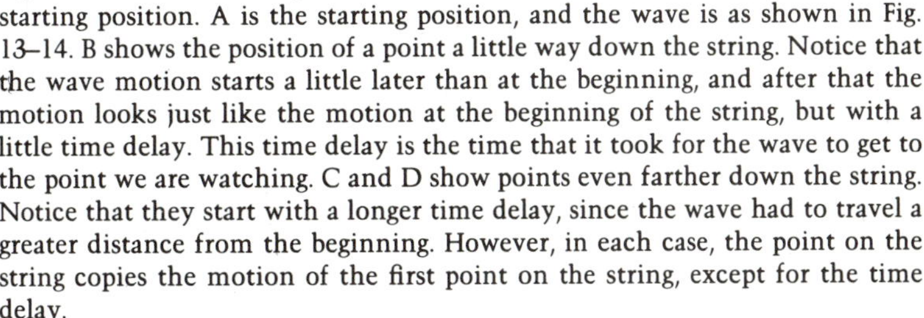

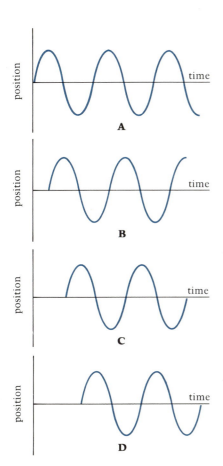

Fig. 13–15 Position of points on the string at various distances from the starting position. A is the starting position, and the wave is as shown in Fig. 13–14. B shows the position of a point a little way down the string. Notice that the wave motion starts a little later than at the beginning, and after that the motion looks just like the motion at the beginning of the string, but with a little time delay. This time delay is the time that it took for the wave to get to the point we are watching. C and D show points even farther down the string. Notice that they start with a longer time delay, since the wave had to travel a greater distance from the beginning. However, in each case, the point on the string copies the motion of the first point on the string, except for the time delay.

time. Eventually we go so far that we see the position of the end point one full cycle ago. That means we have reached another point that is a maximum height. Of course, this maximum corresponds to the maximum of the end point one full cycle ago. If we go further down the string, we find another maximum corresponding to the maximum of the end point two full cycles ago. We find maxima periodically down the string.

The distance on the string over which the pattern repeats itself is called the wavelength. The wavelength is related to the period because the wavelength is the distance for which the delay time is exactly one period (see Fig. 13–16).

Since the wave moves through a distance of one wavelength in a time of one period, the speed (distance/time) is the wavelength period. Since the frequency is 1/period, we have

$v = \lambda f$ *speed = wavelength × frequency.*

We have already seen that the speed is determined by the properties of the string. Once we choose the string with a given tension, we have no control over the speed. We can normally control the period (or frequency) at which we move the end of the string. But once we do that, the wavelength becomes determined.

The only other property of the wave that we can control is how far up and down each other point on the string can move. This property is called the amplitude of the motion.

All of these considerations hold for sound waves too, except that sound waves are longitudinal waves instead of transverse waves. In this case, to get our harmonic waves started, we might put a piston at the end of the tube. The piston could then be moved back and forth on the end of a spring to produce the simple harmonic motion. This would alternately produce areas in which the pressure is higher than average and areas in which the pressure is lower than average (see Fig. 13–17).

We can generate pressure waves at any frequency. However, our eardrums do not respond to frequencies that are too high or too low. The audible frequency range is from about 20 Hz to about 20,000 Hz. The actual range of frequencies that can be heard varies

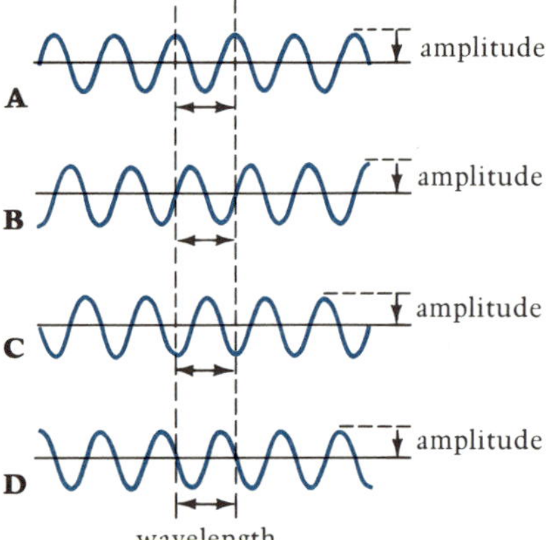

Fig. 13–16 Pictures of a string, driven by simple harmonic motion at different times. At different times we see the same wave pattern, shifted over by the distance that the wave has traveled in the time between pictures. In this figure, the time between successive pictures is a quarter of the period. On any of the pictures we can find the wavelength by measuring the distance for the wave to go through one full cycle. The amplitude of the wave is the maximum height that the wave can reach.

1 Hz = 1 Hertz = 1 cycle per second

with the individual and with age. Some people can be quite sensitive to high frequencies, and may often hear hums from signals that were not intended for human ears (such as the horizontal sweep frequency on a television set). Dogs hear higher frequencies than humans, which is how we can have dog whistles that are audible to dogs but not their masters. Frequencies higher than about 20,000 Hz are referred to as ultrasonic.

The human ear is generally very good at distinguishing between sounds of different frequencies. Different musical notes are at different frequencies. For example, concert A is at 440 cycles per second. The "A" one octave higher is at 880 cycles per second. (Each octave corresponds to a doubling of the frequency.)

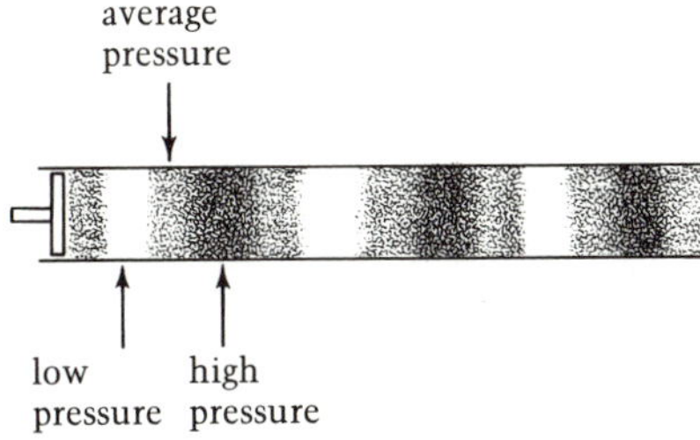

Fig. 13–17 As we move a piston at the left end of the tube back and forth, we get regions with alternating higher and lower pressure than the average.

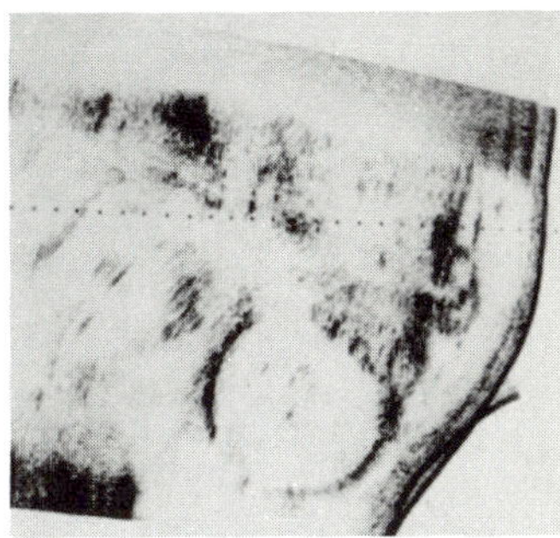

Fig. 13–18 Ultrasonic reflections from this fetus reveal its growth and position months before its birth.

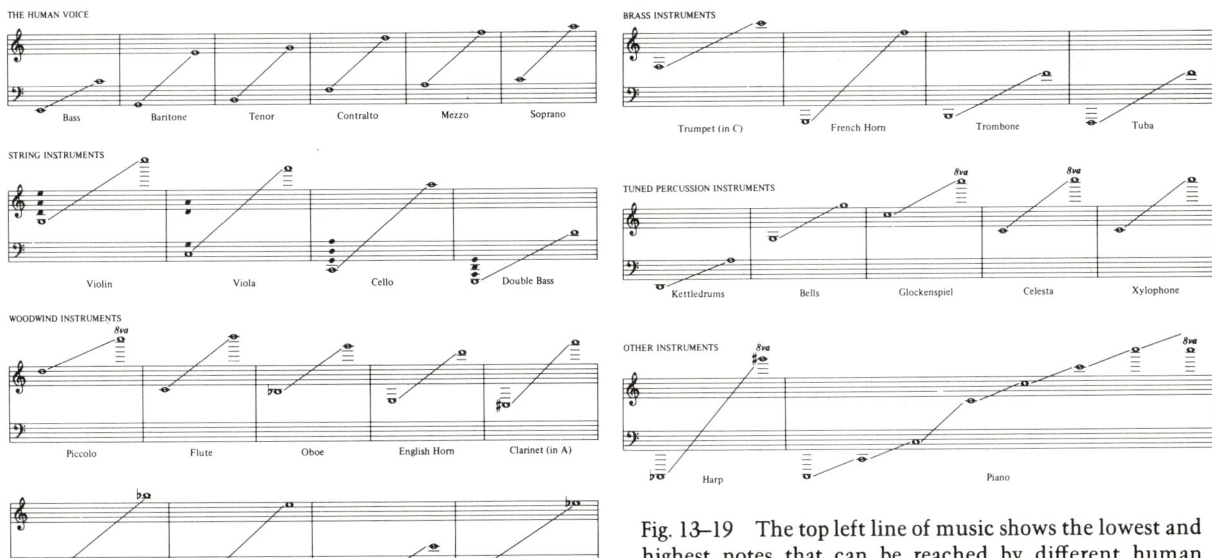

Fig. 13–19 The top left line of music shows the lowest and highest notes that can be reached by different human voices. The remaining lines show the ranges for a variety of instruments.

13.3 *Polarization*

We have already seen that to send waves down a string, we move the string's end perpendicular to the direction of the string. We can move the string up and down; we can move it left and right; we can move it at some slanted direction. All of these directions are at right angles to the string itself, but they are different from each other. (Notice that this choice does not exist for longitudinal waves on a *spring*, since the end must be moved parallel to the direction of the spring.) We call the direction in which you choose to make the string vibrate the direction of polarization for the waves. We have two choices to specifying the direction, corresponding to two dimensions.

If we continuously move the end of the rope back and forth along the same line, then the wave is linearly polarized. Another possibility is to move the end of the rope in a circular motion, so that the polarization continuously rotates. At one point the string will be vibrating left-right, and so on. When this happens, the wave is circularly polarized.

Let's look at the linearly polarized case. We can make a "filter" that allows only waves polarized in a certain direction to pass through. For the case of string, this need consist only of a wooden board with a long slot that is oriented in the direction of the polarization direction we wish to pass. This is shown in Fig. 13–20. Waves parallel to the slot get through. Waves perpendicular to the slot do not.

If we send a wave that is neither parallel nor perpendicular to the slot, only some of the wave gets through. However, the wave that gets through is different from the wave that came in. First, it is weaker, by which we mean that the amplitude is smaller. Also, the

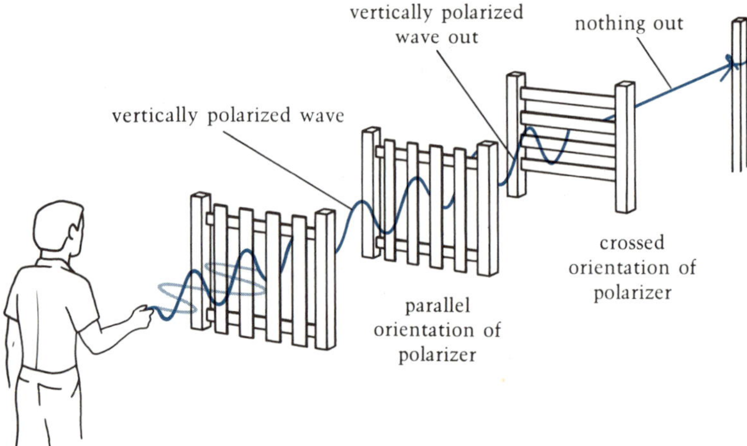

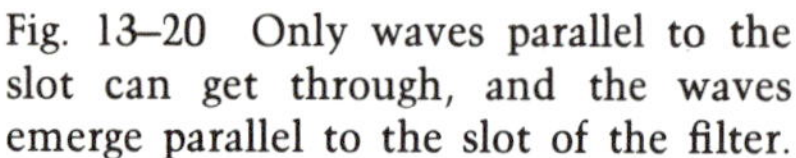
Fig. 13–20 Only waves parallel to the slot can get through, and the waves emerge parallel to the slot of the filter.

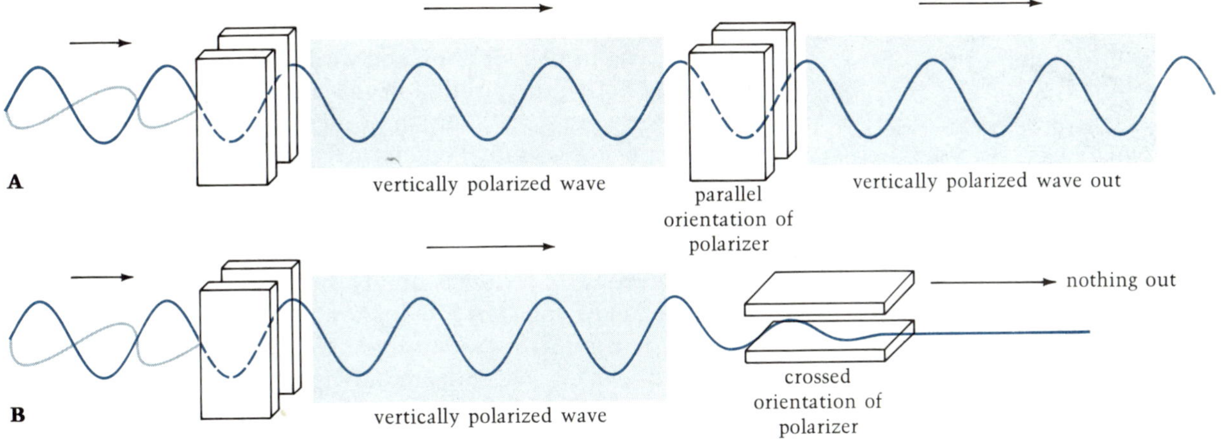

Fig. 13–21 If two filters pass polarizations that are perpendicular to each other, then no wave can get through the combination of the two filters.

direction of polarization has changed. The wave emerges polarized parallel to the slot.

We can demonstrate polarization further with a very curious result. Suppose we use two filters along the rope, the first is up-down and the second left-right. We then send some wave polarized in some direction down the rope. When it reaches the first filter, only the up-down part will get through, and the wave that comes out will be polarized up-down. When this up-down wave encounters the second filter, none of it will get through, since up-down cannot get through left-right. Therefore, if we use two filters oriented at right angles to each other, nothing gets through no matter what the polarization of the incoming wave (see Fig. 13–21).

13.4 *The Doppler Effect*

Our ears can discriminate very well between notes of different frequencies and can also detect changes in frequencies. One thing that you might notice when a car approaches is that, as it passes by, the steady tone of the engine seems to shift lower in frequency (pitch).

This change of frequency caused by relative motion was pointed out by Christian Doppler in 1842. The Doppler effect is a general property of all wave signals, and applies equally well to light waves. We can more easily see how the effect comes about for mechanical waves, such as sound waves, which move through some medium, than we can for light. For sound waves, we must distinguish between the effects caused by the motion of the observer and the motion of the source. We'll treat the two effects separately.

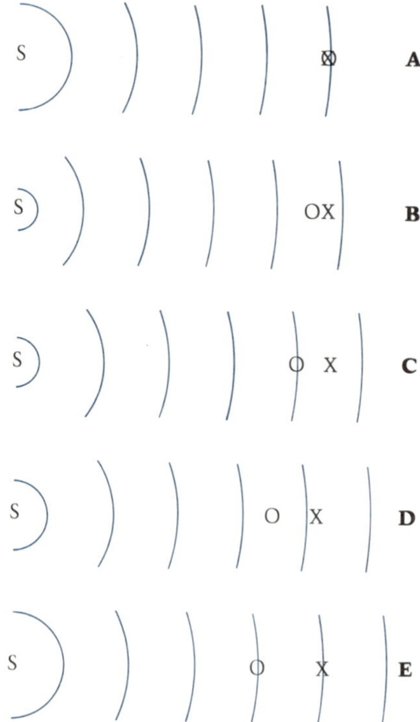

Fig. 13–22 Doppler effect due to a moving observer. Sound waves are emitted by the source, *S*, which is stationary. These waves spread out in circles. In this figure, we watch the progress of these waves in frames taken a quarter of a period apart. A. The waves are to be observed by two observers, *O* and *X*, who start at the same point. B. While *X* stays stationary, *O* starts to walk toward the source. C–D. Observer O has already come into contact with another wave, while *X* is still between waves. E. By the time the wave reaches *X*, a second wave has reached *O*. Therefore, in the time interval shown here, *O* will count more waves than *X*. Therefore, *O* will conclude that the frequency is higher than the value found by *X*. A higher frequency corresponds to a shorter wavelength. If *O* had been walking away from the source, then *X* would have encountered more waves in any given time interval, and *O* would measure a lower frequency than *X* would.

Figure 13–22 shows what happens when the source of sound is fixed and the observer is moving toward the source. The source gives off waves in all directions, and we can keep track of the waves' progress by noting the position of the high point, or crest, of each cycle. As the wave spreads out, these crests form circles centered on the source of the sound. The distance between adjacent crests is exactly the wavelength of the wave.

Now suppose that you want to measure the frequency of waves. All you have to do is time the interval between the passage of successive crests. This time is simply the period, and the frequency is 1/period. Now you start moving toward the source. That means that you are running into the waves. Not only are the waves moving toward you, but also you are moving toward the waves. Therefore, your speed with respect to the waves is increased, and you cover the distance between crests more quickly. This means that you encounter more waves per second than if you were standing still. The frequency appears higher. If you were listening to music coming from the source, all the notes would sound too high in pitch. The effect is just the opposite if you are moving away from the source. Then you will be moving away from the approaching wave crests and you will encounter them less often, making the frequency seem lower.

If you are standing still and the source is moving toward you, as shown in Fig. 13–23, the net result is pretty much the same, but the details are a little different. The waves still spread out as circles. However, now each wave crest is emitted at a different point, so the circles are no longer centered at the same point. If the source is moving toward you, then each wave crest is emitted closer to you than the previous one. Therefore the spacing between wave crests is smaller than the original wavelength. This means that the wavelength appears to be shorter. Since the frequency is the speed of sound divided by the wavelength, a shorter wavelength means a higher frequency. Thus, when the source moves toward you, the frequency of the sound appears higher. If the source is moving away from you, then the waves will be farther apart and the frequency will be lower.

If the source and observer are both moving, we must combine these two effects. Notice that in the cases we chose, the source and observer were moving either directly toward or directly away from each other. Although we chose the easiest cases to visualize, there is still a Doppler shift for motions in other directions. The Doppler shift is a result of the source's and observer's moving apart or together. If the motion is not along the line joining the observer and the source, then the Doppler shift depends only on that part of the motion that tends to bring the source and the observer closer

A small Doppler shift for motion perpendicular to the line joining the source and observer shows up only in special relativity when the relative speed of the source and observer is close to the speed of light.

together or farther apart. The Doppler effect is greatest when the motion is along the line joining the source and observer, and disappears for the instant that the motion is perpendicular to the line joining the two. We often refer to that part of the motion that brings the source and observer closer together or farther apart by the name *radial motion*. If we think of an observer at the center of a circle, radial velocity is the motion that would be along a radius on the circle.

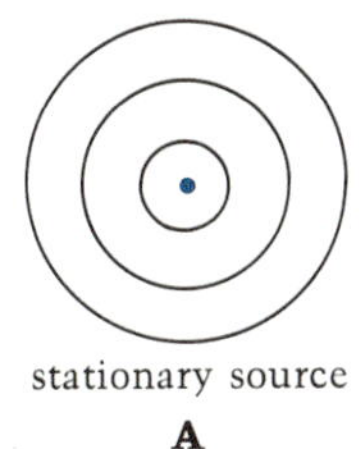

stationary source

A

moving source

direction of motion of source

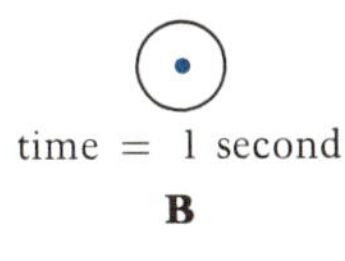

time = 1 second

B

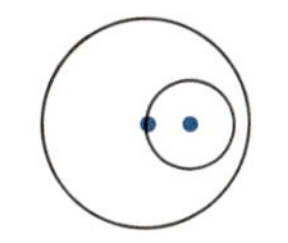

time = 2 seconds

C

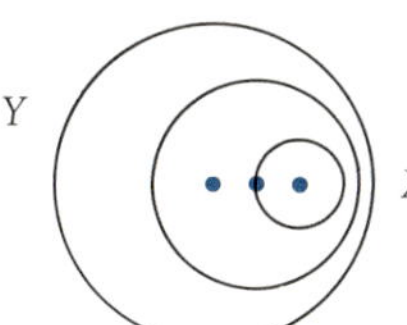

time = 3 seconds

D

Fig. 13–23 We illustrate the Doppler effect by considering a source that gives off a pulse every second. A. For a stationary source, we see that the pulses simply keep expanding about the source. B–D. Let us consider times 1 second, 2 seconds, and 3 seconds after the first pulse is emitted. In B, 1 second after the pulse was emitted, we see a sphere about the point where the source was at time zero. In C, 1 second later, the first pulse has continued to expand about the original point, but a later pulse is expanding about a different point, since the source has moved. In D, we see the state of three spheres, each emitted one second after the other and each expanding about a different point. If we look at the moving source from point X, the pulses appear closer together than they would for a stationary source, so we would measure a shorter wavelength than if the source were at rest. From Y, we would measure a longer wavelength. From Z, the wavelength would be longer than for a source at rest but not as much longer as it is from Y. This is shown in a water tank in Fig. 13–10A.

It is often easier to talk about the wavelength shift, the *difference between the observed wavelength and the rest wavelength*, than to talk about the new wavelength. When the radial velocity is much less than the wave speed, this gives the following relationship for the Doppler shift:

$$\frac{\Delta\lambda}{\lambda_0} = \frac{v}{c}$$

$$\frac{\textit{wavelength shift}}{\textit{rest wavelength}} = \frac{\textit{radial velocity}}{\textit{wave speed}}.$$

The result may be easily remembered, since it merely states that the wavelength change is the same fraction of the original wavelength as the velocity of motion is of the wave speed.

Notice that the amount of shift depends on the actual wavelength of the signal. Let's look at a specific example. Suppose the radial velocity is 10 percent of the wave speed. This means that all of the signals will be shifted in wavelength by 10 percent. For example, a signal whose rest wavelength is 1 meter will be detected with a wavelength of 1.1 meters. We get the 1.1 by saying that the shift is 10 percent of 1 meter, or 0.1 meter, and then adding the shift onto the original wavelength. If the rest wavelength is 2 meters, then the shift is 0.2 meter, so the signal is observed with a wavelength of 2.2 meters. If the objects were moving together instead of apart, all the wavelengths would be shorter. The signal

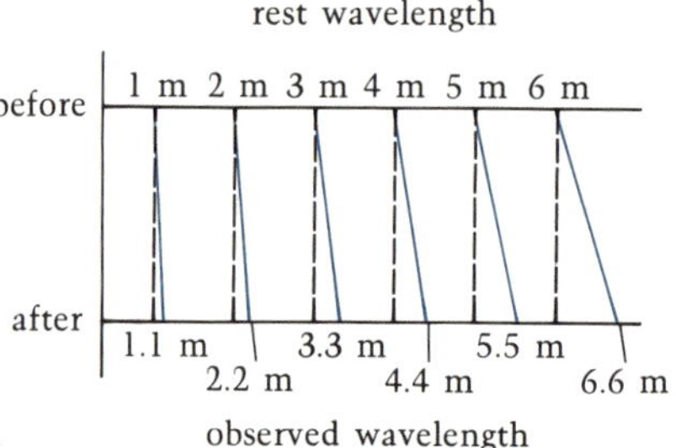

Fig. 13–24 Effect of a 10 percent Doppler shift to longer wavelength. The observed wavelength of any wave will be 10 percent longer than the rest wavelength of that wave. That means that the increase in the wavelength will depend on the original wavelength. The slanted lines connect the rest wavelength to the observed wavelength. The vertical dashed lines show where the 1-, 2-, 3-, 4-, 5-, 6-m markers are on the observed wavelength scale.

whose rest wavelength is 1 meter would be observed with a wavelength of 0.9 meter (1.0 meter − 0.1 meter). The signal whose rest wavelength is 2 meters would be observed at 1.8 meters (Fig. 13–24).

Key Words

wave pulse
elastic
mechanical wave
medium
transverse wave
longitudinal wave
compressions
rarefactions
sound waves
supersonic
shock wave
sonic boom
decibel
periodic wave
simple harmonic motion
harmonic wave
period
frequency
wavelength
amplitude
audible frequencies
ultrasonic frequencies
polarization—linear and circular
Doppler effect
wavelength shift

Questions

1. Would a band marching in rows fit our definition of a wave?
2. Suppose we painted one spot on a string red. Describe what happens to the red spot as a wave pulse, like the one shown in Fig. 13–4, passes by.
3. In a violin, would you expect the wave speed to be faster in the highest pitched string or the lowest pitched string?
4. If you made a musical instrument that used two strings with the same length and the same tension, how would you choose the strings to make sure that they had different pitches?
5. Give two examples of a longitudinal wave.

6. Suppose we put red paint on a spring at one spot. Describe what happens to the red spot as a longitudinal wave pulse passes by.
7. You see a flash of lightning and then hear a clap of thunder 9 seconds later. How far away was the flash?
8. Will you hear a sonic boom when a supersonic aircraft is coming toward you?
9. What evidence is there that metals are not perfectly rigid?
10. The period of a particular harmonic wave is 10^{-3} s. What is the frequency (cycles/s) of this wave?
11. Suppose we put red paint on a spring at one spot. Describe the motion of the red spot as a simple harmonic transverse wave passes by.
12. What is wrong with the following procedure for finding the period of a transverse wave on a string? Pick any point on the string at any time and note its height above or below the zero point. Then see how much time expires before that point is next at the same height above or below the zero point.
13. What is the wavelength of a sound wave whose speed is 331 m/s and whose frequency is 440 cycles/s (440 Hz)? (b) What is the wavelength of a sound wave whose frequency is 880 cycles/s (880 Hz)?
14. Suppose we double the frequency of a particular sound wave. (a) What happens to the speed of the wave (assuming it is still traveling through the same medium)? (b) What happens to the wavelength of the wave?
15. Why can't a longitudinal wave be polarized?
16. Suppose a wave is moving along a string, with an up-down motion. It encounters two filters in a row, each filter arranged so that it only passes left-right waves. No signal gets through. (a) If you could only rotate one filter, which one (first) or (second) would you rotate to get some signal through? (b) How does the strength of the signal depend on the angle through which you rotate this filter? Explain.
17. Suppose we have two filters on a string. One only passes up-down polarizations and the other only passes left-right polarizations. What types of polarizations could get through this filter combination?
18. If you were moving toward someone playing an "A," would the note sound sharp or flat? (Sharp corresponds to a slightly higher frequency and flat corresponds to a slightly lower frequency.)
19. You are moving toward a source of sound at a wavelength of 1 meter. You observe the note to have a wavelength of 0.99 meter. (a) What wavelength would you observe for a note emitted by the same source at a wavelength of 2 meters? (b) What wavelengths would you observe for these two notes if you were moving away at the same speed as before? (c) What wavelengths would you observe if you were moving toward the source, but at twice the speed as before?
20. A Doppler shift is often most obvious when a source of sound moves right past you. Why?

CHAPTER 14

Standing Waves

14.1 *Interference—When Waves Get in Each Other's Way*

Now we know how a wave travels from one place to another. In each of the cases that we looked at in the last chapter, there was only one source for the waves. However, what do you observe at the center of a rope if both ends are being shaken and pulses come at you from both directions? Or, what if you are in a rowboat in the center of a lake and ripples are coming at you from various directions?

When two (or more) waves are in the same place at the same time, we say that they *interfere* with each other. We call the pattern of light and dark that results an interference pattern. By "interference," we really mean that the waves are in the same place at the same time and apparently get in each other's way. The result is given by what is called the principle of superposition: *The net wave disturbance at any given time and place is the sum of all the wave disturbances at that particular place at that particular time.* Nothing could be simpler; we just add the waves together.

As an example, let's look at two identical waves that are moving down a string in the same direction (as shown in Fig. 14–1). Let's suppose that each maximum of one wave occurs at the same time

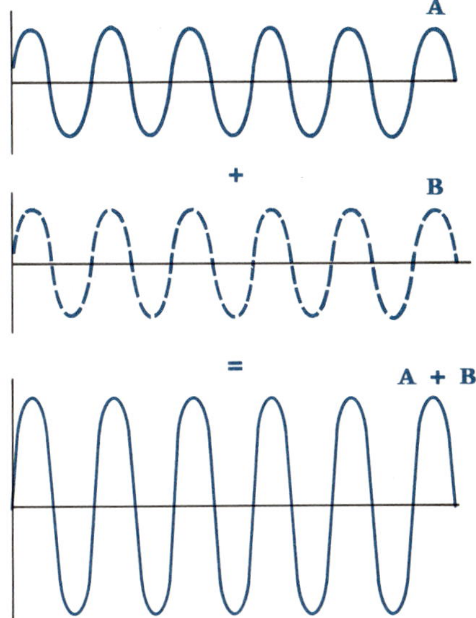

Fig. 14–1 Adding together two waves that are in phase. By "in phase," we mean that the waves have the same frequency, and the high point of one wave occurs at the same place as the high point of the other wave. In the case of this figure, the two waves that we are adding together (A and B) are identical. To find the resulting wave (A + B), we just add the height of the two waves at each point. Since the two waves are identical, we end up with a wave that is twice as high as each of the ones we started with. When two waves add together to produce a wave whose height is greater than the height of either of the waves that we started with, we call the interference between the waves *constructive*.

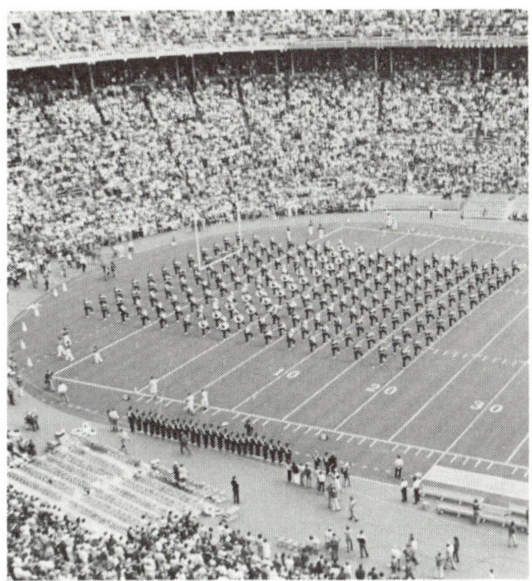

Fig. 14–2 Individuals in a marching band are in phase with each other.

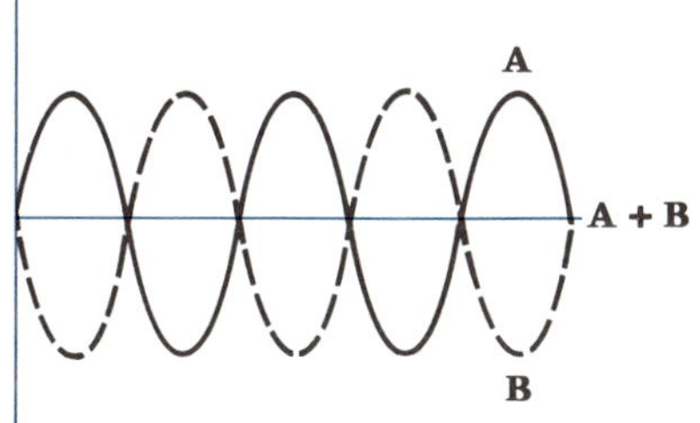

Fig. 14–3 Two waves that are completely out of phase. The high point of one wave falls in the same place as the low point of the other. When we add these two waves together, we get zero at every point. We have complete *destructive interference*.

and place as each maximum of the other, that is, the waves are in phase. Waves in phase are like a marching band with all its players in step.

If we want to know the wave that results at any given time from the superposition of two waves, we simply take a picture of the two individual waves at that time, as shown in Fig. 14–1, and add them together. By adding, we mean that at every position we add the height of one wave at that position to the height of the other wave at that position. In the case of identical waves that are in phase, we get a wave that, at every point, has twice the height of each of the original waves. The resulting wave has the same frequency as the original wave, but has twice the amplitude. When two waves add this way to produce a wave whose amplitude is greater than the amplitude of either original wave, we say that the interference is constructive. By "constructive," we mean that the interference has actually resulted in the building of a stronger wave. Constructive interference results when the waves are in phase.

Now let's look at what happens when the two waves are out of phase. When they are completely out of phase, the high point of one occurs at the low point of the other (Fig. 14–3). Again, at any given time, we can find the resulting wave pattern by adding these two waves together. You can see from the figure that one wave always has a value that is the negative of the other wave. This means that at any point the two waves add to zero. They cancel each other exactly. This is destructive interference. By "destructive," we mean that the two waves combine to produce a wave

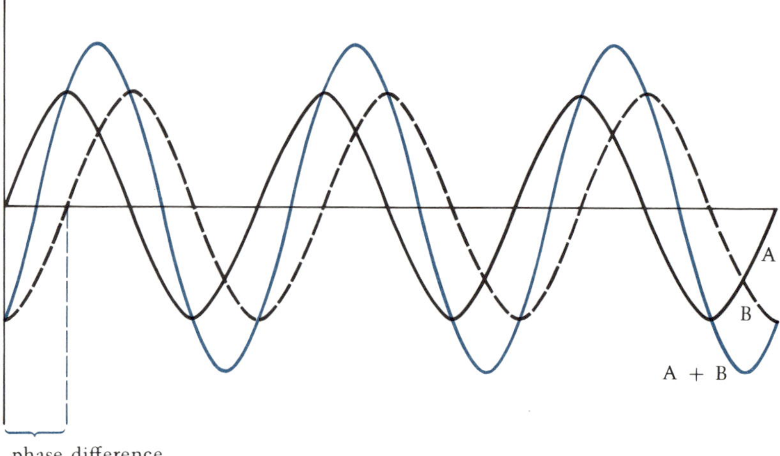

Fig. 14–4 These waves are partially out of phase. Wave B lags behind wave A by some amount, called the phase difference. When we add these two waves together, we get a wave whose amplitude is stronger than that of either A or B, but not as strong as when the two waves were exactly in phase. Notice that the peak of the new wave falls between the peaks of the waves A and B. As an exercise, you might try tracing waves A and B onto a piece of paper. Then, at several points along the wave, use a ruler to measure the height of A and the height of B. Add these two heights together, and that will be the height of the resultant wave at that point. Plot a point that high above the axis. When you have enough points, see if they follow the curve we have drawn in for the sum of A and B.

whose amplitude is less than that of the original waves. In this particular case, the amplitude is zero. This is a very interesting and possibly surprising result. *It is possible to have two waves meet in such a way that no wave results.*

Of course there are also cases in between, in which waves are not exactly in phase or completely out of phase. Some examples of these intermediate cases are shown in Fig. 14–4.

Now that we know how to add waves together, let's look at two waves traveling along a string in opposite directions (Fig. 14–6). We'll have to look at the pattern along the string at a few different instants of time. We can start looking at a time when the two waves exactly cancel each other along the entire length of string; that is, the two waves are completely out of phase so that all points on the string are at zero height at that particular time. We look again a little time later, so that one wave has moved slightly to the left and the other has moved slightly to the right. Since they have moved in opposite directions, there will no longer be exact cancellation over the whole string. The two waves still cancel at some points along the string but not at others. The points at which the waves always cancel are regularly spaced, one-half wavelength apart, along the string. Midway between these two points are points at which the interference pattern has a maximum, but the height of the maxima is greater than the height of the original waves, as shown in the figure.

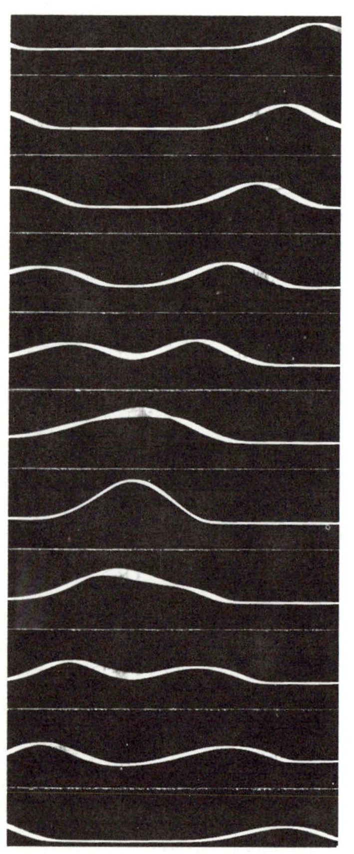

Fig. 14–5 Two pulses passing each other in opposite directions. Each frame shows the pulses at a slightly later time than in the previous frame. At any time, the total wave disturbance is the sum of the two pulses, even when they overlap. Notice that the two pulses retain their identities. We can tell this since the two pulses have different shapes and we can tell which is which.

Now we can let the waves move a little farther along. When we do the addition, we find that there are still points, spaced a half wavelength apart, for which the waves cancel. Moreover, these are exactly the same points for which the cancellation occurred in the previous picture. The zero points always occur at the same places. We call these zero points **nodes**. In between the nodes, the maxima are called **antinodes**. Note that in this case the antinodes are higher than in the previous picture.

We can keep watching the waves slide over. Eventually we reach the point when the two waves line up exactly. The resulting wave then still has the nodes and antinodes in the same places. However, the antinodes have grown stronger, and are now at a height equal to the sum of the heights of the peaks of the two waves. The antinodes never get higher than this. If we let the waves slide over farther, the height of the antinodes gets smaller again. If you were to watch the string for a while, you would see the resulting wave get larger and smaller, but it would not *appear* to be moving along the string, since the nodes are always in the same place. For this reason, we have a **standing wave pattern**.

Everything we have said about interference has been illustrated with transverse waves on a string. However, interference is a phenomenon that is common to all waves, and the analysis for waves on a Slinky, or sound waves in air, or water waves in a tank, would all be the same. In each case we could produce standing waves, and the waves would have the same property.

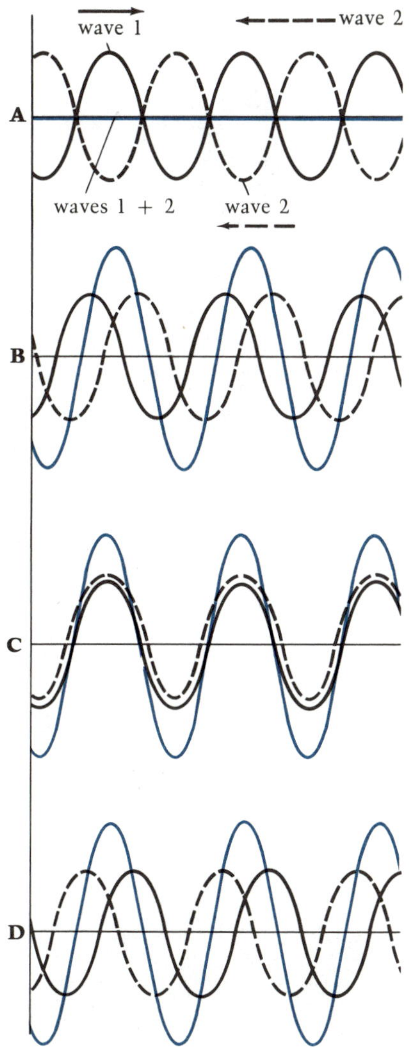

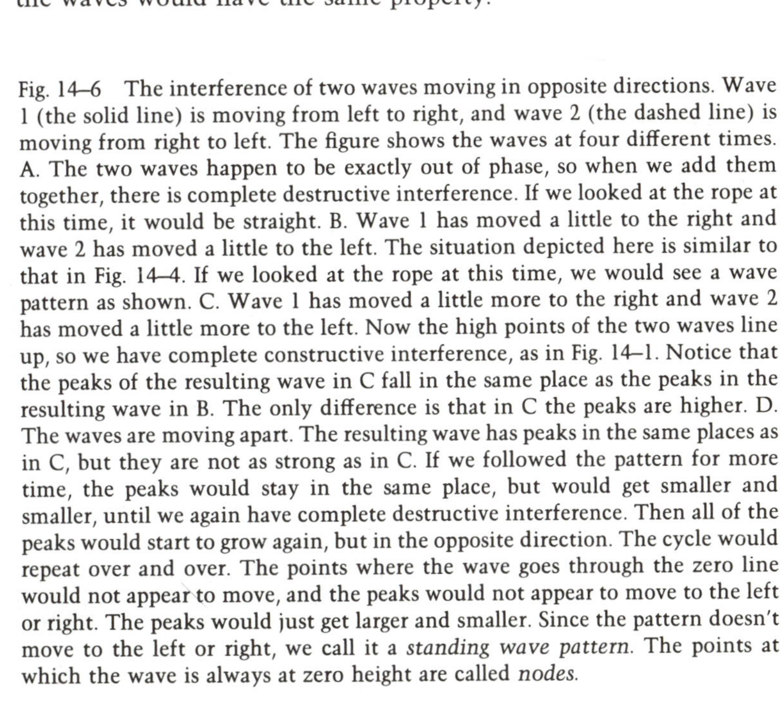

Fig. 14–6 The interference of two waves moving in opposite directions. Wave 1 (the solid line) is moving from left to right, and wave 2 (the dashed line) is moving from right to left. The figure shows the waves at four different times. A. The two waves happen to be exactly out of phase, so when we add them together, there is complete destructive interference. If we looked at the rope at this time, it would be straight. B. Wave 1 has moved a little to the right and wave 2 has moved a little to the left. The situation depicted here is similar to that in Fig. 14–4. If we looked at the rope at this time, we would see a wave pattern as shown. C. Wave 1 has moved a little more to the right and wave 2 has moved a little more to the left. Now the high points of the two waves line up, so we have complete constructive interference, as in Fig. 14–1. Notice that the peaks of the resulting wave in C fall in the same place as the peaks in the resulting wave in B. The only difference is that in C the peaks are higher. D. The waves are moving apart. The resulting wave has peaks in the same places as in C, but they are not as strong as in C. If we followed the pattern for more time, the peaks would stay in the same place, but would get smaller and smaller, until we again have complete destructive interference. Then all of the peaks would start to grow again, but in the opposite direction. The cycle would repeat over and over. The points where the wave goes through the zero line would not appear to move, and the peaks would not appear to move to the left or right. The peaks would just get larger and smaller. Since the pattern doesn't move to the left or right, we call it a *standing wave pattern*. The points at which the wave is always at zero height are called *nodes*.

14.2 *Reflections—The End of the Line*

So, we can produce standing waves by sending waves of the same frequency down a string in opposite directions. To do this by vibrating both ends in the appropriate way takes twice as much work as is necessary. We can simply rely on the reflections at the end of the rope to get the wave traveling down a string in the direction opposite to the original wave. Up to this point we have just been talking about waves going down a string and never coming back. However, this is not the case in real life.

Let's look at what happens when a pulse reaches the end of a string (Figs. 14–7 and 14–8). When a pulse gets to the tied-down

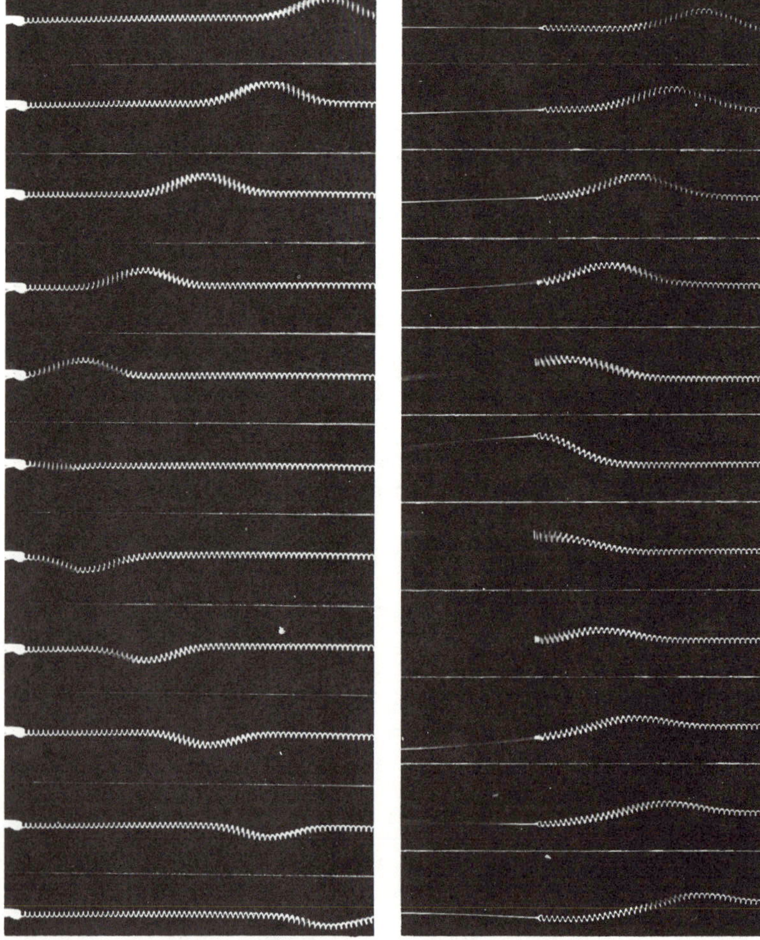

(left) Fig. 14–7 Reflection of a pulse from a fixed end. Each frame shows the pulse at a slightly later time than in the previous frame. The pulse enters from the right and is reflected from the fixed end at the left. Notice that the reflected pulse is flipped. This is because the reflected pulse and the incoming pulse must exactly cancel at the fixed end so the end can really stay fixed.

(right) Fig. 14–8 Reflection of a pulse from an end which is free to move. (In this case the end is tied to a light string.) Since the end is free to move and respond to the incoming pulse, the reflected pulse leaves with the same sense as the incoming pulse. That is, the incoming pulse is up and the outgoing pulse is up.

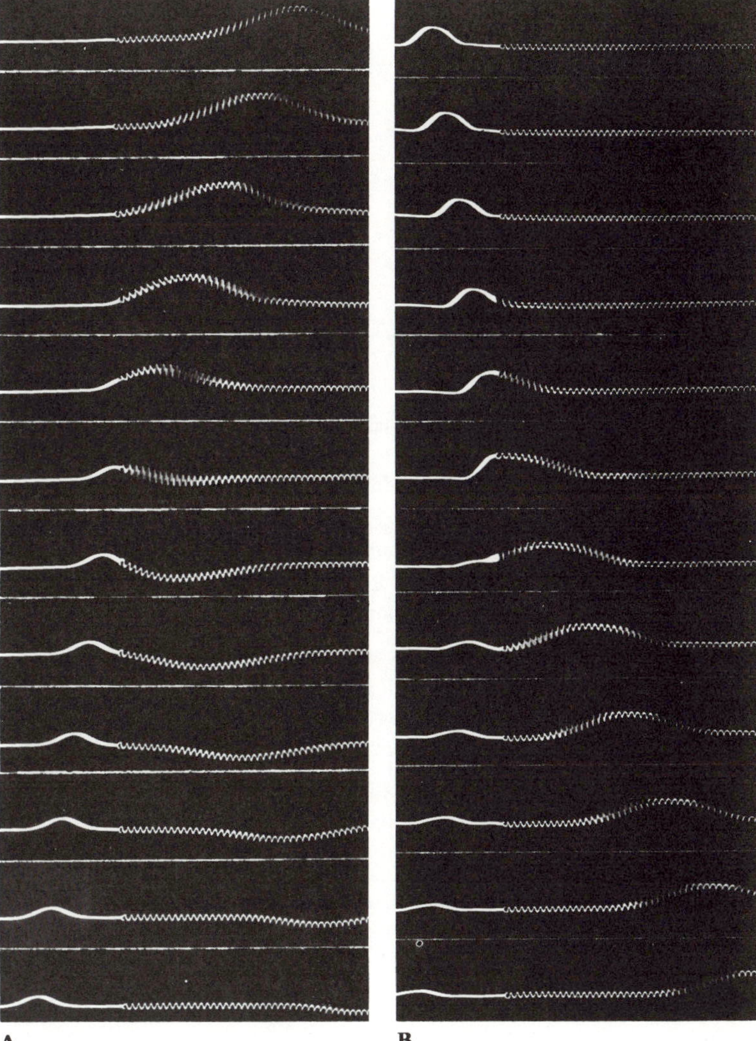

Fig. 14–9 Waves passing from a spring of one weight to a spring of another weight. In each frame of each series, the pulses are shown at a slightly later time than in the previous frame. A. The pulse moves from the light spring on the right to the heavy spring on the left. Notice that, as well as a pulse that is transmitted to the heavy spring, there is also a reflection that goes back in the light spring. The reflection is flipped from the original pulse. B. The pulse moves from the heavy spring on the left to the light spring on the right. As in A, there is a transmitted pulse and a reflected pulse, but this time the reflected pulse is not flipped.

end, it is reflected back. If the end of the string cannot move, the reflected wave is upside down. If the end is not-tied down, the reflected wave is not flipped.

So far we've seen that waves get reflected from a string end whether it is tied down or free. We get a reflection back from any point where the waves pass from one region to another region where their speeds are different. For example, if we attach two ropes with different masses to each other, we get a reflection from the point where the two strings are joined (Fig. 14–9). The reflected wave may be flipped or not depending on which is the more massive string.

Notice that the reflected wave is not as strong as the incoming wave. In addition to the reflected wave, there is one that is transmitted (that is, keeps going in the original direction), taking up some of the energy. This is quite amazing. It seems strange that part of the energy can be reflected and part transmitted when there is a transition from one medium to another. You certainly couldn't think of that happening with a ball. When you throw a ball into a tank of water, the ball does not divide into two parts, one going into the water and one bouncing back at you. (However, it is rumored that there have been cases in which baseball players hit the ball so hard that it came apart, the cover being caught and the core going over the fence.) The ability to bounce off nonrigid points of connection, and to divide into a reflected and transmitted part, is something that clearly distinguishes waves from simple particles.

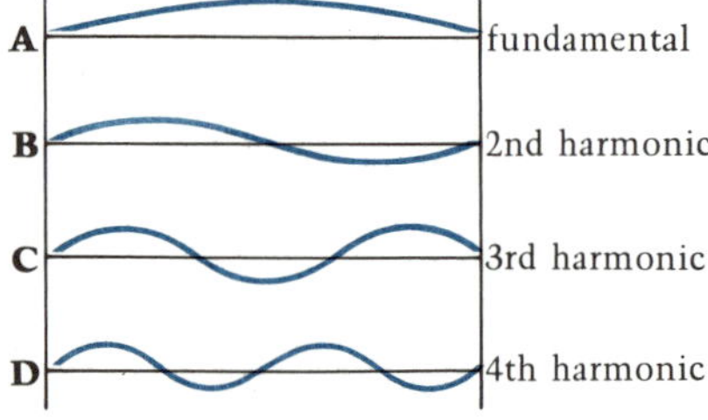

Fig. 14–10 Standing wave patterns on a string with fixed ends. Since the fixed ends cannot move, there must be a node at each end. A shows the fundamental, or first harmonic. Since the nodes are always a half wavelength apart, the wavelength of the fundamental must be twice as long as the string. B shows the second harmonic (or first overtone). The wavelength for the second harmonic is equal to the length of the string. For the third harmonic (in C), the string is $1\frac{1}{2}$ wavelengths long, and for the fourth harmonic (in D) it is 2 wavelengths long. This series continues on and on, each time adding one more node to the string. Note that this figure only shows each rope at one point in its cycle. A half cycle later the up points will be down and the down points will be up, but the nodes will still be in the same places. When we say that there is an antinode between the nodes, it doesn't matter whether the rope happens to be up or down. All that matters is that it is at its maximum possible distance from the zero line (the line joining the nodes).

14.3 *Standing Waves on a String*

The reflection of waves off the ends of a string gives us a way to get waves of the same frequency moving in opposite directions on a string. In fact, it is not even necessary for us continuously to drive one end of the string up and down. We can start a wave going down the string and then let the reflections off both ends take care of the rest. This is essentially what we do when we pluck a string. Our standing wave pattern will satisfy a few conditions. First, the nodes will be half a wavelength apart. Second, each fixed end of the string must have a node, because a fixed end cannot move. Third, each free end will have an antinode. Now let's see what happens when we apply these conditions to a string with both ends tied down and a string with one end tied down and one end free.

First, let's look at the string that has both ends tied down. We have seen that there is a node at each end of the string. Only certain wavelengths are allowed by our conditions. One possibility is that there are no other nodes except those at the ends. This case is shown in Fig. 14–10. In this case, since the nodes are always a half wavelength apart, the length of the string must be equal to half the wavelength of the wave, or the wavelength must be twice the length of the string. Once the wavelength is known, we can find the frequency. We call the oscillation with one node at each end only the fundamental oscillation mode for the string. The oscillation frequency for the fundamental mode is the fundamental frequency.

The fundamental mode is not the only way that we can have standing waves on the string. Another possibility is to have a node in the middle in addition to the nodes at the ends. In this case the nodes are half as far apart as in the fundamental mode. The wavelength is therefore half the wavelength of the fundamental, which is the same as saying that the frequency is twice that of the fundamental. This mode is called the second harmonic or the first overtone. By a harmonic we simply mean a multiple of the fundamental. In that case the fundamental can be thought of as the first harmonic.

Higher modes are also possible. There can be one in which the string is divided into thirds by the nodes. The frequency of this third harmonic will be three times that of the fundamental. This process of division can go on, giving higher and higher harmonics with shorter and shorter wavelengths and higher and higher frequencies. The only requirement for a given wavelength is that the distance between nodes must fit on the string an integral number of times. The series of wavelengths that satisfies this criterion is called the harmonic series for that string.

So we see that the string can support standing waves for certain natural frequencies. These frequencies arise from the condition that the ends of the string must be fixed. In other words, the selection comes from the fact that we are constrained by what the ends of the string will do. If we place different constraints on the ends, we will get a different series of modes, but we will nonetheless get a series of modes.

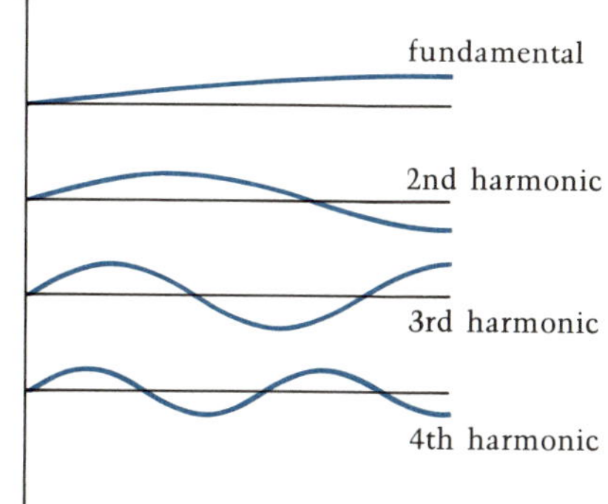

Fig. 14–11 The four longest wavelength modes for a string with one end fixed and one end free. The fixed end must have a node and the free end must have an antinode (a point of maximum distance either above or below the zero line). In the fundamental, there is only one node, so the rope must only be a quarter of a wavelength long. This means that the wavelength for the fundamental in this case is four times the length of the rope. In the second harmonic the rope is three-quarters of a wavelength long, in the third harmonic it is $1\frac{1}{4}$ wavelengths long, and so on.

Suppose we want to change the fundamental frequency of a given string, something that is a question of practical interest for many musicians. Each of the strings has a different fundamental frequency. However, suppose you are playing on one string. How do you change its fundamental frequency? One way is to change the speed of the waves on the string by changing the tension. The greater the tension, the faster the waves travel. In this case, the wavelength will still be the same for the fundamental mode, but the frequency corresponding to that wavelength will change (since *frequency = speed/wavelength*). When you tune a guitar by turning pegs, you increase or decrease the tension on the string. Another way to change notes is to change the length of the string. Guitarists use their fingers to shorten the vibrating part of their strings. A shorter string has its ends, and thus the two nodes, closer together. The wavelength of the fundamental mode becomes shorter. A shorter wavelength means a higher frequency.

Next let's look at the case of a string that is fixed at one end and free at the other end. Now our conditions are slightly different. The fixed end must have a node and the free end must have an antinode. For this string, the fundamental mode is one in which there is only one node—at the fixed end. The first overtone occurs when there is one node in addition to the one at the fixed end. These modes as well as the higher ones are illustrated in Fig. 14–11. Can you draw the modes for a string that is free at both ends?

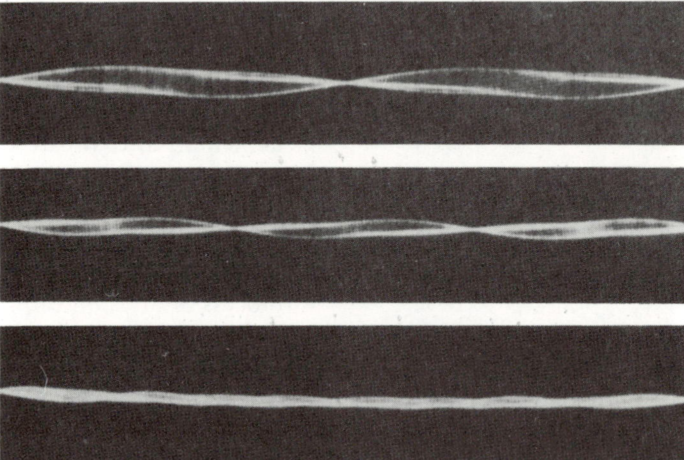

Fig. 14–12 When we play the string of a musical instrument, we don't simply get one harmonic or another. Instead we get a combination of modes, as can be seen in these photographs.

Given all of these ways in which a string can actually vibrate, how does it decide in any given situation how it will actually vibrate? In part it depends on how the string is started in vibration. This could be by plucking or bowing the string. The exact form of the oscillations will depend on where the string is plucked, how hard it is struck, etc. For example, for a violin string, the fundamental frequency gives the note that you are actually playing. But much of the quality of the sound comes from the combination of harmonics that also contribute to the sound. The fact that two violins playing the same "A" don't sound the same comes from differences in the overtones that accompany the fundamental.

Set a guitar or violin string vibrating, and touch its center midpoint lightly with the tip of a handkerchief. This damps (shuts off) the fundamental, and leaves the first harmonic. Once you hear the first harmonic, set the whole string in vibration again. Careful listening will reveal that you can hear both the fundamental and the first harmonic, and have been doing so all along. By touching the string one-third, one-quarter, etc., of the way along, you can listen to the higher harmonics as well. All are present when you listen to an ordinary note.

14.4 *Resonance*

So far we have been talking about getting the string to vibrate by striking it once. It is also possible to be continuously driving the string, such as when we repeatedly move the free end up and down.

A

B

Fig. 14–13 Standing waves on a drumhead. In two dimensions the standing wave patterns can be a little more complicated, but the ideas are the same. For the drum, there must be a node all the way around the edge, where the skin is attached. In A, we see the fundamental mode. In B, we see one possibility for the second harmonic, where there is a node along one diameter of the circle. It is also possible to have a second harmonic in which the node is in a circle about the center. Compare these patterns with Fig. 14–10.

Fig. 14–14 The New York Philharmonic's tympanist tuning his drum by tightening the drumhead.

We start the vibrations slowly, so that the frequency of the motion is below the fundamental frequency of the string. The motion is disorganized. No coordinated pattern will appear on the rope and it is hard to keep the motion going.

As we increase the frequency of motion, we reach the fundamental frequency. You will note that as we reach that frequency, the motion will suddenly lose its disorganized appearance, and will oscillate in the fundamental mode. It is also easier to keep the motion going in this mode. It seems that at this frequency it is easier to keep the energy flowing into the string.

As we increase the frequency of vibrations, the string goes back to its disorganized motion, and it again gets harder to keep the motion going. Eventually you reach the second harmonic and things revert to their easier, more organized fashion. By a similar process, other harmonics occur as you go higher in frequency. Every time you move the end of the string at a frequency corresponding to one of the frequencies at which standing waves can occur (or one of the natural frequencies of the string), it becomes very easy to keep the motion going and to put energy into the string. At all other frequencies the motion of the string looks disorganized and you get the feel of doing a lot of work with little result. This phenomenon, by which a system responds most favorably to energy pumped in at one of its natural frequencies, is called resonance.

Resonances are very important in almost all branches of physics. One way to learn about some physical system, be it a solid, a crystal, a nucleus, or a subnuclear particle, is to find out its resonant, or natural, frequencies.

Resonances are also important on a larger scale. In your television, your channel selector allows you to pick the frequency at which the electronic circuit in your set will resonate. Then, when it receives a signal of that frequency, your TV responds to it alone, producing a nice clear picture. You may also notice certain resonances in driving your car. For example, many cars give a very smooth ride except at one particular speed. At that speed, something may be vibrating at a frequency that corresponds to a resonance in the frame of the car. Although the car is going along smoothly as you accelerate, when you reach the speed at which the resonance occurs, you will suddenly notice the whole car start to vibrate. As you pass that speed, the vibration will stop.

In this section we have been discussing a vibrating string, which is a one-dimensional object. So the waves we have been talking about are traveling in one direction only. However, everything that we said applies to more complicated systems. For example, a vibrating drumhead is a two-dimensional analogy to a vibrating string. Figure 14–13 shows what happens when we drive a drum-

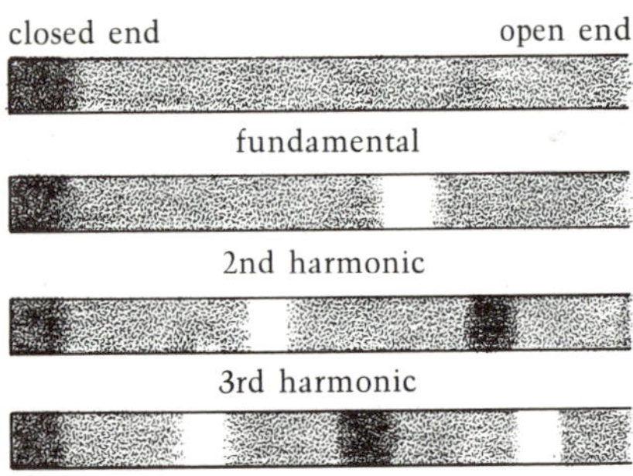

Fig. 14–15 Standing waves in a tube with one end open and one end closed. At the open end the pressure must always equal the atmospheric pressure, since the air at the end of the tube and the atmospheric air are always in contact. In this diagram, we have lightly shaded the air that is approximately at atmospheric pressure (node). The darker shading corresponds to pressure higher than atmospheric (antinode), and no shading corresponds to pressure lower than atmospheric (also antinode). Remember that the pressure wave is strongest where the pressure is either higher or lower than the atmospheric pressure (the antinodes). Note that, just as we did in the figures of the standing waves on ropes, we have only shown one part of the cycle. If we waited half a cycle, the lightly shaded areas will remain the same (since these are around the nodes), and the dark shaded and nonshaded areas will be interchanged. In the fundamental mode there will be an antinode at the closed end (either high or low pressure, but in this case we have drawn a high-pressure shading), and a node at the open end. This is similar to the fundamental mode on a rope with one tied end and one free end. The wavelength of the fundamental will be four times the wavelength of the tube. In the second harmonic, we will have an antinode at the closed end, and another two-thirds of the way down the tube.

head at certain resonant frequencies. For the drumhead, the rim is stationary. Therefore, there must be a node all the way around. In the fundamental oscillation, the center of the drumhead moves only up and down. As you go to higher frequencies, however, the motion can get a little more complex than on a simple string.

14.5 *Standing Sound Waves*

Our analysis of waves on a string can carry over to sound waves in a pipe. We must remember that a sound wave is a longitudinal wave, with pressure variations. When we have a pressure node, it doesn't mean that the pressure is zero. It means simply that the pressure is always equal to its normal value at that point. When there is no node, the pressure is slightly higher or lower than this normal pressure.

Just as our string could have a fixed end or a free end, a pipe with sound waves can have an open end or a closed end. At an open end, the pressure must always be equal to the pressure just outside the tube. There can be no change in the pressure at this end. *So the open end of a tube always has a pressure node.* Since molecules cannot move through a closed end, *the closed end of a tube always has a pressure antinode.* (Remember—a pressure node means no variations in pressure; an antinode means the maximum variation both above and below normal.) We can use these rules to figure out the types of standing wave patterns that are allowed in various pipes, just as we did for standing waves on a string.

Figure 14–15 shows the fundamental mode for a tube with one open end and one closed end, as might be the case for a drinking glass. The fundamental oscillation occurs when there is a pressure node at the open end, an antinode at the closed end, and no other nodes in the tube. Remember that the distance between nodes is half a wavelength. In this fundamental oscillation, we don't even get from one node to another. In the next harmonic, there is an additional node between the two ends. The wavelength for this oscillation is shorter than for the fundamental, which means that it has a higher frequency. Of course, this series can continue on, and we can get higher and higher frequencies.

When you strike one of these tubes, most of the sound comes at the fundamental frequency. But again, the overall impression and quality of the sound depend on the presence of overtones. You can make a musical instrument by filling a series of glasses with different amounts of water, but the overtones are insubstantial. An organ works in essentially the same way. Air is forced over an opening in each pipe; the other end can be either open or closed,

Fig. 14–16 When a pipe organ is played, standing waves are created in pipes of different sizes. This organ is in Alice Tully Hall, Lincoln Center, in New York City.

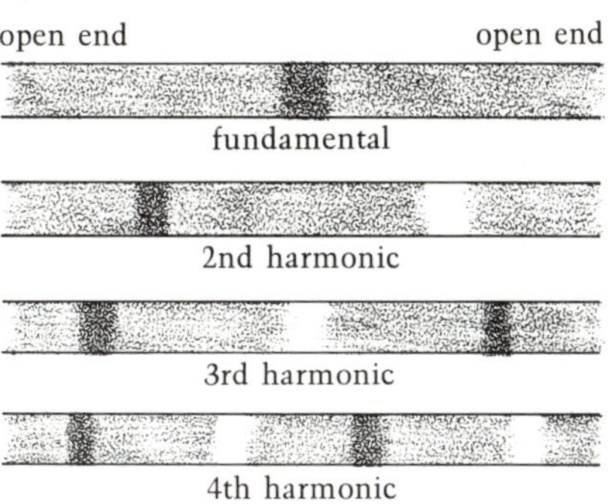

Fig. 14–17 Modes in a pipe with two open ends. There must be a pressure node (pressure always equal to outside pressure) at each end. Compare these to the rope with two fixed ends, shown in Fig. 14–10. In the fundamental, there is a pressure antinode in the center, and the wavelength is twice the length of the tube. In the second harmonic, there are two antinodes (the dark region and the light region), and the wavelength is equal to the length of the tube.

depending on the construction of the organ. We have just analyzed the case of a closed-end pipe.

What about an organ pipe with an open end—that is, with both ends open? At each open end we must have a pressure node, so the oscillations of this system will be equivalent to those of the string with two fixed ends (Fig. 14–17). For the fundamental, there will be only the nodes at the two ends. In this case the length of the pipe is half the wavelength of the wave. This means that the wavelength is twice the length of the pipe. Of course, higher harmonics, with higher and higher frequencies, are also possible.

To start the standing waves in a pipe, we can just start air flowing through the pipe. In this case we hear mostly the fundamental and some variety of overtones, depending on how we send the air through and the exact properties of the pipe. Blowing against a small hole in the end of an organ pipe starts a swirling motion in the air on the other side of the hole. The swirling motion starts the vibrations (Fig. 14–18). This is why you blow across the top of a soda bottle, and not directly into it, to create a sound. Notice also that wind instruments like the flute are played by blowing across the hole rather than into it.

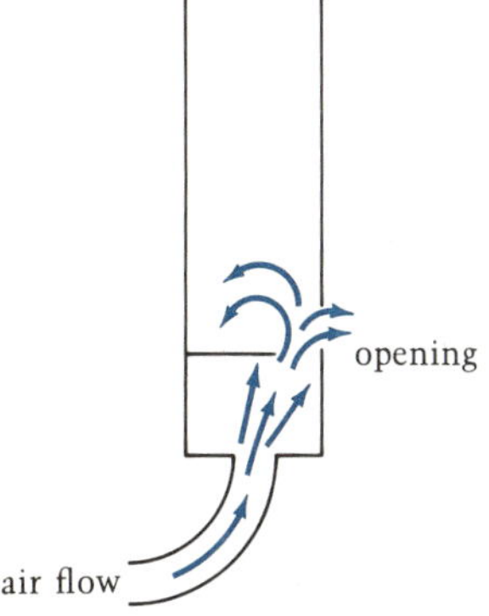

Fig. 14–18 Air flow in an organ pipe. Air is forced in through a hole in the bottom. It then passes by a partial obstruction which forces the air to start circulating rather than flowing smoothly. The circulating air at one end of the pipe starts vibrations in the air in the pipe and a standing wave pattern is set up. There is an opening at the lower end of the long pipe. This means that the pressure at the bottom of the pipe is always equal to the outside pressure, so there is a pressure node at the bottom of the pipe. The standing wave pattern then depends on whether the top end of the pipe is open or closed.

In practical situations the calculations can get quite difficult, and there is often no substitute for a good experiment. For this reason, when the interior of what is now known as Avery Fisher Hall in New York, where the New York Philharmonic plays, was renovated to improve the sound, an audience had to be recruited to fill the hall while an orchestra played. Only tests under such real conditions could prove the success of the redesign. Also, as a test of the reverberation of various parts of the hall, a direct approach was made: a cannon was brought on stage and fired.

In other instruments, such as the oboe, you blow directly in, but your breath does not flow directly into the pipe. Instead, it forces a reed or reeds to oscillate, and it is the oscillation of the reed that starts the standing waves in the instrument.

We can change the fundamental frequency of any pipe by changing the speed of the waves, but this is harder to do than for a string. One way to do so would be to change the temperature significantly. Alternatively, we could change the gas in the tube. For example, if we replaced air in a tube with helium, the speed of the waves would be higher, which would increase the frequency that corresponds to any given wavelength. You will have noticed this effect if you have ever heard someone who has breathed in a little helium. The voice sounds much higher. (Don't try this yourself, since breathing in any gases other than oxygen can be very dangerous.) Still another way to vary the natural frequency of a tube is to change the length; again, this is something that is not as easy to do as with a string.

14.6 *The Nature of a Note*

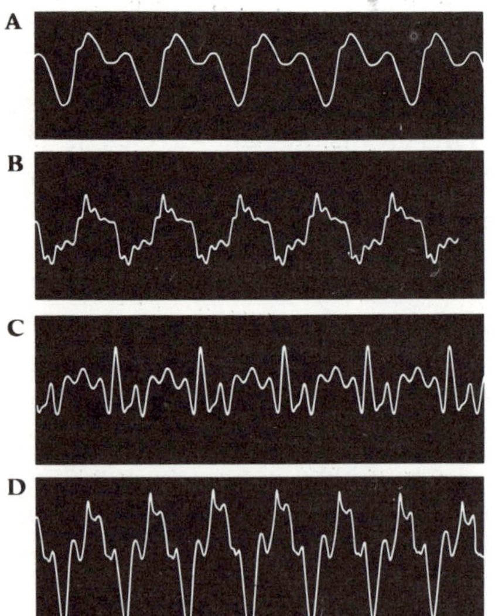

Fig. 14–19 The wave patterns from a flute (A), a clarinet (B), an oboe (C), and a saxophone (D), all playing the same note. The fundamental is the same but the harmonics are very different. The differences give the instruments their respective sounds.

Our ears can distinguish between frequencies—different notes sound different. But all notes of the same frequency don't sound the same. An A on a piano does not sound the same as that A on a violin or a flute. The exact mixture of all of the harmonics of the fundamental gives each its distinctive sound.

If we look at how pressure changes with time for the same note played on various instruments, we can pick out the periodic structure in each (Fig. 14–19). However, the result is not a simple wave. There may be a very complex pattern. The exact nature of this pattern depends on how strong the contribution is from each harmonic. We could use a mathematical technique, called Fourier analysis, that tells us how much of each harmonic is present. To check the result, we could draw (or have a computer draw) the waves corresponding to each of the harmonics. In this plot, the amplitudes of the various waves would correspond to the amount of the harmonic that is present. If we add all of these waves together, we get a pattern that exactly duplicates the original pattern we got from the instrument.

We can reverse this procedure. Rather than analyzing a given sound wave, we can create notes that sound like any instrument we want. This process is called *synthesis.* It is relatively easy to build an electronic circuit that will oscillate at a single frequency, with no harmonics present. For a given note, we would then have to build several circuits. One would correspond to the fundamental

Fig. 14–20 The Moog synthesizer creates waves electronically. It can be used to create complex sounds including many harmonics.

frequency, and the others would correspond to higher and higher harmonics. Then, after analyzing the sound of a violin, we would know how much of each harmonic we need. We would know how loud to make the contribution from each oscillator. In theory, we could then produce a sound wave that would be indistinguishable from that produced by a Stradivarius in the hands of Isaac Stern. Unfortunately (or perhaps fortunately) we can come close, but there are practical limitations that render the match imperfect. Nevertheless, synthesizers, such as those built by Robert Moog, have played an important role in modern music.

14.7 *Beats*

This chapter has thus far been devoted to what happens when we mix together waves of the same frequency. Let's look briefly at what happens when waves of different frequencies are added together. One rule of superposition still holds: at any time and place, the total wave disturbance is the sum of the individual wave disturbances at this place. If we do this addition, we find that the wave that results looks like a high-frequency wave, whose amplitude varies in a wavelike way (Fig. 14–21). The high-frequency wave has a frequency equal to the average frequency of the two waves with which you started. The amplitude of the wave varies with a frequency equal to the difference between the two frequencies that we mixed together.

Suppose the two frequencies only differed by one cycle per second. If these are sound waves being mixed together, you hear a tone

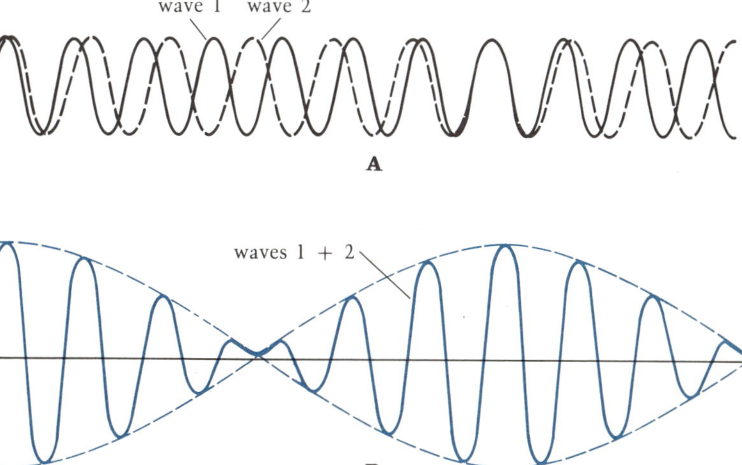

Fig. 14–21 Beats. A. Waves 1 (solid line) and 2 (dashed line) have a slightly different frequency (and therefore a slightly different wavelength). You can see that sometimes the waves appear to be in phase and sometimes they appear to be out of phase. B. This means that the resulting wave, while oscillating with a frequency equal to the average of the frequencies of waves 1 and 2, will seem to get stronger and weaker. That is, the amplitude of the resulting wave will not be constant. The resulting wave is shown in the solid line. Notice that the highest points occur when waves 1 and 2 appear to be in phase. The dashed line keeps track of the changes in the amplitude of the resulting wave. It turns out that this dashed line is also a wave, but with a much longer wavelength than either wave 1 or 2. A longer wavelength means a lower frequency than either wave 1 or 2. In fact, its frequency is equal to the difference between the frequencies of waves 1 and 2. When the frequencies of waves 1 and 2 are close together, we hear a tone that gets louder and softer.

whose frequency is almost equal to the frequency of the two tones. However, the tone varies in strength with a frequency of one cycle per second. You would thus hear a note that gets louder and softer once per second. These variations in loudness are called beats.

It is easy to demonstrate beats on a string instrument. You play a note on one string corresponding to the natural frequency of the next string, and also play that next string. If one of the strings is not tuned right, you will hear a rising and falling sound level. The frequency of this rise and fall will depend on how far out of tune the string is. If the two strings differ by two cycles per second, you will hear two beats per second. As the strings are tuned closer together, the time between beats increases. It is therefore possible to use the beats to get an instrument very well tuned, even if you don't have a good ear for pitch. You don't even have to hear the beats; you can even feel them on the sounding board of the instrument.

Key Words

interference
principle of superposition
constructive interference
destructive interference
nodes and antinodes
phase difference
standing wave pattern
fundamental oscillation mode
fundamental frequency
harmonics
overtones
harmonic series
natural frequencies
resonance
Fourier analysis
beats

Questions

1. At some point in space, two waves of equal amplitude and wavelength interfere destructively. Does that mean that the wave disturbance (due to these two waves together) is always zero at that point?
2. At some point in space, two waves of equal amplitude and wavelength interfere constructively. Does the wave disturbance ever go to zero at that point?
3. Two identical waves interfere constructively at some point. Then the phase of one wave is shifted by half a cycle. What happens to the resulting wave disturbance at that point?
4. When we have standing waves on a string: (a) Is the wave disturbance at the position of a node always zero? (b) At the position of an antinode, does the wave disturbance vary with time?

5. A musical instrument is made to play with two strings of equal length, but the strings have a different composition and are kept at different tensions. As a result, one has a fundamental frequency of 440 cycles/s and the other has a fundamental frequency of 880 cycles/s. How do the wavelengths for the fundamental modes on the two strings compare?
6. Draw a figure showing what each of the strings in Fig. 14–10 will look like at half a cycle later than at the instant shown in the figure.
7. The fundamental mode on a certain string has a wavelength of 4 meters. How long is the string? (Assume that both ends of the string are fixed.)
8. If you have one kettle drum and would like it to be able to sound different notes, what can you do?
9. For a string of given length, would you get a lower frequency fundamental (lower note) by leaving both ends fixed or one end fixed and one end free?
10. What do we mean when we say that someone has a "resonant" voice? How does this idea of resonance compare with the idea of resonance presented in this chapter?
11. With standing sound waves, does the existence of a pressure node mean that the pressure is zero at that point?
12. With standing sound waves, does the existence of a pressure antinode mean that the pressure is always greater than the average pressure at that point?
13. Draw a figure showing the pressure variations in the tubes shown in Fig. 14–15 a half cycle after the time shown in the figure.
14. Why is there always a pressure node at the open end of an organ pipe?
15. If you had an organ pipe with air coming in the bottom end, would you get a lower frequency note for the fundamental if you closed the upper end or left it open?
16. If you beat together two sound waves whose frequencies differ by 1 cycle/s, would you hear a sound wave whose frequency was 1 cycle/s, or would you hear a higher frequency tone that got louder and softer once per second? Explain.

CHAPTER 15

Light—The Wave Side

One of the "great debates" in the history of physics has been over the nature of light. Is a light beam composed of a stream of particles, or is it some form of wave disturbance? The discussion involved such outstanding physicists as Newton and Einstein, and by the early part of the twentieth century it served as the leading edge for one of the great revolutions in physics—the birth of quantum mechanics. (The quantum side of this story will be told in Chapter 23.)

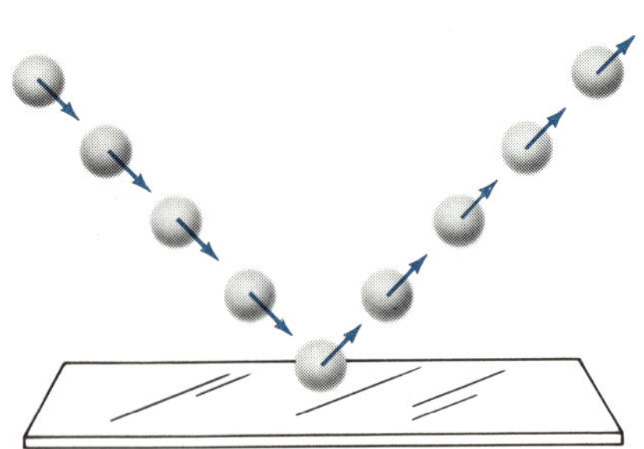

Fig. 15–1 Newton's view of light bouncing off a mirror. In this picture the light consists of tiny particles, each one bouncing off the mirror in the same way as a ball would bounce off the floor.

15.1 *Particle or Wave?*

Early theories of light were formulated in terms of streams of particles. Newton gave this idea considerable weight. His belief in the particle theory was based on the observation that light appeared to travel through the air in a straight line. He also reasoned that the reflection of light off a mirror could be explained in the same way as the bouncing of a ball from a wall (Fig. 15–1). Finally, Newton worked out a particle theory that was able to explain refraction, the bending of light that takes place when light passes from air to water or to glass. His theory required that the light speed up when passing from the air to the glass. (We now know that the opposite is true, but we'll consider this further in Chapter 17.)

Fig. 15–2 Refraction. Light rays are bent when passing from the air to the glass and again when passing from the glass to the air.

(above left) Fig. 15–3 The Dutch scientist Christian Huygens (1629–1695).

(above right) Fig. 15–4 Thomas Young (1773–1829) demonstrated interference for light in 1801.

(above center) Fig. 15–5 Augustin Fresnel (1788–1827).

The wave theory of light is attributed to the seventeenth-century Dutch physicist Christian Huygens. However, Newton's ideas about particles predominated until the wave theory was revived in 1801 by the British physicist Thomas Young, who studied the interference of waves. This work was furthered both theoretically and experimentally by the French physicist Augustin Fresnel. Fresnel's experiments showed, among other things, that light does not always travel in a straight line. It can bend around a corner. (Fresnel also developed a thin type of lens that has recently gained widespread popular use.) By the mid-nineteenth century, the wave theory was gaining momentum, especially when Foucault was able to show that, contrary to Newton's explanation of refraction, the speed of light is less in water than in air.

Even greater acceptance of the wave theory came with the theory linking electricity and magnetism that the Scottish physicist James Clerk Maxwell advanced in the 1860s and 1870s. Maxwell predicted the existence of electromagnetic radiation, waves of varying electric and magnetic fields. Maxwell predicted that the speed of his waves would be the same as the speed of light, which suggested that electromagnetic waves and light waves are really the same phenomenon. Maxwell's work was particularly important in establishing a wave theory of light because, even though experiments seemed to be favoring the wave theory over the particle theory, it was important to find out what is actually "waving" in a light wave. According to Maxwell, as light waves move along, electric and magnetic fields are getting stronger and weaker in a periodic pattern. The clincher for the wave theory seemed to come in 1887 when Heinrich Hertz was able to send and receive electromagnetic waves in the form of radio waves. This direct detection of electromagnetic waves was experimental confirmation of Maxwell's predictions. The acceptance of the wave theory of light was probably never greater than at that time.

However, the particle theory was not quite dead. It was to make a dramatic comeback. In fact, the seeds for that comeback were sown in the very experiment in which Hertz confirmed the predictions of Maxwell's theory. We will save the rest of this story for Chapter 23 when we discuss the origins of the quantum theory. In this chapter we will concentrate on those ways in which light behaves as a wave.

One of the reasons that the wave nature of light is not always apparent is that the wavelength of light is very small compared to most of the things in our everyday experience. Waves are most evident when they encounter things their own size. By their own size, we mean things whose size is similar to that of a wavelength. How small is the wavelength of light? First of all, we must realize that by "light" we generally mean that part of radiation to which

It is traditional to express light wavelengths in angstrom units, *with the symbol Å, after the nineteenth-century Swedish physicist Ångstrom. One angstrom is 10^{-10} meters, so the wavelength range of light is 4000 Å to 6600 Å. In SI units, only multiples of 1000 are recognized as appropriate units. Therefore, many people now measure the wavelength of light in* nanometers *(abbreviated nm). One nanometer is 10^{-9} meter, so 1 nm = 10 Å. (Many traditionalists—including most astronomers—continue to use the angstrom as a wavelength unit. Chemists are also resisting the change from angstroms, because a hydrogen atom is about one angstrom across, which makes the angstrom a convenient unit.)*

our eye responds. The only difference between light and other electromagnetic radiation is the wavelength. The wavelength of light ranges from 4×10^{-7} meters to 6.6×10^{-7} meters. This means that the wavelength range of light is 400 nm to 660 nm.

This wavelength range corresponds roughly to the wavelength range over which the sun puts out most of its energy. It is also the range in which the sun's energy gets through the earth's atmosphere. These facts are not coincidences. Our current ideas of biological evolution say that natural selection will favor those species equipped with receptors that are most sensitive to the part of the spectrum where the sun is brightest.

The whole of the visible part of the sun's light, taken together, is called white light. It was Newton who first showed that sunlight was actually composed of many colors of the rainbow. We now associate different colors with different wavelengths. Red is at the long wavelength end of the visible spectrum. The short wavelength end is the violet.

Although it is usually hard to "see" the wave nature of light, we are surrounded by its effects. You are taking advantage of the wave properties of light every time you put on a pair of Polaroid sunglasses. The telephone company takes advantage of it when they send telephone conversations through glass fiber cables, as they are doing increasingly. A photographer benefits from the wave nature of light through the antireflection coatings on camera lenses. An astronomer uses it in measuring the velocity of a distant galaxy. A surgeon uses it in repairing a damaged eye with a laser. We shall see more about these applications in the following chapters.

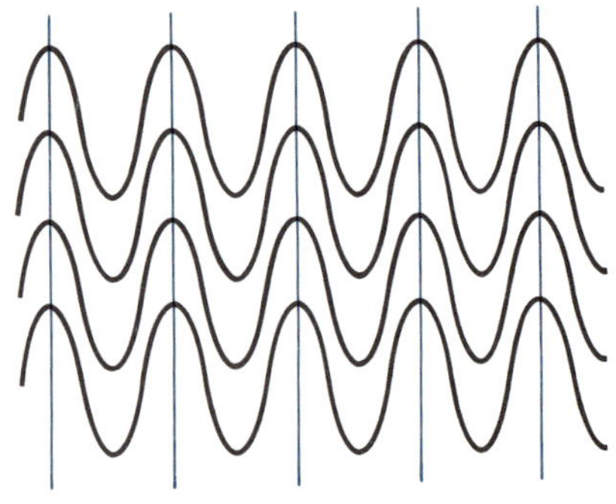

Fig. 15–6 Wave fronts. If we have a series of waves moving along, we can draw a series of lines, each line connecting a peak on one wave to the nearest peaks on the neighboring waves. These lines, called *wave fronts* (shown here in blue), then keep track of the positions of the rows of peaks of the waves coming along. Once we know where the wave fronts are, it is not really necessary to draw in the wavy lines. The wave fronts help us keep track of the progress of the wave. Notice that the wave fronts are each separated by one wavelength.

15.2 *Light-Wave Propagation*

We need some convenient way of picturing light waves as they travel through space. At any given time we can "take a picture" of the location of the waves and note the positions of the high points—*crests*—of the waves. We can represent these positions by lines, which we call wave fronts (Fig. 15–6). Wave fronts are one wavelength apart. Sometimes, we may also wish to indicate the direction in which the wave fronts are traveling. In these cases, we draw an arrow in that direction. When we do this, we are representing a light ray (Fig. 15–7).

The representation in terms of wave fronts is useful only when the light we are considering consists of only one wavelength. We call such light monochromatic (single color). Monochromatic light is really an idealization. In any real experiment there will always be some range of wavelengths, but this range can be so small that the wave behaves almost as if it were monochromatic.

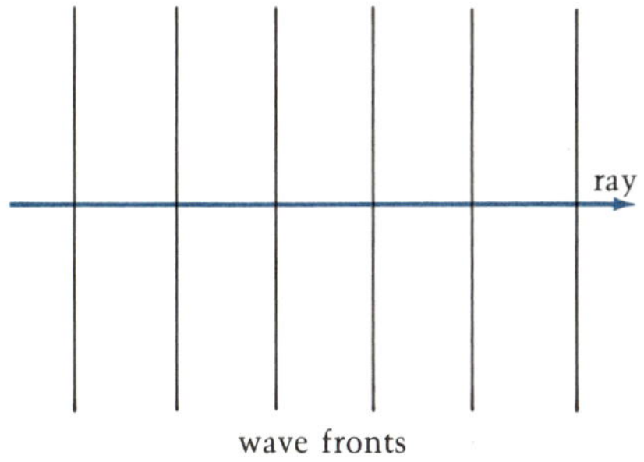

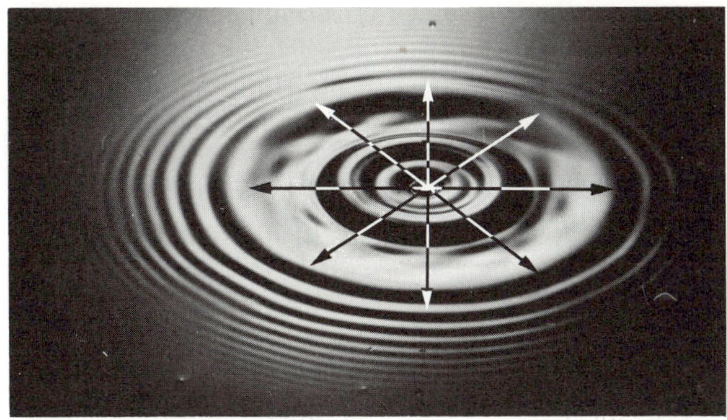

(above, left) Fig. 15–7 A *ray* helps us keep track of an advancing series of wave fronts. It is drawn in the direction in which the wave is traveling.

(above, right) Fig. 15–8 A series of waves moves outward. They were caused by drops of falling water.

There is another requirement if our analysis is to be simple. Let's look at the light coming from a normal light bulb. The light is being emitted simultaneously at many different points of the bulb, and each point is unrelated to the others. We call the light bulb an incoherent light source. We can see that it makes no sense to locate the wave fronts coming out of an incoherent source. If, on the other hand, a light source puts out light with a well-defined wave pattern, then we refer to the source as coherent. For our discussion in this chapter we will always assume that we are dealing with coherent, monochromatic sources. (A laser, which will be discussed in Chapter 24, is an example of such a source. For many purposes it doesn't matter how a laser works—it only matters that it is a coherent, monochromatic source.)

When a coherent light source is very small compared to the wavelength of the light, we refer to it as a point source. (Actually, there are a variety of definitions of a "point source," depending on the application. A star, for example, is so far away that it can be considered as a point source for many applications.) The wave fronts moving outward from a point source move out in circles (really wave-front spheres, if we are considering three dimensions). We refer to these wave fronts as circular (or spherical) waves. When we have such waves, the rays point radially outward from the source. When you are very far from a point source, the wave fronts reaching you will be almost straight lines parallel to each other (Fig. 15–9). When the wave front is not curved and all of the

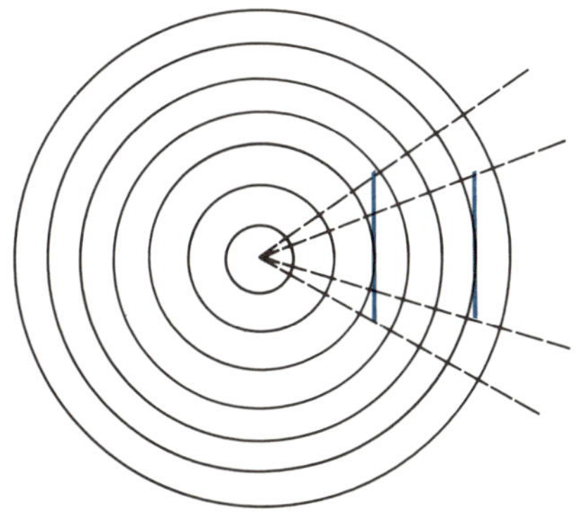

Fig. 15–9 The two screens in this diagram (shown in blue) are the same size, but are different distances from the source of the waves. Notice that some of the light which would hit the first screen would not hit the second screen. The dashed lines outline the rays that will strike the two screens. The closer screen will be hit by a larger range of angles than the farther screen. Notice also that the curvature is more evident at the closer screen than at the farther screen. If we had another screen much farther to the right (away from the source), the wave fronts would arrive virtually as straight lines rather than as curves.

rays are parallel to each other, we call the wave a plane wave. (This is because the wave fronts are flat—planes.) The wave in Fig. 15–7 is a plane wave.

We have always associated wave motion with a transfer of energy from one place to another. Light fits into this picture. You can tell that light contains energy by its effect on the materials it strikes—light falling on an object heats it up (Fig. 15–10).

Fig. 15–10 One advantage of a lens or mirror is that it can produce brighter images than we might see with our naked eye. This is because the lens intercepts the light coming in over a large area and converges it to a small area. Here the concentrated sunlight is used to burn a newspaper.

How does the amount of energy that reaches us depend on our distance from the light source? Let's look at one spherical wave front as it spreads out. The total amount of energy contained in that wave never changes. However, the energy is spread over a larger and larger area. Consider two collectors, the same size, but one twice as far as the other from the light source. When the wave reaches the first detector, the energy has been spread out over a sphere of a certain size. When the wave has reached the outer collector, it will cover twice the radius it did before, which means that it will be spread out over four times the surface area. (The surface area depends on the square of the radius.) Therefore, the outer detector receives only one-quarter of the energy that the inner detector receives. We say that the intensity of the wave, the amount of energy striking each square centimeter of the collector each second, is proportional to $1/\text{distance}^2$ (this is known as the inverse square law). For example, Venus is 0.7 times as far from the sun as is the earth. That means that the amount of radiation it receives from the sun is $1/(0.7)^2$ or about twice that received by the earth. Similarly, people in the foreground of a flash picture are often overexposed while people in the background are underexposed. (See Fig. 15–11.)

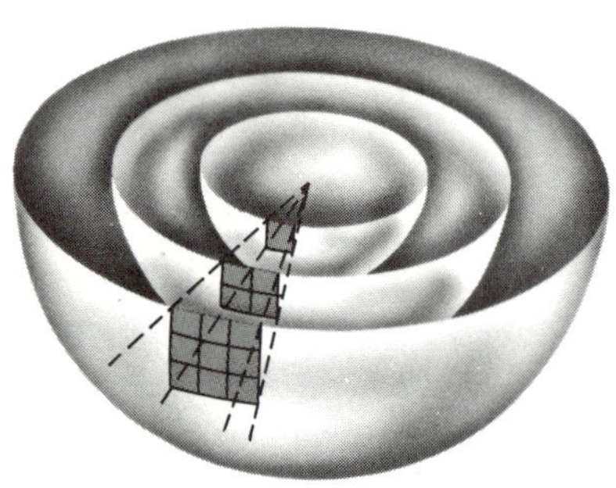

Fig. 15–11 Intensity in a spreading light wave. On each of the imaginary shells we have shaded the area that would be hit by some group of light rays (dashed lines). The second shell has twice the radius of the first and the third shell has three times the radius of the first. The same amount of energy would strike each of the shaded areas, but notice that, as you get farther away, the energy is spread out over a larger and larger area. When you go two times as far away, the area is four times as large. When you go three times as far away, the area is nine times as large. This means that the intensity of light falls off as the area that the same amount of energy hits gets larger.

15.3 *Interference of Light Waves*

15.3a YOUNG'S EXPERIMENT

In 1801 Thomas Young pointed out that if light is a wave, then light waves should be able to interfere with each other. But it was hard to observe interference patterns. Because the wavelength of light is very short, it was not practical to set up standing waves. Instead, Young used two coherent point sources of light and let their radiation interfere (Fig. 15–12).

Since until recently it has been relatively hard to get two coherent light sources, we usually use a *double slit* arrangement to demonstrate interference. When a plane wave from behind reaches the slits, each slit acts as a point source of light. The light spreads out in circular waves. With such an arrangement, Young was able to show

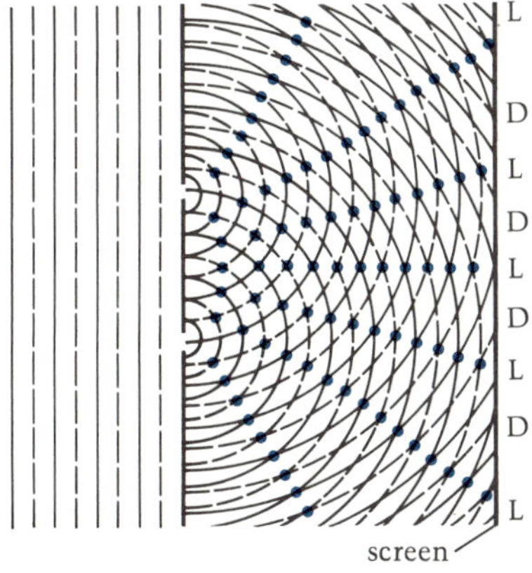

Fig. 15–12 Plane wave fronts come in from the left. For clarity, we have used the solid lines to represent the actual crests, or high points, of the wave. The dashed lines represent the low points of the wave. This is an important point because the wave is just as strong at the high point as at the low point. "High" and "low" just refer to pointing in opposite directions. The solid lines are one wavelength apart, as are the dashed lines. The solid lines and dashed lines are a half wavelength from each other. As the waves pass through the two slits, they spread out in a circular pattern from each slit and then interfere with each other. (When a plane wave reaches two slits not aligned parallel to its direction of motion, each slit then functions as a point source.) Wherever a high point meets another high point (solid line meeting solid line) we have constructive interference. Similarly, wherever a low point meets another low point (dashed line meeting dashed line) we also have constructive interference. Points of constructive interference have been marked with blue dots. Wherever a high point meets a low point there is cancellation—destructive interference. (This would be where the solid lines cross the dashed lines.)

If we place a screen in the pattern, we will see a light spot on the screen (L) wherever there is constructive interference and a dark spot on the screen (D) wherever there is destructive interference.

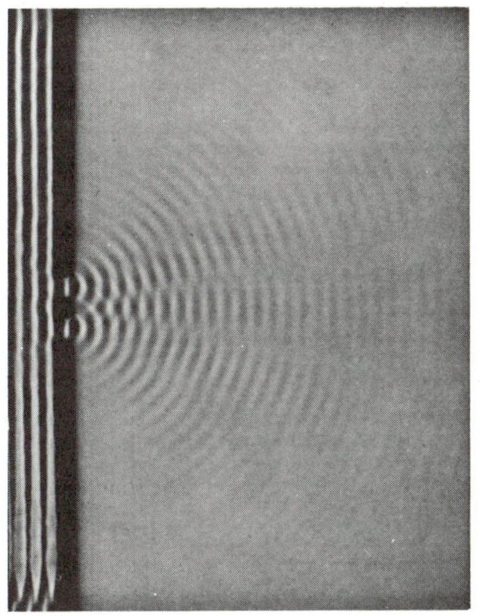

Fig. 15–13 Interference seen in water waves passing through a pair of slits.

Fig. 15–14 Interference of light passing through a pair of slits.

that you can observe interference in light. We generally refer to this setup, with a light source, a double slit, and a screen, as Young's experiment.

The circular wave fronts spread out from each of the point sources. At any given time and place there will be two waves present, one from each of the sources. In order to find the total wave disturbance, we must add up the effects of the two waves. If we are at a point where the wave crests meet, a stronger wave results. We have constructive interference—the light is brighter there. If we are at a point where the crest of one wave meets the bottom of the other wave, we have destructive interference, and there will be no resulting intensity. It is dark. In Fig. 15–12 we have marked those points at which there is complete constructive interference. We see that these points tend to lie along certain directions. When we take another picture of the waves a half cycle later, the points at maximum move out half a wavelength, but they still lie along the same lines. When we place a screen in this pattern, we should see alternate light and dark lines.

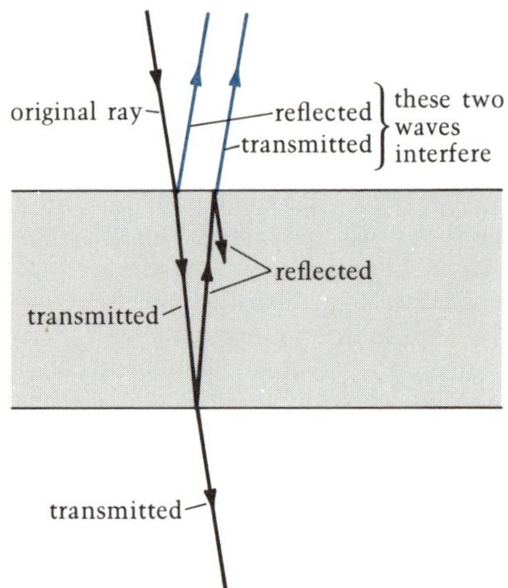

Fig. 15–15 Interference from a thin oil film. The incoming light splits up with some light being reflected and some transmitted. When the transmitted light strikes the opposite surface, some of it is transmitted and the rest is reflected back. When the reflected part again hits the upper surface, some of it is reflected back and some is transmitted. The part that is transmitted can then interfere with the part of the original ray that was reflected off the top surface.

The exact spacing corresponding to light and dark may change by half a wavelength, depending on possible additional phase shifts when the light is reflected.

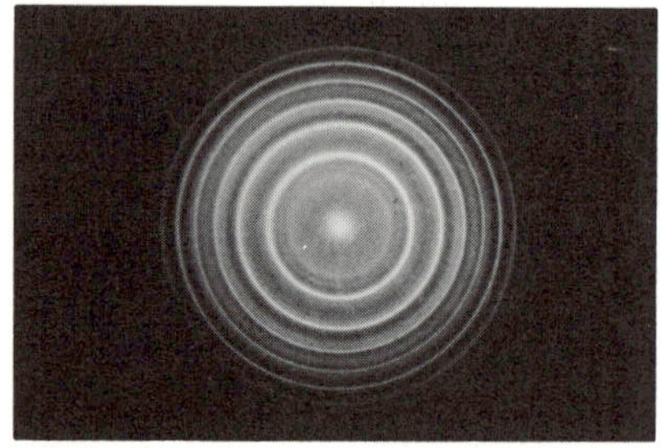

Fig. 15–16 The interference pattern between layers of thin films.

How can we predict where on the screen the light and dark regions would be? First, whether a point will be light or dark depends on how much farther it is from one slit than from the other. If the path difference (the difference between the distances from each slit to a given point on the screen) is zero, or an exact multiple of the wavelength (so that a crest from each slit arrives at the same time), then we will get a bright spot.

15.3b INTERFERENCE AND THIN FILMS

Although Young's experiment is a very clear demonstration of the wave nature of light, you are more likely to run across interference when you see the beautiful patterns in the light reflected off thin films of oil or other substances (Fig. 15–15). When a ray of light falls on the surface of a thin film of oil, some of the light is reflected and some passes through the surface. The part that passes through the front surface reaches the back surface of the oil where some of it is again reflected while the rest continues through.

Now some light is moving back toward the top surface and some of that will be sent right through the surface. This light will now interfere with the part of the incoming wave that was originally reflected from the top surface. One of the interfering waves has now traveled farther than the other, so the crest of one wave will not necessarily match up with the crest of the other. If the total path back and forth through the film is exactly equal to one wavelength, the crest of one wave will line up with a crest from the other wave. The two light rays will constructively interfere. The same occurs if the path length through the film is two wavelengths or three, or the wavelength times any integer. If there is an extra half wavelength left over, though, then there will be destructive interference. You will see dark instead of reflected light.

Thus, when white light is incident on such a film, only certain colors appear to be reflected. These are the ones for which the thickness of the film is right for the wavelengths to produce constructive interference. Which colors appear also varies with the angle at which the light enters, since the distance the light travels through the film increases as you tilt the beam away from the vertical. As a result, the reflection off a thin film is prettier for white light than for light of any given particular color. For monochromatic light, you will see alternating light and dark rings. For white light, rings of different colors result.

Besides producing pretty color effects, this type of interference can be very useful. For example, the surfaces of most camera lenses are coated so that they do not reflect too much light. The thickness of the coating is chosen to produce destructive interference for the reflected light. The method really only works perfectly for one color (wavelength), but multiple layers can be used effectively to cut down reflection at a wide range of wavelengths.

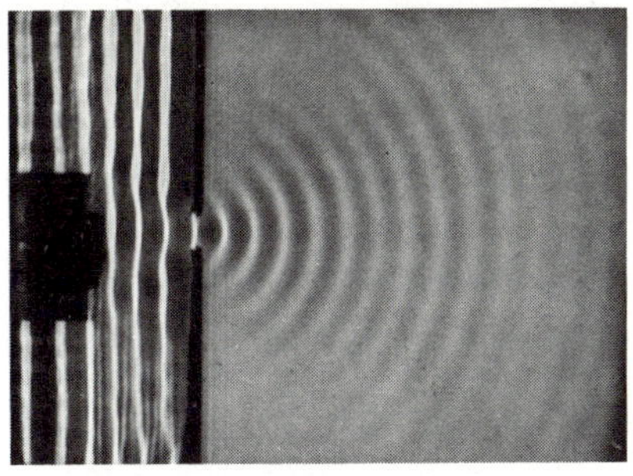

Fig. 15–17 Diffraction of waves in a shallow tank of water. Plane waves enter from the left and strike the barrier. The hole in the barrier serves as the source for an expanding wave front (no longer a plane wave). However, because of the effects of interference, this spreading wave front appears stronger in some directions than in other directions.

15.4 *Diffraction*

We have thus far seen that waves have some very unusual properties that clearly distinguish them from particles. Waves can transport energy without transporting any matter; they can interfere with each other, producing in some places a very strong signal and in others a very weak or even zero signal; when they reach a boundary they can be partially transmitted and partially reflected. We now encounter another property—the ability of waves to bend around a corner. This phenomenon is called diffraction.

You can see diffraction of a wave in a tank of water. Plane waves come along the side of a barrier, but as soon as they reach the edge, a spherical wave spreads out around the barrier. It is almost as though the corner of the barrier suddenly acts as the point source for a spherical wave.

We can visualize the process in terms of a geometric representation that was proposed by Huygens, whom we mentioned in Section 15.1. Huygens said that each point on a wave front acts as a point source for circular waves (Fig. 15–18). At any given time, we can draw several representative points along the wave front and say that these will then emit circular waves. After some time, the circular waves will have spread by some amount. We then draw a line that connects the fronts of all these new little waves, and this line gives you the new position of the wave front. As Fig. 15–18 shows, the method works very easily for the case of plane waves. We can also see that if we block part of a plane wave, we are left with the circular wave pattern going around the wall. The basic idea of Huygens's presentation is that a wave does not simply sweep along; it must be regenerated from point to point.

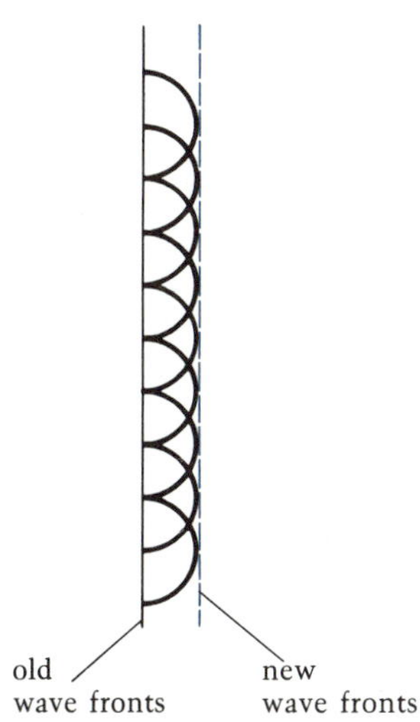

Fig. 15–18 Huygens's construction of wave fronts. In this diagram, each point on the old wave front serves as a point source for spherical waves that spread out. To find the position of the new wave front, we just connect the forward edges of these spherical waves.

You might say that if we are really to take this idea seriously, we shouldn't merely draw in the position of the wave front. We should, in this view, really consider each point on the old wave front as a real source for spherical waves. In this case, we have to work out the interference pattern that results from all these spherical waves. In some regions there is complete destructive interference—no light—and in other regions there is constructive interference—bright zones. We can actually see these alternating light and dark bands if we look at monochromatic light projecting the image

A

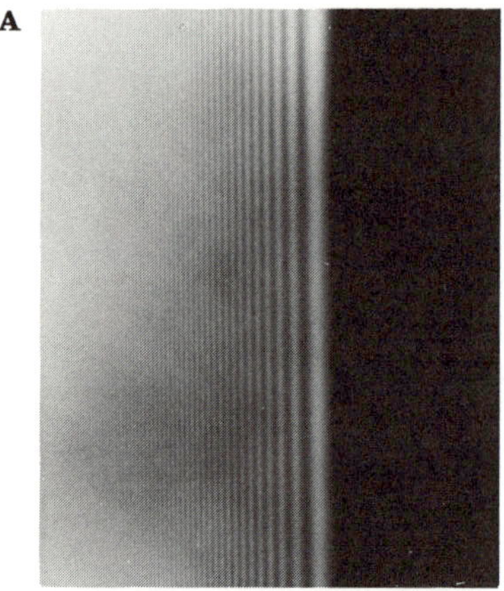

B

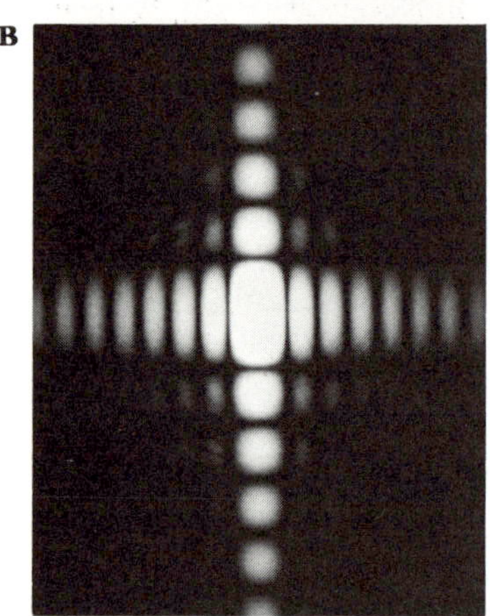

C

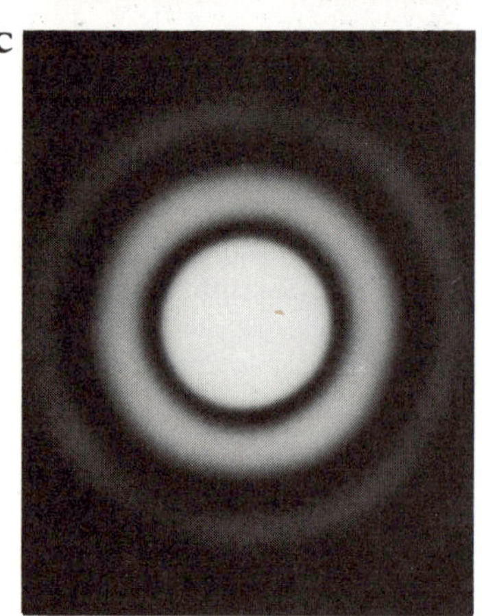

Fig. 15–19 Diffraction patterns due to: A. A straight edge. B. A rectangular aperture. C. A circular aperture.

of a sharp edge, such as a razor blade, on a screen (Fig. 15–19A). The shadow of the blade is not sharp. A zone of alternating dark and light bands appears in between the extended dark and light regions.

The effects of the phenomenon of diffraction are often with us. One of the most common examples comes in the question of resolution. By "resolution," we mean the ability to distinguish between (resolve) two objects that are close to each other. For example, when you are on a long straight highway at night, notice the headlights of approaching traffic. It is very easy to see that cars close to you have two headlights. However, you really cannot tell that faraway cars have two headlights, and until the car gets close enough, you cannot be sure if you are looking at a car or a motorcycle. When a car is at the distance where we can first make the distinction, we say that the headlights are barely resolved.

Why can't you see that there are two headlights even when the car is far away? The answer is diffraction. Assume that each headlight is a pinpoint of light. As the light passes through your pupil, diffraction causes the light to spread out. A diffraction pattern forms on the retina of your eye, rather than a pinpoint image. The light is smeared out. Two blurry images form from the two headlights; the size of each is determined by the relative size of the wavelength of the light and the pupil opening.

15.5 *Diffraction Gratings and Spectroscopy*

Let's take another look at Young's experiment, which has light coming through two slits. We now know that the real pattern we see results from the combined effects of diffraction and interference. The addition of diffraction does not change the positions of the bright spots. However, their relative brightness follows the wide diffraction pattern of a single slit.

What happens if we have more than two slits? If all the slits are identical and the spacings between them are the same, then the bright spots will still be in the same place as when there were only two slits. The only difference will be that as we increase the number of slits, each bright area will become narrower and better defined, and the dark regions in between will grow wider.

If we want to get very sharp images, we need a large number of slits. Such a device is called a diffraction grating. Diffraction gratings come in two sorts: If the light is transmitted through many little slits, we have a transmission grating. If the light is reflected off many tiny mirrors, we have a reflection grating.

We normally build a diffraction grating into a device that allows us to examine the spectrum. When we examine the spectrum with only our eyes, the device is called a spectroscope (Fig. 15–20). For

Gratings can be made by taking a very flat piece of glass and having a machine rule lines on the surface of the glass with a very sharp point. The point is usually a diamond, so that it will retain its sharpness. A high-quality grating may have as many as 10,000 lines per centimeter! (By comparison, a phonograph record has fewer than 100 grooves per centimeter.) The ruling of gratings is a difficult art.

A new way of making gratings is to form an interference pattern with an extremely narrow spacing, and to allow the pattern to fall on photographic film. A grating made in this way is called a *holographic grating*.

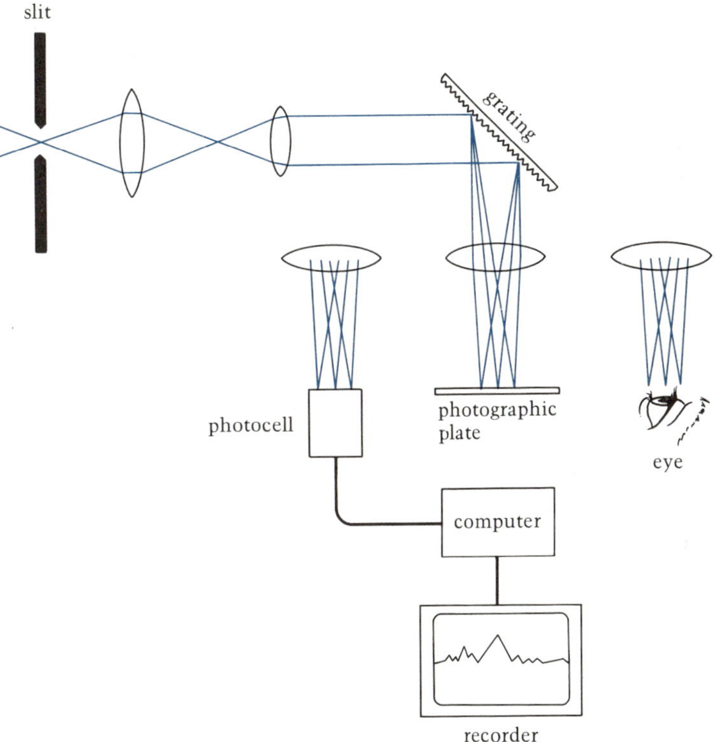

Fig. 15–20 A spectrograph. The light from the object being studied enters through a small slit and then passes through a series of lenses to produce a beam of parallel rays, which then strike a grating, breaking the light into its component colors. This light is then focused on a photographic plate to record the spectrum. If, instead of the photographic plate, we use our eye to look at the light, as shown on the right, we call the device a spectroscope. If, as on the left, we use a photocell, which converts the light into electric pulses, and then record those pulses, the device is called a spectrometer.

most studies, we want a permanent record of the spectrum for further study. We can do this in a spectrograph, in which the spectrum is projected on a photographic plate. Sometimes we measure the intensity of light at each wavelength electronically, and call the device that does so by the name spectrometer. The field of study of the spectrum is called spectroscopy. Many physicists are engaged in spectroscopy of one type or another.

In the early 1800s, William Wollaston in England and Josef Fraunhofer in Germany independently spread out the light of the sun into its component colors. They noticed that there were cer-

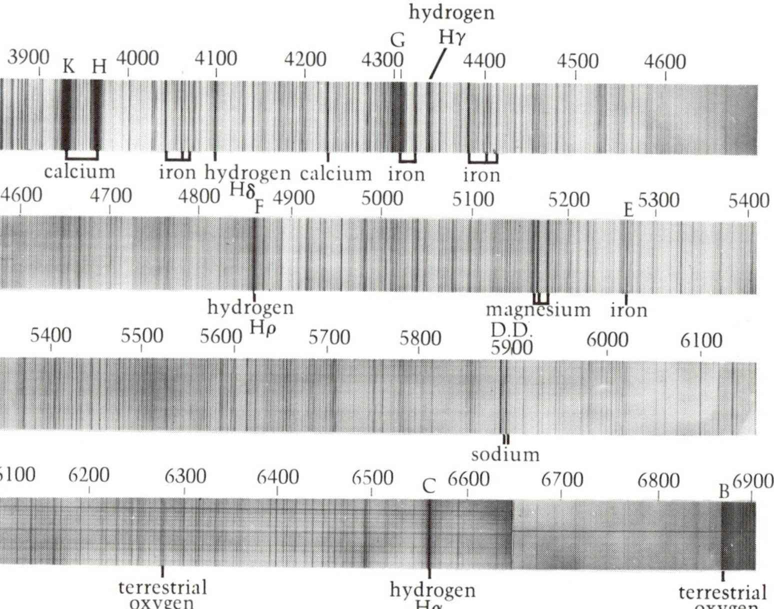

Fig. 15–21 The solar spectrum, taken with the 13-foot spectroheliograph at the Hale Observatories.

tain gaps in the band of bright light, certain colors that were missing (see Fig. 15–21 and Plate 17). These gaps appeared as dark lines across the spectrum and were therefore called spectral lines. We now know that these lines result from the presence of atoms of the various chemical elements in the atmosphere of the sun. At certain wavelengths, these atoms absorb a little of the energy of the sun's light, so the parts of the spectrum where they do so appear darker than other parts.

In 1859, Gustav Kirchhoff, a German chemist, observed a similar effect in his laboratory when he passed bright light through a container of gas. Kirchhoff noted that the wavelengths at which absorption takes place depend on the particular gas in the container. Each chemical element has its own set of spectral lines, as unique to that element as your fingerprints are to you. We now know that if we make the atoms hot enough, they will emit (rather than absorb) radiation at these particular wavelengths. Whether we have bright emission lines (as these are called) or absorption lines (the dark lines), the wavelengths are the same for each type of atom. The reasons for these distinctions did not become clear until the development of quantum mechanics in the first part of the twentieth century, and we will leave further discussion for later in this book. The important point for now is that we can identify an element by its spectral lines.

Spectroscopy has many applications. We can tell what elements or molecules are present in a given sample by passing a light through the sample and studying the spectrum that results. We can even tell how much of a given element is present, since the more we have, the stronger the absorption lines will be. We can, for example, monitor the level of various pollutants in the earth's atmosphere. Moreover, by studying the relative strength of one absorption line with respect to another, we can also learn about the temperature in some inaccessible region, such as the interior of a blast furnace.

15.6 *Polarization of Light*

If you have ever used a pair of polarized sunglasses, then you will know that light certainly exhibits the effects of wave polarization. If we have polarizing filters for light, we can duplicate all the effects that we saw when we discussed polarization of waves on a string in Section 13.3. For example, we can pass light through two successive filters (as illustrated in Fig. 15–22). As we rotate one filter, we see that there is one orientation for which no light gets through. When the maximum amount of light gets through, the polarization directions of the two filters are the same. When no light gets through, the polarization directions of the two filters are at right angles to each other.

The fact that we can demonstrate polarization in light waves means that *light is a transverse wave.* In the case of light, the things that determine the polarization, the things that have some direction associated with them, are the fields that make up the electromagnetic radiation. We'll discuss this point further in Section 22.5.

Polarizing filters are at their best in screening out light when that light is polarized. However, in our normal experience we have light

Fig. 15–22 A. When two polarizing sheets are held so that they pass light polarized in the same direction, most of the light gets through. B. When one of the sheets is rotated so that it now blocks the polarization that is passed by the other, no light gets through.

A

B

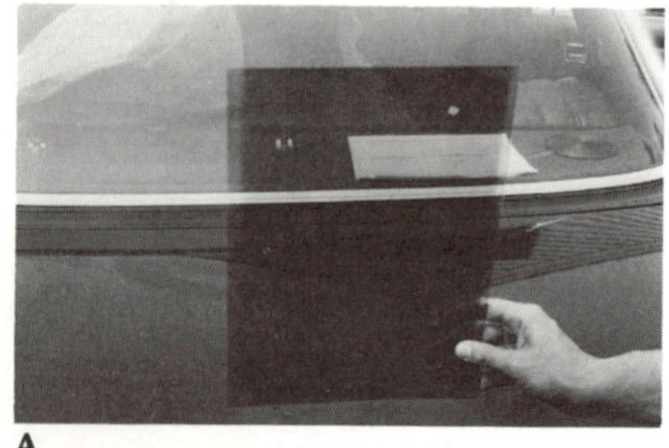
A

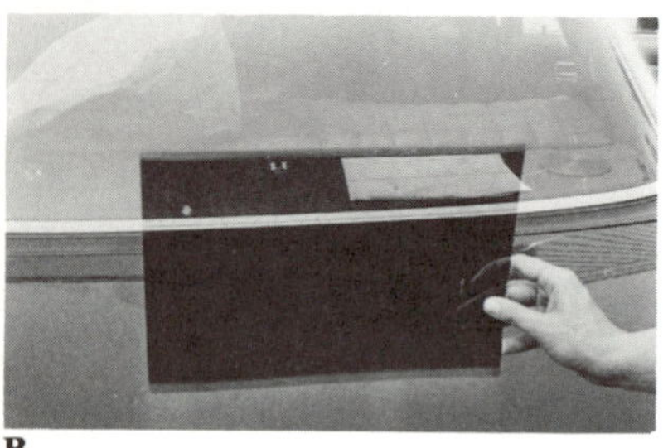
B

Fig. 15–23 Glare from the hood of a car is partially polarized. A. This filter is oriented so that the glare is not blocked. We see the reflection of the house. B. When the filter is rotated by 90°, most of the glare is blocked. We cannot see any reflection.

Fig. 15–24 The blue sky is darker when seen through the horizontal pair of sunglasses than when seen through the vertical pair of sunglasses because the sky itself is polarized. Notice that since one of the glasses is held horizontally and the other held vertically, the polarizing filters are crossed in direction and no light passes through both of them.

that is polarized up-down traveling side by side with light that is polarized left-right and light that is polarized in various other directions. Such a mixture of polarizations is called unpolarized light. Polarizing sunglasses are so effective because the process in which light is reflected to cause "glare" tends to cause the reflected light to become at least partially polarized in one direction. The filters in the sunglasses preferentially cut out some of this reflected light.

The sky is another place where reflected light becomes polarized. When the sunlight is reflected by the atoms in the atmosphere (causing the blue sky, since blue light is reflected much more efficiently than red light), the reflected light becomes polarized. If you have a pair of polarizing sunglasses, take them off and then look through them at a deep-blue patch of sky halfway around the sky from the sun. Then rotate the sunglasses. Notice that the sky appears to get lighter and darker, indicating that the light you get from the blue sky is polarized to some extent.

15.7 *Doppler Shifts in Light*

We have seen that light exhibits the interference, diffraction, and polarization phenomena that we associate with a wave. Another wave property that we discussed in Chapter 13 is the Doppler shift, a change in the observed wavelength of a wave resulting from movements of the source of the waves or of the observer. A Doppler shift for light also occurs. However, the details of the theory are a little different. Again, it doesn't matter if the Doppler shift comes from the motion of the source, from the motion of the observer, or from some combination of the two. We always get the same result. If the speeds involved are much lower than the speed of light, then the result is the same as for sound waves when the speed is much less than the speed of sound. The fractional change in the wavelength of light is equal to the velocity divided by the speed of light. So, if something is moving at 0.01 percent the speed of light, all the wavelengths will increase by 0.01 percent.

A certain terminology has grown out of the application of the Doppler shift to light. If the source of radiation is coming toward us, the light will be shifted toward shorter wavelengths. The short wavelength end of the visible part of the spectrum is the blue-light end. We therefore say that Doppler shifts that shorten the wavelength are blueshifts. On the other hand, Doppler shifts that lengthen the wavelength are redshifts.

Some of the most striking Doppler shifts are encountered in astronomy, because astronomical objects can move at considerable speeds. For example, in the early part of this century it was found that galaxies in all directions from ours have redshifts, which

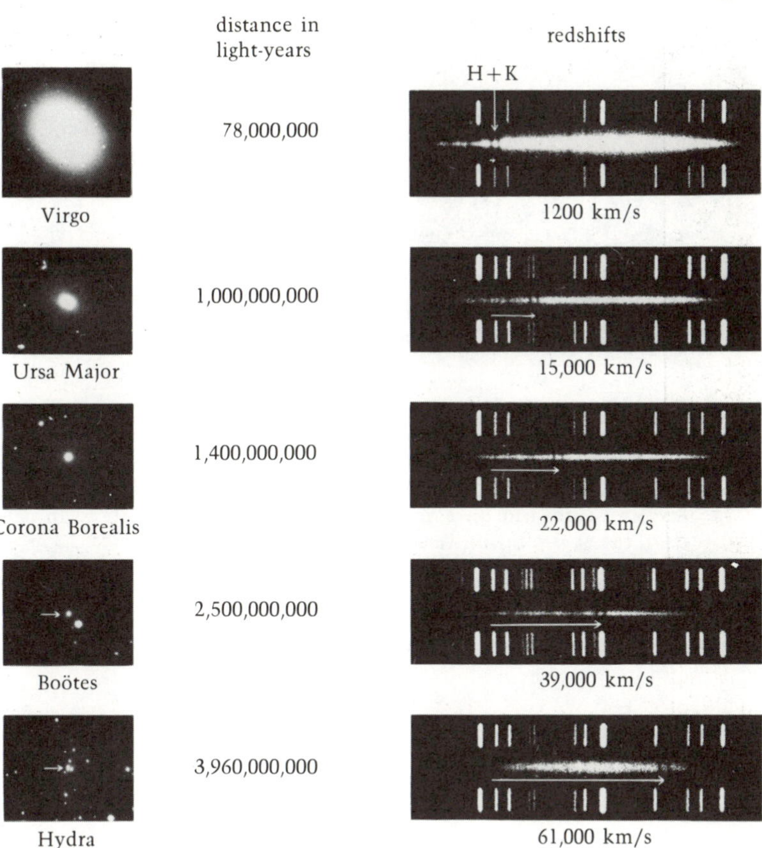

Fig. 15–25 Photographs and spectra of five galaxies. Each of the galaxies shown in the left column is in a cluster of galaxies in the constellation listed. The distance to each galaxy is given next to the picture. Notice that as the galaxies get farther away they get fainter. In each spectrum, the top and bottom row have nothing to do with the galaxy, but are for purposes of calibration. The spectrum of the galaxy is a bright band with a few dark lines. The longer wavelengths are to the right. In this figure, we look at the redshift (shift to the right) of the pair of lines marked "H + K" as the galaxy we look at is farther away. The amount of the redshift is shown with an arrow. Under each spectrum is the speed that would be required to produce the observed redshift of the lines.

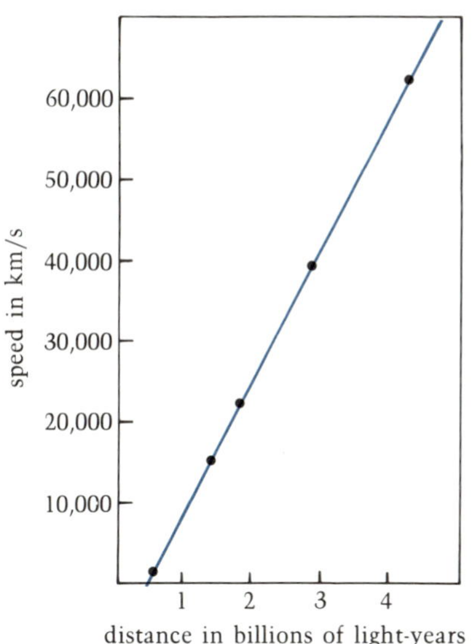

Fig. 15–26 Each point represents one of the galaxies in Fig. 15–25, and we make a graph of distance and speed. Notice that the points fall very close to a straight line. The relation is called Hubble's law.

means that all of the galaxies are moving away from ours. In 1929, the astronomer Edwin Hubble found that the amount of redshift is proportional to the distance of the galaxy from ours. The galaxies farther away are moving away faster. This relation between redshift and distance is known as Hubble's law. In Chapter 29 we will discuss the implications of Hubble's law for the structure of the universe. For the moment, let us simply say that Hubble's law establishes that the universe is expanding, with all of the groups of galaxies moving apart from the other groups of galaxies. (See Figs. 15–25 and 15–26.)

Twenty years ago, astronomers discovered a new class of objects that have the largest redshifts of all. Because the objects, which were first detected by their strong radio emission, appeared almost (quasi-) point-like (stellar), they were called "quasi-stellar radio sources." This has been contracted to *quasars*. We now think that the quasars are galaxies with especially bright cores, which enables us to see them at especially great distances. Studies of the quasars and the distant ordinary galaxies are being used to explore the origin of our universe and to predict its future.

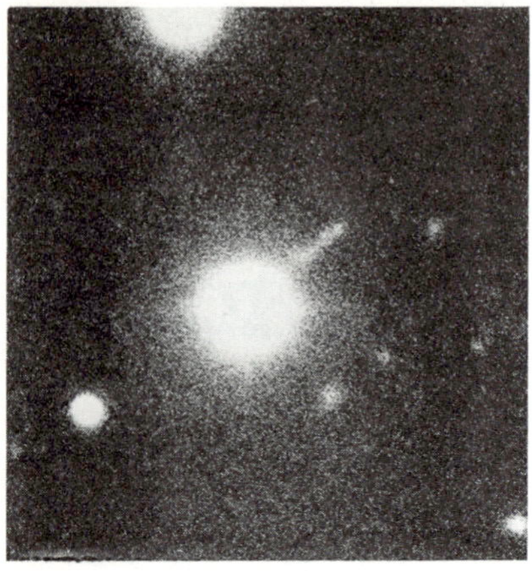

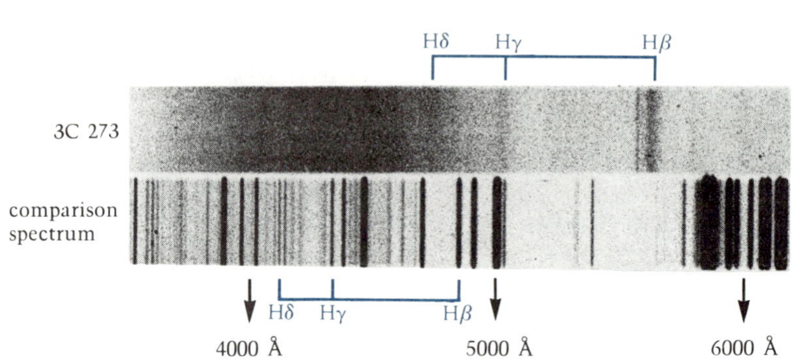

Fig. 15–27 A. Enlarged portion of 200-inch photograph of the quasi-stellar radio source 3C 273. The object looks like any other thirteenth magnitude star, except for the faint narrow jet visible out to about 20 seconds of arc from the quasar. B. Spectrum of the quasar 3C 273. The lower spectrum consists of hydrogen and helium lines, and serves to establish the scale of wavelengths. The upper part is the spectrum of the quasar. The Balmer lines Hβ, Hγ, and Hδ in the quasar spectrum are at longer wavelengths than in the comparison spectrum. The redshift of 16 percent corresponds to a distance of two billion light-years in the expanding universe.

Key Words

refraction
Maxwell
electromagnetic radiation
white light
visible spectrum
wave fronts
light ray
monochromatic
incoherent
coherent
point source
plane wave
intensity
inverse square law
Young's experiment
diffraction
resolution
spectroscopy
diffraction grating
  transmission grating
  reflection grating
spectroscope
spectrograph
spectrometer
unpolarized light
spectral lines
  emission lines
  absorption lines
blueshift
redshift
Hubble's law

Questions

1. Briefly describe two pieces of evidence that show that light is a wave.
2. Why don't you see interference effects from light entering a room from two different windows?
3. Why would an ordinary light bulb not be a good source of light if you want to do Young's experiment?

4. Two stars have the same intrinsic brightness. However, one star is ten times as far away as the other. How bright do they appear to us relative to each other?
5. In Young's experiment, if we move the screen farther from the two slits, what happens to the spacing between the dark fringes?
6. Suppose we set up two versions of Young's experiment. In one we use light with a wavelength twice that used in the other. In the experiment with the longer wavelength, the slits are twice as far apart as in the other. How does the spacing between bright lines compare in the two experiments?
7. The first bright line away from the center of the screen in Young's experiment is 1000 wavelengths of the light away from the nearer slit. How far is it (in terms of the wavelength) from the farther slit?
8. Draw a figure like Fig. 15–15, showing how two waves could emerge from the bottom interfering destructively, so that there is no transmitted light. (*Hint:* Extend one of the waves shown.)
9. Why isn't it possible to make a single coating that would be nonreflecting for all visible wavelengths?
10. At some wavelength a given coating is just the right thickness so there is no reflection. Suppose we now increase the thickness of the coating by half a wavelength. What happens to the reflection?
11. Why don't we see diffraction of light around the corner of a window?
12. A certain object is moving away from us at 0.001 percent of the speed of light. (a) At what wavelength would we observe a line that was emitted at 700 nm? (b) If two lines were emitted at wavelengths that are 100 nm apart, how far apart will they appear to be in wavelength? (c) How would your answers to parts (a) and (b) change if the object were moving toward us instead of away from us?
13. As the object emitting the radiation moves faster and faster away from us, does the spacing (difference in wavelength) between any two spectral lines increase, decrease, or remain the same? Explain.
14. If you are making a pair of polarized sunglasses, does it matter which direction, forward or backward, the filters are put in?

CHAPTER 16

Holography

A photograph may be very pretty, but something is missing. The picture has no depth, no third dimension. When we look out the window, on the other hand, we have the impression of a third dimension. If we move our heads from side to side, our perspective changes somewhat. We can even look around nearby objects to see what is behind them. And we can focus near and far.

Why is a window on the world better than a photograph of the same view? Can we get around the two-dimensional limitation of the camera? The answer to this limitation is holography, a growing technology. Its very name, based on the Greek word *holos*, which means "whole," implies that we get the whole view.

16.1 *Lensless Photography*

We'll discuss lenses and how they work in the next two chapters, but you don't have to know about lenses here. Holography doesn't use them.

When we take a picture with a camera, we capture the view that the camera's lens "sees." Once we snap the shutter, we have fixed that view in time and space. To see the view from a location 20 centimeters to the left, we have to move our camera and take another snapshot.

But the light from the entire view outside has been coming through the whole window all the time. The camera intercepts only a small fraction of this light. How can we capture the information that is in *all* the light waves that pass through the window?

Dennis Gabor, a British scientist born in Hungary, described his idea of how to do so over thirty years ago, in 1947. He suggested that to capture all the information carried by the light waves that come through the window at one instant, it is not enough to know only the intensity at every point. (We get this intensity from the amplitude—the height—of each wave.) We must also know the phase of each wave. Gabor also proposed detecting the phase information by using the phenomenon of interference, which we discussed in the previous chapter.

Let us consider a particular example to illustrate Gabor's idea. Figures 16–1 and 16–2 show a model outside a "window"; we put a photographic plate at the "window" to capture the information. Since we want to record the phases of light waves, it makes it much easier if we start out with waves that (1) are all the same wavelength (that is, are monochromatic), and (2) are all in phase to begin with (that is, are coherent). By coherent, we mean that all the waves are marching in step, with all the peaks in amplitude coming together and all the valleys coming together as well.

To get color holograms, we would have to use three monochromatic beams. Most holograms at present are in only one color.

Now let us illuminate our model with coherent light. The light that hits it will bounce back toward the film. Since some parts of the model are farther away from the film than other parts, the reflected waves have to travel different distances and so are no longer in phase when they reach the film. But if we simply expose the film to reflected light in this way, the entire film will be fogged (exposed). (After all, light is reflected by the model in all directions, and film does not ordinarily care about the phase of the light waves that hit it.)

Gabor's brilliant idea involves splitting the original coherent light beam into two parts. Half the beam is sent at the model, but the other half (the reference beam) is directed without obstruction at the film. Thus, when the waves reflected by the Lincoln model reach the film, they interfere with the reference beam, and the film

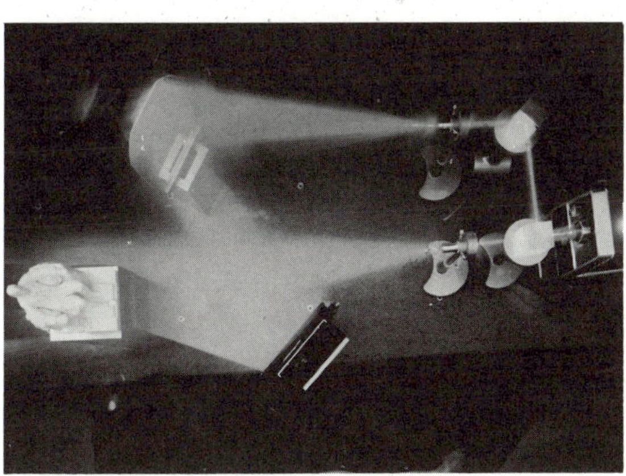

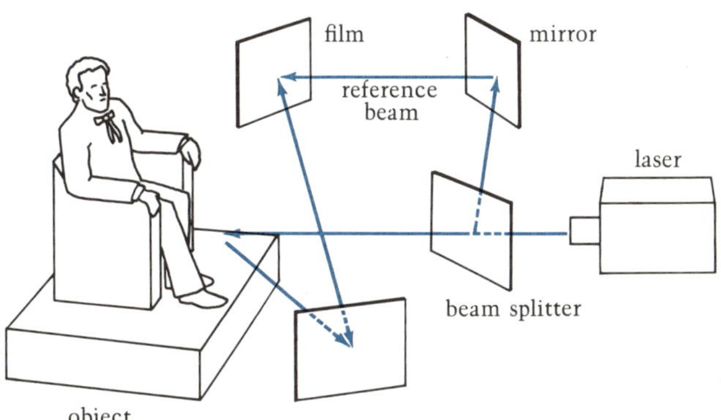

Figs. 16–1 and 16–2 Making a hologram. Light from a laser strikes a partially silvered mirror, which reflects some of the light and lets the rest through. The part that gets through is called the *reference beam* and is directed toward the film. The reflected part strikes the object and then goes toward the film. This part of the beam interferes with the reference beam, producing an interference pattern on the film. This is the hologram.

Our original beam, B, and our reference beam, R, interfered to make a hologram, H. We might say that B + R = H. Now we take H and pass R through it, giving H − R = B, our original beam.

We shall discuss the principles and details of lasers in Chapter 24. Here we need only know that they provide a bright, coherent, monochromatic beam.

records the interference pattern. The interference pattern provides all the information we need about the phase of the light waves.

In principle, the interference pattern that results on the film contains all the information about objects that are in the view of the film (and that are lighted by the beam). We can develop the film, and then at any later time we can illuminate it with light of the same wavelength as the original light. This recreates the outgoing light beam that would have been present if the original reflected beam had not been stopped by the film. Thus when we look through the illuminated film from the far side, we are not able (either in principle or in practice) to tell that the film is there at all. We have recreated our window. We are able, for example, to move our heads from side to side and see the effects of our changing perspective. We can focus at different distances. We have a *hologram*. Since the wave front that hit the film in the first place has been recreated, another name for the process is wave-front reconstruction.

Gabor's idea lacked only the technical means to put the process into operation. No source of coherent light bright enough to make the process work was known at that time. With the invention of the laser in 1960, however, this situation changed. Holograms became a reality when two University of Michigan scientists used a laser to make one in 1962. Gabor was awarded the Nobel Prize for physics in 1971.

16.2 *Practical Holography*

Early holograms were weak and feeble recreations of the original object, but the three-dimensionality of the image was unmistakable. One limitation was that the intensity of the lasers was relatively low and the films were not sufficiently sensitive to compensate for the faintness of the images. So the object being photographed had to be perfectly still for hours, which is why inanimate objects like chess pieces were used. And since we are working with interference, which depends on tiny spatial differences of the order of the wavelength of light, any jiggling of the object or holographic setup prevents the making of an interference pattern and thus of the holograph. Thus large concrete tables were and are often used, anchored directly to bedrock so that they are

Fig. 16–3 The hologram made in Fig. 16–1 can be viewed by shining a beam of coherent light on the hologram and then looking at it from the other side. The image of the object will appear on the opposite side of the hologram from your eye. It is not necessary to use a laser for the light source.

Fig. 16–4 A closeup view of a transmission hologram is very unimpressive. (Hologram by Tung Hon Jeong)

The pattern we see when we look at a hologram bears no resemblance to the picture stored there. The information that allows the picture to be reconstructed is stored in tiny interference lines; the larger patterns of whorls and stripes that we see (Fig. 16–4), are usually from dust on the plate or from other unimportant features. There is really no way of telling what information is stored in a hologram without illuminating it to reconstruct the image.

One reason that the films were not sufficiently sensitive is that special films that had exceptionally high resolution (roughly, the ability to detect detail) had to be used. These films are much slower (require much more light) than the films we use in our everyday cameras.

free of the vibration that takes place resulting from both people walking and from cars and trucks on nearby streets. For these reasons, the best holograms are made in the still of the night, when few people or vehicles are around.

The best images are viewed when the hologram is illuminated by a laser identical to the one that was used to make the hologram in the first place. (The playback beam does not have to be as purely monochromatic as the original beam, so lasers don't always have to be used for playback.) Since the early laser beams were narrow, only small objects could be recorded, and the resulting view was limited in scope. Still, we could see around the side of an object, or see something else that an object in front was blocking.

As time went on, better lasers and faster films were developed, which allowed shorter exposure times and larger objects to be used. Holograms are now made that, although they are made with laser light, can be viewed with ordinary light-bulb light, so-called white-light holograms. One advantage of white-light holograms is that since no laser is involved in viewing, it is impossible for a viewer accidentally to look directly into a laser, something that would be dangerous.

At first, holograms had to be viewed by looking through them. This type is called transmission holograms (Fig. 16–5). Ways were

Fig. 16–5 Two views through a transmission hologram that has been made into a cylinder and illuminated inside by monochromatic light (actually, yellow sodium light rather than laser light). Notice how we are seeing the car from a different angle when we look at the hologram from a different angle. This effect would not exist in looking at an ordinary photograph from a different point of view. (Hologram by Tung Hon Jeong)

A

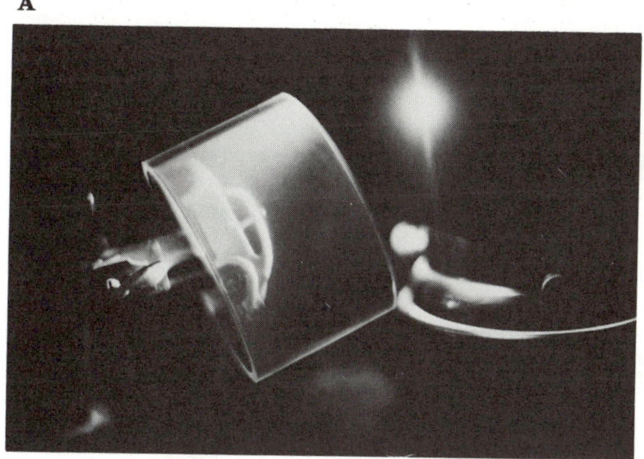

B

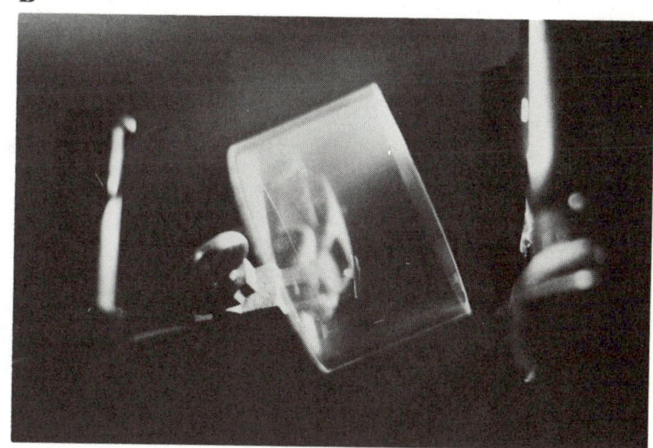

A

B

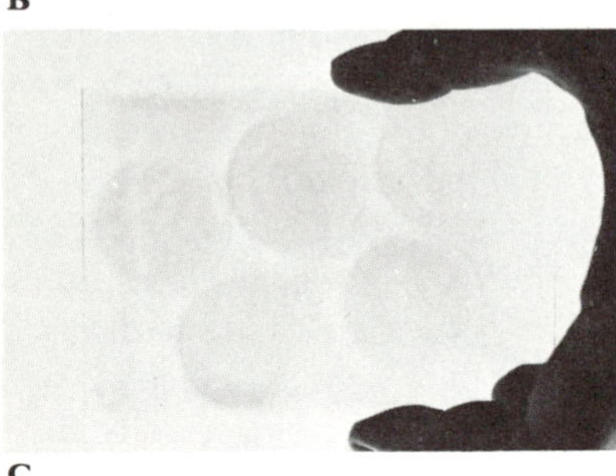
C

Fig. 16–6 A. A reflection hologram looks as though it has depth. B. But a side view shows that it is actually an image in a glass plate. C. When held up with a light behind it, it appears mostly transparent. (Hologram by Tung Hon Jeong)

developed of making holograms that could be viewed with reflected light. These reflection holograms (Fig. 16–6) could be put on a table top or hung on a wall. As time went on, the quality of the image has been continually improved by a variety of technical modifications in the method, none of which, however, altered the basic principles that we discussed.

All the holograms we have discussed were monochromatic; even those viewed in white light give a monochromatic image. But black-and-white film preceded color film as well. Since color film really consists of three monochromatic images one against the other in the film, we can presumably make full-color holograms by superimposing three monochromatic holograms. Methods of making such full-color holograms are being studied, but this ability is still beyond us.

One interesting property of a hologram is that since all light reflected from the object can hit any given point of the film, each point on the film retains information about the entire scene. Thus if we break a hologram in half, each half still contains all the information about the appearance of the scene. Even a small piece of a hologram is enough to see the entire holographic view (although the resolution gets worse as we use smaller and smaller pieces).

16.3 *The Uses of Holography*

The obvious use of a hologram—to put a picture on the wall that looks like a three-dimensional view through a window—is but one possibility. Holograms have many prospective uses of great practical benefit, and the list grows rapidly as people explore this exciting field of physics.

For example, the three-dimensional nature of the view can be used to store a lot of information in a limited space. If all the pages of a book for example, were stored on a single hologram—which we

Fig. 16–7 Each individual bit of a hologram contains an entire image, as we see in these two views. Each shows an image projected on a table from the one bit of the hologram that is illuminated (bright spot) by the laser at the bottom. (Hologram by Tung Hon Jeong)

A

B

can do by changing the angle of the beams slightly for each page—then by simply moving our heads to the side a bit, we could see whatever pages we wanted. This technique would certainly result in a saving of time for telephone operators, or for others who continually have to refer to different pages of reference works.

If we could get inside the human body (perhaps with fiber optics) and make a hologram of the inside of the heart or lung, we could then have a three-dimensional view at our disposal that we could study and refer to later on. This capability, when developed, will surely lead to advances in health care.

The interference pattern recorded in the hologram is very sensitive to changes in the distance of the object from the film. So a car manufacturer can monitor how tightly an engine is bolted down (Fig. 16–8). Flaws in the structure of a steel part show up clearly. By making a double exposure, a loudspeaker manufacturer can test how a speaker is vibrating (Fig. 16–9). The laser light falling on the speaker remains in phase between the two exposures, so changes in the position of the speaker cone are revealed. Holography will be used aboard Space Shuttle to observe crystals grow.

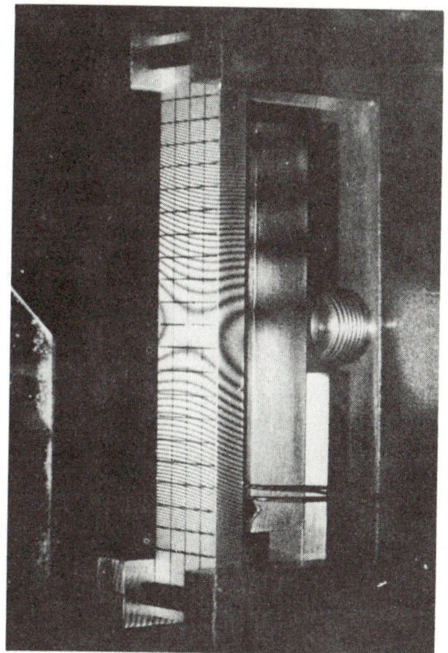

Fig. 16–8 The dark lines show the displacement in the aluminum bar induced by stress.

A

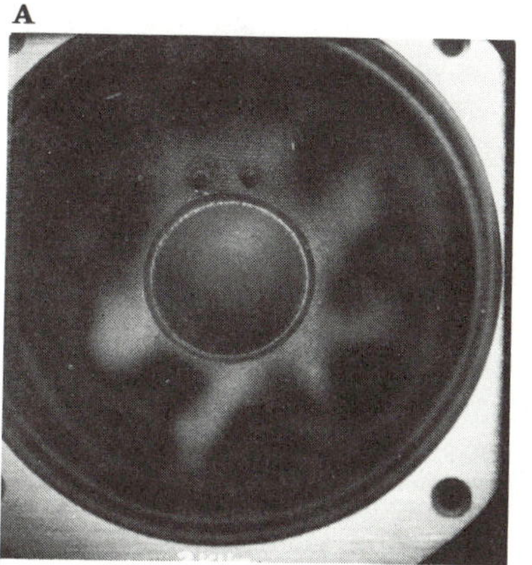

B

(above) Fig. 16–9 A pair of holograms used to study loudspeaker cones by a hi-fi manufacturer. We see how the modes and reflections in the cone of the loudspeaker change shape as a function of frequency. White areas are nodes, which do not move. The black rings show differences of deflections of the speaker cones by 320 nanometers (12 millionths of an inch), a half wavelength of the laser light. We see the cones at 3 kHz (left) and 10 kHz (right).

A

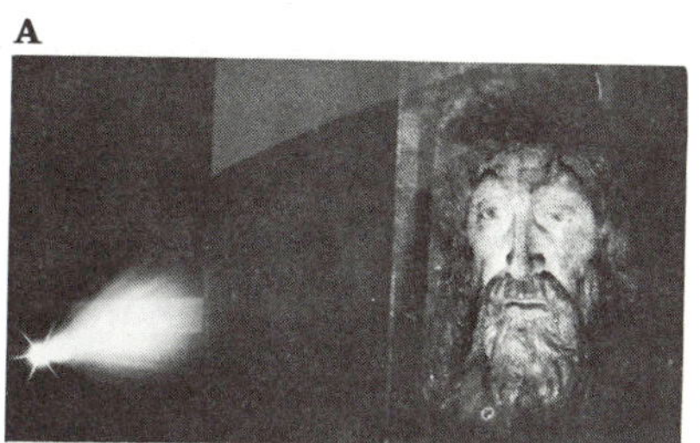

B

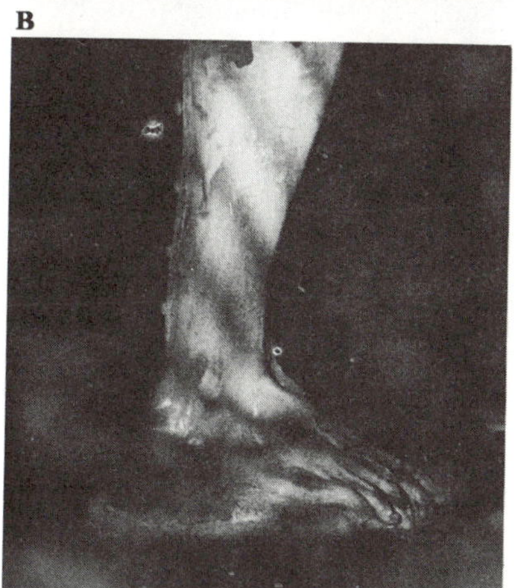

(left) Fig. 16–10 A. A photograph of a hologram of Donatello's statue *John the Baptist,* a fifteenth-century statue studied in Venice, Italy. B. Holography can be used for archival purposes, and can also be used to find faults. This is a double-exposure hologram of the statue's leg taken with a half-hour interval, during which the statue was placed in a humid environment. The pattern of dark fringes is mostly regular, and shows that the leg tilts by a small amount (0.3 millidegrees). The jog in one of the fringes outlines an internal structure added long after the statue was made.

One limitation with our modern movies is that colors in film change with the passing of time. (You may have found that some of your old color snapshots or slides have faded.) Even the master prints of old movies are fading, no matter how carefully the studios try to protect them. But when we develop ways of recording color images holographically, the information that determines the colors will be in interference patterns that are fixed in the film; this information is now in dyes that fade with time. Even the films that already exist could be transformed to holograms in order to preserve their colors; holograms, in this way, would make better two-dimensional movies, quite apart from their capabilities for making three-dimensional ones.

Although for making a movie it would be desirable to be able to record a scene and play it back at the same wavelength or combination of wavelengths, for some purposes we might want to record at one wavelength and play back at another. For example, we might one day be able to make holographic x-ray pictures by using x-rays to make the image, and to play back the images in white light. In this way, we might even be able to take advantage of the fact that x-ray wavelengths are extremely short, by magnifying the image greatly in the playback. (Difficulties in focusing x-rays prevent our using x-rays in a microscope in ordinary fashion; we do not have to focus the x-rays to make a hologram.) Or we might even be able to make a hologram itself using sound waves instead of light waves; this might be important under water or for medical purposes. Then we could view the hologram and use our eyes to "see" the otherwise invisible object. Electron waves, such as those used in an electron microscope, are another possibility, because waves typical of electrons have very short wavelengths. This possibility was in Gabor's mind from the beginning.

There is nothing sacred about the use of photographic film to record a hologram. Any surface that can retain an image is usable; a gelatin that shows the effect of light and then hardens permanently is often used. Alternatively, easily erasable surfaces are used to make holographic memories for computers. And there is no reason in principle why a fine-grained television system couldn't transmit holographic images.

A

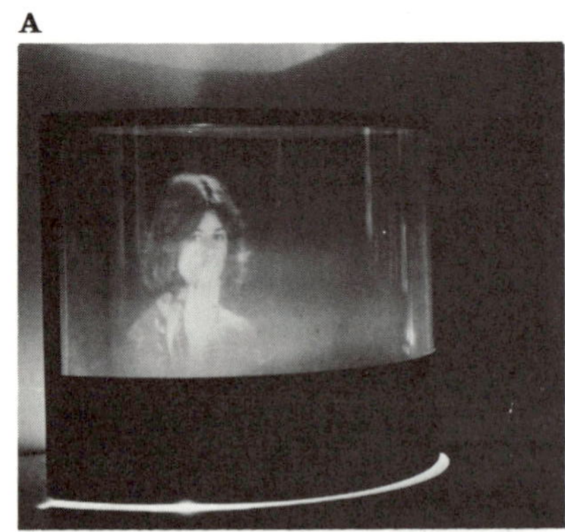

B

C

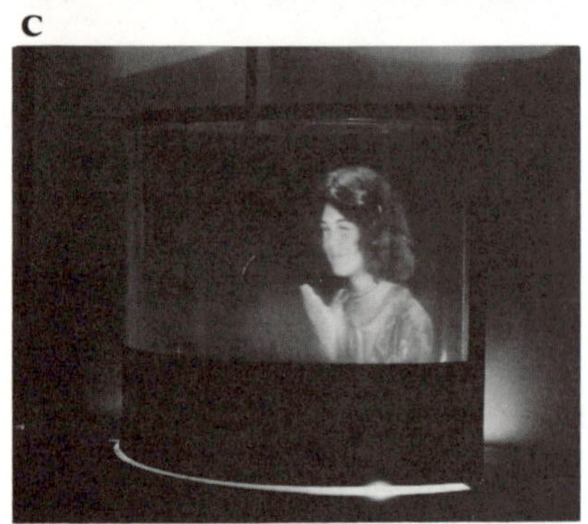

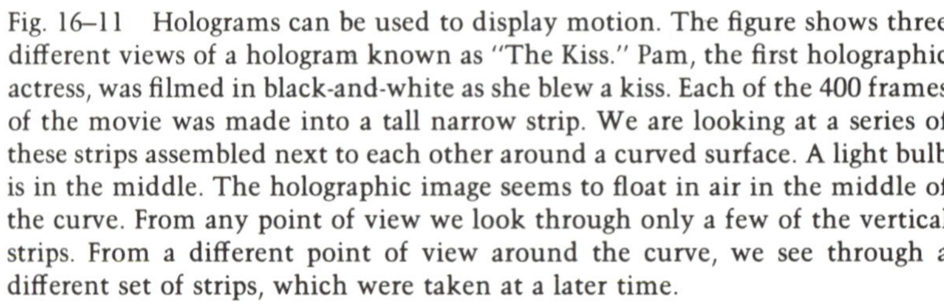

Fig. 16–11 Holograms can be used to display motion. The figure shows three different views of a hologram known as "The Kiss." Pam, the first holographic actress, was filmed in black-and-white as she blew a kiss. Each of the 400 frames of the movie was made into a tall narrow strip. We are looking at a series of these strips assembled next to each other around a curved surface. A light bulb is in the middle. The holographic image seems to float in air in the middle of the curve. From any point of view we look through only a few of the vertical strips. From a different point of view around the curve, we see through a different set of strips, which were taken at a later time.

Key Words

holography
reference beam
wave-front reconstruction
white-light hologram
transmission hologram
reflection hologram

Questions

1. Why is a monochromatic light important in making a hologram?
2. In making a hologram, why do you split one beam into two beams rather than just using the beams from two different lasers?
3. In white-light holograms, for which part of the process can we use the white light?
4. You and a friend have some information that you would like to keep secret. You make a hologram of this information, and to protect the secret, you cut the hologram in half, so that you and your friend each get half. That way, if one half should fall into "enemy" hands, they still won't know what's on the other half. What is wrong with this reasoning?

CHAPTER 17

Optics

Fig. 17–1 Large reflecting telescope at the Cerro Tololo Inter-American Observatory, in Chile. The mirror in this telescope is 4 meters in diameter.

When you look in a mirror to comb your hair in the morning or when you use a large telescope, you are taking advantage of the reflection of light. Light has bounced off a mirror. When you look through a pair of glasses or binoculars, take a picture with your camera, or watch a football game through the eye of a television camera, you are taking advantage of the refraction of light: light is bent or refracted, when it passes from air to glass (or from any substance to any other). The study of reflection and refraction and their applications is called optics.

In our discussion of optics we will be dealing almost entirely with situations in which we can ignore the effects of diffraction. We will thus usually be able to treat all the light waves as though they travel in straight lines, although the directions of these lines may change at the boundaries between one substance and another. When we can ignore the effects of diffraction in this manner, and keep track of the light by drawing the rays (as defined in Chapter 15), we say that we are dealing with geometrical optics. This term comes from our using the light rays to construct simple geometric figures that help us analyze the behavior of the light.

17.1 *Reflection and Refraction*

When you throw a ball straight at a wall, it bounces straight back at you. If you throw a ball so that it strikes the wall at an angle, the ball

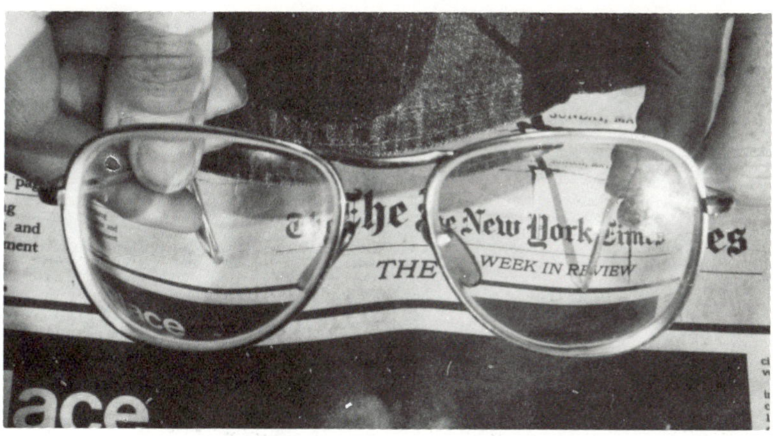

Fig. 17–2 Eyeglasses can change our view of things.

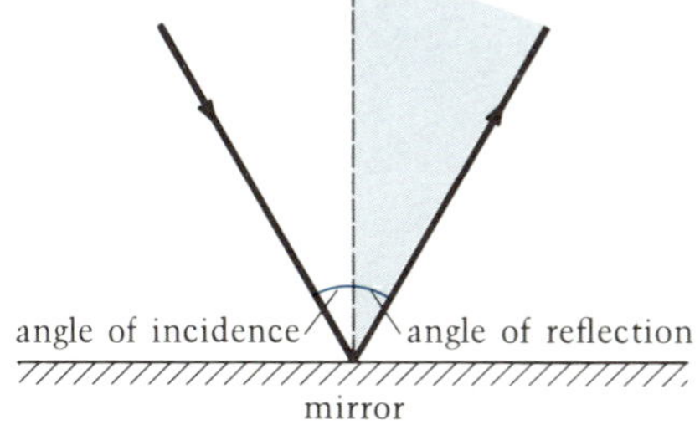

Fig. 17–3 The angle of incidence is the angle between the incoming light ray and a line drawn perpendicular to the surface of the mirror (dashed line). The angle of reflection is the angle between the outgoing light ray and the line drawn perpendicular to the surface of the mirror. The angle of incidence equals the angle of reflection.

does not come directly back to you. Instead, it bounces off at the same angle with the wall as it had going toward the wall.

Light behaves in much the same way when it is reflected off a surface. If we draw an imaginary line perpendicular to the surface, the angle that the incoming ray makes with that line is called the angle of incidence. Similarly, the angle that the outgoing ray makes with the imaginary line is the angle of reflection (Fig. 17–3). The law of reflection says simply that *the angle of incidence is equal to the angle of reflection.*

The law of reflection makes it easy to keep track of how light bounces off a flat mirror, but what happens if the mirror is curved, as in Fig. 17–5? In this case we can consider a very small area of the

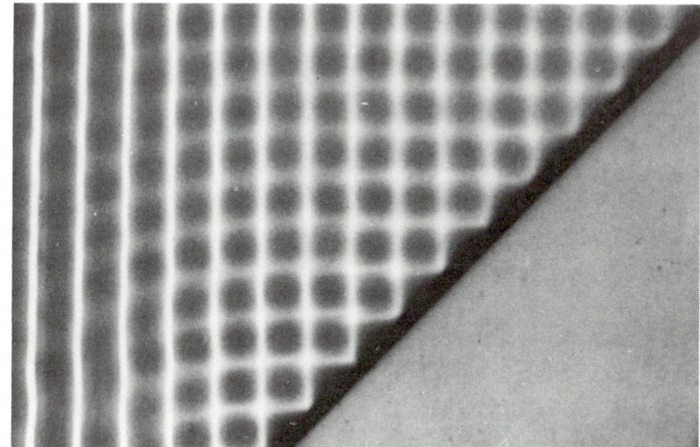

Fig. 17–4 We can see what happens to the wave fronts when a wave is reflected by looking at this picture of reflected waves in a water tank. The plane wave fronts come in from the left and are reflected off the barrier, forming plane wave fronts moving off to the top.

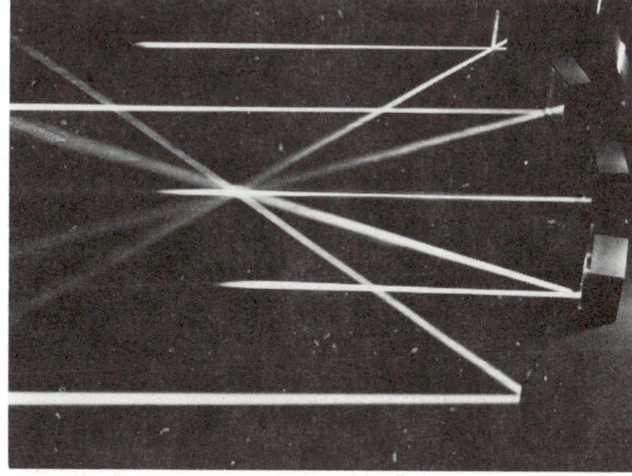

Fig. 17–5 Five parallel beams of light, coming from the left, strike five small plane mirrors. The mirrors are individually oriented so as to reflect the beams to a common focus. These individual plane mirrors could be replaced by a single large curved mirror of suitable shape to provide the same focusing.

Fig. 17–6 A recreation of Archimedes' setting a ship on fire in 212 B.C. by using many plane mirrors to focus the sun by approximating a parabola.

A seagoing observer sees such a small piece of the ocean that it appears flat, even though it is part of the curved surface of the earth. Similarly, a given light ray is affected by such a small part of a curved surface that we can consider the light wave to be reflected from a flat surface at the point of reflection.

mirror surrounding the place where the light ray hits. That small area of mirror behaves almost as though it were flat.

When light passes from a vacuum into glass (or any other material), it slows down. The ratio of the speed of light in a vacuum to the speed of light in any material is called the index of refraction of the material. The index of refraction of the vacuum is 1, and that of air is very close to 1, which is another way of saying that light is not slowed much in air. However, the index of refraction of some glasses is about 1.5.

index of refraction $= c/v_{\text{light in a material}}$.

Example: Since light travels 3×10^8 m/s in vacuum, it travels 2×10^8 m/s in these glasses, since their index of refraction is 1.5.

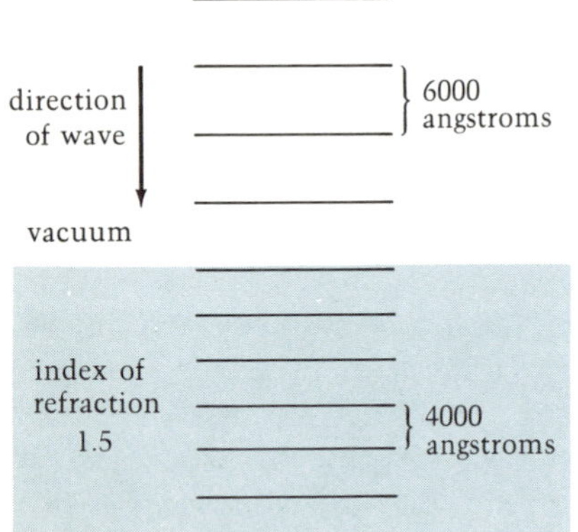

Fig. 17–7 Wave fronts going from one medium to another, showing the change in wavelength. In this case, we show the waves moving from a vacuum to a material with an index of refraction of 1.5. The waves move through the vacuum at 1.5 times the speed with which they move through the material. This speed change is accompanied by a corresponding change in wavelength. (The frequency stays the same.) If we put in light with a wavelength of 6000 Å in the vacuum, it will have a wavelength of 4000 Å in the material.

We have already seen that the speed of a wave is equal to the wavelength times the frequency. When light slows down, its frequency does not change, but its wavelength gets shorter. The wavelength changes by the same fraction that the speed changes, so the index of refraction also gives us the ratio of the wavelength in a vacuum to that in the material.

$$\textit{index of refraction} = \frac{\textit{wavelength (in a vacuum)}}{\textit{wavelength (in a material)}}.$$

Example: Light with a wavelength of 6000 Å in a vacuum has a wavelength of 4000 Å in glass that has an index of refraction equal to 1.5. The higher the index of refraction of a material, the shorter the wavelength of light in that material (Fig. 17–7).

Figure 17–8 shows how the fact that the wavelength changes at a boundary between two substances causes the light to change direction. Suppose that we are looking at waves going from air to water.

Fig. 17–8 Explanation of refraction. As wave fronts pass from one medium to the next, they can bend but they cannot break. A given wave front must continue in one medium from the previous medium. In this case the wavelength is longer in the top medium and shorter in the bottom medium. The only way that we can match up the wave fronts is to tilt one set with respect to the other.

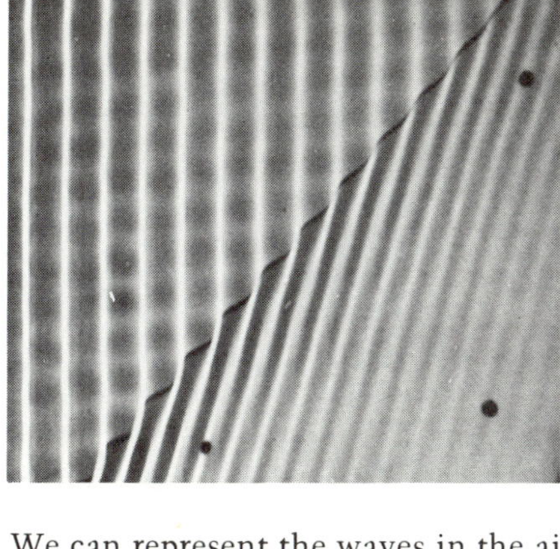

Fig. 17–9 Photo of refraction in a water tank. The two regions have different depths of water, resulting in different wave speeds in the two regions. Notice that each wave crest is bent, but unbroken, as it crosses the boundary.

We can represent the waves in the air by wave fronts separated by an amount that corresponds to the wavelength in air. In the water, which has a higher index of refraction than air, we represent the waves with wave fronts that are closer together, because the wavelength is now shorter.

We can choose some angle of incidence for the incoming waves and orient our wave fronts accordingly. As any given wave front passes from air to water, there is a time when it is partially in both. The crest of a wave must be continuous, so there is no break in it as it goes from the air to the water. However, in the water the wave fronts must be closer together than in air. How can we arrange to have the wave fronts unbroken but still change the spacing at the right place?

To see how to solve this puzzle, take a piece of paper with equally spaced ruled lines. Then take another piece of paper with equally spaced lines that are closer together. The way to make each line in one paper run into a corresponding line in the other is to tilt one paper with respect to the other. The greater the difference in spacing between the lines, the more you have to tilt the paper.

Waves behave the same way. In order for the waves to satisfy all of the conditions at the boundary, the wave fronts must tilt a little going from one medium to the next. The amount of tilt (which for waves is called *refraction*; see Fig. 17–10) depends on the difference

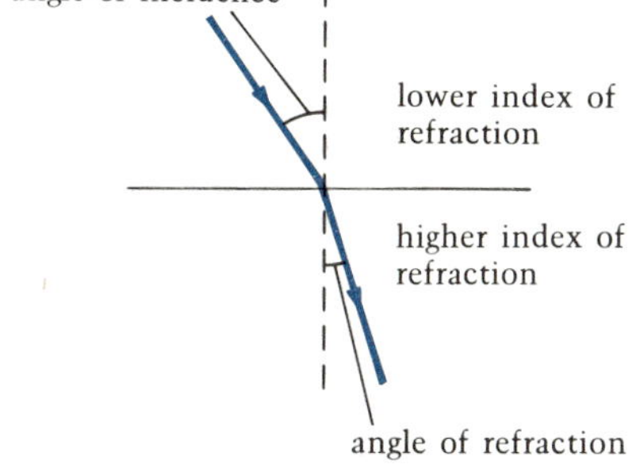

Fig. 17–10 The law of refraction. The *angle of incidence* is defined as the angle between the incoming light ray and a line drawn perpendicular to the surface between the two materials. The *angle of refraction* is defined as the angle between the emerging light ray and the same perpendicular line. The relationship between these two angles was investigated experimentally in the early 1600s by Willebrod Snell, a Dutch mathematician and astronomer, and the mathematical law describing the bending of light is called *Snell's law.* When the light passes from a material of lower index of refraction to one of higher index of refraction, it bends toward the dashed line. When it passes from a material of higher index of refraction to one of lower index of refraction, it bends away from the dashed line.

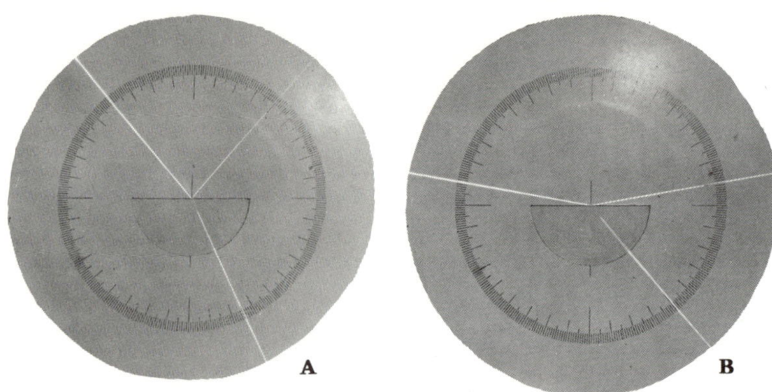

Fig. 17–11 A beam of light is incident from the left in both parts of this figure, and splits into a reflected and a refracted beam. We can see how the angles at which the reflected and refracted beams go off are different for incident beams of different angles.

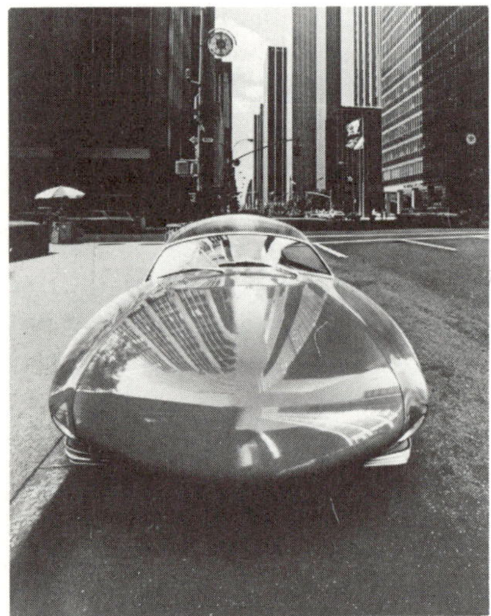

Fig. 17–12 When a car-wax company wants you to admire the shine, they may show you a reflection in the hood of a car arranged so that you are looking at a glancing angle.

in the wavelengths. We get a larger tilt when the change in index of refraction is larger.

The principles of reflection and refraction plus a little geometry are all we need to know in order to do geometrical optics. Most of geometrical optics involves light bouncing off or passing through objects. Although detailed answers to some problems require geometric or trigonometric calculations, we can get a feel for most of the results by simply looking at the paths that certain light rays take through various optical systems. Such an analysis is called ray tracing. As we will see in following sections, there is a trick to choosing the right rays to trace that simplifies the problem at hand.

Before looking at some of these applications, let us consider one additional feature of reflection and refraction. When a wave passes from one medium to another, some of the wave is reflected and the rest passes through to be refracted (Fig. 17–11). If we like we can bias things in favor of either reflection or refraction. For example, we can take an ordinary piece of glass and put a very reflective coating on it, so that almost all of the wave is reflected. (An ordinary mirror is a piece of glass with a coating behind it to make it more reflective than, say, a window pane.) Or we can get more reflection by glancing the light off the surface at a low angle, like skipping stones across the surface of a lake. Even x-rays, which penetrate surfaces, glance off if they are incident at a sufficiently low angle. The Einstein Observatory, the first telescope able to make images using x-rays, uses a set of concentric cylinders to focus the x-rays. The x-rays are incident on the glass insides of the cylinders at an angle of less than 1°.

Another phenomenon occurs when light goes from a medium with a high index of refraction to one with a low index of refraction. When we shine light almost perpendicular to the surface, some light is reflected back and some gets through. As we increase

Light, Optics, and Lasers

Plate 50 When a narrow beam of white light, as we see entering from the left, passes through a prism, we get a spectrum. Each wavelength of light is refracted by a different amount by the prism. (Courtesy Deutsche Bank)

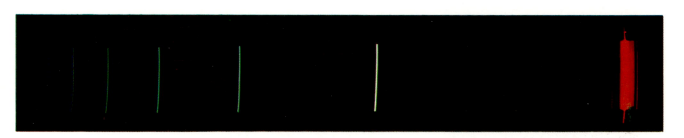

Plate 51 The Balmer series of hydrogen extends across the visible part of the spectrum. The red H alpha line, at the right, appears wide because it is overexposed. The lines appear here in emission. (Courtesy Alfred Leitner, Rensselaer Polytechnic Institute)

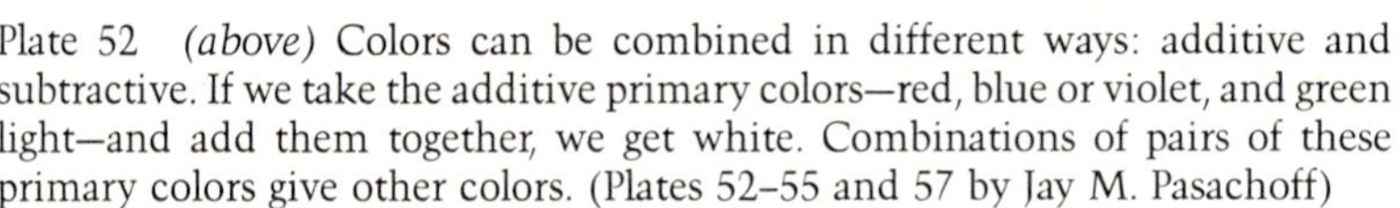

Plate 52 *(above)* Colors can be combined in different ways: additive and subtractive. If we take the additive primary colors—red, blue or violet, and green light—and add them together, we get white. Combinations of pairs of these primary colors give other colors. (Plates 52–55 and 57 by Jay M. Pasachoff)

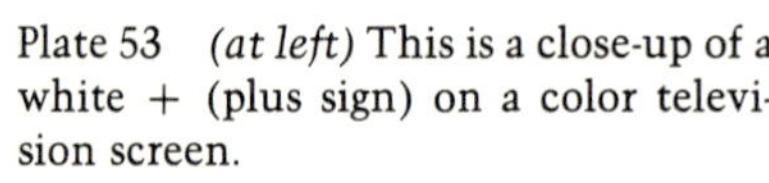

Plate 53 *(at left)* This is a close-up of a white + (plus sign) on a color television screen.

Plate 54 Subtractive color works opposite to additive color. Here we see a photograph of three filters, each the opposite of an additive primary. Cyan (bluish) is the opposite of red, magenta (purplish) is the opposite of green, and yellow is the opposite of blue. By opposite, we mean that it blocks the other completely. We are looking through the filters at a white background.

Plate 55 The process of painting uses subtractive colors. All the other colors can be mixed from blue, yellow, and red. The pure colors can be lightened with white. Mixing all three subtractive colors together gives black.

A B

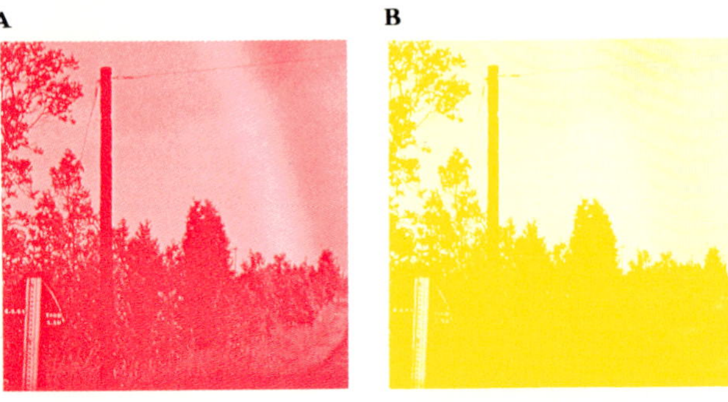

C

D

Plate 56 shows how color printing in magazines and in books like this one uses tiny dots of color in an additive process. In A–C we see three plates in each of the three colors. Part D shows all three printed over each other with black added as well. From a sufficient distance, the eye does not resolve the individual dots.

Plate 57 Rainbows are formed when sunlight enters a droplet of water, bounces around inside, and exits. Because the light of different colors is bent differently, a spectrum is formed. The primary rainbow here, the brighter one, comes from a single reflection inside the raindrops. The secondary rainbow comes from a double reflection.

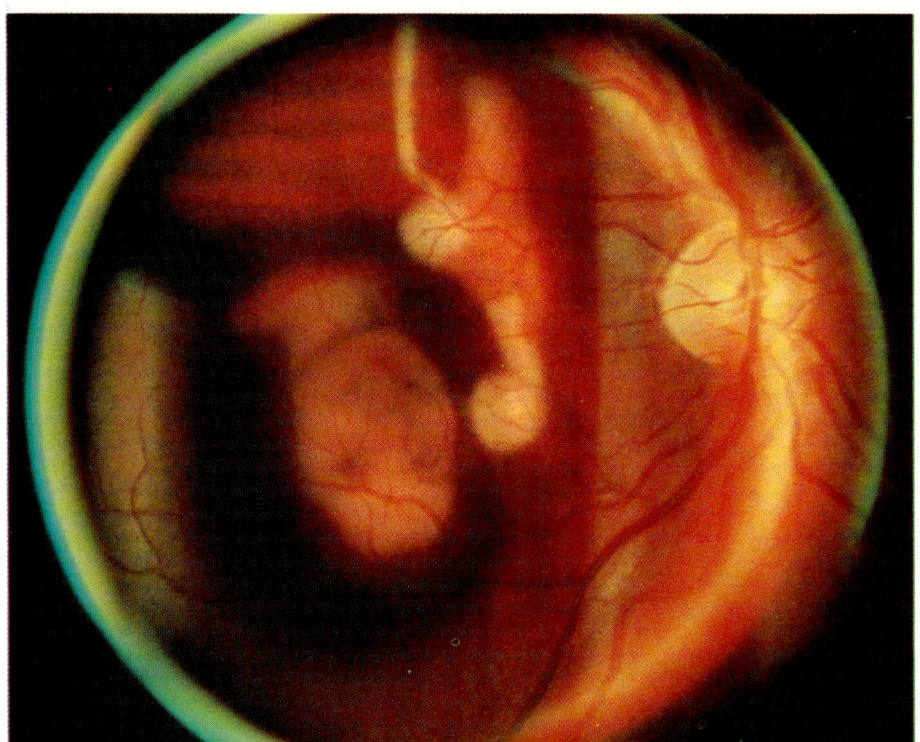

Plate 58 *(at left)* The eye acting as a lens. The photo shows the image of a girl using a telephone on the eye's retina. The image is inverted, as the laws of optics show. It was taken with a special camera that looked through the eye's pupil. The yellow region at the right is the location where the blood vessels and nerve fibers leave the eye. (Photo by Lennart Nilsson)

Plate 59 *(below)* A mirage. We see two horizons with a "lake" in between to lure desert travelers. The mirage is an optical illusion caused when light from the sky and from the mountains is refracted in the hot air above the desert. The "lake" appears to recede as you move forward. Such mirages are called Fata Morgana after King Arthur's stepsister Morgan le Fay, who used mirages to exert her power. (Photo by Emil Schulthess)

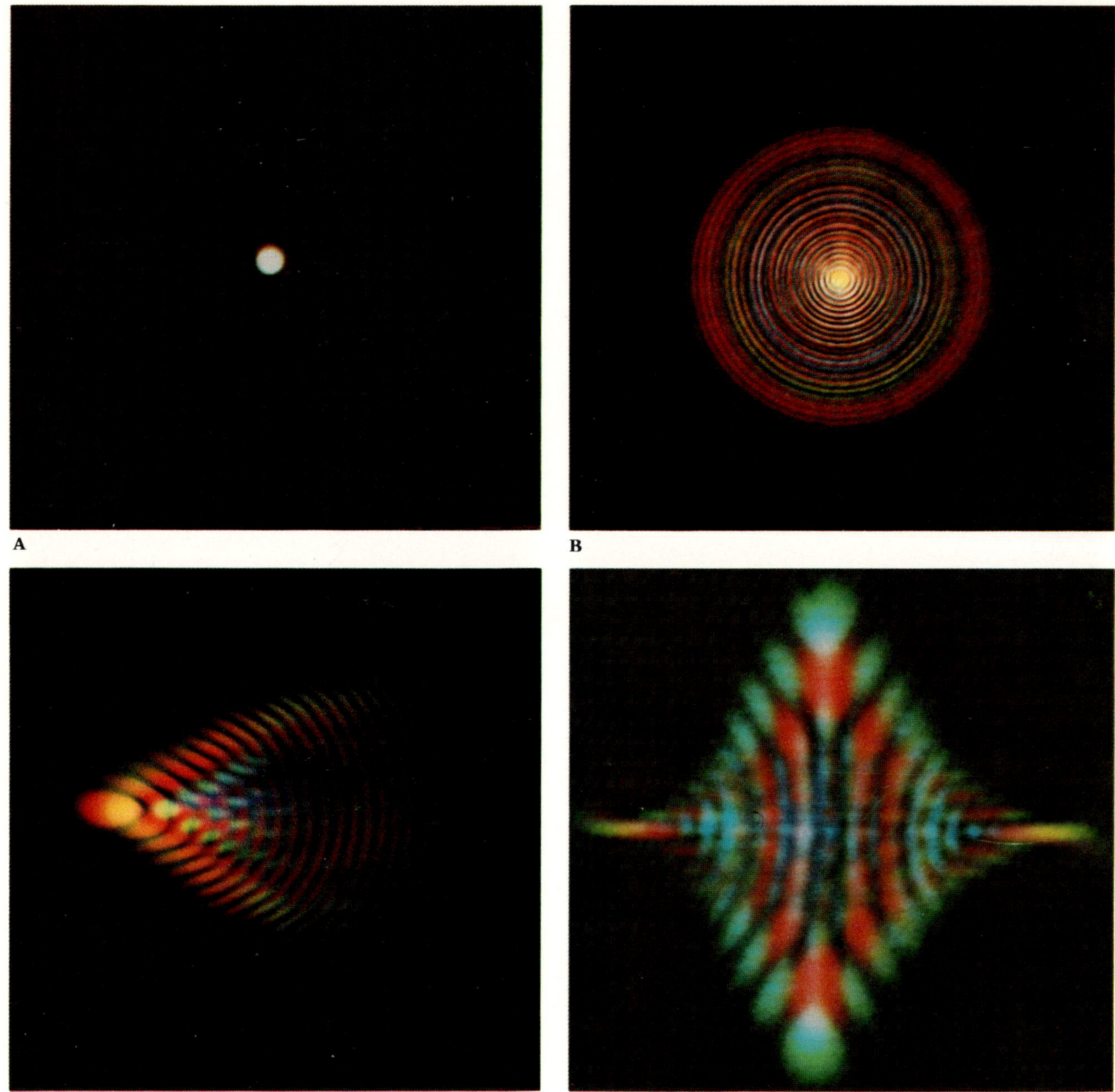

Plate 60 Aberrations in lenses, shown as greatly enlarged images of a point source. A. A perfect image. B. Spherical aberration. Light passing through the lens at different distances from the axis is focused at different distances, producing a blurred image. C. Coma. For off-axis rays, light passing through the lens far from the axis is focused in a different part of the focal plane than light passing near the center. The image appears as a spot of light with a spreading tail. D. Astigmatism. This aberration occurs for objects that are not on the axis of the lens. Light that passes through the lens top and bottom is focused a different distance behind the lens than light that passes through the lens left and right. (Photos by Norman Goldberg, *Popular Photography*)

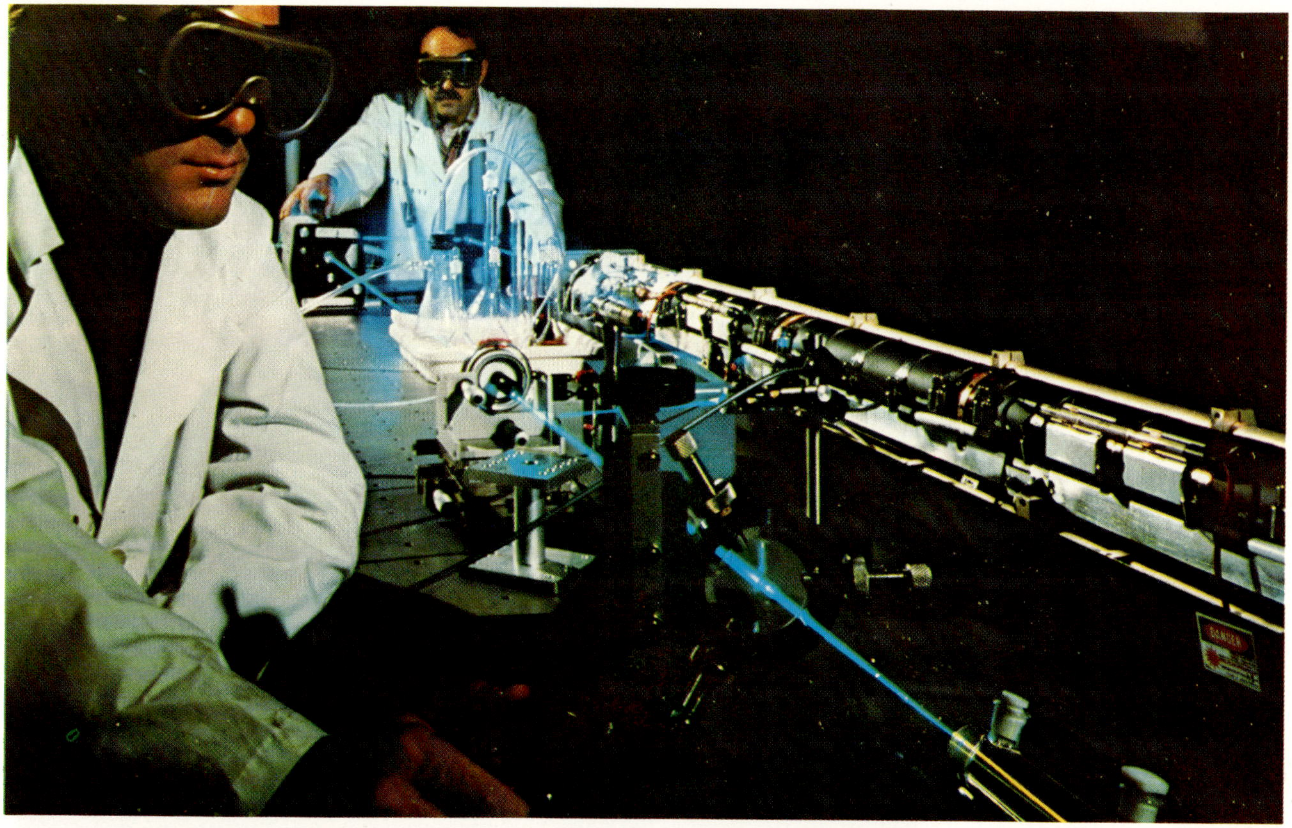

Plate 61 *(above)* A laser in use on a laboratory test bench. The ability of lasers to provide a narrowly focused and coherent beam of a single color is finding many uses in science and industry, as we describe in Chapter 24. (Photo by Bell Labs)

Plate 62 *(at left)* A diffraction pattern set up on the surface of a videodisk, some of the new technology that is beginning to provide additional entertainment and technology directly to homes. The fact that a diffraction pattern is set up shows that the details of the surface of the disk approximate the wavelength of light in size. (John Nijst, Phillips Research Laboratories)

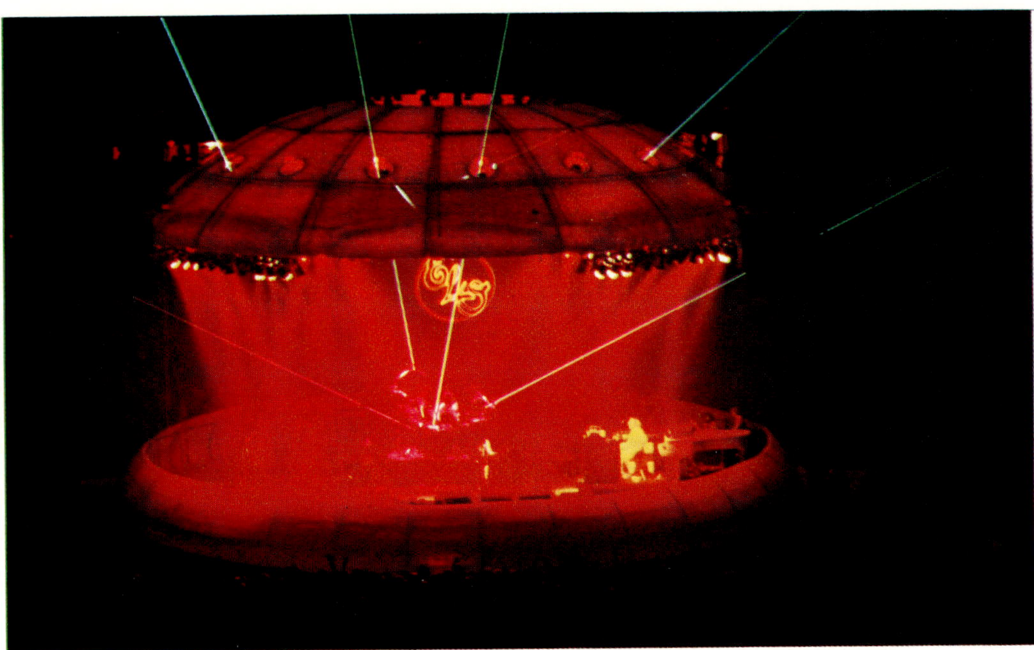

Plate 63 Note the narrow laser beams at this Electric Circus rock concert light show. (Courtesy LaserMedia)

Plate 64 Here we see a natural lens: droplets of water on a leaf making enlarged images of the leaf's veins. (Photo by Ernst Haas)

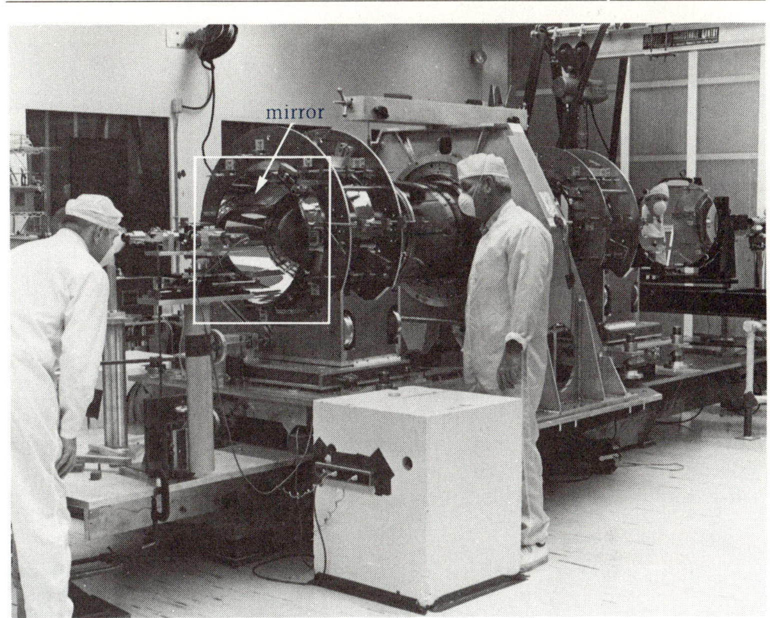

Fig. 17–13 Grazing incidence at work. Outlined is one of the grazing incidence mirrors as it was installed in NASA's Einstein Observatory to focus x-rays from celestial objects. The arrow points to the inside of a cylinder that is polished to the shape of part of a parabola. In Chapter 29, we will see how this spacecraft has been put to use.

the angle of incidence, the angle of refraction increases too. However, since the light is passing from a high-index medium to a low-index medium, the angle of refraction will be greater than the angle of incidence. Eventually we reach an angle of incidence for which the angle of refraction is 90°. At this angle, the light is bent so much that it runs parallel to the surface (Fig. 17–14). The angle of incidence at which this happens is called the critical angle. When we go beyond the critical angle, none of the light gets out at that point. It is all reflected, a phenomenon that is called total internal reflection because an incident ray traveling at the critical angle will remain within the medium.

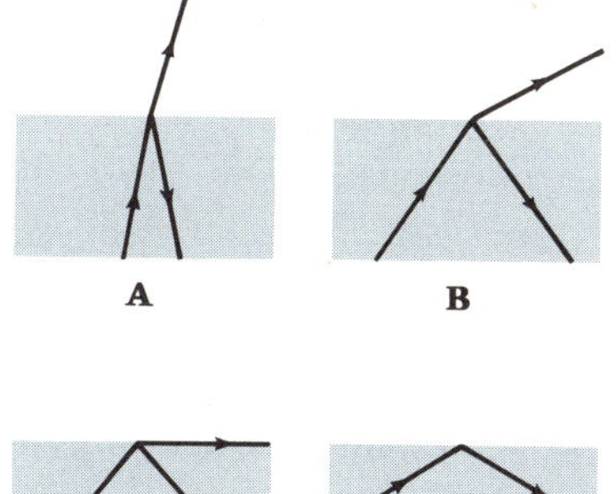

Fig. 17–14 Total internal reflection. The medium on the bottom has a higher index of refraction than the medium on the top. In A, the light comes in from the lower left. Some of the light is reflected and some is transmitted. Notice that the transmitted ray bends over farther than the incident ray. (This is because the index of refraction is higher below than above the surface.) In B we let the angle of incidence get larger. Notice that the transmitted ray is bent over even farther. In C, we have increased the angle of incidence so much that the transmitted beam goes right along the surface. We say that we have reached the *critical angle* for the incident ray. In D, we go beyond the critical angle, and there is no more transmitted beam. We have only reflection. This is total internal reflection.

17.2 *Fiber Optics*

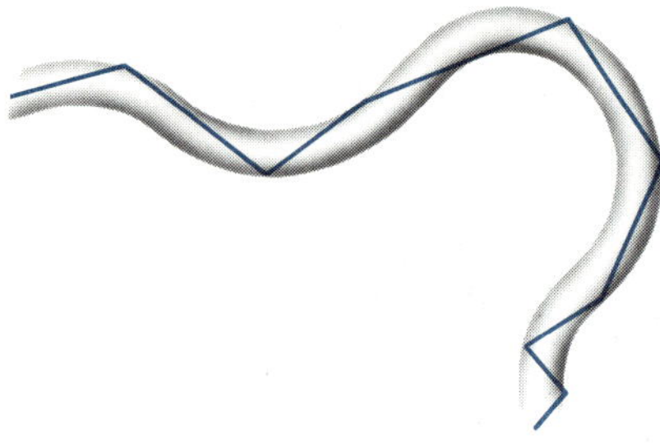

Fig. 17–15 Fiber optics. Using total internal reflection, we can keep a light beam inside a glass fiber. As the light moves down the fiber, it bounces off the walls. We can bend the fiber (and the light) around corners at will.

The phenomenon of total internal reflection has been used to make very thin filaments, thinner than a human hair, inside which light can travel (Figs. 17–15 and 17–16). These filaments are called *optical fibers*, and the field of research is called fiber optics. Since the light stays within a fiber even when the fiber is bent, doctors use fiber optics to see inside the human body. They can even insert a fiber through a vein to photograph the inside of a patient's heart. Such fibers can also be used to carry telephone messages more efficiently than copper wires, and for cases where telephone traffic is heavy, the fibers can be made much more cheaply than copper cables. Fiber optics, by making more channels available, by more readily allowing video images to be transmitted, and by cutting costs, may soon lead to a revolution in communications.

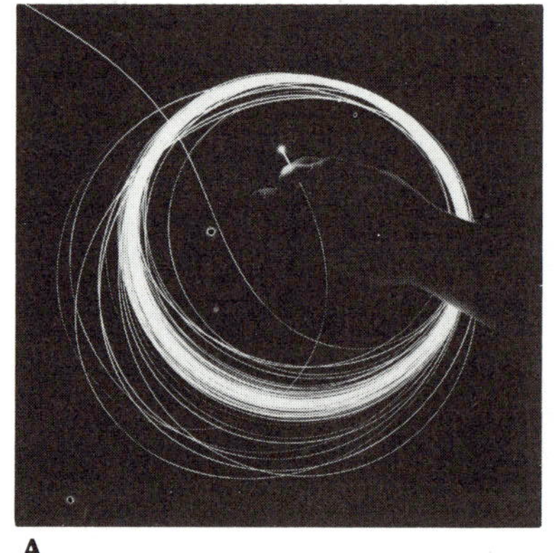

A

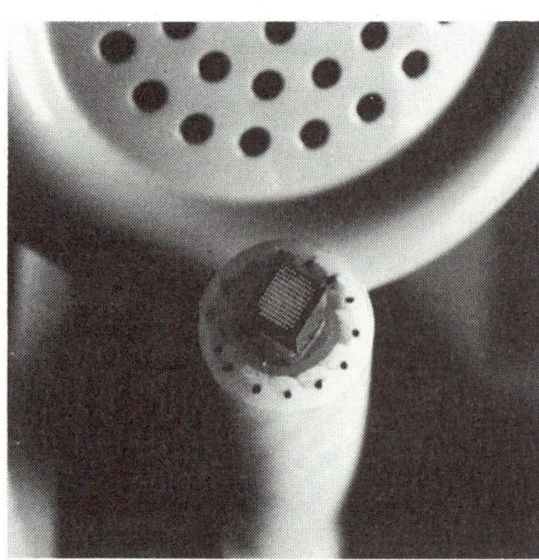

B

Fig. 17–16 A. A single optical fiber with light shining through it. B. A set of optical fibers in a telephone wire. A cable like this can carry 40,000 voice channels.

17.3 *A Look at Images—Mirror, Mirror on the Wall*

If you look directly into a mirror, you see an image of yourself. If you look at an angle, you see an image of something else. The law of reflection—the angle of incidence equals the angle of reflection—shows us why this happens. But the image actually appears to be behind, or inside, the mirror. If people stand behind you while you are looking directly into the mirror, you will see their images behind yours. How can light reflected off a plane surface produce such a realistic image?

Let's start by looking at something simple, a light bulb, for example, as seen directly by you and as seen in the mirror. The important thing to remember is that the only information that your eye gets comes from the light rays that enter it. The eye has no way of knowing where the light rays actually came from. It knows only the

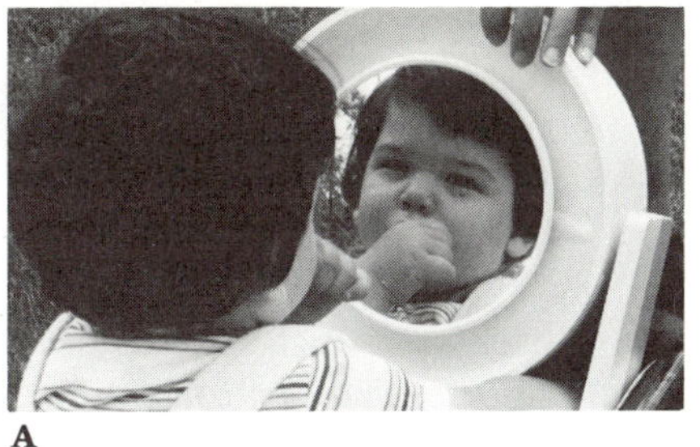

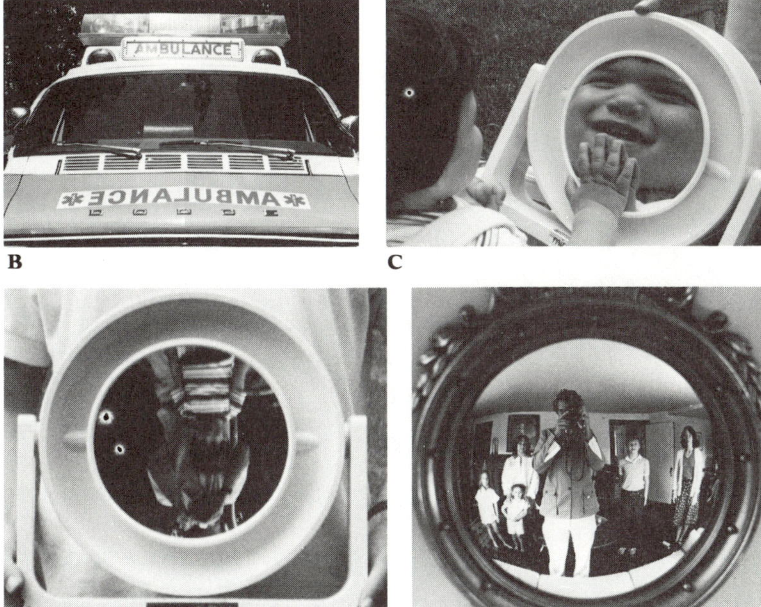

Fig. 17–17 A. Images are reversed from left to right on a plane mirror. The baby is sucking his left thumb but the baby's image is sucking his right thumb. B. The word "ambulance" is reversed so that it will be seen normally when viewed in a rear-view mirror. C. When observed from within the focal length of a concave mirror, the image looks enlarged and right-side up. D. When observed from outside the focal length of a concave mirror, the image looks smaller and upside-down. E. The image from a convex mirror shows a wide field of view.

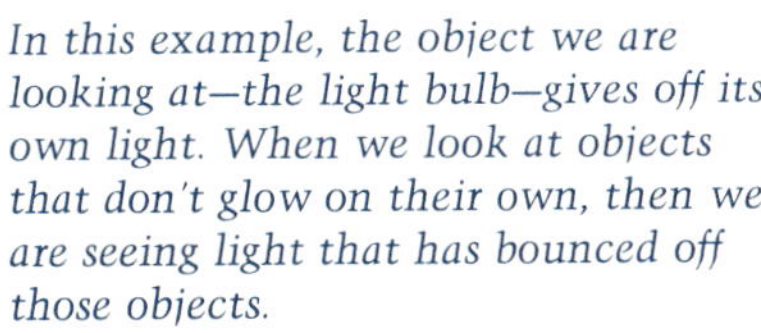
In this example, the object we are looking at—the light bulb—gives off its own light. When we look at objects that don't glow on their own, then we are seeing light that has bounced off those objects.

directions from which they are coming. In Fig. 17–18, you can compare the light rays reaching you directly from a light and those reaching you when the light rays are bounced off the mirror. Notice that, for the two cases shown, the light-ray pattern reaching your eyes is exactly the same. Therefore, there is no reason for the eye to record anything different in the two instances, even though they are different physical situations.

Notice that in both cases the rays leaving the light are spreading out, which we call diverging. In the first case, with no mirror, we

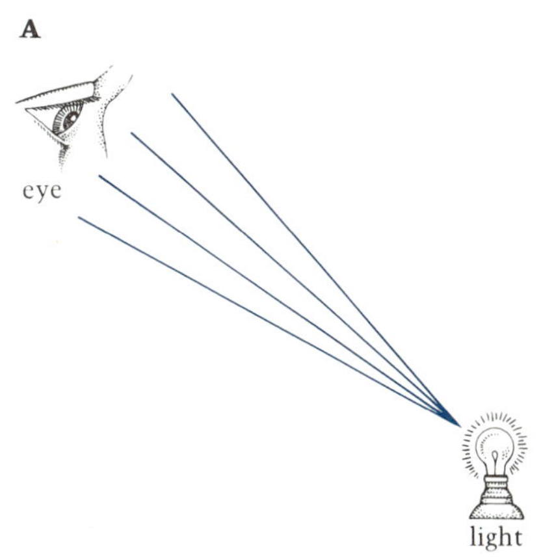

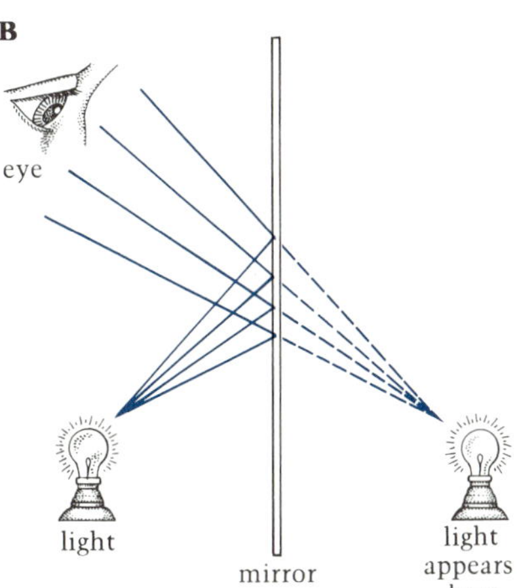

Fig. 17–18 A. The rays of light coming directly from a bulb to your eye. B. The light in a different position with a mirror also in place. Notice that the light rays reaching your eye are the same in the two pictures, and the eye has no way of knowing that the rays were reflected in B. Based on the information reaching your eye, both scenes look the same.

A

B

Fig. 17–19 This pair of pictures in funhouse mirrors shows the distortion caused by curvature. The mirror in A stretches out the images of people, while the mirror in B compresses them.

find the location of the light by tracing the rays back to the point where they cross. (Our brains do this automatically, a skill that newborn infants have to learn.) In the second case, with the mirror, we trace the rays back to where they *appear* to cross. The point where all the lines cross or appear to cross is called the *image* point. This is where you will "see" the light.

Although we are tracing the rays back as though they had all traveled in straight lines, they really did not. The real rays never passed through the image point. For this reason, we call the image in the mirror a virtual image. The word "virtual" is another way of saying that if you look behind the mirror, you will find no image there.

With a plane mirror, the image that you see appears as far behind the mirror as the object is in front of the mirror. Since reflection preserves all angles—the angle of incidence equals the angle of reflection—the image appears to be the same size as the object. There is no magnification.

The situation gets more interesting when we consider a curved mirror. If you have ever looked in the mirror in an amusement park fun house (Fig. 17–19), you know that a curved mirror can distort the image. Some parts appear larger and others smaller, because the curvature of the mirror changes the angles at which some of the rays reach our eyes.

We can analyze why this happens by looking at a simple curved mirror, one in which the mirror is a section of the surface of a sphere, as shown in Fig. 17–20. The imaginary line from the center of the mirror, drawn so that the mirror is symmetric about this line, is called the optical axis (or, simply, *axis*) of the mirror. It comes out perpendicular to the center of the mirror.

Let's look at what happens when rays come in parallel to the axis. (This situation corresponds to a source located very far away and on the axis.) If a ray comes right along the axis, it is reflected back out along the axis. To see what happens to rays parallel to the axis but not right on the axis, we can imagine that the mirror is not exactly a piece of a sphere but is made up of a lot of little flat mirrors glued together to form something close to a sphere. We

Fig. 17–20 Light rays striking a spherical mirror. All of these rays are coming in parallel to the axis of the mirror (the dashed line). These rays bounce off the mirror and all cross at the same point, the *focal point*.

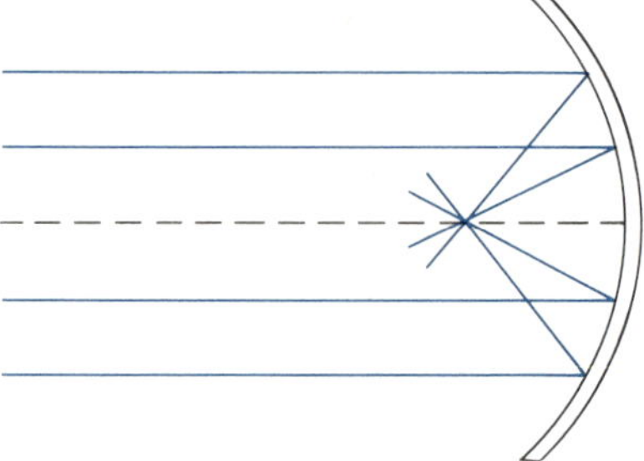

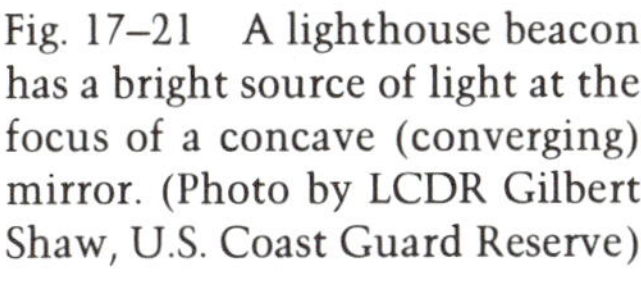

Fig. 17–21 A lighthouse beacon has a bright source of light at the focus of a concave (converging) mirror. (Photo by LCDR Gilbert Shaw, U.S. Coast Guard Reserve)

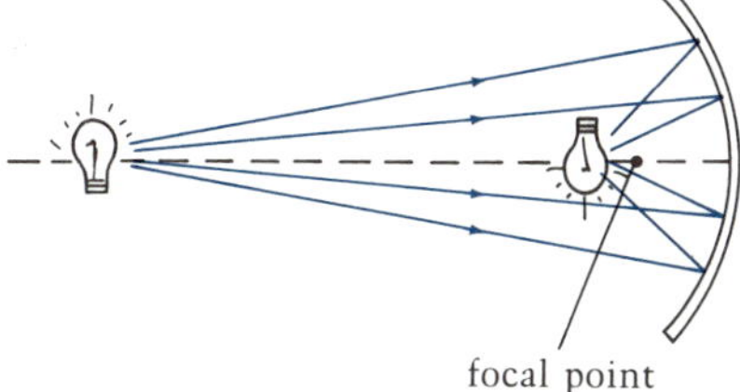

Fig. 17–22 All of the light rays that strike the mirror after leaving the light bulb come back together again. At the point where they cross, there is an image of the light bulb. In this case, we say that the image is real, since the light rays actually do cross. If we put a screen at that point, we would actually see a projected image of a light bulb. Notice that the image is not at the focal point.

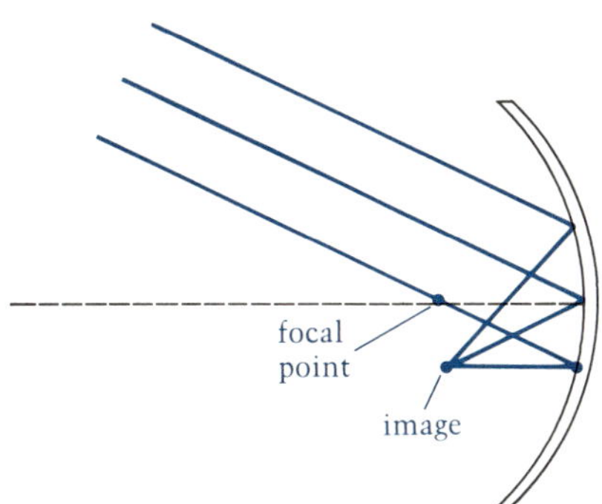

Fig. 17–23 If parallel rays come in from a distant object that is not on the axis of the mirror, the rays will still come together at a point, but that point will not be the focal point. The image will still be the same distance away from the mirror, but now will be off the axis.

then find that the rays are reflected toward the optical axis. The rays begin to approach each other—converge. For this reason we call such a mirror a converging mirror. If the mirror is just a small piece of a sphere, then we will find that all the rays cross the axis at about the same point. We call this point the focal point (or *focus*) of the mirror. *The focal point of a mirror is the point on the axis at which rays coming in parallel to the axis will all cross after reflection.* The distance from the focal point to the mirror itself is called the focal length of the mirror.

We have already seen that an image formed by a mirror appears to be at the place where the set of rays that originated at some particular point meet again. In the case of the plane mirror, although the rays never meet again in the forward direction, we can trace their paths backward along straight lines until they appear to meet. In the case where the rays actually do meet, we have a real image instead of a virtual image. If a curved mirror is focusing a light bulb, for example, you can place a screen at the image point and actually see a sharp bright spot—the image of the light bulb (Fig. 17–22).

The image of a distant object located on the axis of the mirror is at the focal point. What if the object is located far away, but is off the axis? Its rays still reach us parallel to each other, but make some angle with the axis of the mirror. As shown in Fig. 17–23, these rays also come together at a point. That point will be at a distance from the mirror equal to the focal length, but not on the axis. If the object is above the axis, the image will be below the axis. If the object is to the left of the axis, the image will be to the right of the axis.

If we look at the images of many distant objects, with each object located at a different distance above, below, or to the sides of the axis, we find that the images all fall on the same plane. This plane is

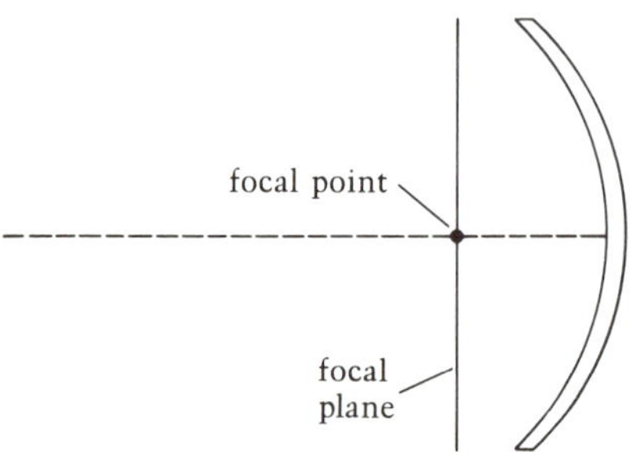

Fig. 17–24 The images of distant objects will be a distance from the center of the mirror equal to that of the focal point but will be off the axis of the mirror, as shown in Fig. 17–23. The points that make up all the possible images for distant objects fall on a surface called the *focal plane*.

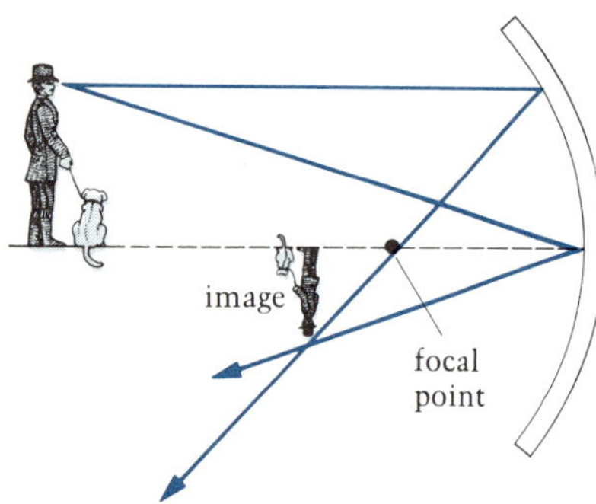

Fig. 17–25 Ray tracing and mirrors. If we can find where any two rays from the same point cross after striking the mirror, then we have the point where all the rays cross. So we trace two easy rays. One is the ray that leaves the object and travels parallel to the axis of the mirror. After striking the mirror, this ray will pass through the focal point. The other ray is one that strikes the mirror right on the axis. This ray will bounce off the mirror making the same angle with the axis going out that it did on the other side coming in. Wherever these two rays cross, you will have the image.

at a distance from the center of the mirror equal to the focal length, and is called the focal plane (Fig. 17–24).

If we want to find the image of an object that is not very far away, we find the image point by point. For example, if you are the object (and you are standing up straight with your arms down), we need only find the image of your feet and your head. The rest of your image must fall in between. Since the image of each point is the place where *all* the rays leaving that point come together, it is necessary to find out only where any two rays cross. The rest of the rays must cross there too. This is the idea behind ray tracing. For any point, we choose two rays that are easy to trace and see where they cross. This gives us the image of that point. For example, we know that a ray that comes in parallel to the axis of the mirror will be reflected back through the focal point. A second easy case is that of a ray that strikes the mirror right at the mirror's center, where the axis crosses the mirror. This ray will be reflected off with the same angle, as though it had been reflected off a plane mirror (Fig. 17–25).

Example: Let's find your image as formed by a spherical mirror. The details depend somewhat on how far you are from the mirror, so we'll look at a few cases. We'll assume that you are standing so that your feet are on the axis and your head is above it. This way, we know where the image of your feet will be—always on the axis—so we only have to find the image of your head. We can assume that the rest of your body will fall in between.

Let's start with the case in which you are close to the mirror (Fig. 17–26A); that is, you are closer than the focal length. In this case, your image appears behind the mirror and is a virtual image. You appear taller, which means that this mirror acts as a magnifier. As you walk backward from the mirror, your image moves farther behind the mirror. As the image gets farther away, it gets larger and larger. Eventually, when you reach the focal point, the rays all leave the mirror parallel to each other, which means that they will never cross. For this situation, there is no image. Because we think (imprecisely) of parallel lines approaching each other when they have gone an infinite distance, we sometimes speak of an image "at infinity."

When you are farther from the mirror than the focal point, the rays cross and there is a real image. As shown in Fig. 17–26B, however, your image is *inverted,* that is, you appear upside down. Other than that, you would appear quite real to any observer.

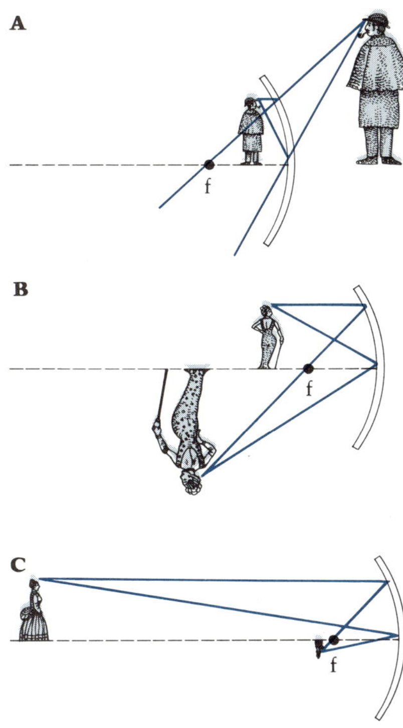

Fig. 17–26 Finding your image in a mirror, using the rays shown in Fig. 17–25. A. You are closer to the mirror than the focal point (*f*), so the image will be virtual (behind the mirror), upright, and larger than you. B. You are between one and two focal lengths from the mirror. The image is real, upside-down, and larger than you. C. You are more than two focal lengths from the mirror. The image is real, upside-down, and smaller than you. In general, whichever is farther from the mirror, you or the image, will be larger.

17.4 *Images by Refraction*

We can apply all the ideas that we introduced in the last section to see how refraction plays a part in forming images. Actually, apart from the plane mirrors and occasional magnifying mirrors in our bedrooms and bathrooms, we are probably more familiar with images formed by refraction than reflection. Certainly anyone who wears eyeglasses or contact lenses is. The same applies to anyone who looks through any type of eyepiece, whether it is on a pair of binoculars, a microscope, or a simple magnifying glass.

We can see the effects of refraction on an image by looking at an object immersed in a tank of water. Our discussion of refraction at the flat surface of the water will have many similarities with that of reflection from a plane mirror. Suppose you are looking down at a fish from above the surface of the water. Figure 17–27 shows that you see a virtual image of the fish closer to the surface of the water than the fish actually is. If you try to judge the depth of the fish and

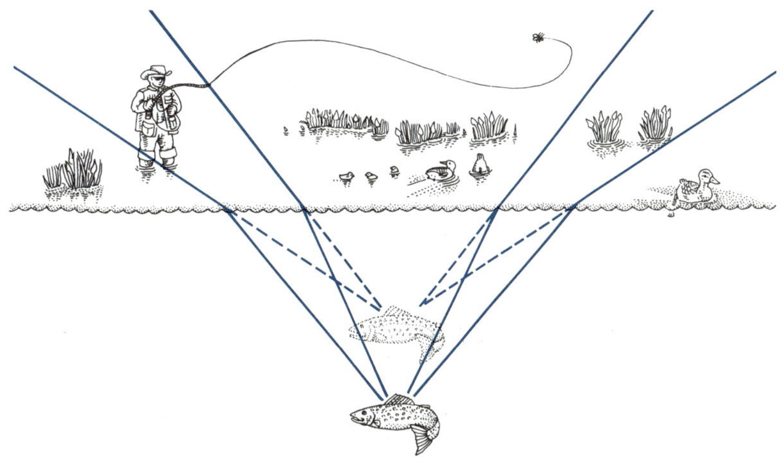

Fig. 17–27 The image of a fish under water. We look at all the rays leaving the fish. When they reach the surface, they bend away at larger angles from the vertical, since the index of refraction is greater in the water than in the air. Any observer seeing the rays would "trace" them back, along the dashed lines, and "see" an image of the fish that is closer to the surface of the water than the fish itself. This is why a stream always looks shallower than it actually is.

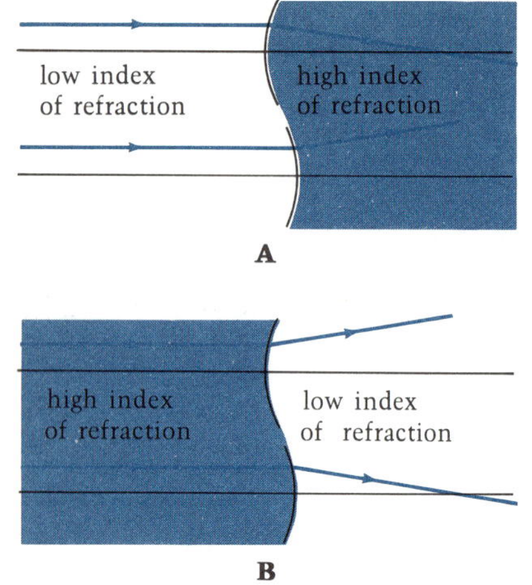

Fig. 17–28 Refraction of light at a curved surface. A. The light is going from a medium of low index of refraction to one of high index of refraction. B. The situation is reversed.

reach in, you will come up empty-handed. Remember, your eyes pay attention only to the angle at which the rays enter. They know nothing about how these angles might have been changed along the way. If you want to take a picture of the fish, you should aim the camera at the image of the fish, but if you want to spear the fish, you would have to use your knowledge of optics to figure out where the fish really is.

When we consider refraction at a curved surface, we again have two possibilities. If you are on the inside of the curve, we say the surface is concave; if you are on the outside of the curve, we say that the surface is convex. In addition to the direction of curvature, we must also know whether the light is traveling into a medium of higher or lower index of refraction. Figure 17–28 shows how rays traveling parallel to the axis are bent in each of these four cases. Notice that in two cases the rays converge and in two cases they diverge.

A piece of transparent material with one of these curved surfaces on each side makes a lens. (Actually, one side of a lens can be flat.)

MIRAGES

Many of you have looked out over hot dry land and seen a shimmering that looks like the glistening surface of a pond. However, as you walk toward the sight, you never seem to get closer and eventually it disappears. This little trick of nature is called a mirage. Mirages can be quite convincing. For example, in his 1906 trip toward the North Pole, the explorer Robert E. Peary "discovered" a beautiful range of mountains in the Arctic. When another team came to explore these mountains, a few years later, they again "saw" them. But when they approached the mountains, the mountains turned out to be a mirage.

A mirage is an example of refraction of light by the earth's atmosphere. The atmosphere is acting as a lens. The index of refraction of air depends on the density of the air, which in turn depends on the temperature (as we saw in Chapter 11). When light passes from air of one temperature to air of another temperature, it bends. The greater the change in temperature, the greater the bending.

There are often large temperature changes near the ground, so light passing near the ground can be subject to large bending. Thus mirages usually appear close to the ground. The exact nature of the mirage depends on how the temperature changes in a given situation. (Examples are shown in Plate 59 and Figs. 17–30, 17–31, and 17–32.)

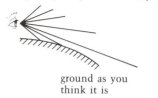

A

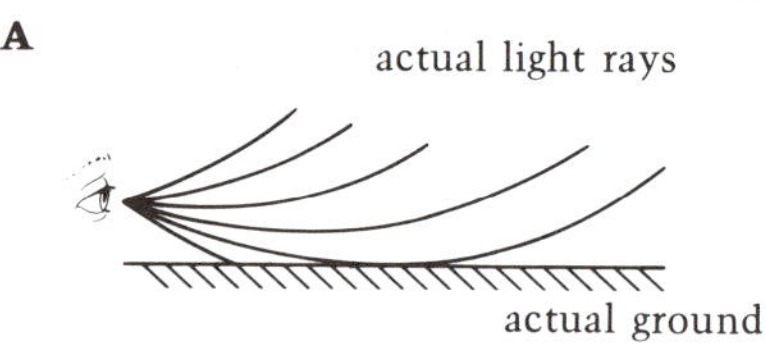

B light rays as you think they came

ground as you think it is

Fig. 17–30 One example of how a mirage is formed. A. The air gets cooler as we get higher above the ground, so the rays of light are bent. B. Your eye thinks all the rays came in straight, and the image you see is determined by tracing the rays back along straight lines. In the image, the ground appears to curve away. Different temperature distributions result in different effects.

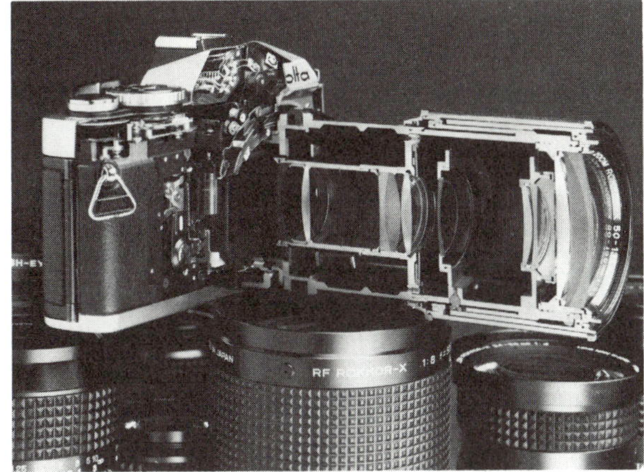

Fig. 17–29 Modern camera lenses like these involve several optical elements of different shapes made of different kinds of glass.

In our examples, we will be working with thin lenses, by which we mean that the lenses are so thin that we can ignore the distance between the two surfaces (other than the effects of the curves) in finding images.

In normal use, we place our lenses in air or a vacuum, so the lens always has a higher index of refraction than the surrounding medium. In making a lens, we can choose any combination of convex and concave surfaces (Fig. 17–29).

Once the light from an object reaches the eye, the apparent location of the image depends on tracing the rays back in a straight line. You see a distorted version of reality, or even something that has little to do with reality.

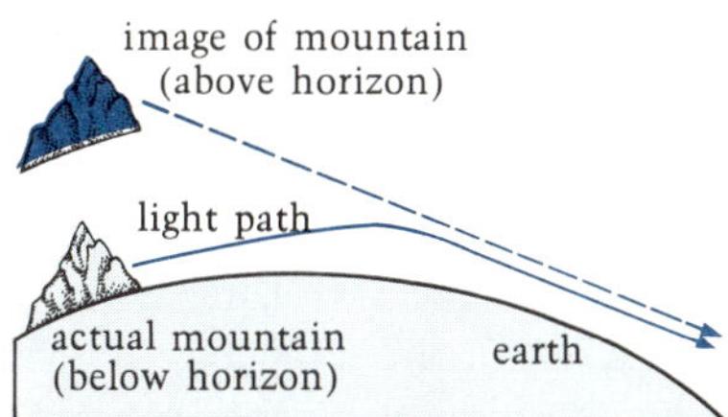

Fig. 17–31 When the temperature variations in the atmosphere are right, light can be bent "around" the horizon. When the light ray reaches your eye, there is no information on how its path has been bent. The image of the mountain appears along a straight line (dashed line) running away from your eye at the angle which the true light ray entered your eye. The temperature conditions for such a mirage usually occur in arctic regions, and such a mirage is therefore called an arctic mirage. In an arctic mirage you see an object that really exists; however, you see it in the wrong place.

A

B

Fig. 17–32 A mirage in which a puddle of water seems to be shimmering in the roadway. We even think we see the reflections of cars in it. Part A was taken from a crouching position. In B, taken a few seconds later from a standing position, the mirage has disappeared because of the slightly different perspective.

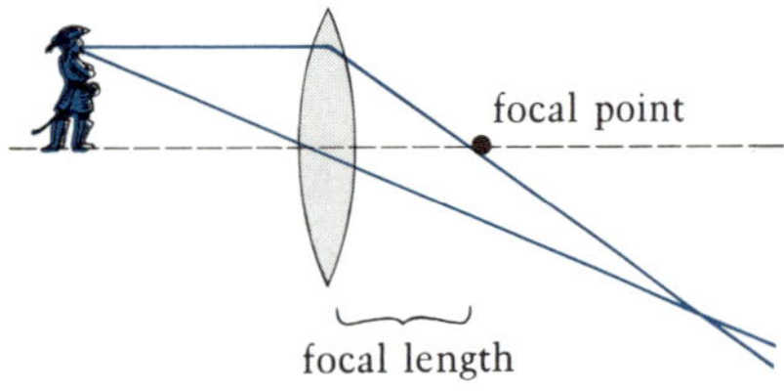

Fig. 17–33 Ray tracing with thin lenses. As in the case of a mirror, we find the two easiest rays to trace. The first one is one that is parallel to the axis of the lens. This will pass through the lens and be bent so that it passes through the focal point of the lens. The other ray passes through the center of the lens, virtually undeflected (in a straight line). The reason for this straight-line path is shown in Fig. 17–34.

A double converging lens (shown in Fig. 17–33) illustrates how a lens forms an image. The lens is called "doubly converging" because each of the two curved surfaces tends to bend light toward the axis. Notice that although the surfaces are curved in opposite directions, at one surface the light is going from air to glass and at the oppositely curved surface the light is going from glass to air, so that both surfaces have similar effects.

First let us see what happens when we look at a distant object. For this case, all the light rays come in parallel to the axis of the lens. All the rays are brought together—converge—at a point on the other side of the lens. This is the focal point of the lens, which has a definition similar to the definition of the focal point of a mirror. For a lens, however, the focal point is on the opposite side of the lens from the object. For an object that is far away but not on the axis of the lens, the light reaches the lens with the rays parallel but making some angle with the axis. These rays are also brought together at a distance behind the lens equal to the focal length. However, the image formed by these rays is not on the axis. If we look at a variety of distant objects, as we did for a mirror, all their images will lie in a plane. This is the focal plane for the lens.

What about objects that are not far away? We can again find their images by ray tracing (Fig. 17–33). As we did with mirrors, we find where all the rays that left a point come back together, and we do this simply by finding where any two of the rays come back together. One ray that is easy to trace is the one that comes parallel to the axis of the lens. It will be bent just enough to pass through the focal point. The other ray we trace is the one that passes through the center of the lens. As Fig. 17–34 shows, this ray passes through the lens undeflected.

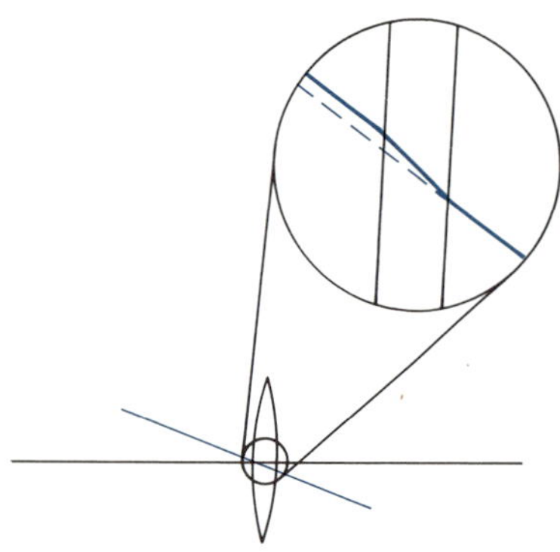

Fig. 17–34 A ray passing through the center of a thin lens. At the center of the lens the front and back surfaces are parallel to each other. This means that however much the light ray is bent in entering the lens, it bends back by exactly the same amount on leaving the lens. The emerging ray is parallel to the continuation of the incoming ray (dashed line), but is shifted over slightly. The thinner the lens, the less chance it has to shift over, and when we consider a lens to be very thin, we can ignore the amount of shift. We can effectively say that the ray passes through the center of the lens in a straight line.

Example: We will try to find your image as created by this thin lens (Fig. 17–35). As with the mirror, the exact situation depends on how far you are from the lens. Let's start with the case in which you are more than two focal lengths from the lens (Fig. 17–35C). Again we find the image of your head and assume that we can fill in the rest. The rays leaving your head actually come together so this is a real image. If you had a miner's light on top of your head and placed a screen at the image point, you would see the image of the light as a bright spot of light on the screen. The image is inverted; that is, you appear upside-down. The image is also closer to the lens than you are and is smaller than you.

As you walk closer to the lens, your image moves farther away and gets larger. When you reach the point at which your distance from the lens is equal to twice the focal length, then the image falls at the same distance behind the lens as you are in front of it, and your image is the same size as you are. As you continue to move

The distance of the focal point from the lens is called the focal length.

closer, your image moves farther away from the lens than you are and becomes larger than you. The image keeps moving away and getting larger until your distance from the lens is exactly equal to the focal length. Then all the rays leaving any point on you emerge from the lens parallel to each other and will never cross. Your image is then "at infinity."

What happens when you move closer than one focal length? Now the rays we are tracing do not cross. There is no real image. Instead, there is a virtual image, which is defined by tracing the rays backward to the point where they intersect. This image is farther away from the far side of the lens than you are from the near side, and it is larger than you are. However, a screen placed at the distance of the virtual image shows nothing projected on it, since no light rays actually reach that point.

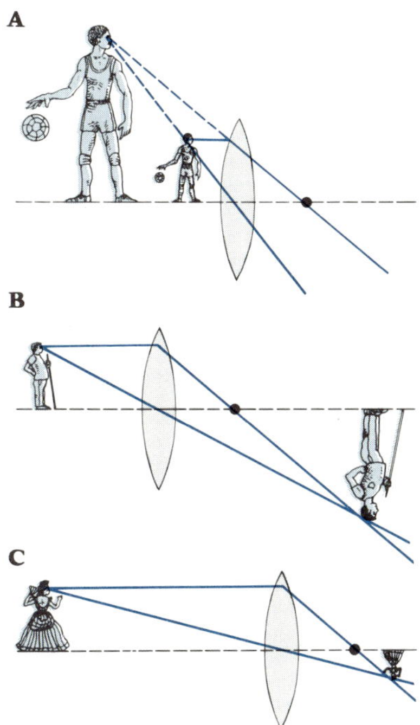

Fig. 17–35 Finding your image in a lens. We are tracing the two rays shown in Fig. 17–33. A. You are closer to the lens than one focal length. Your image is virtual, that is, the rays never cross. To find the image, you must trace the rays back. Your image is behind you, and since it is farther from the lens than you are, it is larger than you. It is also upright. B. You are between one and two focal lengths from the lens. Your image is real (and on the other side of the lens), upside-down, and larger than you. C. You are farther than two focal lengths from the lens. Your image is real, upside-down, and smaller than you are.

17.5 *Colors and Refraction*

We have been discussing the index of refraction of a given material as though it were a constant. However, the index of refraction of any material is different for different wavelengths of light. This variation occurs because the speed of light in a material is different for different colors. When light is refracted, the different colors are bent by different amounts.

We see this phenomenon clearly when we use a prism (Fig. 17–36 and Plate 50). The simplest prisms have a triangular shape. Light passing through is bent at each of two surfaces, and light of different colors emerges at different angles. When white light passes through the prism and falls on a screen, we see a progression of colors from red, which is bent the least, to violet, which is bent the most. The range of colors, the spectrum, is beautiful to behold. Spectroscopy, the study of the spectrum, has many practical applications, such as analyzing the constituents of a gas from afar (studying pollution, for example). We have already discussed spectra and how we study them using gratings. Prisms were used long before gratings; Newton studied the nature of light by passing white light through a prism, thus separating its component colors: red, orange, yellow, green, blue, indigo, violet. (We can most easily

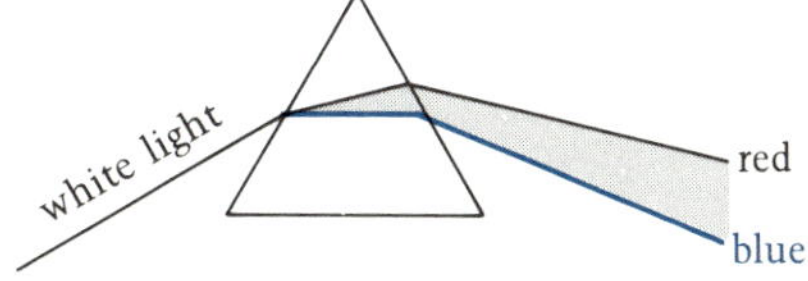

Fig. 17–36 A prism. White light enters, but the index of refraction is different for different colors, and the different colors leave the prism at different angles.

remember this list by thinking of their initials as a name: Roy G. Biv.) In general, the resolution of prisms is not as good as the resolution of gratings of comparable size, but there are still cases in which prisms are useful.

The bending of different colors by different amounts results in one of the most beautiful of natural phenomena—the rainbow. In a rainbow, refraction takes place as light passes through water droplets. The details of the formation of a rainbow are quite complicated, and physicists still argue over some of the subtleties. The problem is that the light is not simply refracted by the raindrops. It is also reflected, including internal reflections within the raindrop. As a result, a rainbow has a more complicated structure than the spectrum cast by a simple prism.

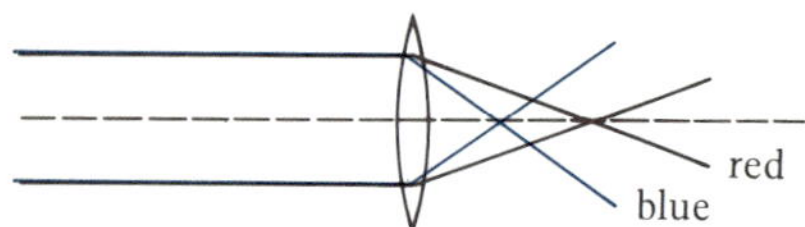

Fig. 17–37 Chromatic aberration. Different colors are focused at different distances from the lens because the index of refraction is different for different wavelengths of light.

Unfortunately, the bending of different colors by different amounts sometimes causes difficulties. What applies to prisms and raindrops also applies to lenses. When white light comes in parallel to the axis of a lens, the different component colors are bent by different amounts. Thus the focal length of the lens is slightly different for different colors (Fig. 17–37). The different colors of any image converge in different places. The effect is not a very large one. For plain glass, the index of refraction may change by a few percent from red to violet, so the focal lengths for different colors may vary by a few percent. However, the effect can be sufficient to blur an image slightly and reduce the clarity of the picture.

The variation of focal length with color is one of several problems that can occur in a real optical system (as opposed to an ideal one that we might use for simple calculations). Such variations from the "normal" or ideal operation of an optical system are called aberrations. The particular problem discussed above, in which different colors are imaged at different points, is called chromatic aberration.

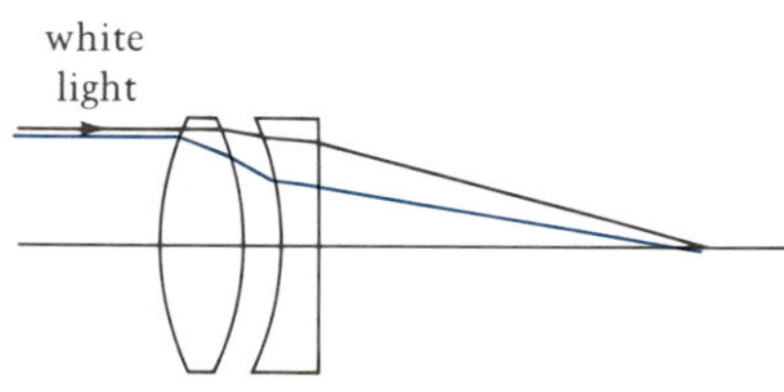

Fig. 17–38 Correcting for chromatic aberration. Two lenses are used and the lenses are made out of different types of glass so the change in the index of refraction for different wavelengths is not the same for the two lenses. As the light passes through the first lens, it is broken up into different colors. However, the second lens brings the colors back together again. This lens combination is called an *achromat.*

We usually want to eliminate chromatic aberration, or at least reduce its effects as much as possible. We can do this by using a series of several lenses, each made out of a different material so that each has a different index of refraction. Because the indices of refraction vary differently with wavelengths for the different materials, the lenses can be chosen so that the chromatic aberration of one lens is offset by that of the other lens. Such a combination is called an achromat. A two-lens achromat can correct perfectly for chromatic aberration at only two wavelengths. However, if properly designed, it can also reduce chromatic aberration greatly at other wavelengths. If still better performance is required, the achromat must be made with more lenses (Fig. 17–38).

Chromatic aberration only applies to lenses and not to mirrors because the law of reflection holds equally well for all colors. However, the other common aberrations affect lenses and mirrors alike

As long as there is not too much curvature, a piece of a sphere and a piece of a parabola of the same focal length differ only slightly. In fact, if you are grinding a parabolic mirror (a pastime of many amateur astronomers), you first grind and polish a piece of glass to produce a spherical mirror, and then make some very small adjustments to convert your mirror into a better approximation of a parabola.

(Fig. 17–39). Spherical aberration arises because a mirror or lens shaped as part of the surface of a sphere does not act as a perfect focusing device for rays all arriving parallel to the axis. For example, for reflection, the perfect shape that focuses parallel rays is a parabola rather than a sphere.

Another aberration is called coma. Coma occurs when rays that come along the axis of the lens have a different focal length than those that approach at an angle. Coma results in a focal "plane" that is slightly curved. If you project such an image on a flat screen, small points of light have images that look like fiery comets. This appearance gave this type of aberration its name.

Another aberration is called astigmatism, which results when light that approaches a lens at a certain distance above or below the center of the lens is not focused in the same place as light that approaches at the same distance left or right of the center of the

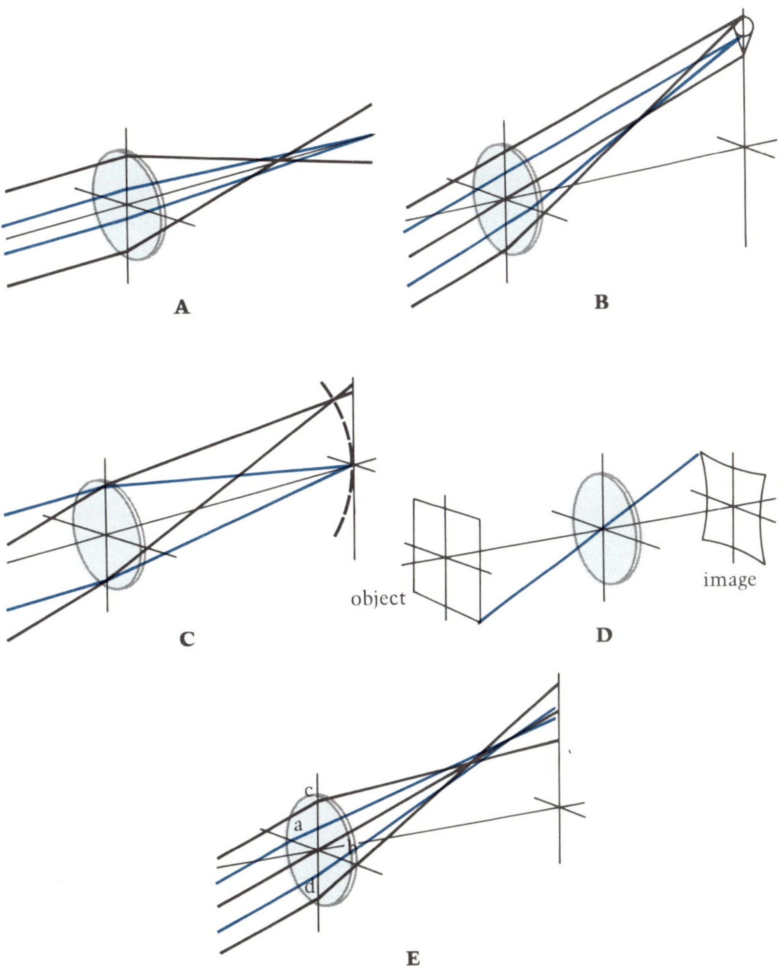

Fig. 17–39 Aberrations in lenses. A. Spherical aberration. Light passing through the lens at different distances from the axis is focused at different distances, producing a blurred image. B. Coma. For off-axis rays, light passing through the lens far from the axis is focused in a different part of the focal plane than light passing near the center. The image appears as a spot of light with a spreading tail. C. Curvature of field. The focal "plane" is no longer a plane, but a curved path. D. Distortion. E. Astigmatism. This aberration occurs for objects that are not on the axis of the lens. Light that passes through the lens top and bottom (*c*,*d*) is focused a different distance behind the lens from light that passes through the lens left and right (*a*,*b*).

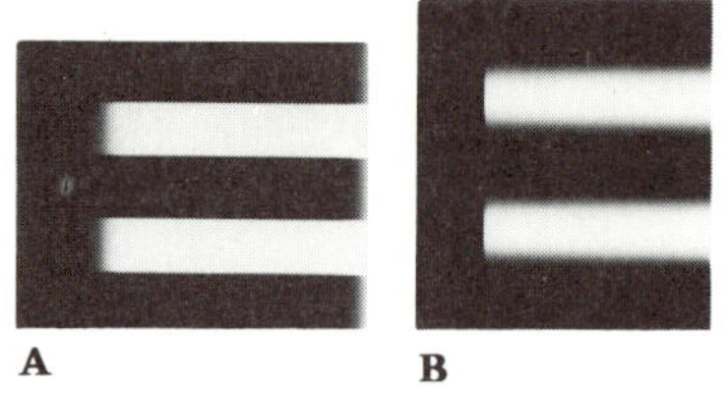

Fig. 17–40 Astigmatic images are blurred in some particular direction. These views through an astigmatic lens are blurred horizontally (A) and vertically (B).

lens. Correction for astigmatism must be done with a lens that is not symmetric about the axis; such a lens looks different in the up-down direction than in the left-right direction. The other aberrations are usually corrected or reduced by the use of multiple lens systems.

Modern lens makers for the leading camera companies use computers to design complicated series of optical elements that can correct all aberrations quite well over a wide range of image distances. Some lenses can even "zoom" in and out, changing their focal lengths while retaining a constant focus.

In the next chapter, we'll see how these various optical elements are put together into useful optical systems.

Key Words

reflection
refraction
optics
geometrical optics
angle of incidence
angle of reflection
law of reflection
index of refraction
ray tracing
critical angle
total internal reflection
fiber optics
diverging
virtual image
optical axis
converging
focal point
focal length
real image
focal plane
concave
convex
lens
mirage
spectrum
aberrations
chromatic aberration
achromat
spherical aberration
coma
astigmatism

Questions

1. Which aspects of the wave nature of light are important for geometric optics, and which aspects of the wave nature of light do we ignore? (*Hint:* An equivalent question is to ask which aspects of geometric optics depend on the wavelength of light used.)
2. The image of a light bulb in a flat mirror appears to be 10 meters away from you. How far did the light from the bulb actually travel in striking the mirror and reaching your eye?
3. The speed of light in a glass block is 1.5×10^8 m/s. (a) What is the index of refraction of this material? (b) Light whose wavelength in air is 600 nm passes into the glass block. What is its wavelength in this material?

4. The index of refraction of a given material is 1.2. (a) What is the ratio of the wavelength of light in vacuum to that in the material? (b) What is the ratio of the speed of light in vacuum to that in the material?
5. Light passes from air, into a piece of glass, and back out into the air again. When the light emerges into the air, how does its wavelength compare with the wavelength it had just before it went into the glass?
6. Light passes from air into glass. Of wavelength, speed, and/or frequency, which do (does) not change?
7. Light is traveling through air and reflects off a window. How does the angle of reflection depend on the wavelength of the light?
8. We are shining light at a surface at the critical angle. If we want to tilt the light beam and still have total internal reflection, should we make the angle of incidence greater or smaller?
9. We are sending light down an optical fiber and want to bend the fiber. Do you think that sharp turns or gradual turns would be better? Explain your answer.
10. A slide projector projects the image of a slide on a screen. Is this image a real image or a virtual image?
11. When you look at your image in a mirror, your left and right appear reversed, but up and down do not. Why do you think this is?
12. Imagine that we have a mirror that is curved outward instead of inward (that is, convex instead of concave). In this case, rays parallel to the axis would be reflected away from the axis, so that if you traced them back through the mirror they would also meet at a focal point behind the mirror. (a) Draw a diagram analogous to Fig. 17–20 showing the paths of rays coming in parallel to the axis of such a mirror. (b) Draw a ray-tracing diagram to find the image of a person two focal lengths from such a mirror.
13. If we put a light bulb in the focal plane, but not at the focal point, of the mirror shown in Fig. 17–23 (that is, the bulb is not on the axis), draw a diagram to show what happens to the light rays that leave the bulb and strike the mirror.
14. For the mirror shown in Fig. 17–26, under what conditions is the image larger than the object?
15. You look in a curved fun house mirror, like the one in Fig. 17–26, and see an upright image of yourself. Is that image real or virtual?
16. In Fig. 17–27, assume that the index of refraction of the water is greater for blue light than for red light. Will the fish appear closer to the surface of the water if viewed in blue light or red light?
17. We are tracing rays to find an image in a mirror. (a) What happens to the rays that strike the mirror while traveling parallel to the axis? (b) What happens to the rays that strike the mirror at the point where the optical axis and the mirror intersect?
18. We are tracing rays through a thin lens. (a) What happens to the ray that strikes the lens while traveling parallel to the axis? (b) What happens to the ray that passes through the center of the lens?

19. In Fig. 17–35, which situation would correspond to using the lens as a simple magnifying glass?
20. Light from the sun bends continuously as it passes through the atmosphere, although the effect is noticeable only when the sun is low in the sky. What property of the atmosphere makes it bend?
21. Why would chromatic aberration cause a blurred image even if you were only taking a black-and-white picture?
22. How could you tell if a certain lens had a correction for astigmatism?
23. Of the aberrations discussed in this chapter, which cannot take place with a mirror?

CHAPTER 18

Optical Instruments

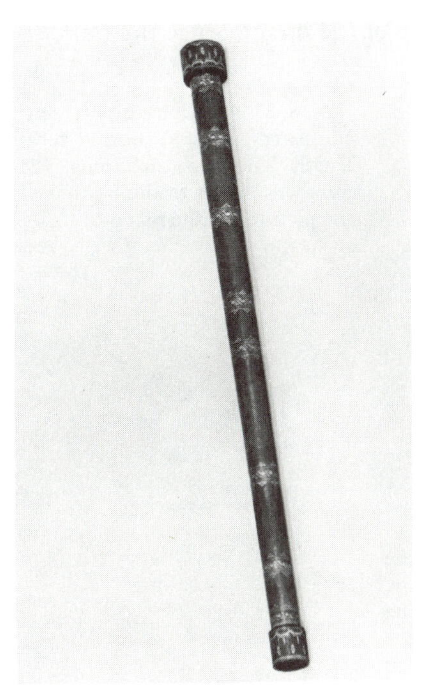

Fig. 18–1 One of Galileo's original telescopes.

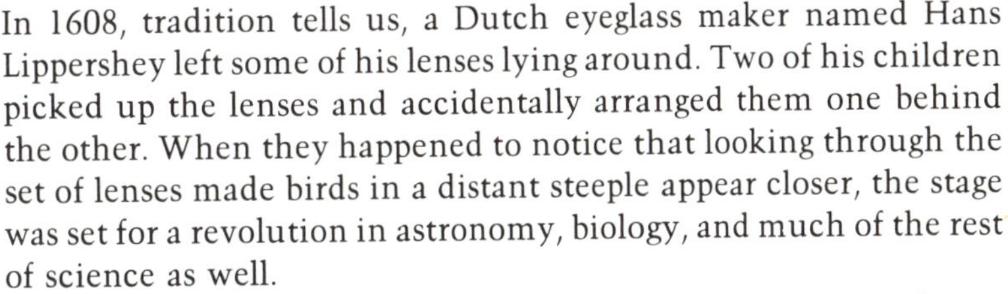

In 1608, tradition tells us, a Dutch eyeglass maker named Hans Lippershey left some of his lenses lying around. Two of his children picked up the lenses and accidentally arranged them one behind the other. When they happened to notice that looking through the set of lenses made birds in a distant steeple appear closer, the stage was set for a revolution in astronomy, biology, and much of the rest of science as well.

Following his children's discovery, Lippershey put together a simple telescope. Its fame spread quickly. In Italy, the following year, Galileo heard of the telescope, figured out for himself how to put one together, and turned it toward the sky. Since that time, telescopes have been important tools for studying the universe.

In this chapter, we apply the principles of geometrical optics that we have just discussed to telescopes and some of the other optical instruments that are now in use.

18.1 The Simple Camera

Let us first discuss a straightforward optical instrument, a simple camera. A simple camera consists of a single converging lens with a flat piece of film behind the lens. We'll start by photographing objects that are very far away, so the light rays from any point come in parallel to each other. (A star is a truly distant object, although a distant landscape will do.) When an object is very far away, the

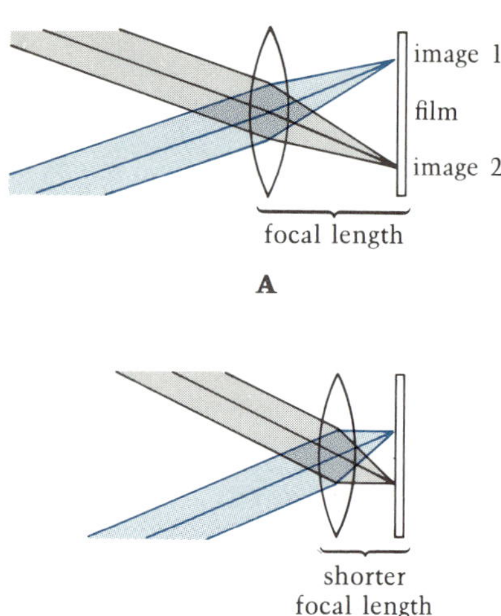

Fig. 18–2 Image formation in a simple camera. In this case we are looking at the images of two objects, one above the axis and one below the axis. The objects are so far away that all of the rays arriving from each object are parallel to each other. For each bundle of rays, we have marked the ray that passes through the center of the lens. This ray is undeflected. As the light from any one bundle passes through the lens, the rays are bent together, and meet one focal length beyond the lens. This is where the image of the distant object will be, and this is where we place the film. The only difference between A and B is that the focal length in B is less than in A. This means that the rays in A come together sooner than in B. Thus the images of the two objects are closer together in B than in A. The longer the focal length, the more spread out the image will be.

image is located one focal length behind the lens, so we place the film there.

Figure 18–2 shows the light arriving from two different objects. (We can also think of the two points at which the light originates as different parts of the same object, perhaps your head and your feet.) The angle between the two bundles of rays is not changed by the camera lens. However, the farther the two beams travel between the lens and the film, the farther apart the images get. Cameras with longer focal lengths place the images farther apart on the film. Thus, with a longer focal length, your head and feet will appear farther apart. The image will appear larger.

Enough light must get in to expose the film, so we must also worry about the brightness of the image. Two factors determine the image brightness. One is fairly obvious—the size of the lens opening. The larger the lens opening, the more light we can get in. The amount of light that gets in depends on the area of the lens. The area is proportional to the square of the diameter (d) of the lens. For example, if you double the diameter of a lens, you let in four times the light (Fig. 18–4). The brightness also depends on the focal length (f) of the lens. A larger focal length means a larger image, in which case the light is spread out over a larger area. There is less light available for any given piece of the image. Since the image is

Fig. 18–3 A telephoto lens, left, compared with a shorter focal length lens. On each lens, the numbers closest to the bottom of the picture are the f-stops.

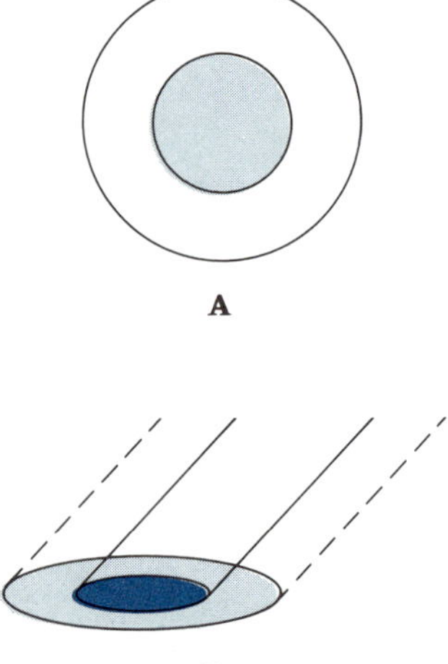

Fig. 18–4 Factors that determine image brightness. A. The shaded circle has half the diameter of the larger circle. The area of the shaded circle is therefore one quarter the area of the larger circle. This means that four times as much light passes through the larger circle as passes through the smaller circle. More light passing through means a brighter image. B. Here we let the same amount of light fall in two spots, one twice the diameter (four times the area) of the other. In the larger spot, the light is more spread out, so any part of the spot is not as bright as the smaller spot. From A we can conclude that more light will get into a larger lens opening (in proportion to the square of the diameter), and from B we can conclude that a smaller, more concentrated image will be brighter. The smaller image is obtained with a smaller focal length.

Fig. 18–5 Lenses of the same focal length but with different focal ratios have openings of different diameters.

A typical camera lens for normal use with a 35-mm camera (that is, a camera that uses film that has a width of 35 mm) might have a focal length of about 50 mm, while a telephoto lens might have a focal length possibly four times that, 200 mm. If our film size is smaller, as for a Kodak Instamatic, a lens for normal use has a shorter focal length than the 50-mm focal length used for the larger film.

spread out in two dimensions, the brightness depends on the square of the focal length.

The image brightness therefore is proportional to the square of the ratio d/f. The ratio (f/d) of the focal length to the diameter is called the focal ratio, or *f-number* of the lens. These numbers are conveniently printed on a camera as f-stops. The f-stops are usually listed as 1.4, 2, 2.8, 4, 5.6, 8, 11, etc. Since the brightness depends on the square of the focal ratio, the brightness of the image varies as 2, 4, 8, 16, 32, 64, 128. Each number in this series of squared numbers is twice the previous number; each succeeding f-stop therefore corresponds to a factor of 2 in brightness (doubles the brightness). We can also vary the amount of light that gets into the camera by varying the exposure time. Camera shutter speeds do so, and are normally marked by factors of about one-half ($^1/_{30}$, $^1/_{60}$, $^1/_{125}$, $^1/_{250}$ second). A change of one notch on the shutter speed dial exactly compensates for a change of one f-stop.

What is actually happening when you change the f-stop of your camera? You are not changing the focal length of the lens. You are, rather, changing the diameter of the part of the lens that you are using. The lens contains a diaphragm, an opaque piece that has a hole whose size can be varied. When you change the f-stop, you are really changing the size of the hole. When the hole gets larger, the f-number gets smaller.

Although so far we have discussed only the images of objects that are very far away, the same general ideas about image size and brightness apply to objects that are relatively close to us. In this case, the image is farther away from the lens than one focal length. We have to move the lens farther from the film if we want the image plane still to fall on the film (Fig. 18–6). When you adjust the distance on the focus ring of your camera, you are moving the lens for this purpose.

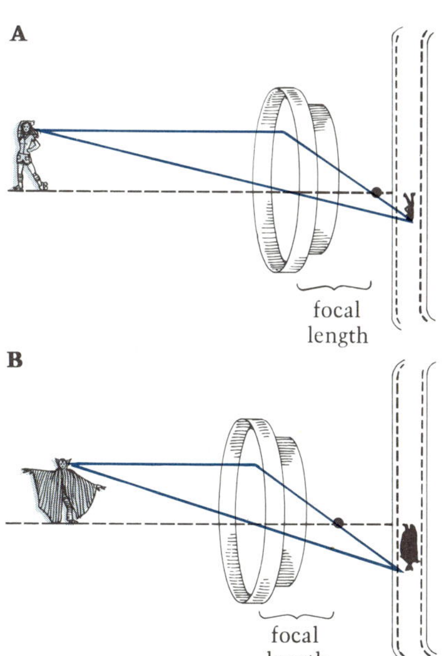

Fig. 18–6 Changing the focus of a camera. Since the focal length of the lens is fixed, as the distance between the object and the lens changes, we have to change the distance between the lens and the film, so that the image always falls on the film. A. Here the camera is properly set for the subject at some distance. Note that we have demonstrated the location of the image by tracing two rays from the subject's head. B. The subject has walked closer to the camera, so the image falls farther behind the lens than in A. To keep the image on the film, we have to move the lens farther from the film.

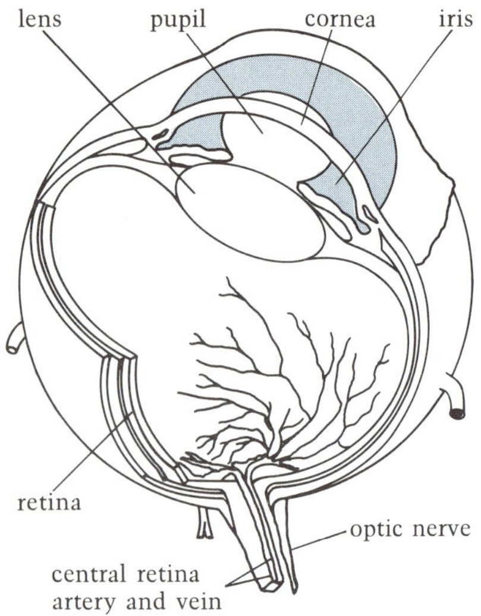

Fig. 18–7 The human eye.

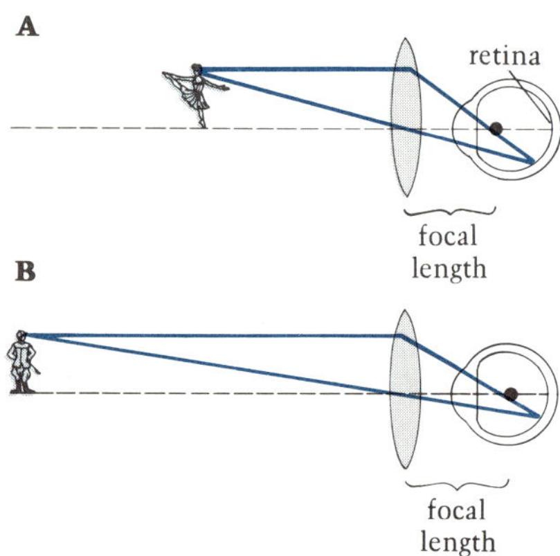

Fig. 18–8 How the eye adjusts to subjects at different distances. Unlike the camera, the eye has a fixed distance between the lens and the retina. This means that as the subject distance changes, the only way to keep the image on the retina is to change the focal length of the lens. This is done by the muscles of the eye. A. Here we show the proper imaging for the subject at some distance. B. The subject is farther away. If the eye did not adjust from A, the image would fall in front of the retina. However, the muscles allow the lens shape to change slightly so the focal length in B is longer than in A, and the image is still on the retina.

18.2 *The Eye*

18.2a FOCUS

Your eye is like a simple camera (Fig. 18–7). It has a *lens,* which works in much the same way as a camera lens (although the eye's *cornea* also refracts light). There is one difference, however. With a camera, when you want to look at an object at a distance that is not far away, you change the distance between the lens and the film by moving the lens in and out. In the eye, the lens does not move. To focus on objects at different distances, the eye muscles exert pressure on the lens, changing its shape. These changes in shape of the lens result in a change in the focal length. The process of changing the eye's focal length is called accommodation (Fig. 18–8).

Like the diaphragm in a camera, the *pupil* of the eye controls the amount of light that gets in by limiting the amount of the lens that is used. When the light is dim, the pupil dilates and has a large opening. Some people whose eyes suffer from certain aberrations, such as astigmatism, find that the condition is worse in dim light because more of the lens is then used. The more of the lens you use, the worse the aberrations become. You cannot consciously control the opening that your eye chooses. With a camera, you can compensate for a different size opening by changing the exposure time. Your eye's exposure time is "built-in" rather than adjustable. The eye's exposure time is about $\frac{1}{30}$ second, so your pupil must automatically open and close accordingly.

The *retina* in the eye takes the place of the film in the camera. The retina contains the tiny sensing elements that detect the light and transmit the signals from the eye to the brain along the optic nerve. The lens must adjust itself so that it forms an image on the retina. If the light from distant objects is focused in front of the retina, then we say that the person is nearsighted. Nearsightedness

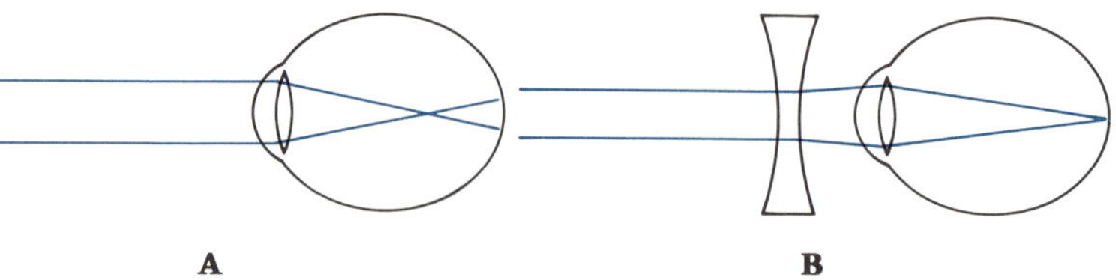

Fig. 18–9 Nearsightedness. A. In the nearsighted eye, two parallel rays are brought together in front of the retina. Images then fall in front of the retina instead of on it. B. This is corrected with a lens placed in front of the eye. This lens causes the parallel rays to diverge slightly, so it now takes a little longer for them to come back together, and the image falls on the retina.

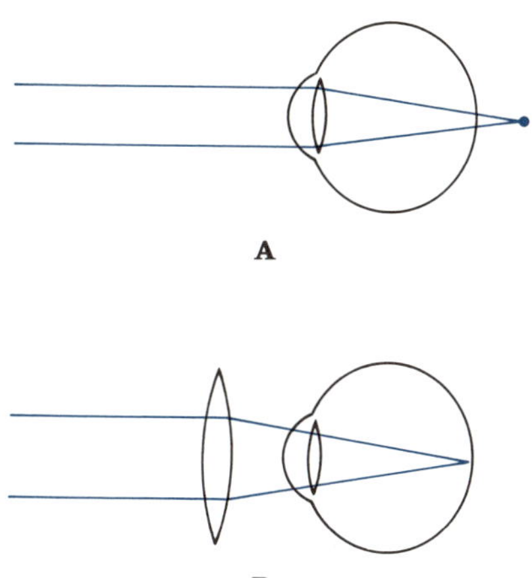

Fig. 18–10 Farsightedness. A. We see that two parallel rays are brought together behind the retina, so images will be behind the retina instead of on it. B. This is corrected by placing a converging lens in front of the eye. This starts to bring the rays together a little sooner, so that now the image will fall in the right place.

is corrected by the use of a diverging lens (Fig. 18–9). When the image of distant objects falls behind the retina, then the person is farsighted. Farsightedness is corrected with the use of a converging lens (Fig. 18–10). Even for persons whose vision is normal, the lens cannot produce a clear image on the retina when the object is too close to the lens. The closest distance at which you can have an object and still get a good image is called the near point. The near point for a person changes with age. A typical value is about 25 cm, but a child might have a near point of about 7 cm and an older person's may be out at about 200 cm.

18.2b COLOR

Sometimes—probably in a physics lab—we may have light of a pure color to work with; we call it *monochromatic light.* Monochromatic light consists entirely of waves of one wavelength. However, in real situations, no light is perfectly pure, and we usually observe some range of wavelengths.

The light from most of the things we see contains many wavelengths. On the retina of the eye are small receptors called the color cones, also called simply *cones,* which detect the colors and feed the information through the optic nerve into the main part of the brain. The other type of receptor on the retina is the rods (rods and cones are shown through a microscope in Fig. 18–11), which are more sensitive than the cones. The rods are more effective in dim light, but are not sensitive to colors. Thus, in dim light we do not perceive colors very well; try this in a fairly dark room or outside at night. Your eye contains perhaps 120 million rods and 5 million cones; the general run of colors your brain perceives corresponds to the divisions of the spectrum. But the eye and brain can be made to think that all the colors are in a picture even when they are not. If a red beam of light and a green beam of light are allowed to fall together on a white piece of paper, we perceive a yellow region in the overlap. Our brain cannot tell that we are seeing red and green instead of light of a wavelength that corresponds to yellow. Commonly, we think that three colors, primary colors, are sufficient to recreate the effect of the entire range of color. We think that the eye's color cones come in three different types, one for each color sensitivity, although no theory of how the brain perceives color is generally accepted.

There are two opposite ways of composing primary colors into other colors: *additive,* where light of different colors is superimposed, and *subtractive,* where paints or dyes are mixed. Color television is an example of an additive process and a slide or movie film is a subtractive process.

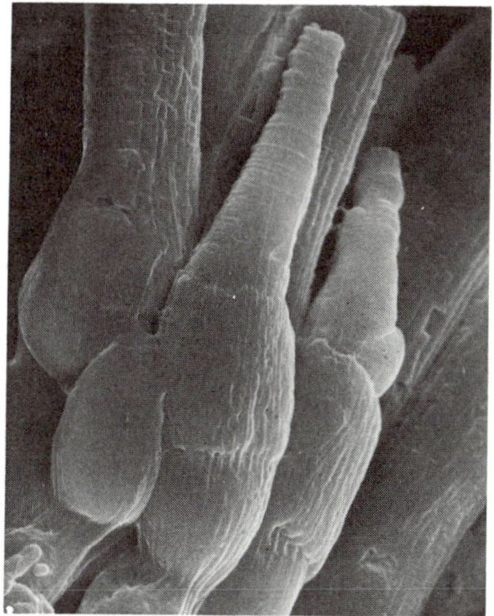

Fig. 18–11 The rods and cones in the eye of a mudpuppy can be distinguished by their shapes in this picture from a scanning electron microscope. (Courtesy of E. R. Lewis, Y. Y. Veezy, and F. S. Werblin)

The eye's visual purple *(rhodopsin) is a pigment (a substance sensitive to a limited range of color) whose structure varies when exposed to light. The visual purple affects the electrical conductivity of the membrane that supports it. Biologists in laboratories can insert probes to measure such changes. When the brain senses the changes, it interprets them in terms of the intensity of light falling on the visual purple.*

The way that the different primary colors add can be shown on a chromaticity diagram, where hue (the exact shade or tint) and saturation (the degree to which the color is pure, without having any white light mixed in) can vary. Intensity, the degree of brightness, is the other value that must be considered. Ideally, one can come closer to making the full range of colors with an additive process than with a subtractive one, so the colors on television have the potential of being truer (if your set is well adjusted) than the colors on film.

The way that colors subtract when mixed is shown in Plate 54. Whereas all the colors in an additive process make white when combined, all the colors in a subtractive process make black when put together. Plate 56 shows a close-up of a printed color picture to show how the illusion of varied colors is created by printing many dots of three colors plus black (so-called four-color printing). This is an additive process. A close look at an apparently white area on a color TV screen reveals that red, blue, and green dots are present (Plate 53).

Color film is composed of three layers of dye, plus other layers for various purposes; the colors in the three layers reflect (for a print) or transmit (for a slide) their complements. The films thus contain yellow, magenta, and cyan dyes, which give us blues, greens, and reds, respectively. This subtractive process makes the colors we perceive.

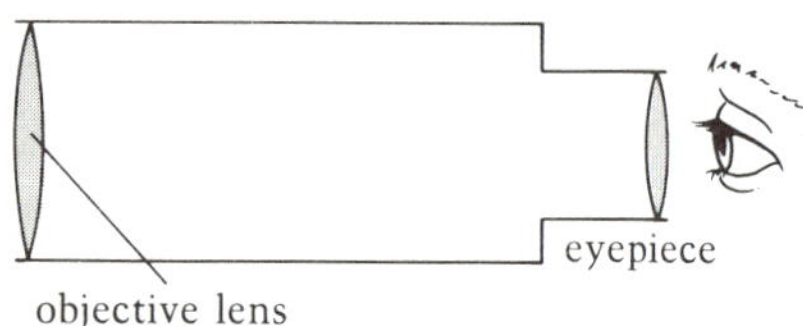

Fig. 18–12 A simple telescope. The objective lens forms an image which is then inspected with the aid of an eyepiece lens.

18.3 *Telescopes*

Cameras and eyes are examples of optical systems with only one lens. (Often a camera lens may consist of a series of individual lenses to reduce aberrations, but we still treat it as one lens; Fig. 18–12.) Telescopes and microscopes employ more than one lens.

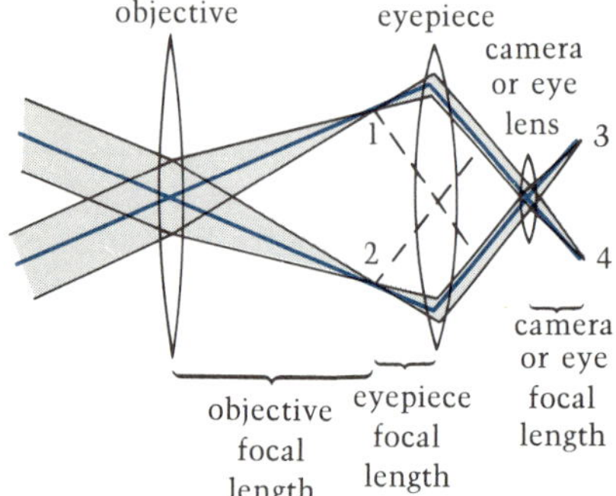

Fig. 18–13 Image formation in a simple telescope. We start with two objects far away and follow the bundle of rays through the telescope. One bundle comes from an object above the axis of the telescope and the other comes from an object below the axis of the telescope. Each bundle starts out as parallel rays, but these rays are bent to come together one focal length (of the objective) behind the objective. That means we get images of the two objects at points 1 and 2. The rays then continue on, spreading out again. However, if we place the eyepiece so that it is one focal length (of the eyepiece) away from 1 and 2, then the rays from 1 all emerge from the eyepiece parallel to each other and the rays from 2 also all emerge parallel to each other. These two bundles of parallel rays now make a greater angle with each other than the two incoming bundles of parallel rays. This is where the magnification of a telescope comes from. The final step is for the camera or eye lens to bring the rays in these two bundles back to two points, 3 and 4.

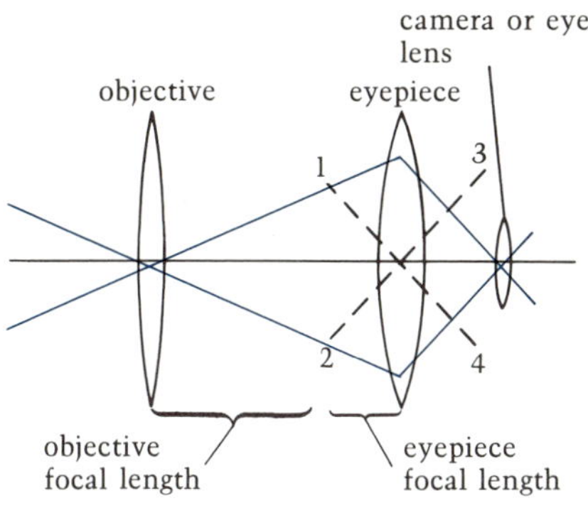

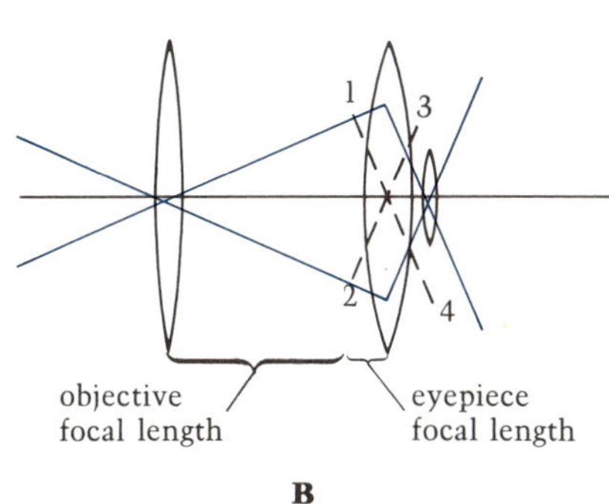

Fig. 18–14 In this figure we trace through one of the rays from each of the objects in Fig. 18–13, to see the effect of changing eyepieces on the final image size. A. Here we reproduce the situation in Fig. 18–13. Notice the dashed line drawn from the image at point 1 to the center of the eyepiece. All rays leaving 1 and passing through the eyepiece will emerge parallel to this dashed line. Similarly, rays emerging from point 2 and passing through the eyepiece will be parallel to the dashed line from 2 through the center of the eyepiece. B. Now we change to an eyepiece of shorter focal length. This means that the two bundles of rays emerge from the eyepiece making a greater angle with each other than in A. Thus the images on the film (or retina), at points 3 and 4, are farther apart with the shorter focal length eyepiece. When the image points are spread out more, the image is larger and we can get more detail. Of course, the image will not be as bright, since it is spread out over a larger area.

To find the image in a system with more than one lens, simply take the lenses one at a time. We use ray tracing to find out where the first lens forms an image. This image then becomes the object for the second lens, and so on.

When Galileo demonstrated his telescope in Venice in 1610, the lens through which the light entered his tube was a few centimeters in diameter. This lens, which the light strikes first, is called the objective lens, or simply the *objective*, of the telescope. Galileo's second lens was smaller and at the opposite end. This eyepiece was used to examine the image formed by the objective. The focal length of an eyepiece lens is generally shorter than that of the objective.

The operation of a simple telescope is illustrated in Fig. 18–13. Galileo used a concave eyepiece but we more commonly use a convex one, as is shown, or a "compound" eyepiece made of several lenses in a series. We trace the rays from two distant objects through the telescope in much the same way that we did with the camera. The images of the two objects are formed one focal length behind the objective. The rays now continue beyond this point. Let's follow one set of rays through the eyepiece. We place the eyepiece at a distance from the objective's image equal to the focal length of the eyepiece, so that all rays leaving a point emerge from the eyepiece parallel to each other. To find the angle they make with the axis of the telescope, we simply use our rules of ray tracing. We draw a line from the image formed by the objective through the center of the eyepiece. This ray is undeflected. Therefore, all the other rays from the same image point must emerge from the eyepiece parallel to this one.

The angle at which the rays emerge from the eyepiece is greater than the angle at which they enter the objective. The ratio of these angles is just the ratio of the focal lengths of the lenses. The ratio of the angles is the amount of magnification provided by the telescope. For example, if the objects are 1° apart and the ratio of the focal lengths is 10, then the objects will appear to be 10° apart. We can change the amount of magnification by simply changing the eyepiece. As we use eyepieces with shorter focal lengths, allowing us to move the eyepiece closer to the image formed by the objective, the magnification increases.

To observe the rays leaving the eyepiece, we use either our eye or a camera. If we are using a camera, each bundle of parallel rays is imaged at a different point on the film. Since the rays emerge from the eyepiece at a greater angle than that at which they entered the objective, they come into the camera at a greater angle than if the telescope were not there. Thus the images will be farther apart on the film; if the two points we have been considering are parts of a single object, then the object's image will appear bigger on the film.

18.3a THE USES OF TELESCOPES

Can we keep increasing the magnification as much as we want by using eyepieces of shorter and shorter focal lengths? Not really. There are practical limitations that determine how much magnification we can use. For example, suppose we are looking at a point of light, such as a star. When the light passes through the objective, there is some spreading due to diffraction. The larger the objective,

Fig. 18–15 Space Telescope, a 2.4-m diffraction-limited telescope scheduled for launch into earth orbit by NASA in 1983. Because it will be in space, the limits on resolution that the earth's atmosphere place on ground-based telescopes will not apply. Also, the sky above it will be significantly darker than the sky over ground-based telescopes, even when the latter are at the best nighttime sites. Thus Space Telescope will be able to study much fainter objects, and will see about seven times farther into space (with equivalent detail) than a ground-based telescope.

USING A TELESCOPE IN SPACE

Much of astronomy represents a marriage of analysis, using the laws of physics, with technology (often optical), based on other laws of physics. Here we describe a typical piece of astronomical research.

For years astronomers have traveled to remote mountaintops to get the clearest, darkest skies with the best seeing. But the sky is always clear above the earth's atmosphere, it is much darker than it ever gets on earth, and the stars shine steadily. Within the last few years, the opportunities to observe from space have increased.

In 1978, one of the most powerful and versatile telescopes yet to be launched was sent into an orbit 23,600 km above the earth. It contained a 45-cm (20-inch) primary mirror made of beryllium, which is strong for a given weight. A reflecting coating covers the beryllium.

This telescope is used to study the ultraviolet part of the spectrum, and is known as the International Ultraviolet Explorer (IUE). It was built jointly by NASA, the European Space Organization, and the British Science Research Council. The telescope provides light to two spectrographs.

IUE is operated from a control room at NASA's Goddard Space Flight Center in Greenbelt, Mary-

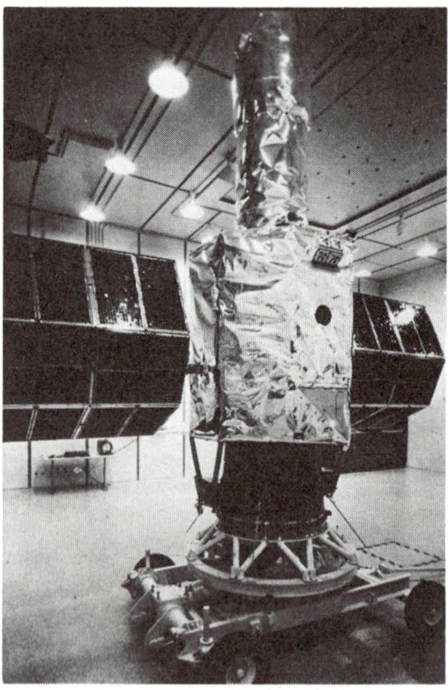

Fig. 18–16 The International Ultraviolet Spacecraft contains a 45-cm telescope. It is in synchronous orbit, and is supplying data about the ultraviolet spectra of all types of astronomical objects to astronomers from all over the world.

Fig. 18–17 The control room for IUE at NASA's Goddard Space Flight Center contains video readouts for the scientists and technical staff. An image of a bright star is on the screen in this view.

land (near Washington, D.C.), and a twin control room in Europe. The satellite's high orbit is such that IUE circles the earth once a day. Since the earth also turns once a day, the satellite always remains in the same spot overhead (a synchronous satellite, as we discussed in Chapter 7). The scientists on the ground can be in constant communication with it.

IUE is open to scientists from countries around the world, who come as guest investigators. Let us say that we have studied a certain type of star with conventional ground-based telescopes and want to extend our observations into the ultraviolet. We apply for observing time. When our proposal is accepted, we travel to the IUE control room.

Once upon a time, astronomers had to sit out in the cold to use a telescope, and this is still necessary for certain types of observations. But we are deep inside a building, and computer screens and telephones are our only links to the outside. The telescope operator types into the computer the position of the star we want to observe.

The message from the computer goes to one of NASA's radio telescopes and is beamed up to IUE overhead. The spacecraft rotates, and soon we see a video picture of a part of the sky. We identify our star. Then the telescope operator presses a button to start our exposure of the spectrum.

With electronic devices such as those on IUE, there is no need to wait for film to develop. Within minutes after our exposure is over, it is radioed to earth and appears on a screen before us.

We press a few buttons on the computer, and all of a sudden the TV screen shows a graph of the spectrum. We ask the computer to magnify the image; a few seconds later the improved view of the spectrum is displayed. Now we press another button to have a "hard copy" printed out on paper. The spectral lines that we were looking for show nicely. We move on to the next star.

When we go back home at the end of our run, we take with us the hard copies we made at the control room. Reels of computer tape that carry the data in digital form for later analysis on our own computer soon follow. It will be months before we get it all "reduced," and make our conclusions. The final publication takes a year, and we get to see the paper in print just as we return home from our next observing run with the International Ultraviolet Explorer.

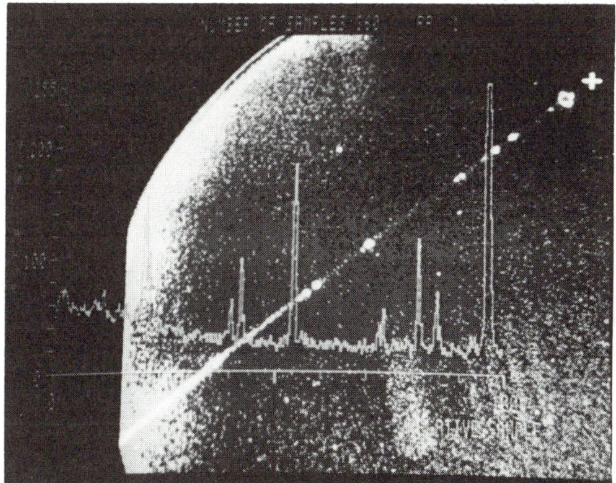

Fig. 18–18 The interactive computer system of IUE can show the spectrum (diagonal streak) within minutes after an exposure is completed. The bright dots on the streak are emission lines at wavelengths less than 200 Å. The Lyman alpha line of hydrogen is the bright spot near the + sign at upper right. The computer has also printed out a graph of the spectrum for easy reference on the video screen. Color coding helps the visibility of the data.

the less the spreading, but diffraction is always present to some degree. Diffraction blurs the image of the star slightly. There is usually no point in magnifying the blur. Another factor that limits the useful resolution of a telescope is the earth's atmosphere. When light passes through the atmosphere, it is refracted slightly. However, the atmosphere is continually churning about, causing slight changes in refraction. These changes cause the image of a star to dance around and apparently vary in brightness, which produces the "twinkling" of starlight. The dancing around of the image is called seeing by astronomers. When using a telescope, it generally does not pay to magnify your image any larger than the seeing will allow. For most large telescopes, the seeing rather than the diffraction limits the useful resolution. At a good observing site, seeing fluctuations may be about 0.2 arc second (that is, $\frac{1}{5}$ of $\frac{1}{3600}$ of 1 degree) on the best nights. Improved seeing is one reason why observatories are usually built at high altitudes. If you get above the atmosphere, then the starlight does not have to travel through the air on the way down and seeing problems disappear. Space Telescope, the 2.4-m telescope scheduled for a 1983 launch, will be our first large diffraction-limited telescope in space. Its resolution will be better than that of the best ground-based telescopes by a factor of 7.

The telescope serves another purpose, which is often much more important than the magnification. It collects light. The objective is larger than the pupil of your eye, and so much more light falls on the objective than falls on your naked eye. The telescope brings this light together at your eye, which then receives much more light than it could without the telescope. As we have already seen, the amount of light that gets in depends on the surface area of the objective, which depends on the square of the diameter. For exam-

(below, left) Fig. 18–19 The 1.0-m refractor at the Yerkes Observatory, the largest refractor in the world, was built in 1897.

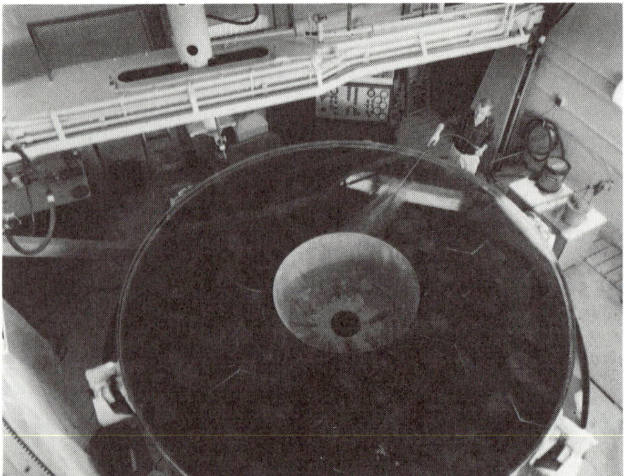

(below, right) Fig. 18–20 The mirror of the largest reflector telescope at Kitt Peak National Observatory is 4 meters across. This view is of the original mirror blank, made of fused hexagonal quartz ingots, before it was coated with aluminum.

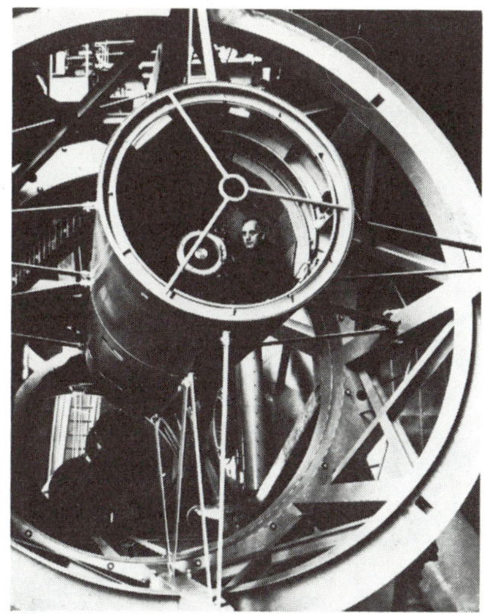

Fig. 18–21 The astronomer Edwin Hubble in the focus of the 5-meter reflecting telescope on Palomar Mountain. Only a few telescopes in the world are large enough for anyone to sit at the "prime focus" without blocking the incoming light. Secondary mirrors are used to reflect and refocus the light to more convenient observing locations.

ple, if you use a telescope whose diameter is 5 meters, and your pupil's diameter is 5 mm, the ratio of the diameters is 10^3. This means that the brightness of objects is increased by a factor of 10^3 squared, or 10^6—a factor of one million! And if we take an hour-long exposure, that gives us 10,000 times more light than we can receive in the $\frac{1}{30}$-second "exposure" that an eye makes. These calculations show why a telescope allows us to view and photograph objects that we could not even imagine seeing with the naked eye.

18.3b TYPES OF TELESCOPES

So astronomers like to have larger and larger telescopes. From Galileo's time on, telescopes did get larger. At first, all telescopes used an eyepiece and an objective *lens*. Such a telescope is called a refracting telescope. However, refracting telescopes have some disadvantages. They suffer from chromatic aberration, since they consist entirely of lenses. Some correction can be made by the insertion of additional lenses, but the problem does not disappear completely, and each additional lens inserted cuts down the amount of light that gets through, since some light is reflected away at the surface of each lens and a bit of light is lost in transmission. Further, bigger lenses are more difficult to make, since you need a large piece of glass with no defects inside it. And as the glass gets heavier, it becomes more difficult to support at the rim. For all these reasons, the largest refractor ever built is the 1-meter telescope at the Yerkes Observatory in Williams Bay, Wisconsin. It was constructed in 1895.

The largest telescopes in the world are reflecting telescopes, also called *reflectors*. In these telescopes, the objective lens is replaced with a curved mirror (Fig. 18–22). The principle of such a telescope is similar to that of the refractor in that the mirror forms an image that is then examined with an eyepiece. All of the considerations regarding image size, brightness, and focal length still apply.

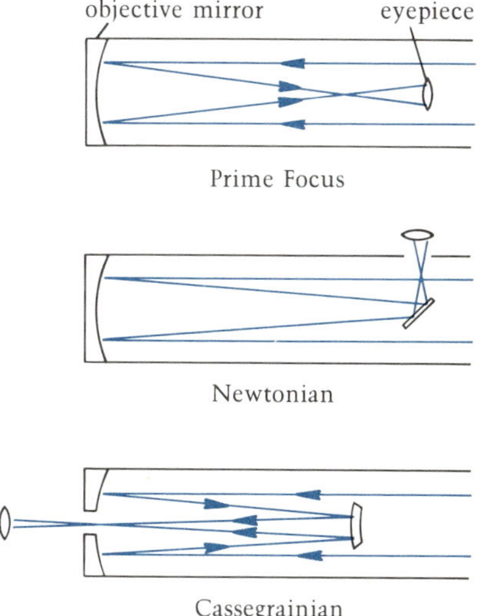

Fig. 18–22 Three arrangements for reflecting telescopes.

Reflecting telescopes have many advantages. The reflection of light is not color dependent, so there is no chromatic aberration. Since light does not penetrate the surface of the mirror, the quality of the inside of the mirror does not matter much. Also, since light does not pass through a mirror, the mirror can be supported completely from behind against the pull of gravity. Mirrors are usually made by grinding and polishing a piece of glass or ceramic into the desired shape and then placing a reflecting coating on the front surface.

Until very recently, the largest reflector was the 5-meter (200-inch) telescope that was opened on Palomar Mountain in 1949. Telescopes can't be used until the mirror shape has reached equi-

Fig. 18–23 The 5-m Hale telescope at Palomar Mountain, pointing to the zenith.

librium after being exposed to the cool night air. The mirror itself was made out of Pyrex at the Corning Glass Works; Pyrex represented a new material with a lower coefficient of expansion than had been used previously. Recently, several 4-meter telescopes have been built in the United States, in Chile, and in Australia. Their mirrors are of ceramics or of quartz, which have lower coefficients of expansion than Pyrex. And the Soviet Union even built a 6-meter reflector.

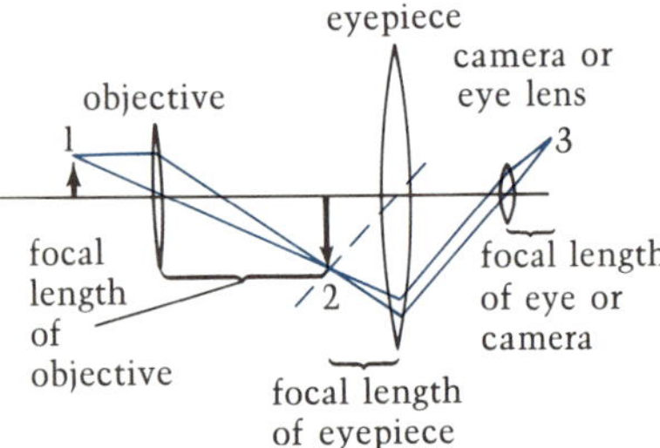

Fig. 18–24 Image formation in a simple microscope. We are looking at the arrow under the microscope, and following the location of the image of the head of the arrow. We start with the rays leaving point 1. We place the arrow closer than two focal lengths (of the objective) in front of the objective so that the image of the arrow as formed by the objective will be larger than the objective. We find the image of the arrowhead by tracing the usual two rays, one through the center of the lens and one parallel to the axis and then through the focal point. These two rays come together again at point 2. We now inspect this image with an eyepiece, in much the same way that we used the eyepiece of a telescope. Light emerges from the eyepiece as a parallel bundle until it is brought together by a camera or eye lens, to a final image at point 3. The camera or eye will only bring the rays together in the right place if it is focused at "infinity." To do this with your eyes, you must relax the eye muscles. For this reason, it is easier to cover one eye, leaving both eyes relaxed, than to try to close one eye and keep the other open.

18.4 *Microscopes*

The microscope is very much like the refracting telescope, except that the goals of a microscope are somewhat different. A microscope is used to magnify, to produce a large image of something that is close at hand. It still employs an eyepiece and an objective, and we can trace the rays through a microscope in the same manner as we did through a telescope (Fig. 18–24).

The object to be studied is placed between one and two focal lengths below the objective. We get a real image that appears larger than the object. The closer we place the object to one focal length from the objective, the larger the image becomes. However, as the image gets larger it also gets farther away from the objective, requiring a longer and longer microscope.

We then view the image with the eyepiece. We can again treat

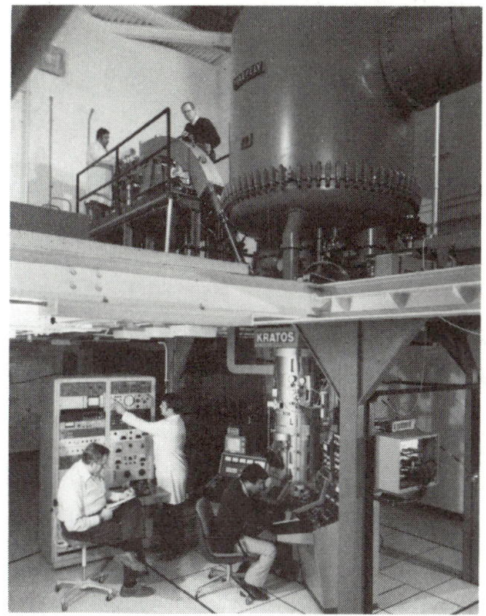

Fig. 18–25 This giant electron microscope at the Argonne National Laboratory near Chicago is used to study the fundamental properties of materials, especially those under consideration for solar, fusion, and fossil-fueled energy systems. Its energy of 1.2 million electron volts enables its electrons to penetrate surfaces more deeply than those from electron microscopes of lower energy.

the image of the objective as though it is a real object and trace rays from it through the eyepiece. Again we can see that if we use an eyepiece with a shorter focal length, then the rays emerge from the eyepiece at a greater angle. This produces greater magnification.

Since most of the things we look at under a microscope don't give off their own light, we must provide some illumination. Depending on what we are looking at, this illumination can be either in the form of light passed through the object from behind (which often requires examining only a thin slice of the object) or light directed at the object from some angle and then bounced back into the microscope. In either case the light that comes into the microscope has been scattered by the object. The resolution of a microscope—the smallest detail that we can see—is limited by diffraction. If we magnify too much, we no longer see the details of what we are looking at. We see, instead, the diffraction pattern. For this reason, work at higher and higher resolution requires the use of shorter and shorter wavelengths to illuminate the object being studied. Blue light provides almost twice the detail of red light. If we can go to even shorter wavelengths, such as ultraviolet radiation, then the resolution improves further. Technical problems still prohibit us from using x-rays.

When we study quantum mechanics, we'll see that we can think of particles as acting like waves under certain circumstances. High-energy electrons have the equivalent of a very short wavelength, and therefore have less severe diffraction effects than visible light. This is the reason why the highest resolution pictures that we get come from electron microscopes. In these microscopes, a beam of electrons is directed at the target and then the scattered electrons are detected in a microscope-like arrangement. Obviously there must be differences, since our normal lenses would stop the electrons rather than focusing them. Electron microscopes have even succeeded in seeing objects as small as atoms and molecules.

Fig. 18–26 Ordinary salt crystals observed with a scanning electron microscope, magnified 100 times. (© 1977 David Scharf from the book *Magnifications*, Schocken Books, Inc.)

Key Words

simple camera
focal ratio
f-stops
accommodation
nearsightedness
farsightedness
near point
color cones
rods
primary colors
hue
intensity
objective lens
eyepiece
magnification
seeing
refracting telescope
reflecting telescope

Questions

1. Consider two cameras. One camera has a lens with twice the diameter of the lens of the other camera. The camera with the larger lens also has a focal length that is twice the focal length of the other camera. (a) How do the exposure times necessary to get the same image brightness compare for the two cameras? (b) How does the detail that we can resolve compare on the two cameras?
2. Two lenses have the same focal length, but one is twice the diameter of the other. Which lens would you use if you wanted to get a brighter image for the same exposure time?
3. Two lenses have the same diameter, but one has twice the focal length of the other. (a) Which lens would you use if you wanted to get a brighter image for the same exposure time? (b) Which lens would you use if you wanted to take pictures with the greatest detail?
4. Why do we sometimes call a camera lens with a very long focal length a telephoto lens?
5. Although we often speak of "a camera lens," the lenses on most good cameras really consist of a few individual lenses. Why is this?
6. (a) Why are lenses with very small focal ratios usually quite expensive? (b) When taking a picture, why would you want to cover part of the lens, if the light level allowed it?
7. Compare the basic optical features of the eye and the camera. What are the similarities, and what are the differences?
8. When we talk about nearsightedness and farsightedness, what do the "near" and "far" refer to?
9. Estimate how far away you would have to hold a dime so that it would just appear to cover the full moon.
10. The sun is about 400 times farther from the earth than is the moon. During an eclipse of the sun, the moon just manages to cover the sun. What can you conclude from this about the relative diameters of the sun and moon?
11. What do we mean by monochromatic light?
12. What is the main advantage that a telescope with a large objective has over a telescope with a small objective?
13. Suppose you were using a telescope to look at Jupiter and wanted a larger image. How would you change things?
14. For a telescope, when the image of the objective is exactly one focal length of the eyepiece in front of the eyepiece, why is it important that you have your eye focused at infinity?
15. List the relative advantages and disadvantages of reflecting and refracting telescopes.
16. How does changing the objective in a microscope allow you to change the magnifying power? (In many microscopes, there are a series of objectives that can be rotated into place.)

PART V

Electricity and Magnetism

When we talk about electricity, we are usually referring to the wealth of its practical uses that help us with our daily lives. However, for the physicist, electricity is more than just a source of conveniences, or a way to make our equipment run. Electricity is a result of one of the few fundamental ways in which things can interact with each other.

Although we will begin our discussion by treating electricity and magnetism separately, we will later see that they are really part of the same fundamental force—the electromagnetic force. We have already encountered another one of the fundamental forces—gravity. The four fundamental forces of nature are the gravitational force, the electromagnetic force, and two other forces related to nuclear and elementary particle physics (these we will discuss in Chapters 26 and 28).

Of the four known forces, the electromagnetic force has two important distinctions:

1. It is responsible for most of our everyday phenomena. The book you are reading, the chair you are sitting in, and your own body as well, are all held together by the electromagnetic force. As we discussed in Chapter 3, whenever you touch something, the force between you and the object is really the electromagnetic force. In addition, light originates in a bulb and radio waves originate in an antenna by the electromagnetic force. The electromagnetic force also allows us to detect that light with our eyes, and to receive the radio waves with our radio.

2. It appears that the electromagnetic force is the one that physicists understand best. Even at the most fundamental levels, we seem to understand the origin of the force, and can make very detailed predictions about the results of experiments.

CHAPTER 19

Electricity at Rest

If you have turned on a light to read this book, or plan to watch television when you have finished reading this chapter, you are certainly taking advantage of conveniences and luxuries brought about by applications of our knowledge of electricity. Before we see how these things work, we begin our discussion by looking at the sources of electricity.

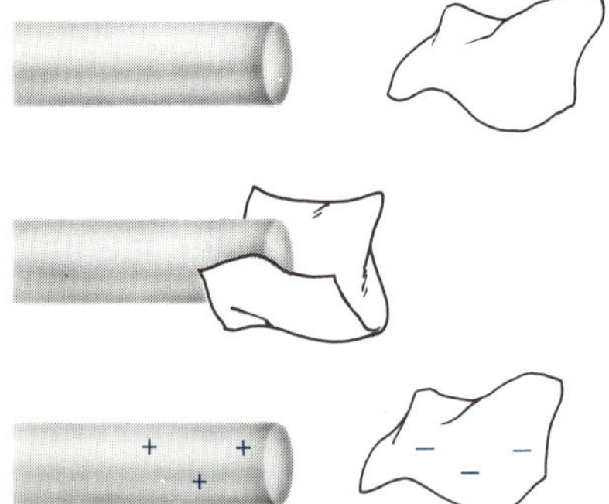

Fig. 19–1 Charging a rod by rubbing it with a cloth. Both the rod and cloth start out neutral. When the rod is rubbed with the cloth, some charge is transferred. When the two are brought apart, the rod and cloth have equal and opposite charges. Whether the rod or the cloth has the positive charge depends on the particular materials used for the rod and the cloth.

19.1 *Electric Charge*

When we studied gravity, we saw that the mass, or more properly, the gravitational mass, was a measure of the ability of an object to exert or feel gravitational forces. Similarly the electric charge of a particle is the measure of its ability to feel or exert electric forces.

There are two types of electric charge. The idea that two types of charge exist goes back to the early eighteenth-century work of Charles François Du Fay, who did a series of experiments involving electric effects brought about by rubbing two objects together. Even the ancient Greeks knew that amber, when rubbed, acquired the ability to pick up small objects. (In fact, the Greek word *electron* means amber.) When we bring together two glass rods hung on strings and both rubbed with silk (Fig. 19–1), the rods move apart. However, when we rub a plastic rod with fur and bring it near a glass rod that has been rubbed with silk, the glass rod and the plastic rod tend to move together. Rubbing a glass rod with silk produces one of the types and rubbing a plastic rod with fur produces the other.

We conclude that the two glass rods rubbed with fur have the same kind of charge as each other, and that two of the same kind of charge—we say "like charges"—repel each other. On the other hand, unlike charges attract each other. We refer to the two types of charges as positive and negative. These names are totally arbitrary, and we owe the terminology to Benjamin Franklin, whose contributions to our understanding of electricity go far beyond flying a kite in a thunderstorm. (Franklin, apart from all his other claims to fame, is also noted as the first great American physicist.)

Fig. 19–2 In the eighteenth century, Benjamin Franklin showed that lightning was a form of electricity, and studied the types of electric charge.

Franklin asserted that charge is neither created nor destroyed, but merely transferred from one body to another. We now call this idea the law of conservation of charge. This law says that in any process, the total charge on hand at the end must be the same as that in the beginning. We can shift charge around, but there can be no net gain or loss of charge. (By contrast, we saw that heat is not a quantity that is conserved in itself, but one that can be created or destroyed when work is done.)

Ideas on how charge is distributed in matter developed, as did general ideas on the structure of matter, through the late nineteenth and early twentieth centuries. In 1897, J. J. Thomson of the Cavendish Laboratory in Cambridge, England, took an important step. Thomson was studying the behavior of beams of charge in tubes that were like simplified versions of a television picture tube. He found that these beams bend in a magnetic field as though they consist of a stream of identical particles. We now call these particles electrons.

Fig. 19–3 Sir J. J. Thomson, left, discussing a new type of vacuum tube on a visit to the United States.

It was not until 1909, however, that Robert Millikan, an American physicist, demonstrated that there is a fundamental charge by showing that all the charges he could measure were integers times this fundamental unit of charge (Fig. 19–4). Another way of stating

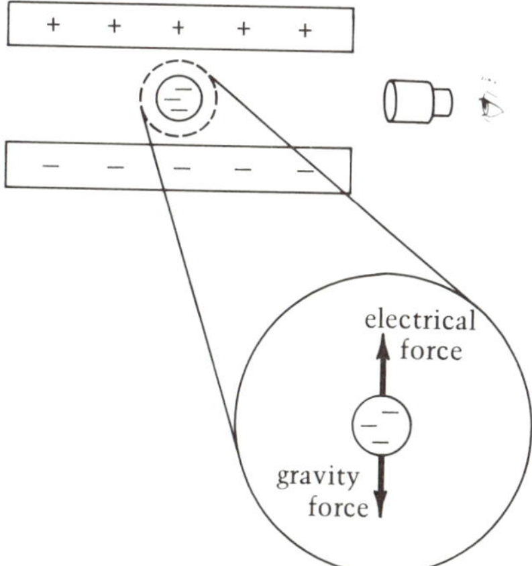

Fig. 19–4 A schematic view of Millikan's oil-drop experiment. Oil drops are placed between two charged plates. Each drop has a different charge, but for many drops the electrical force will just balance the gravitational force. By controlling the strength of the electric field, the drop can be made to move up and down. The motion is studied through an eyepiece, and by measuring the speed up and down, the observer can determine the charge on the drop.

Fig. 19–5 Millikan's oil-drop experiment. The picture is reproduced from his book *The Electron* (University of Chicago Press, 1917).

Later in this book, we shall return to the theory that the "fundamental particles" are composed of quarks, which have charges of $\frac{1}{3}e$, $-\frac{1}{3}e$, and $\frac{2}{3}e$. So far, individual quarks have not been detected, and may not even be detectable. All charges we can deal with directly are positive or negative integer multiples of e.

Other charged objects such as ions*—atoms that have lost one or more electrons—can also serve as charge carriers.*

this fact is to say that the charge is quantized; by this, we mean simply that something comes only in integral multiples (that is, multiplied by 1, 2, 3, and so on) of some basic quantity. The fundamental amount of charge is designated e, and all other charges are positive or negative integers times e. The charge on the electron is $-e$. Millikan's experiment involved measurement of the charges on many small drops of oil, and his experiment is known as Millikan's oil-drop experiment. (Almost every physicist has done some version of this experiment in a college course and has come away with great respect for anyone who can sit for hours on end following oil drops in a microscope!)

We now know that matter is made up of a variety of fundamental particles, each having a charge that is some integral multiple of e. The charge on the proton is $+e$. The charge on the neutron is zero. An atom has an equal number of protons and electrons. Thus, even though an atom contains a number of charged particles, it is neutral. The neutrality is an important feature of the structure of matter.

When we study electricity, we often deal with charge that is moving from one place to another. Charge moves easily through some materials, called conductors, and hardly at all through others, called insulators. Most familiar conductors are metals, although certain solutions, such as salt water, are also good conductors. Most conductors have a large number of electrons that are free to move around. Even though a conductor has these free charges, the conductor is still neutral. For every electron that moves around, there is a proton that stays in place. Most materials fall neatly into either the category of conductor or that of insulator. However, there is also a class of materials called semiconductors, which we will deal with in Chapter 25.

If we have a charged rod, we can use it to induce a charge—that is, to cause a charge to appear—on another object without touching that object. Let's assume that our rod has a negative charge, as shown in Fig. 19–6. If we bring the rod near a conductor (a metal ball, for example), then the free electrons in the metal will be repelled by the negative charge of the rod. These free electrons will go to the side of the metal object farthest from the rod. Even though the metal object is still neutral, the side farthest from the rod will have a negative charge and the side nearest the rod will have a positive charge. If we start with two metal balls in contact, the electrons can flow from one to the other. When we separate the two balls, one will have a positive charge and the other will have a negative charge. The process is called charging by induction (because the charging is *induced* by the presence of the charged rod). The earth is a very good conductor, so you can also charge a metal

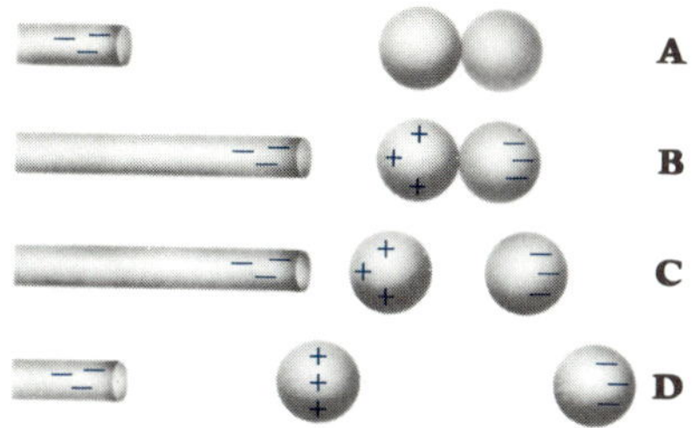

Fig. 19–6 Charging by induction. A. A negatively charged rod is brought near two conducting balls that are in contact with each other. B. As the charged rod comes closer, some of the positive charge in the balls is attracted toward the rod and some of the negative charge on the balls is repelled away from the rod. Note that at this point one ball has a net positive charge and the other has a net negative charge, but together they are still neutral. C. The balls are now separated. D. When the rod is removed, the charges cannot run back to their original places. We are left with one positively charged ball and one negatively charged ball. This process is often carried out using the earth (which can hold a lot of charge) as one of the balls. When the conducting ball is in contact with the earth, we say that it is grounded.

object by induction if it is in contact with the earth. Then you separate the object from the earth.

Since the entire earth is a good conductor (although a very large one), anything that is in electrical contact with the earth is in electrical contact with all other objects that are in electrical contact with the earth. Objects that are in electrical contact with the earth are said to be grounded.

19.2 *Coulomb's Law*

We learned a lot about gravity, another fundamental force, by looking at the force between two particles. Now let us evaluate the electric force between two charged particles.

Like the gravitational forces, the electric force is an inverse-square-law force. This idea was first proposed in the eighteenth century by Joseph Priestly, following a suggestion from Ben Franklin that charged particles floating in a charged cup did not seem to respond to an electric charge on the cup. Newton had already showed, a hundred years earlier, that a particle inside a spherical shell could not feel the *gravitational* force of the shell because the attraction of all of the parts of the shell would exactly cancel. Priestly reasoned that, in at least this case, the electric force must behave similarly to gravity—as an inverse-square law.

The inverse-square-law force for electricity was confirmed experimentally by Charles Coulomb. (There is now some question as to what degree Coulomb's experimental results were a product of his firm belief in the inverse-square-law force instead of being unbiased.) Coulomb and other experimenters found that the electrical force that one particle exerts on another is given by

Coulomb's law is

$$F = \frac{kq_1 q_2}{r^2}$$

where k is a constant, q_1 and q_2 are the charges, and r is the distance between the particles.

$$\textit{force} = \frac{\textit{constant} \times \textit{charge on particle 1} \times \textit{charge on particle 2}}{(\textit{distance between particles})^2}.$$

This force law for electricity is called Coulomb's law.

Notice that the force is proportional to the charges of both particle 1 and particle 2. The electric force obeys Newton's third law, which means that the force that particle 1 exerts on particle 2 has the same strength as the force exerted by particle 2 on particle 1. If the charge on *either* particle is zero, the force is zero. *We can have an electrical force only when both particles are charged.*

If we are using SI units, then we measure charge in coulombs *(abbreviated C). The charge on the electron is -1.6×10^{-19} C.*

We already know that forces have directions associated with them; the electric force is no exception. The electric force acts along the imaginary line joining the two particles, just as is the case

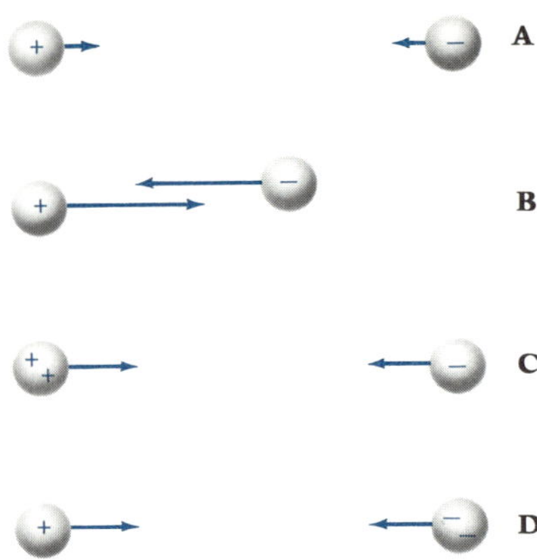

Fig. 19–7 Coulomb's law. A. Here we represent the forces that two balls exert on each other when one has a positive charge and the other has a negative charge. In each case, the arrow on a given ball is the force that that ball feels due to the presence of the other ball. Notice that the force on one ball is equal and opposite to the force on the other ball. B. If we halve the distance between the balls, the force on each is now four times as great as it was in A. C. We move the balls back to the original distance, but double the charge on the positively charged ball. The force on each ball is now twice as great as it was in A. D. Here it is the negatively charged ball that has twice the charge, but the forces are still the same as in C.

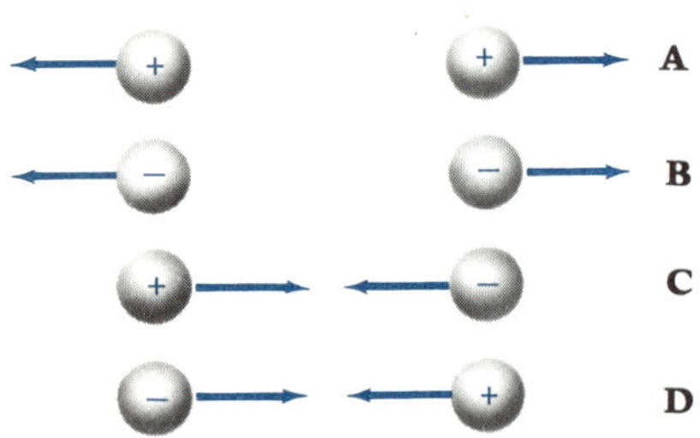

Fig. 19–8 Effect on the sign of the charges in Coulomb's law. A. We have two positively charged balls. Each one feels a force directed away from the other. We call such a force repulsive. B. We have two negatively charged balls. The force is still repulsive. C–D. We have two oppositely charged balls, and in each case the force is attractive, that is, each ball is pulled toward the other.

with the gravitational force. However, the direction of the force depends on the signs of the two charges (that is, whether they are positive or negative). If the two charges have the same sign, either both plus or both minus, then the particles will push each other away. The force is repulsive. If the two particles have charges of opposite sign, one plus and one minus, then the force is attractive. An attractive force acts in the direction that would bring the two particles together (see Figs. 19–7 and 19–8).

If we want to find the force that *several* charges exert on another charge, we simply calculate the force that each one exerts and then add all of these forces (as vectors) together. We wind up with the resultant force (Fig. 19–9). This is the same procedure that we follow when answering a similar question for gravitation.

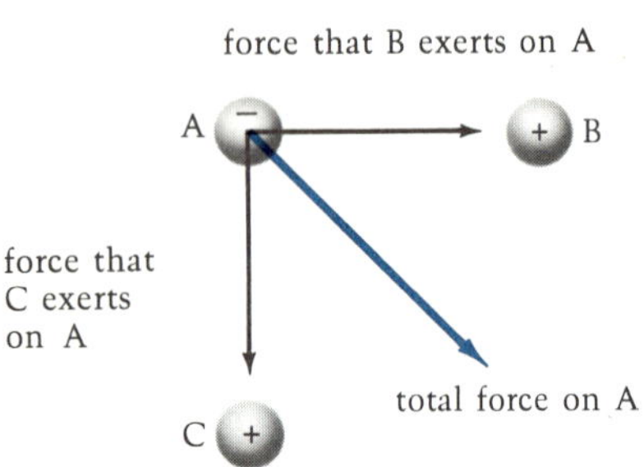

(at left) Fig. 19–9 To find the total force on ball A, we take the (vector) sum of the forces that B exerts on A and C exerts on A. If there are even more balls exerting forces on A, we also add their individual effects in to get the total force on A.

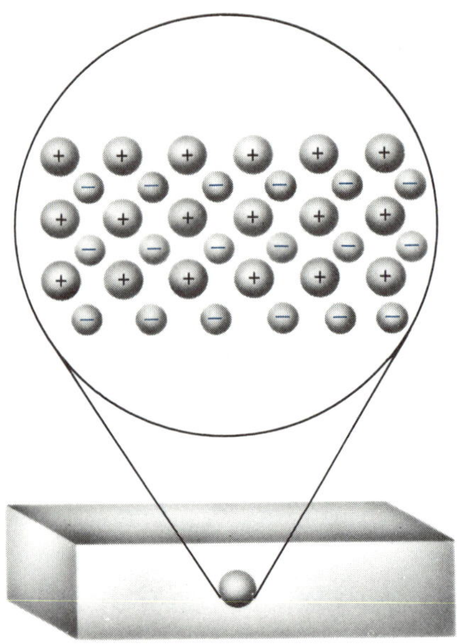

(at right) Fig. 19–10 The neutrality of matter. Even though matter consists of a large number of charged particles, we generally find an equal amount of positive charge and negative charge.

Since the gravitational and electric forces seem to behave in the same way, we might ask about the relative strength of these two forces. The electric force is much stronger. For example, the electric force between two protons is about 10^{36} times greater than the gravitational force between the protons! With such a strong electrical force, why are the motions of the planets, stars, and galaxies governed by gravity and not by electricity? The answer lies in the ability of the positive and negative charges to cancel each other's effects. Although the earth has many charged particles, it has equal numbers of plus and minus charges, and so is electrically neutral (Fig. 19–10). Planets don't have electric charges between them. Gravity, on the other hand, has only one type of mass. There is nothing to cancel it out. (Only science fiction writers have written of "antigravity.") The gravitational effect of all the little particles adds up bit by bit, and the collective effect can be enormous.

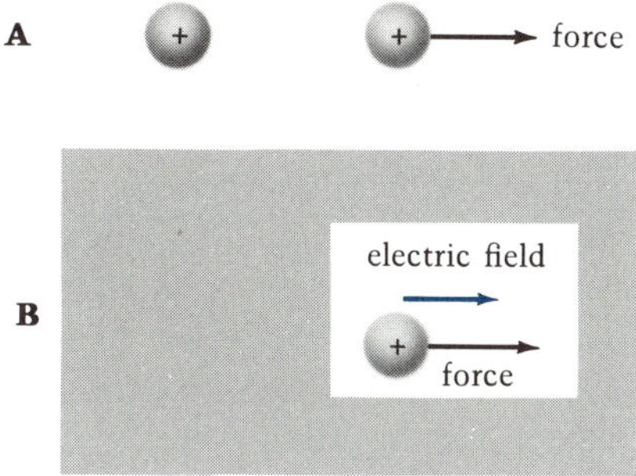

Fig. 19–11 Forces and fields. A and B show two different ways of looking at the same situation. In A, we directly write down the force on the right ball due to the left ball. In B, we "block out" the left ball, and leave a window so we can see the right ball. Even though we cannot see the other ball, we can still see that it has an effect, and we describe that effect by saying that there is an electrical field at the position of the ball that we can see through the window. Then, knowing the strength and direction of the electric field, as well as the charge on the ball we can see, we can find the force on that ball.

19.3 *Electric Fields*

Instead of referring back each time to the prime object causing a force (the earth for the earth's gravitational force, for example, or a charged ball for a particular electric force), we often find it convenient to work out the effect of the source of the force at each point of space. We can then think as though there is a "field of force" at each location. This electric field has a direction associated with it, just as the electric force does. The direction of the electric field is in the same direction as the force felt by a positively charged particle.

The idea of an electric field is a convenience for doing calculations. Once we calculate the electric field at a point, we can find the force on any new charged particle placed at that point without having to calculate the field over and over (see Fig. 19–11).

We can make useful drawings of electric fields that show both the strength and direction of the field and how these vary from place to place. Our drawing uses what are called electric field lines or lines of force. The rules for making such drawings are quite simple:

1. *The lines we draw should point in the direction of the electric field at each point.* We can think of each field line as though it is a chain of tiny arrows, each representing the direction of the electric field at its location.
2. *The stronger the field, the closer together we draw the field lines.*
3. *Electric field lines can only begin and end at charged particles.*

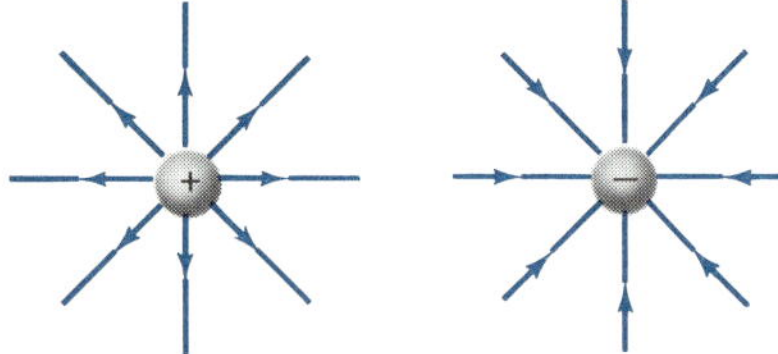

Fig. 19–12 Electric field lines near a single positive charge and near a single negative charge. Notice that the electric field points away from the positive charge and toward the negative charge, since these are the directions in which the positive charge would feel a force. In each case the electric field is stronger the closer you are to the charge. This is represented by the fact that the field lines get closer together as you get closer to the charges.

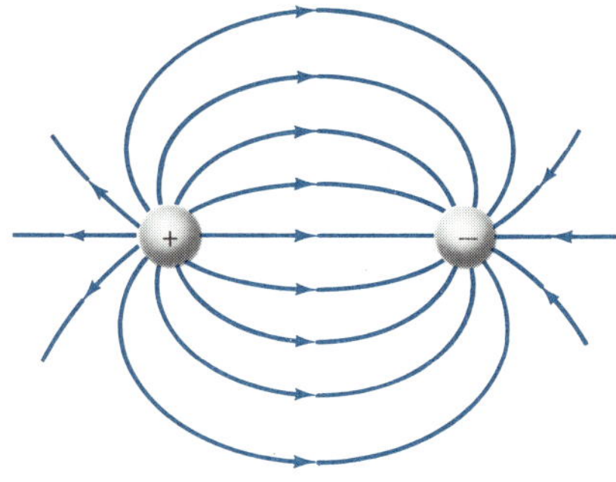

Fig. 19–13 The electric field due to an electric dipole. A dipole consists of a positive charge and a negative charge separated by some distance. Notice that the field lines go from the positive charge to the negative charge.

The simplest electric field is that created by a point charge (Fig. 19–12). Let's say that the charge is positive. In this case a positive charge would be repelled, and so the electric field points outward. Notice that the field lines get farther apart as we move farther from the charge, indicating that the field gets weaker as we get farther away. If we used a negative charge instead of the positive charge, everything would be the same except that all the arrows would point into the charge instead of away from it.

Figure 19–13 shows the field pattern produced by two charges of opposite signs separated by some distance. This arrangement of

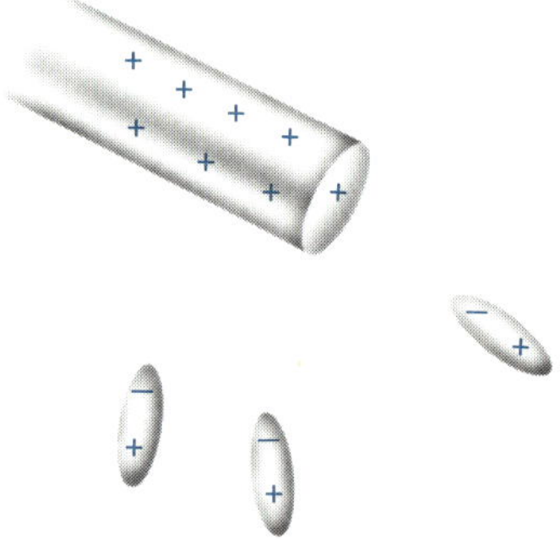

Fig. 19–14 The behavior of electric dipoles near a charged rod. In this case the rod has a positive charge. The negative ends of the dipole are drawn toward the rod and the positive ends are pushed away from the rod, so the dipoles end up with their negative ends pointing toward the rod.

Fig. 19–15 The electric field around unit charges. (Here the tip of the charged rod acts as a unit charge.) A. The field around one unit charge. B. The field around two unit charges of opposite polarity. C. The field around two unit charges of the same polarity.

A

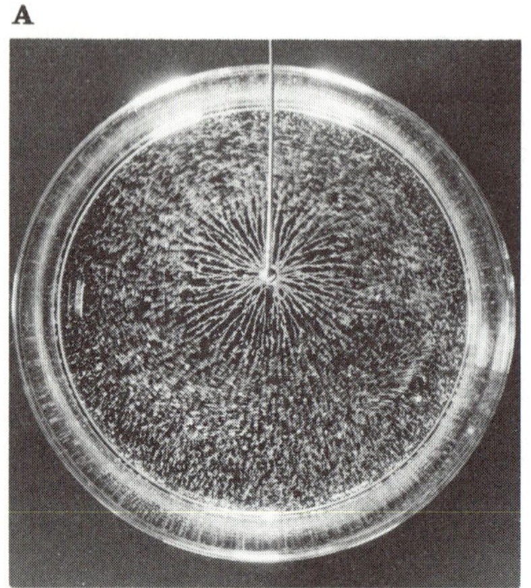

B

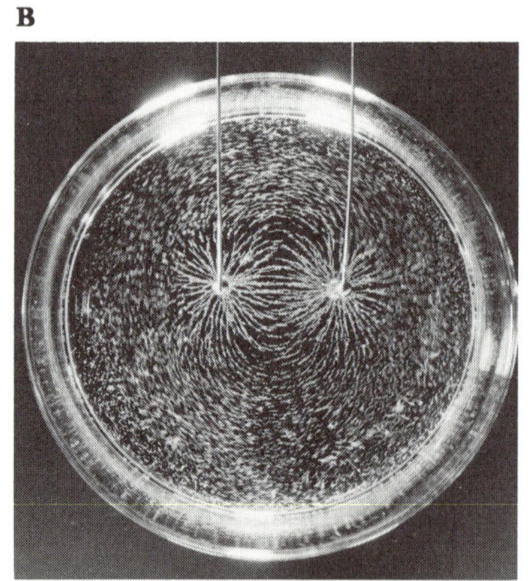

C

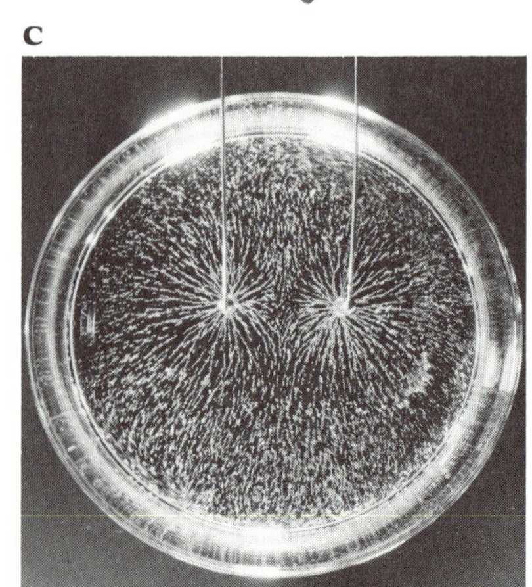

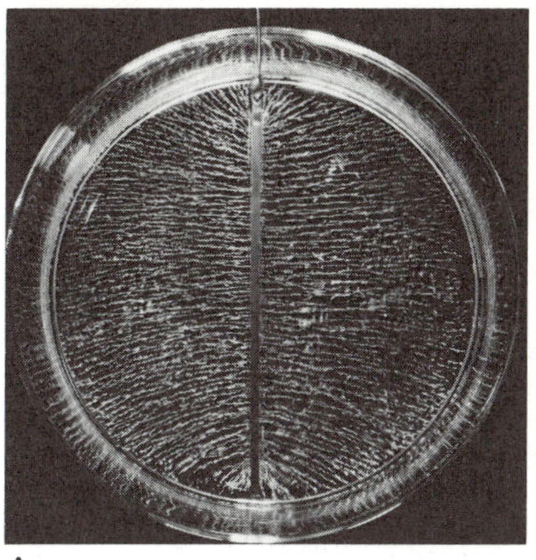
A

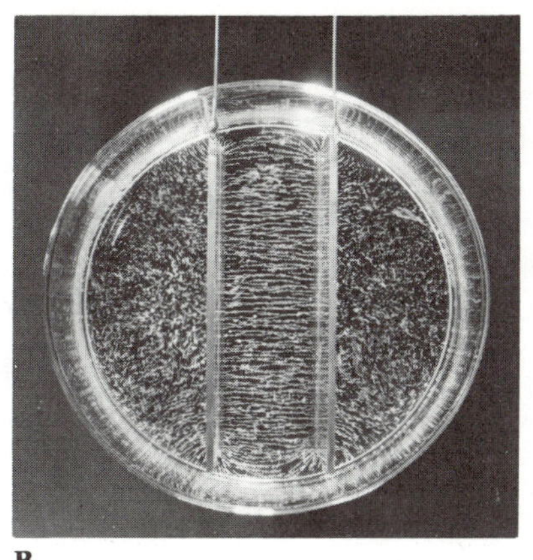
B

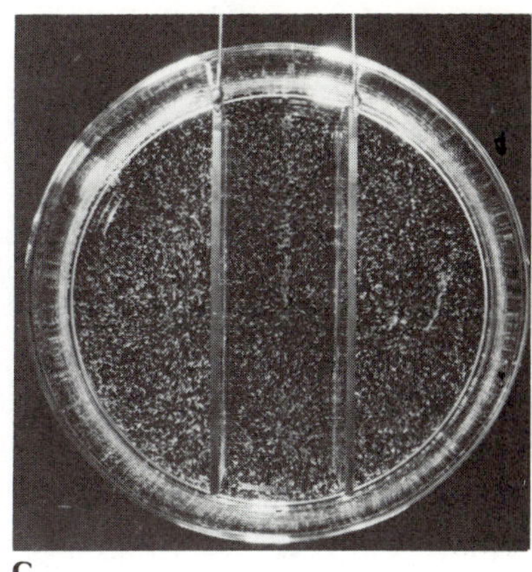
C

Fig. 19–16 The electric field around charged bars. A. The field around one charged bar. B. The field around two bars with opposite charge. C. The field around two bars with the same charge.

charges is called an electric dipole (the prefix *di-* is from the Greek for "twice") since it has a positive end and a negative end. If the charges are of equal strength but of opposite sign, then the total charge of the dipole is zero. However, since the charges are not in the same place, the electric field caused by the dipole is not zero. Dipoles have important effects since although most of matter is neutral—an equal number of positive and negative charges—the positive and negative charges are not always in the same place. Thus they can create an electric field and have an effect on other charged particles (see Fig. 19–14).

19.4 *Electricity and Energy Conservation*

The concept of an electric field is especially useful when we want to calculate how a charged particle will move. We have found that our study of the motion of a particle in a gravitational field could be greatly simplified by applying the law of conservation of energy. Since there are many similarities between the electric force and the gravitational force, we might expect that we can do the same with electric fields. In fact, we can. The application of the law of conservation of energy to electric fields is quite useful.

A charged particle in a uniform electric field behaves in many ways like a ball in the earth's gravitational field. For example, if a proton starts at rest, it will accelerate in the direction of the electric field. However, for a particle "falling" in an electric field, the acceleration depends on the mass of the particle and on the charge of the particle. All particles do not "fall" at the same rate. Further, negative and positive particles fall in opposite directions.

If a charged particle is moving when it encounters an electric

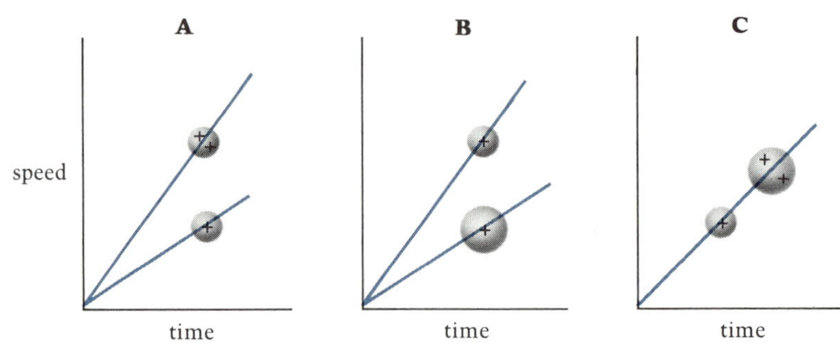

Fig. 19–17 Effects of charge and mass on the motion of particles placed in a constant electric field. In each case, the particles start at rest and we graph their speed at various times. A. In this case, we compare two balls of equal mass, but with one having twice the charge of the other. The ball with the greater charge feels a greater force, and since both balls have the same mass, the greater force will produce the greater acceleration. At any time, the ball with twice the charge will be moving at twice the speed of the other ball. B. In this we compare two balls with equal charges, but with one ball having twice the mass of the other. The electrical forces felt by the two balls will be the same, since their charges are the same, but the ball with the larger mass will have a lower acceleration. At any time, the ball with twice the mass will be traveling at half the speed of the other ball. C. Here we have two balls, one of which has twice the charge *and* twice the mass of the other. The ball with the greater charge feels a greater force, but this is exactly balanced by the greater mass, so both balls have the same acceleration, and will have the same speed at any time.

field, then the problem is similar to that of projectile motion near the surface of the earth. The velocity of the particle in the direction perpendicular to the electric field does not change. However, the motion along the electric field will change (Figs. 19–17 and 19–18).

If we take a proton and push it some distance against an electric field, we will have to do some work. After we let go, the proton will accelerate and pass its starting position with a kinetic energy equal to the amount of work we did pushing it.

Since we can recover the work done against an electric field, we can define an electric potential energy in much the same way we defined a potential energy for gravity. Again, we are mainly interested in *changes* in the potential energy. *If we move a charged particle in an electric field from one place to another at constant speed, then the electric potential energy changes by the amount of work done in moving the charge.* Recall that the work done is the force × distance, where we only consider the component of the force that points along the direction of motion. We know that the force is just the electric field strength times the charge, so (1) the greater the electric field strength, the greater the change in potential energy; and (2) the greater the charge, the greater the change in potential energy (see Fig. 19–19).

It is useful to define a quantity that represents the change in potential energy of a particle *per unit of charge* when we move the particle from one place to another. We call this quantity the electric potential. Just as with potential energy, we are only interested in the changes in potential from one place to another. *The potential difference between two points is the work required, per unit*

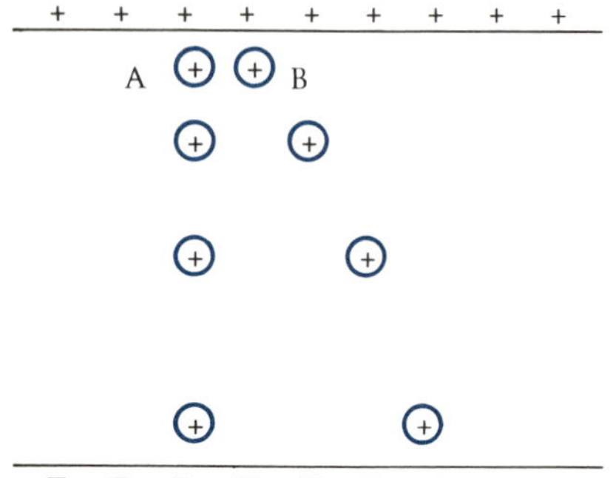

Fig. 19–18 Projectile motion in an electric field. This diagram shows two identical balls, A and B, at four different times after they are released in an electric field. Both will accelerate toward the negative plate, but we have started ball A from rest and have given ball B an initial speed to the right. This motion is not changed and ball B continues to move over to the right the same amount in each time interval. The progress of both balls toward the negative plate is the same.

THE VOLT

We could express potential difference in terms of energy per unit charge. However, this ratio comes up so often that we give it a special name. The SI unit for potential difference is the *volt*. ("Volt" is named after the Italian scientist Alessandro Volta, 1745–1827, who is known for his work on potential differences. It is abbreviated "V.") The potential difference between two points is sometimes referred to as the *voltage difference* (or, simply, the *voltage*) between two points. It is important to remember that we are always measuring the difference.

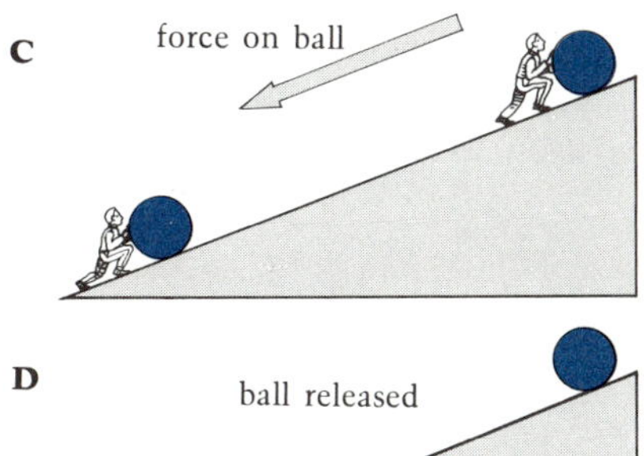

Fig. 19–19 Work against an electric field and against a gravitational field. In A, the person starts on the left and wants to push the positively charged ball to the right. The electric field exerts a force to the left, so the person must exert a force to the right to move the ball. In moving the ball from left to right, the person will have to do some work. However, the energy lost can be recovered. If the ball is released on the right, as in B, it will arrive back at the original position on the left, but will have acquired some kinetic energy. Its kinetic energy will be equal to the amount of work the person did in moving the ball from left to right. The situation is repeated in C and D, except that we have substituted gravity for the electric field, to show that the situation is quite similar. In this case, the person wants to push the ball up a ramp. To do this, the person must exert a force to counteract that of gravity, and will have to do some work to get the ball to the top of the ramp. However, if the ball is released, as in D, the energy will be recovered in the form of the kinetic energy of the ball when it reaches the bottom of the ramp.

charge, to move a particle from one point to the other. The electric potential difference between two points depends only on the location of the two points and not on the path taken in traveling between them.

work = potential difference × charge

Suppose we want to know how much work we must do to move a particular charge from one point to another. All that we need to know is the potential difference between the two points and the charge on the particle. The potential difference tells us the amount of work that we must do for each unit of charge, so if we multiply work per unit of charge by the number of units of charge, then we have the total work required. (If you know how much work is required to carry one bucket of water up a hill, then ten times as much work is required to bring ten buckets of water up the hill.)

Suppose we want to move our proton opposite to the direction of the electric field. We have to exert a force to oppose the electric force, so as we push the proton along we are doing work. As we push the proton, its potential energy increases. It's like pushing a ball uphill. Therefore, if you push a charge opposite to the direction of the electric field, you move it from a lower potential region to a higher potential region.

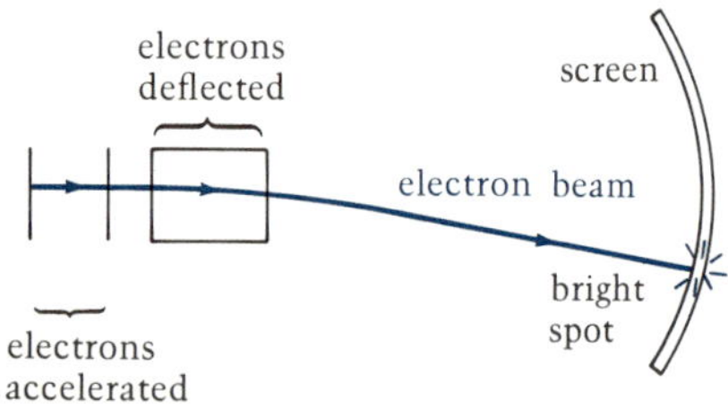

Fig. 19–20 Basic operation of a TV picture tube. We need high-speed electrons so that when they strike the screen, their energy will go into making a bright spot. The electrons are accelerated by letting them "fall" through a region with a large potential difference, as large as 30,000 volts, as they go from the negative plate to the positive plate. (The electron beam is deflected, to move the bright spot around the screen, by magnetic processes, which will be discussed in Chapter 21.)

If we let the proton go, it accelerates in the direction of the field. *A proton "falls" from a high potential to a low potential.* (Of course, a negative charge goes the other way.) By the time that the proton accelerates back to its original position, it acquires a kinetic energy equal to its charge times the potential difference through which it has "fallen."

We give a special name to the energy that a proton or electron acquires when it passes through a potential difference of 1 volt—the electron volt (abbreviated eV). The electron volt is a unit that is a convenient size for atomic physics.

In a television picture tube, we want to create a stream of high-energy electrons. The high-energy electrons eventually strike the screen, and when the screen absorbs the energy, it glows. To produce the electrons, a piece of wire is heated up so that some electrons are driven off. If an electric field is present, the negatively charged electrons accelerate, acquiring an energy equal to the charge on the electrons times the voltage difference through which they accelerate. In a color TV set, this potential difference may be as much as 30,000 volts, so the electrons reach the grid with an energy of 30,000 electron volts (written 30 keV, where 1 keV $= 10^3$ eV; see Fig. 19–20).

19.5 *The Oscilloscope*

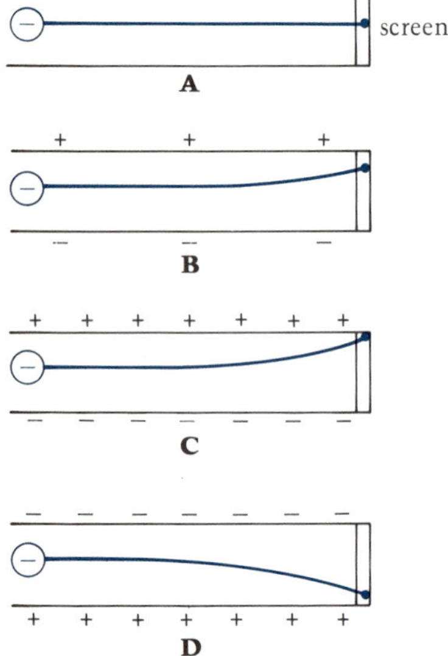

The oscilloscope, a device that displays potential differences on a screen, is one of the most important tools of the experimental physicist. An oscilloscope allows us to monitor and automatically graph the variations in the potential difference between any two points in a piece of equipment. For most experimental physicists, having an oscilloscope in a laboratory is taken for granted, and they would be hard-pressed to carry out their experiments if deprived of this valuable tool.

We start by making a beam of electrons, and accelerating that beam, as described above for the TV picture tube. The beam then strikes a screen, and the screen glows briefly wherever the beam

Fig. 19–21 Operation of an oscilloscope. An electron beam starts on the left and strikes the screen on the right. The screen glows wherever the beam strikes. The position where the beam strikes the screen is controlled by changing the electric field in the region through which the beam passes. In A, there is no vertical field, and the beam strikes the center of the screen. In B, there is a small field and the beam strikes above the center of the screen. In C, there is a stronger field and the beam is deflected more. In D, the field goes the other way and the beam strikes the bottom of the screen.

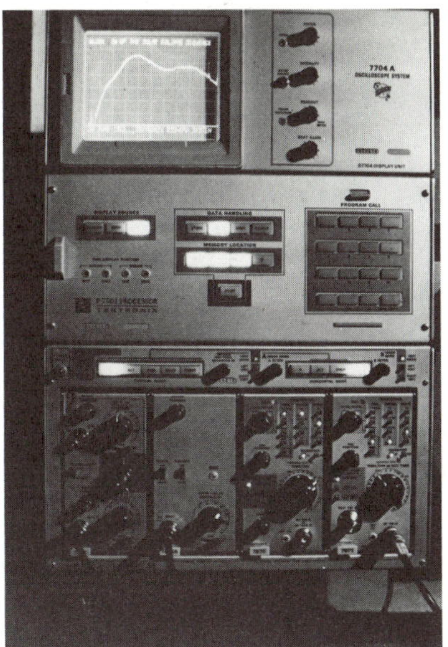

Fig. 19–22 An oscilloscope with a display on the tube from an electron beam inside it. The frequency of the signal guiding the bending of the beam from left to right is half that guiding the bending up and down.

strikes. The beam passes through two sets of parallel metal plates before it reaches the screen. One set of plates creates an electric field in the up-down direction, and the other creates an electric field in the left-right direction. We change these electric fields by varying the potential differences between the pairs of plates. The greater the potential difference, the greater the electric field.

If there is no potential difference between the plates of each pair, the beam strikes the center of the screen. If we want the beam to strike above the center of the screen, we must put the upper plate at a higher potential than the lower plate. (Remember, the negatively charged electrons bend their paths toward the higher potential plate.) If we reverse the potential difference, the beam strikes below the center. We use the other pair of plates in a similar manner to control the left-right position of the beam (see Fig. 19–21).

Doctors use equipment that works in this manner to examine your heart, as part of making an electrocardiogram. The electric impulses associated with your heartbeat cause the potential difference between your wrist and your chest to vary with each heartbeat cycle. Medical personnel attach a wire from your wrist to the upper plate and one from your chest to the lower plate. Since the potential differences are small, the oscilloscope also has amplifiers to make the potential difference large enough to have an effect on the beam. To create a continuous monitoring of your heart, as in an intensive care unit, an oscilloscope-like device is also used. The potential difference between the left plate and the right plate is made to increase with time, so that the beam sweeps across the screen from left to right. When the beam gets all the way to the right the cycle is repeated. (The equipment is carefully constructed and used in such a manner so that you don't get a shock.)

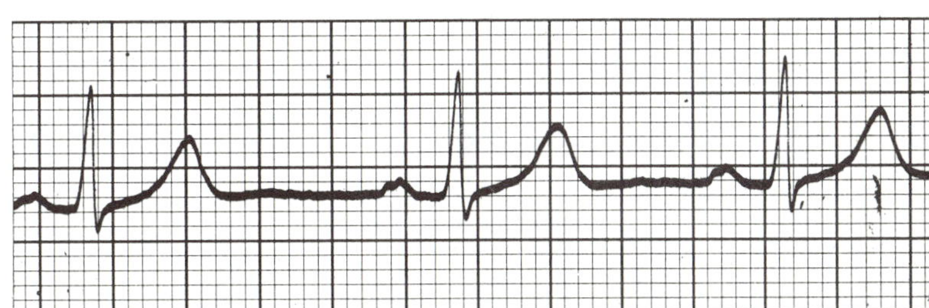

Fig. 19–23 Electrocardiogram of the heartbeat of one of the authors.

19.6 *Capacitors—Storage for Charge*

We are often faced with situations in which we would like to have a surge of electricity for a short period of time. For example, if you use an electronic flash with a camera, you would like to be able to

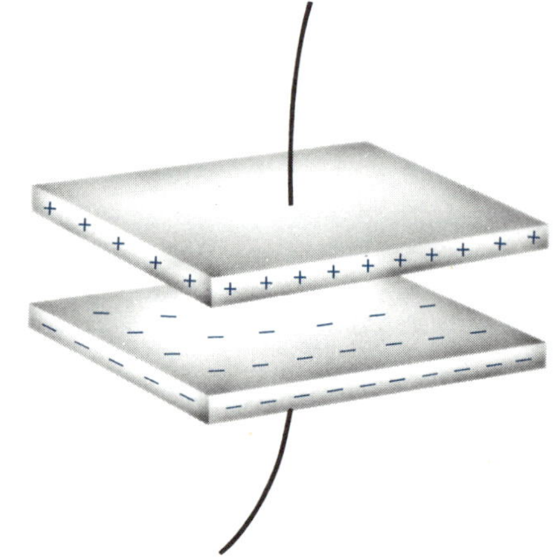

Fig. 19–24 A capacitor. This capacitor consists of two parallel rectangular plates made out of a conducting material. A positive charge sits on one plate and an equal amount of negative charge sits on the other plate.

hold charge in the flash unit until you take a picture, at which time you want a surge of current and a bright flash. A device that we can use to store charge is called a capacitor.

The simplest kind of capacitor consists of two parallel plates of metal, and is called a parallel-plate capacitor (Fig. 19–24). When we say that we are "storing charge," we don't really add charge to the capacitor. We simply take some electrons from one plate and transfer them to the other plate. As long as we keep the plates separated, the electrons cannot run back. If we connect the plates with a wire, the electrons turn back to their original plate. (The parallel-plate arrangement was first devised by Ben Franklin.)

Since we have to do work to charge a capacitor, we can think of the capacitor as something that stores energy as well as charge. We must do a certain amount of work to charge the capacitor. Then, when we discharge it, we get that energy back. Since charging a capacitor creates an electric field, we can think of the energy as being stored in the field, a way of thinking that is more general than its specific application to the case of a capacitor. It takes a certain amount of energy to set up any electric field, and we can think of that energy as being stored in the field.

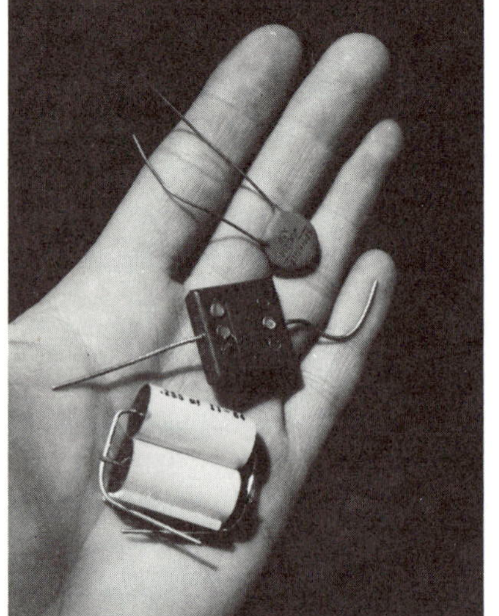

Fig. 19–25 Various types of capacitors.

19.7 *High Voltage*

There are often occasions when we want to create large potential differences, for example, to accelerate charges. One device for doing this is the electrostatic generator, also known as the Van de Graaff generator, illustrated in Fig. 19–26. Charge is carried on an insulating conveyor belt up to the inner surface of a large conducting sphere. The charge spreads out evenly on the sphere and the

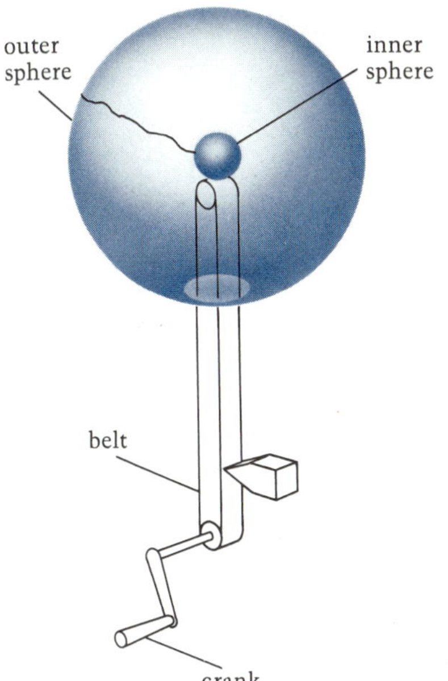

Fig. 19–26 The electrostatic generator. As you turn the crank, charge is rubbed onto the belt at the bottom and is carried to the small sphere on the top. A small wire then carries it to the outer sphere. No matter how much charge is on the other sphere, it produces no net force on the charge within, and we can keep adding more and more charge to the outer sphere, creating a large potential difference between the sphere and the ground.

Fig. 19–27 The electric charge this man picks up from a Van de Graaff generator spreads through his body and makes individual strands of his hair repel each other.

belt continues to bring up more charge. Usually the belt is turned by hand or by a small motor. The more charge we put on the sphere, the higher the resulting potential.

If you put your hand on the sphere while it is charged, enough charge is transferred to you to raise you to the same potential as the sphere. This charge spreads out over your body, and because each of the hairs on your head then carries a little charge, the hairs repel each other. In the process of trying to get as far from each other as possible, the hairs stand on end. (The hairs stand up best on a dry day. On a humid day, the water vapor in the air carries away some of the charge.)

There is a limit to how high we can make the potential in an electrostatic generator. Eventually a spark will jump from the generator to another object. When the electric field becomes too high, the air molecules become ionized, that is, some electrons are torn away from the rest of the molecule. Ionized gases are very good conductors of electricity.

For air between the plates, a spark can jump across a gap of 1 meter if you get a potential difference of 3×10^6 volts across the gap, or will jump across 1 mm with a potential difference of 3×10^3 volts.

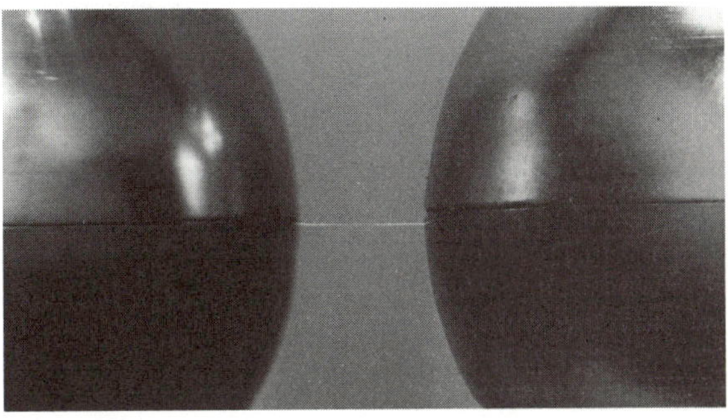

Fig. 19–28 When the potential difference between two objects gets large enough, a spark will jump. In this case, one sphere is part of an electrostatic generator and the other is grounded.

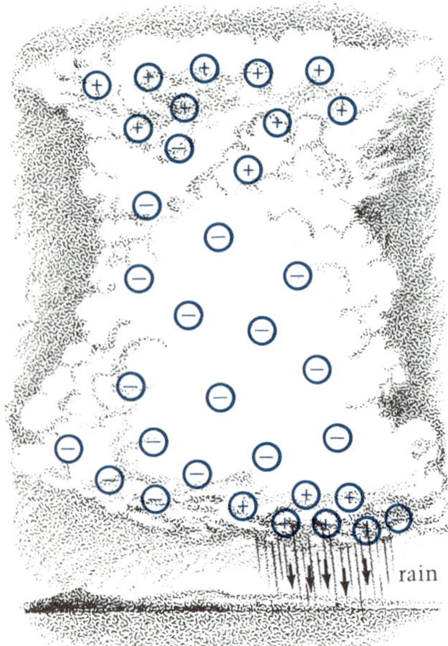

Fig. 19–29 Charge distribution in a thunderstorm. The top half of the cloud will be generally positively charged and the bottom half of the cloud will be generally negatively charged. One exception is a small positively charged region near the bottom of the cloud where the rain is falling. The top of the cloud is usually more than 10 km above the ground. Lightning may occur between the upper and the lower parts of the cloud, between two clouds, or between the cloud and the ground.

Fig. 19–30 Lightning is electrical energy jumping between regions that have different electrical potentials.

When the breakdown occurs, a lot of charge suddenly flows across the gap. Usually it flows across where the distance is shortest. The small region around the flowing charge is heated and glows momentarily. The sudden pressure change due to the sudden temperature increase causes a sound wave, which propagates away from the spark. You hear a crackling sound.

Lightning is a form of breakdown in the earth's atmosphere. The tremendous convection currents that run up and down inside thunderheads act like giant electrostatic generators, transferring charge around the cloud. Eventually, a sufficiently large potential difference builds up between two clouds or between the clouds and the ground. The air breaks down, allowing a spark to jump. We can think of the cloud and earth as a giant capacitor that is storing a lot of energy. When the spark jumps, that energy is released in the form of light and sound. Lightning usually strikes the tallest object in a given area since that allows it to take the shortest path, which creates a larger electric field for a given potential difference. Because of this, never seek refuge under a tree in a thunderstorm (Figs. 19–29 and 19–30).

If the spark jumps a kilometer, then the potential difference must be 3×10^9 volts to give an electric field of 3×10^6 volts/m, certainly a large potential difference!

Key Words

electric charge
positive
negative
law of conservation of charge
electrons
quantized
Millikan's oil-drop experiment
conductors
insulators
charging by induction
grounded
Coulomb's law
coulomb
repulsive
attractive
electric field
electric field lines
lines of force
electric dipole
electric potential
potential difference
voltage difference
electron volt
oscilloscope
capacitor
parallel-plate capacitor
electrostatic generator (Van de Graaff generator)

Questions

1. Two objects 1 meter apart are given enough charge so that the electric force between them happens to be just as strong as the gravitational force between them. When the same two objects are 2 meters apart, which (if either) is stronger, the gravitational or the electric force?
2. If our bodies contain so many charged particles, why are we not electrically attracted to (or repelled from) each other?
3. Based on our discussion on the inherent strength of the electric force, when we charge a rod by rubbing it, do you think that we transfer a large number (comparable to the number of atoms in the rod) of charges or a small number?
4. Suppose we start with a neutral hydrogen atom and break it into an electron and a proton. We started with a neutral particle and ended up with two charged particles. Does this process conserve charge? Explain.
5. To deduce the quantization of charge, do you think that it is necessary to observe a particle whose charge is exactly e?
6. A dish of pure water is a good insulator. However, if you stick your hand in the water, shortly thereafter the water becomes a good conductor. Why?
7. If you want to charge an object by induction, why must that object be made out of a conductor?
8. How does the force exerted by a proton on a proton compare with the force exerted by a proton on an electron at the same distance?
9. The force between two particular charges is 2 N. (a) If we double the size of one charge, what is the force? (b) If we double the size of both charges, what is the force?

10. If we put a 1-coulomb charge at some point and it feels a force of 100 N, what is the strength of the electric field at that point? (b) What force would a 2-coulomb charge feel at that point?

11. What happens to an electric dipole when we put it into a uniform electric field?

12. If an electric dipole is neutral, how can it be affected when it is placed in an electric field?

13. Suppose a dipole is made of some positive charge at one end and equal negative charge at the other. The dipole is placed in an electric field so that the dipole is parallel to the field lines. However, the electric field is not constant in strength and is stronger at the positive end than at the negative end. What do you think will happen to the dipole?

14. Does the potential difference between two points depend on which of the two points we designate as the zero of potential?

15. Suppose you have an electric field in your room, pointing from one wall to the opposite wall. If you double the strength of the electric field, what happens to the potential difference between the two walls?

16. One TV set uses a 30,000-V picture tube and another uses a 24,000-V picture tube. Which one would you expect to give you a brighter picture?

17. (a) If we increase the potential difference between two plates in an oscilloscope, what happens to the electric field between the plates? (b) What happens to the deflection of the beam?

18. Suppose you were given a black box with two connectors, and you wanted to find out how the potential difference between the connectors changed with time. How would you use an oscilloscope to find out?

CHAPTER 20

Electric Circuits

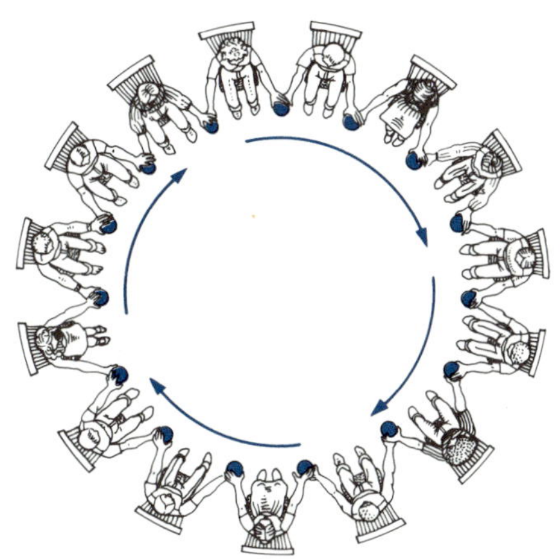

Fig. 20–1 In this drawing, each student has a ball. At regular intervals, the students pass the balls to the left. At any one time, each student has one, and only one, ball. However, if you were to watch any one student, you would see a number of balls passing by. This flow of balls past any point in the ring is equivalent to the flow of current in a wire.

When we think about electricity, we think primarily about the kind that we buy from the power company. This electricity flows through the wires in our homes and furnishes us with energy to run our refrigerators, our lights, and our televisions.

Try turning on a light switch. See if you can detect any delay between the throwing of the switch and your perception of the light coming on. Even though the electricity must flow through a substantial distance from the light switch to the light, it seems to cover the distance almost instantaneously. This little experiment tells us something about the flow of electricity through the wires in our houses. It is fast. In this chapter we take a closer look at what happens when you throw the switch.

20.1 *Current in a Conductor*

As we saw in the last chapter, a conductor is a material in which charge is free to flow when an electric field is present. Free charges can be present even though the conductor is neutral. An analogy is a group of students sitting in a circle, each holding a blue ball. The students pass the balls around. We would say that the balls flowed clockwise. But when we are done, each student is still left holding one ball. If the balls each had one unit of negative charge and the students each had one unit of positive charge, the group would start neutral and remain neutral, but charge could flow (Fig. 20–1).

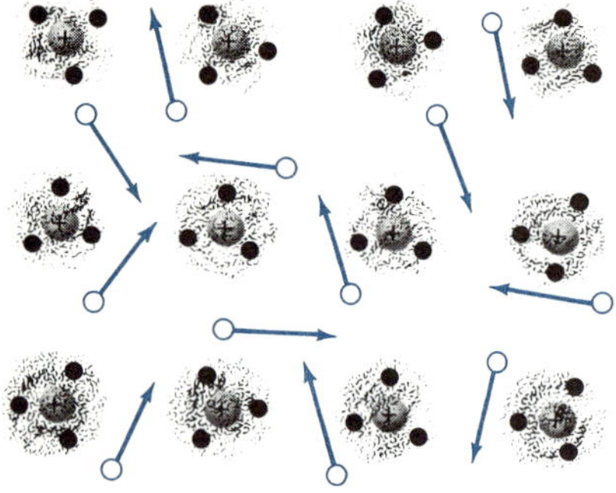

Fig. 20–2 Free electrons in a conductor. In this solid metal, the atoms are arranged in some regular arrangement. The nucleus of each atom is indicated by the +. Around each nucleus are a certain number of electrons that are bound to the nucleus (filled circles). These are called bound electrons. However, each atom has also contributed one electron to a pool of free electrons (open circles). These electrons are free to move all about among the stationary atoms.

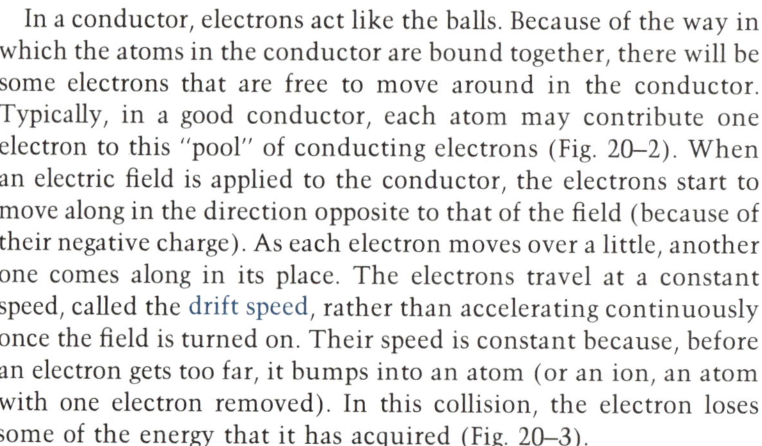

In a conductor, electrons act like the balls. Because of the way in which the atoms in the conductor are bound together, there will be some electrons that are free to move around in the conductor. Typically, in a good conductor, each atom may contribute one electron to this "pool" of conducting electrons (Fig. 20–2). When an electric field is applied to the conductor, the electrons start to move along in the direction opposite to that of the field (because of their negative charge). As each electron moves over a little, another one comes along in its place. The electrons travel at a constant speed, called the drift speed, rather than accelerating continuously once the field is turned on. Their speed is constant because, before an electron gets too far, it bumps into an atom (or an ion, an atom with one electron removed). In this collision, the electron loses some of the energy that it has acquired (Fig. 20–3).

If the charges move so slowly in response to an electric field, why does a light come on almost as soon as you throw the switch? The charges in any part of the wire flow as soon as they feel the electric field. The light doesn't have to wait for electrons from the switch to reach it. Once we turn on an electric field in the wire, charge starts

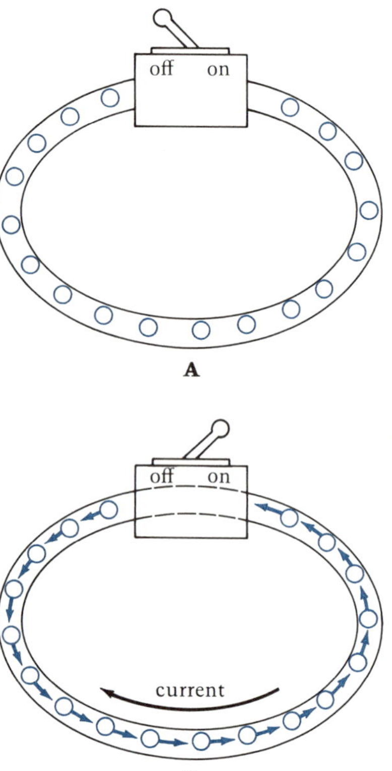

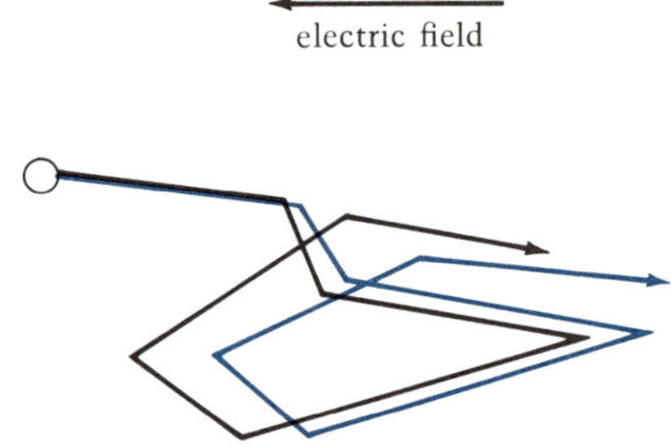

Fig. 20–3 The drift of electrons. The black line represents the path of a free electron through a metal when there is no electric field applied to the metal. The sudden changes in direction are when the electron runs into stationary atoms and rebounds. If there is no electric field, the electron will wander aimlessly about. The blue line shows what happens when an electric field is applied. Remember, the electric force on the electron is in the direction opposite to the electric field, since the charge on the electron is negative. Because of this force, in each piece of the electron's motion it ends up a little farther to the right than it would have if the field were not present. This means that even though the electron still bounces around a lot, it drifts slowly to the right. It is this drift that produces the current in a wire.

Fig. 20–4 The effect of throwing a switch. A. When the switch is *off* there is no electric field in the wire and the electrons don't flow in any particular direction. B. Once the switch is moved to *on* (assuming you are hooked up to a power supply) there is an electric field felt everywhere in the wire, almost instantaneously. This means that electrons start to flow everywhere in the wire, and we get a current.

In copper at room temperature, the drift speed is about 10^{-3} cm/s. At this rate it would take electrons three days to go 3 meters from the wall switch to the light.

to flow everywhere in the wire. (Actually, the field cannot spread out faster than the speed of light, but this is still much faster than the drift speed; see Fig. 20–4.)

The rate at which charge flows through a wire is called the current. We measure the current as the amount of charge passing through a section in the wire per unit of time. We normally measure current in coulombs per second. The unit of current is called the ampere (or *amp*), named for André Marie Ampère, a French physicist who did extensive experiments on flowing currents and magnetism in the early nineteenth century.

1 ampere = 1 coulomb/second

It is convenient to think of the flow of current as a flow of positive charges. By convention, we take the direction of the current to be in the opposite direction to that in which the electrons are physically flowing. In fact, we often talk as if there were indeed positively charged particles flowing in the direction of the current. Doing so saves us from the trouble of having to keep track of a minus sign in calculations (Fig. 20–5).

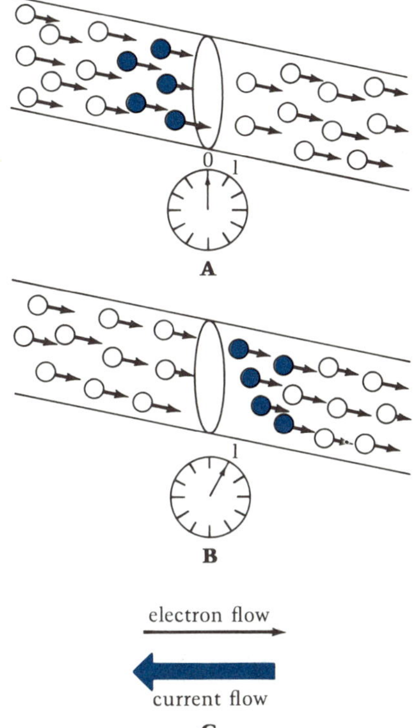

Fig. 20–5 Current in a wire. The circles represent electrons flowing through a wire in the indicated direction. Suppose we want to find the current flowing past any section of the wire (shaded region). We want to see how many electrons flow past our viewing point in one second. In A (when the clock is at zero), the filled circles represent the electrons that will flow across the shaded region in the next second. In B (with the clock at 1 second), the filled circles indicate the electrons that flowed through in the past second.

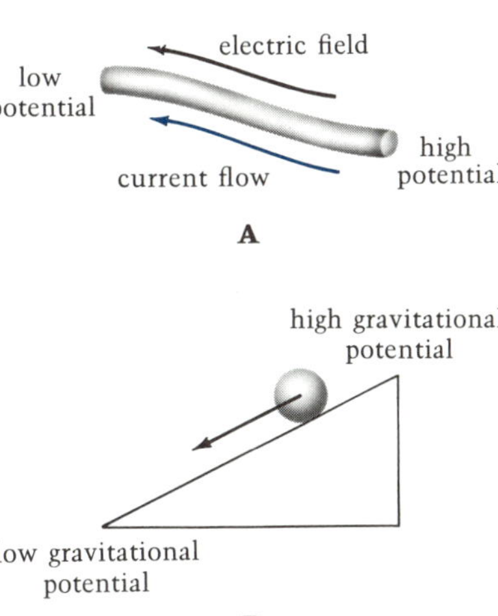

Fig. 20–6 A. Current flows through a wire in the direction of the electric field. This is from the end of high electric potential to low electric potential. B. The same reasoning applies to gravity. A ball will roll from a region of high gravitational potential to a region of low gravitational potential. This is just another way of saying that the ball will roll from the top of a hill to the bottom.

If there is an electric field in a wire, then there must be a potential difference between the ends of the wire. So, instead of talking about the current flowing in response to the electric field, we can say that a current flows through a conducting wire whenever there is a potential difference between the ends of the wire. Since positive charges flow from regions of high potential to regions of low potential, the current flows from high potential to low potential. (Of course, the electrons are flowing the other way; Fig. 20–6.)

The size of the potential difference determines how much current will flow through a wire. The greater the potential difference between the ends of the wire, the faster the charges will flow. This

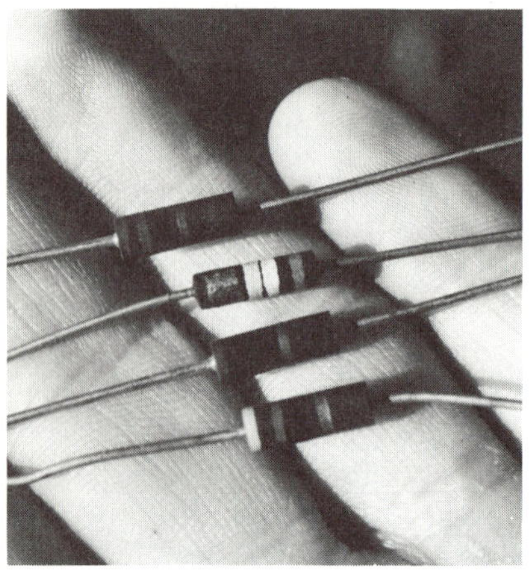

Fig. 20–7 Several resistors. The bands on the resistors are in different colors. Through a code they tell the resistance of the resistor.

statement is equivalent to saying that a ball will roll faster down a steep hill than down a gentle one. Charge moving faster means that more current is flowing. We normally talk about the ability of wire to limit, or resist, the flow of current when a potential difference is applied. The resistance is the measure of this ability. Resistance is defined by

$$\text{resistance} = \frac{\text{potential difference}}{\text{current}}.$$

If the resistance is large, then a small current will flow when a potential difference is applied. If the resistance is low, then a large current will flow when the same potential difference is applied. A wire used for the purpose of providing resistance is called a resistor (Fig. 20–7).

The resistance of a wire depends on the material out of which the wire is made. For example, the more free electrons there are in the material, the greater the current will be for a given potential difference. And a greater current means a lower resistance. The ability of a material to resist the current flow also depends on how the atoms are arranged in the material, since it is collisions with the atoms that slow down the flow (like a commuter trying to get

Fig. 20–8 The rush-hour crowd at Grand Central Terminal in New York City as viewed by the rushing commuter.

through the rush-hour crowd at Grand Central Terminal; Fig. 20–8). For any given material, we define a quantity called the resistivity that is a measure of the ability of that material to resist the flow.

If we want good current flow, we pick a material with low resistivity. Of the metals with low resistivity, copper is the most plentiful. This has led to a situation where there is more copper under the streets of New York City (in wires) than in any copper mine in the world (Fig. 20–9). Besides the resistivity of the material, the

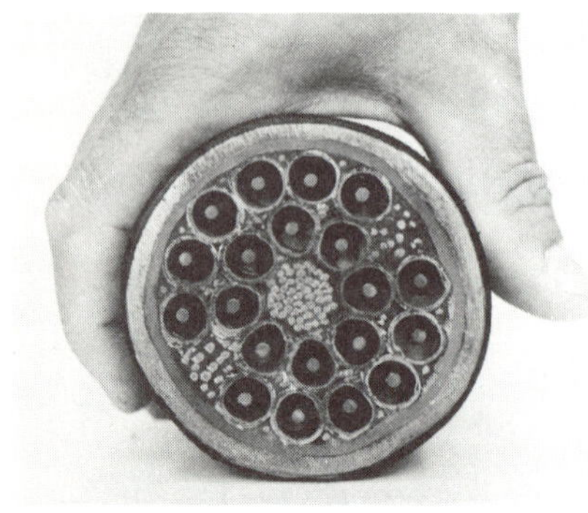

Fig. 20–9 These copper wires in a telephone cable can carry 18,000 telephone calls simultaneously.

Fig. 20–10 The effect of using a thicker wire on resistance. Notice that any section of the thicker wire (right) will have more electrons to cross a surface in a given time. This means that more current will flow for the same electric field applied to the same wires. The thicker wire will have more current, so it must have a lower resistance.

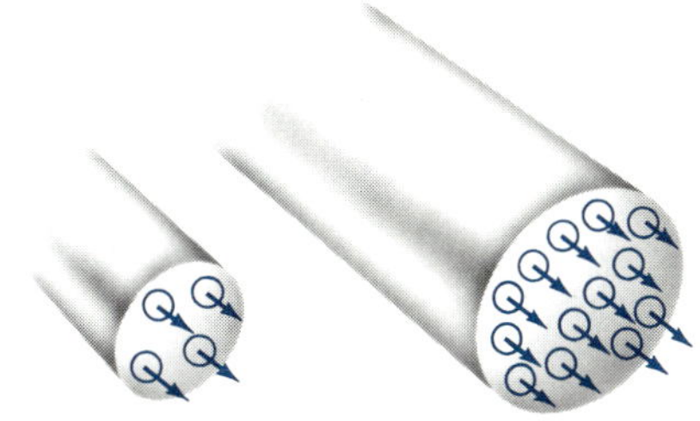

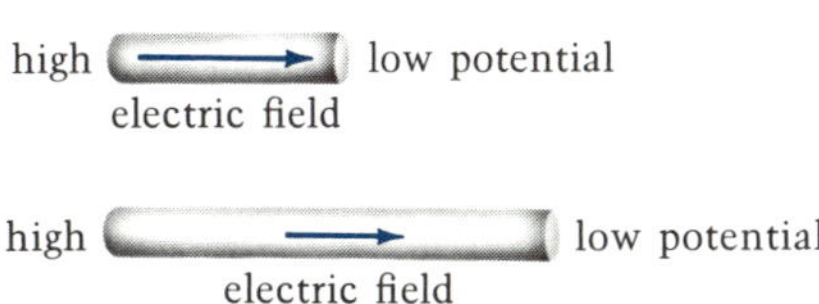

Fig. 20–11 The effect of wire length on its resistance. In the two pieces of wire above, we've applied the same voltage. There will then be an electric field pointing from the high potential to the low potential. However, when the ends of the wire are farther apart, the electric field will be weaker. Less current will flow, so we say that a longer wire has a greater resistance.

The unit of the resistance is the ohm—1 volt produces a current of 1 ampere through a resistance of 1 ohm.

resistance of a wire will depend on the size of the wire. If we use a nice fat piece of wire, then there will be more room for the current to flow, and a greater current flow, just as you can get more cars down a four-lane highway than a one-lane highway (Fig. 20–10). The resistance will also be greater if the wire is longer (Fig. 20–11).

When we measure the resistance of wires of various substances, we find that the resistance for many materials does not depend on the potential difference between the ends of the wire. For example, if we double the potential difference, we double the current, so the resistance (the ratio of the potential difference to the current) remains the same (Fig. 20–12).

Materials for which the resistance does not depend on what voltage we apply or how much current is flowing are said to obey Ohm's law, after the German physicist Georg S. Ohm (1787–1854). We sometimes refer to resistors that obey Ohm's law as ohmic resistors. There are certain materials for which Ohm's law does not hold. These materials are called nonohmic resistors. You will sometimes see the expression $V = IR$ (where V is the voltage, I is the current, and R is the resistance) referred to as Ohm's law. Actually this expression is simply the definition of resistance. Ohm's law tells us that we can use this expression with a constant value of R.

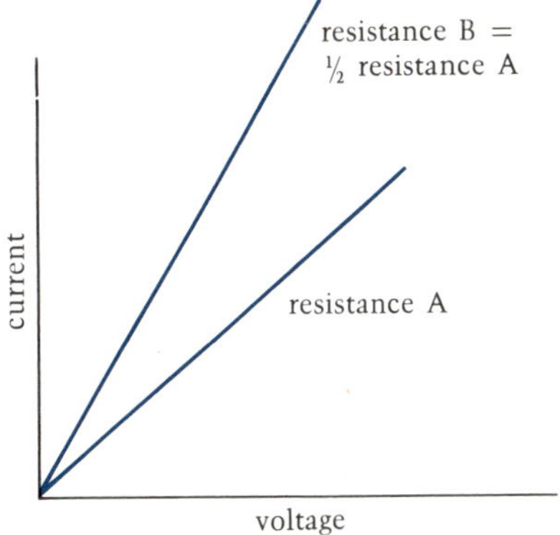

Fig. 20–12 A graph of the current at various voltages, for two different resistors. Resistor B has half the resistance of resistor A. This means that at any voltage, twice as much current will flow through resistor B as will flow through resistor A. In both cases, since the graph of current at various voltages is a straight line, we call the resistors *ohmic*.

20.2 *Circuits*

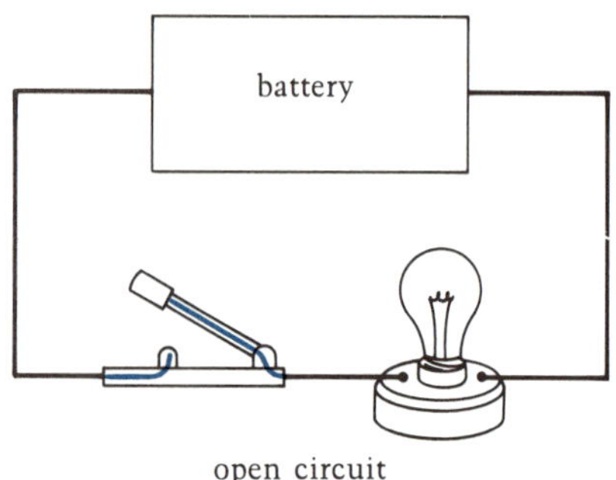

open circuit

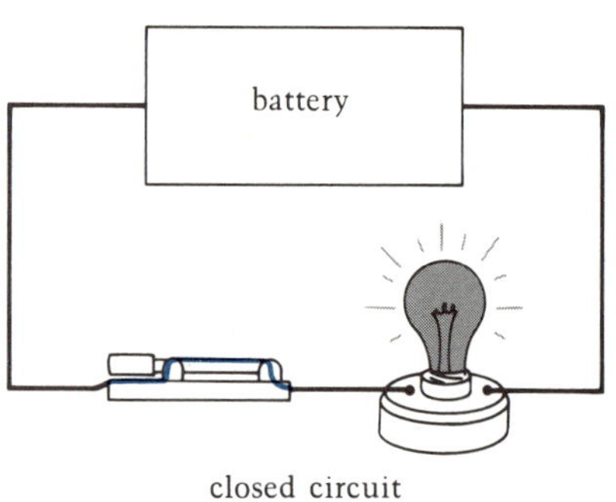

closed circuit

Fig. 20–13 An open circuit and a closed circuit. In an open circuit, there is no complete loop for the current to flow around. No current will flow. In a closed circuit, there is a complete loop and the current will flow through that loop. That is why the light bulb is on in the closed circuit and off in the open circuit. We normally use a switch to control whether a circuit is open or closed.

If we want current to flow through a wire, we must provide a potential difference between the ends of the wire. Any device that provides a potential difference to allow current to flow is called a source of electromotive force (emf). The first such device was the battery, invented in 1800 by Volta. In a battery, the energy from chemical reactions is used to provide the potential difference. Other sources of emf may be the generators of your local electric company, or, increasingly, devices that produce a potential difference when exposed to sunlight.

When we connect a wire so that one end touches one terminal of the battery (or any other source of emf) and the other end of the wire touches the other terminal, then we say that we have an electrical circuit. We can break our wire at various places and put a variety of electrical devices in the circuit. As long as there is a path that charge can follow to get from one terminal to the other, then we have a closed circuit. If the circuit is set up so that there is no such path, then we have an open circuit and no current will flow (Fig. 20–13). Often we will build a closed circuit and then cut the wire in a convenient place to insert a switch. The switch can convert the circuit from an open circuit to a closed one or from closed to open. When you turn on your light switch, you are closing the circuit that starts at the electric company generator, goes through the light, and returns to the generator.

Since an ordinary light bulb has resistance, it is used as a resistor in these examples.

DC AND AC CIRCUITS

When a battery is the only source of emf in a circuit, then the current will flow steadily in one direction. We call such a circuit a direct current (DC) circuit. However, when the source of emf is a generator, such as that used by the electric company, the potential difference between the main wires in your house reverses direction at brief, regular intervals. First one wire is at the higher potential and then the other wire is at the higher potential, and then the process repeats over and over (60 times per second in the United States, Canada, and Mexico). As a result, the current sometimes flows in one direction and sometimes flows in the opposite direction. Such a circuit is called an alternating current (AC) circuit. Our discussions in this chapter apply mainly to DC circuits. However, many of the properties of DC circuits can be applied to AC circuits, and we are even able to use your house wiring as an example in this chapter. AC circuits have additional features that depend on the magnetic effects of changing currents, which we will discuss further in Chapter 22.

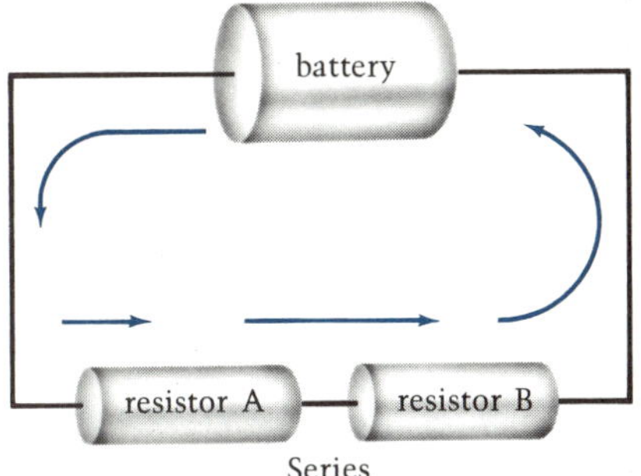

Series

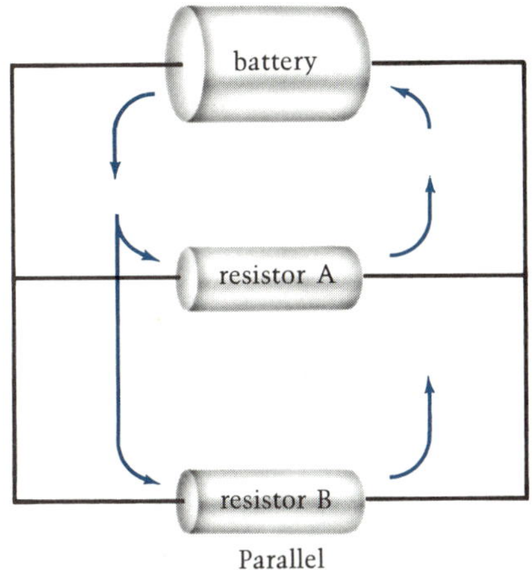

Parallel

Fig. 20–14 Resistors in series and parallel. When the two resistors are connected in series, as in the diagram on the top, then any current flowing through resistor A must also flow through resistor B. The sum of the voltage drop across A and that across B must equal the voltage supplied by the battery. When the resistors are in parallel, as in the figure on the bottom, the current has a choice of going through either resistor, and more current can flow than when the resistors are in series, or than when you have only one resistor or the other. Since both resistors are connected directly across the terminals of the battery in this figure, the voltage drop across A must be the same as that across B. This is true even in more complicated circuits where the two resistors are not directly attached to the battery. The important thing is that for two resistors in parallel, the ends of the two resistors are directly connected.

Since we can put many resistors in a circuit, we would like to see what happens when we put two resistors together. There are two possible ways in which the two resistors can be connected: (1) We can connect them so that the current will first flow through one resistor and then through the other. When we do this, we say that the resistors are in series (Fig. 20–14). (2) We can connect the resistors so that the current has a choice of going through one resistor or the other. When we do this, we say that the resistors are in parallel.

Let's look at the series hookup first. Since the current must pass through each of the resistors in its trip around the circuit, it will lose energy in each of them. Since all the current must pass through the retarding effects of each resistor, the two resistors together have a greater resistance than either of the resistors alone. The circuit behaves as though there were a single resistor whose resistance is equal to the sum of the resistance of the two that are in series. *The total resistance of two resistors in series is the sum of the individual resistances.*

In the parallel case, the current that is flowing has a choice of routes. It will thus be easier for this current to get through the two resistors in parallel than if only one of the two resistors had been there. *The total resistance provided by two resistors in parallel is less than the resistance of either of the two resistors.* The following analogy illustrates a similar situation to the parallel case. You are on a crowded freeway during the rush hour, with three lanes of traffic. All of a sudden, a one-lane service road is opened up next to the highway. (The service road is in parallel with the main road.) Now, some of the traffic gets off the main road and onto the service road, thereby allowing all the cars to get through more easily. It doesn't matter how bad the quality of the service road is. The same thing occurs when we put two resistors in parallel. No matter how high the resistance of the second one, it still provides an alternative path for some of the charge (see Fig. 20–15).

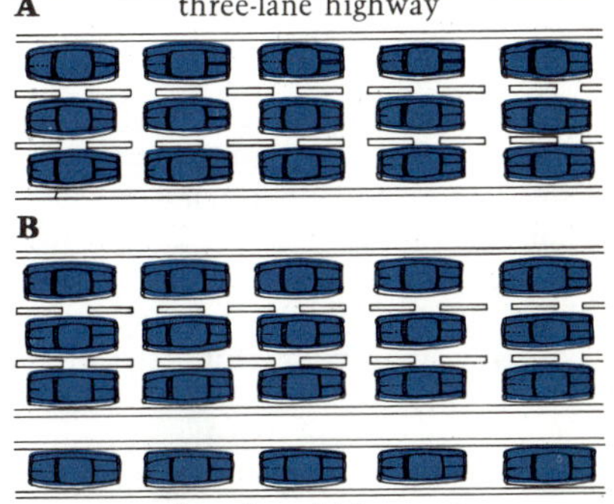

Fig. 20–15 We can relate the flow of current to the flow of cars on a highway. In A we see the cars moving along a three-lane highway. If we add a service road, as in B, even if it is only one lane, then more cars can flow past any point in a given time. In the same way, adding a wire in parallel to another, even if the added wire has a very high resistance, decreases the resistance of the whole combination.

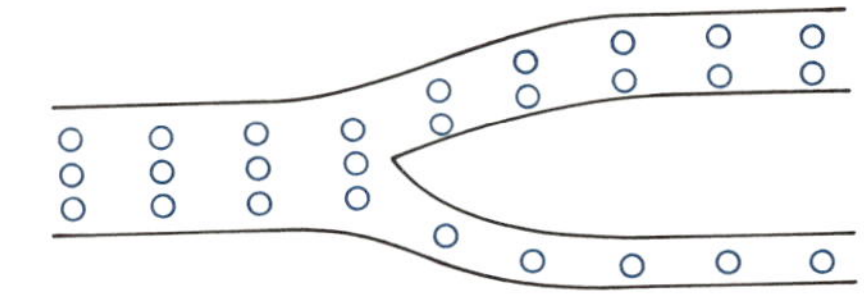

Fig. 20–16 When a wire divides, more current will flow through the wire with the lower resistance.

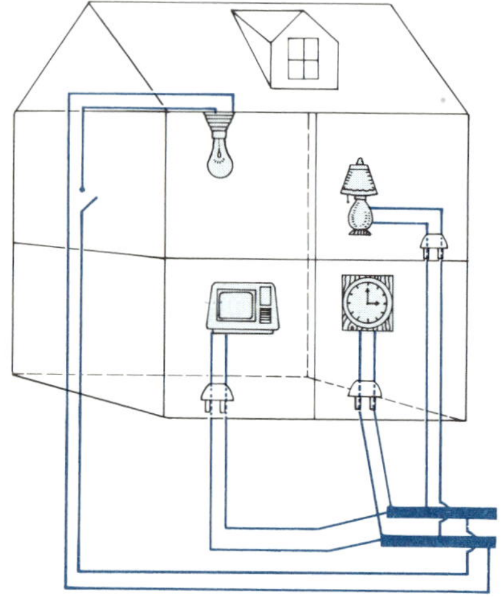

It is not hard to calculate how much of the current flows through each of the parallel branches of the circuit. To do so we make use of the fact that the potential difference between any two points does not depend on the path taken between the two points. If we choose as our two points the places where the ends of the resistors meet, we see that the potential difference across the two resistors must be the same (Fig. 20–16). More current flows through the path of lower resistance and the current divides so that

$$\frac{\textit{current in resistor 1}}{\textit{current in resistor 2}} = \frac{\textit{resistance of resistor 2}}{\textit{resistance of resistor 1}}.$$

The wiring in your home is a good example of things connected in parallel. Two wires run through the wall with the maximum potential difference maintained at some fixed value by the electric company, usually 120 V in the United States, Canada, and Mexico

(above) Fig. 20–17 Supplying electricity to a house. The electric company tries to keep a fixed maximum voltage difference between the two wires running through the walls of your house. You have access to those two wires via the wall outlet. All electrical appliances are connected across the two wires, so they each operate on the same voltage. The appliances are attached in parallel, so the current passing through a particular appliance passes through no other. Also, the appliances with the lower resistance will "draw" more current.

Rather than running one long pair of wires through your house, each room may be on a different circuit. This means that the wires in the wall of any room are connected to the two main lines and each circuit in the house is in parallel with the other circuits in the house. Current can flow through any or all of the circuits. The lines coming into the house have to be large since they must carry the current demanded by all the individual room circuits.

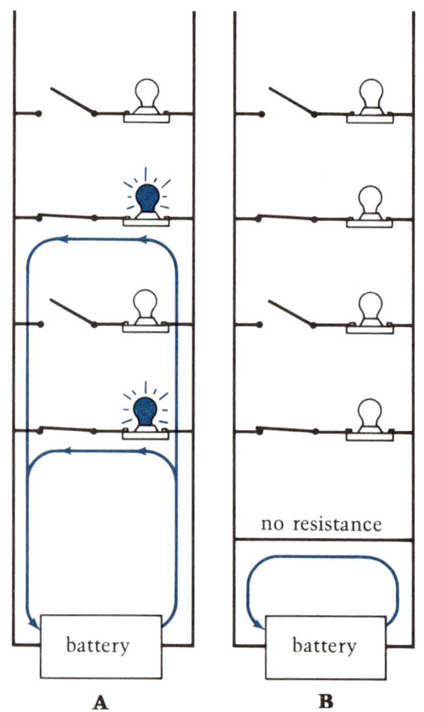

(at right) Fig. 20–18 The effect of a short circuit. A. Here we have a normal circuit. The light bulbs are all connected in parallel. Current will flow through any light bulb that has a closed switch. No current will flow through any light bulb that has an open switch. B. This is a short circuit. In this case there is a wire of zero resistance (actually very low resistance) placed in parallel with the light bulbs. Since this wire has much less resistance than the light bulbs, virtually all of the current passes through this wire and none through the light bulbs. Also, since the resistance is so low, a lot of current will flow through the short circuit. Note that the shorting wire can be placed anywhere, not just close to the battery. For example, it could be placed in parallel between the third and fourth light bulbs and the result would be the same.

(and 210–230 V elsewhere). Your outlets are set up so that one wire from the outlet is connected to one of the main wires and the other wire is connected to the other main wire. When you plug something into the outlet, you are connecting it from one main wire to the other. All the things that you plug in are in parallel, because each goes independently from one main wire to the other. Every time you plug something in and turn it on, it is equivalent to adding another resistance in parallel. In order to accommodate the demand for current through this new resistor, more current must flow into the two main wires (Fig. 20–17).

Another common example of parallel circuits is the so-called short circuit. Suppose that we have an appliance plugged into the outlet, as shown in Fig. 20–18, and the two wires come into contact. Current can now flow from one wire to the other without flowing through the appliance. This accidental contact provides a shortcut (a "short circuit") for the current to return to the source of emf. Also, since the appliance may have a large resistance and the resistance of the short circuit is very small, almost all the current flows through the short circuit. Essentially, a short circuit connects one main power line in your house directly to the other. Plugging in an appliance with faulty wiring is thus the same as connecting the two main lines with a resistor of almost zero resistance. A very large current will flow in that case.

20.3 *Energy in Circuits*

Since current flows through a wire from a point of high potential to one of low potential, the potential energy of each charge decreases as it moves. The principle of energy conservation tells us that this lost potential energy should show up as kinetic energy. However, the gain of kinetic energy is very short-lived, since the electrons are forever bumping into the atoms in the conductor. Any kinetic energy the electrons gain between collisions is quickly imparted to the atoms. As a result of the energy absorbed from the electrons, the conductor heats up. This is called joule heating (Fig. 20–19).

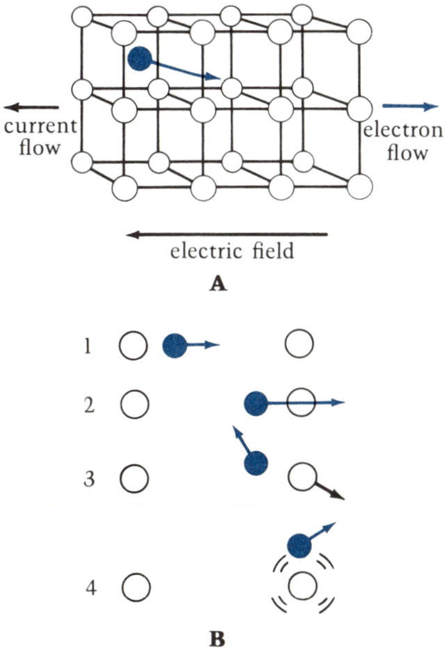

Fig. 20–19 Joule heating. A. The open circles represent the positive nuclei of atoms arranged in the regular pattern in a solid. The filled circle represents a free electron moving through this pattern. B. Here we follow this electron for part of its trip. In B1 it starts with a small velocity to the right. Under the influence of the electric field, that velocity increases. However, it soon strikes a stationary atom (B2) and bounces back (B3). Eventually the electron starts moving to the right again under the influence of the electric field. The atom that was struck does not go too far, since it is held in place by its neighbors. However, the atom will vibrate back and forth (B4). This vibration shows up as heat in the solid. In this way the electrons lose some of the energy they gained from the electric field as heat in the conductor. This is the reason for joule heating.

Fig. 20–20 An electric toaster uses joule heating.

Capturing this waste heat and using it is proving to be a major source of energy saving for businesses.

Sometimes the effects of joule heating are quite useful. For example, electric heaters and toasters work by passing a large current through several wires (Fig. 20–20). As the wires get very hot, they begin to radiate energy. The same thing happens in an incandescent light bulb. A current passes through a small wire that gets hot enough to give off a white light (Fig. 20–21).

Sometimes the heat can have unwanted effects. Often large electrical machinery must be cooled with air conditioning, or by water passed through tubes in the machinery, to remove the heat produced by the currents flowing through the wires. Computers, for example, can generate tremendous amounts of heat, and it is very important to have a computer well ventilated and probably air-conditioned. It seems paradoxical that first you spend a certain amount of money on electricity to make your machinery go and then you have to spend an equal or even greater amount of money on air conditioning to get rid of the heat caused by your original purchase of electricity. It's almost like paying for the same energy twice.

If the wires in the walls of a house get too hot, then the walls might catch fire. The wires will get too hot when they carry too much current. To prevent excessive current, we can use a fuse, which takes advantage of joule heating. The fuse is a very short wire through which all the current in the circuit must pass. When the current gets too large, the fuse wire, made of a material with a low melting temperature, heats up so much that it melts. Since there is then a break in the circuit, current no longer flows and the wall wires never reach the overheated point. When a fuse blows, it means that you are trying to draw too much current through that particular wire. The situation is not remedied by simply replacing

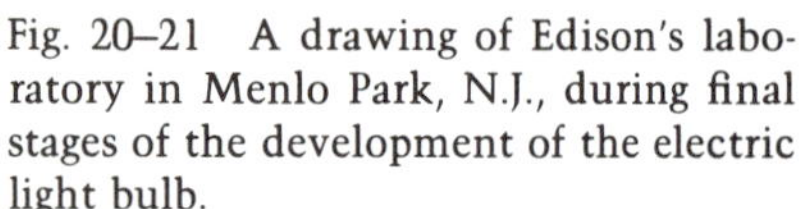

Fig. 20–21 A drawing of Edison's laboratory in Menlo Park, N.J., during final stages of the development of the electric light bulb.

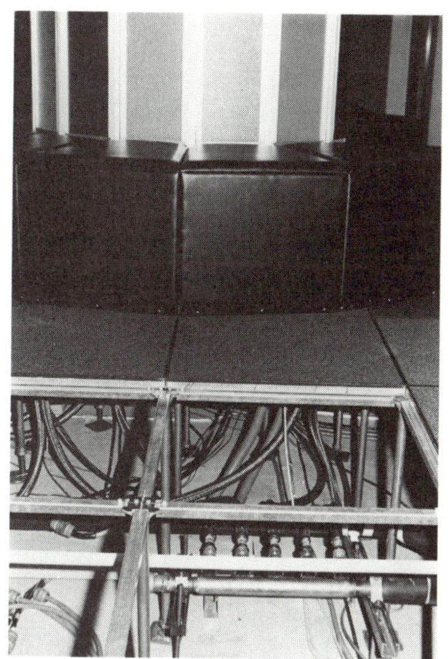

Fig. 20–22 An extensive system to cool the Cray-1 computer enables it to be made very compact. Because signals cannot travel faster than the speed of light, its compact size allows computations to be made very quickly. An overall view of the computer appears in Fig. 1–9.

the fuse. You must also reduce the amount of current that you are asking that wire to carry by turning off certain appliances, moving them to other circuits, or fixing a short circuit. (See Fig. 20–23.)

One disadvantage of fuses is that once they burn out, they must be replaced. For this reason, most newer houses have circuit breakers instead of fuses. Circuit breakers are switches in which the magnetic effect of a large current forces the switch to be thrown. (We'll see more about how this procedure works in the next chapter.) With a circuit breaker, once you have restored the circuit to a safe operating level, all you have to do to restore the current is to throw the switch.

The energy losses in a circuit must be made up somehow if the current is to continue to flow. This is where the source of emf comes in. In maintaining a constant potential difference, the source of emf must do work to keep the charges moving. As a result, flashlight batteries eventually run down. Similarly, the utility

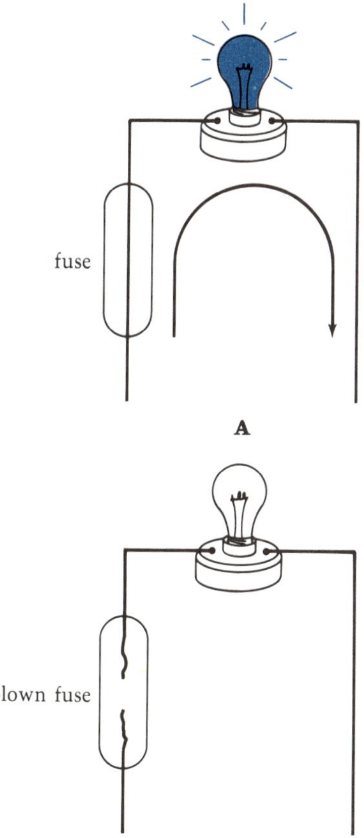

Fig. 20–23 A fuse can protect against a circuit's drawing too much current. The fuse is placed in series with the circuit that is to be protected, as in A. Should the current in the circuit reach an unsafe level, the fuse will overheat and blow, as in B. In a blown fuse, there is no longer a wire connecting one end to the other and no current will flow. Fuses are rated in terms of the maximum current that they will allow to flow before the fuses blow.

company has to do work to maintain the potential difference between the wires in your walls.

Let's follow the current around a simple circuit, one that has only a battery and a resistor. We start at the high-potential side of the battery. The current then flows to the high-potential side of the resistor and loses energy as it passes through the resistor, emerging at the low-potential side. It next arrives at the low-potential side of the battery. Then, as the current passes through the battery, it returns from the low-potential side to the high-potential side, which means that it gets back the energy it lost in passing through the resistor. It is now back where it started, ready for another go-round (Fig. 20–24).

We can think of a mechanical analogy to this electrical circuit. Start with a block at the top of a tilted piece of wood. The block starts sliding down but the friction is so great that the block slides at a constant rate. As the block slides down, the gravitational potential energy is converted into heat, because of the friction. The block then arrives at the bottom of the ramp. If we want the block to get

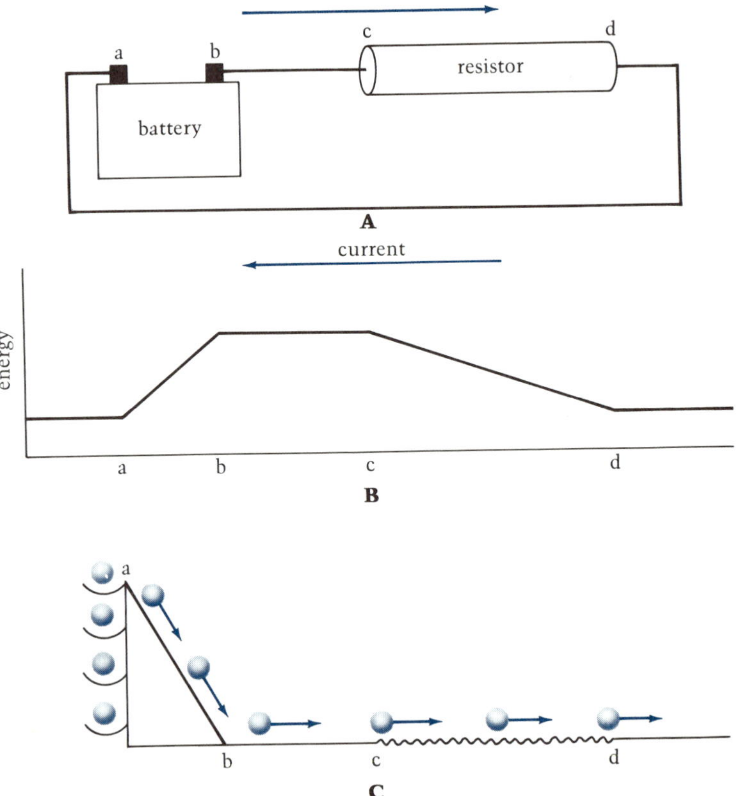

Fig. 20–24 Energy changes in a circuit. In A, we show a simple circuit with a battery and a resistor. We look at the energy at four different points, a, b, c, and d. The energy changes are graphed in B. The energy starts low at point a, but is increased as we go through the battery. There is no change in energy going from b to c (if we assume that the connecting wires have no resistance). In going through the resistor from c to d the energy falls again, until we get back to a. C shows a mechanical analogy to this. In this case we will look at the kinetic energy of some balls. After being raised up by a lift, they start at a, and roll down the hill to b, where they now have a high speed and a high kinetic energy. There is no change in energy from b to c, if there is no friction, but the region between c and d has friction, so the energy decreases as the ball goes across. It finally arrives at d with no kinetic energy, and must be taken back up the lift at a. In this analogy, the ride up the lift and down the hill is like going through the battery (getting a new boost), and the part of the track with friction is like a resistor.

to the top of the ramp to start again, we must lift it to the top. The work to do this lifting is equivalent to the work done by the battery in raising the potential energy of the charge as it passes through the battery.

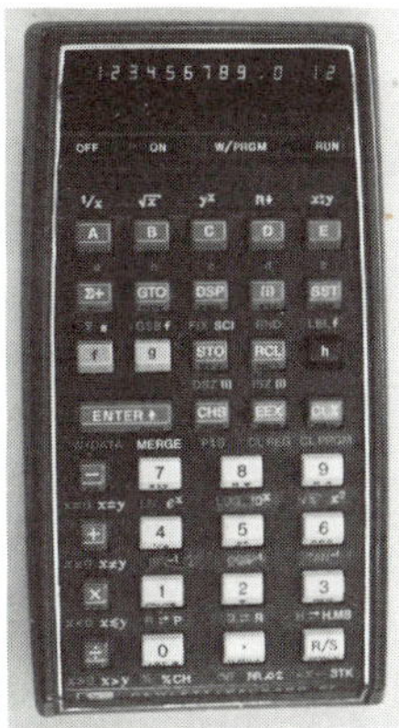

Fig. 20–25 An electronic calculator. Besides doing the indicated functions, this one can also be programmed.

20.4 *Microcircuits*

Electronic calculators (Fig. 20–25) that add, subtract, multiply, and divide cost less than students paid ten years ago for mechanical devices for multiplying and dividing. The past twenty-five years have seen a real revolution in making electronic circuits smaller and smaller, and also in reducing their cost greatly. At the heart of even a fairly sophisticated calculator is a tiny "chip"—only a few millimeters on a side—that contains all of the circuitry to interpret the commands punched in from the various buttons, perform all the required calculations, and signal the display light to light up in the pattern that you read. (We are amazed that a calculator can fit in our pockets. It's even more amazing to realize that most of the size is to accommodate the buttons, digital display, and battery. The actual "works" of the calculator are much much smaller than the pocket calculator itself.)

Fig. 20–26 Before the transistor was invented, large vacuum tubes, like this one, were quite common.

To give you an idea of how far we have come, we can compare our new tiny "microcomputers" with the first large electronic computer, the ENIAC, which was built in the 1940s (Fig. 20–27). ENIAC cost a few million dollars, while the microcomputer costs a few hundred. ENIAC occupied 30,000 times the space, but was 20 times slower. ENIAC also had less storage capacity and took a fan-

Fig. 20–27 ENIAC, built in the 1940s, was huge. (Smithsonian Institution photo no. T53192)

tastic amount of power to keep it going. Also, there was only one ENIAC, while you can buy as many microcomputer chips as you need for a given project.

Fig. 20–28 A transistor (center) compared in size to a wood screw and a vacuum tube.

20.4a INTEGRATED CIRCUITS

The first step in the direction of miniaturization was the development of the transistor (which we'll discuss in Chapter 25). However, even the transistors that we might normally buy and wire together are still a few millimeters in any dimension (Fig. 20–28). Therefore, another stage of shrinking beyond the normal transistor was necessary. This shrinking was accomplished by stopping use of individual components—such as transistors and resistors—that were then connected by wires. The new idea involved building the whole circuit into a single tiny board so that all the circuit elements, including the connections between the elements, were part of the board.

These single-unit circuits are called integrated circuits (ICs). The "backbone" of these circuits is a wafer of very pure silicon, to which some impurities are added to get the desired semiconducting effect. Then, by using special techniques to cut into the surface, to add different impurities to different regions, to put insulating layers where necessary, and, finally, to add conducting metals connecting all the right places, integrated circuits that serve a multitude of purposes are made. A *mask* that covers the regions where a desired material stays on the wafer is an important part of the process. Once the mask covers these parts of the wafer, acid removes the covering of the other parts.

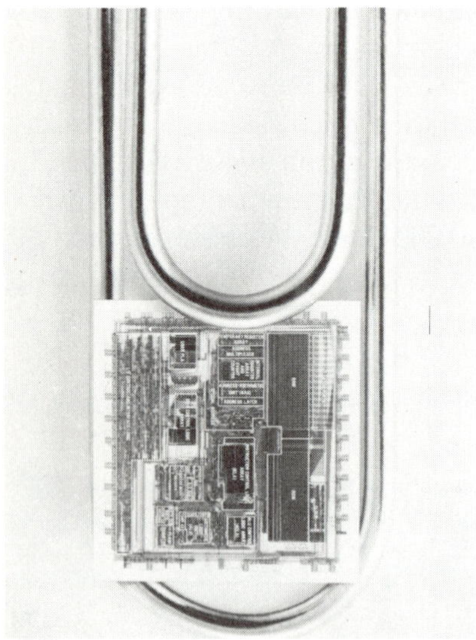

Fig. 20–29 A one-chip computer containing thousands of transistors, compared in size to an ordinary paper clip.

In practice, many duplicates of the same circuit will be made on the same chip, which allows for a form of mass production. After the processing is finished, the last step is to cut the wafer into tiny chips, each one identical to the others. Each has a large number of circuit elements on it. Finally, each chip is placed in a package, which can easily be plugged into place in the device for which the chip is made.

With the complexity that has developed in integrated circuits—by now we can make a whole computer on one chip—even the production of the masks is quite a job. The circuit design is done mostly by computer. In the days before integrated circuits, an engineer wishing to test a circuit design would collect the necessary components and wire them together in a crude fashion to see if they behaved as predicted. This is no longer possible because of the complexity, and also because some of the particular elements in the circuit are designed in such a way that it is not practical to produce them as single elements that can be wired together. Now the response of a proposed circuit to various inputs is tested by computer.

One of the most amazing things about integrated circuit technology is that it keeps on improving. There is no sign yet that the rate of advancement is slowing down. Circuits keep getting more complex and the chips keep getting smaller and cheaper. At this point it seems like a major part of the challenge is to put this new technology to efficient use.

20.4b THE POCKET CALCULATOR

Fig. 20–30 The digits in this calculator display are made up of various combinations of short lines. There are seven different lines, any of which can be turned on for any particular digit.

Even though the techniques of integrated circuits provide us with the capability of concentrating a fantastic number of circuits in a very small area, there has to be more to the pocket calculator than the chip. Actually, given the miniaturization provided by the chip in a calculator, we are hard-pressed to match this small size in the other functions. In a calculator, we must have a way of entering commands and numbers. This is done via a keyboard, but the buttons of the keyboard must be connected to the "heart" of the calculator. We must have a way of interpreting the various commands and then performing the required operations, including storing numbers for later use. Finally, the result of a calculation must be displayed, usually through either an array of red glowing light-emitting diodes or through liquid-crystal devices that reflect light shined on certain parts of their surfaces.

Despite appearances, a calculator is very active whenever it is turned on, even when it is not calculating. An oscillating crystal inside acts as a clock, sending out pulses 250,000 times per second. All the calculator operations are synchronized to this signal. This synchronization means that the calculator does not have to be looking at all of the buttons at the same time. Instead, it looks at the buttons in sequence, to see which ones are depressed. The buttons are scanned so rapidly that, no matter how short a time your finger spends on a given button, the scanning will catch it. Calculations are also synchronized to the internal clock. Although it may take over 10,000 clock cycles to do a simple addition, this is still only a few hundredths of a second!

The lights showing the answer also operate in a sequential fashion. In a typical calculator, each digit is actually made up of seven straight segments. All the numbers are made of some combination of these segments. To avoid the expense of having a separate wire from each segment to the chip, the individual segments are also turned on and off one after another. Each segment may be on less than 10 percent of the time but the segments flicker so fast that the eye cannot detect the flicker.

20.5 *Safety and Electricity*

In any situation in which you are dealing with electricity, there is always some risk of injury from electric shock. You can "get a shock." Each year, too many people are seriously injured because they did not take the proper precautions in dealing with electricity. Also, electrical accidents often turn out to be worse than they should because people don't know how to give proper assistance to a victim. Most accidents involving electricity can be avoided with a little knowledge of how electricity flows and what its effects on the body are, in addition to using common sense in recognizing dangerous situations.

Fig. 20–31 A reading lamp near a bathtub can be very dangerous because water with your body salts dissolved in it is very conductive and the plumbing is a good ground.

Injury in the event of electric shock comes in two forms. Burning caused by joule heating is one problem. Also, parts of the nervous system can be disrupted, since nerves work by transmitting weak electrical signals. Both of these types of injury occur in response to a current flowing through the body at a strength that is too large for the body to handle. This explains why shock from an electrostatic generator of several hundred thousand volts might sting for only a second or make your hair stand on end, while touching a faulty lamp and getting a 120-volt shock from the wiring can be much worse. The electrostatic generator has very little charge to flow, so the current never gets very large. For the wiring in your house, however, the electric company keeps the current coming. The current you can draw through your body from home wiring is limited only by the fuses in the circuits. These fuses normally limit the current to 15 amperes, but a current of 0.1 amperes or even less (especially if it flows through the heart) can be fatal.

For current to pass from one part of your body to another, there has to be a potential difference between these points. As long as there is no potential difference, you are safe. A bird can sit on a high-voltage wire because it touches only one wire at a time. The potential change in a few millimeters of wire is small. However, if you touch something else at a different potential while you are holding onto the high-voltage wire, then there is a potential difference and current will flow.

Most shocks result when you touch something at a high potential with one part of your body while another part of your body is

In laboratories with high-voltage equipment, you will often find scientists walking around with one hand in their pockets. They do this so they don't inadvertently grab hold of something grounded with one hand while the other hand is touching the high voltage.

grounded. The current then flows from the high potential through you to ground. The grounding usually occurs when you touch a pipe or radiator, since the plumbing in a building is usually grounded. The "ground" provides a path of virtually no resistance down to the ground.

If you do get a potential difference across your body, then current will flow. The amount of current depends on the voltage applied and the resistance of your body. Most of the resistance is in getting through your skin into your body. If your skin is immersed in salt water, the resistance becomes very low because the ions (atoms that have lost an electron) dissolved in the water make the water a good conductor. Even distilled water, which is normally a bad conductor, becomes a good conductor as your body salts are dissolved in it. Never touch anything electric when you are wet, such as when you are in the bath.

Once inside, the resistance is only a few hundred ohms from one side of the body to the other. However, if your skin is very dry, that will add a few hundred thousand ohms to the resistance.

If you come across the victim of an electrical accident, it is possible to give some help. However, the first thing you should do is to make sure that you don't become the next victim. If the other person is still touching a "hot" wire and you touch the other person, then you will receive the same shock. So first try to remove the reason for the electrical shock: perhaps pull the plug. Then it is important to give the victim artificial respiration, since many victims of electrical shock stop breathing because the electricity affects the nerves that control breathing. Don't give up if the victim doesn't come around right away. There have been many cases of victims of lightning strikes who, although they did not breathe on their own for several hours, were eventually revived and completely regained their good health.

Key Words

drift speed
current
ampere
resistance
resistor
resistivity
Ohm's law
ohmic resistors
nonohmic resistors
electromotive force
battery
electrical circuit
closed circuit
open circuit
switch
direct current circuit
alternating current circuit
series
parallel resistors
short circuit
ammeter
voltmeter
joule heating
fuse
circuit breakers
integrated circuits

Questions

1. The drift speed of electrons in a wire is very small, but when you turn on a light switch, the light goes on almost immediately. How can this be?
2. A group of free electrons flow from left to right in a wire. In which direction is the current flowing?
3. A group of protons in a plasma (ionized gas, composed of protons and electrons) flow from left to right. In which direction is the current flowing?
4. When an electric field is applied to a particular plasma, the protons move from left to right and the electrons move from right to left. Is there a current flowing in such a situation?
5. When you pick up a neutral piece of wire and carry it across the room, is a current flowing from one side of the room to the other?
6. What effect will each of the following have on the resistance of a piece of wire: (a) make the wire longer and thinner (b) make the wire shorter and wider (c) change the potential difference between the ends of the wire.
7. A resistor with a high resistance and a resistor with a low resistance are both connected across the terminals of a battery. (a) How does the potential difference across the two resistors compare? (b) How does the current in the two resistors compare?
8. What are the similarities between replacing a given resistor with one of the same length but greater width and adding another resistor in parallel with the first resistor?
9. Is a short circuit an open or closed circuit?
10. What are the similarities between replacing a given resistor with one twice as long and adding another resistor in series with the first resistor?
11. You plug a toaster and an iron into the same circuit in your house, using two different outlets. (a) Is the potential difference across the iron the same as the potential difference across the toaster? (b) Is the current through the iron the same as the current through the toaster? (Assume that the toaster and iron have different resistances.)
12. You overload a circuit and a fuse blows. You replace the fuse and then move one appliance to a different outlet on the same circuit. Does moving the appliance in this way prevent the fuse from blowing again?
13. Suppose you have a few appliances connected to the same line. Why will a short circuit in only one appliance be sufficient to make the fuse blow? (Draw a circuit diagram to illustrate your answer.)
14. List three examples in which joule heating in wires produces desirable

results. List three examples in which joule heating produces undesirable results.

15. An ammeter has a fuse in it to prevent it from passing too much current. A student uses the ammeter in a circuit and the fuse blows. The student replaces the fuse with an identical one, throws the switch, and the fuse blows again. The student then asks the laboratory instructor for a larger fuse. What has the student done wrong?
16. How does the resistance of a 100-watt light bulb compare with that of a 200-watt light bulb?
17. List three examples of equipment that makes use of microcircuits.
18. How is a microcircuit chip different from a standard circuit that performs the same functions (other than in size)?
19. Why is the danger of electrical shock greater when your skin is wet?
20. What are the two ways in which an electrical shock can cause injury?
21. Why can a bird sit with both feet on a power line and not be injured? (*Hint:* How does the resistance from one of the bird's feet, through the bird's body, to the other foot compare with the resistance in the part of the wire between the bird's feet?)

CHAPTER 21

Magnetism

Fig. 21–1 The arrow in the compass points north when it is free of external disturbances.

A huge magnet picks out the metal from the garbage at a recycling center. A small magnet picks up nails that have dropped on the floor. A compass needle points north. An aurora, the northern or southern lights, lights up the sky with beautiful pulsating arcs of color. All these phenomena are descended from the discovery by the Greeks two thousand years ago that certain stones from a region known as Magnesia attracted pieces of iron. From the name of the region, we get the name magnetism.

Magnetic effects play important roles on many different scales. For example, the magnetic field of the earth, besides its importance in navigation, serves as our shield from a stream of high-energy particles from the sun that would be lethal if they hit us. The earth's shield is not totally effective, and variations in this stream of particles, this solar wind, seem to affect our weather and climate. Where the magnetic shield allows the sun's charged particles to get closest to the earth, at the polar regions, we have the spectacular displays of auroras. The rocks of the earth are magnetized, which helps us to trace billions of years of geologic history during which the continents have drifted (continental drift) over the surface of the earth. The sun's magnetic field varies with an eleven-year cycle, during which varying numbers of sunspots pop out through the solar surface.

Much of the early activity in the study of magnetism concerned itself with the behavior of particular materials that were discovered to be magnetic. William Gilbert began the scientific study of mag-

Fig. 21–2 An aurora over New York City, with the George Washington Bridge in the foreground.

netism in the sixteenth century. We now know that all magnetism is just a result of moving charges, an idea that had its start in 1820. In that year, in Denmark, Hans Christian Oersted showed that a current flowing in a wire could influence the needle of a compass. Immediately upon hearing of Oersted's results, Ampère set up a number of experiments to test the relationship between magnetism and flowing charges. He showed that a wire with a current could attract bits of metal, and that wires with currents flowing in them could repel or attract each other, depending on the direction of the currents. Based on these results, Ampère proposed that all magnetic phenomena actually arise from currents.

Fig. 21–3 William Gilbert, physician to Queen Elizabeth I, began the scientific study of magnetism. This plate from his 1600 book on magnetism shows a blacksmith making a permanent magnet. He is beating a glowing iron bar while holding it in the north (septentrio) to south (auster) direction, thus making it magnetic in agreement with the earth's magnetic field.

Fig. 21–4 A. When the switch (at left) is open, no current flows in the coil (top). The compass aligns with the earth's magnetic field. B. When the switch is closed, current flows through the coil and a magnetic field is generated. The compass needle now points in the direction of this field, which is stronger than the earth's field.

A

B

21.1 *Magnetic Fields and Forces*

Einstein's original childhood scientific inspiration, as we mentioned, was the realization that the compass needle always sought north. No doubt he realized to some extent that there must be an invisible field present.

For an electric force between two stationary charged particles, we have seen that the force depends only on the charges of the two particles and the distance between them. However, for a magnetic force the situation is a little more complicated. The magnetic force also depends on how fast and in what direction each charged particle is moving. Because of these additional factors, it becomes even more important in discussing magnetic forces to divide the calculation into two parts. First we calculate a magnetic field, and then we calculate the force that the field exerts on a particular particle. In this section, we'll look at the force that a particle feels when it moves through a magnetic field. In the next section, we'll see where magnetic fields come from.

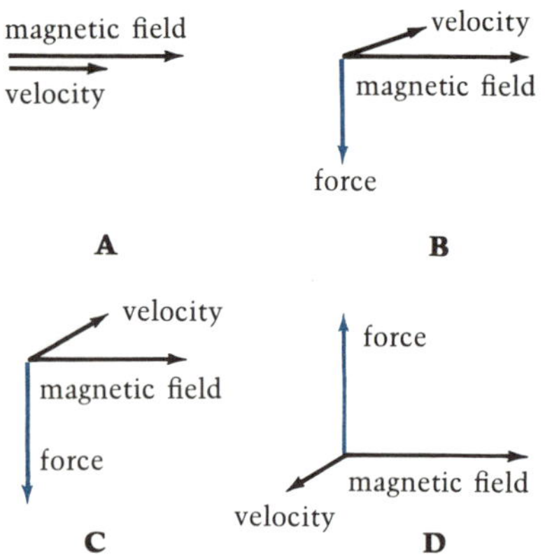

Fig. 21–5 Magnetic force on a positively charged particle. A. The direction of motion is the same as the direction of the magnetic field, so there is no force. B. The direction of motion is neither parallel to nor perpendicular to the magnetic field, and there is a force, at right angles to both. C. The direction of motion is perpendicular to the magnetic field and the force is maximum, but is still in the same direction as it was in B. D. The direction of motion is reversed from C, so the direction of the force is reversed. If the charge on the particle is negative, all forces are reversed in direction.

We can draw pictures of the magnetic field lines, in the same way that we draw pictures of electric field lines. The direction of the field line at any point is the direction of the magnetic field. The stronger the magnetic field, the closer together the field lines are drawn.

When a particle moves through a magnetic field, *the magnetic force is proportional to the strength of the field, to the charge of the particle, and to the speed of the particle.* The dependence of the magnetic force on the strength of the field and the charge of the particle are familiar from our treatment of electric fields, but the dependence of the magnetic force on the speed of the particle is a new concept. It not only depends on the speed of the particle but *the force also depends on the direction of the motion of the particle with respect to the direction of the field. When the motion of the particle is parallel to the field, the force is zero. When the motion is perpendicular to the field, the force is at its maximum. The magnetic force is perpendicular to both the direction of motion and to the magnetic field.* This is illustrated in Fig. 21–5.

Suppose a positively charged particle enters the magnetic field, and moves at a right angle with the field lines. The force on the particle (shown in Fig. 21–7) is upward, so the path of the particle is bent upward. However, now the direction of motion has changed.

MEASURING MAGNETIC FIELDS

The SI unit for magnetic field strength is the tesla, *where 1 tesla = 10^4 gauss.*

We measure magnetic fields in units called gauss. The earth's magnetic field, the field that makes compass magnets point north, is about half a gauss. An ordinary toy magnet produces about 100 gauss. The strongest steady field produced in a terrestrial laboratory is about 300,000 gauss.

Fig. 21–6 Scientists at the National Magnet Laboratory at MIT rejoice as the superconducting magnet in the foreground achieves the world's record for continuous magnetic field of over 30,000 gauss (30,000 times more powerful than the average field of the earth). The magnet is a hybrid containing both a superconducting core and a water-cooled part with ordinary conductivity. (Courtesy Robert J. Weggel, Francis Bitter National Magnet Laboratory)

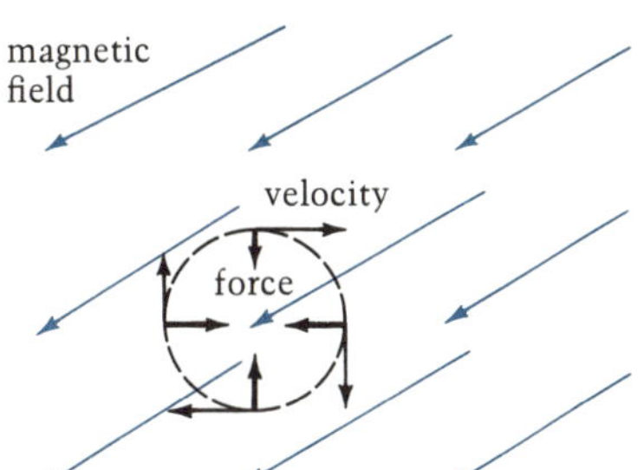

Fig. 21–7 When a particle travels perpendicular to a constant magnetic field, its path will be bent around in a circle. At any point on the circle, the force is perpendicular to the direction of motion, exactly what we need for circular motion.

The particle is still going at right angles to the field lines, but it is now going in a different direction. As a result, the direction of the force changes slightly, as shown in the figure. The process keeps repeating itself, and the particle makes a complete circle, and then goes around again.

Since the force is always perpendicular to the direction of motion of the particle, the magnetic field does no work on the particle. (Remember, work depends only on the part of the force that is directed along the direction of motion.) The particle's speed does not change at all as it goes around in the circle; it arrives at the starting point with the same speed as it started. It keeps going around in the same circle.

If the particle also had some motion parallel to the magnetic field, the force would not affect the part of the motion parallel to the field, since the force is always perpendicular to the field. As a result, the particle will still circle around the field lines, but because of its additional motion parallel to the field lines, the actual path of the particle will be a spiral (Fig. 21–8).

William Gilbert's description of the earth as a giant magnet is seen in Fig. 21–9. We now know how the earth's magnetic field

Fig. 21–8 If a particle has some motion parallel to magnetic field lines, then that motion will not change. Thus the particle will go around in circles, but will also drift along the direction of the field lines (either way, depending on the initial motion). The resulting path is a spiral.

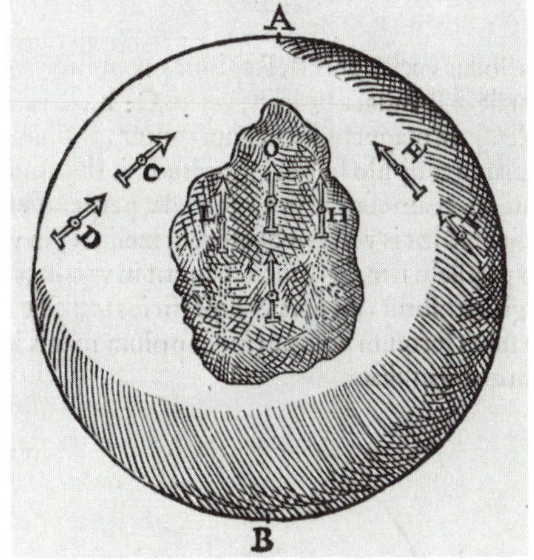

Fig. 21–9 In his 1600 book *De Magnete,* William Gilbert provided this view of the earth as a magnet.

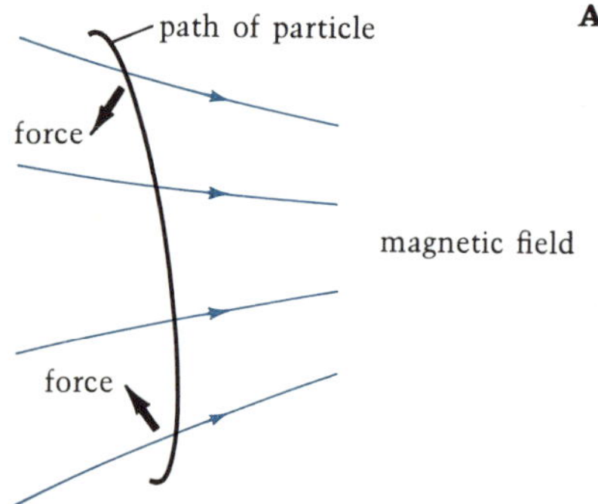

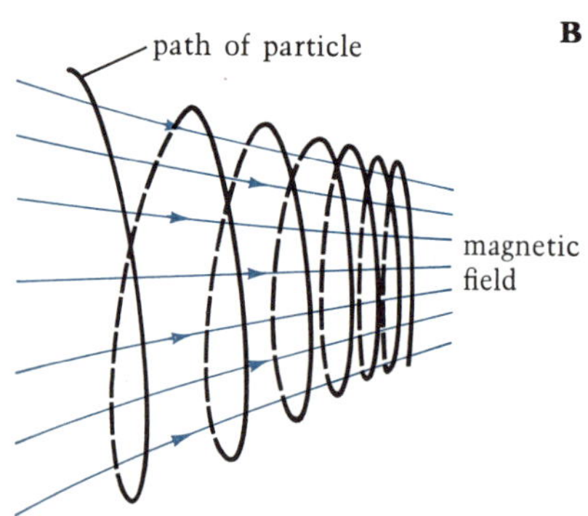

Fig. 21–10 The magnetic mirror effect. A. When a particle moves from a region of weaker magnetic field to one of stronger magnetic field, it will be moving to a region where the field lines are closer together. Since the field lines are not parallel, in addition to the force producing the circular motion the particle will experience a slightly backward force. This will slow the particle down. B. The resulting path of such a particle is shown, with the spirals getting closer together as the particle slows down. Eventually the particle will stop and move back the other way.

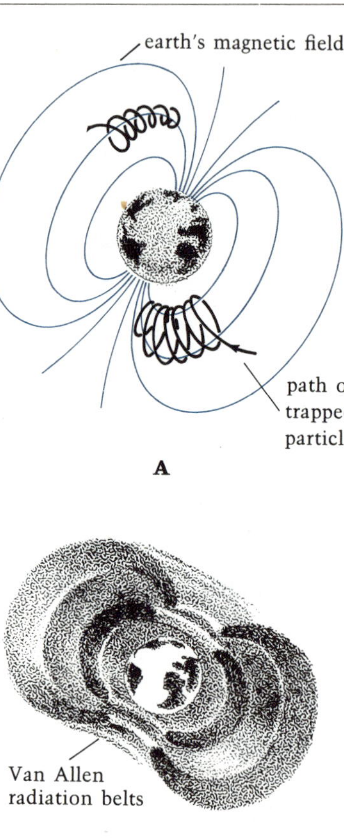

Fig. 21–11 A. The earth's magnetic field can trap charged particles that come in from outer space. These particles then spiral back and forth from magnetic pole to magnetic pole. In this way the magnetic field protects us from these charged particles. When a large number of these particles get close to the earth's surface, they cause the air to glow, producing the aurorae. These are most commonly seen near the magnetic poles, since it is at those places that the particles are closer to earth. B. The trapped particles fall mainly in two belts, known as the Van Allen radiation belts, named after James Van Allen, who discovered them, and for the fact that they can emit large amounts of radiation.

keeps charged particles trapped. The particles continually spiral around the magnetic field lines, from north pole to south pole and back again. The particles turn around because as the particles get closer to the poles, the field lines start to come together. As shown in Fig. 21–10, the increased strength of the magnetic field slows the particle down. Eventually the particle turns around. We say that we have a magnetic mirror.

The charged particles in the earth's magnetic field radiate energy. These charged particles are generally found in two regions, called the Van Allen belts. These belts of charged particles were discovered in January 1958 by the first American earth satellite, with a detector whose operation was under the direction of James A. Van Allen of Iowa State University (see Fig. 21–11).

Radiation from particles spiraling around magnetic fields also plays an important role in radio astronomy, the study of the universe with radio waves instead of light waves. In many regions of our galaxy, high-energy electrons are trapped by strong magnetic fields and spiral around the field lines. They emit radiation as they

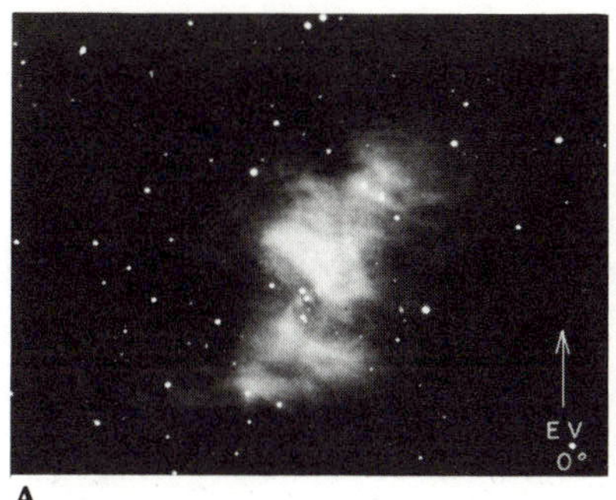

A

B

C

Fig. 21–12 The Crab Nebula in the constellation Taurus photographed in polarized light. Note the different appearance of the nebula as the polarizing filters are rotated through 90°. EV stands for "electric vector," and indicates the direction of the electric field passed by the polarizing filter.

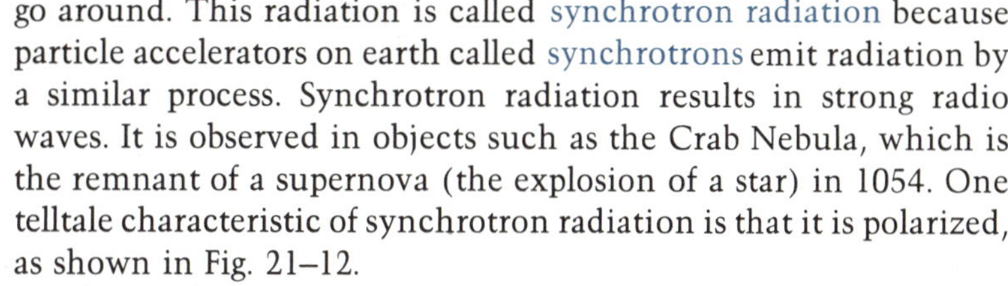

go around. This radiation is called synchrotron radiation because particle accelerators on earth called synchrotrons emit radiation by a similar process. Synchrotron radiation results in strong radio waves. It is observed in objects such as the Crab Nebula, which is the remnant of a supernova (the explosion of a star) in 1054. One telltale characteristic of synchrotron radiation is that it is polarized, as shown in Fig. 21–12.

What we learn about magnetic mirrors on earth or in space may lead to the solution of our pressing energy problem. We can, theoretically, generate energy by fusing hydrogen into helium. But to get hydrogen nuclei together, we have to overcome their tendency to repel each other. A high temperature, millions of kelvins, is necessary to overcome the repulsion. We need to control a plasma, a hot gas of charged particles. Plasmas cannot be restrained in ordinary containers because of their high temperatures. However, by shaping a magnetic field properly, one can make a "magnetic bottle" so that the particles of a plasma are contained without touching any walls. The particles merely spiral back and forth along the field lines (Fig. 21–13). Massive nuclear fusion projects are under way (Fig. 21–14) in both the U.S. and the U.S.S.R.

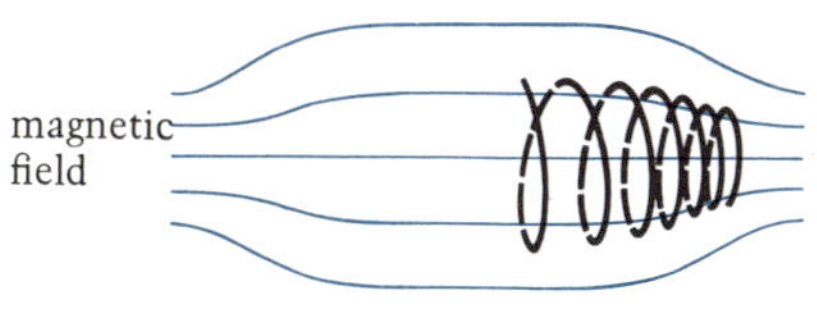

Fig. 21–13 A magnetic bottle. This is a magnetic field arrangement with a magnetic mirror at both ends. In theory, charged particles could be trapped here for long periods of time, running back and forth without ever striking the walls of a container.

Fig. 21–14 Buried inside this mass of machinery is the Princeton Large Torus, one of the major efforts to use magnetic bottles to restrain plasmas for nuclear fusion research. Toruses—doughnut shaped—are used because the plasma can flow through continuously in them without having to be reflected. Temperatures of 75 million degrees have been achieved, but not at high enough densities or for long enough times to send more energy out than it takes in. The devices are called Tokamaks, the Russian name. A 1980 law set up a $20-billion U.S. national fusion plan calling for an engineering Tokamak to be in operation by 1990, and for commercial fusion power to be demonstrated between 1995 and 2000. This represents a speeding up of the prior schedule by twenty years.

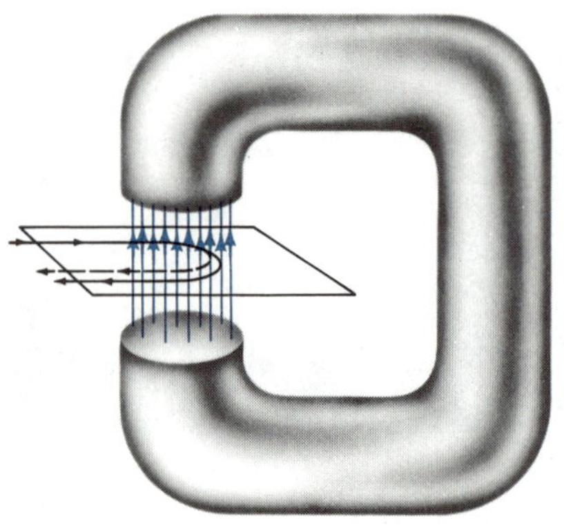

Fig. 21–15 We can use a magnetic field to separate particles of the same charge but different masses. In this case, the path of the high-mass particle is the solid line and that of the low-mass particle is the dashed line. The two particles originally go in with the same energies and along the same path.

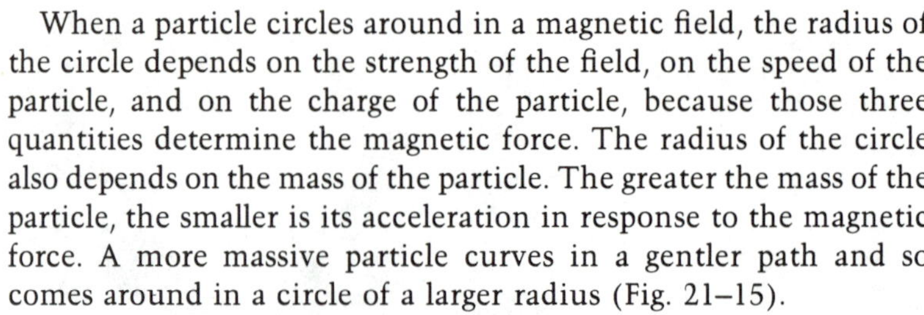

When a particle circles around in a magnetic field, the radius of the circle depends on the strength of the field, on the speed of the particle, and on the charge of the particle, because those three quantities determine the magnetic force. The radius of the circle also depends on the mass of the particle. The greater the mass of the particle, the smaller is its acceleration in response to the magnetic force. A more massive particle curves in a gentler path and so comes around in a circle of a larger radius (Fig. 21–15).

By measuring the radius of the circle in which particles spiral, we can accurately measure the mass of particles whose charge is known. An arrangement for making such measurements is called a mass spectrograph. In general, particles are first accelerated through some difference in electrical potential. They then enter the region of the magnetic field, where they go around one-half circle. The more massive particles emerge farther from the entry point, since they go around in circles of larger radius. This is one technique used to make accurate measurements of pollution levels. Mass spectrometers have landed on Mars to analyze the soil in a search for life (and did not detect any traces of organic material) and plunged into Venus's atmosphere to analyze the chemical makeup.

21.2 *Magnetic Moments*

Now that we have looked at the forces on individual charges in a magnetic field, we can look at the force on a wire that carries a current. We can think of the current simply as positive charges moving through the magnetic field. To find the force on the wires, we add up the forces on all the moving charges. Figure 21–16 shows how the force on a current-carrying wire depends on the orientation of the wire.

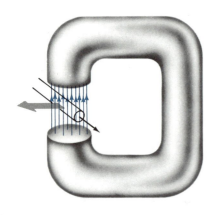

Fig. 21–16 Force on a current-carrying wire. The force is perpendicular to the wire and to the magnetic field.

An interesting case occurs when we look at the forces acting on the current flowing through a closed loop of wire. To make the situation easy to analyze, let us take the loop to be rectangular in shape. We'll start with the rectangle formed by the loop making some angle with the magnetic field, as illustrated in Fig. 21–17. The forces are in opposite directions for wires on opposite sides of the rectangle, because the currents in the opposite sides are flowing in opposite directions. Thus there is no net force on the loop. However, if the plane of the loop is not perpendicular to the magnetic field, two of the forces pull in a way that tends to rotate the loop—they exert a torque. The loop rotates until the rectangle is perpendicular to the direction of the magnetic field. The greater the current in the wires, the greater the torque on the loop. Also, the larger the loop, the greater the torque (Fig. 21–18). These features hold true even if the loop is not rectangular.

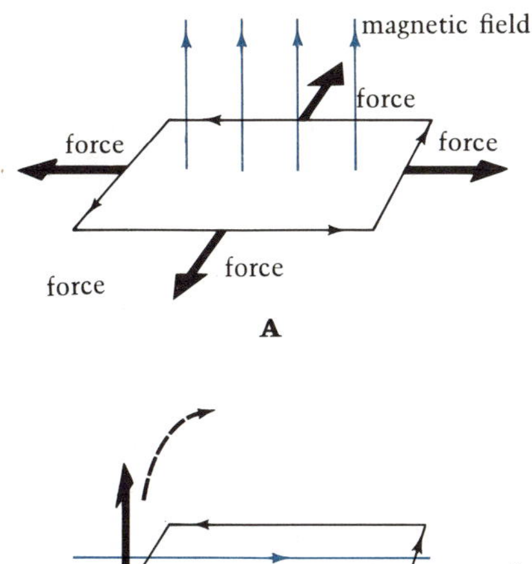

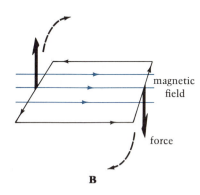

Fig. 21–17 Torque on a loop carrying current. A. The magnetic field is perpendicular to the loop. There is a force on each wire, but the forces in opposite directions cancel, and forces in opposite directions are all along the same plane so there is no tendency to rotate. B. When the magnetic field is at any angle not perpendicular (shown here parallel) to the plane of the loop, the forces on two sides cancel (in fact, when the loop is parallel to the magnetic field, these forces are zero). The forces on the other two sides are equal and opposite, but these two forces do not pull along the same line and the loop tends to rotate in the direction shown. It will stop rotating when the magnetic field is perpendicular to the plane of the loop.

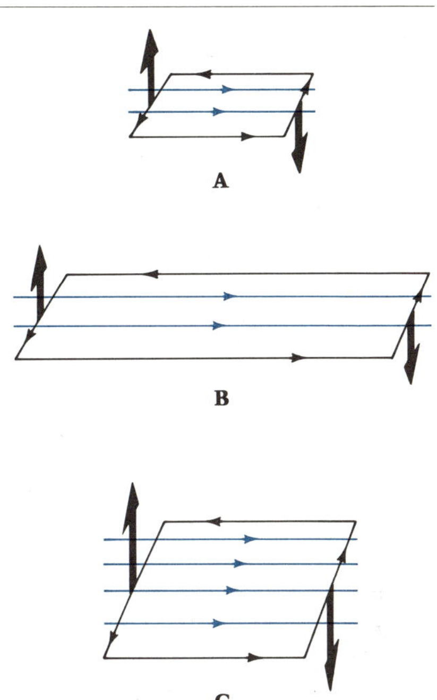

Fig. 21–18 Effect of loop size on torque. A. This shows a comparison circuit. B. We make the loop longer along the direction of the magnetic field. The forces are the same as in A, but they are pulling farther apart so there is more torque. C. Now we make the loop larger in the direction perpendicular to the field. The two forces increase, thus increasing the torque.

We define a quantity for the loop, called the magnetic moment. The greater the magnetic moment, the greater the torque on the loop. The magnetic moment is a vector, that is, it has a direction to it, as illustrated in Fig. 21–19.

A magnetic field tends to twist current loops until their magnetic moments line up with the magnetic field. We can think of a compass needle as consisting of many little magnetic moments. When placed in a magnetic field, these moments tend to line up with the field lines. In the earth's magnetic field, the compass needle turns until its total magnetic moment lines up with the magnetic field lines (see Fig. 21–20).

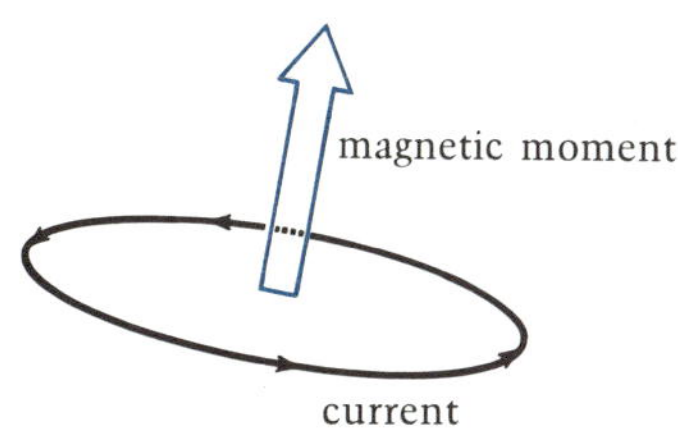

(at left) Fig. 21–19 The magnetic moment of a loop of current.

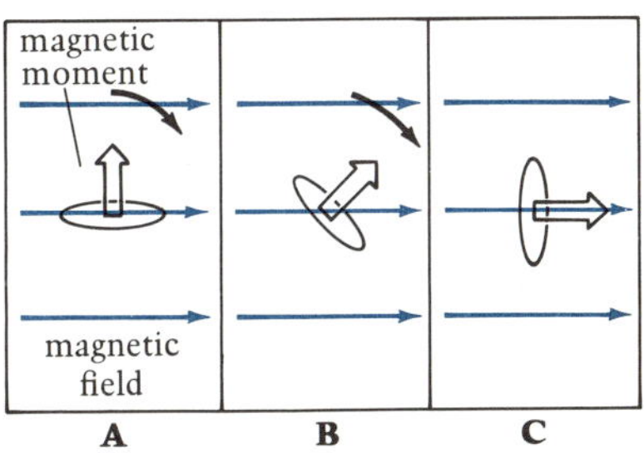

(at right) Fig. 21–20 The alignment of a dipole with a magnetic field. A. The magnetic moment is perpendicular to the direction of the field and experiences the maximum torque. B. The torque is slightly less, but is still present. C. Now the magnetic moment is lined up with the magnetic field, and there is no torque.

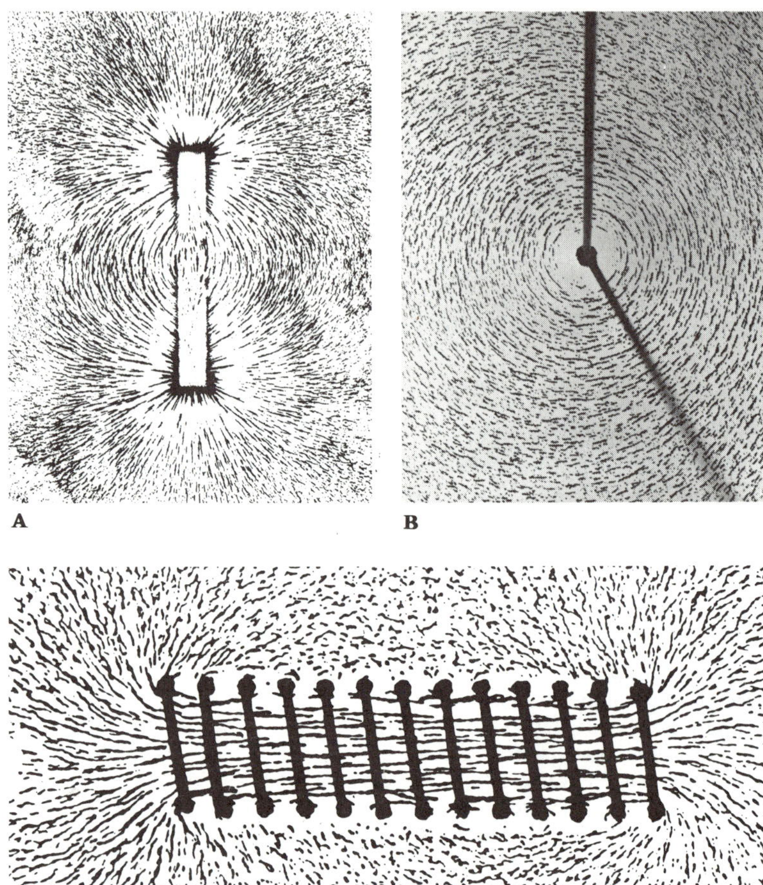

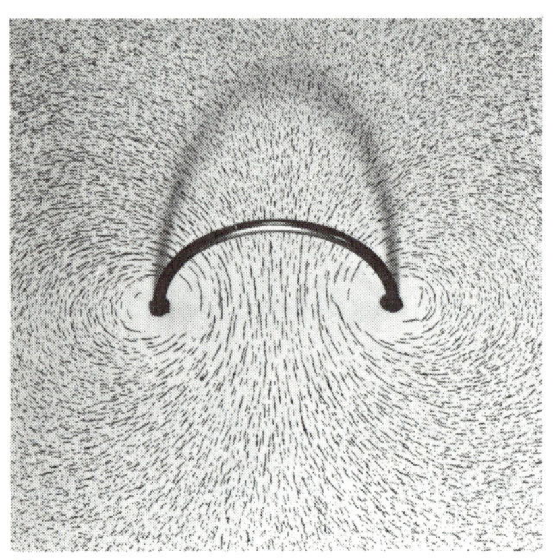

Fig. 21–21 Iron filings trace out the magnetic field lines of (A) a bar magnet, (B) a wire carrying a current, (C) a loop carrying a current, (D) several loops carrying current. The arrangement in D is called a solenoid.

We can take advantage of the twisting nature of magnetic moments to trace out a magnetic field pattern. All we need is a good supply of little magnetic moments. Small iron filings are very good for this. If we place a magnet under a piece of paper and sprinkle the iron filings on top of the paper, they line up with the magnetic field lines to provide a clear picture of the magnetic field pattern (Fig. 21–21).

21.3 Where Magnetic Fields Come From

When we studied electricity, we saw that *charges* are the fundamental source of electric fields. For magnetism, *moving charges*, or *currents*, are the fundamental source of magnetic fields.

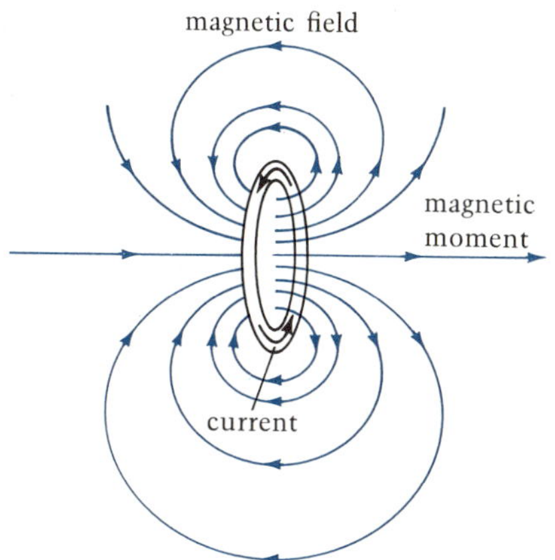

Fig. 21–22 The magnetic field produced by a loop of wire carrying a current.

Let's look at the magnetic field set up in a few simple cases. For a long straight wire, the field lines are circles centered at the wire. The magnetic field gets weaker as you get farther from the wire.

Figure 21–22 shows the magnetic field created by a loop of wire with a current flowing in it. Notice that the magnetic field at the center of the loop points in the direction of the magnetic moment of the loop. In fact, the magnetic moment of a loop is useful in describing both the field caused by a loop and the response of the loop to a field.

21.4 *Magnetic Poles*

In all the magnetic field line patterns that we have drawn in this chapter, there is one important distinction between magnetic fields and electric fields. While electric field lines start and end at charges, *magnetic field lines neither start nor end. They go around in complete loops.* Magnetic fields do not arise from simple stationary particles. Charges must be moving to create a magnetic field (Fig. 21–23).

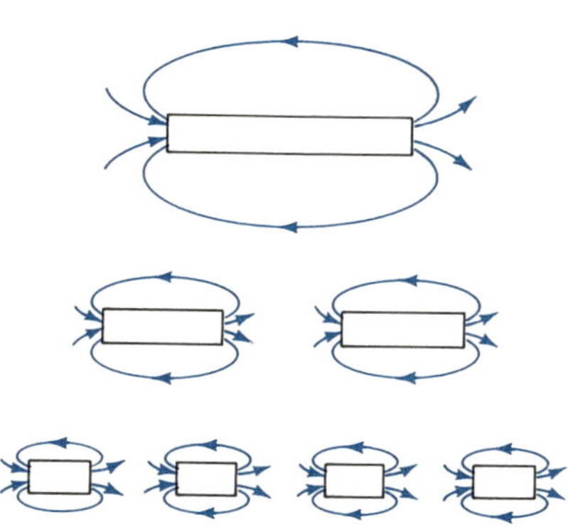

Fig. 21–23 No matter how many times we slice a magnet in half, we still end up with two dipoles. It seems that we can never get to the point where we have a magnetic monopole.

We know of no single particle that could create a magnetic field in the same way that a proton or electron can create an electric field, but we have a name for such a particle anyway: a magnetic monopole. (By "monopole," we mean that the particle could create the field from a single point.) In 1976, the world of physics was excited when a group of physicists announced that they had found a particle that might be a magnetic monopole. They had flown a balloon high in the atmosphere to study cosmic rays, charged particles that reach the earth from space. They found traces of one particle that they could not explain in terms of known particles, and suggested that it was the long-sought magnetic monopole. However, since that time, alternative explanations have been proposed. Also, it is usually difficult to say anything meaningful about a single event—the case would be more convincing if more such particles were found. At this point, most physicists think that the 1976 report of the discovery of a magnetic monopole was in error and retain their belief that magnetic monopoles have not yet been found.

Since there are no magnetic monopoles, the simplest magnetic field is that produced by a current loop. We have already seen what the magnetic field of a current loop looks like—along the axis, the field points in the direction of the magnetic moment. Because of this directionality, we can tell one side of the loop from the other, as though there were a front and back. We normally refer to opposite sides as poles, so we call a current loop a magnetic dipole and

The magnetic moment of a current loop points from the south pole of the dipole to the north pole of the dipole.

the field produced by the loop a magnetic dipole field. The names we give to magnetic poles go back seven hundred years. We have our choice of any two names, and sometimes use + and − . But mostly the two poles are called *north* and *south*.

Let's look at the force that one current loop exerts on another. Assume that the loops are lined up so that their magnetic moments are parallel, as shown in Fig. 21–24. We can then figure out the direction of the force that the magnetic field of one loop exerts on the current of the other loop. As we see in the figure, the force is attractive—each loop is pulled toward the other. That means that when the loops are placed so that the north pole of one loop is next to the south pole of the other, the two loops will attract each other. If we turn one of the loops around, we will find that the loops now repel each other. When the loops are placed so that the north pole of one is nearest the north pole of the other (or the south pole of one is near the south pole of the other), the force between the loops is repulsive. All of our natural magnetism arises from many little magnetic moments acting together.

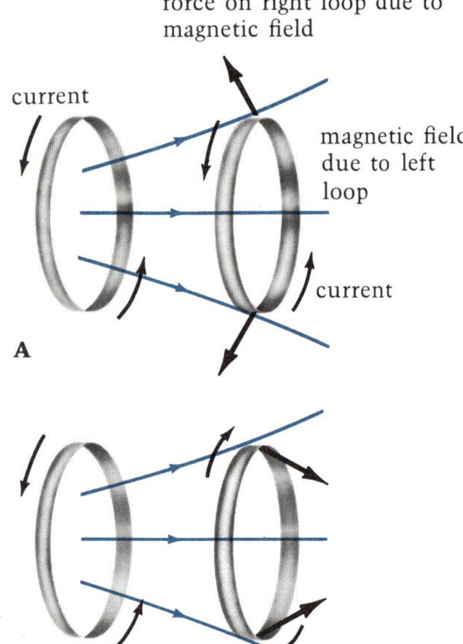

Fig. 21–24 The force that one current loop exerts on another. A. The magnetic field lines are those due to the current in the left loop. This is the magnetic field that will exert a force on the right loop. The heavy arrows represent the force on the right loop due to this field. The up-down parts of this force cancel from opposite sides of the loop, but there is still a residual force to the left. B. If we reverse one of the currents, the loop on the right is pushed to the right. From this we can see that when the currents in the two loops flow in the same direction, the loops are attracted toward each other, and when they flow in opposite directions, the loops are forced away from each other. The same thing happens with bar magnets.

21.5 *Natural Magnetism*

Different materials have different types of magnetic properties, and we have three categories of magnetism:

1. Ferromagnetism. This type is so named because the most common example is provided by iron (*ferrum* is "iron" in Latin). Ferromagnetic materials are strongly attracted to a magnet. Even after a magnetic field is removed, ferromagnetic materials may remain magnetized permanently.
2. Paramagnetism. Paramagnetic materials show no permanent magnetic effects, but they are weakly attracted to a magnet.
3. Diamagnetism. Diamagnetic materials are repelled by either pole of a magnet.

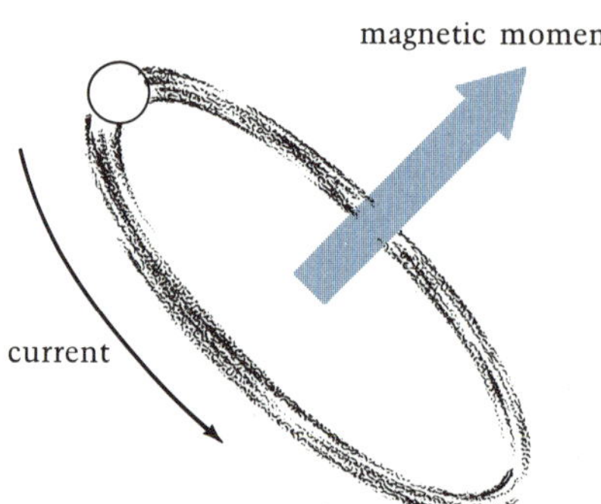

Fig. 21–25 The magnetic moment of an orbiting electron.

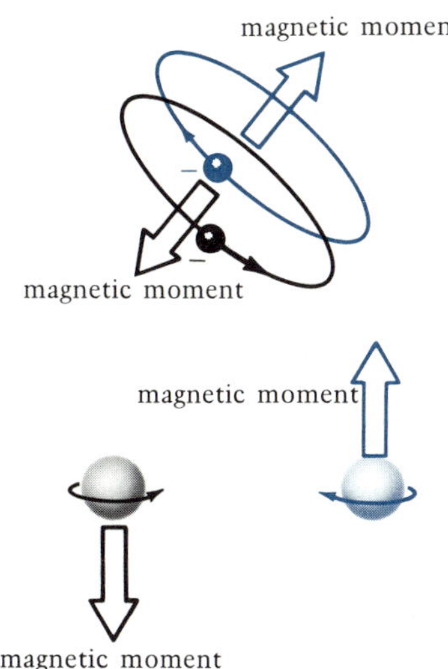

Fig. 21–26 The magnetic moments of two electrons orbiting in opposite directions or spinning in opposite directions cancel.

The source of all these magnetic phenomena is the fact that atoms and molecules can have magnetic moments. In some elements these magnetic moments occur naturally, while in other elements they are induced by the presence of magnetic fields. On the atomic scale, magnetism comes from the motions of the electrons within the atoms. In the classical picture of the atom, when an electron goes around the nucleus in a circular orbit, we have a charge going around in a circle. This is the same as having a current flow around the circle. The orbiting electron behaves just like a current loop, and has a magnetic moment (Fig. 21–25).

Each electron also has a magnetic moment that is intrinsic to the electron, and has nothing to do with the orbital motion of the electron. In many atoms the electrons are arranged in such a way that all these magnetic moments cancel each other, and the atom is left with no net magnetic moment. This cancellation comes from the existence of pairs of electrons that orbit in opposite directions and spin in opposite directions (Fig. 21–26).

Ferromagnetic atoms have permanent magnetic moments. However, in any sample of a material made of such atoms, there is a large number of atoms. In general, the magnetic moment of any atom in the sample can point in any direction, and all the magnetic moments are randomly pointed. As many point one way as point the opposite way and the sample as a whole has no magnetic moment.

Curie temperature for iron = 1043 K. (Room temperature = 300 K.)

In iron (and other ferromagnetic materials), once some of the magnetic moments start to line up, each atom begins to exert forces on its neighbors. This causes all the atoms to line up. The result is that the magnetic moments line up over a very large region. Even if the magnetic field is then turned off, the atoms keep each other in line and we have a permanent magnet. This alignment will take place as long as the temperature is not too high. The temperature above which alignment will not take place is called the Curie temperature. (See Fig. 21–27.)

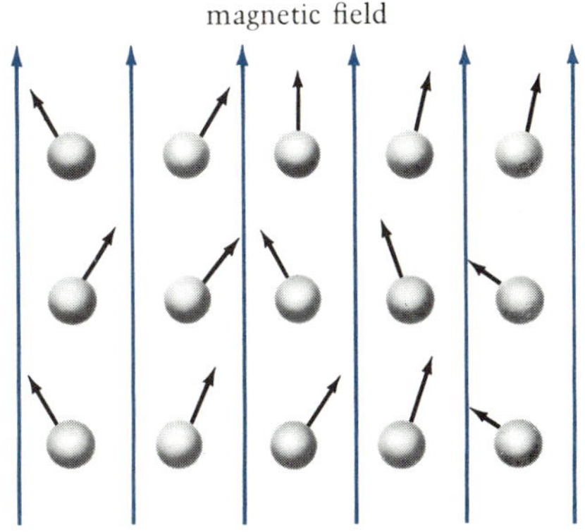

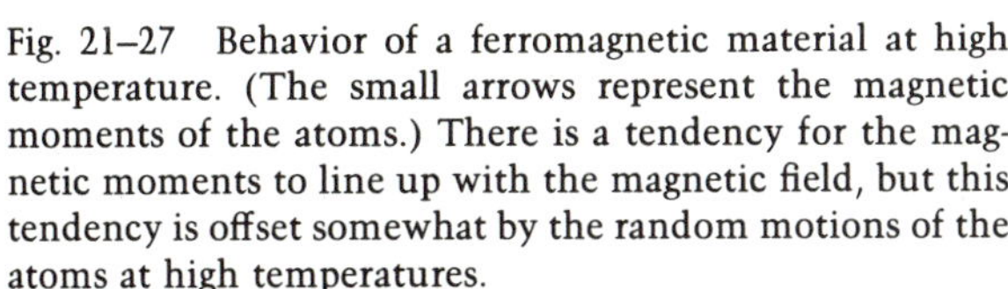
Fig. 21–27 Behavior of a ferromagnetic material at high temperature. (The small arrows represent the magnetic moments of the atoms.) There is a tendency for the magnetic moments to line up with the magnetic field, but this tendency is offset somewhat by the random motions of the atoms at high temperatures.

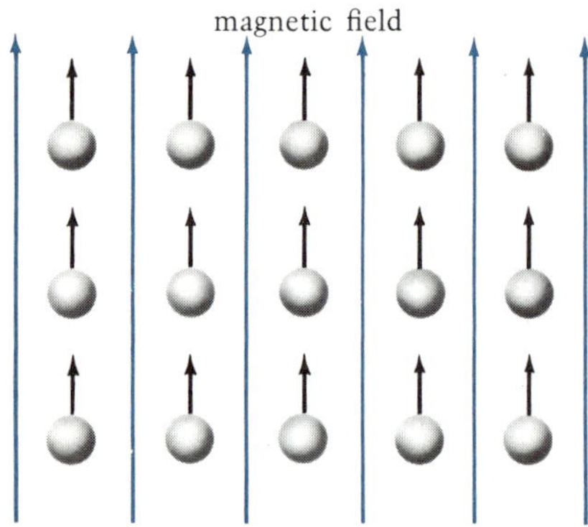

Fig. 21–28 When the temperature is low enough, the magnetic moments of ferromagnetic materials will line up with the magnetic field. Even if the magnetic field is removed, the magnetic moments will tend to hold each other in place, and we will have a permanent magnet.

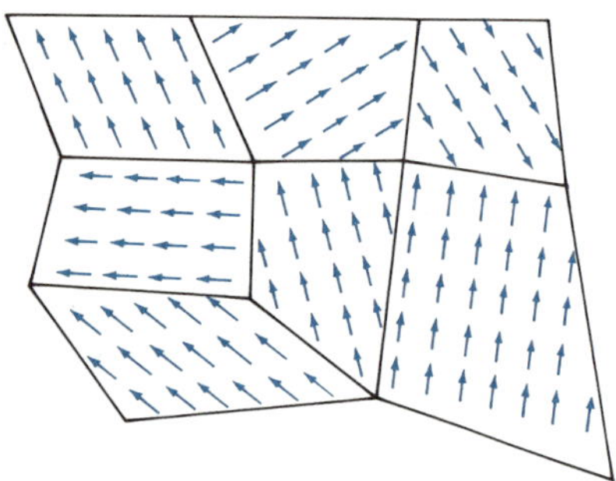

Fig. 21–29 Domains in a ferromagnetic material. Within each domain the magnetic moments are aligned, but the direction can change from domain to domain.

In general the dipole moments don't line up over the entire sample of the material; instead the material divides itself into smaller regions, called domains. Within each domain, most of the magnetic moments are aligned. However, from domain to domain, the direction of alignment will be different. Therefore, in a large sample of a ferromagnetic material, there may be no net magnetic moment, since the different domains tend to cancel each other out.

However, when a magnetic field is applied, the field causes most of the domains to line up in the same direction. This gives the material a very large magnetic moment. An electromagnet is generally made by placing a ferromagnetic material in a coil of wire. The ferromagnetic material makes the magnetic field produced by the coil of the wire much stronger. (See Figs. 21–28 to 21–31.)

You can easily make an electromagnet. Take an iron nail and wrap a coil of wire around it. Then use a battery to pass a current through the wire coil. Notice that the nail picks up paper clips. When you take the nail out of the coil, the magnetic field of the coil alone is not strong enough to pick up paper clips.

When you remove the nail from the coil, you will notice another interesting effect. The nail will still be able to pick up the clips. A ferromagnetic material will "remember" its magnetic alignment, making a permanent magnet. Even after the magnetic field is removed, the domains stay lined up in the same direction. They will remain aligned unless something is done to destroy the alignment. For example, heating the metal above the Curie temperature would destroy all of the alignment, even within the domains. Even below the Curie temperature the alignment can be disturbed. For example, hitting the magnet with a hammer will scramble the magnetic domains and destroy the alignment. Dropping a magnet will sometimes destroy its magnetic properties.

21.6 *Magnetism in the Solar System*

The earth's magnetic field has been studied for a long time because of its importance to navigation. Generally, the field looks like that of a dipole, but the magnetic poles do not coincide with the poles that mark the axis of rotation of the earth. The magnetic axis of the earth is tilted with respect to the rotation axis. We also find that the magnetic field lines point away from the earth in the southern hemisphere and toward the earth in the northern hemisphere. The north pole of the earth's magnetic field actually lies in the southern hemisphere.

The magnetic field of the earth is not exactly that of a perfect dipole; it has some irregularities. We think that the irregularities are due primarily to large deposits of magnetic materials under

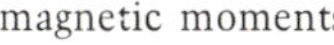

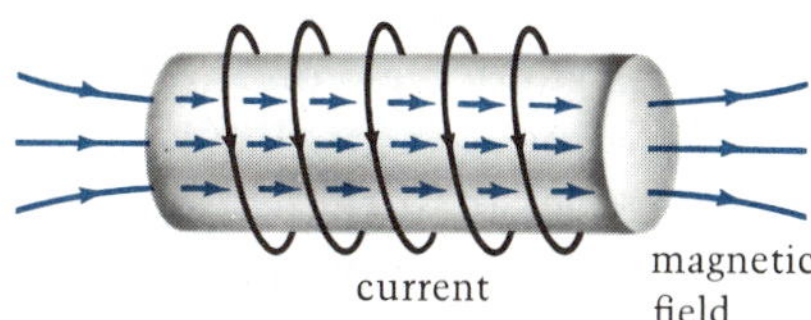

(above, left) Fig. 21–30 An electromagnet. We place a rod of ferromagnetic material inside a coil of wire. When the current flows, the magnetic moments in the material line up and we have a magnetic field that is much stronger than that of the coil of wire alone.

(above, right) Fig. 21–31 This giant superconducting electromagnet (shown here being transported from Argonne National Laboratory to the Stanford Linear Accelerator Center) can create a magnetic field 36,000 times as strong as that of the earth.

various parts of the earth's surface. There is also evidence that the earth's magnetic field changes with time. We can trace how the magnetic poles have moved. For example, the north magnetic pole has recently been moving about 8 kilometers each year. There is even evidence that, on a longer time scale, the earth's magnetic field has completely reversed its direction. Reversals may have happened several times in the past few million years. The evidence for the reversals comes from studying the patterns of magnetic materials in rocks that were formed eons ago.

The mechanism that causes the earth's magnetic field is poorly understood. The field apparently comes from the earth's liquid core of conducting materials—iron and nickel. Somehow, the earth's rotation causes large circulating currents within the core, which creates the magnetic field.

Astronomers have recently been able to examine other bodies in the solar system for magnetic fields. The Apollo lunar landings provided an excellent opportunity to search for a magnetic field on the moon. No general field was found, confirming the idea that the moon's interior is not molten.

One of the big surprises of Mariner 10's 1974 mission to Mercury was the discovery that Mercury possesses a small magnetic field. It has only about 1 percent of the strength of the earth's magnetic field, but the existence of a magnetic field at even that low level was quite unexpected. Scientists had thought that Mercury, like the moon, had never been heated enough to form a molten iron core, but even if a molten core had formed, it would have cooled by now because of Mercury's small size. Also, we believe that the earth's rapid rotation is important in creating the earth's magnetic field. However, Mercury's period of rotation is long (58 days). Thus Mercury's magnetic field tells us that we may not understand the existence of the earth's magnetic field even as well as we had previously thought.

Jupiter, however, has a very strong magnetic field (Fig. 21–32). The existence of the field had been suspected for several years, because Jupiter is a very strong emitter of radio waves. The most

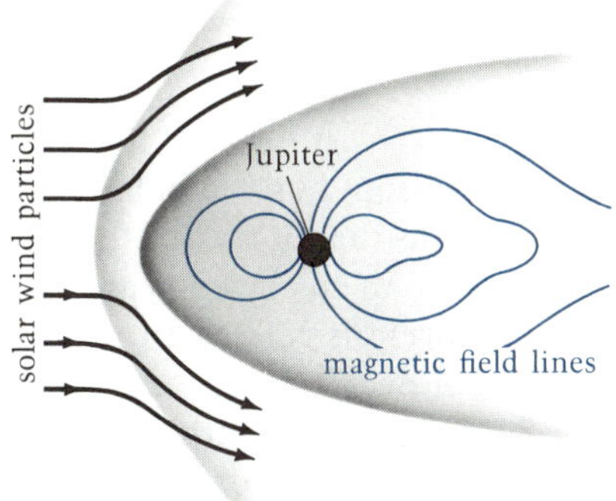

Fig. 21–32 The magnetic environment of Jupiter. The blue lines represent the magnetic field lines. The shaded region is the region over which charged particles are kept trapped by Jupiter's magnetic field. The charged particles in the solar wind come in from the left, and are deflected by the strong magnetic field.

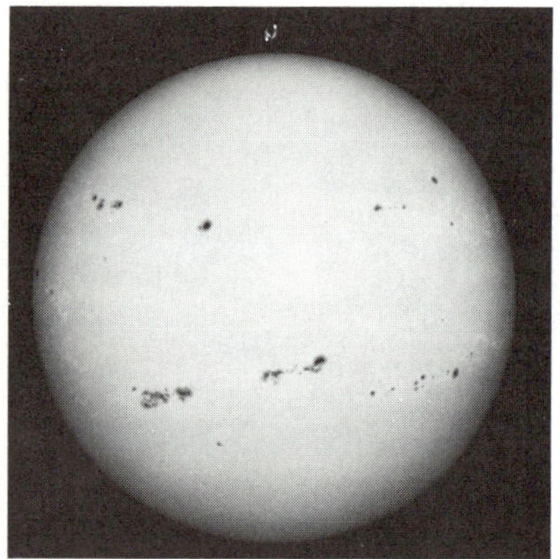

Fig. 21–33 Sunspots in 1979, maximum of the latest solar cycle.

plausible mechanism for such radiation was that Jupiter had belts of trapped charged particles similar to our Van Allen belts. Such belts would require the existence of a magnetic field on Jupiter. As a result of space missions to Jupiter, we now know that Jupiter's magnetic field is at least ten times stronger than the earth's magnetic field, even stronger than we had expected. Jupiter's magnetic field is so strong that it begins to deflect the solar wind at a distance of 7×10^6 kilometers from the planet! (Saturn's magnetic field is also much stronger than the earth's, although not as strong as Jupiter's.)

Very strong magnetic fields are found in some regions of the sun, namely, in the sunspots (Fig. 21–33). Sunspots are the regions that appear as dark spots on the surface of the sun. Actually, these regions are quite bright but are not quite as bright as the surrounding regions. Only in contrast do they appear darker. Sunspots normally occur in pairs. In each pair, one sunspot has the polarity of the north pole of a magnet and the other has that of the south pole.

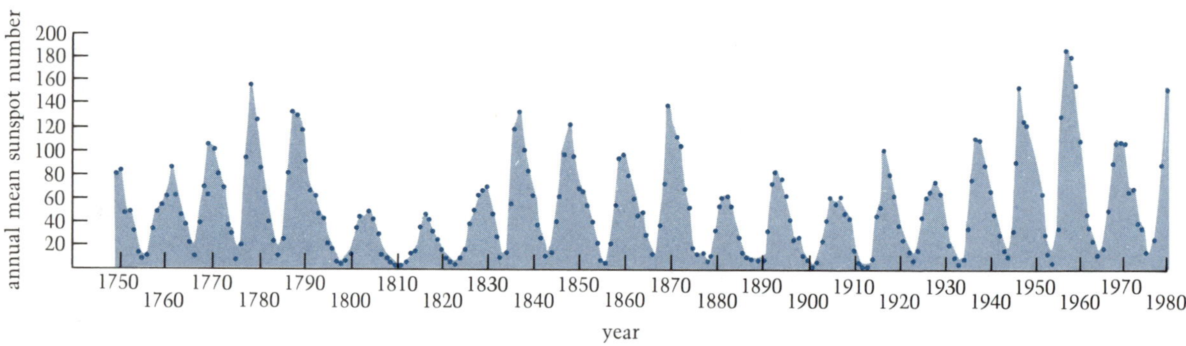

Fig. 21–34 The number of sunspots on the sun varies with a cycle of about 11 years.

The number of sunspots on the sun varies in an eleven-year cycle (Fig. 21–34). We now believe that this cycle is related to regular variations in the sun's magnetic field. For every eleven-year cycle, the sun's magnetic field flips over. The north and south magnetic poles switch places! The changeover comes a year or two after a maximum in the number of sunspots.

Maxima of the sunspot cycle are usually accompanied by increases in the numbers of solar flares and in the intensity of the solar wind. On earth, we see spectacular aurora displays and experience radio blackouts, CB skip, and surges in power lines. It even appears that solar activity affects the weather on the earth by changing upper atmospheric conditions.

Fig. 21–35 The 1980 solar eclipse observed from India.

The solar magnetic field extends far above the sun's everyday surface. The beautiful shapes of streamers in the sun's corona, best visible at total solar eclipses (Fig. 21–35), trace out the shape of the magnetic field.

Key Words

magnetism
solar wind
continental drift
gauss
magnetic mirror
Van Allen belts
synchrotron radiation
synchrotron
plasma
mass spectrograph
magnetic moment
monopole
pole
magnetic dipole
magnetic dipole field
ferromagnetism
paramagnetism
diamagnetism
spin
Curie temperature
domains
electromagnet
permanent magnet

Questions

1. Explain the following statement: All magnetism, even that of a bar magnet, arises from moving charges or currents.
2. Consider two protons. One has a speed of 100 m/s parallel to a given magnetic field. The other has a speed of 10 m/s perpendicular to the same magnetic field. Which one feels the greater magnetic force?
3. An electron is traveling in a circular path perpendicular to a constant magnetic field. (a) How does the speed of the electron change? (b) Is any work being done on the electron by the field?
4. How do we know that the particles in the Van Allen belts are charged particles?
5. Describe the path of a particle trapped in the Van Allen belts.
6. Is any work done on a particle while it is being reflected in a "magnetic mirror"?
7. What can you hold in a "magnetic bottle" that you cannot hold in a glass bottle?
8. Suppose you have a gas with two different molecules. Explain how you could use a mass spectrograph to separate these two different molecules. Under what conditions is such a separation possible?
9. Describe what happens to a current loop placed in a magnetic field. Explain your answer both in terms of the actual current loop and in terms of the magnetic moment of the loop.
10. How would you detect the difference between "magnetic north" and "true north" for your location?
11. Why do you think that magnetic compass readings taken inside many buildings may be misleading?
12. A needle of a compass will line up with the earth's magnetic field, but iron filings on a piece of paper usually will not. Why? (*Hint*: Why is the compass constructed with a pivot?)

13. Explain what we mean when we say that the fundamental unit of magnetism is the dipole.
14. Use the fact that currents circulating in the same direction attract each other to explain why the north pole of one bar magnet is attracted to the south pole of another bar magnet.
15. Since all atoms have electrons circling the nucleus, why aren't all materials ferromagnetic?
16. Why do you think that exposing a mechanical wristwatch to a strong magnetic field is bad for the watch? (This is not true for all watches.)
17. Why is dropping a permanent magnet bad for the magnet?
18. Why did we suspect that Jupiter has a strong magnetic field even before we sent space probes to Jupiter?
19. Why is the existence of a magnetic field on Mercury surprising?
20. Which regions on the sun's surface tend to have strong magnetic fields?
21. List three ways in which the sunspot cycle affects us on earth.

CHAPTER 22

Changing Fields: Motors to Maxwell

22.1 Induction

The early 1830s saw the discovery of a very important feature of magnetic fields. The British physicist Michael Faraday and the American physicist Joseph Henry, working independently, found that changing magnetic fields could cause a current to flow in a wire. The current flows even when no other source of electromotive force is present. Since the current is induced to flow by factors external to the circuit, this phenomenon is called induction.

Figure 22–1 shows induction in a circuit that consists of a coil of wire and an ammeter. When we bring a bar magnet toward the coil, the meter shows that a current is flowing. If we stop moving the magnet, even when the magnet is inside the coil, no current flows. As we pull the magnet back out, a current flows again, but in the direction opposite to the original current. In these cases, the current flows because of the motion of the magnet. As the magnet gets closer to the coil, the magnetic field at the coil gets stronger. As the magnet gets farther from the coil, the magnetic field at the coil gets weaker.

Fig. 22–1 A moving bar magnet can induce an electric current in a coil.

Fig. 22–2 A hand-cranked generator and light. The light is in the lower left. A. When the generator is not being cranked, the light is off. B. When the generator is being cranked, the light is on.

Another demonstration of induction is the familiar situation shown in Fig. 22–2, a coil of wire attached to a light bulb. The coil is placed in a magnetic field, and turned end over end with a crank. As we turn the coil, the bulb starts to glow. The faster we turn the crank, the brighter the bulb glows. (Many bicycles have headlights that operate this way, with the turning motion coming from a wheel.)

We'll take a closer look at these and other examples later, but first let's see what is happening when a current is induced to flow. For a loop of wire in a magnetic field, the induction is related to a quantity called the magnetic flux. The magnetic flux is proportional to the number of magnetic field lines that pass through the loop (see Fig. 22–3).

The magnetic flux depends on three things: (1) *The strength of the magnetic field.* The stronger the field, the closer together the field lines are and the greater the number of lines that pass through the loop. (2) *The size of the loop.* The larger the loop, the greater the number of field lines that pass through the loop. (3) *The orientation of the loop relative to the field lines.* You can see this by holding up a sheet of paper and rotating it. The sheet appears very small when you view it edge on but it appears large when you see it face on. When the magnetic field lines approach the loop of wire edge on, the loop presents a very small target for the field lines and very few get through. When the magnetic field lines approach the

A

B

C

Fig. 22–3 Flux through a loop. A. This shows the effect of changing the magnetic field strength. On the right the field is stronger, so more field lines pass through the loop and the magnetic flux is greater than on the left. B. This shows the effect of changing the size of the loop. More field lines flow through the larger loop on the right, so the flux is greater. C. This shows the effect of tilting the loop. Fewer field lines pass through the tilted loop and the flux is therefore lower.

loop of wire face on, the loop presents a large target and many field lines get through.

Whenever the flux through a loop changes, a current is induced to flow in the loop. It doesn't matter what caused the change in the flux—a change in the field strength, a change in the size of the loop, or a change in the orientation of the loop relative to the field. The circuit behaves as though a source of emf has been placed in the circuit. The strength of the induced emf is described by Faraday's law: *The strength of the induced emf is proportional to the rate at which the flux changes.* Faraday's law means that the emf depends on how fast the flux is changing, and not on the total amount by which it changes.

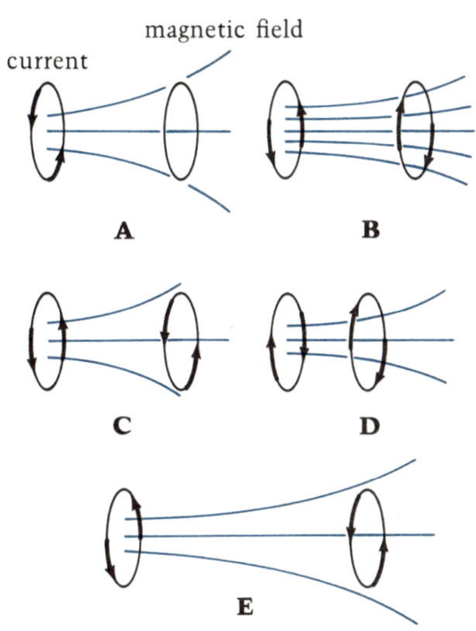

Fig. 22–4 Induced current in a loop. In each case we start with a current flowing in the loop on the left. The magnetic field shown is that due to the current in the loop on the left. In each picture we do something to change the flux through the second loop and look at the current induced in the second loop. A. Initial position. B. Increase current. C. Decrease current. D. Move closer. E. Move apart.

Which way around the loop does the induced current flow? The answer is actually included in the mathematical statement of Faraday's law, but is also given as a separate rule known as Lenz's law. Lenz's law says that *the induced current will flow in a direction such that the magnetic field created by that induced current opposes the change in the flux.*

We can illustrate the basic ideas in Faraday's law by looking at two identical loops, lined up as shown in Fig. 22–4. We'll refer to one loop as the primary loop. This loop has a current flowing in it. We will use the primary loop to create changes in the flux through the other loop, the secondary loop.

1. *Increase the current in the primary loop.* The magnetic field created by the primary loop also increases, which means that the flux from the primary loop that passes through the secondary loop increases. The increasing flux induces a current in the secondary loop. The current flows in the direction that leads to the formation of a magnetic field that tends to decrease the flux through the secondary loop. The magnetic field created by this induced current must be in a direction opposite to that of the magnetic field created by the first loop. For this to occur, the induced current in the secondary loop must flow in the direction opposite to that of the current flowing in the primary loop.

Remember that the induced current flows only while the current in the first loop is changing. If the current in the first loop levels out at some constant higher value, the induced current dies out. It can only oppose the change while the change is actually taking place.

2. *Decrease the current in the primary loop.* Then the magnetic field in the secondary loop decreases. The induced current flows in the direction that increases the magnetic field through the second loop. The induced current in the secondary flows in the same direction as the current in the primary.

3. *Move the primary loop closer to the secondary loop.* The magnetic field through the second loop increases, increasing the flux. The induced current in the secondary must flow in a direc-

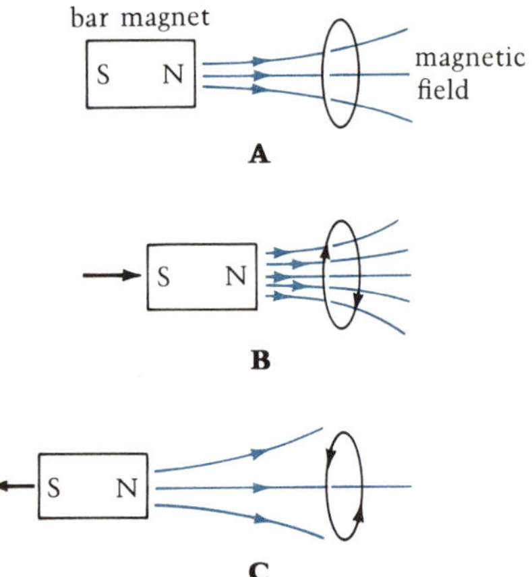

Fig. 22–5 Current induced in a loop by moving a magnet closer to (B) and away from (C) the loop. The magnetic field is that due to the magnet. Compare this situation with that in Fig. 22–4 D and E. The magnet just replaces the loop on the left as the source of a magnetic field.

tion that tends to decrease this flux. It must create its own magnetic field in the opposite direction. Thus the induced current in the secondary flows in the direction opposite to that of the current in the primary.

4. *Move the primary farther from the secondary.* The magnetic field, and thus the magnetic flux, through the secondary, decreases. The current in the secondary is then induced to flow in the direction that strengthens the magnetic field. It thus flows in the same direction as that in the primary. Moving the secondary instead of the primary would have the same effect. Either way, when we move the loops farther apart, the field gets weaker.

In these four situations, the secondary loop did not know anything about the primary loop except that it felt the magnetic field produced by the primary loop. That magnetic field might also have been produced by a simple bar magnet. Moving the bar magnet closer to and farther from the secondary loop would have produced induced currents in the same way that moving the primary loop did (Fig. 22–5).

We get an interesting effect if we make a coil with many loops in it, like a spring. If we pass a magnet through the coil, there is a flux change in each loop of the coil, so there is an induced emf in each loop of the coil. The total emf induced in the coil is the sum of the emfs induced in the individual loops. (The situation is like connecting a lot of batteries together in a row.) The more loops you have, the greater the emf.

Fig. 22–6 Thomas Alva Edison, developer of DC electrical systems.

22.2 Electricity Goes to Work

Before electricity can work for us, we must get it into our homes and factories. It is not efficient to put a battery in every electrical device you have or to put a large battery in your basement to supply electricity for the whole house. It is much more efficient to generate electricity in centralized locations and then transmit this electricity to the various users.

In Chapter 20, we discussed the difference between alternating current (AC) and direct current (DC) circuits. Thomas Edison proposed and started a system in which DC electricity was produced and distributed. However, we generally now use AC electricity, a method developed by Edison's rival Nikola Tesla. AC won out primarily because, for most applications, we like to have our electrical appliances operate on relatively low voltages. (This is a safety consideration.) However, we want a low current for the most efficient transmission. (A low current leads to low losses from joule heating in the wires.) But to transmit a lot of power at a low current, you need a high voltage. So we want to transmit electricity with a high voltage and use it with a low voltage. As we will see in this chapter, it is easier to make the necessary conversion with alternating than with direct current. We'll also see that it is very straightforward to generate electricity in the form of alternating current.

Advances in the last two decades have revived interest in DC power transmission systems.

We can take advantage of induction to generate alternating current: a simple AC generator, shown in Fig. 22–7, contains a rectangular loop of wire in a uniform magnetic field. The loop is broken at a convenient point and two wires come out of the region where the magnetic field is located. The two wires are connected so that current can flow in one wire, around the loop, and then out the other wire.

As we have already seen, the flux through the loop depends on the tilt of the loop with respect to the magnetic field. If the loop is perpendicular to the field lines, then the flux is maximum. If the loop is parallel to the field lines, the flux is zero. If we start to turn the loop around, the flux changes continuously and so an emf will always be induced in the loop. In fact, as we rotate the loop, the induced emf is not constant. Instead, the induced emf oscillates. It sometimes causes a current to flow one way in the loop and sometimes the other way.

Each time we go through one full turn of the loop, the induced emf goes through one full cycle. Just as with wave motion, the number of cycles that it goes through in one second is called the frequency of the output of the generator. The normal frequency for generating power in the United States is 60 cycles/s, which we now call 60 Hertz.

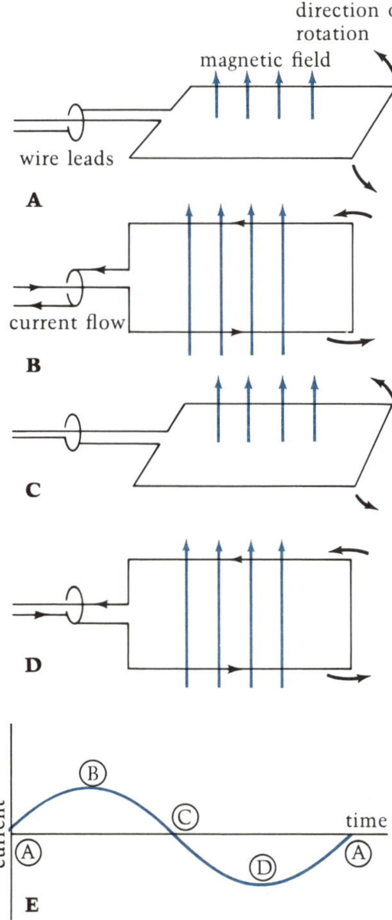

Fig. 22–7 A simple generator. The magnetic field is provided by some large magnet not shown in this drawing. As the loop is turned, the flux of the field, through the loop, changes. The flux is maximum in A and C, and zero in B and D. However, the *rate* at which the flux is changing is maximum in B and D and zero in A and C (in much the same way that a ball thrown up in the air has zero speed when it reaches its highest point). Thus, the induced emf is greatest in B and D and zero in A and C. In A, no current is flowing. In B, we see a current flowing in the indicated direction. In C, there is no current, and in D, the two leads of the loop have reversed position, so the current is flowing in the opposite direction. In E, we show the actual current flowing at various times, with the points A, B, C, and D marked. Notice that the situation returns to A after one full cycle.

We can increase the peak (maximum) emf by using a larger loop, by turning the loop faster, or by using a stronger magnetic field. We can also increase the voltage by using more than one loop, since the same voltage is generated in each loop and the voltages add up. A generator normally has a large number of loops in the form of a coil.

A generator turned by a flow is called a turbine.

If you turn the crank of a generator attached to a light bulb, you will notice that your arm gets tired very quickly. You have to do work because the magnetic field exerts a force on the induced current, and that force is always opposite to the direction in which you are cranking. The fact that it is opposite is guaranteed by Lenz's law. The induced current always opposes the change in the flux.

Of course, it shouldn't surprise us that we have to do some work to turn the crank of a generator. After all, we are getting electrical energy out. It would violate the law of conservation of energy to get energy out of something without putting any energy or work in. In actuality, *a generator is merely a means of converting mechanical work into electrical energy.* In any generator, we need some source of work to keep the generation of electricity going. Even in the ideal case, with no friction or losses, we can only get out an amount of energy equal to the work we put in.

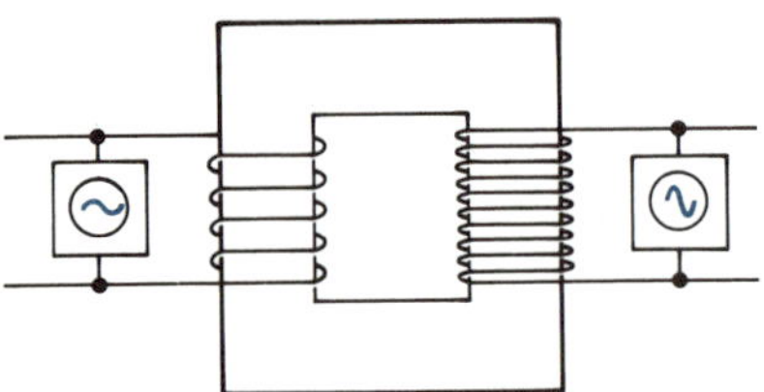

Fig. 22–8 A transformer. Any current flowing through the loop on the left will cause a magnetic field in the magnetic material, and this field will be transmitted, through the magnetic material, to the loop on the right. If the current in the loop on the left is allowed to change, the flux through the loop on the right will change and a current will flow through that loop. Since the loop on the right has more turns than the loop on the left, it will have a greater voltage than the loop on the left. The squiggles in the boxes at left and right show what you would see in a oscilloscope screen if you hooked an oscilloscope across the leads coming in and those going out. If we use the transformer as shown here, then it is a step-up transformer. If we were to hook it up in reverse, it would be a step-down transformer.

There are several ways of doing large amounts of work on a generator. One way is to connect the shaft of the loops to a paddle that is turned by flowing water, as in hydroelectric generating plants. For example, the waters of Niagara Falls are used to generate large amounts of energy.

Another way to get the required work is to use steam. Water is boiled to form steam under high pressure. The steam can be directed to flow at high speed past the blades of a turbine, which turns the loops of a generator at the other end of a shaft. Of course, you must somehow create the steam from the water. In this case, the work or energy you must supply is in the form of heat. The heat may be generated by burning a conventional fuel such as coal, or it may be generated in nuclear reactors, the heat being the by-product of the nuclear reactions. (This is a point that is often missed. When we talk about getting electricity from nuclear reactors, what we are really getting is heat. We still have to use a generator to convert that heat into electricity.)

Once we generate the electricity and transmit it to the location where it is to be used, we are still faced with the problem of converting the electricity to the correct voltage. The potential differences in the transmission lines are in the range of thousands of volts. For home use we reduce the voltage to about 120 volts. However, there are many occasions for which we want to get the voltage higher or lower for a particular application. For example, in an automobile engine, we start with a low voltage from the battery but have to get up to almost 20,000 volts to make a spark jump in the spark plugs. With alternating currents, such voltage conversions are easy. We use a device called a transformer, which takes advantage of the fact that an alternating current in a loop of wire produces an alternating magnetic field. The changing magnetic fields, in turn, induce an emf in another loop of wire, from which we let the output current flow.

Since the output of a battery is DC, we first convert it to AC before using the transformer.

A schematic diagram of a transformer is shown in Fig. 22–8. The transformer consists of two coils of wire. The power is put into one coil, the primary coil, and the output power is taken from the secondary coil. Usually, an iron core passes through each of the coils in order to increase the magnetic field produced by the coil.

Now let's apply an alternating current to the primary coil. The current produces a magnetic field inside the coil and thus a magnetic field inside the iron. This magnetic field alternates in direction, just as the current does. The magnetic field changes in the same way throughout the piece of iron, so there is a changing magnetic field inside the second coil. Faraday's law tells us that an emf is induced in each loop of the secondary coil. The total emf in the second coil is proportional to the number of turns of the wire in this second coil. We find that the output voltage and the input voltage are simply related by

$$\textit{output voltage} = \textit{input voltage} \times \frac{\textit{no. of turns in secondary}}{\textit{no. of turns in primary}}.$$

If the number of turns in the secondary coil is greater than the number of turns in the primary coil, then the output voltage is greater than the input voltage. In this case, we say that the transformer is a step-up transformer. When the number of turns in the secondary coil is less than in the primary coil, the output voltage is less than the input voltage, and we have a step-down transformer. Any transformer can be used as either a step-up or a step-down transformer. It merely depends on which coil you use as the primary and which you use as the secondary.

Now that we have generated the electricity and transformed it to the correct voltage, we want to use this electricity. If we want the electricity to do mechanical work for us, we use a motor. We can think of an AC motor as a generator running backward. We have a coil of wire in a magnetic field. This field is usually supplied by a permanent magnet. The coil of wire generally has an iron core in it, to increase the magnetic moment of the coil (Chapter 21) when a current flows through it (Fig. 22–9).

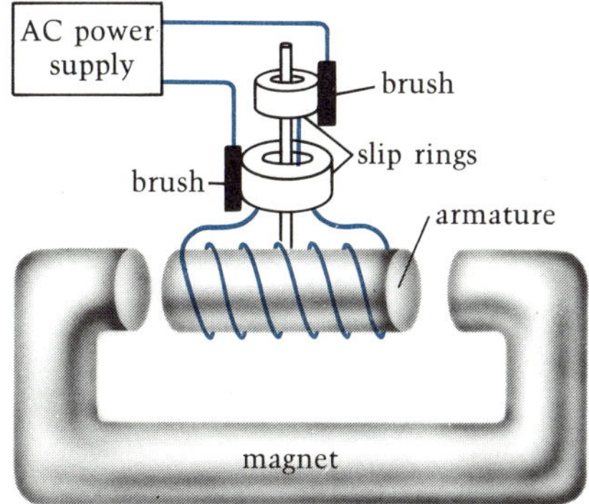

Fig. 22–9 A motor. AC electricity is supplied to the motor through two brushes in contact with slip rings. This arrangement allows electrical contact even when the rings are turning. Wires from the two rings carry the current to a coil of wire around an iron core. The wire coil plus core form the armature. The AC current in the wire causes a changing magnetic moment in the armature, and the armature rotates around in the magnetic field of a large permanent magnet.

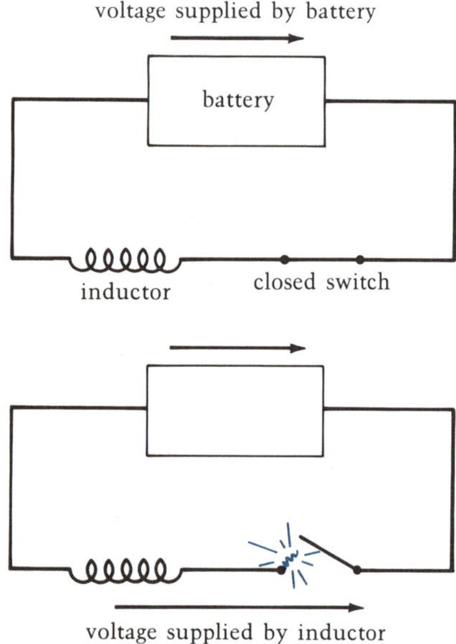

Fig. 22–10 *Opening a switch in a circuit with a large inductor can be dangerous.* When the switch is opened, there is a sudden change in the current, meaning that there is a sudden change in the magnetic field in the inductor. Since the change is so fast, there will be a very large induced voltage across the inductor, and this may be large enough to cause a spark to jump across the switch.

No matter what type of motor you use for a given job, it is interesting to follow the transfer of energy. Suppose your electrical power plant is a hydroelectric plant at Niagara Falls. The plant takes the mechanical energy of the falls and converts it into electrical energy. The voltage is stepped up and fed along transmission lines to your home. As the power leaves the main lines to enter your home, it is transformed to a low voltage. You might then use it to run a motor. The net effect is to take the mechanical energy in Niagara Falls and convert it into the kinetic energy of the motor shaft. The electricity has merely served as a means of getting that energy from one place to another. It is certainly more efficient than digging a canal from Niagara Falls to your house.

22.3 *Inductance*

When we looked at two loops to demonstrate the effects of Faraday's law, we saw that a changing current in the first loop produces a changing magnetic field which, in turn, induces an emf and thus makes a current flow in the second loop. We might also ask about what happens to the first loop. After all, the current flowing through the first loop creates a magnetic field that also exists inside the first loop. Therefore, when the current in the first loop changes, the magnetic flux through the first loop is also changing. Doesn't this mean that there should be an emf induced in the first loop to oppose the change in this flux? The answer is yes. This is called self-inductance, or, simply inductance. A device that possesses this property is called an inductor. (This is like saying that a device with resistance is a resistor.)

The inductance of a single loop of wire might not be very noticeable. However, the inductance of a coil is proportional to the square of the number of turns (because of the effect that each turn has on the others). Also, placing an iron core in the coil increases the inductance, since you get larger magnetic fields.

An inductor placed in a circuit resists changes in the current flowing through it. If the current is increased, while that increase is taking place an emf is induced in a direction that tends to make current flow the opposite way, in order to oppose the increase. If there is a decrease in the current, the induced emf tries to boost the current back to the original value. This emf only exists while the change is actually taking place.

Circuits containing large inductors present certain safety problems. If a current is flowing in a circuit with an inductor and you

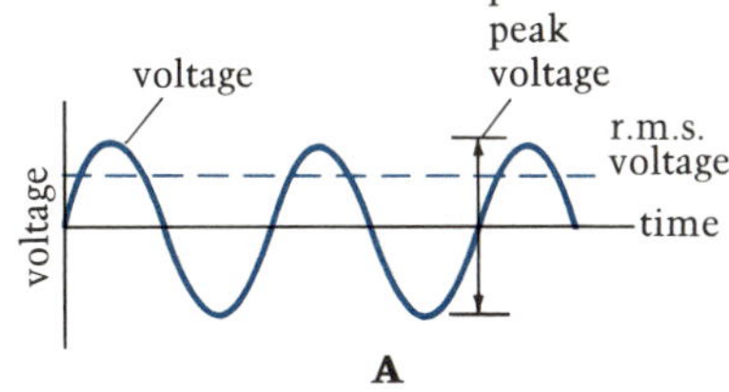

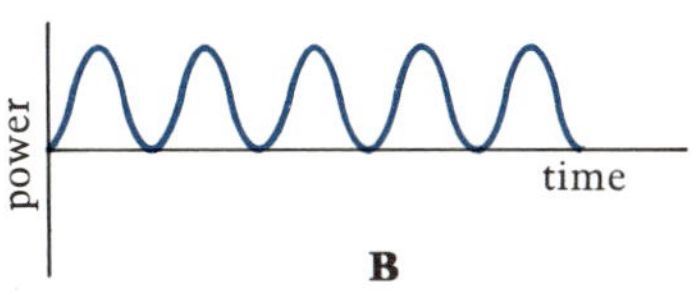

Fig. 22–11 Important quantities in an AC circuit. A. How the voltage changes with time. The peak-to-peak voltage is the maximum difference between the highest and lowest point. The r.m.s. voltage corresponds to the average power delivered by the circuit. B. Even though the voltage goes back and forth, the power delivered is always positive, although it does change in strength.

The effective value of the voltage or current is sometimes called the r.m.s. *value (for "root-mean-square," having to do with the way in which it is calculated).*

suddenly open a switch, the voltage induced by the inductor because of the sharp drop in current may be enough to make sparks fly across the gap in the switch (Fig. 22–10).

22.4 *AC Circuits*

When we apply an alternating voltage to a circuit, we get an alternating current, oscillating with a certain frequency determined by the frequency of the generator. The potential difference between any two parts of the circuit oscillates between some maximum value and the negative of that value. The difference between the peak value and its negative is called the peak-to-peak voltage (Fig. 22–11) in the circuit. However, we are normally interested in how much power is being delivered to some device or appliance. Furthermore, it is the average power that counts. The voltage that corresponds to the average power turns out to be the peak voltage divided by the square root of 2 ($\sqrt{2}$ is about 1.4), called the effective voltage. When we talk about the 110-volt circuits in our houses, this is the effective voltage. The peak voltage is actually about 155 volts. In the same way, for a 100-watt light bulb, 100 watts is the average power it puts out.

22.5 *Maxwell's Equations and Radiation*

The culmination of the classical theory of electricity and magnetism came with Maxwell's presentation in the 1860s and 1870s of four basic equations that describe the properties of electric and magnetic fields. The equations are known as Maxwell's equations. Actually, of the four equations, it is probably proper to say that three and a half were not deduced by Maxwell. For example, one of Maxwell's equations is known as Gauss's law, and describes the basic relationship between charges and electric fields. A second equation is similar to the first except that it is for magnetic fields and that it includes the fact that there are no magnetic monopoles. A third equation is Faraday's law, which we discussed earlier in this chapter. Finally, there was a law describing the relationship between currents and magnetic fields, known as Ampère's law.

Fig. 22–12 James Clerk Maxwell. The largest volcano in the solar system, recently mapped on the planet Venus, is named after him.

Maxwell's great contribution was to point out that Ampère's law needed an important modification. Faraday's law tells us that a changing *magnetic* field can produce an *electric* field (and thereby

Fig. 22–13 Readers of this text may not have to know Maxwell's equations, but physics majors like this one have to understand them.

an induced emf). Maxwell reasoned that a changing *electric* field should be able to produce a *magnetic* field. It was this addition that brought the whole package together into a unified theory of electricity and magnetism.

Maxwell realized that these four equations contained *all* the information you need to solve any problem in electricity and magnetism. The trick is always in seeing how to apply them to a given situation.

Maxwell even went beyond presenting the four equations. He showed that combining them resulted in an equation that described the propagation of waves—a wave equation. This theoretical prediction of electromagnetic waves was confirmed in 1887 when Heinrich Hertz, in Germany, succeeded in transmitting and receiving electromagnetic waves. Hertz's experiment was a great triumph for Maxwell's theory. We now take the transmission and reception of electromagnetic waves, like those that bring radio and television programs, for granted.

We have already talked about the properties of electromagnetic waves in terms of light in Chapter 15. But what *are* electromagnetic waves? Actually, electric and magnetic fields are moving through space. No material is actually moving. The fields at one location in space influence the field at the next location, and so on.

In the case of electromagnetic waves, the important point is that changing electric fields induce magnetic fields and changing magnetic fields induce electric fields. The induction must work both ways for the waves to exist. The wave begins to travel at some point, such as an antenna, when we create both changing electric and magnetic fields there. The changing electric field induces magnetic fields at neighboring locations and the changing magnetic fields induce electric fields farther away, and so on.

A simple electromagnetic wave is illustrated in Fig. 22–15. At any given point in space, we find an electric field and a magnetic field.

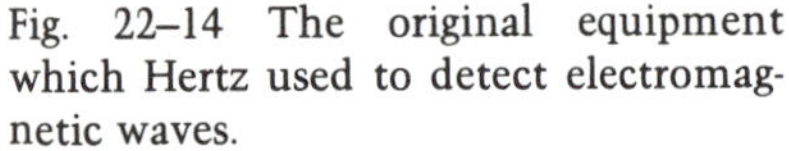

Fig. 22–14 The original equipment which Hertz used to detect electromagnetic waves.

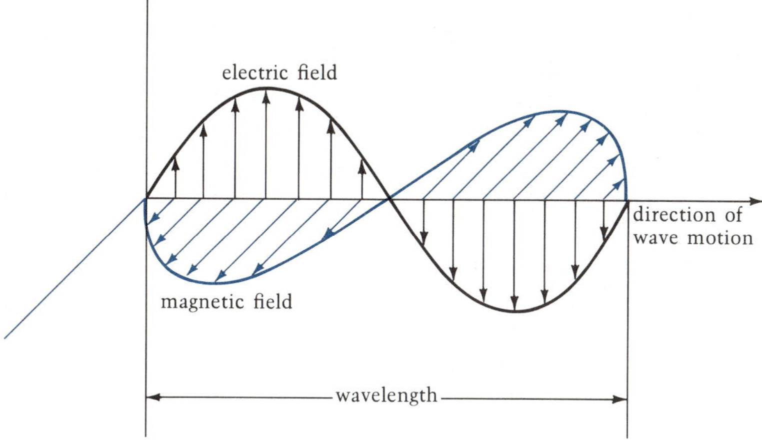

Fig. 22–15 The fields in an electromagnetic wave at some instant of time. Notice that the electric field is perpendicular to the magnetic field and both are perpendicular to the direction of motion. In fact, when we talk about the polarization of an electromagnetic wave, we are talking about the direction of the electric (or magnetic) field. For example, a polarizing filter might only pass radiation whose electric field was up-down. Notice also that the positions where the electric field is strongest are also the positions where the magnetic field is strongest, and both fields are zero at the same places. In this figure, the fields go through one full cycle, and the distance shown is one wavelength.

The electric field is perpendicular to the magnetic field and both fields are perpendicular to the direction in which the wave is traveling. If we look at another point along the direction of travel, we find a similar situation, but the electric and magnetic fields have different strengths than at the first point. If the electric field is stronger, then the magnetic field is also stronger.

Suppose we could take snapshots of the fields at a series of points along a line in the direction of travel of the wave, and see what the fields look like at each instant. We would find that both fields increase, go through a maximum, decrease, go through zero, then start to increase in the opposite direction, go through a maximum in that direction, and turn back toward zero. This pattern repeats in a regular cycle. The strongest electric field and the strongest magnetic field occur in the same place. The distance between successive maxima of the fields is the wavelength.

If we stayed at any point in space and watched the wave pass by, we would see the fields go through a cycle. They would get stronger and weaker and turn the opposite way. Thus we see that electromagnetic waves behave just like the other types of waves we discussed in Chapters 13 and 14. We can even classify the electromagnetic waves as transverse, since the disturbances—in this case, the electric and magnetic fields—point perpendicularly to the direction in which the wave is moving. When we talked about polarization of electromagnetic waves in Chapter 13 we were really talking about an alignment of the electric (or magnetic) fields.

One big difference between the electromagnetic waves and the other waves we have discussed is that *electromagnetic waves do not need a medium to travel through.* A changing electric field can induce a magnetic field and so on, even in a vacuum. The possibility of waves in a vacuum was not fully appreciated in Maxwell's time, and physicists postulated the existence of the ether (as discussed in Chapter 8) to explain the propagation of electromagnetic waves. It was a surprise when the Michelson-Morley experiment ruled out the existence of the ether.

Maxwell's theory placed no restriction on possible wavelengths for the electromagnetic radiation, and we have now succeeded in producing and detecting a wide range of wavelengths. The whole range of electromagnetic radiation, from very long to very short wavelengths, is called the electromagnetic spectrum. We tend to divide the electromagnetic spectrum into different regions (Table 22–1; Fig. 22–16). These divisions are made on the basis of the various techniques that are used to create and detect radiation in different parts of the spectrum. However, there is really nothing basically different about these regions. They merely represent electromagnetic waves with different wavelengths. Waves with different wavelengths may interact with matter differently, but all these waves must still obey Maxwell's equations.

TABLE 22–1 Regions of the Electromagnetic Spectrum

Wavelength range (meters)		*Name*
greater than 10^{-3}	greater than 1 mm	radio
10^{-3} to 7×10^{-7}	1 mm to 700 nm	infrared
7×10^{-7} to 4×10^{-7}	700 nm to 400 nm	visible
4×10^{-7} to 10^{-8}	400 nm to 10 nm	ultraviolet
10^{-8} to 10^{-10}	10 nm to 0.1 nm	x-ray
less than 10^{-10}	less than 0.1 nm	gamma ray

Fig. 22–16 The electromagnetic spectrum.

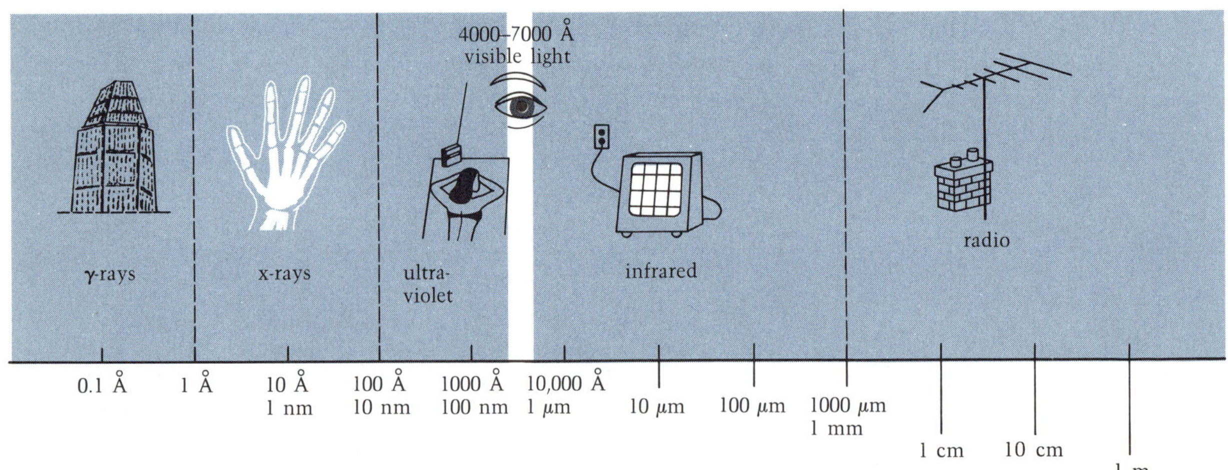

Key Words

induction
magnetic flux
Faraday's law
Lenz's law
primary loop
secondary loop
frequency
peak
hydroelectric
turbine
transformer
step-up transformer
step-down transformer
motor
self-inductance (inductance)
inductor
peak-to-peak voltage
r.m.s. value
Maxwell's equations
wave equation
electromagnetic spectrum

Questions

1. In your own words, describe what is meant by magnetic induction.
2. In each of the following cases, tell what is changing to cause a current to flow through a loop in a circuit: (a) You bring the north pole of a magnet closer to the loop. (b) You place a bar magnet inside the loop (pointing along the axis of the loop) and then rotate the magnet end over end. (c) You bring a second loop, with a current flowing, closer to the first loop. (d) You leave the second loop in one place, but decrease the current in the second loop.
3. What are the advantages of AC electricity over DC electricity?
4. Sometimes, under very heavy loads, the generators of the electric company will slow down to slightly lower than 60 cycles per second. (a) What effect will this have on electric clocks? (b) What can the electric company do to make sure the clocks are right at the end of the day?
5. Turning the crank of a generator attached to a light bulb will increase the frequency of the electricity, but it will also make the bulb burn brighter. Why will the bulb be brighter?
6. List one use each for a step-up and a step-down transformer.
7. Suppose you have two transformers. One will take a 10-V signal and convert it into a 100-V signal. The other will take the 100-V signal and convert it back into a 10-V signal. Can the two transformers be constructed identically? Explain.
8. In what ways can we think of a motor as a generator running backward?
9. Suppose you have a current flowing through a circuit with a large inductor. You suddenly open a switch in the circuit. Explain what happens and why.
10. In talking about the average power that you get from some appliance, why is it important to use the r.m.s. voltage as opposed to the peak-to-peak voltage?

11. In what ways is a circuit with an inductor and a capacitor like a pendulum or a spring?
12. Why was the existence of a medium for electromagnetic radiation (the ether) postulated after Maxwell's theory was presented?
13. (a) In what ways are light waves and radio waves similar? (b) In what ways are they different?
14. Suppose a pulse of electromagnetic radiation passes by you. How will your neighborhood of space be different during the passage of that pulse?
15. If changing magnetic fields could produce electric fields, but changing electric fields could not produce magnetic fields, do you think that electromagnetic waves could exist? Explain your answer.
16. What was Maxwell's main contribution to "Maxwell's equations"?
17. Of infrared, visible, ultraviolet, and x-ray, which part of the electromagnetic spectrum covers the greatest range of wavelengths?

PART VI

Quanta and All That

The last part of the nineteenth century saw the culmination of the age of what we call classical physics. We can think of this era as having begun with Newton, although his work, in turn, was built on that of several major predecessors. By the end of this era, it seemed that there were complete theories of mechanics, electricity and magnetism, and thermodynamics. The successes of classical physics were many, and there was even a feeling at the end of the nineteenth century that the role of the twentieth-century physicist would be to come up with applications of well-founded theories and to refine experiments a little more. People spoke in terms of measuring one more decimal place. . . .

This complacency was shattered on several fronts. We have already seen that problems concerning the intertwined nature of electric and magnetic fields started Einstein on what was to become the theory of relativity. In Chapters 8 and 9 we saw how the concepts introduced by Einstein were a great departure from the classical ideas, and caused a complete rethinking of our ideas on the nature of space and time.

Around the same time, another revolution was taking place—the quantum revolution. By the time it was finished, it had completely altered our ideas about the nature of radiation and the structure of matter. The new ideas were so radical that they also rekindled philosophical debates that are still raging. Even the discoverers of these new ideas were not entirely happy with them right away. However, there is an important feature to keep in mind. Despite the sometimes esoteric-sounding ideas and philosophical discussions that go along with the quantum theory, the theory has a strong foundation in experimental fact. It had been the inability to explain certain well-established experimental results that drove physicists to the quantum theory. Even after the quantum ideas were introduced, they underwent several modifications. However, during the early years and continuing right through the present time, the string of experimental successes—phenomena explained and phenomena predicted—has been varied and impressive.

The ideas of quantum mechanics have had such a profound effect on physics and have represented such a departure from the past that we mark the birth of what we call "modern physics" from the time of the introduction of both the quantum theory and the theory of relativity. We teach these subjects in courses under the heading of "modern physics" and physicists even talk about them as very new ideas. We often seem to forget that many of the most important events took place some fifty to seventy-five years ago, and that most of the pioneers of the quantum theory are no longer active.

The designation as "modern" simply serves to emphasize the importance that physicists place on the fundamental role of these ideas. Also, it is probably fair to say that many of the current avenues of research that are truly modern are based on ideas that had their start in the quantum revolution.

We'll start this section by seeing how Maxwell's theory of radiation failed to explain some very basic experimental results. The solution to this problem revived the old debate about whether light is a particle or a wave. We'll then see how the classical theory failed to explain the observed properties of atomic structure, and how the ideas of quantum mechanics and wave mechanics grew out of the attempts to explain atomic structure. In Chapter 24, we'll see how the ideas of quantum mechanics apply to a device called the laser. In Chapter 25, we'll extend the ideas of atomic structure to talk about the structure of solids, and try to explain why different solids have very different electrical properties. Finally, in Chapter 26, we'll launch into the realm of a new entity, one that is even smaller than the atom—the nucleus. In Chapter 27, we'll see how our ideas of the nucleus can be applied to our "energy crisis."

CHAPTER 23

Quantum Theory and Atomic Structure

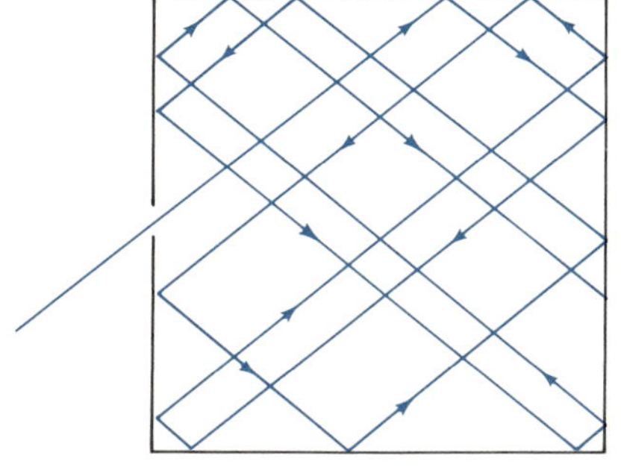

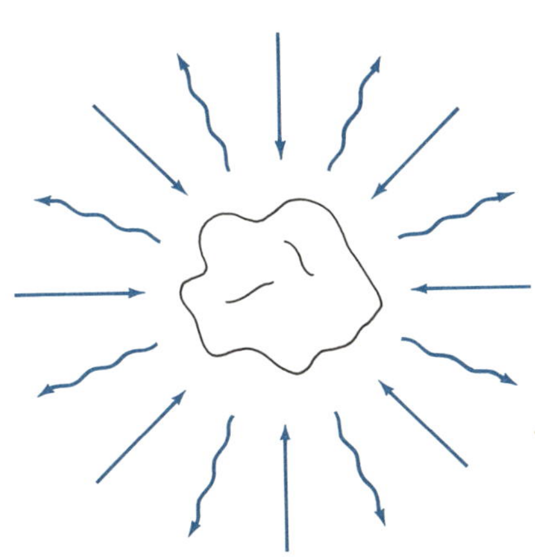

23.1 Black-Body Radiation—Classical Theory Fails

The first hint that classical radiation theory had problems came with its failure to explain the spectrum of radiation emitted by objects called black bodies. A black body is simply something that absorbs all the radiation that hits it. (A black piece of paper absorbs essentially all the radiation from the visible region of the spectrum that hits it, but a black body does this completely, and absorbs all the radiation from other parts of the spectrum as well; Fig. 23–1.)

Black bodies also emit radiation (Fig. 23–2). If they did not, as they absorbed radiation they would get hotter and hotter. When a black body is in equilibrium with its surroundings—that is, neither gaining from nor giving to its surroundings any net amount of

(above) Fig. 23–1 A cavity with a very small hole in it comes very close to being a black body. Remember, a black body is something that absorbs all electromagnetic radiation that hits it. In this case we see that light coming in the hole might get bounced around quite a bit inside the cavity, but the hole is so small that it will not get out again, and will eventually be absorbed in one of its collisions with the wall.

(below) Fig. 23–2 Even though a black body absorbs all the radiation that hits it, it does give off radiation of its own. In equilibrium, it will give off as much energy in a second as it absorbs in a second. However, it will give it off with a different spectrum than it absorbs.

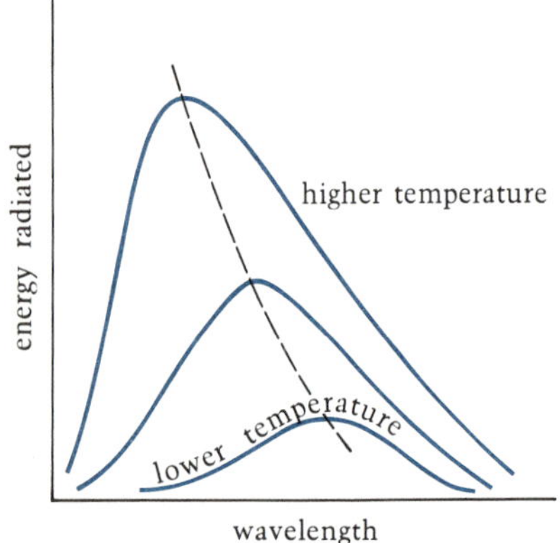

Fig. 23–3 The black-body spectra of objects at different temperatures. The higher the temperature, the shorter the wavelength at which most of the radiation is given off. Also, higher temperature objects give off more energy than lower temperature objects at any wavelength. In this diagram, the top curve is for the highest temperature and the bottom curve is for the lowest temperature.

energy, which happens when it is at the same temperature as the surroundings—it will absorb exactly as much energy as it radiates. This does not mean that a black body must emit radiation at the same wavelengths as those at which it absorbs. It means, rather, that the *total* energy that it emits must be the same as the total energy that it absorbs. The black body is free to choose how much energy it will radiate at each wavelength to accomplish that.

In discussing a black body, we usually talk about its spectrum. To describe its spectrum, we specify the amount of radiation given off at each wavelength. By the year 1900, the properties of black-body spectra were known experimentally (Fig. 23–3). For example, if you place an iron poker in the fire, it begins to glow as it gets hot. This glow starts as a red, turns yellow as the poker heats up, and eventually turns blue-white. The apparent color change occurs because, as the object gets hotter, it gives off a greater fraction of its radiation at shorter wavelengths. The wavelength at which the black body gives out the most energy (where the spectrum *peaks*) becomes shorter as the object gets hotter. There is a simple relationship between the temperature of the black body and the wavelength at which the spectrum peaks:

$$\textit{wavelength of peak} \times \textit{temperature} = \textit{constant}.$$

For example, if you double the temperature of an object, the wavelength at which its spectrum peaks is halved.

As we heat a black body, the total amount of radiation given off increases quickly. In 1879, Josef Stefan found that the total energy radiated by a black body is proportional to the fourth power of the temperature. For example, if you double the temperature of an object, the total energy output increases by a factor of 16 ($2^4 = 2 \times 2 \times 2 \times 2 = 16$).

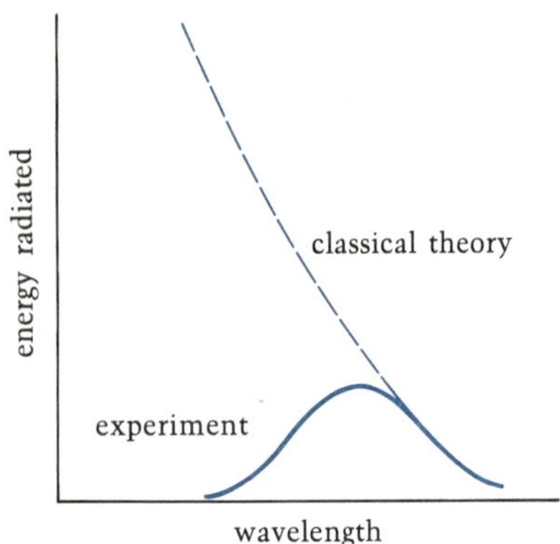

Fig. 23–4 The classical theory of radiation agreed well with experimental results at long wavelengths, but predicted much too much energy at short wavelengths.

When classical radiation theory was used to calculate what the spectra of black bodies would be, the resulting calculations did not agree with the experiments. The theoretical results indicated that the intensity of radiation would continually get stronger as the wavelength got shorter (Fig. 23–4). This predicted behavior was called the ultraviolet catastrophe (the ultraviolet used to signify the shorter wavelengths) since it predicted that an infinite amount of energy would be emitted at very short wavelengths.

The situation changed in 1900 when Max Planck, a German physicist, suggested a formula that fit all the experimental black-body curves. At first Planck could not justify the formula, but within a few years he managed to derive it on theoretical grounds. In his derivation, he had to make an unusual assumption. He assumed that the energy of something he was considering could take on only a certain set of values. (For the purposes of this book, let us not go into the details of which energy this is.) The values had to be integers multiplied by a value of energy. Furthermore, Planck said that the lowest energy possible for each frequency must be proportional to the frequency; that is,

$$energy = constant \times frequency.$$

The constant in this equation is called Planck's constant, and is designated by the symbol h. Planck was able to find a value for this constant by considering the value that was necessary for it to make his theoretical curves agree with experimental ones. Once this constant is set, Planck's formula predicts the correct amount of energy at any given wavelength for a black body at any temperature.

Even though it was nice to be able to derive a black body's spectrum, there was still some dissatisfaction, even on Planck's part. He had succeeded in producing the right formula, but in order to do it he had to make an assumption that he could not really justify—the assumption that the energy could take on only a certain set of values and not values in between. Why couldn't the energy take on any other value? Nobody knew.

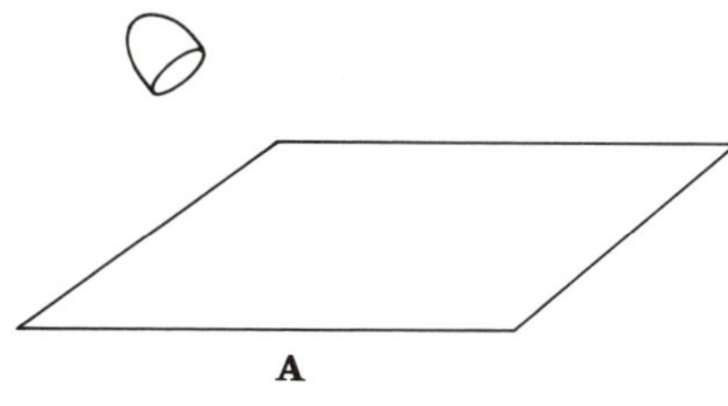

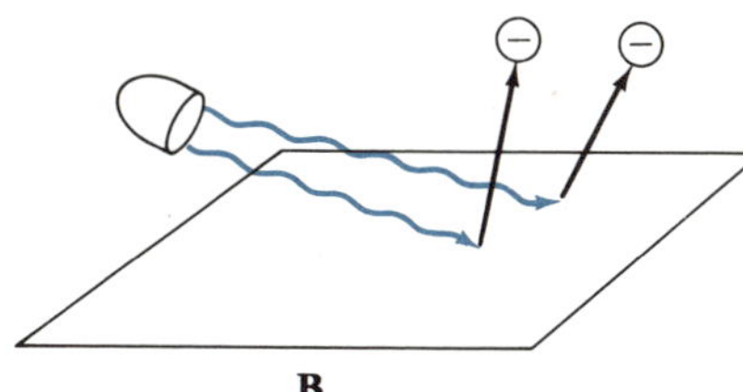

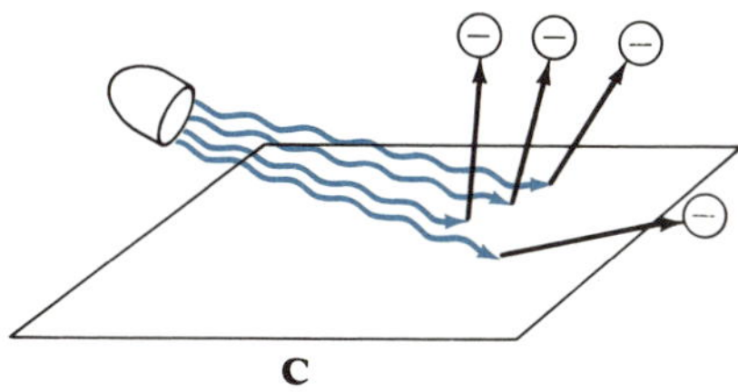

Fig. 23–5 The photoelectric effect. When we shine a light on a metal surface, electrons will be given off by that metal surface. Changing the intensity of the light does not change the energy with which the electrons are ejected from the surface, but does change the number of electrons ejected from the surface.

23.2 *Photons*

As we have already seen, the great triumph of Maxwell's theory had come in the laboratory of Heinrich Hertz, where Hertz was able to transmit and receive electromagnetic waves. It is ironic that in the very experiment in which he confirmed Maxwell's prediction, Hertz made another discovery that was to play a major role in the undoing of classical radiation theory. He discovered the photoelectric effect. In Hertz's apparatus, the waves he received caused a spark to jump across a gap. Hertz found, quite accidentally, that when light was shining on the gap, then the spark could jump across a larger gap than it could in the dark. Somehow, the presence of light made it easier for the sparks to jump.

Although Hertz did not follow up on this discovery, it was pursued by others. By 1900 a number of curious facts had emerged about the photoelectric effect. When light fell upon a clean metal surface, particles were emitted that had the same properties as the electrons recently discovered by Thomson (Fig. 23–5). The number of electrons emitted from the surface was proportional to the intensity of the light, as might be expected. However, contrary to expectations, the energy of the electrons did not depend on the intensity of the light; a stronger light would lead to more electrons, but not more energetic ones.

The explanation of the photoelectric effect was proposed by Einstein in 1905. Einstein said that energy comes only in bundles of specific size. In his picture, light is absorbed or emitted in little packages, or quanta (singular form: quantum) of energy. A quantum of energy is called a photon. The energy of each photon is given by Planck's relation

$E = hf$ *energy of photon = Planck's constant × frequency.*

It was a surprise that Planck's constant should appear in this new context. Light, after all, was thought to be a wave because it can interfere, as we saw in Young's experiment (Chapter 15). Only later was it realized clearly how the work of both Planck and Einstein led directly to our current understanding that light also acts like a set of particles.

Einstein's 1905 theory held that when a photon strikes a metal surface, it knocks an electron out. The energy of the photon goes into the energy of the electron, but the electron immediately loses a little energy in the act of leaving the metal. This lost amount of energy is the same for all the electrons in a given type of metal.

The energy of a photon is very small. A 100-watt light bulb gives off over 10^{19} photons in one second!

The remaining energy shows up as the kinetic energy (energy of motion) of the electron. If we increase the intensity of the light (keeping the wavelength fixed), the *number* of photons is increased, but the energy of each photon remains the same. As a result, more electrons are emitted, but each of these electrons still has the same energy as did the electrons ejected by a lower intensity beam (Fig. 23–7).

Remember: the photons are in the light, and the electrons are emitted by the surface when the light hits it.

Einstein noted that a straightforward test of his idea could be made. The experiment had to be repeated for light of different frequencies, to see if the energy increased in the prescribed fashion as the frequency got larger. The experiment turned out to be more difficult than anticipated, and it was another decade until Millikan presented results supporting Einstein's idea.

"I spent ten years of my life testing that 1905 equation of Einstein's and, contrary to all my expectations, I was compelled in 1915 to assert its unambiguous verification in spite of its unreasonableness since it seemed to violate everything we know about the interference of light." Robert Millikan

Arthur Holly Compton, in 1923, provided another demonstration of the particle nature of radiation. Compton analyzed the collisions between radiation and electrons. He reasoned that if the radiation acted like a particle, then he could apply conservation of energy and momentum, using the standard techniques that are applied to particles. In doing this, Compton calculated that the difference between these energies should show up as a shift in the frequency of the radiation after the collision. After calculating this shift, Compton performed an experiment to measure the shift of an x-ray (a very short-wavelength photon) scattered by an electron. The experimental result agreed with the prediction, serving as an-

Fig. 23–6 Max Planck and Albert Einstein.

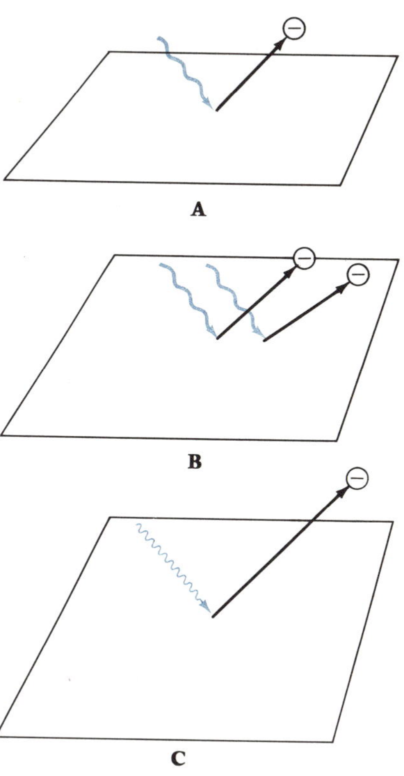

Fig. 23–7 Explanation of the photoelectric effect. A. In Einstein's explanation, each electron was ejected by one photon. B. If we increase the intensity of the light, we are increasing the number of photons, so more electrons are ejected, but they have the same energy since each one is ejected by a photon of the same energy. C. If we increase the energy of the photon (decrease the wavelength), the energy of the ejected electron is increased.

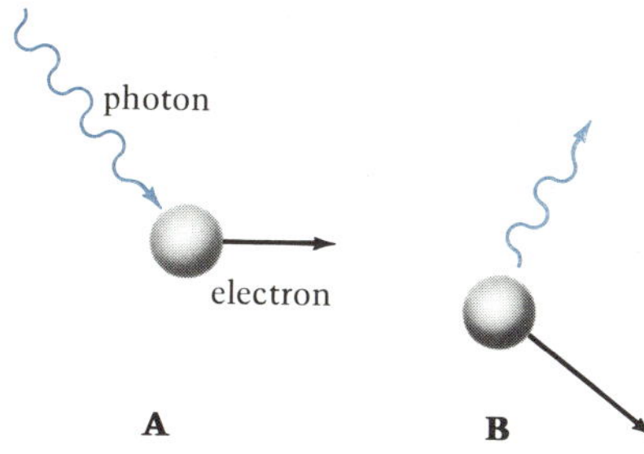

Fig. 23–8 Compton scattering. A. Here we see an electron and a photon before a collision. B. After the collision the electron has gained some energy, so the photon must lose some energy. The lower energy photon has a longer wavelength than the original photon.

other confirmation that radiation behaves as though it were made up of particles. (See Fig. 23–8.)

In Chapter 15, we discussed the long-standing debate over whether light is a wave or a particle. We saw that despite Newton's endorsement of the particle theory, most of the evidence favored the wave theory. This evidence in favor of the wave theory included the interference of light and the diffraction of light, and, finally, the theoretical explanation provided by Maxwell. Now it may seem that we are turning around and saying that light is a particle, and the theory that light is a wave is all wrong. Certainly, the results we discussed in this chapter leave little doubt that, in many situations, light behaves like a particle. The black-body radiation spectra, the photoelectric effect, and Compton's scattering cannot be explained without the photon picture. However, refraction, diffraction, interference, and the existence of radio waves cannot be explained without the wave picture.

Which is it? Wave? Particle? Both? Neither? Radiation is certainly not a wave in the classical sense, in that it does not always exhibit wave-like properties. By the same token, it is not a group of particles

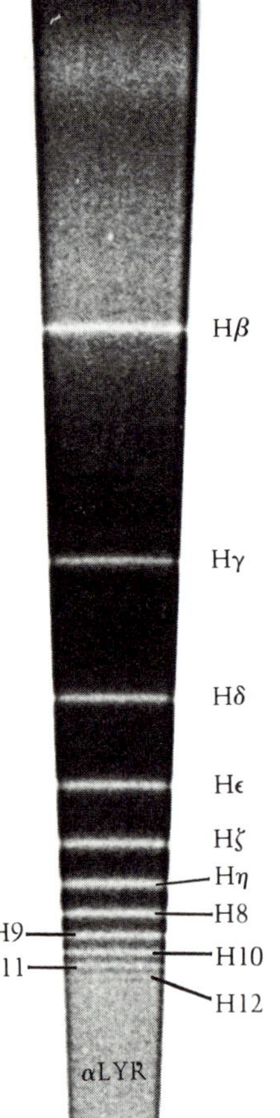

Fig. 23–9 The Balmer series shows nicely in this spectrum of the star Alpha Lyrae. The film was not sensitive to the red color of H alpha, which would appear as far to the top of H beta as H12 is to its bottom. (Courtesy of D. Chalonge and L. Divan)

in the classical sense, in that it does not always exhibit particle properties. It seems as if we have to make room for a new entity. *Light is neither a wave nor a particle but exhibits either wave or particle properties, depending on the experiment you are doing.* This explanation may seem like a bit of a cop-out in that we have avoided making a choice. But we *have* made a choice. In the "wave-particle" ballot, we have inserted a write-in choice. Admittedly, our answer takes a while to get accustomed to. As we'll see in the following section, the idea that something can be both a wave and a particle will grow on us.

23.3 *Problems in Atomic Structure*

While the difficulties with black-body radiation and the photoelectric effect were being explored, there was a parallel set of developments on another front—the structure of the atom—that played an equally important role in the development of the quantum theory. Many of these developments originated in the search for the origin of the *spectral lines* that we discussed in Chapter 15. The set of wavelengths at which an atom or molecule of a specific substance can emit or absorb energy is unique to that substance.

23.3a MODELS OF THE ATOM

The first step in interpreting the spectral lines of various elements was to look for regularities in the wavelength patterns. Johann Balmer, a Swiss schoolteacher, found in 1885 that the wavelengths of one series of lines in the spectrum of hydrogen could be represented by a simple formula. The series of lines is now called the Balmer series, and is prominent in the optical part of the spectrum (Fig. 23–9). Balmer's formula was generalized to the rest of the hydrogen spectrum in 1908. The important feature is that there is a simple formula for the wavelengths that contains only small integers (1, 2, 3, etc.).

The atom, in Thomson's model, would be like a plum pudding, with hard bits distributed throughout.

The challenge then for the theoreticians was to come up with a model of the atom from which these formulae could be directly derived. After Thomson discovered the electron, he proposed a model of the atom in which the positive charge was distributed

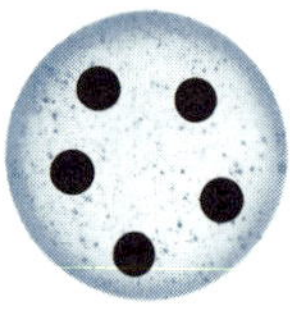

Fig. 23–10 Thomson's plum-pudding model of the atom. The solid circles are the electrons, and the larger area is the rest of the atom, which has a positive charge. In this picture of the atom, the positive charge was uniformly distributed throughout the atom (the pudding) and the negative charges were placed at certain spots within the atom.

Fig. 23–11 Ernest Rutherford.

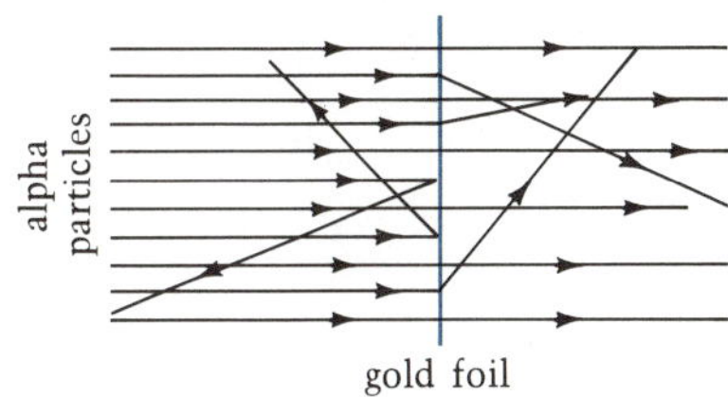

Fig. 23–12 Rutherford scattering. Alpha particles are shot at a target of a thin gold foil. If the plum-pudding model of the atom were correct, the alpha particles would just be slowed down slightly in passing through the pudding, but would not change directions very much. However, when researchers did the experiment, many of the particles passed right through but others suffered strong deflections. This experiment suggested that the positive charge of the atoms is concentrated in one small area. If an incoming alpha particle strikes that area, then it can suffer a strong deflection. Otherwise it will pass through.

evenly over the whole atom, with the negative charge carried by electrons that were implanted in the positive charge. Thomson tried unsuccessfully to come up with an arrangement that was stable. (By "stable," we mean that the electric forces within the atom would hold it together; Fig. 23–10.)

No alternative model appeared until 1911, when Ernest Rutherford at Cambridge University's famous Cavendish Laboratory reported the amazing results of an experiment carried out by two of his students. They had shot a beam of particles known as alpha particles at a target made out of gold foil. (Rutherford had correctly concluded that alpha particles were actually helium atoms that had lost electrons.) The object of the experiment was to observe how the beam passed through the foil, and from that get information on the structure of the gold atoms. Rutherford's students expected, if Thomson's model was correct, that the alpha particles would undergo only very small deflections. However, they found that most particles were not deflected, but some were deflected by large amounts (Fig. 23–12).

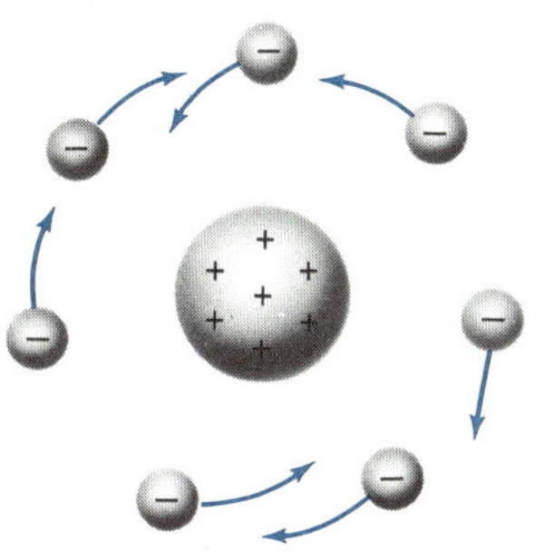

Fig. 23–13 The nuclear model of the atom. As a result of the Rutherford experiment, this model was proposed, in which the positive charge (and mass) of the atom is concentrated in the nucleus, and the electrons are in orbit around the nucleus.

Based on these results, Rutherford proposed a new model of the atom—the nuclear model. In this model, the positive charge is concentrated in the center of the atom, in the nucleus, and the electrons are in orbit about the nucleus. The electrons are held there by the electrical attraction of the positively charged nucleus for the negatively charged electrons. Whenever an alpha particle hits the nucleus, it is liable to be deflected severely. (See Fig. 23–13.)

In Rutherford's nuclear model, the hard bits of an atom are concentrated at its center.

Rutherford's model certainly seemed to be preferable to Thomson's, but it still had many inadequacies. One came from the fact that an electron in orbit is always accelerating (because the direction of its motion is always changing, as discussed in Chapter 6). Classical radiation theory says that accelerating particles must radiate, which means that they will be losing energy. Thus, in the classical theory, the orbiting electron loses energy and travels in a spiral path into the nucleus (Fig. 23–14). But, on the contrary, atoms appear to be quite stable, so something had to be wrong with the classical theory as applied to Rutherford's model. In addition, the model still did not explain the origin of spectral lines.

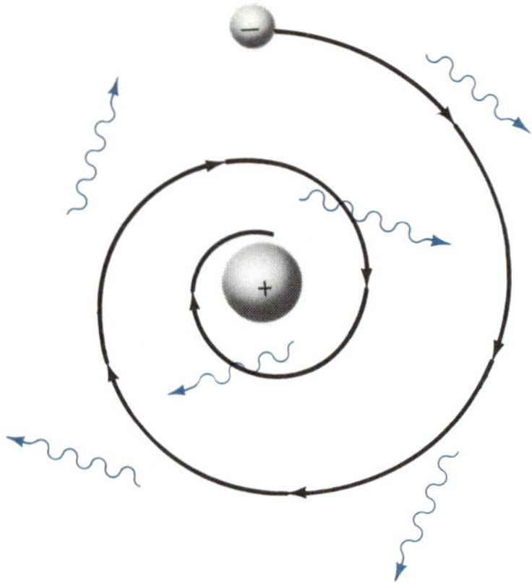

Fig. 23–14 A problem with the nuclear model of the atom was that classical radiation theory predicted that circling electrons would radiate energy away. As the energy of the electrons decreased, the orbits would get smaller and smaller until the electrons fell into the nucleus.

23.3b THE BOHR ATOM

In 1913, the Danish physicist Niels Bohr proposed a way out of this dilemma. Bohr started with the Rutherford picture of the atom but, just like Planck, he made some additional assumptions in order to make the theory agree with the observations. Although the orbits of electrons could have been circles or ellipses, he chose to analyze circular orbits.

Bohr's first assumption stated that *there are certain specific orbits in which electrons are allowed to stay. In these orbits, which are called stationary orbits, they do not radiate away all of their energy and spiral into the nucleus.* Bohr then assumed that *an atom could only radiate when an electron transferred from one stationary state to another.* Furthermore, when the atom did make such a transition, *the radiation would come off in a single bundle*—a photon, as proposed by Einstein—*whose energy was just equal to the difference between the energy of the electron in the initial and final stationary states.* Since the energy is equal to Planck's constant times the frequency,

Planck's constant × *frequency of photon*
= *energy difference between initial and final states.*

Fig. 23–15 Albert Einstein and Neils Bohr.

Next Bohr calculated the energies of the stationary states in hydrogen. He first proposed the correspondence principle, which says that when the electron is not too close to the proton, *the results of the quantum theory calculations must agree with the classical calculations.* The correspondence principle seems to be a reasonable requirement, since we know that the classical laws seem to hold on the large scale.

Based on his assumptions of the existence of stationary orbits, the value of the energy of emitted radiation, and the correspondence principle, Bohr went on to show theoretically how the energy of the stationary states is related to the angular momentum of the electron. He showed, in particular, that *the angular momentum is quantized.* He was then able to calculate the energies of the stationary states. The energies of these stationary states (sometimes referred to as energy levels) are shown in Fig. 23–16.

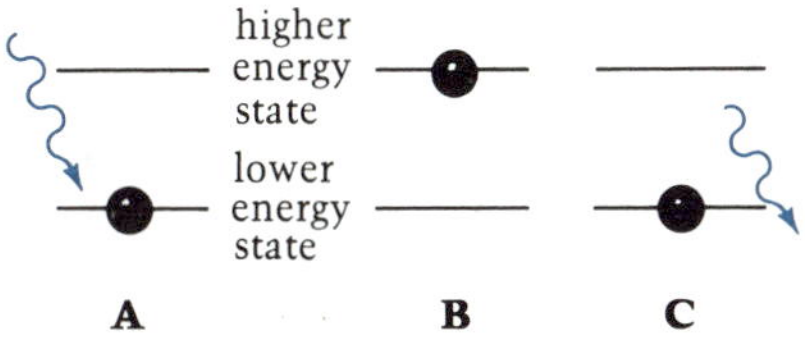

From the energies, Bohr could use Planck's formula ($E = hf$) to calculate the wavelengths at which hydrogen could emit or absorb radiation. The predicted wavelengths agreed with the observed values. In the Bohr atom, the energies of the various states depend on a single number, called the principal quantum number, which can only take on values that are integers.

The Bohr atom turned out to be quite a successful model. To some extent, the structure of other atoms could also be calculated, although there was really no complete technique for handling atoms with more than one electron.

In 1914, an experiment by J. Franck and Hertz provided another success for the Bohr theory. The experiment showed that when electrons collided with atoms, they could only lose a certain set of amounts of energy (we say, certain discrete energies) to those atoms.

higher energy state

lower energy state

A B C

Fig. 23–16 Energy levels and radiation in atoms. In this figure we look at an atom with two energy levels. A. We start with the atom in the lower energy state, and a photon approaching the atom. B. If the energy of the photon is just equal to the energy difference between the two levels, then the atom can absorb the photon and the electron will jump to a higher level. C. Eventually the electron will jump down to the lower state, and will emit a photon whose energy is equal to the energy difference between the levels.

So we see that Bohr's theory had gone a long way toward allowing a treatment of atomic structure and atomic spectra. It was confirmed by further experiments and generalized theoretically (by considering elliptical orbits, for example). But how much did it explain? Bohr had been forced to introduce some postulates to get a result that agreed with experiment, much as Planck had been forced to do. Of course, Bohr's work did include the quantization of radiation—photons—that had already been explained by Einstein. The problem still left was to explain why the stationary states existed.

Also, the Bohr theory did not explain why certain spectral lines were stronger than others. Although the Bohr atom was a great step forward, it must still be regarded as a stopgap measure rather than as a true explanation of atomic structure. We now think of the Bohr theory, along with its early generalizations, as the old quantum theory. It was not until ten years after Bohr proposed his atomic picture that the new quantum theory got its start.

23.4 *Matter Waves and Wave Mechanics*

The first hint of the new theory came in 1924, when a French graduate student, Louis de Broglie (pronounced approximately "duh-broy"), made a very simple observation. He noted that physicists seemed to be stuck with a theory of radiation in which light had certain properties of waves and certain properties of particles, and asked why the same could not be true of the electron. When we have a beam of electrons in a tube, their particle properties are apparent, but maybe when we get down to the scale of atoms,

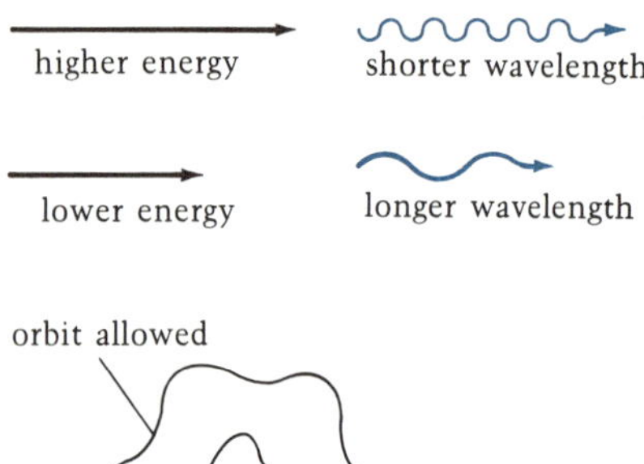

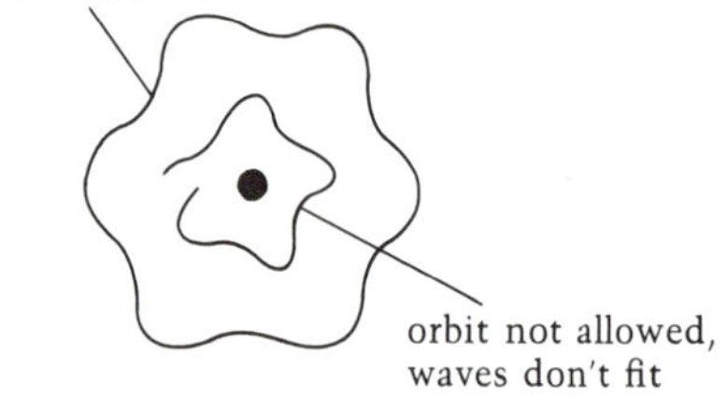

Fig. 23–17 Electron waves, as proposed by de Broglie. A higher energy (faster) electron corresponds to a shorter wavelength and a lower energy corresponds to a longer wavelength. An orbit for a particular electron in an atom is allowed only if you can fit an integer number of wavelengths around the atom. In this way, an electron orbit that is allowed is something like a standing wave.

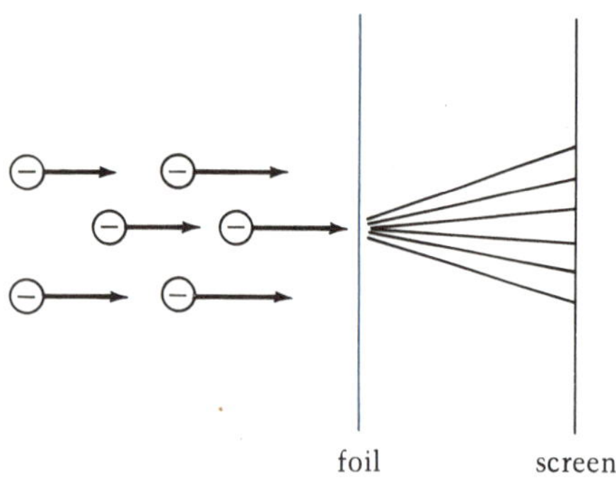

Fig. 23–18 Electron diffraction experiment. A beam of electrons strikes a foil target and the electrons are then deflected toward a screen. If the screen has a photographic plate, and we develop the plate to see where the electrons struck, we will see a pattern that is just like an interference pattern for light. This proves that electrons can act like waves.

electrons begin to take on wave properties. To make any test of these ideas, de Broglie first had to assign a frequency and a wavelength to a traveling electron. He was guided by the known results for photons. The wavelength of an electron turns out to be extremely short, only a few angstroms. It depends on the electron's momentum. As the momentum increases, the wavelength gets even smaller.

Because the electron's wavelength is so short, the wave properties of the electron are not normally obvious. After all, the effects of any wave are most apparent when we are dealing with objects comparable in size to the wavelength. And in our everyday world, objects are much bigger than the wavelength of an electron.

Using his idea, de Broglie was able to explain Bohr's stationary states. He said that electron orbits were like standing waves (Fig. 23–17). However, de Broglie's ideas were most unorthodox and hardly formed a complete theory. It is said that his professors were reluctant to accept de Broglie's thesis as suitable for a degree, and awarded the degree only after a favorable recommendation from Einstein. In any case, if de Broglie's matter waves were to have any real meaning, two major problems remained. One was the experimental problem of verifying the wavelike nature of the electron. The other was the problem of completing the theory to describe the properties of such waves. In such a theory, the structure of the hydrogen atom would be but one application out of many.

Experimental proof arrived in 1927, when Bell Telephone Laboratories scientists were studying what happened when electrons were scattered by crystals. The electron scattering patterns for several crystals looked very much like diffraction patterns. These patterns were characteristic of waves with wavelengths as predicted by de Broglie. The very fact that a diffraction pattern was observed could be explained only if the electrons behaved like waves in the scattering process. This experiment confirmed the wave nature of the electron as surely as Young's interference experiments with light demonstrated the wave properties of light (Fig. 23–18).

By the time this experimental work was done, considerable progress had been made on the theoretical side. The period around 1926 was a time of astonishingly rapid development and of extensive exchanges of ideas. Many physicists of the time have recounted the excitement that attended every calculation and discovery, as a group of young physicists worked in close correspondence with each other, especially in Copenhagen, Denmark, and in Göttingen, Germany. These groups of researchers were not sure where these new ideas of quantum theory and matter waves were taking them, but they were determined to pursue them as vigorously as possible.

A

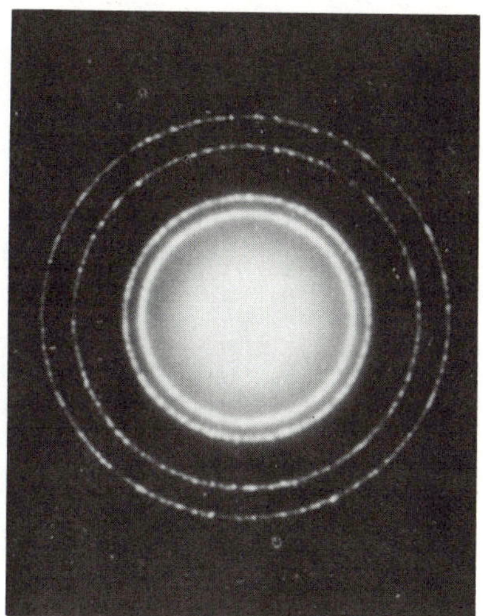
B

Fig. 23–19 Diffraction patterns produced by scattering from an aluminum foil target. A. The pattern is produced by x-rays. B. The pattern is produced by electrons whose wavelength is calculated to be close to the wavelength of the x-rays used in A.

23.5 *Quantum Mechanics*

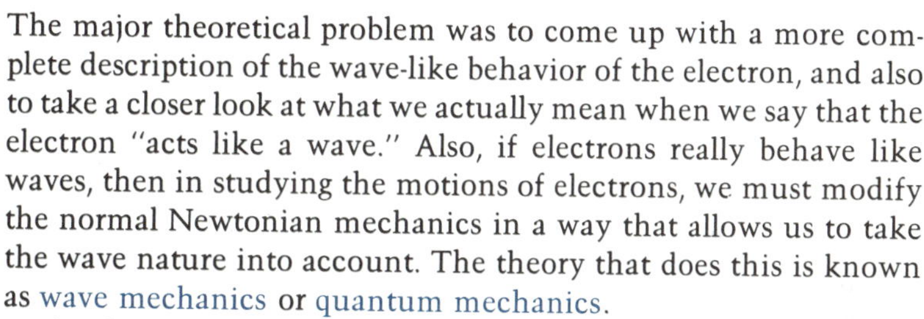

The major theoretical problem was to come up with a more complete description of the wave-like behavior of the electron, and also to take a closer look at what we actually mean when we say that the electron "acts like a wave." Also, if electrons really behave like waves, then in studying the motions of electrons, we must modify the normal Newtonian mechanics in a way that allows us to take the wave nature into account. The theory that does this is known as wave mechanics or quantum mechanics.

Fig. 23–20 Erwin Schrödinger.

The first real step in this area was taken in 1925 in Germany, when Erwin Schrödinger (pronounced "shro-ding-gur") presented an equation that described the wave-like motion of the electrons. This general equation, now known as the Schrödinger equation, could, in principle, be solved for any problem to predict how an electron would behave.

The Schrödinger equation looked very much like the equations in classical physics that describe the propagation of waves, such as sound waves, waves on a string, or electromagnetic waves. However, in the classical cases, the wave equations (the shorthand term for "equations that describe waves") were describing the variations of some particular quantity at different points in space and time. For example, with sound waves we deal with the pressure variations, with waves on a string we deal with the amplitude of the displacement on the string, and with electromagnetic waves we deal with the variations of electric and magnetic fields. What is it that varies when we talk about an electron wave? Schrödinger's equation is written in terms of a mathematical expression called the wave function, and this is the quantity that varies from place to place and from time to time.

But what is this wave function? How does knowing the wave functions tell us anything about the electron? The wave function turns out to be related to the probability of finding the electron at a given place at a given time.

When we deal with electrons as waves we cannot predict exactly where they will be. We can only know the probability of finding them in various places. This situation is very different from the

PROBABILITY

What do we mean by "probability"? It's not really such a strange idea. We all make decisions based on probability every day. For example, suppose you have to drive to school, to be there at 9:00, and you want to decide what time to leave your house. You rely on your past experience in getting to school to tell you how much time to allow. Of course, it doesn't take exactly the same amount of time every day. The conditions are always a little different, and you don't always hit all the traffic lights the same way. If you kept records of how long it took you for the last hundred trips you made (Fig. 23–21), you might find that eighty times it took between 55 minutes and 65 minutes, five times it took you between 50 and 55 minutes, and five times you really breezed through in less than 50 minutes. In addition, five times it took between 65 and 70 minutes, and five times (about which you would care to forget) you were stuck in bad traffic jams and it took almost 90 minutes. You could then use all this information to decide how much time to allow yourself. Past experience tells you that if you leave at 8:10 you have a chance to make it on time, but you would have to have some good luck. If you leave at 8:00, you can feel pretty confident that you will get there within five minutes of 9:00. If you want to be very confident of getting there by 9:00, you might leave at 7:55. When you actually leave depends on how important it is for you to be there by 9:00. For example, if you will fail the next time you are late,

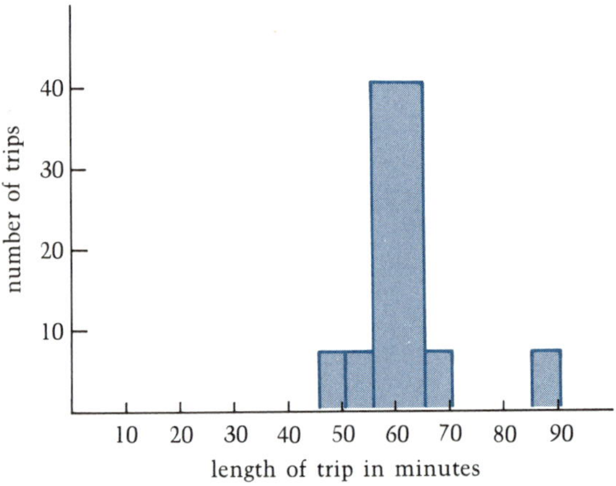

Fig. 23–21 Record of trips. For 100 trips, we graph the number of trips with length falling in each five-minute range. For example, 40 trips took between 55 and 60 minutes and 40 trips took between 60 and 65 minutes, while 5 trips took between 50 and 55 minutes.

you may feel that you have to leave at 7:30 to allow for traffic jams, even though they are rare (and, of course, you never know—there may be a two-hour jam). The lesson is that past experience tells us that we cannot predict exactly how long it will take you to reach school on any given day. It can only tell you the probability that the trip will take a certain amount of time. That is why weather forecasters now give probabilities that it will rain rather than absolute predictions.

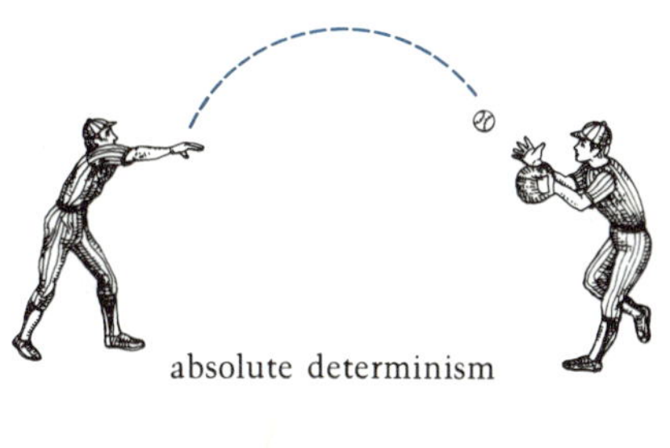

Fig. 23–22 When one baseball player throws a ball to the other, the laws of physics tell us exactly where the ball will go, and the fielder knows where to go to catch the ball. However, for an electron in an atom, we do not have such absolute determinism. We can only describe its position by a probability wave that gives us the probability of finding the electron at different positions.

definite answers that we get from Newtonian mechanics. In Newtonian mechanics, if we know where a particle starts, what its initial velocity is, and all the forces that act on that particle, we can predict *exactly* where it will be at any given time. We call this situation absolute determinism. It seems to work quite well on the large-scale world. However, when we descend to the subatomic world, it no longer applies. (It is this point that has led to many philosophical debates; Fig. 23–22).

Of course, we can do experiments and measure positions of things even on a subatomic scale. For example, we can measure how far an electron beam is bent as it passes through an electric field. How does such a measurement fit in with the probability notions? A beam of particles is made up of a large number of particles. Even though each electron may have its path bent by a *slightly* different amount, and although we cannot tell where any *particular* electron will end up, we can calculate where many will end up *on the average*. We find that the large majority of electrons fall very close to the average value. Although individual electrons can deviate from the average value, the farther we stray from that value, the less probable it is to find an electron.

Therefore, when we speak of an electron moving along, we are really talking about the motion of the probability wave that represents the electron. Such a concentration of the probability in one place (like a single pulse on a string) is called a wave packet (Fig. 23–23). Physicists have found that these quantum mechanical wave packets have many features in common with classical waves, such as those on the string.

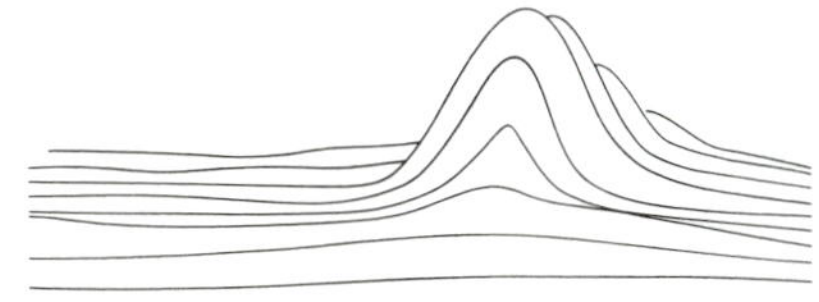

Fig. 23–23 If we plot the probability of finding a particular particle, we get what is called a wave packet. The particle is most likely to be found in the regions where the wave packet is highest above the zero level.

23.6 Uncertainty

Heisenberg had been working on another way of formulating quantum mechanics. In Heisenberg's picture, which was different in appearance from Schrödinger's but really amounted to the same thing, the emphasis was not on individual variables, such as energy or momentum, but only on the outcome of experiments—on observable quantities.

There is another interesting consequence of the wave nature of particles. It is called the uncertainty principle, and was first enunciated by Werner Heisenberg in 1927. Heisenberg's uncertainty principle says that *we cannot measure both the position and momentum of a particle with aribitrary accuracy.* There will be some uncertainty in each quantity. Morever, if you try to measure one more accurately, the other will be determined with less accuracy.

Fig. 23–24 Werner Heisenberg.

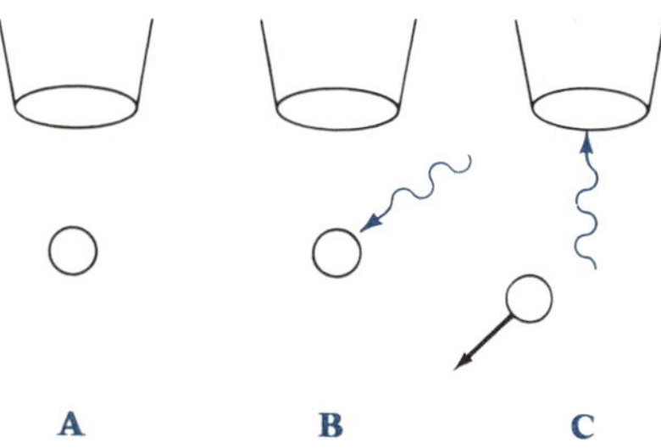

Fig. 23–25 The Heisenberg microscope. This is actually just a thought experiment, but illustrates the impossibility of absolutely determining the position and momentum of a particle, in this case the electron. If we want to observe an electron, we bounce some light off the electron and then look at the light in our microscope. However, the photon will give the electron a boost in momentum (as shown in C). The more accurately we wish to know the position of the electron, the shorter the wavelength of the photon that we use. However, a short-wavelength photon has a very high energy and will therefore give the electron a bigger kick. Therefore the more we know about the position of the electron, the less we know about its momentum.

In fact, if you take the product of the uncertainty in both of these quantities you find that

uncertainty in momentum × *uncertainty in position*
is greater than (Planck's constant)/2π.

Heisenberg illustrated the principle by talking about observing an electron through an imaginary microscope. If you want to locate the electron as accurately as possible, you must use light of a very short wavelength, or else the diffraction of the light will give you a large uncertainty in the position of the electron. However, shorter wavelength light carries more momentum, so in making a very accurate measure of the position of the electron, the light hits the electron and changes its momentum, giving you a large uncertainty in determining the momentum. You can't win. (See Fig. 23–25.)

However, the uncertainty principle goes beyond this description. It is not simply a property of microscopes and how we measure things. It is a consequence of the fact that electrons and other particles propagate like waves.

If the probability wave picture of the electron is correct, we should not be talking about electrons being in definite orbits or levels within the atom. Instead, we should talk about a smeared-out cloud of probability for each electron in the atom. However, the probability is not evenly smeared out throughout the atom. There is still a very high probability of finding the electron close to a particular fixed orbit, and each orbit will have some average energy. When we observe many photons that come from a particular transition in a group of atoms, the average energy of the photons is equal to the difference between the average energies of the two levels involved. In this sense, many physicists still find it conve-

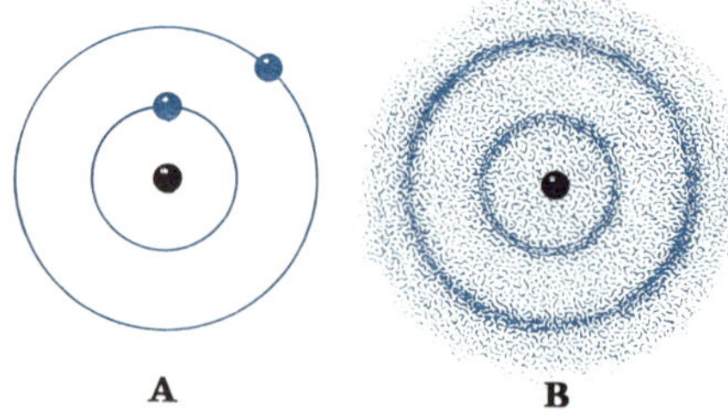

Fig. 23–26 A. Energy levels in an atom. B. The probability cloud for electrons in the atom. The probability is greatest for finding the electrons in the allowed energy levels, but there is still a small chance of finding them elsewhere.

Despite his many contributions to the development of quantum theory, Einstein did not like the probability interpretation of the wave function. "God does not play dice with the universe," he said.

nient to talk in terms of energy levels and orbits, even though they know that this is only a shorthand for what is really going on within the atoms. (See Fig. 23–26.)

23.7 The Pauli Exclusion Principle and Atomic Structure

By "state," we mean energy level in its most detailed sense, including not only the main level as defined in the Bohr atom but also subdivisions that depend on such factors as the spin orientation of the electron and the angular motions of the electron in its orbit. No two electrons can have all these properties in common.

When we try to figure out how electrons actually distribute themselves in atoms, we must also take into account the effects of the spin of the electrons. In 1927, Wolfgang Pauli considered the spin of the electrons in the framework of quantum mechanics. He came up with a very important idea, known as the Pauli exclusion principle, which says basically that *no two electrons can occupy exactly the same state.*

The Pauli exclusion principle means that if you are building an atom, each electron has to seek out a state that is not already occupied. You can get two electrons into the same basic energy level only when the two electrons are spinning in opposite directions (since the opposite spins mean that the electrons are not really in the same state).

Fig. 23–27 Wolfgang Pauli (left) and Neils Bohr watching a top spin.

As a result, the exclusion principle tells us that as we add more electrons to an atom, they have to go into higher and higher energy states. When we work out the details of the energy levels, we find that they group together, into shells. Because of the exclusion principle, each shell has a maximum number of electrons, which is equal to twice the number of energy levels within the shell (for there can be two electrons per energy level, one spinning each way; Fig. 23–28).

Shells that have their full complement of electrons are very stable, and are called closed. Atoms try to participate in chemical reactions, that is, form compounds with other atoms, that leave each atom with all shells filled. Atoms in the resulting molecules exchange or share some of their electrons in such reactions. When the electrons arrange themselves in an atom, they often fill inner shells before starting on the outer ones. The outermost shell, called the valence shell, is the one least likely to be full. The number of

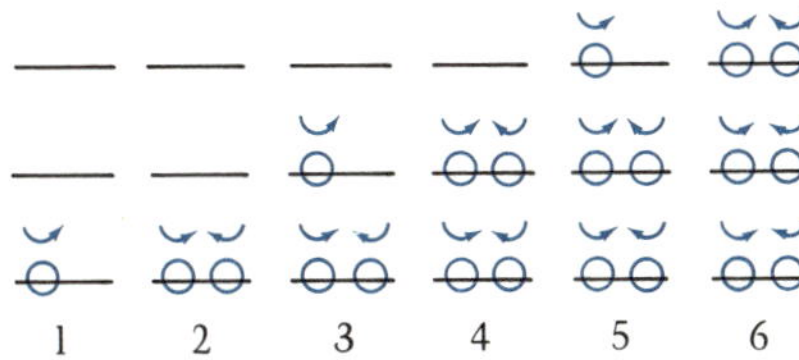

Fig. 23–28 The Pauli exclusion principle. This figure shows the lowest energy arrangements of one through six electrons. The exclusion principle says that no two electrons may occupy the same state. We can place two electrons in the same level, if they are spinning in opposite directions. This means that the second electron *may* be in the same level as the first, but the third *must* start a new level, and so on.

valence electrons in an atom determines the atom's chemical properties, and leads to the existence of the periodic table of the elements (Appendix 2).

23.8 *The Triumphs of Quantum Mechanics*

Quantum mechanics was not really thoroughly developed in the flurry of excitement around 1926. During the 1930s the quantum theory was applied to specific cases and had many successes in explaining atomic structure and a variety of other phenomena. The

---

PERILS OF MODERN LIVING, by H. P. Furth

A kind of matter directly opposed to the matter known on earth exists somewhere else in the universe, Dr Edward Teller has said. . . . He said there may be anti-stars and anti-galaxies entirely composed of such anti-matter. Teller did not describe the properties of anti-matter except to say there is none of it on earth, and that it would explode on contact with ordinary matter.

San Francisco Chronicle.

Well up beyond the tropostrata
There is a region stark and steller
Where, on a streak of anti-matter,
Lived Dr Edward Anti-Teller.

Remote from fusion's origin,
He lived unguessed and unawares
With all his anti-kith and kin,
And kept macassars on his chairs.

One morning, idling by the sea,
He spied a tin of monstrous girth
That bore three letters: A. E. C.
Out stepped a visitor from Earth.

The "A.E.C." was the Atomic Energy Commission.

Then, shouting gladly o'er the sands,
Met two who in their alien ways
Were like as lentils. Their right hands
Clasped, and the rest was gamma rays.

Note: An antimacassar is the doily that we used on the arms and backs of sofas and chairs to protect them from hair oil (of which Macassar was a brand). Obviously, an anti-antimacassar must be a macassar.

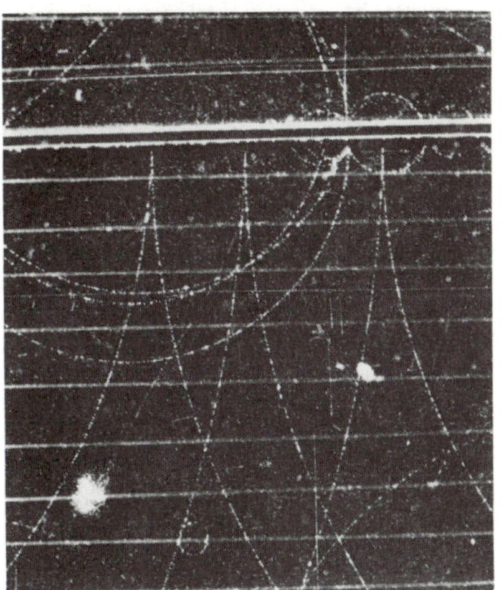

Fig. 23–29 Cloud chamber tracks of electron-positron pairs.

one thing that Heisenberg and Schrödinger had not done was to come up with a way of expressing the ideas of quantum mechanics in a way that was also consistent with the special theory of relativity, and Paul A. M. Dirac succeeded in this task in 1928. An interesting consequence of Dirac's work was the prediction of another particle, similar to the electron but with certain properties, such as the charge, reversed. This particle was called the positron, and its discovery in 1932 was a triumph for quantum theory.

The positron and electron have the interesting property that if one of each comes close enough together they can annihilate each other, converting all of their mass into energy ($E = mc^2$). We say that the positron is the antiparticle to the electron; it was the first example of antimatter.

Considerable theoretical refinement followed the development of the basic ideas of quantum mechanics. The most complete the-

TABLE 23–1 Time Line of the Quantum Revolution

1900	Planck's explanation of black-body radiation. Investigations of photoelectric effect by Lenard.
1905	Einstein's explanation of the photoelectric effect.
1907	Einstein's use of quantization to explain heat capacities of solids.
1909	Quantization of charge—Millikan oil-drop experiment.
1911	Nuclear atom proposed by Rutherford.
1913	Bohr atom proposed.
1914	Franck-Hertz experiment confirms energy quantization in atoms.
1914–1916	Millikan's experiments verify Einstein's explanation of photoelectric effect.
1916	Wilson-Sommerfeld generalization of Bohr theory to elliptical orbits.
1923	Compton confirms particle properties of light in scattering.
1924	De Broglie proposes that electrons can act as waves.
1925	Wave mechanics developed by Schrödinger; alternative but equivalent theory developed by Heisenberg. Pauli exclusion principle.
1927	Heisenberg uncertainty principle; Davisson and Germer electron-diffraction experiment.
1928	Development of relativistic quantum mechanics by Dirac, including prediction of positron.
1932	Discovery of neutron by Chadwick (Chapter 26). Discovery of positron by Anderson.

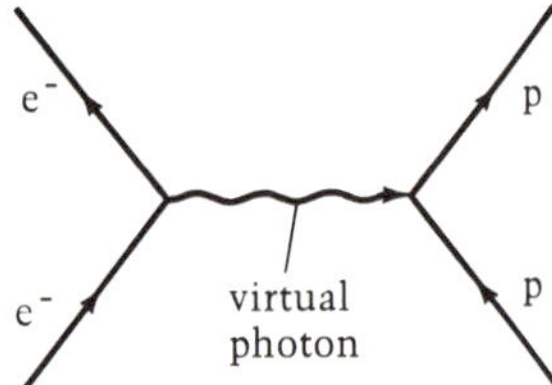

Fig. 23–30 Feynman diagram for electron-proton scattering. In this schematic drawing, the electron comes in from the lower left, emits a virtual photon (wiggly line), and changes its direction. The proton enters from the lower right, absorbs the virtual photon, and changes its direction. For the observer, the virtual photon is never seen.

In 1965 Richard Feynman, Julian Schwinger (then at Harvard), and Sin-Itiro Tomonaga shared the Nobel Prize in physics for their development of QED.

ory to date of the quantum description of electricity and magnetism is known as quantum electrodynamics (QED). Much of the pioneering work on QED was done in the late 1940s by Richard Feynman, a Caltech professor who is one of the strongest advocates of the idea that physical theories should be simple and beautiful. In his paper on QED, Feynman suggested a simple pictorial way of showing what goes on when one charged particle exerts a force on another. These pictures, known as Feynman diagrams, serve as a guide to the physicist who wants to calculate the details of a given process. (The calculations are much more complicated than the diagrams.)

One of the important features of QED, as illustrated in various Feynman diagrams (Fig. 23–30) is that when a charged particle exerts an electric force on another charged particle, the "force" is actually carried by a photon emitted by one particle and absorbed by the other particle. We cannot directly observe the photons that carry the electric force. These photons are therefore called virtual photons, similar to our use of the word "virtual" in optics to signify an image that isn't really there. QED has been subjected to a variety of exacting tests, and it has had remarkable success, although a few theoretical problems still remain.

23.9 *The Dual Nature of Light and Particles*

We have come a long way in this chapter. We started with light as a wave and with an electron that was a particle, and we now see that neither always acts like either. Each one possesses certain properties that sometimes make it act like a particle and certain properties that sometimes make it act like a wave. This double identity is called the wave-particle duality. The behavior of an electron in a given experiment depends on the experiment you are doing. Wave and particle properties do not manifest themselves at the same time. It is also important to remember that once you have said that particles can behave like waves, then the existence of energy levels is not so hard to understand. The conditions for energy levels arise whenever you confine the particle waves to a small region, exactly the same way that standing waves arise on a string only for certain wavelengths.

These ideas must sound a little strange, and it certainly takes time to get used to them. Indeed, Einstein didn't like them at all. In trying to accept the ideas of quantum mechanics, we must look at things very objectively, ignoring our previous prejudices about the way things are. When we do this, then the quantum theory stands up to the tests we make as well as does any theory we accept as valid. Quantum mechanics has by now been confirmed over and over in laboratories around the world.

Also, such things as the wavelength assigned to an electron are not merely theoretical ideas whose only place is in the calculations of atomic structure. For example, because the wavelengths of electrons are so short compared with those of light, electrons are not diffracted as much as light when they go through apertures. We have already seen in Chapter 19 that the shorter the wavelength of light you use to illuminate the object being studied in a microscope, the better the resolution you have. For this reason, when exceptionally high resolution (and magnification) is required, light beams are no longer useful in a microscope. They are replaced by electron beams, and we have the electron microscope!

Key Words

black body
equilibrium
ultraviolet catastrophe
Planck's constant
photoelectric effect
quantum
photon
spectral lines
Balmer series
alpha particles
nuclear model
nucleus
stationary orbits
correspondence principle
energy levels
Bohr atom
principal quantum number
old quantum theory
matter waves
wave mechanics (quantum mechanics)
Schrödinger equation
wave function
probability
absolute determinism
wave packet
uncertainty principle
shell
closed shell
valence shell
valence electrons
positron
annihilate
antiparticle
antimatter
quantum electrodynamics (QED)
virtual photons
wave-particle duality

Questions

1. Which of the ideas discussed in this chapter do you consider "revolutionary"? List three.
2. List three ways in which the classical theory of black-body radiation failed to predict the results of experiments.
3. (a) List those pieces of evidence that lead us to conclude that light is a wave. (b) List those pieces of evidence that lead us to conclude that light is a particle.

4. What is the ultraviolet catastrophe?
5. Which has more energy, a photon in the infrared or one in the ultraviolet? (Explain your answer.)
6. In the photoelectric effect, why doesn't stronger light produce more energetic electrons?
7. The Balmer formula correctly describes the wavelengths of a series of hydrogen lines. Why do we say that the existence of the formula did not constitute a real theory?
8. If the mass of atoms were not concentrated in small volumes at their centers, how would the results of Rutherford's scattering experiments be different?
9. (a) What assumptions did Bohr make in describing the atom? (b) Using these assumptions, what was he able to derive?
10. (a) List those pieces of evidence that lead us to conclude that electrons are particles. (b) List those pieces of evidence that lead us to conclude that electrons are waves.
11. One electron is traveling twice as fast as another. Which has the longer wavelength?
12. When we talk about the electron as a wave, what quantity actually exhibits the wave-like behavior?
13. Why can we control a car and predict where it will go without worrying about the uncertainty principle?
14. Which aspects of quantum mechanics are related to philosophical problems of determinism? How might a hidden-variables theory be more appealing in a philosophical sense?
15. How does the Pauli exclusion principle influence the structure of atoms?
16. What is meant by the "wave-particle duality" of matter?
17. As you heat a black body, (a) what happens to the wavelength at which the greatest amount of radiation is given off? (b) what happens to the total amount of radiation given off?
18. One black body is at 100 K and another is at 1000 K. (a) Which one gives off more energy at a wavelength of 5 m? (b) Which one gives off more energy at a wavelength of 500 nm? (Assume that both objects are the same size.)
19. In Chapter 12 we discussed the greenhouse effect. In what ways are the properties of black-body radiation important in the operation of the greenhouse effect? (*Hint*: In what part of the spectrum does radiation from the sun peak and in what part of the spectrum does radiation from the earth peak?)
20. If we double the temperature of an object, what happens to the wavelength at which its radiation peaks?
21. A given star appears blue. Do you think that this star is hotter or cooler than the sun? Explain your answer.

22. (a) If we double the frequency of a photon, what happens to the energy? (b) If we double the wavelength of a photon, what happens to the energy?
23. What are the similarities between the question of the wave/particle nature of light and the question of the wave/particle nature of the electron?
24. In what way did Einstein's explanation of the photoelectric effect incorporate Planck's work on black-body radiation?
25. What determines the energy of electrons ejected in the photoelectric effect?
26. What determines the number of electrons ejected per second in the photoelectric effect?
27. In what ways did the observations of atomic spectra contradict the ideas of classical physics?
28. (a) What features of atomic spectra was the Bohr atom able to explain? (b) What features of atomic spectra was the Bohr atom not able to explain?
29. What assumptions did Bohr have to make to get his model for the atom?
30. Why are the wave properties of the electron difficult to observe?
31. List three situations in everyday life in which you make decisions based on probabilities, rather than a certainty of what will happen.
32. Suppose we have one electron that is contained within a room and another that is contained within a breadbox. Which one has a larger uncertainty in momentum according to the uncertainty principle?
33. Some people have said that chemistry is just applied atomic physics. What information in this chapter tends to support that idea?
34. List the refinements that took place in quantum theory after the work of Bohr.
35. What is a virtual photon, and what is its importance in electromagnetic theory?

CHAPTER 24

Lasers

Fig. 24–1 Hooray for . . . Searchlights and laser beams crisscrossing across the night sky at the dedication of the new sign on Mount Lee above Hollywood, California. The narrower beams are the laser beams.

We live in an age where laser light shows are common and lasers light up the sky. Within the past few years, lasers have come into widespread use in science, medicine, manufacturing, and in many other fields. Laser stands for "light amplification by stimulated emission of radiation." Although the name is long, the mechanism is straightforward. We saw in Chapter 16 how lasers have brought holograms from a theoretical principle to a practical reality. In this chapter, we shall see how lasers work and how they have been put to practical use.

24.1 *The Discovery of Masers*

In the preceding chapter we described how electrons in atoms occupy certain energy states. The lowest energy state is called the ground state and the other states are called excited states. When an atom is excited, one or more of its electrons move into states of higher energy. In the Bohr model of the atom (which is a sufficient model for most of the purposes of this chapter even though it is no longer our best understanding), exciting an electron corresponds to moving it into an orbit of larger radius (Fig. 24–3).

We can excite the atom either by hitting it with other atoms or with radiation.

When left alone in an excited state, an electron will eventually drop back down to the ground state. Since it is jumping from a higher energy level to a lower one, it spontaneously (by itself) emits

Fig. 24–2 Luke battles Darth Vader. © Lucasfilm Ltd. (LFL) 1980. All rights reserved. From the motion picture: *The Empire Strikes Back*, courtesy of Lucasfilm, Ltd.

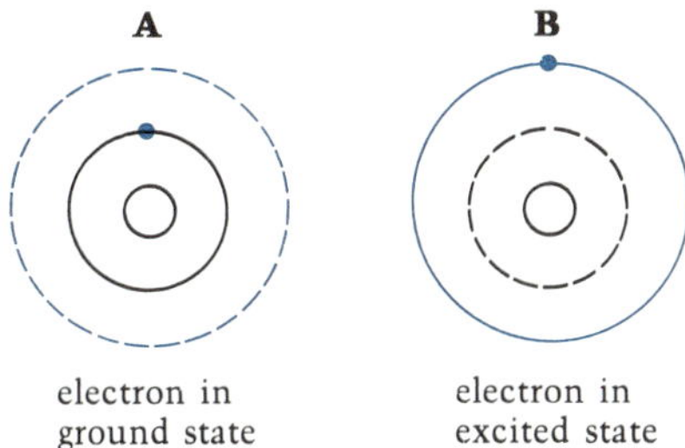

Fig. 24–3 The ground state is the lowest energy state in an atom. Any higher energy state is an excited state. In this drawing, the state that the electron is actually in is depicted with a solid line and the other state with a dashed line. A. Electron in ground state. B. Electron in excited state.

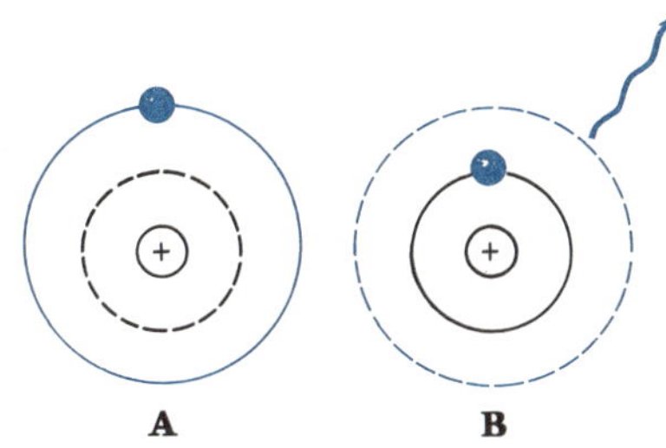

Fig. 24–4 Spontaneous emission. A. The electron is in the excited state. B. After some time, the electron jumps to the ground state and a photon is emitted.

energy as it does so, and this process is called spontaneous emission. To get to the excited state in the first place, it had to absorb energy. The process of spontaneous emission, in which the electron emits a photon and drops to a lower energy level, is just the reverse of the process of absorption, in which a photon is absorbed and the electron jumps to a higher energy level. Remember, absorption can take place only if the photon has exactly the amount of energy required to get the electron from the lower state to the higher state. Similarly, when the emission process takes place, the photon emitted has an energy equal to the difference between the energy of the upper level and that of the lower level (Fig. 24–4).

It might seem that an atom will interact with radiation either by absorption or spontaneous emission, and that these are the only possibilities. However, in 1917, Albert Einstein realized that there was a third possibility: stimulated emission. This process gets its name because the emission is made to happen—is stimulated—by radiation passing by, rather than happening spontaneously. Stimulated emission takes place only from excited atoms. The emission can only be stimulated by a photon whose energy is exactly equal to the energy of the photon that will be emitted, that is, the incoming photon must be tuned to the energy difference between two energy levels. Before the stimulated-emission process, we have only one photon of that energy—the photon that is going to stimulate the

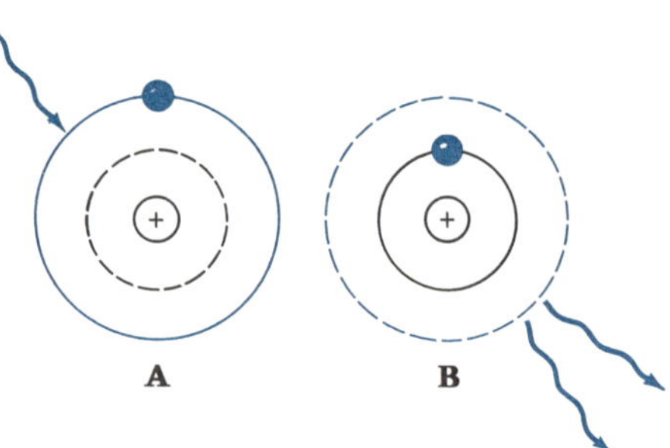

Fig. 24–5 Stimulated emission. A. The electron is in the excited state. A photon approaches the atom, and the energy of the photon is equal to the energy difference between the excited state and the ground state. The photon induces the electron to jump from the excited state to the ground state, emitting a photon in the process. B. Now we have two photons of the same energy instead of the original one.

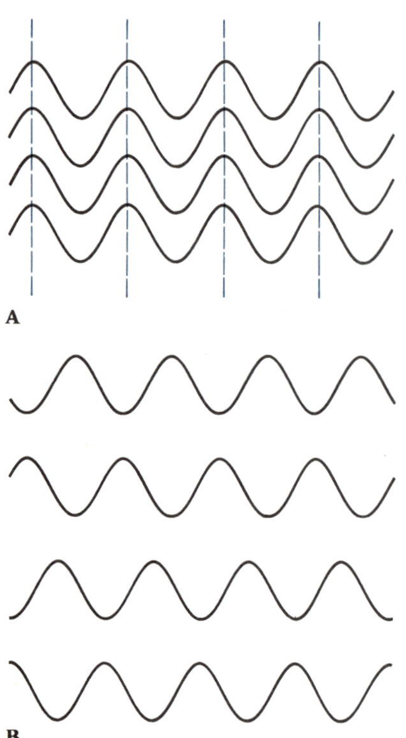

Fig. 24–6 A. Coherent waves are in step with each other. B. With incoherent waves there is no relationship between one wave and another.

emission. After the emission, we have two or more photons of that energy—the original one and the emitted ones. Since we have increased the number of photons at a given energy, stimulated emission can be used to amplify—strengthen—the original radiation passing through a material (Fig. 24–5).

Further, since the emitted photon is in phase with the original photon, all the stimulated emission can be made coherent, as we discussed in Chapter 15; that is, all the waves go up and down together. That lasers produce coherent radiation is one of their most important properties; it is the property we use in holography. (Ordinary light is *incoherent;* that is, the waves vibrate randomly—see Fig. 24–6.)

Let us now consider a large number of atoms, as shown in Fig. 24–7. If we could get a large fraction of them to the same excited state, then even a weak stimulus could get all the atoms to jump down to a lower energy state (perhaps the ground state). As a result, the coherent radiation given off would be much stronger than the weak radiation that we put in. It doesn't matter for this purpose how the atoms get excited in the first place. In fact, there are several ways to raise them to a higher energy state: the atoms can be excited by special kinds of collisions with other atoms, or by shining another wavelength of light on them.

Normally, when we simply heat a gas, liquid, or solid, the atoms

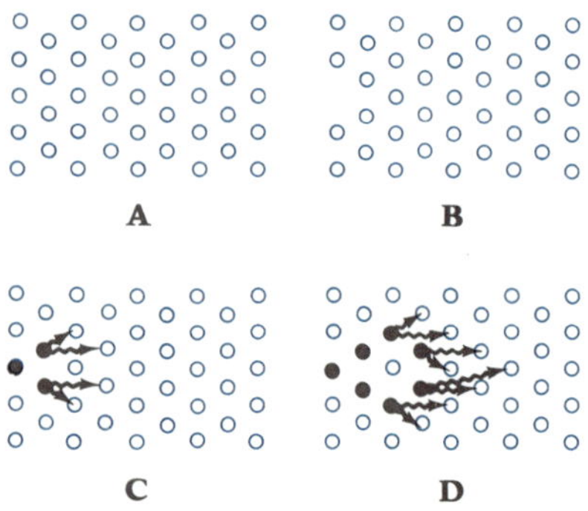

Fig. 24–7 Stimulated emission in a maser. In this drawing the open circles represent atoms in excited states and the dots represent atoms in the ground state. A. We start with all the atoms in the excited state. (In a real maser not all, but a large fraction, will be in that state.) A photon with an energy equal to the energy difference between the excited and ground states enters from the left. B. The photon strikes an atom, inducing it to emit a photon. Now there are two photons with the same energy. C. The two photons strike atoms, inducing them to drop to the ground state. We now have four photons. D. Each of these photons induces the same process in an atom, again doubling the number of photons. This process continues as the number of photons grows.

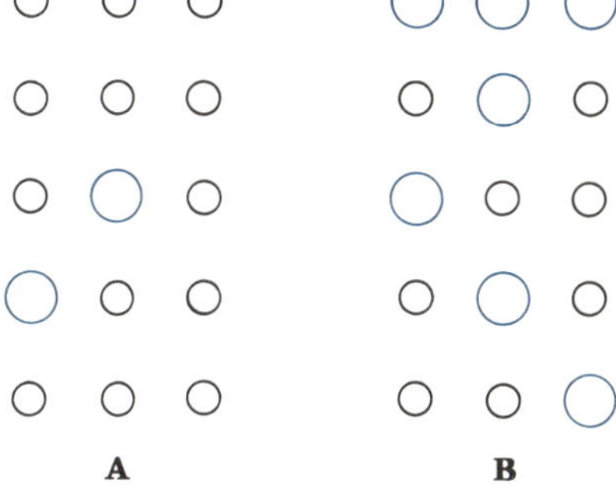

Fig. 24–8 In this drawing the large circles represent atoms in the excited state and the small circles represent atoms in the ground state. A. At a low temperature, most of the atoms will be in the ground state. B. At a higher temperature, more of the atoms will be in the excited state.

Fig. 24–9 In order to get a maser or laser we must artificially prepare a sample with an extraordinary number of atoms in the excited state. This condition is called a population inversion, since the fraction of atoms in the ground and excited states is inverted from the normal situation.

distribute themselves among the ground state and several excited states depending on how hot we heat the medium (Fig. 24–8). If we heat it more, then more of the atoms are changed from lower energy states into higher energy states. But the process is a gradual one, and the number (the population) of atoms at a particular energy level always declines as we go from lower energy states to higher ones. But we can choose some particular level (say, level 6) and put many more atoms at that level than at lower levels. Then when the medium is stimulated, the amount of stimulated emission is particularly strong. Having more atoms in a high-energy state is the inverse of the normal situation, in which most atoms are in the lowest energy states, so we say that we have a population inversion for our atoms. We concentrate on inverting the relative populations of two particular levels, say, making the population of level 6 much greater than the population of level 3 (Fig. 24–9).

The basic way of using this principle to amplify radiation was first worked out in 1947 and 1948 by Charles Townes and Arthur Schawlow in the United States and simultaneously by a group of Soviet scientists. (Townes and the leaders of the Soviet group were awarded the 1954 Nobel Prize in physics for this research.) They worked in the short end of the radio region of the spectrum; the wavelengths in this band—from about 1 cm to 30 cm—are called

A

B

Fig. 24–10 Charles Townes and Arthur Schawlow, the inventors of masers, shown in the mid-1960s after masers had led to the invention of lasers. A. Townes in his lab. (Photo by Howard Sochurek © National Geographic Society) B. Schawlow playing with a laser for a classroom demonstration. He is bursting a colored balloon inside a clear one by shooting a laser beam of that color at it. The beam passes through the clear balloon, but its energy is absorbed by the colored one. (Photo by Jack Fields © National Geographic Society)

"microwaves," because they are relatively short for radio waves. Since Townes and the others amplified microwave radiation by applying the principle of stimulated emission, they made the first maser, "microwave amplification by stimulated emission of radiation."

One disadvantage of the early masers was the need to operate them at very cold temperatures, close to absolute zero. Some of these masers were used in radio telescopes to amplify the very faint radio signals that come from some stars and galaxies in outer space. Keeping the masers surrounded by liquid helium in order to maintain a temperature of 4 K made them very difficult to use.

Shortly after the construction of the first masers, astronomers were astounded to find that certain clouds of gas between the stars give off radiation that shows the effects of a population inversion among the atoms that make up the clouds. The different interstellar molecules are excited by several different processes, sometimes by collisions and sometimes by radiation. It turned out that these interstellar clouds were themselves masers, so masers were not such extraordinary things after all.

24.2 *The Discovery of Lasers*

Many masers were made before anybody succeeded in making a similar device that worked in the visible part of the spectrum. In 1957, Gordon Gould, a graduate student at Columbia, was working on the problem and hit upon an idea that he thought could provide

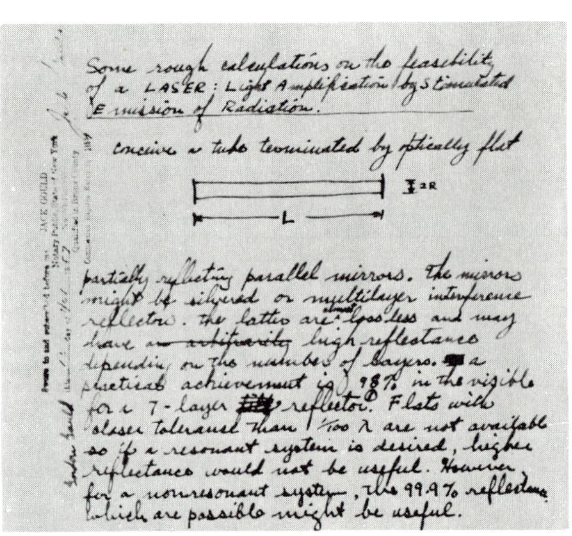

Some rough calculations on the feasibility of a LASER: Light Amplification by Stimulated Emission of Radiation.

Conceive a tube terminated by optically flat

2R

L

partially reflecting parallel mirrors. The mirrors might be silvered or multilayer interference reflectors. The latter are almost lossless and may have ~~an arbitrarily~~ high reflectance depending on the number of layers. a practical achievement is 98% in the visible for a 7-layer reflector. Flats with closer tolerance than 1/100 λ are not available so if a resonant system is desired, higher reflectance would not be useful. However for a nonresonant system, the 99.9% reflectance which are possible might be useful.

JACK GOULD
Notary Public, State of New York

Fig. 24–11 The notarized page from Gordon Gould's original notebook in which he advanced the idea of a laser cavity. The notary, though also of the name Gould, was no relation.

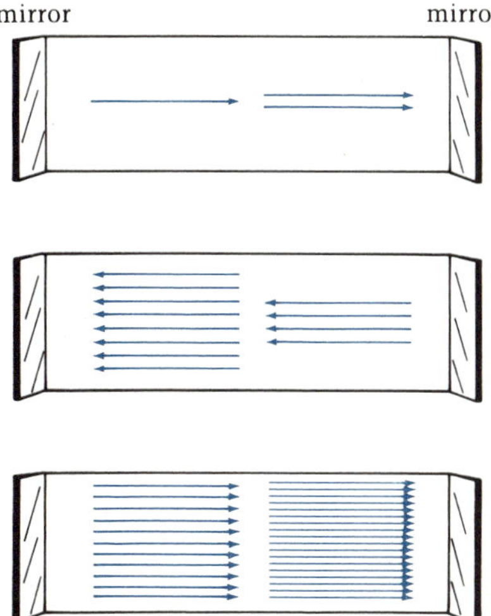

Fig. 24–12 The use of mirrors in laser design. In each pass through the cell, the increase in the number of photons is not very great unless the cell is made very long. However, if we put mirrors at the ends, the radiation will pass back and forth, increasing in strength each time. In a real laser, the mirror on one side will be made only partially reflecting, so some of the light can get out, giving us our laser beam.

the solution. He went down to his neighborhood candy store in the Bronx, and had the page of his notebook on which he wrote his idea notarized to establish his legal priority. Townes was also working on the problem and came up with a similar idea at about the same time. (Just who discovered what, and when, has been the subject of lengthy court battles over patent rights, and we make no attempt to resolve the delicate distinctions.)

These new ideas led to the construction of what was at first called an "optical maser" but was soon called a laser, for "light amplification by stimulated emission of radiation." Less than two years after the theoretical principles were first put forth, Theodore Maiman at Hughes Aircraft built the first working model.

In interstellar space, masers cover huge regions, and the stimulated emission has a lot of room to develop. On earth, however, the container for the atoms (we call it the cavity) is small, and the new design used mirrors to make the space effectively bigger. The light bounces back and forth many times between the mirrors (Fig. 24–12) building up the amount of stimulated emission. Usually the cavity is a cylindrical tube with mirrors at both ends; the mirror at one end is made partially transparent so that the laser light can escape. The stimulated emission is almost all traveling in the direction of the original radiation, so most of it comes out through the partially transparent end rather than through the sides.

Different materials can be used inside the cavity. At first, solid materials, such as rubies, were used (these were industrial rubies, not of gem quality). Later, ways were found to use gases, such as a helium-neon mixture, giving us helium-neon lasers in particular and gas lasers in general. By now, scientists have succeeded in making almost any kind of material they choose undergo laser action. Even a Jell-O laser has been built. (See Fig. 24–13.)

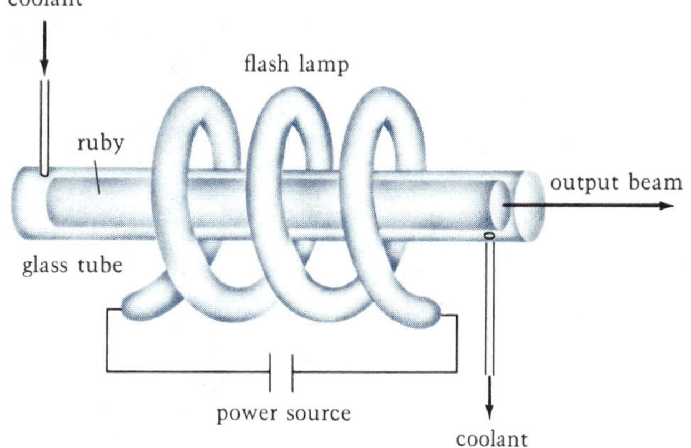

Fig. 24–13 A ruby laser. A synthetic ruby crystal is cut so that the ends are parallel. The ends are polished; one is covered with a completely reflective coating and the other with a partially transmitting coating. A flash lamp is used to excite the electrons, by providing energy, up to a state from which they can be stimulated to jump down. The entire apparatus is cooled to reduce the movement of the atoms and thus lower the number of collisions of atoms that might hit the excited atoms before they can emit radiation.

When we look at a piece of paper that is illuminated with a laser, we see a pattern of light and dark called a speckle pattern. *The speckles are caused by the coherent light interfering with itself (as was discussed in Section 15.3) as it bounces off the paper (see Fig. 24–14).*

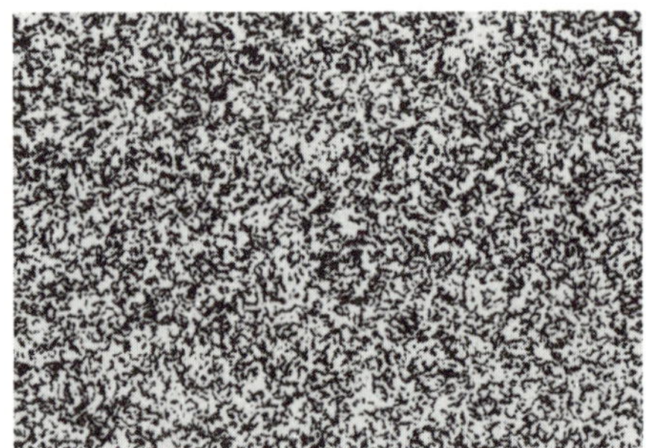

Fig. 24–14 A close view of a laser beam hitting a surface, photographed with a very short exposure, shows this speckled pattern. (Courtesy of James C. Wyant, University of Arizona)

Each different material emits laser light at different characteristic frequencies. Helium-neon lasers, for example, emit red light at a very specific wavelength. Certain types of tunable lasers have been developed that can emit laser light at any wavelength in a given range.

Some lasers can emit laser light only in short bursts, while other lasers can give off radiation continuously. The power in a burst can be much higher than the power in a continuous laser, so both types have advantages.

Fig. 24–15 James Bond faces a laser in *Goldfinger*.

24.3 *The Uses of Lasers*

The laser was a product that was quickly put to a wide variety of uses. Within three years of the first working laser, a laser appeared in a James Bond movie. We have already mentioned that amplification and coherence are important laser properties. Because stimulated emission strongly tends to be emitted in the direction of the stimulating photon, we can focus a laser beam with extreme precision.

Laser beams can be made so strong and focused so well that we

Fig. 24–16 This laser beam traveled more than 27 miles over London, striking Big Ben. The $\frac{3}{8}$-inch beam, produced by a 60-watt argon laser, spread to over 10 feet in diameter.

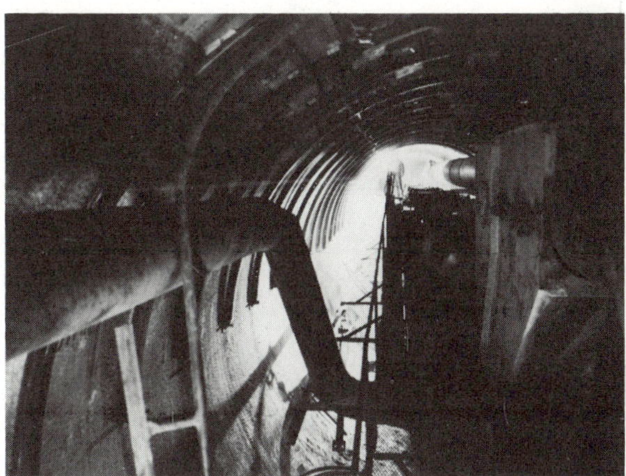

(left) Fig. 24–17 Measuring the distance from the earth to the moon by laser technology—to within a few centimeters.

(right) Fig. 24–18 The straight line marked out by the laser beam keeps a huge tunneling machine boring straight ahead through solid rock in this irrigation tunnel under construction. The laser beam can be seen traveling down the left side of the tube. Lights on the console of the tunneling machine tell the operator if the machine is off course. The tunnel was straight to a centimeter after 2 km. (Photo by Hughes Tool Company © National Geographic Society)

can bounce them off the moon. A laser beam sent by bouncing the light off a 2.5-meter telescope mirror (Fig. 24–17) spread out only to the size of a house when it reached the moon. Several of the Apollo missions to the moon carried reflecting mirrors that bounce back light in the direction that the light came in, and we can receive this reflected light back on earth. By carefully timing the round trip of the light, we can tell the distance to the moon to within a few centimeters, a factor of a hundred or more better than the accuracy of prelaser measurements. Now we can refine our knowledge of the moon's orbit.

The amount of information that can be carried per second on a beam of electromagnetic waves increases as the frequency increases. Since light has much higher frequencies than radio waves, much more information can be carried by a light beam than by a radio beam. Thus by modulating the beam in much the same way that we modulate radio transmissions (that is, by causing slight variations of a basic beam), we could carry many television channels or telephone conversations simultaneously. Optical fibers, as discussed in Chapter 17, are now being used to carry laser beams that carry telephone and television signals.

Light travels in a straight line, so a focused laser beam makes an exceptionally accurate ruler. Lasers were used, for example, to lay out the tunnels of the Bay Area Rapid Transit (BART) system in California.

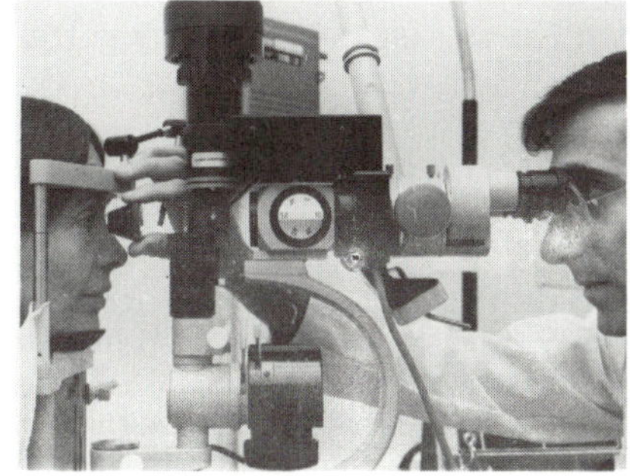

Fig. 24–19 A corneal contact lens is being applied to the cornea of this woman's eye by an argon laser.

Surgeons use laser beams to focus light at the tip of a detached retina, welding the retina back to the inner surface of the eye without need to cut into the eye. And surgeons also use lasers as knives that cut through human skin and tissue cleanly while heat from the laser closes the blood vessels that were cut, thus minimizing bleeding.

A laser beam can be focused so that it delivers all its energy at a small area. This energy, turned to heat, can burn through a block of wood or weld together arteries or other parts of the body.

Laser cutting works in industry too. Lasers can be used to cut through many layers of cloth simultaneously, in order to shape, at one time, the material for fifty pairs of pants. And on a Detroit assembly line, small holes in pieces of plastic or metal can often be cut better with a laser than they can with a metal drill. Further, lasers have been used to make holes in diamonds, which metal won't drill into at all. (The diamonds are then used to draw out wire into long strands.)

Fig. 24–20 The worker in the background is using a laser to grind metal.

Optical physicists use laser beams to line up series of lenses and sometimes to enable direct ray tracing. The lenses in telescopes and cameras are also routinely aligned with lasers when they are set up. A new telescope system (Fig. 24–21) is now using laser signals to keep six 1.8-meter (72-inch) telescopes perfectly lined up in order to make the equivalent of a 4.5-meter (176-inch) telescope. This Multiple Mirror Telescope (MMT), built on an Arizona mountain by a joint Harvard–Smithsonian–University of Arizona effort, opened for observation in 1978.

Biologists use a laser's ability to make ultrashort pulses of radiation. A laser that produces pulses that are only 0.3 picoseconds (0.3×10^{-9} seconds) has been used to study the excitation of rhodopsin, the eye's visual purple, which we discussed in Chapter 18. The reaction that results when light hits rhodopsin takes place on that time scale and could not be studied before lasers were available.

Chemists use lasers to excite specific reactions in order to study particular isotopes of atoms or molecules. One reaction that has been studied is that by which nitric oxide (NO) and ozone (O_3) make nitrogen dioxide (NO_2) and molecular oxygen (O_2). This process is important in the production of smog from automobile exhausts; it also plays a role in the chemistry of the ozone layer that surrounds the earth and protects us from the sun's harmful ultraviolet radiation. Other studies with laser picosecond pulses have shown how excited molecules bend and twist and how they participate in reactions.

Solid-state physicists use lasers too, as we shall see in the next chapter. For example, because we can accurately focus a laser beam,

Fig. 24–21 The Multiple Mirror Telescope, a new telescope design erected on Mount Hopkins in Arizona, is a joint project of the Harvard-Smithsonian Center for Astrophysics and the University of Arizona. It contains six large mirrors each of which is 1.8 m across. Their pointing is controlled with lasers so that each mirror focuses its light at the same location. The ensemble contains the same surface area to collect starlight as a 4.5-m telescope, which makes the MMT the third-largest optical telescope in the world.

A

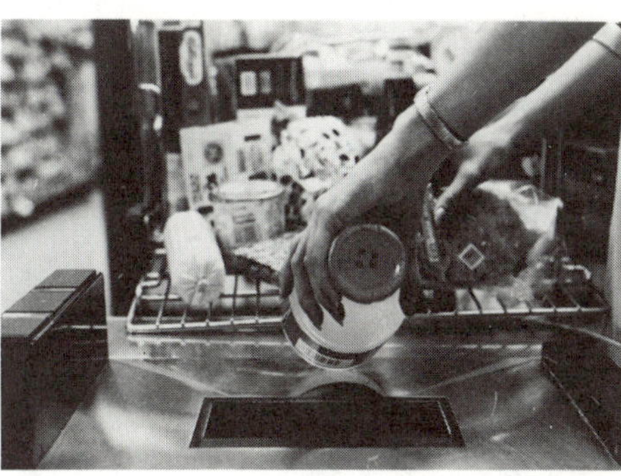

B

Fig. 24–22 A. Universal Product Code on a cereal box. B. A laser-scanning system automatically reads the Universal Product Code symbol on the label. The weak laser beam is directed upward through the window on the counter.

we can use it to draw very small circuits, some of which now appear on tiny chips.

Lasers are reaching the consumer market with increasing frequency. Most packages you buy in your supermarket now contain Universal Product Codes (Fig. 24–22), which can be read quickly by bouncing a laser beam off them. The information that is read is fed into a computer, which automatically puts the price into the cash register (avoiding the chance of mispunching) and also keeps track of sales for inventory.

The use of lasers in energy production through fusion was the subject of the color essay between pages 88 and 89.

We are about to see laser video disks introduced. These disks will have both sound and pictures on them, and will provide movies and educational material that can be played back on devices that hook up to home TV sets. The playback devices will cost seven hundred dollars at first, but the price will eventually drop. The disks, a type of record, have signals on them in the form of hills and valleys and are read by bouncing a laser beam off them. When you realize that a TV picture changes 30 times a second, you see that an hour's television really contains 30 images per second $\times$ 36,000 seconds/hour $=$ 100,000 images. We could put an entire encyclopedia on one disk, including animated illustrations!

24.4 *Laser Safety*

Laser beams concentrate so much energy into a small region that they can injure an eye in a fraction of a second. Thus we must be very careful not to look into a laser beam, nor even at its reflection from a shiny surface. Eye protection is mandatory for work in the laboratories where lasers are in use.

Key Words

laser
ground state
excited state
spontaneous emission
stimulated emission
incoherent
population
population inversion
maser
cavity
helium-neon lasers
tunable lasers
speckle pattern

Questions

1. What is the difference between a laser and a maser?
2. What do we mean when we say that an atom is "excited"?
3. In the process of stimulated emission, after one photon strikes an atom, how many photons leave the atom?
4. In the process of stimulated emission, how does the energy of an outgoing photon compare with the energy of the incoming photon?
5. Why is laser light coherent (that is, with the waves emerging from the laser in phase)?
6. Why will a population inversion produce strong radiation when stimulated?
7. What is the purpose of the mirrors in a laser?
8. In a laser, why do we want to make one of the mirrors less than 100 percent reflecting?
9. Describe three different uses of lasers.
10. What role might lasers play in using fusion as an energy source?

CHAPTER 25

Solid-State Physics

The technology associated with solid-state physics—transistor radios and pocket calculators, for example—has become so widespread that we almost take it for granted. The technology involves special properties of materials that are in their solid state (rather than being in liquid or gaseous states). These materials show a variety of fascinating effects that are still being explored extensively. The field of study is known as solid-state physics.

When we analyze gases, we usually treat the individual atoms or molecules of the gas as though they were small billiard balls bouncing off each other. Many of the properties of gases can be explained with this simple model. This classical picture, with certain modifications, is generally useful even for liquids. However, the atoms in solids are so close together that we cannot ignore quantum mechanical effects.

25.1 Structure of Solids

When we cool a liquid to form a solid, the actual way in which the atoms come together to form the solid depends on how we cool the liquid. The atoms try to arrange themselves in a pattern that requires the least amount of energy (just as, given a chance, a ball will roll downhill). If we cool the liquid quickly, the atoms do not have enough time to achieve this optimum arrangement before their

Fig. 25–1 Quartz crystals are not only chemically interesting, but also beautiful.

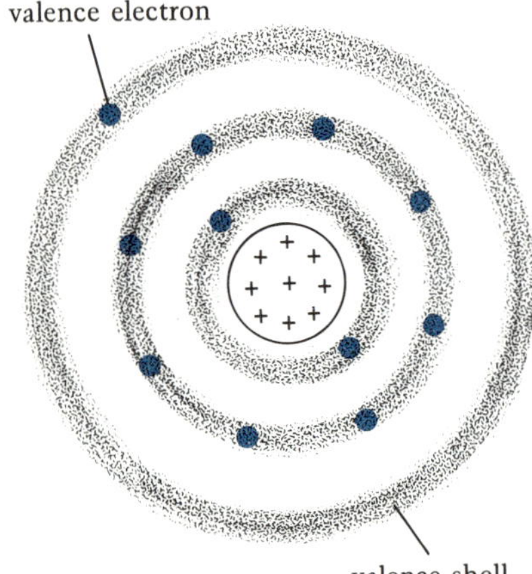

Fig. 25–2 The chemical properties of an element are determined by the number of electrons in the outermost, or valence, shell.

positions are, loosely speaking, frozen in. The resulting solid is called an amorphous solid, because its large-scale structure is quite irregular. Glass is a common example of an amorphous solid. The study of amorphous solids is just beginning to play a very important role in solid-state physics.

When we cool down our liquid slowly, then the atoms have the time to arrange themselves in the most stable possible manner. In this situation, each atom tries to bond itself to as many nearby atoms as possible. A very regular geometric pattern—a crystal—is formed by these atoms. We can think of a crystal as a single entity, like a giant molecule.

The chemical properties of an element are determined by the number of electrons in its outer—or valence—shell. If the valence shell is full, then the atom is very unlikely to react since it is quite content the way it is. Certain gases—helium, neon, argon, krypton, xenon, and radon—have valence shells that are filled and are chemically unreactive. (See Fig. 25–2.)

These gases are known as noble gases *because their unreactive tendencies were thought to resemble the aristocracy's.*

25.2 *Conductors and Insulators*

We generally think of an electrical conductor as a material that has electrons that are free to flow; these electrons are not all bound to individual atoms so they can flow through the crystal. In this picture, sodium is an example of a good conductor because it has one valence electron. When a large number of sodium atoms come together in a crystal, the individual atoms are only too happy to give up this electron. All these orphan electrons are then free to roam through the crystal. The free electrons in this case behave like a gas of electrons, moving around with random speeds and in random directions (Fig. 25–3). When an electric field is applied, these

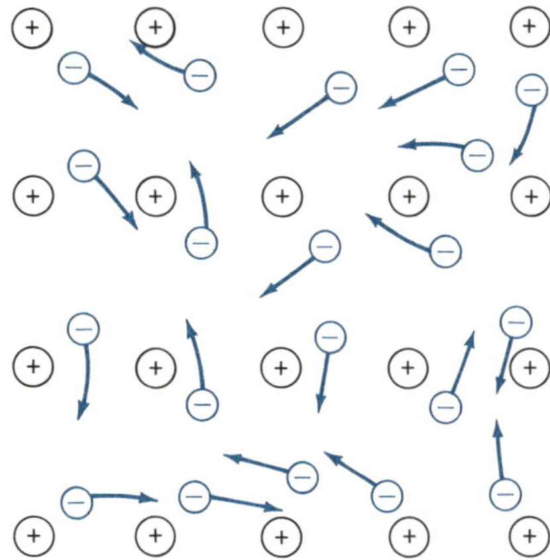

Fig. 25–3 In a metal, the valence electrons from each atom are free to move all around the crystal, and are shared by all the atoms. This figure shows the free electrons in a metal for which each atom contributes one valence electron.

free electrons acquire a drift in a direction aligned with (but pointed in the opposite direction to, because of the negative charge) the direction of the electric field.

Plausible as this model seems, it fails to explain the conducting properties of various materials, or the way in which these properties change when the temperature of the solid is varied. In addition, this picture of conductors does not give accurate results when we calculate the heat conduction properties or the heat capacity of these materials.

The problem is that the above picture takes no account of the wave nature of the electrons. To explain the properties of conductors, and why some materials are good conductors while others are quite poor conductors, we must apply the rules of quantum mechanics. One effect of quantum mechanics is that the electrons cannot move through the lattice in just any old manner. As in an atom, they are confined to certain energies.

Another important quantum-mechanical effect comes from the fact that the electrons must obey the Pauli exclusion principle. Even if we could cool a gas of electrons to absolute zero, we could not get all the electrons into the lowest energy level. Only two electrons will fit in each energy level. Therefore, even at absolute zero, there would be electrons in metals that are in relatively high energy levels. These energy levels would not normally be reached in a classical gas until the temperature got as high as 100,000 K! (See Fig. 25–4.)

You might feel a little more comfortable about the Pauli exclusion principle in solids if you could experience its effects. It turns out that you can. Take a quarter and try to squeeze it. (This is an update on penny-pinching.) You will find that metals are pretty hard. What gives the quarter the ability to resist your squeezing? It is a result of the exclusion principle. The electrons in the high energy levels exert a tremendous pressure. When you squeeze the quarter, you are trying to force the electrons together and they just won't go. If you are not impressed by their ability to withstand your squeezing, just think of the pressure that a steel girder on the ground floor of the Empire State Building must withstand.

To understand the properties of conductors, we must still calculate the energy levels of the electrons within the solid. We must solve the Schrödinger equation (the wave equation) for an electron

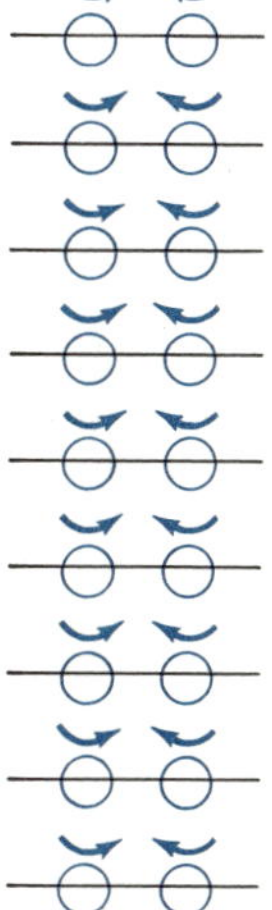

Fig. 25–4 The effect of the Pauli exclusion principle in a metal. The exclusion principle says that we can only put one electron in any state. We can put two electrons, with opposite spins, in any energy level within a metal, but can put no more. Therefore, the shared valence electrons from the atoms can end up in very high energy levels. The pressure caused by having so many high-energy electrons is responsible for the hardness of the metal.

moving among the ions that form the crystal. When we do this calculation, we find that the allowed energies for electrons are grouped in bands (Fig. 25–5). In the simplest picture, the bands represent a range of allowed energies that electrons traveling through the crystal can have. The bands are separated from each other by energy *gaps.* These gaps represent energies at which electrons cannot move freely through the crystal. Each band actually contains a number of energy levels, but each level can only have two electrons.

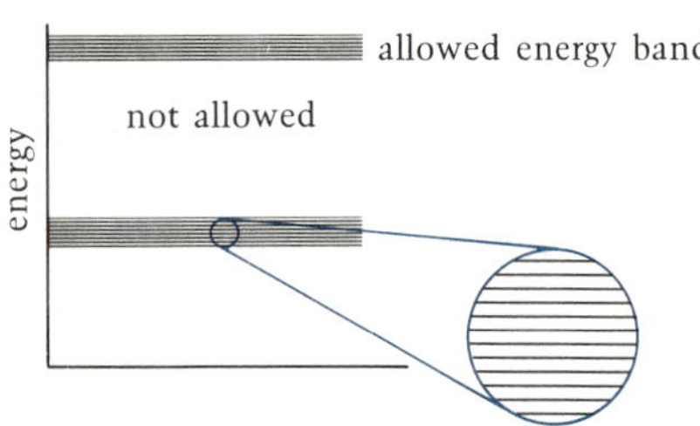

Fig. 25–5 When we look at the energy levels for the "free" electrons in a solid, the levels group together into bands. A close look shows that each band is made up of many closely spaced energy levels.

The band structure in a solid determines whether the solid will be an insulator or a conductor. The bands will be filled up to a certain level by the electrons within each atom. The highest band in which electrons are still predominently attached to their atoms are found is called the valence band. This is the band in which the valence (outermost) electrons from each atom will be located. These are the electrons that are the possible conductors of electricity. However, in order for an electron to conduct, it must get up to a slightly higher energy so that it is free of the grip of its atom. Think of yourself walking through a crowded forest, not making very good progress as you have to weave your way through the trees. If you could only get 50 feet up in the air, as with a James Bond rocket backpack, you would be able to zoom along with no interference from the trees. The same problem is faced by a would-be conducting electron. The lowest band in which an electron can be free of the crowd of atoms is called the conduction band.

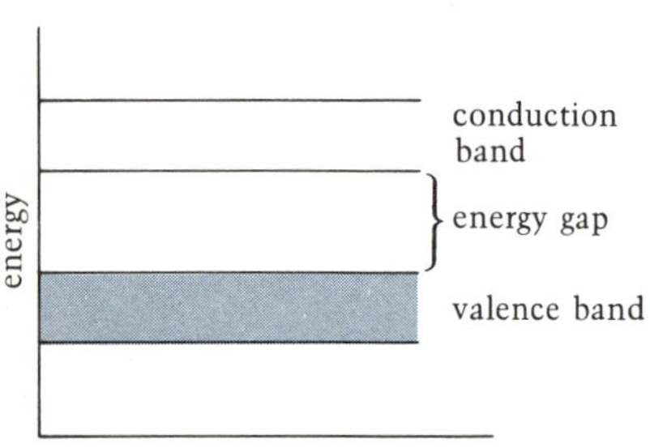

Fig. 25–6 The valence band is the highest energy band in which the valence electrons are normally found when they are bound to their atoms. The conduction band is the lowest energy band in which the electrons are free to move about the solid. This figure shows a good insulator. The valence band is filled, but there is a large energy gap between the valence band and the conduction band, so no electrons can get up to the conduction band.

Let's first look at the case when the valence band is full, that is, when there are no more available energy levels. Then, a valence electron must jump (increase its energy) into the next higher band to be free. If the energy gap between the valence band and the conduction band is too large, then the electron will not be able to make that jump. Such a material will not be a good conductor of electricity, and is called an insulator. The band theory of solids tells us that *an insulator is a material in which the valence bands are filled and the energy gap to the conduction band is too great for the valence electrons to jump* (Fig. 25–6).

What happens in a conductor? There are actually two possibilities. One is that the valence band is not completely filled. Then an electron in the valence band can get free of its atom by simply jumping to a higher energy level within the same band. This jump requires a small amount of energy, and many electrons can therefore make that jump (Fig. 25–7).

Another situation in which a conductor results stems from the fact that the size of the energy gaps between the valence band and the conduction band are different for different materials. In some materials this gap can be quite small. It can even disappear, when the valence band and conduction band overlap. In this latter case, the electrons can move very easily from the valence band to the conduction band. Therefore, the band theory tells us that *we have a conductor when*

(a) *the valence band is not filled, so electrons can move to higher states in the valence band and be free, or*

(b) *when there is no energy gap between the valence band and the conduction band, so electrons can easily make the transition from the valence to the conduction band.*

The band theory correctly explains many properties of solids. It can be used, for example, to calculate how properties vary with temperature, how heat is conducted, and the heat capacity of various solid materials.

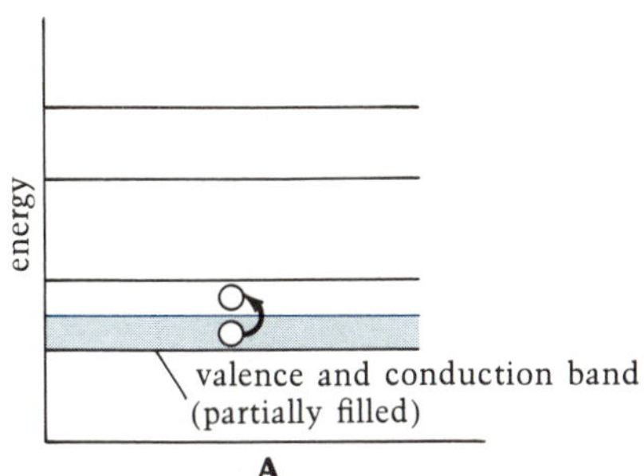

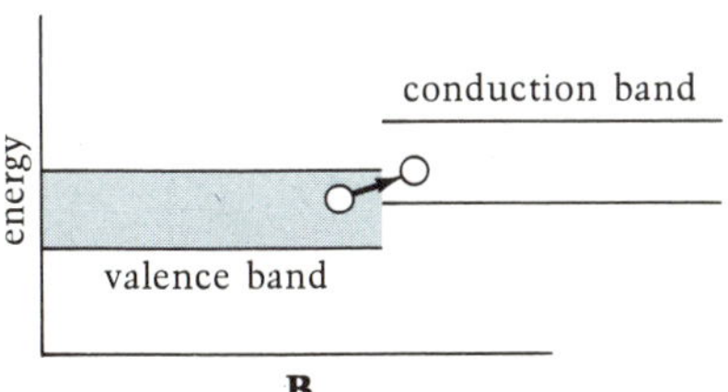

Fig. 25–7 This drawing shows two ways in which we can get a good conductor. A. The valence band is only partially filled. (The colored region is that which is filled.) This means that the upper part of the valence band can serve as a conduction band. An electron in the valence band doesn't need much additional energy to get to the conduction part of the band. B. The valence band and conduction band overlap slightly, so an electron can easily go from one to another.

25.3 *Semiconductors*

There is a case in between "conductor" and "insulator." A material with intermediate properties is called a semiconductor (Fig. 25–8). In a semiconductor,

(a) *the valence band is filled, and*

(b) *although there is an energy gap between the valence band and the conduction band, that energy gap is not very large.*

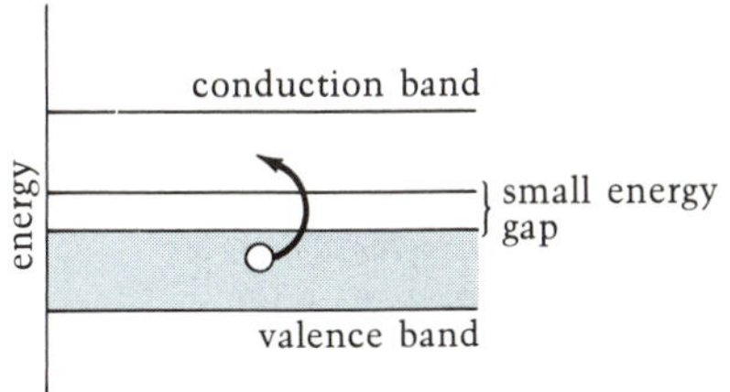

Fig. 25–8 In a semiconductor the valence band is filled, but there is only a very small energy gap between the valence band and the conduction band. Enough electrons can get the required boost to cross this small gap, and the material will conduct under the right conditions.

Examples of solids that are semiconductors are silicon and germanium. The gaps between their valence and conduction bands are sufficiently small that the normal thermal energy of the solid at room temperature is enough to knock a few electrons into the conduction band. These electrons conduct a current, but, as the name "semiconductor" implies, a semiconductor does not conduct as well as a real conductor. After all, a semiconductor has fewer free electrons than a conductor.

Students filling a row of chairs while waiting to buy concert tickets can serve as an analogy to the flow in semiconductors. When the person at the left end gets up, an empty chair (a "hole") is left. When the second person moves to sit in the empty chair, the second chair from the left is now empty. We can say that the "hole" has moved right, and so on.

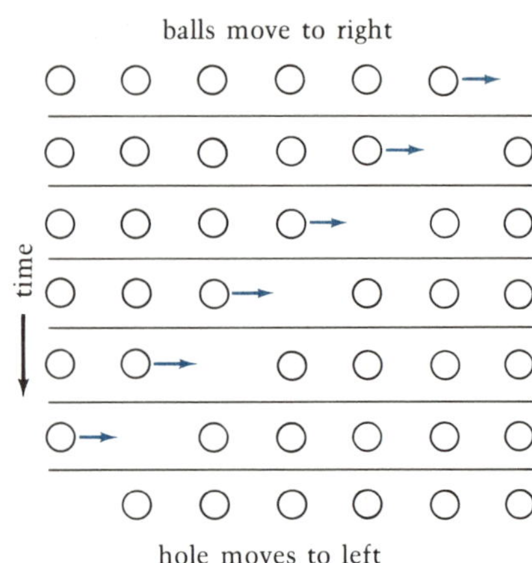

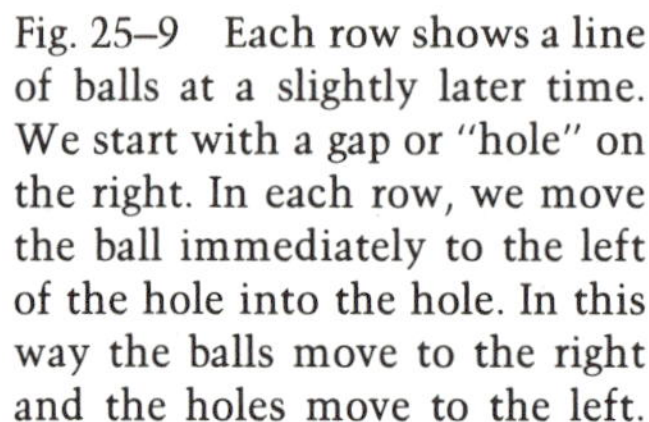

Fig. 25–9 Each row shows a line of balls at a slightly later time. We start with a gap or "hole" on the right. In each row, we move the ball immediately to the left of the hole into the hole. In this way the balls move to the right and the holes move to the left.

When an electron jumps into the conduction band, it leaves behind a hole. When an electric field is applied, this electron moves in the direction opposite to the electric field. As the other electrons move to fill in the hole, the hole "moves" in the direction of the electric field. Thus, we can think of the flow of current in the presence of an electric field as being either in the form of negatively charged electrons moving opposite to the field direction or positively charged "holes" moving in the same direction as the field. Solid-state physicists often find it convenient to talk about the holes as though they were as real as the electrons (Fig. 25–9).

The semiconductors that we have been talking about are called intrinsic semiconductors. By "intrinsic," we mean that the semiconducting property arises naturally for these particular materials when they form solids. If this were all there were to semiconductors, they might still be interesting, but their uses would be limited.

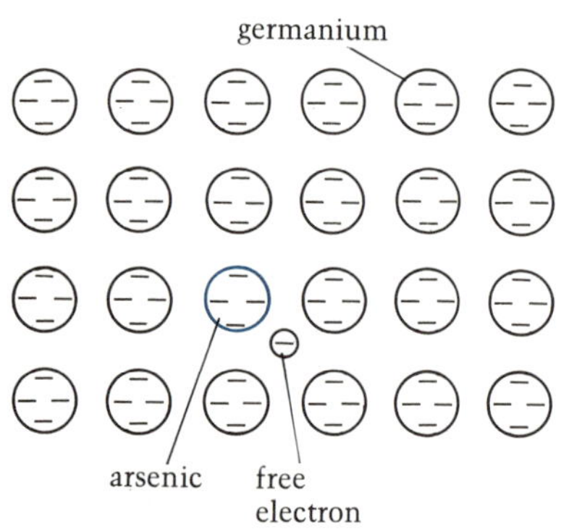

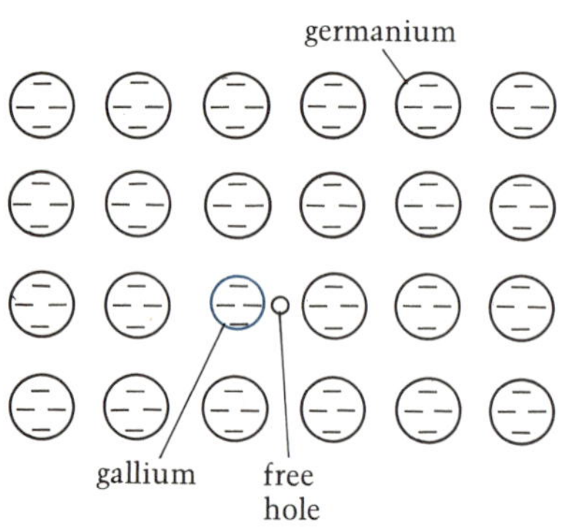

Fig. 25–10 Impurity semiconductors. In each case, we "dope" a crystal of germanium (four valence electrons) with an impurity. The impurity occupies a site where one of the germanium atoms would normally be. A. The impurity is arsenic, with five valence electrons. In order to fit into the germanium crystal, the arsenic must only use up four of its valence electrons. The fifth is left over and is free to conduct electricity. The arsenic is called an n-type donor, since it donates the negatively charged electron. Germanium doped with arsenic is an n-type impurity semiconductor. B. The impurity is now gallium, with only three valence electrons. In order to fit into the crystal of germanium, the gallium must go into debt by one electron, creating a hole, which is free to conduct electricity. The gallium is a p-type donor, since it contributes a free positively charged hole. Germanium doped with gallium is called a p-type impurity semiconductor.

After all, relatively few elements are intrinsic semiconductors, and in any case, we have very little control over their properties.

A much larger class of semiconductors is made artificially. These impurity semiconductors are made by mixing a small amount of impurity into a normal crystal. This process is called doping. By using different crystals and different doping materials, and by varying the amount of doping we use, we can produce semiconductors that have a wide variety of properties.

How does this doping work? One possibility is to use a doping material that has one more valence electron than the material of the crystal. For example, arsenic has five valence electrons and germanium has four valence electrons. If we add a small amount of arsenic to germanium (that is, use arsenic as a doping material on a germanium base), the basic structure of the crystal will be determined by the germanium. Wherever an arsenic atom ends up, it will only have to use up four of its valence electrons to join the crystal. This will leave one extra electron, which is weakly bound and easily makes its way into the conduction band. The effect of the arsenic is to add one more electron that can normally be accommodated in the valence band. We call the arsenic a donor, since it donates excess *electrons*. The resulting material is called an n-type semiconductor (the *n* standing for "negative").

Alternatively, we can dope the germanium with gallium, which has only three valence electrons. In this case, we have added a hole instead of adding an electron. However, the effect of this hole is to provide a means for the valence electrons to move around by jumping into the hole. The situation is equivalent to adding carriers of positive charge to the material, so we call this a p-type semiconductor. (See Fig. 25–10.)

You might wonder what is so exciting about having a piece of germanium with some extra electrons or missing some electrons. The uses of a simple, single semiconductor are indeed quite limited. The interesting effects take place when we put two different semiconductors together. The boundary between two different semiconductors is called a junction. The simplest junction to analyze is the one between a p-type semiconductor and an n-type semiconductor. This type of junction is called a diode ("di-" meaning "two").

n-type p-type

A

high voltage + – low voltage

B

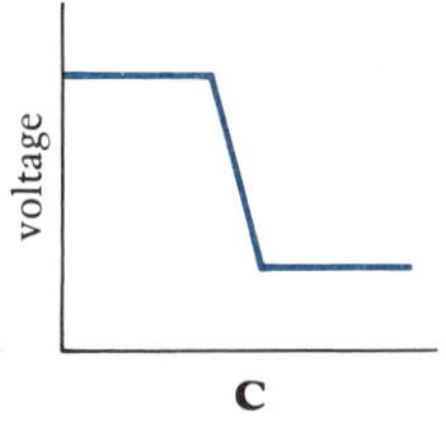

C

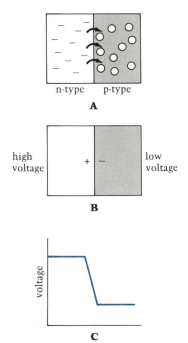

Fig. 25–11 A diode. We put an n-type and a p-type semiconductor together. A. Some of the electrons near the boundary will rush in to fill the holes on the other side of the boundary. B. The result is that the n-type side is left with a deficiency of electrons and a net positive charge, while the other side has a negative charge. C. As a result of the charge separation, the n side is at a higher voltage than the p side. All of this is just if we leave the diode alone.

In a diode (Fig. 25–11), although both semiconductors are neutral, the n-type has a high concentration of *free* electrons, which spread out into the p-type. The result is an excess of positive charge on the n-type and of negative charge on the p-type. As a result, a potential difference arises across the junction.

If we connect the p-type side to the positive terminal of a battery,

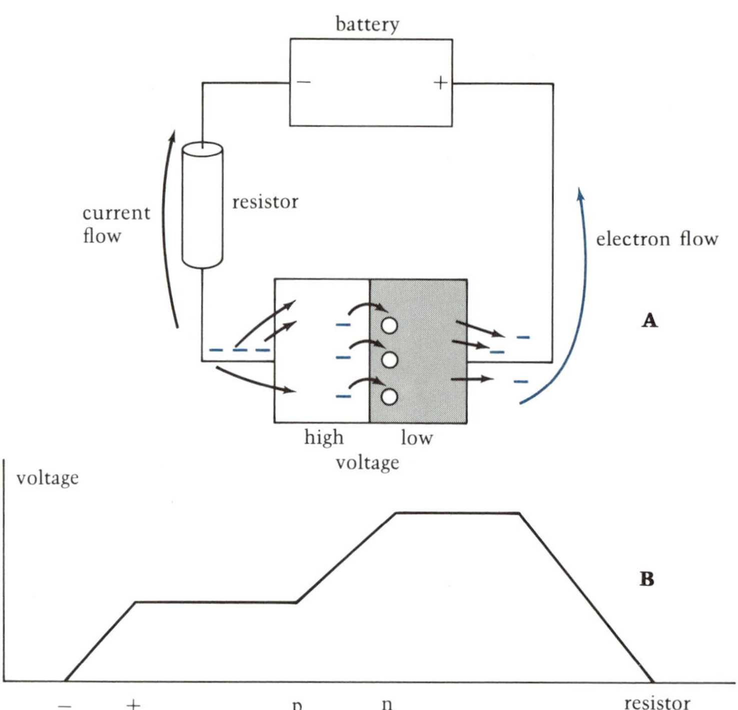

Fig. 25–12 We now put the diode in a circuit with a resistor and a battery. A. If the diode is connected so that it wants to make current flow in the same direction as the battery, current will flow. B. The graph shows the voltage change as we go around the circuit. The voltage increases as we go from the negative to positive terminal of the battery and increases more in going from the p to the n side of the diode. The voltage then drops back down as the current flows through the resistor.

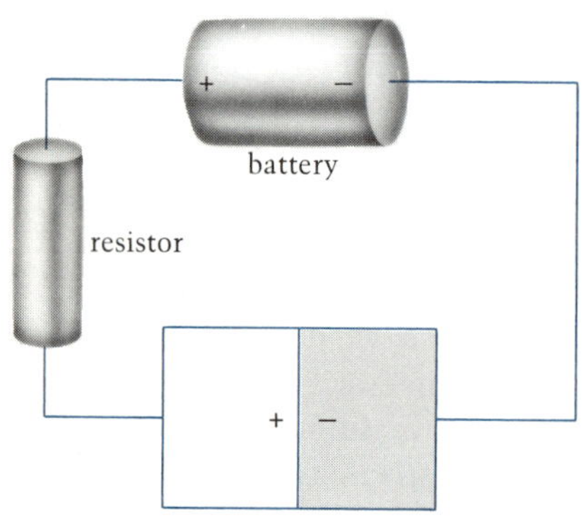

Fig. 25–13 Now the diode is connected the opposite way. Its voltage difference wants to counteract that of the battery and no current will flow in the circuit. (In real cases, a very small current will flow.) If we substitute a battery with a greater potential difference, the battery can overcome the retarding effect of the diode. Current will flow through the circuit, but in the "backward" direction as far as the diode is concerned.

current flows (as shown in Fig. 25–12). We define this direction as the forward direction. When the battery is connected in the reverse direction (as in Fig. 25–13), very little current flows because the potential difference across the junction opposes that across the battery. If the battery has a large enough voltage, it overpowers the potential difference at the junction and we say that the diode "breaks down." A large current flows in the reverse direction. For practical purposes, we can think of the diode as conducting electricity in only one direction (as long as the voltage does not get high enough for breakdown). Thus a diode is good to use in a circuit in which we want current to flow in only one direction. For example, if we pass the current in a normal AC circuit through a diode, then only half of every cycle will get through, as shown in Fig. 25–14.

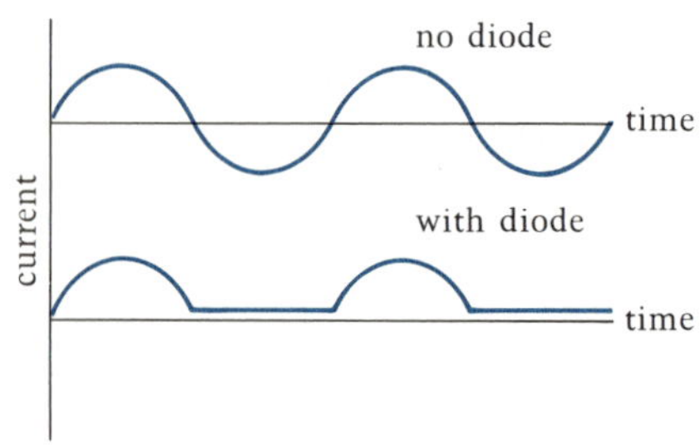

Fig. 25–14 As long as we don't apply too great a voltage, a diode will only allow current to pass in one direction. If we put in some AC current, as shown above, the current will be normal in one direction, but will be zero for half the time when we put a diode in the circuit.

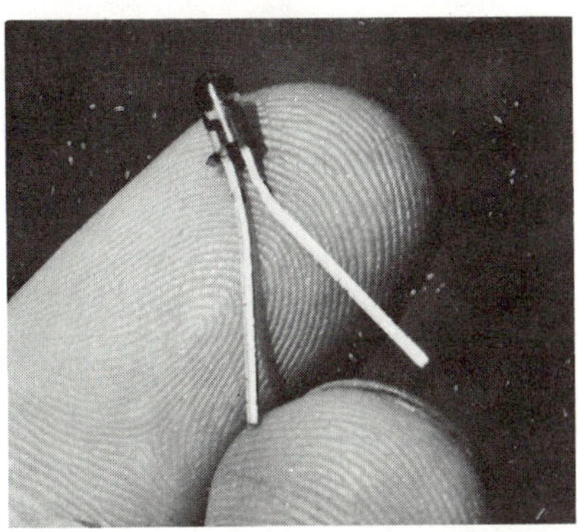

Fig. 25–15 A light-emitting diode.

Sometimes an electron in a diode will go from the conduction band to the valence band. The result is one fewer free electron and one fewer hole. In the process a little bit of energy is given off. In semiconductors such as silicon and germanium, the energy that is liberated shows up as heat in the crystal. However, in some semiconductors (gallium arsenide, for example) the energy is given off as radiation. Most of the radiation is in the infrared, but some is visible and is seen as light (usually red in appearance). A diode with these properties is called a light-emitting diode (LED). LEDs are used for the number displays in many pocket calculators and wristwatches because they take very little electricity to keep glowing and because they don't heat up like ordinary light bulbs.

A diode capable of the reverse process—absorbing light and creating an electron-hole pair—is called a photodiode. Photodiodes can be used in detecting a beam of light. Thus they can tell when a beam of light is broken, indicating that someone, or something, has passed through. They are therefore used in burglar alarms and in some automatic door openers.

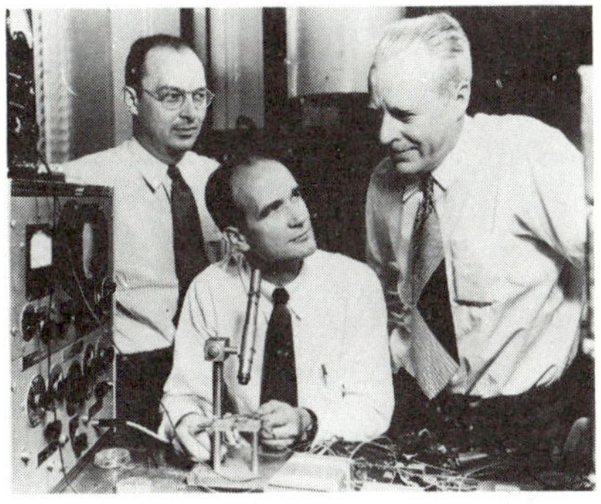

Fig. 25–16 John Bardeen, William Shockley, and Walter H. Brattain (left to right), inventors of the transistor, for which they received the 1956 Nobel Prize in physics.

25.4 *Transistors*

The great revolution in the use of semiconductors came with the invention of the transistor in 1948. A transistor consists of three pieces of semiconductor material arranged in a sandwich. If the middle piece is n-type, then the outer two pieces are p-type. This arrangement is called a pnp transistor. The opposite arrangement, called an npn transistor, is also possible. In any situation, either can be used, but we will talk about pnp transistors. (If an npn is substituted, all the voltages must be reversed; the general idea is the same.) Another important feature of the construction of a transistor is that the center section, the base, must be very thin—less than 100 nm thick. (See Fig. 25–17.)

The three layers are designated (in order) the emitter, the base, and the collector. The emitter is much more heavily doped than the other two sections, and therefore serves as the primary source

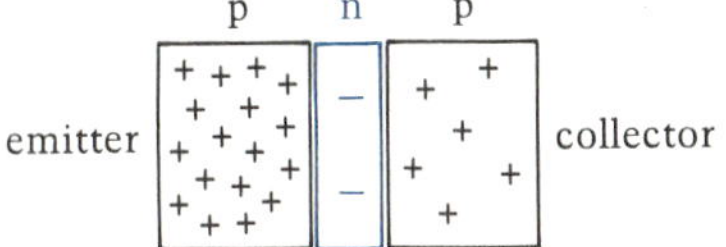

Fig. 25–17 Two transistors. The upper one is a pnp transistor and the lower one is an npn transistor. In the p sections of semiconductor there is an excess of holes that are free to conduct electricity, even though the whole piece is electrically neutral. Similarly, in the n sections, there is an excess of free electrons, even though the piece is neutral. In each case the base is least heavily doped and the emitter is most heavily doped. The base is very narrow, even narrower relative to the other sections than depicted here.

for charge carriers (which are electrons in the case of an npn transistor). In normal operation, connections are applied to all three so that the emitter has the lowest potential and the base has the highest potential. As Fig. 25–18 shows, we can regulate the flow of electrons from the emitter to the collector by regulating the rate at which we remove electrons from the base. For every electron that we remove from the base, we get a much larger number of electrons flowing from the emitter to the collector. The ratio of these currents can be anywhere from ten to a thousand. This is the basis for using the transistor as an amplifier, a transistor amplifier (Fig. 25–19).

The amplification comes from the fact that the ratio of collector current (the current leaving the collector) to base current (the current leaving the base) remains constant. Let's say that the ratio is fifty, which is typical for a transistor amplifier. Then, if you make a small change in the base current, the collector current changes by an amount fifty times as great. We change the base current by changing the voltage that we apply between the emitter and the base. The changes in voltage must be small in comparison to the voltage that we have used to establish the basic potential

current flow
electron flow
–
low
potential
+
high
potential
e
b
c
A

–
+
B

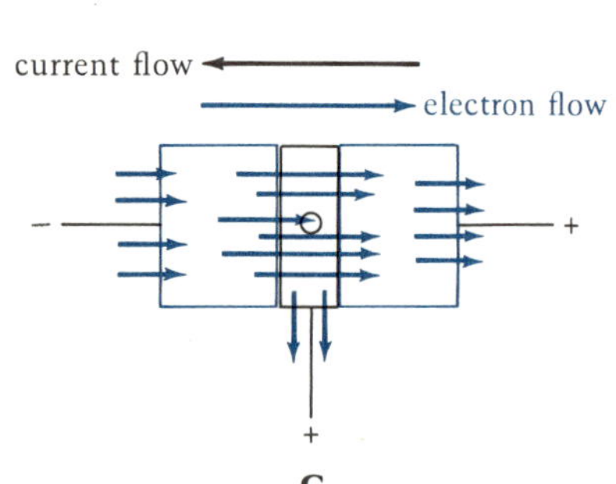

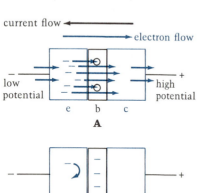

Fig. 25–18 Operation of a transistor. A. Suppose we just connect our npn transistor with no base connection and with the emitter at low potential and collector at high potential. The free electrons in the emitter will then start to flow to the collector, across the very narrow base. However, some will be trapped by the free holes in the base. B. There will then be a net negative charge on the base, so electrons will be repelled as they approach the base, and the flow of current will stop. C. We now make a connection to the base, and put it at a high potential, drawing some of the excess electrons off the base. Then many more electrons can flow across the base from the emitter to the collector. For every electron we pull off the base, away from a hole, fifty electrons may get across before one of them runs into the hole and has to be pulled out. We can control the electron flow from the emitter to the collector by controlling the rate at which we take electrons out of the base. Notice that the actual current flow is in the direction opposite to that of the electron flow.

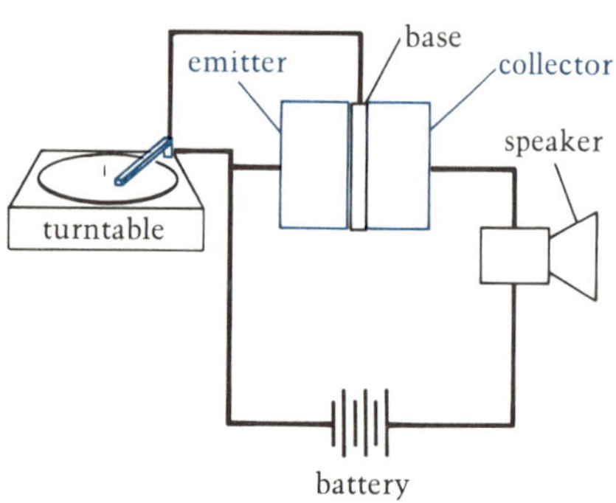

Fig. 25–19 A transistor amplifier. We have a speaker and battery hooked up across the emitter and collector. However, unless we draw electrons out of the base, no current will flow. In this case the electrical signals from the turntable determine the rate at which electrons are removed from the base, and regulate the flow of current to the speaker. Small changes in the current flowing through the phonograph can produce large changes in the current flowing through the speaker. In a real transistor amplifier, better quality is obtained by using several transistors, often using combinations of pnp and npn transistors.

Fig. 25–20 The decreasing size of electronic components. Vacuum tubes (at left) were replaced by transistors (right center) starting in 1948. The integrated circuit (far right) contains 22 transistors and other components. It helps generate the musical dial tones in push-button telephones. To note the size of a vacuum tube, see Fig. 20–26. The latest chips contain the equivalent of tens of thousands of transistors.

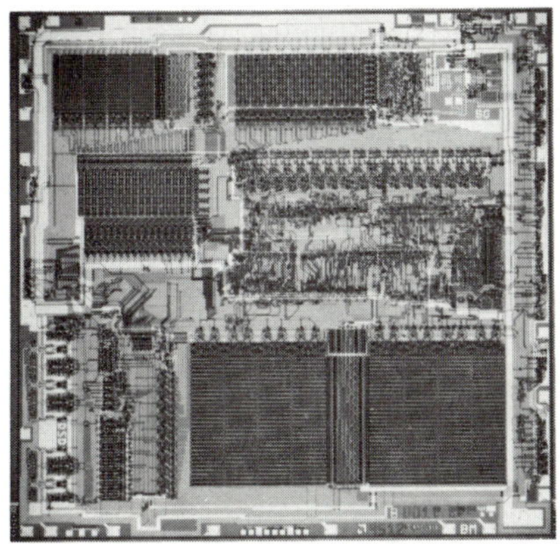

Fig. 25–21 A close-up view of an unusually large chip, about 0.6 cm on each side.

difference between the emitter and the base. For example, we might have a battery maintaining the proper operating voltage difference between the emitter and the base. To this battery voltage, we then add a much smaller voltage that varies with time. (This varying voltage might be produced by the radio signals picked up by the antenna of your transistor radio. These signals are very weak, so the voltage in the antenna is very small.)

These weak signals then cause slight variations in the base current. The variations are a factor of fifty stronger (for our example) when they show up as variations in the collector current. We have amplified the original signal.

If we need more amplification, we work in stages. The signal from the first amplifier is put into a second, into a third if necessary, and so on. Eventually, the signal from the radio antenna is amplified enough to produce sounds in a loudspeaker.

Originally, all radios and televisions used vacuum tubes to amplify signals. Such tubes are expensive to make, require high voltages, and use a lot of electricity. They also heat up and so need ventilation. Further, vacuum tubes are much larger than transistors that perform the same function, and tubes break more easily. At present, the only tube in a new television set is the picture tube, the tube that contains a beam of electrons that put the picture on the face of the tube. All the rest of the circuitry is composed of solid-state devices such as transistors and diodes. These are usually relatively inexpensive to make, run at lower voltages, and consume less electricity. They don't heat up as much as tubes, and tend to last a long time. Their only disadvantage is that if they are overloaded with too high a voltage or current, they can burn out very quickly.

25.5 *The MOSFET*

Consumer industries now sometimes keep up with the latest advances in technology. The hi-fi industry certainly does so. Ads for the latest hi-fi component scream about MOSFETs, and how they make the Rolling Stones sound like they're in your living room. What are MOSFETs, and what good are they?

The type of transistor that we have been discussing thus far in this section is called a bipolar transistor because the motions of both electrons and holes is important in its operation. In another type of transistor, only one type of charge carrier is involved. This charge carrier can be either electrons or holes, depending on the doping. The result is called a field-effect transistor (FET).

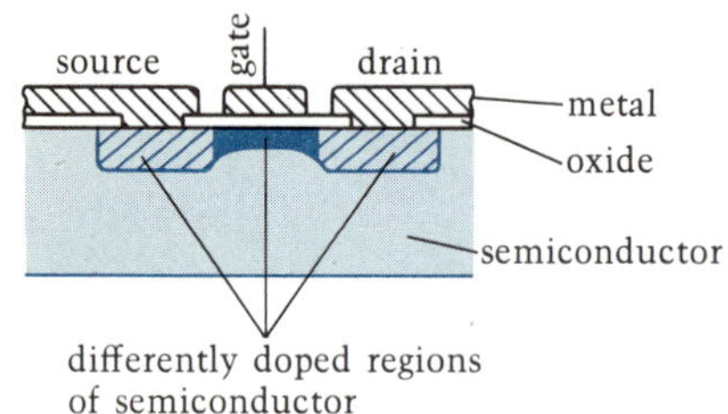

Fig. 25–22 Structure of the MOSFET. The electric fields produced by the gate regulate the flow of current between the source and the drain.

A typical FET is illustrated in Fig. 25–22. Instead of having an emitter, a base, and a collector as in a bipolar transistor, the three sections of an FET are the source, the gate, and the drain. The flow of charges from the source to the drain is controlled by varying the electric field set up by the gate, hence the name "field effect." The FET shown has a three-layer construction—a top layer of conductor (metal), a middle layer of oxide, and a bottom layer of semiconductor. This particular arrangement is called a MOSFET. MOSFETs have the advantage that they can be made somewhat smaller than the normal bipolar transistors. The smaller size is particularly useful when we are interested in making circuits that are as small as possible. One major advantage of the MOSFET is that the particular set of three layers that it has makes it very suitable for inclusion in the microcircuits discussed in Chapter 20. For a hi-fi, we take advantage of the fact that MOSFETs can take higher power levels than ordinary transistors without distortion.

Fig. 25–23 H. Kamerlingh Onnes (seated) and J. D. van der Waals with the helium liquefier in their laboratory in Leiden, The Netherlands, in 1911.

25.6 *Superconductors*

The study of superconductors is one of the most exciting areas of solid-state physics. *Superconductors are materials that conduct electricity with no apparent resistance.* The phenomenon of superconductivity was discovered in 1911, when the Dutch physicist H. Kamerlingh Onnes found that, when it was cooled to very low temperatures, less than about 4.2 K, mercury suddenly lost all its resistance. Since that time, superconductivity has been observed in several materials, but it has always been observed at low temperatures (Fig. 25–24).

Only certain materials become superconductors at low tempera-

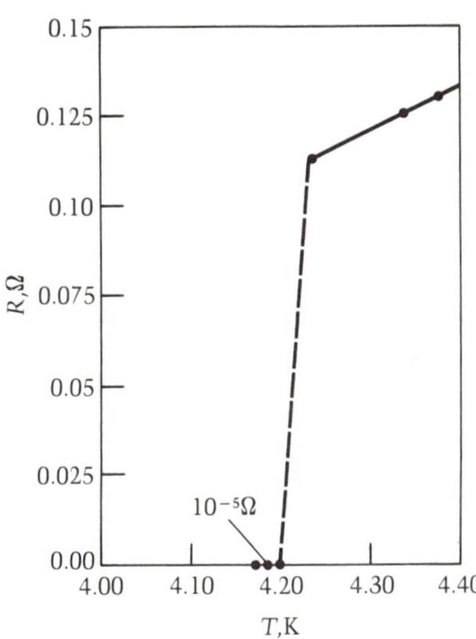

Fig. 25–24 Plot by Kamerlingh Onnes of resistance of mercury versus temperature showing sudden decrease at the critical temperature $T = 4.2$ K. The resistance, R, is measured in ohms, designated by the symbol Ω.

tures. For them there is a temperature, called the critical temperature, at which the behavior suddenly switches from the normal conducting behavior for that material to superconducting behavior. The value of the critical temperature can be influenced by the presence of a magnetic field. The greater the field, the lower the critical temperature, so it appears that magnetic fields tend to inhibit superconductivity. In fact, for very high fields, no superconductivity is observed. Also, the magnetic field lines near a superconductor go around the superconducting material. It appears that magnetic fields cannot exist inside a superconductor.

The phenomenon of superconductivity was not understood until a theory was proposed in 1957 by John Bardeen, Leon Cooper, and J. Robert Schrieffer that explained many of the properties of superconductivity. (The three were awarded the Nobel Prize in 1972, making Bardeen the only person to receive the physics prize twice.) Their theory, known as the BCS theory, involves a mechanism by which electrons begin to travel in pairs at low temperatures. The two electrons in a pair are not directly attracted to each other. In fact, the normal electric force between them is repulsive. Instead, one electron interacts with the crystal of atoms, and the crystal then interacts with another electron. The net effect of this electron-crystal-electron interaction is to produce an apparent attraction between the two electrons. Once the electrons are paired, they can be accelerated by electric fields, but the pair does not lose energy in collisions with the atoms in the crystal, while a single electron does.

It is interesting that the best superconductors are not the best normal conductors. Copper, for example, is a very good conductor, meaning that the individual electrons in the conduction band have relatively few collisions with the atoms in the crystal. Because of the lack of collisions, the electron-crystal-electron interaction required for superconductivity does not occur.

The uses of superconductors are obvious. Once a current starts going around in a superconducting ring, it continues to circulate without any loss of energy. Currents have been observed to circulate in this manner for years without diminishing in strength. Also, since there is no resistance, there is no joule heating. At research institutions, large amounts of electricity are used to run huge magnets. When the accelerator magnets are made superconducting, there will be a tremendous saving in electricity. No electricity will be then required to maintain the magnetic field, and no cooling will be required to remove the heat generated in normal wires. The changeover is imminent.

There is, however, one major drawback. In order to make a material a superconductor, you must get it very cold. Therefore, in

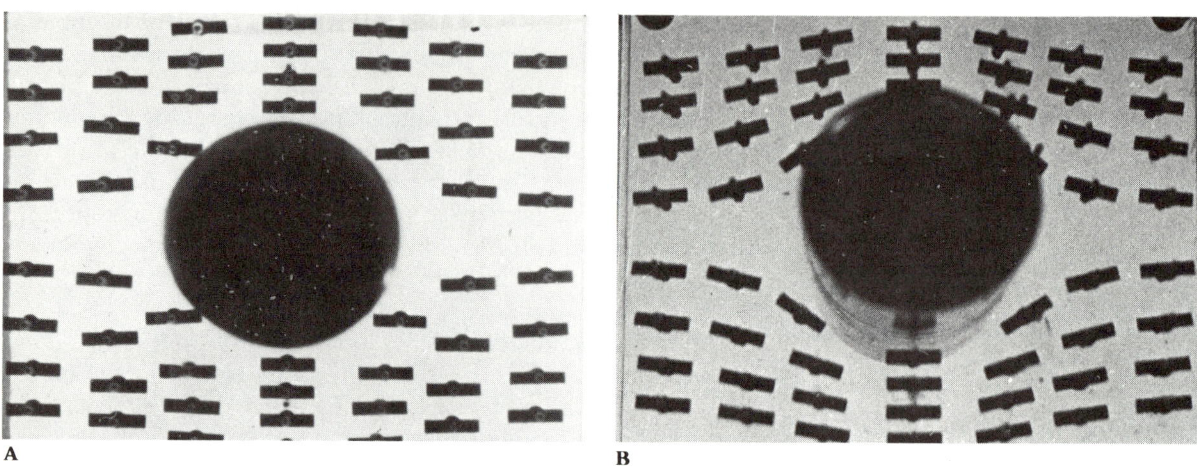

Fig. 25–25 Magnetic field in a normal conductor and in a superconductor. The strips act like little compasses showing the direction of the magnetic field. A. A normally conducting tin cylinder, seen here in cross section, does not interfere with the magnetic field. The strips go straight through without being deflected. B. The tin is now cooled so that it is superconducting. The magnetic field is now deflected so that it passes around the superconductor. (Courtesy A. Leitner, Rensselaer Polytechnic Institute)

order to have superconducting coils on the magnets, a large investment is needed in refrigeration equipment, and some power will be required to run that equipment.

Physicists are trying to find materials that have superconducting properties at higher and higher temperatures. Currently, the highest temperature superconductors operate at is about 30 K. This is considerably higher than 4 K, but is still much lower than room temperatures. The development of a room-temperature superconductor would allow tremendous savings in energy consumption, but before you get your hopes too high, we must point out that solid-state physicists are not even sure that such a material can exist. Nevertheless, it's nice to think about the possibility.

Key Words

solid-state physics
amorphous solid
noble gases
bands
valence band
conduction band
semiconductor
hole
intrinsic semiconductor
impurity semiconductor
doping
donor
n-type semiconductor
p-type semiconductor
junction
forward direction
reverse direction
light-emitting diode (LED)
photodiode
transistor
pnp transistor
npn transistor
emitter
base
collector
transistor amplifier
bipolar transistor
field-effect transistor (FET)
source
gate
drain
MOSFET
superconductors
critical temperature
BCS theory

Questions

1. What is the difference between an amorphous solid and a crystal?
2. Give one example of an amorphous solid and one example of a crystal.
3. Why are noble gases noninteracting chemically?
4. Why are solid metals usually good conductors? (*Hint:* Metals usually have one, two, or three valence electrons.)
5. What would happen to the chair you are sitting on if the Pauli exclusion principle were not in effect?
6. What is the relationship between energy bands in solids and energy levels in atoms?
7. What is the relationship between the valence band and the conduction band in a solid?
8. What determines whether a solid will be a good insulator or a good conductor?
9. Metals are good conductors of electricity, but they are also good conductors of heat. What is the relationship between these two properties? (Think of how heat might be transmitted through a metal.)
10. What type of band structure will give rise to a semiconductor?
11. In a particular semiconductor the electrons are moving from left to right. (a) In which direction are the holes moving? (b) In which direction is the current flowing?
12. How does doping a material with an n-type impurity improve the conducting ability of the original material?
13. Describe the behavior of a diode (a) when a voltage is applied in the forward direction, (b) when a voltage (less than the breakdown voltage) is applied in the reverse direction, and (c) when a voltage (greater than the breakdown voltage) is applied in the reverse direction.
14. Compare the operation of a light-emitting diode and a photodiode.
15. What are three advantages of transistors over vacuum tubes?
16. How are transistors useful in the manufacture of a computer?
17. What are the advantages of the MOSFET over the standard bipolar transistor?
18. What happens when a superconductor is placed in a magnetic field?
19. Why don't we use superconductors in all of our electrical transmissions today?

CHAPTER 26

Inside the Nucleus

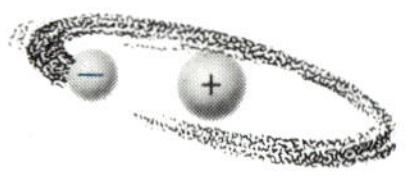

hydrogen
atomic number 1

helium-4
atomic number 2

helium-3
atomic number 2

Fig. 26–1 The atomic number of an element is the number of protons in the nucleus of the element.

No sooner did some of the answers appear to the riddles of the structure of the atom than the new quest began to understand the structure of the nucleus. Although atoms are small, nuclei are smaller yet. A typical hydrogen atom is about 0.1 nm (10^{-10} m) across. Rutherford found that a nucleus has a radius of about 10^{-15} m, a factor of 10^5 smaller than an atom. Since the volume of a sphere depends on the cube of the radius, then the volume of an atom is $(10^5)^3$, which is equal to 10^{15} times that of a nucleus! When we start talking about nuclei, those tiny atoms begin to seem enormous.

In this chapter, we shall see that the structure of the nucleus presents a whole new set of challenges to the physicist. To meet these challenges, a variety of new laboratory techniques has been established.

26.1 *Building a Nucleus*

26.1a THE BUILDING BLOCKS

The charge of a nucleus, which is positive, is provided by protons. When we talk about a given element, it is really the electrons orbiting the nucleus that give the element its *chemical* properties. Since atoms are normally neutral, this number of electrons outside

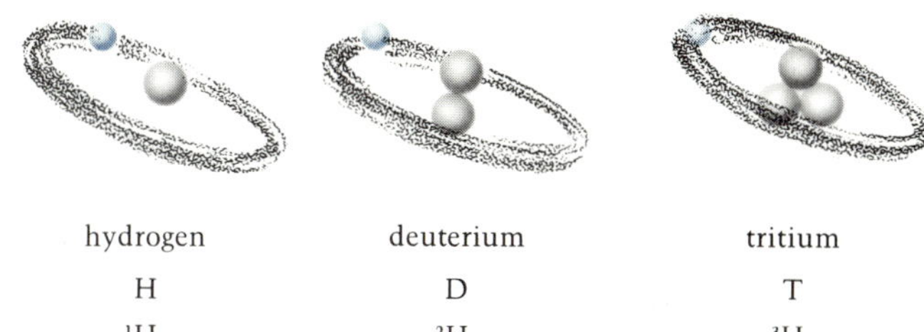

Fig. 26–2 The three isotopes of hydrogen. Notice that each only has a single proton. The only difference is the number of neutrons.

the nucleus must equal the number of protons in the nucleus. Therefore, *the number of protons in the nucleus determines the properties of a given element.* The number of protons in a nucleus is called the atomic number, and is given the symbol Z. For example, hydrogen, the simplest element, has one proton in its nucleus, so the atomic number of hydrogen is 1. Helium has two protons in its nucleus, and its atomic number is 2. (See Fig. 26–1.)

As early as 1920, Rutherford postulated the existence of a new particle in the nucleus. This particle has approximately the same mass as the proton, but has no electric charge. Since the new particle is electrically neutral, it is called a neutron. This particle was eventually discovered by Sir James Chadwick in England in 1932.

It is possible for nuclei of a given element to have different numbers of neutrons. Nuclei of the same element (that is, nuclei with the same number of protons) but with different numbers of neutrons are called isotopes of the same element. For example, the simplest form of hydrogen consists of a single proton. Another form of hydrogen still has one proton, but also has one neutron. This isotope is so important that it even has its own name—deuterium. An additional isotope of hydrogen is called tritium, which consists of a single proton and two neutrons. But most isotopes do not have different names.

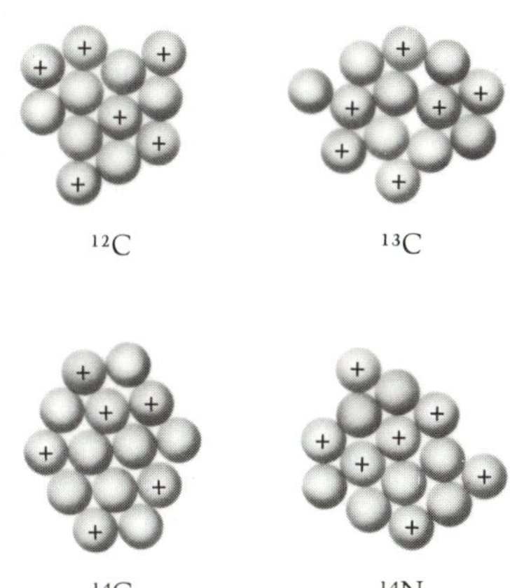

Fig. 26–3 Each of the three isotopes of carbon has six protons. The number in the upper left of each symbol gives the total number of protons and neutrons. Notice that carbon-14 and nitrogen-14 each has 14 nucleons. The carbon-14 has six protons (which is what makes it carbon) and eight neutrons; the nitrogen-14 has seven protons (which is what makes it nitrogen) and seven neutrons.

Note that even though we give them different names, and even though they have different numbers of neutrons, ordinary hydrogen, deuterium, and tritium are all hydrogen, because they each have a single proton. Thus they each have a single electron and very similar chemical properties (see Fig. 26–2).

We sometimes refer to both protons and neutrons as nucleons, since they are the principal particles in the nucleus. The masses of the proton and neutron are almost identical, so the mass of the nucleus depends on the total number of nucleons, which is called the mass number of the nucleus (symbol A). For example, $A = 1$ for normal hydrogen and $A = 2$ for deuterium.

Nuclei are usually designated by giving the chemical symbol for the element (Appendix 2), with the mass number written to the upper left of the chemical symbol (Fig. 26–3). For example, normal hydrogen is ^1H, deuterium is ^2H, and tritium is ^3H. The most abundant form of carbon has six neutrons (and carbon is the element that has six protons), so this form is designated ^{12}C. There is also an isotope with six protons and seven neutrons—^{13}C—and one with six protons and eight neutrons—^{14}C.

26.1b THE GLUE

Now that we know what goes into a nucleus, we also want to know what holds the nuclear particles together. Keeping the nucleus together is not an easy task, since the protons all have positive charges and so tend to repel each other. The electric forces within a nucleus try to burst the nucleus apart. The force binding the nucleus cannot be provided by gravity, since we have already seen that the gravitational force between two protons is much smaller than the electrical repulsion. The nucleons must be held together by a different force—the nuclear force. Actually, there are two nuclear forces, the strong force and the weak force. The weak force is inherently weaker than the electric force, so it does not play an important role in keeping the nucleus glued together. As we'll see later, the weak force does play an important role in certain nuclear reactions.

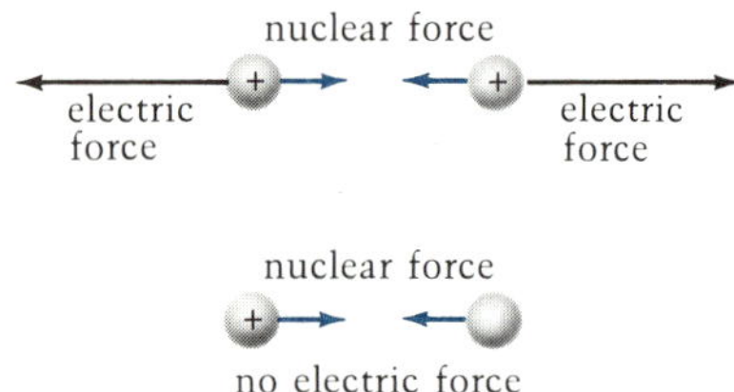

Fig. 26–4 The nuclear force between two protons is the same as the nuclear force between a neutron and a proton (or between two neutrons). However, two protons will also exert an electric force on each other, while a proton and a neutron or two neutrons will not exert an electric force on each other.

It is the strong nuclear force that keeps the nucleus bound together. Protons and neutrons are both subject to the nuclear force. In fact, extensive experiments have shown that *the nuclear force between two protons has the same strength as that between two neutrons or between a proton and a neutron.* We can simply think of the strong force as a force between nucleons, without regard to the charge of the nucleon.

The neutron plays an important role in a nucleus. (After all, it would seem inelegant to have a particle that serves no useful purpose.) We cannot build a nucleus out of protons only (except for hydrogen, which is a single proton) and have it stick together. However, when we add a neutron to a nucleus, we get all the advantages of increasing the strong force attraction without the disadvantage of increasing the charge of the nucleus, since the neutron has no charge. Therefore, the role of the neutron in the nucleus is to provide additional sticking power without increasing the electric repulsion. (There are, however, limits to the number of neutrons that we can put together; see Fig. 26–4.)

Let's take a closer look at some of the properties of the nuclear force. We know that when two protons are close together—as close together as they are in a nucleus—the strong force between them must be stronger than the electric repulsion. If not, the protons would fly apart. At large distances, the electric force must be stronger than the nuclear force. If it were not, then the protons in this book would all be attracted to each other, and the book would shrink to virtually nothing in size. (See Fig. 26–5.)

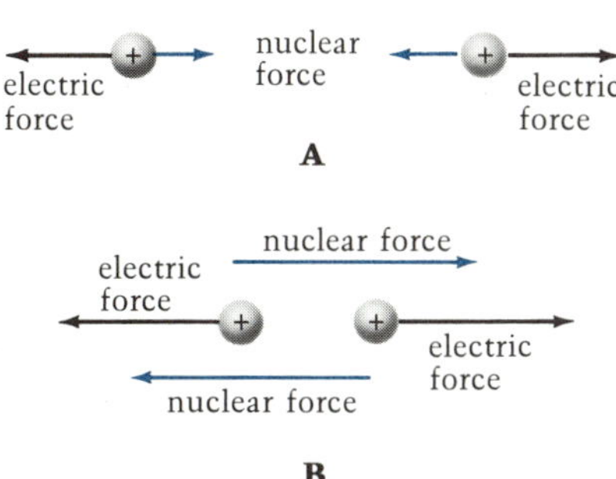

Fig. 26–5 A. When two protons are far apart, the nuclear force between them is much weaker than the electric force between them. As we move the protons closer together, the electric force increases, but the nuclear force increases even more. B. When the protons are very close together, the nuclear force is much stronger than the electric force. We call the electric force a long-range force and the nuclear force a short-range force.

Thus the strength of the nuclear force drops off with distance much faster than the electric force does. Since the strong force is only effective when the nucleons are very close together, we call it a short-range force.

The Japanese physicist Hideki Yukawa first theoretically explained the range of the strong force in 1935. Yukawa reasoned that if the electromagnetic interaction is carried by a particle—the photon—then the strong interaction must also be carried by a particle. He called this particle a meson. Yukawa said that the short-range nature of the strong interaction could be explained if the meson had some mass. The more massive the meson, the shorter the distance over which it could be transferred, and the shorter the range of the force. Yukawa calculated that this meson should have a mass about 200 times that of the electron, which is about 1/1800 that of the proton. The name meson is from the Greek for "in between."

In 1946, Yukawa's meson was discovered in showers of cosmic rays, high-energy charged particles that arrive from deep space; it had a mass about 40 percent greater than predicted. This particle is called the (pi) meson, or pion. Actually, there are three types of pions: neutral (π^0), positively charged (π^+), and negatively charged (π^-). Charged pions carry plus or minus one unit of fundamental charge. Other types of mesons exist, each having a higher mass than pions have.

Yukawa's ideas have been very fruitful in describing some of the basic features of the strong force. His theory works particularly well when the protons are not too close together. However, there are many further complications, and theoreticians are still working on theories involving the exchange of several mesons between interacting particles. Despite progress in this area, there is still much that we do not know about the nuclear force. Nuclear physicists are faced with a twofold problem—finding the way in which nuclei are arranged or put together, and finding information about the strong interaction. Nuclear physicists must figure out how to apply the rules of quantum mechanics to the many particles that make up the nucleus in order to discover the structure of various nuclei. Their work is especially hard because we do not really know how the nuclear force behaves when particles are very close together.

26.1c THE BUILDING CODE

Now we have the materials to make up the nucleus—neutrons and protons. We also have the "glue" that holds the nucleus together in

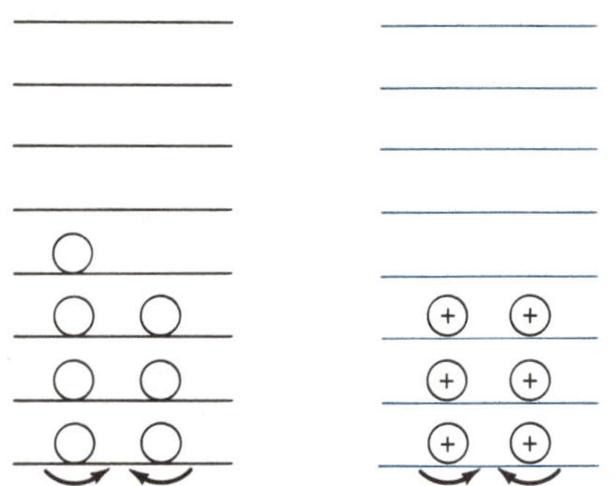

Fig. 26–6 When we put neutrons or protons in a nucleus, they obey the Pauli exclusion principle. We can only put two neutrons (spinning in opposite directions) in each neutron energy level and two protons (spinning in opposite directions) in each proton energy level. The nucleus shown in this figure is carbon (six protons) -13. The spacing between neutron (left) and proton (right) energy levels is slightly different because of the electric repulsion among the protons.

the strong interaction (the pions). Our next step is to figure out how it all goes together.

We must first realize that the rules of quantum mechanics still apply. Each "particle" in the nucleus is really a wave packet. Since each particle is confined within the nucleus, the waves bounce around inside and standing waves result. As in the case of the electrons orbiting the nucleus, the nucleons fit into the nucleus only in allowed states. These states have particular energies, so we talk about nuclear energy levels in much the same way that we talked about atomic energy levels for the electrons.

Like the electron, the proton and the neutron obey the Pauli exclusion principle. No two protons in the nucleus, or no two neutrons in the nucleus, can be in exactly the same state. Since the definition of a "state" includes the spin direction, we can place two protons in the same energy level as long as they have spins in opposite directions. Since there are only two opposite directions, we cannot fit three protons on the same energy level (Fig. 26–6).

The properties of a nucleus are often very complicated because there can be a large number of nuclear particles. The situation is not hopeless, however. While the number of particles involved and the unknown properties of the nuclear force have made it virtually impossible to make detailed calculations of nuclear structure, a number of very fruitful results have been found. Certain patterns have emerged that have given rise to models of nuclear structure.

One picture that has been particularly successful is the so-called shell model, proposed in the late 1940s. It had been observed that certain nuclei seemed excessively reluctant to take part in various nuclear reactions. This reluctance was reminiscent of the long-known fact that atoms such as helium, neon, and argon were reluctant to undergo *chemical* reactions. These "noble" gases were explained by the fact that they have only filled shells of electrons and therefore have no reason to interact chemically with other atoms. The idea of filled shells was carried over to the nucleus. It was assumed that nuclei that were reluctant to react had shells that were filled. By knowing the number of nucleons in such nuclei, one could work out the shell structure. The number of nucleons for the reluctant atoms are called magic numbers. (See Fig. 26–7.)

Even though the shell model is successful in explaining many features of the spectra of nuclei, many properties still cannot be

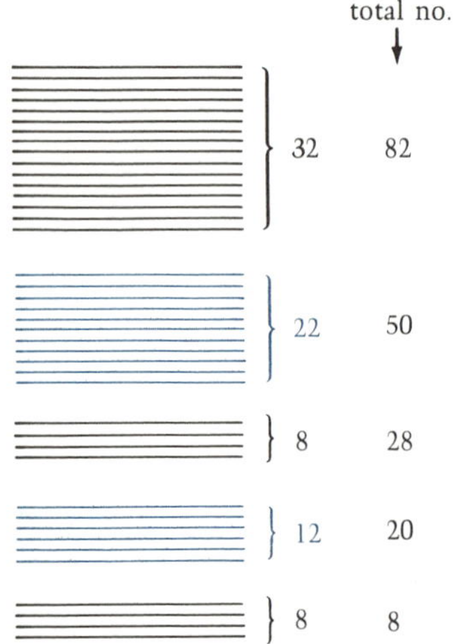

Fig. 26–7 Nuclear shell model. Each line represents an energy level in a nucleus for a nucleon. (The energy separations are not drawn to scale.) The levels fall into groupings called *shells*. The Pauli exclusion principle limits two of any type of nucleon (proton, for example), to each level. The maximum number of protons in each shell is shown at the right, with a running total on the far right.

Fig. 26–8 Maria Goeppert-Mayer and her daughter. Mayer and J. Hans Jensen received the 1963 Nobel Prize in physics for their discoveries about the shell model of the nucleus.

explained with it. Certain effects depend not only on the energy levels of the nucleons but also on the arrangement of the nucleus as a whole. Another model takes this behavior into account. This *collective model* considers the collective, or group, properties of the nucleons in a nucleus. The nucleons that fill a closed shell are treated as we would treat a drop of liquid in an ordinary fluid. Such a property as resistance to changing the shape of the nucleus turns out to be very similar to the classical behavior of a drop of water. For this reason, the collective model is also referred to as the liquid-drop model (Fig. 26–9).

Often the shell model and the liquid-drop model must be applied simultaneously to understand observed phenomena. It seems quite strange that we need two such different models to operate at the same time. After all, one of the models reflects the essence of the quantum-mechanical results—shells and energy levels. The other model seems to say that despite all of this quantum-mechanical behavior, there are still some properties that can be calculated from classical physics. Sometimes, these properties are even simulated with real drops of liquid in the laboratory.

We must keep in mind, however, that despite the success of these models in predicting many properties of nuclei, they are only models rather than a fundamental theory. However, while the slow search for a more fundamental understanding takes place, the models help us on the road to that theory. They are also practical aids in predicting the results of experiments that are important for applications of nuclear physics.

26.2 *Mass, Energy, and Stability of Nuclei*

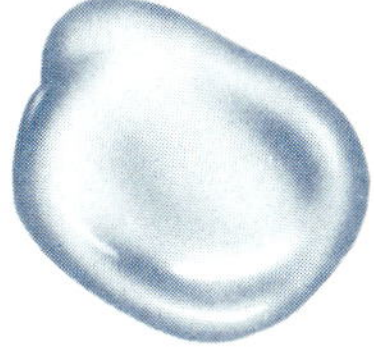

Fig. 26–9 In many situations, nuclei behave like drops of water.

Let's say that we want to take apart a deuterium nucleus (a proton and a neutron, called a deuteron). We would have to pull the proton and neutron apart with a force that is at least as strong as the nuclear force that holds the two particles together. Since we are exerting a force, we are doing some work as we move the proton and neutron apart. The amount of work that we must do in separating a nucleus into its component protons and neutrons is called the binding energy of the nucleus. The greater the binding energy, the harder it is to pull the nucleus apart.

One way to tell how much binding energy a nucleus has is to measure its mass. The mass of a nucleus, when measured, is slightly *less than* the mass of its component parts. For example, the mass of the neutron plus the mass of the proton is about 0.1 percent greater than the mass of the deuteron. How can the whole be less

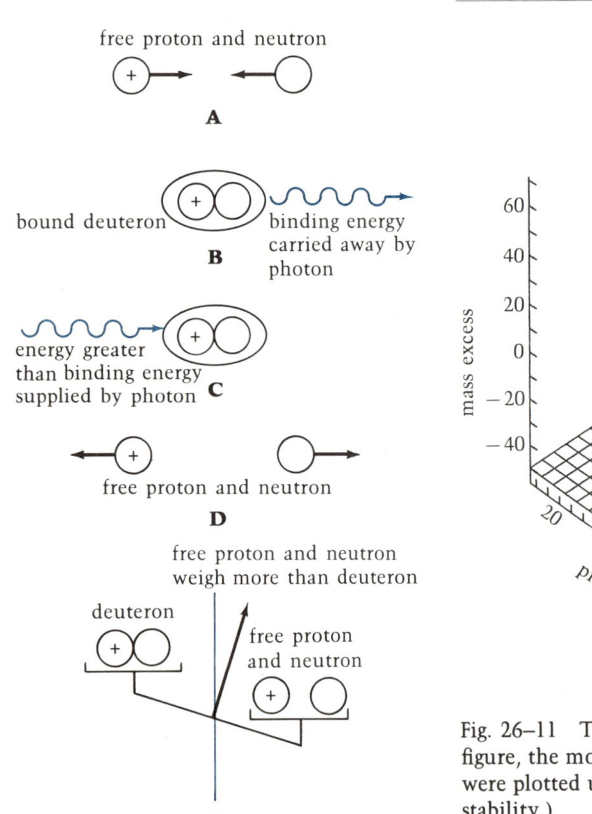

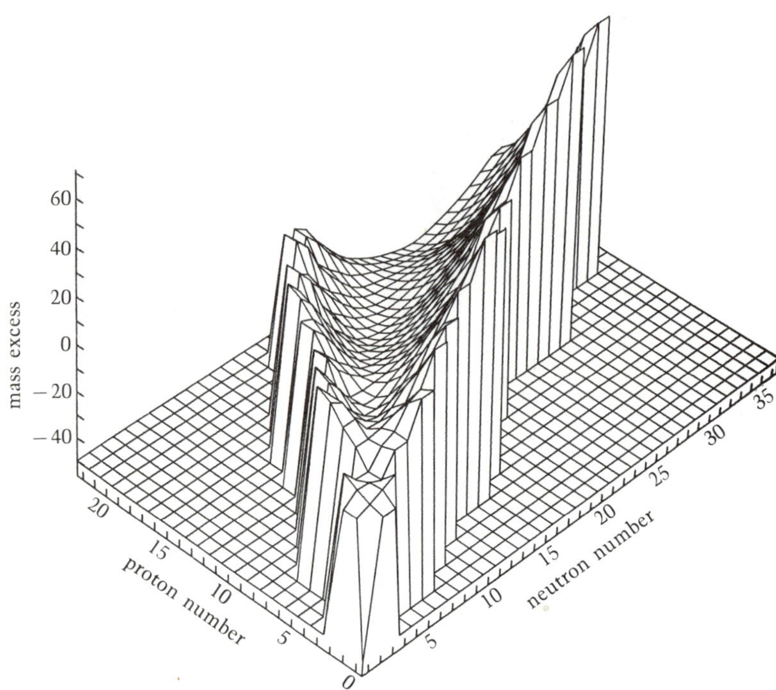

Fig. 26–11 The more excess mass a nucleus has, the less stable it is. In this figure, the most stable nuclei are in the valley running up the middle. (If this were plotted upside down, instead of a valley we would have the peninsula of stability.)

Fig. 26–10 Binding energy of a deuteron. A–D. When a free proton and neutron come together to form a deuteron (A), some excess energy is given off (B). This is the binding energy. If we want to break the deuteron apart, we must supply at least this much energy (C), and we get back a free neutron and proton (D). The total energy of the bound deuteron plus the binding energy must equal the energy of the free proton and free neutron. E. For this reason, the energy (and therefore the mass) of the deuteron is less than that of the free proton and neutron.

than the sum of its parts? Where did this 0.1 percent of the mass go? When the deuteron was formed, this amount of mass was converted to energy, according to Einstein's formula, $E = mc^2$. The energy that is given off in forming the deuteron is equal to the binding energy. It is the same as the amount of energy that you must put back if you are to separate the neutron and proton again (Fig. 26–10).

The greater the binding energy of a nucleus, the more stable that nucleus is. A stable nucleus is one that is hard to pull apart. If a nucleus is unstable, it will transform itself into a more stable nucleus—a nucleus with a greater binding energy. For this reason, we cannot arbitrarily combine protons and neutrons in any quantity. Only a limited number of nuclei are stable. A graph that shows the number of neutrons and the number of protons in nuclei is shown in Fig. 26–11. For nuclei with fewer than about 20 protons, the stable forms have approximately as many neutrons as protons, perhaps one or two more neutrons. However, as we go to larger numbers of protons, we need more and more neutrons to keep the nucleus from flying apart as a result of the electric repulsion of the protons.

The figure also shows that for a given number of protons, the nucleus is stable for only a limited range of neutrons. The stable nuclei on our graph form a long, narrow band. Because of its shape, we call this band the peninsula of stability.

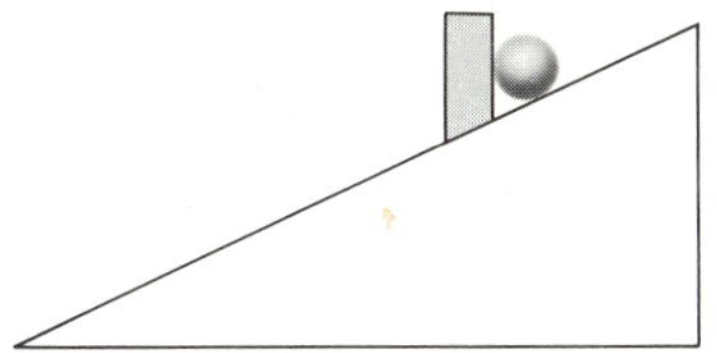

Fig. 26–12 The ball would like to roll downhill but there is a barrier to prevent it. If the ball can somehow get past the barrier, it will roll down the hill.

Of course, just because one nucleus has a lower binding energy than another does not mean that there will be a transformation from one to the other. There would have to be a means for that transformation to take place. This situation is analogous to a ball sitting on top of a ramp. The ball would like to get to the bottom, and will do so as long as there is some path available to it. However, if we place a block in the path of the ball, the ball stays above the block. If the ball is to get down the ramp, it must find a way over, around, or through the block. The same is true for nuclei. If one is to be converted into another, a path for that conversion must exist. (See Fig. 26–12.)

26.3 *Radioactivity*

There are actually a number of ways in which one nucleus can be converted into another. (The original nucleus is traditionally called the parent and the resulting nucleus the daughter.) Several such processes are lumped together under the name radioactivity. Radioactivity was discovered by Henri Becquerel in Paris in 1896, when he noticed that a small amount of uranium salt had fogged a photographic plate. This discovery was well before Rutherford's discovery of the nucleus.

Fig. 26–13 Marie Curie.

Many scientists were immediately fascinated with radioactivity and the associated transmutation of one element into another. Marie Curie, a student of Becquerel's, took up the project and discovered other radioactive elements besides uranium. Mme. Curie, working with her husband Pierre, discovered both radium and polonium. The Curies and Becquerel won the 1903 Nobel Prize in physics for their work on radioactivity. Within a few years, Mme. Curie managed to isolate a sample of radium, for which she was awarded a second Nobel Prize, this one in chemistry.

It was quickly noted that the number of particles emitted by a radioactive sample decreased with time. After a reaction the daughter may not be radioactive. As time goes on there are fewer radioactive nuclei in the sample, and the rate of emission of particles decreases. We normally specify the rate at which a given sample decays as its half-life, the time that it takes for half of the nuclei in any sample to decay into something else. For example, suppose that the half-life of a given nucleus is one year. After one year we have half parent and half daughter. After another year, only a

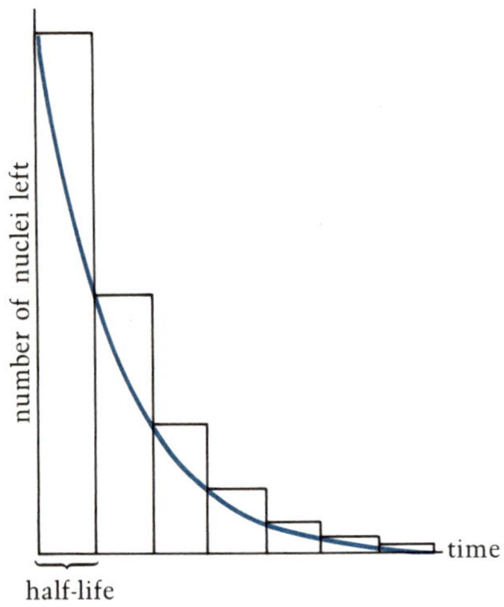

Fig. 26–14 A graph of the number of nuclei left in a radioactive sample after each half-life. After each half-life there are half the number left as at the beginning of the half-life-long time interval.

quarter of the parent nuclei remain (that is, half of the remaining half), and so on (Fig. 26–14).

Early studies of radioactivity indicated that three different types of rays were emitted. They were classified according to their ability to pass through various types of matter: alpha, beta, and gamma. We now know that the first two of these are really particles while the third is indeed an electromagnetic ray.

26.3a ALPHA DECAY

Alpha particles are helium nuclei—two protons and two neutrons bound together. An alpha particle is a particularly stable nucleus. Let us consider a parent nucleus that emits an alpha particle. The daughter nucleus will then have two protons fewer and two neutrons fewer than the parent nucleus. An alpha decay will occur spontaneously only if a more stable system forms in the process.

Even though the alpha decay of a given nucleus may be allowed by considerations of stability, the actual emission of the alpha particle is not an easy process because the alpha particle is held tightly in the nucleus. In order for it to escape, it must get far enough from the rest of the nucleus so that the electric repulsion between the alpha particle and the rest of the nucleus is stronger than the strong-force attraction. Remember the ball that we have sitting at the top of a ramp with a block in its way. We could lift the ball over the barrier, but this would require the addition of some energy. Such additional energy is not available in a nucleus. In the world of classical physics, the alpha particle would be trapped, and the decay could not take place.

However, the alpha particle gets out by a trick that can be played only in the quantum world. It goes right through the barrier! We call this process barrier penetration, or tunnelling. It is something that waves can do, but classical particles cannot. At a barrier, the wave function cannot suddenly drop to zero; it needs a small space

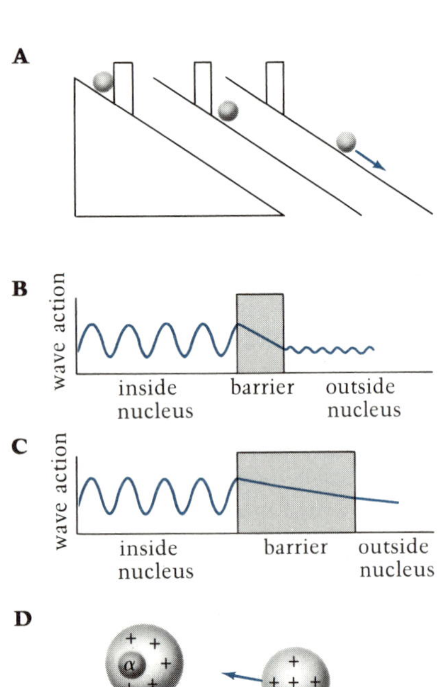

Fig. 26–15 Barrier penetration and alpha decay. A. If the ball can get past the barrier, it will roll downhill. In the classical world, the ball must get over the barrier. B. In the quantum-mechanical world, the alpha particle can get through the barrier. The graph shows the wave function for the alpha particle inside the nucleus, in the barrier, and outside the nucleus. The higher the wave, the greater the probability of finding the alpha particle in a given region. The wave function does not go to zero immediately upon entering the barrier region. It dies out slowly. If the barrier is not too large, there will still be some wave function left for the outside world. C. As the barrier gets larger, the wave function outside the nucleus gets smaller. D. Once the alpha particle gets outside the nucleus, the electric repulsion of the positively charged nucleus and the positively charged alpha particle accelerates the alpha particle.

in which to fall off. As a result, there is a small probability (corresponding to a small wave function) for a particle to be a little bit within the barrier region. If the barrier is very thick, the distance a particle can penetrate is so small that it does not get out the other side. However, if the barrier is narrow, there is a small probability that the particle can actually get through (see Fig. 26–15).

Barrier penetration comes out of the quantum-mechanical calculations—the same calculations that give energy levels in atoms—with no additional assumptions. The calculations involving tunnelling of alpha particles have been quite successful in predicting the properties of the alpha decay of various nuclei.

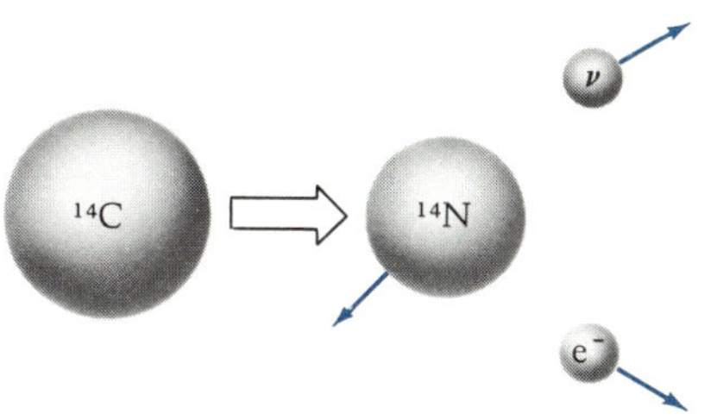

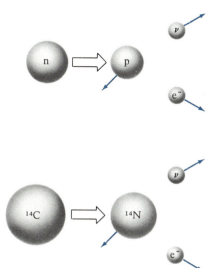

Fig. 26–16 Two beta decays. In each case the net result is to turn a neutron into a proton (plus an electron and a neutrino).

26.3b BETA DECAY

A beta particle is simply an electron or a positron. The positron is the antiparticle of the electron, but for these purposes we can think of it as an electron with positive charge.

The most straightforward example of a beta decay is that of a free neutron. Unlike the proton, which is a stable particle, a neutron outside a nucleus is unstable and will "beta decay." When it does, the result is a proton and an electron and something else. That something else is a mysterious particle called the neutrino (named from the Italian for "little neutral one"). Neutrinos appear to have little or no rest mass. If neutrinos have no rest mass, they travel at the speed of light. Neutrinos are very difficult to detect, but their existence was inferred from analysis of conservation of energy, momentum, and angular momentum in beta-decay data. Because the beta particle does not carry away all of the energy and spin available in the decay, there has to be another particle that is also carrying some of the energy and spin. Although the existence of the neutrino was predicted by Wolfgang Pauli in 1930, and the theory was worked out by Enrico Fermi soon thereafter, neutrinos were not observed experimentally until 1957, when a reactor was available that could produce large numbers of them. (Actually, antineutrinos were observed in this experiment; see Fig. 26–16.)

Detection of neutrinos is difficult because they interact with matter only via the weak nuclear force. The interactions are very infrequent. Any interaction in which a neutrino is either created or destroyed must take place via the weak force. (We saw a neutrino destroyed in Fig. 5–21.) Therefore beta decay is a weak interaction, and usually takes place very slowly. For example, the half-life of the free neutron is about 11 minutes. If we put a billion neutrons on the table, after 11 minutes only half a billion would remain. While this may not seem too long, we must remember that most things on the nuclear level take place in tiny fractions of a

Fig. 26–17 An accurate scale for carbon dating is obtained by comparing the carbon dating scale with a tree-ring time scale.

second. (Some reactions involving the strong force take place in 10^{-23} seconds!)

When the beta decay takes place, an electron and a neutrino are emitted. One neutron in the nucleus has been converted into a proton. The new nucleus has one more proton but one fewer neutron than the old nucleus. However, since the total number of nucleons remains the same, the mass number of the new and old nuclei are the same. An example of a beta decay is

$$^{14}C \rightarrow {}^{14}N + \beta^{-} + \bar{\nu}$$

carbon - 14 → nitrogen - 14 + electron + antineutrino

This decay has a half-life of 5730 years and is very important in what we call carbon dating. The most common form of carbon is carbon-12 (six protons and six neutrons). Carbon-14 (six protons and eight neutrons) decays, so it must have been recently created if

COSMIC GALL, by John Updike

Neutrinos, they are very small.
  They have no charge and have no
    mass
And scarcely interact at all.
The earth is just a silly ball
  To them, through which they
    simply pass,
Like dustmaids down a drafty hall
  Or photons through a sheet of
    glass.
  They snub the most exquisite gas,
Ignore the most substantial wall,
  Cold-shoulder steel and sounding
    brass,
Insult the stallion in his stall,
  And, scorning barriers of class,
Infiltrate you and me! Like tall
And painless guillotines, they fall
  Down through our heads into the
    grass.
At night, they enter at Nepal
  And pierce the lover and his lass
From underneath the bed—you call
It wonderful: I call it crass.

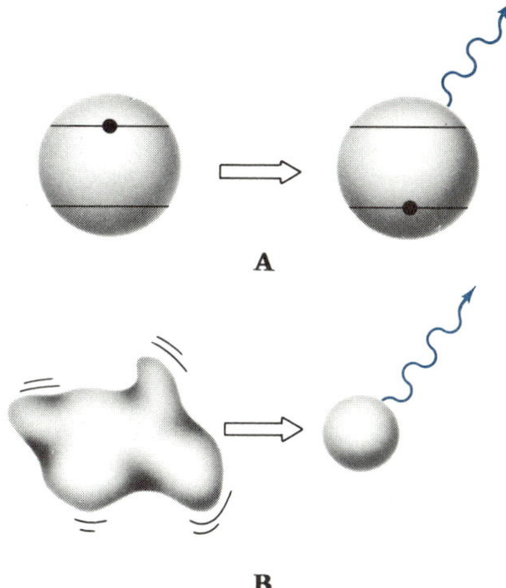

Fig. 26–18 Emission of a gamma ray. The same situation is viewed slightly differently in A and B. A. The nucleus starts in a high (excited) energy level and goes to a lower (possibly the ground) energy level and emits a photon in the process. B. In this case we see the upper energy level as one of high agitation and the lower energy level as one of tranquility. The emission of the gamma ray allows the nucleus to get rid of some excess energy without breaking up.

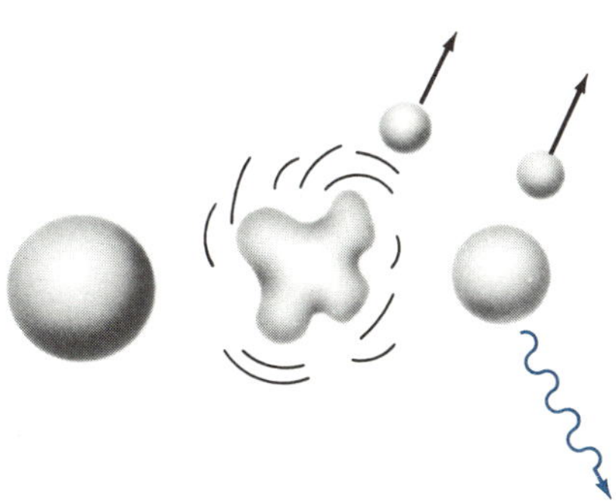

Fig. 26–19 Alpha decay followed by emission of a gamma ray. A. The nucleus before alpha decay. B. An alpha particle is emitted, leaving the nucleus in an agitated state. C. The excess energy is given off in the form of a gamma ray. Remember, a gamma ray is just a high-energy photon.

we can find it. On the earth, cosmic rays continually produce carbon-14 in the upper atmosphere, and then atmospheric circulation mixes it into the lower atmosphere. Both these isotopes of carbon act the same chemically, so when we breathe, in addition to oxygen, we take in some carbon dioxide with a mixture of the two isotopes. When an organism stops breathing, the carbon-14 is no longer replaced and continuously decays away. By noting how much carbon-14 is in a tree ring or in a fossil, in relation to the amount of carbon-12, we can tell how long ago the living thing died. (Since the amount of carbon-14 in the atmosphere varies a bit over the years, the carbon-14 date scale we get from tree-ring studies is more accurate than the carbon-14 date scale we get by assuming that the amount of carbon-14 in the atmosphere long ago is the same as the amount present now.) With the current carbon-14 date scale, we can find the dates of archaeological sites and of anthropological remains, for example.

Beta decay can also take place with the emission of a positron. In this case, one of the protons in the nucleus turns into a neutron (something that cannot happen in free space). Again, this can only occur if the energies involved are favorable.

Actually, the slow rate of weak interactions has a very beneficial effect. The main nuclear reaction by which the sun generates energy happens to be a weak interaction. It proceeds so slowly that the sun's original nuclear fuel is enough to last 10 billion years.

26.3c GAMMA DECAY

Gamma rays are nothing more than photons of much higher energy than visible light. When a gamma ray is emitted by a nucleus, that nucleus does not change into a different nucleus (since the photon carries away neither mass nor charge). Instead, a gamma ray is emitted when a nucleus makes a transition from one energy level to a lower energy level, just as an atom emits a photon when an electron makes a transition from one energy level to a lower energy level. The difference is that the energies involved in nuclei are much greater than those involved in the motion of electrons around the nuclei. As a result, the photons emitted by nuclei have much more energy than those emitted by atoms. Remember, a higher energy corresponds to a shorter wavelength. The shortest wavelengths are gamma rays. (See Figs. 26–18 and 26–19.)

26.3d USES OF RADIOACTIVITY

Radioactivity has many uses. The value of radioisotopes in medical examinations is well established. For example, the radioactive isotope of iodine is treated by the body in the same way as the normal isotope. Injecting a small amount of radioactive iodine into the body allows doctors to study those regions where the iodine collects, such as the thyroid gland. Other medical applications range from the sterilization of instruments to the treatment of certain types of cancer, for radiation can destroy some cancer cells. (Radioisotopes are sometimes implanted directly into a cancerous growth to provide an internal source of radiation to kill the cells.)

Fig. 26–20 Rosalyn S. Yalow received the Nobel Prize in physiology in 1977 for her work with radioisotopes.

26.4 *Nuclear Reactions*

We usually think of any process in which a nucleus undergoes some change (even going from one energy level to another) as a nuclear reaction. The decays that we talked about in the last section are nuclear reactions. Another class of reactions involves two nuclei coming together to form some other combination of nuclei. One of the two initial nuclei may be as simple as a hydrogen nucleus (a proton). In fact, watching what happens when we shoot a proton into a nucleus is one of the ways in which we have obtained a great deal of information about nuclear structure.

In some nuclear reactions, the mass of the final products is less than that of the initial products, meaning that some energy is given off during the reaction. In other cases, the mass of the final products is greater than that of the initial products, meaning that some energy must be provided for the reaction to go. If we want to produce the second kind of reaction, we must find a way of putting energy into the system. An accelerator, for example, can increase the kinetic energy of one of the nuclei.

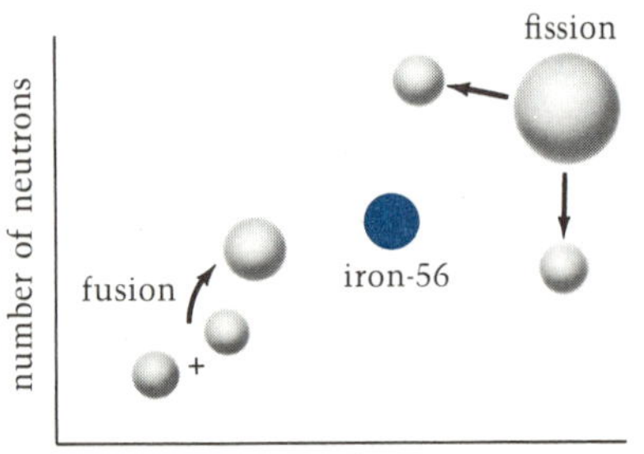

Fig. 26–21 Getting energy out of nuclear reactions. For most elements lighter than iron-56, fusion will give off energy. For most elements heavier than iron-56, fission will give off energy. In each case, you get energy out when the products are closer to iron-56 than the original material.

Even in the case of a reaction that gives off energy, we must put some energy in to get the reaction started. This is necessary because the reaction does not take place until the two nuclei are close enough together so that the attraction caused by the nuclear force will be stronger than the repulsion caused by the electric force. In short, we must put in enough energy to overcome the electric repulsion.

A more common example of this problem occurs for chemical rather than nuclear reactions. Consider what happens when we strike a match. We get heat out of the actual burning of the match. However, we have to do a little work—striking the match against some surface to heat it slightly—in order to get the reaction going. Once the match starts burning, enough heat is generated to keep it going, and we end up getting more energy out than we put in.

We can decide whether a given reaction will give off or require

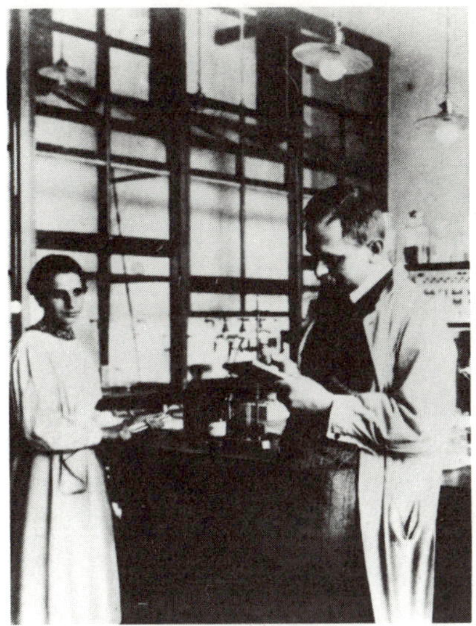

Fig. 26–22 Lise Meitner and Otto Hahn, two of the discoverers that uranium could undergo fission. Their work indicated that a chain reaction, in which the neutrons produced caused additional atoms to undergo fission, could occur. This in turn indicated that large quantities of energy could be released from atoms. (Otto Hahn, *A Scientific Autobiography.* New York: Charles Scribner's Sons, 1966)

extra energy by knowing the binding energies of the nuclei involved. If we look at a graph that shows us the binding energy for various nuclei, we find a very interesting result. For low mass numbers, nuclei become more stable as the mass number increases. For example, sodium-23 is more tightly bound than carbon-12. This trend continues until we reach iron-56 (26 protons, 30 neutrons). After that, the heavier elements become less stable instead of more stable. This means that iron-56 is the most stable nucleus.

To produce nuclear power, we must choose reactions that give off energy. A look at the trends in stability of nuclei tells us that there are two different types of reactions that we can exploit. Either we start on the low-mass side of iron-56 and work up toward heavier nuclei or we start on the high-mass side of iron-56 and work down toward lighter nuclei. In either case we are working toward iron-56, since every time we make a more stable nucleus we get more energy out. If we start with a heavier nucleus, the idea is to break it into pieces. This process is called fission. If we start with lighter nuclei, the idea is to build them into a heavier nucleus. This process is called fusion. (See Fig. 26–21.)

26.4a FISSION

In fission, a nucleus breaks apart into lighter fragments (Fig. 26–23). Once the fragments are separated a little bit, the electric repulsion is stronger than the nuclear attraction and the fragments fly apart. The kinetic energy we get out comes from this electric repulsion. Not every heavy nucleus is likely to break up in this way because there is still the problem of getting the fragments far enough apart for the electric repulsion to take over from the nuclear attraction.

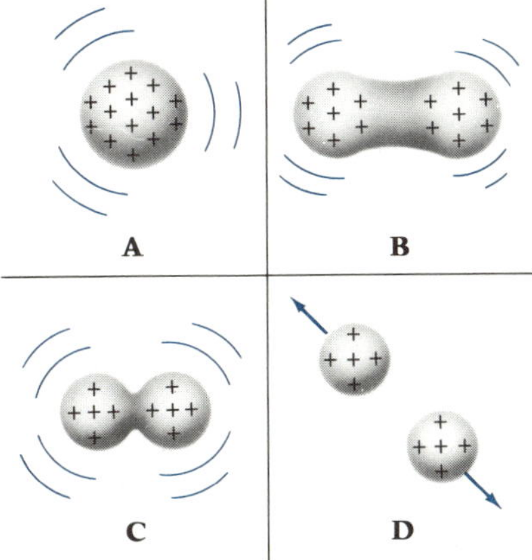

Fig. 26–23 Fission. A. We start with an unstable nucleus. B. The nucleus begins to elongate into two sections. C. The two sections begin to separate. D. Once the two sections are apart, the electric repulsion forces them apart at high speeds.

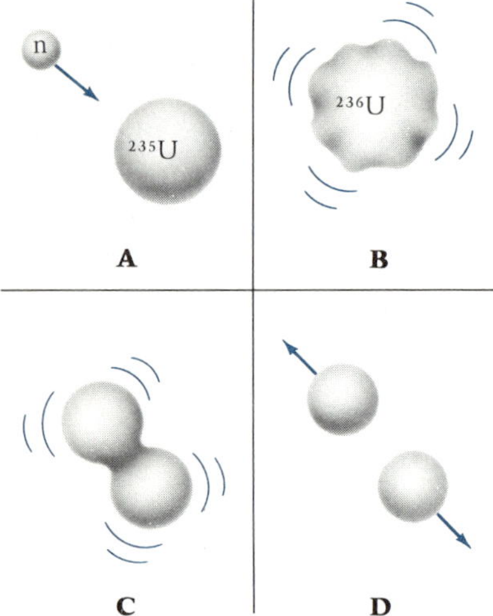

Fig. 26–24 The fission of uranium-235. A. First the uranium-235 captures a neutron and becomes uranium-236. B. ^{236}U is unstable. C. Thus it begins to split apart. D. Once two (or more) pieces are apart, they accelerate away from each other, pushed apart by the electric repulsion between the positively charged fragments.

FISSION AND SUPERSTITION

[*A cautionary verse for parents or children appropriate to the Christmas season.*]

This is the Tale of Frederick Wermyss
Whose Parents weren't on speaking terms.
So when Fred wrote to Santa Claus
It was in duplicate because
One went to Dad and one to Mum—
Both asked for some Plutonium.
See the result: Father and Mother—
Without Consulting one another—
Purchased two Lumps of Largish Size,
Intending them as a Surprise,
Which met in Frederick's Stocking and
Laid level Ten square Miles of Land.

MORAL
Learn from this Dismal Tale of Fission
Not to mix Science with Superstition.

HMK

Usually a fission reaction has to be induced, that is, made to start by some outside process. For example, one important fuel for fission is the isotope of uranium ^{235}U. (Most natural occurring uranium is the isotope ^{238}U, and much effort must be expended to enrich a sample of uranium to contain more ^{235}U than is usual.) If we shoot a neutron at ^{235}U, it can capture the neutron and become ^{236}U. However, this ^{236}U is formed in a very excited state. We can think of this nucleus as a tiny drop of liquid that is vibrating furiously. (The liquid-drop model turns out to be quite useful in studying the properties of fission.) The oscillations are so violent that the drop breaks apart (Fig. 26–24).

The drop breaks up into two lighter nuclei, and also gives off a few free neutrons and some energy. It doesn't always break up in the same way. Various daughter nuclei may be formed, depending on the exact conditions of the excited nucleus before breakup. The important feature is that neutrons are produced. Other ^{235}U nuclei can then capture the new neutrons and themselves undergo fission,

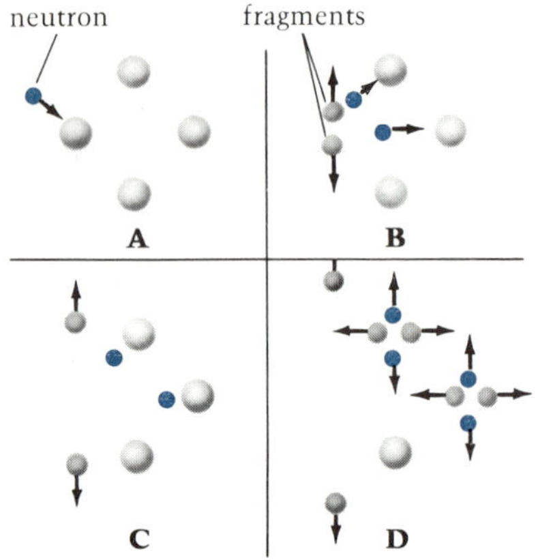

Fig. 26–25 A chain reaction. We start with many nuclei, such as uranium-235. A. A neutron enters and is absorbed by one of the nuclei. B. This nucleus breaks up into two large fragments and several (two or more) neutrons. C. These two neutrons approach other uranium-235 nuclei. D. These nuclei break up producing more neutrons, which are absorbed by other nuclei, and so on.

producing more neutrons, and so on. Thus, once the reaction starts, it can continue without our adding any more neutrons. The process is called a chain reaction (Fig. 26–25). For the chain reaction to continue, enough ^{235}U must be around so that a reasonable number of the new neutrons are captured by ^{235}U nuclei. The minimum amount of material to sustain a chain reaction is called the critical mass. The first controlled chain reaction was produced in 1942 in a former squash court at the University of Chicago by a group headed by Enrico Fermi. We shall discuss the applications of these reactions to energy production in the next chapter.

26.4b FUSION

Fusion is responsible for the fundamental energy production in the sun. The basic effect of the nuclear reactions in the sun is to convert four hydrogen nuclei into one helium nucleus. (This happens in a series of reactions rather than in a single reaction.) In this series of reactions, 0.007 of the mass of the hydrogen nuclei is converted into energy. This is slightly less than 1 percent, but there is so much fuel in the sun that a tremendous amount of energy is generated.

James Bond fans have no trouble remembering .007.

As with other reactions, we usually must supply some energy to get a fusion reaction started. The additional energy must overcome the electric repulsion between nuclei in order to get them close enough together for the nuclear force to take over. One way of adding energy is to fire one of the nuclei at the other in an accelerator. This procedure is often sufficient if we only want to study the reaction, but is very inefficient if we want to get energy out of the reaction. We could also give the particles enough energy by heating them to a high enough temperature. At very high temperatures, the particles have sufficient energy to overcome the repulsion and get close together. In the sun, once a sufficient temperature—10 million K—was reached to get the reactions started, enough energy was released from fusion to keep the temperature high. Because such fusion reactions can be initiated at high temperatures, they are called thermonuclear reactions.

The sun has been shining in this way for about five billion years, and has about another five billion years to go before it exhausts its supply of hydrogen.

26.4c NATURAL, NEWLY MADE, AND "SUPERHEAVY" ELEMENTS

We have now seen that if a nucleus does not like its state, there are many avenues available by which it can be transformed into a more stable nucleus. This means that we cannot go on adding protons and neutrons together in any combination and hope to come up with a nucleus that is stable. In fact there are only about 260 stable nuclei. The largest number of protons in a naturally occurring nucleus is 92, for uranium. To be found naturally, a nucleus must either be stable or must have an extremely long half-life.

In addition to the natural nuclei, some elements have been created in accelerators. These are made by fusion of other nuclei, and since we are building an element heavier than iron-56, the reactions require the addition of a lot of energy. Nuclear physicists have succeeded in making elements with up to 106 protons (Appendix 2), but all of these new elements are very unstable and decay by one means or another. (One exception is plutonium, which has isotopes with very long half-lives.)

An interesting prediction has come of some of the calculations that combine the shell and liquid-drop models. The calculations predict that it may be possible to create much heavier stable nuclei than the nuclei we have already created. These nuclei may have several hundred nucleons, but only particular combinations of neutrons and protons would be stable. When we add these predictions to the graph showing the peninsula of stability, the regions for which these superheavy elements may be produced show up as islands of stability. Nuclear physicists feel that the production of such superheavy elements would be fascinating and would certainly be a triumph for the predictive powers of current nuclear models.

Key Words

atomic number
neutron
isotope
deuterium
tritium
nucleons
mass number
nuclear force
strong force
weak force
short-range force
meson
cosmic rays
(pi) meson/pion
nuclear energy levels
shell model
magic numbers
collective model
liquid-drop model
binding energy
peninsula of stability
parent
daughter
radioactivity
transmutation
half-life
alpha particle
barrier penetration (tunnelling)
beta particle
neutrino
carbon dating
gamma rays
nuclear reaction
fission
fusion
chain reaction
critical mass
thermonuclear reactions
superheavy elements
islands of stability

Questions

1. Suppose you built a scale model of a typical atom, and the model were as big as your room. How large would the nucleus be (approximately)?
2. Why can't we build large nuclei by simply adding more and more protons?
3. Why can't we build large nuclei by simply adding more and more neutrons?
4. Why is the neutron necessary?
5. Consider the following three nuclei:

    i. 2 protons + 2 neutrons
    ii. 2 protons + 1 neutron
    iii. 1 proton + 2 neutrons

    (a) Which two are isotopes of the same element? (b) Which two have the same atomic number? (c) Which two have the same mass number? (d) Which two have the same number of nucleons?
6. Why does the number of protons in the nucleus determine the chemical properties of a given element?
7. (a) How does the strong nuclear force between two protons compare with the strong nuclear force between two neutrons? (b) How does the electromagnetic force between two protons compare with that between two neutrons?
8. Within the nucleus, which force tends to hold the nucleus together and which force tends to push it apart?
9. What is a "short-range" force?
10. What role does the pi meson play in the nuclear force?
11. What are the basic "rules" that protons and neutrons must obey when they form a nucleus?
12. In what ways has the shell model been useful in explaining and predicting nuclear phenomena?
13. In what ways has the liquid-drop model been useful in explaining and predicting nuclear phenomena?
14. Discuss the following statement: "The liquid-drop model and the shell model are not really competing models in explaining nuclear structure."
15. Does a greater binding energy mean that the nucleus is more stable or less stable?
16. As the difference between the mass of the individual components making up a nucleus and the mass of the nucleus increases, what happens to the stability of the nucleus?
17. (a) How much mass would you have to convert into energy to give you enough energy to raise your body by 10 meters? (b) How many protons is this mass equivalent to?

18. For each of the three types of radioactivity—alpha, beta, and gamma—state how the charge, mass number, and atomic number of the final nucleus are related to those of the initial nucleus.
19. Of the three types of radioactivity—alpha, beta, and gamma, (a) which produces the largest change in mass number? (b) which produces the largest change in charge?(c) which provides the least change in charge?
20. Nucleus A "alpha decays" into nucleus B. Which nucleus has the higher atomic number?
21. What is barrier penetration and how does it apply to alpha decay?
22. When an alpha particle leaves a nucleus, does it speed up or slow down after "tunnelling" through the barrier?
23. Of the three forms of radioactivity—alpha, beta, and gamma—which proceeds via the weak interaction?
24. How was the existence of the neutrino deduced?
25. Why are neutrinos very hard to detect?
26. In carbon dating, we actually get an age since the object being studied stopped replenishing its carbon-14. This works well for deciding how long living things have been dead. Under what conditions would the technique work for nonliving things?
27. What is the relationship between gamma rays emitted by a nucleus and the light emitted by atoms when an electron goes from one energy level to a lower one?
28. Why must a proton be accelerated to a high energy if it is to get close enough to a nucleus for a nuclear reaction to take place?
29. What are two ways in which particles can achieve high enough energies for nuclear reactions to take place?
30. We usually get energy out of fission reactions when the initial nucleus is a very heavy one, but we usually get energy out of fusion reactions when the initial nuclei are very light ones. Why is this?
31. What conditions are necessary for a fission chain reaction to start?
32. The mass of the sun is 2×10^{30} kg, and the sun gives off 4×10^{26} joules/s. At this rate how long will it take to convert 0.007 of the sun's mass into energy? Why would you expect this number to be close to the time that the sun could live?

CHAPTER 27

Energy for Tomorrow

If you ask a group of people to tell you what the "energy crisis" is, you will get a variety of answers. Some see it in global terms, asking what type of environment and energy supply we will leave for future generations. Others see it in more personal terms, reflected in the cost of heating their homes or driving their automobiles. Some see it on a ten-year scale; some see it on a hundred-year scale. Others don't agree that there is an energy crisis at all. What we call the "energy crisis" comes out of a complex series of scientific, technological, political, economic, sociological, and psychological problems.

Much of the debate on the energy crisis and what to do about it has been on very emotional grounds. Part of this may be an irrational response to the complexity of the issues involved, part may result from the fact that even "rational" scientists cannot really agree on several of the important "facts."

In short, as we'll see in this chapter, to say simply that we're running out of fuel would be a gross oversimplification. What's more, it is not even true. We *are* running out of inexpensive supplies of the fuels with which we are most familiar and comfortable. We seem to be faced with alternatives that are either very expensive or appear filled with unknown dangers.

Energy decisions eventually fall in the political arena. However, in this chapter we'll look at the background and then take a look at some of the aspects of the problem that involve physics.

27.1 *Fossil Fuels*

Fig. 27–1 A recreation in the Carnegie Museum in Pittsburgh of a carboniferous swamp forest, which after decay, time, and pressure, became coal.

At the present time, the vast majority of our fuel comes from petroleum, natural gas, and coal. These are all called fossil fuels, since they are all the result of decomposition of living matter (Fig. 27–1). As such, the ultimate source of the energy contained in these fuels is the sun, since it was the energy of the sun that originally allowed this living material to form and grow.

The formation of the fuel started with the death of the living vegetable matter—perhaps trees—over 200 million years ago. This material was gradually covered over by other material, and the pressure of the covering layers increased the concentration of the carbon that was released. The material first became peat and eventually was compacted by a factor of twenty to one to become soft coal, known as bituminous coal. This type of coal still contains a lot of volatiles (material that becomes a gas at low temperatures) in addition to the carbon. Bituminous coal burns easily, but gives off a lot of smoke. In some cases, further pressure drove off the volatile materials from the bituminous coal, leaving a hard form of coal, anthracite, which burns more cleanly but is harder to ignite.

Oil and natural gas were formed in more complicated processes involving marine life. Once the fish, crustaceans, and other marine life died and were buried in the mud, the covering layer supplied heat and pressure to convert them into liquid oil and natural gas. (Natural gas is mostly methane, which is a molecule containing one carbon and four hydrogen atoms.) Since marine life was the important first step, many oil and natural gas deposits are in coastal or offshore areas or else are on the sites of former inland lakes. Some areas of the world, like the Middle East, have especially large supplies.

Although coal, oil, and gas are being formed even now, it will be millions of years before these new deposits are ready for use. And oil is even more vital in the petrochemical industry than it is as a fuel.

Fig. 27–2 As a result of the 1973 war between Israel and Egypt, oil supplies were cut off from the Middle East to the West. Here we see the Sinai peninsula in a view from a NASA satellite.

27.1a PETROLEUM AND NATURAL GAS USE

It is now estimated that we have used up about 15 percent of the total supply of petroleum and natural gas on the earth. Even if we could locate and extract all the oil and natural gas supplies, they would still be exhausted in a few years or, at most, several hundred years, depending on how fast we use them up.

However, we do not know where all the oil and natural gas supplies are buried. Also, the oil and natural gas that we have

Fig. 27–3 An oil-drilling head known as a "Christmas tree," on the North Slope of Alaska.

extracted so far have been the easy part. As we search for more deposits, the fuel becomes harder to extract. In addition, conventional drilling techniques do not extract all the oil; only about 30 percent is recovered. A like amount can be brought up with expensive secondary recovery techniques such as pumping water into the wells.

Oil is also found in certain deposits of shale, a fine-grained rock (Fig. 27–4). Estimates place the amount of fuel contained in oil shale at about 100 times the amount in standard crude-oil deposits. However, the techniques for recovering this oil are not yet economical. They may become economically competitive, either by technological advances or by increases in the price of oil from other sources.

The development of natural gas supplies parallels that of petroleum supplies, since they are usually found in the same place. Demand for natural gas in the United States greatly increased in the early part of the 1970s.

The big advantage with the use of oil and gas is that, for the present, they seem to be the least expensive energy sources. Also, because of our experience with them, we feel confident that we can anticipate and control the problems associated with their use. (This may be misplaced confidence as we'll see later.) An additional advantage of natural gas is that it burns fairly clean because the end products of burning methane are carbon dioxide and water (although there is a growing concern over the effects of carbon dioxide).

Oil and gas sources are not without their drawbacks, aside from the dwindling supplies and increased expenses. The combustion of

Fig. 27–4 An outcropping of oil shale.

A

B

Fig. 27–5 A. Smog over Los Angeles, as viewed from Mount Wilson. B. Industrial smoke can be a contributor to smog.

petroleum products still produces a variety of pollutants (Fig. 27–5). There are also problems in getting the fuel to us. Domestic oil and natural gas can be transported by pipeline, but the installation of such pipelines can cause environmental problems. For example, there has been considerable concern about the effects of the Alaskan pipeline which carries oil from Prudhoe Bay to Valdez (Fig. 27–6). Even without considering the spills that occur from the pipeline, environmentalists fear effects on migrating herds of animals because it acts as a barrier, and effects on the nearby environment because of the high temperatures of the oil flowing through the pipe.

Fig. 27–6 The Alaskan pipeline, showing the insulating workpad designed to protect the permafrost soil levels.

Some of the most spectacular environmental problems so far have come from the transportation of oil by tanker. It has now become more economical to transport oil in a relatively small fleet of "supertankers" (one of which appeared in Fig. 4–19) rather than in a larger fleet of small tankers. Unfortunately, it is still not clear that supertankers can be effectively controlled under all of the adverse circumstances that can occur at sea. Their momentum is so great that tremendous distances—up to 15 kilometers—are required to turn or stop a supertanker. There is also a danger—and a history—of oil "spills" in the ocean from offshore drilling platforms (Fig. 27–7). (Actually, most of the oil in the oceans comes not from these spills but from discarded motor oil that is ejected through the sewer systems of large cities.)

There are also problems in shipping natural gas. Tankers are used here too, but the density of gas is so low, compared to a liquid, that it is not efficient to send shiploads of gas. For this reason, the natural gas is liquefied before it is shipped. Under normal shipping conditions, the liquefied natural gas (LNG) must be kept at very low temperatures (−160°C) and at pressures more than ten times that of the atmosphere. In this condition it is highly explosive. As

Fig. 27–7 "Spilled" oil covers the water, pollutes beaches, and kills waterfowl and fish. The oil here was from a runaway well in the Gulf of Mexico.

Fig. 27–8 A ship carrying liquefied natural gas.

(above) Fig. 27–9 Digging coal deep underground.

we write this, there have recently been two major crashes and explosions of LNG tank trucks and one of a railroad tank car, resulting in many deaths.

27.1b COAL

Coal might seem like a very obvious fuel to use, since it is so abundant. Projections suggest that our supply might last for well over a thousand years. (Of course, that's not so long when you consider the hundreds of millions of years it took to build up the supply.) Coal reserves generally occur in two types—either the coal is far below the surface or relatively close to the surface. When it is far below the surface, deep-mining techniques must be used (Fig. 27–9). About half the coal can be extracted from such deposits. (If more were removed, the danger of cave-ins would greatly increase.) For deposits near the surface, strip mining can be used (Fig. 27–10). In strip mining, large amounts of the surface are stripped away with heavy machinery. Strip mining can lead to serious environmental problems even when attempts are made to restore the land after the coal is removed. Coal mining, in any case, is hazardous. The death toll among miners from cave-ins and from black-lung disease is considerable.

Another major problem associated with the burning of coal is air pollution. This is a reason for the trend that had been taking place in recent years for electric generating plants to stop using coal as a fuel. Restrictions on the allowable level of pollutants that can be emitted have become tighter. But the usage of coal has stopped dropping and is now increasing dramatically because of the shortage of oil.

A major source of pollution is the sulfur in coal. When coal is burned, sulfur dioxide is formed; the sulfur dioxide eventually

(at right) Fig. 27–10 Strip mining leaves huge gashes on the landscape unless time and money are spent reclaiming the land.

combines with the water in the atmosphere to form sulfuric acid. Large concentrations of sulfur dioxide in the atmosphere have been linked to a variety of respiratory ailments, and have played an important part in the so-called killer fog that used to occur in London. Coal can be processed so that the sulfur is removed, which raises the cost of coal as an energy source.

An even greater potential problem with burning coal comes from the extra carbon dioxide that this burning introduces into the atmosphere. The additional carbon dioxide will increase the "greenhouse effect" (discussed in Chapter 12), in which the visible sunlight that enters through the atmosphere is transformed into infrared "heat" radiation as it is absorbed and reradiated by the ground. Carbon dioxide in the atmosphere prevents the infrared radiation from passing back outward through the atmosphere, thus warming the atmosphere of the earth. The greenhouse effect has transformed Venus into a planet whose surface is almost 800°C; we don't want that to happen on earth. Even a change in the earth's average atmospheric temperature by a single degree would, by affecting the polar ice caps, change the ocean level substantially and affect coastal cities. The carbon dioxide problem may be the actual limit to the increasing use of coal in our energy future.

The situation with coal is representative of the whole energy problem. Here we have a fuel that is abundant enough to meet our energy requirements easily for many years to come. However, if we choose to mine and burn the fuel in the cheapest ways possible, the effects on the environment, both long term and short term, are enormous. It appears that if we are to mine and use the coal in a safe and ecologically sound way, then the expense will be driven up, even independently of the greenhouse effect. Here the problem is not so much a fuel shortage. It is a question of which fuels to use and how best to use them.

(below, left) Fig. 27–11 A newspaper engraving from 1847 showing a severe London fog. These "killer" fogs disappeared when the use of coal was diminished.

(below, right) Fig. 27–12 The nuclear power plant at Indian Point on the Hudson River, serving New York City.

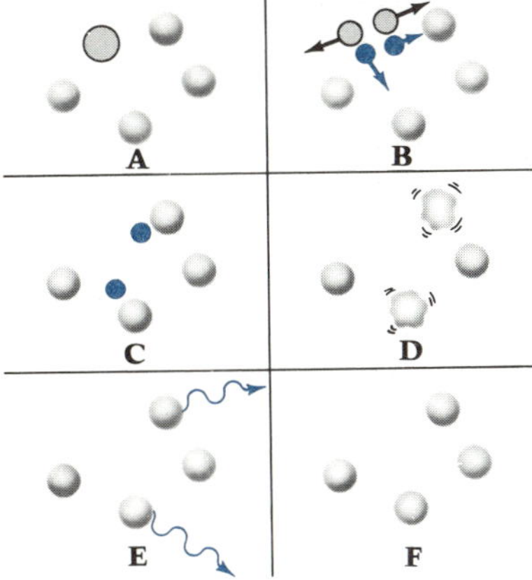

Fig. 27–13 The effect of U-238 in a U-235 reactor. A. In some section we start with one U-235 nucleus (shaded) and several U-238 nuclei (clear spheres). B. After capturing a neutron, the U-235 nucleus undergoes fission, producing some neutrons. C. However, before those neutrons can find more U-235 nuclei to induce more fission, they run into U-238 nuclei and are absorbed. D. This leaves the U-238 nuclei in an excited state. E. Rather than undergoing fission, the U-238 gets rid of the excess energy by giving off a gamma ray. F. The gamma rays escape and there is no additional fission.

27.2 *Nuclear Energy*

A major alternative to fossil fuels is nuclear energy. This is the energy given off in fission or fusion reactions (as discussed in Chapter 26). All the currently operating nuclear power plants and all those planned for the foreseeable future are of the fission type. So let's start our discussion of nuclear energy by seeing how a fission plant operates. It is not generally realized that all we use the fission for is to produce heat (Fig. 27–12). Then that heat is carried away by some coolant, usually water, and is used to produce electricity in much the same way as heat generated in burning coal is used to generate electricity.

Very few nuclei that exist in any appreciable quantity can serve as an appropriate fuel for fission reactors. We call such fuel fissionable material. The most commonly used is the isotope of uranium ^{235}U. However, most uranium is ^{238}U; only 0.7 percent of naturally occurring uranium is ^{235}U. After capturing a neutron, ^{235}U usually breaks up, in the course of which it emits other neutrons. When ^{238}U captures a neutron, the energy it absorbs is too small to break up the nucleus. Thus ^{238}U removes the neutrons that result from the fission of the ^{235}U (Fig. 27–13).

Therefore, to benefit from the energy released in ^{235}U's fission, it is necessary to enrich the uranium sample to produce a much higher fraction of ^{235}U than occurs in nature. This is very difficult and expensive to do, and until very recently the techniques developed during World War II were the only ones employed. The basic idea is first to put the uranium into gaseous form. This is done by forming the compound uranium hexafluoride (UF_6), which is a gas. Recently, there has been an effort made to separate the isotopes

Fig. 27–14 A gaseous diffusion plant for the enrichment of uranium at the Oak Ridge National Laboratory, Oak Ridge, Tennessee.

of uranium by using a gas centrifuge. As the gas is whirled around, the more massive molecules collect near the outer wall of the container and the less massive ones near the inner wall, as shown in Fig. 27–15. Several stages of such separation would have to take place to achieve the desired value of enrichment.

Experiments are also being performed on the use of lasers in isotope separation. Laser enrichment works because the energy levels of the electrons orbiting the nuclei are slightly different in the two isotopes of uranium. Again, uranium hexafluoride gas is used. A laser is tuned in wavelength so that its photons can be absorbed by the atoms of one isotope and not those of the other. After the atoms of the ^{235}U isotope are knocked into an excited state, light from another laser ionizes them (knocks off one electron completely). Since these ions are positively charged, they can be removed with an electric field. After separation of the isotopes by any of the above methods, the UF_6 gas is converted back into metallic uranium.

A

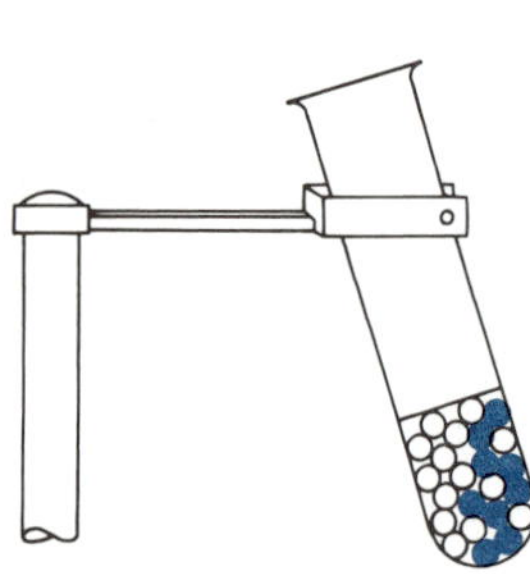

B

Fig. 27–15 Isotope enrichment. We start with a mixture of U-235 and U-238. The uranium is combined with fluorine to make uranium hexafluoride, a gas. ○ = uranium-235 hexafluoride; ● = uranium-238 hexafluoride. A. If we just let the gases stand, the effect of gravity would cause more of the heavier gas molecules (those with the U-238) to be found near the bottom of the container. However, the effect as shown in this drawing is greatly exaggerated, since the mass difference is quite small and the effect of gravity is quite weak. B. We can amplify the effect by providing the equivalent of a very strong gravitational field. This can be done in a centrifuge. As the centrifuge spins around, the heavier molecules are found more in the outer portion than in the inner portion.

27.2a FISSION REACTORS

Fig. 27–16 Enriched reactor fuel.

Once you have a rich enough sample, you can use it for reactor fuel. Remember, a chain reaction can be sustained with the uranium because it takes one neutron to get each ^{235}U nucleus to break up, and in the break-up, more neutrons are produced. These neutrons can then induce fission in other ^{235}U nuclei. However, it is easier for the ^{235}U to absorb slow-moving neutrons (which are technically called slow neutrons), and the neutrons that come out of fission are moving rapidly (fast neutrons). So we need something to slow down (moderate) the neutrons. For a moderator (Fig. 27–18) we would like to have a low-mass nucleus that can carry away a lot of the neutron's kinetic energy after a collision. Thus hydrogen is a good moderator. Water is a cheap and plentiful supply of hydrogen nuclei. A slight drawback to using water as a moderator in a reactor is that hydrogen nuclei can also absorb some of the neutrons, forming deuterium (the hydrogen isotope with a neutron in addition to its proton). When this happens, the neutrons are no longer available to keep the reaction going. One way around this is to use water in which a relatively large fraction of the hydrogen has already been converted into deuterium. This heavy water (HDO instead of ordinary H_2O) is very expensive to produce. Therefore we usually use normal water and accept the loss of some neutrons, since it turns out that there are actually neutrons to spare.

The fission of uranium produces an average of about 2.5 free neutrons per reaction. If all these neutrons are absorbed by other ^{235}U nuclei, then the number of reactions continually increases. A runaway chain reaction would result, and the energy would be released too quickly. (This is something like what happens in an atomic bomb, but in a reactor the concentration of ^{235}U is not as high as in a bomb, so the reactor can't explode. It merely gets hotter and hotter.) You therefore have to control the supply of neutrons.

To control the neutron supply we usually use rods of a good neutron absorber. These rods are called control rods (Fig. 27–19), and are inserted in the reactor however far you choose. To shut down the reactor, you put the rods in all the way. To make the reactor run faster, you pull the rods out a bit. Once the reactor is running, the trick is to keep the rods adjusted so that the reactor runs at a constant rate.

Actually, the use of control rods depends on an important factor,

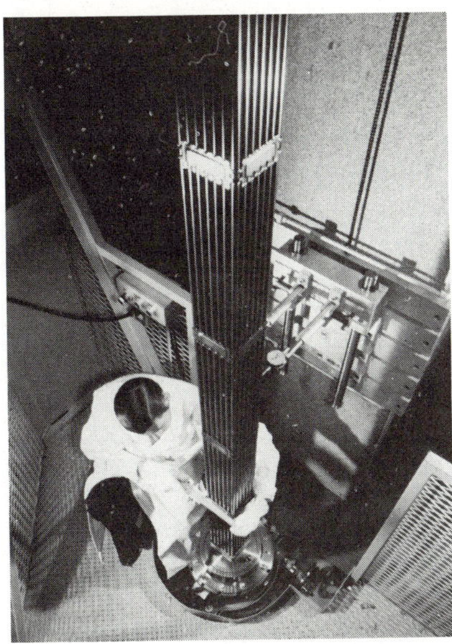

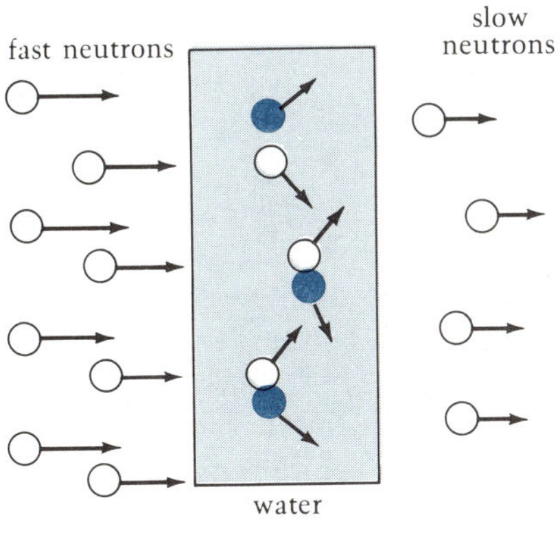

Fig. 27–17 A bundle of uranium-containing fuel rods is being readied for insertion into a reactor.

Fig. 27–18 Use of a moderator in a nuclear reactor. We often wish to have slow neutrons rather than fast neutrons in a reactor. If the neutrons pass through water, they will collide with the water molecules and give up much of their energy to the water, leaving slow neutrons to emerge from the other side. Since the water moderates the energy of the neutrons, we call it a moderator.

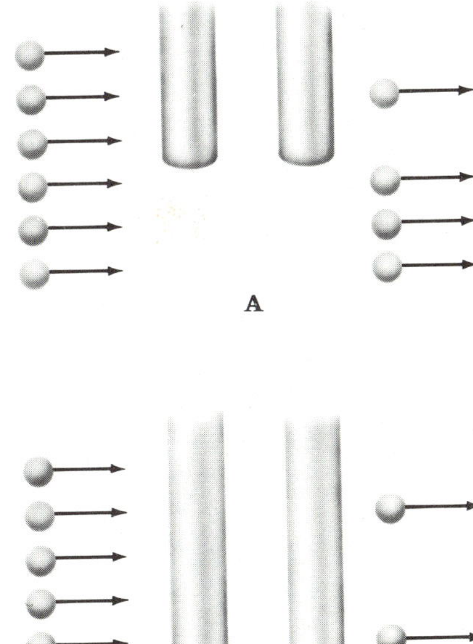

Fig. 27–19 By using control rods that absorb neutrons, we can control the number of neutrons in a reactor. A. Control rods out. B. Control rods in.

namely, that not all of the neutrons emitted in the fission process are released instantaneously. If they were released instantaneously, then a slight increase in the number of neutrons produced per fission would cause the reaction to run away before the control rods could be inserted further. Fortunately, a sufficient number of neutrons are delayed in their release, by up to about 10 seconds, so that we have control over the reactor. This delay comes about because the neutrons are released in decay processes that come after the actual fission.

27.2b BREEDER REACTORS

Alternative fuels to ^{235}U are important because our supply of ^{235}U is quite limited. The rich concentrations of uranium could be used up by the year 2000.

Although ^{235}U is the only naturally occurring fuel that can be used in fission reactors, it is not the only fuel available to us. When ^{238}U absorbs a neutron, it becomes ^{239}U, which then beta decays into ^{239}Np (neptunium), which in turn beta decays into ^{239}Pu (plutonium; Fig. 27–20). (See the list of the elements in Appendix 2.) The ^{239}Pu is radioactive, but its half-life is over 24,000 years. The ^{239}Pu can then serve as a reactor fuel, because it undergoes fission when it captures a neutron.

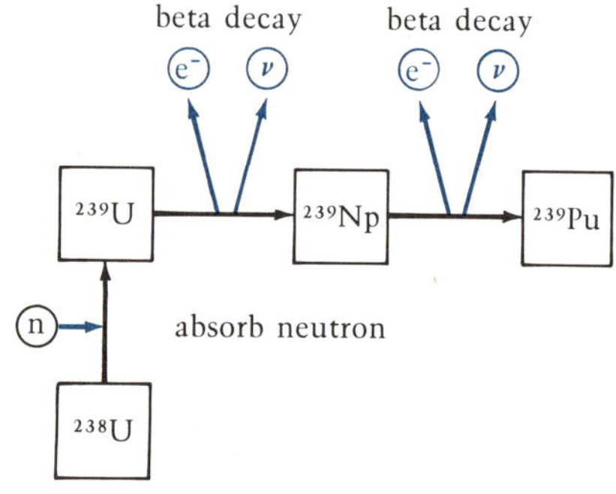

Fig. 27–20 The production of plutonium-239 from uranium-238.

Plutonium has drawbacks as a reactor fuel. One is that there are not as many delayed neutrons in the fission of ^{239}Pu as in ^{235}U, so control rods are not as effective in keeping the reactor rate steady.

The cycle we have just discussed, the conversion of ^{238}U into ^{239}Pu, plays a central role in a type of reactor that can manufacture its own fuel—the breeder reactor. The actual fuel for the reactor is ^{239}Pu. If we produce extra neutrons and have ^{238}U in the reactor, then the extra neutrons can go toward converting the ^{238}U into more ^{239}Pu. The breeder reactor can even produce more ^{239}Pu than it consumes. Then the extra plutonium can be removed and transported to another reactor for use as fuel.

The importance of breeder reactors is that the raw material they use is ^{238}U, which is quite plentiful. No enrichment of the uranium found naturally is required, since it is almost all ^{238}U already. Other cycles can also be used in breeder reactors.

Breeder reactors are still in the experimental stage. Their proponents point out that breeder reactors offer us the best chance for inexpensive power with a plentiful supply of fuel for the foreseeable future. However, as we'll soon see, there are objections to breeder reactors on the grounds of potential hazards and long-term effects on the environment. In any case, other countries are pressing ahead with breeder reactors, no matter what we do in the United States.

In sum, there are a variety of types of fission reactors, some in operation now and others under development, that appear to be capable of meeting our energy needs and for which a large supply of fuel is available. If those requirements were sufficient, there would be no problem. However, we have also said that we want to produce energy in a manner that is safe and not detrimental to the environment. How these requirements apply to fission reactors is controversial. Much of the criticism of fission reactors has centered on issues of safety, environment, and conservation. Let's take a look at these problems more deeply. We can start by taking a general look at the effects of radiation on living beings.

27.2c SIDE EFFECTS OF RADIOACTIVITY

When a radioactive substance decays, it emits some form of high-energy particle—an alpha particle (a helium nucleus), a beta particle (an electron or positron), or a gamma ray (a high-energy photon). When these particles pass through matter they can interact with individual atoms or nuclei, resulting in heating or ionization (knocking electrons off) of the material. After being ionized, an atom or molecule might participate in a chemical reaction that could not have otherwise taken place.

We like to have some standard way of keeping track of how much radiation is given off by various radioactive substances and how much is absorbed by various systems. In discussing the effects that radioactivity can have on biological systems, we often use a quantity called the effective dose, measured in rem, which stands for roentgen equivalent in man.

X-rays are also known as roentgen rays after Wilhelm Roentgen, who discovered them in 1895.

KEEPING TRACK OF RADIOACTIVITY

The unit by which we keep track of radioactivity is the *curie* (abbreviated *Ci*). One curie corresponds to 3.7×10^{10} decays per second, and is independent of the type of particle emitted. Most work in the laboratory deals with radioactivity samples on the order of 10^{-6} Ci (1 microcurie), but certain industrial uses require 10^{6} Ci (1 megacurie). When we consider the amount of energy absorbed by some material, we use the *rad*, which is defined in terms of the amount of energy absorbed by each kilogram of the material. One rad is equivalent to 0.01 joule/kg.

Different particles have different degrees of effectiveness on biological systems. Thus we must take into account the type of radiation (alpha, beta, gamma) and the particular biological system. This gives us the effective dose, measured in rem, the roentgen equivalent in man.

In the final analysis, it is the effective dose, measured in rem, that gives a consistent measure of the danger involved in any situation.

There are many sources of radioactivity in our environment besides nuclear power plants. To keep things in perspective, we should have an idea of how much radioactivity is generally present around us. Some is unavoidable, such as the radioactivity that occurs naturally on the earth and in cosmic rays showering down on us. Comparable to this "natural" level of radioactivity is that which we receive on the average from medical and dental diagnostic x-rays. (Recent findings on biological damage have significantly reduced the use of and the dose in such x-rays.) The dose from these various sources is given in Table 27–1.

When a person is exposed to radiation, the effects can be twofold: those effects experienced by the individual involved, called **somatic effects**; and those that will act on future generations, called **genetic effects**. Somatic effects are divided into two types: the immediate effects of a large dose of radiation, and the long-term effects of accumulated exposure to even small doses.

The short-term effects usually result from ionization damage that kills individual cells. If a sufficient number of cells are killed, **radiation sickness** results. The severity of the sickness and the possibility of cure depend on the severity of the dose. Below 25 rem (spread out over the whole body) there are no visible effects. (The

TABLE 27–1 Average Human Radiation Exposure (in millirem per year)

<table>
<tr><th>Source</th><th>Sea level (New York City)</th><th>1.6 km elevation (Denver)</th></tr>
<tr><td>Natural sources</td><td></td><td></td></tr>
<tr><td>cosmic rays</td><td>32</td><td>52</td></tr>
<tr><td>radiation from the earth (potassium, uranium, thorium, etc.)</td><td>55</td><td>62–133</td></tr>
<tr><td>radioactive elements in the body</td><td>25</td><td>25</td></tr>
<tr><td></td><td>112</td><td>139–210</td></tr>
<tr><td>Other sources</td><td colspan="2"></td></tr>
<tr><td>diagnostic x-rays</td><td colspan="2">70</td></tr>
<tr><td>weapons</td><td colspan="2"><1</td></tr>
<tr><td>power generation</td><td colspan="2"><1</td></tr>
<tr><td>occupational</td><td colspan="2"><1</td></tr>
<tr><td></td><td colspan="2">70</td></tr>
<tr><td>Total</td><td>182</td><td>209–280</td></tr>
</table>

SOURCE: Data taken from S. C. Bushong, "Radiation Exposure in Our Daily Lives," *The Physics Teacher* 15, no. 3 (1977): 135; and Jacob Shapiro, *Radiation Protection* (Cambridge, Mass.: Harvard University Press, 1972), p. 278.

U.S. official limit for workers is 5 rem per year.) Above 600 rem, the central nervous system is damaged, and above 800 rem, death is inevitable.

Long-range effects are not so well understood. It is clear, though, that continued exposure to radiation does lead to increased likelihood of developing cancer. We are only now beginning to collect sufficient data to begin to understand the magnitude of this problem.

Genetic effects result when the radiation damages but does not destroy the DNA molecules involved in reproduction. Such damage can produce mutations. It is still hard to evaluate the rate at which such mutations occur, since some may be recessive and may not appear for several generations.

27.2d SAFETY OF FISSION REACTORS AND OF OTHER ENERGY SOURCES

Fission reactors have obvious advantages over other types of power plants including the availability of fuel. On the other hand, many people have voiced fears about potential dangers of fission reactors. Some of these concerns are legitimate while some others are general reactions to the word "atomic" because of the past use of the atomic bomb. Some of the concerns, in addition, are more a general fear of technology and a yearning to return to a simpler age, and are applied to nuclear energy largely as a ready example.

One fear is that a malfunction would cause the chain reaction to get out of hand and result in an atomic-bomb type of explosion. This is actually impossible, because the fuel used in reactors is not sufficiently enriched to the point that it can explode in a chain reaction. Even if a nuclear reactor did get out of control, it would not turn into an atomic bomb. Rather, if the cooling fluid could not carry the heat away from the reactor fast enough, a meltdown (a rapid heating of the core) can result. The most likely way that a meltdown could occur is from failure of the cooling system, so each reactor is equipped with a backup cooling system. If both systems should fail simultaneously, then the reactor could heat up before it could be shut down.

A meltdown might result in the spread of radioactive materials. To guard against this, reactor cores are surrounded by a double-layered containment vessel. Simulations of a loss-of-coolant accident have shown that protective mechanisms successfully prevented a meltdown and any substantial escape of radioactivity. The situation was unfortunately put to an actual test on March 28, 1979, when something went wrong at the Three Mile Island commercial nuclear reactor in Pennsylvania. The reactor core was uncovered by coolant for one or more periods, which resulted in a partial meltdown of the core.

The situation was serious enough that the President appointed a special commission to look into the matter. The commission was headed by John G. Kemeny, president of Dartmouth College, and contained a cross section of experts and ordinary citizens. Their report showed clearly that we cannot consider the nuclear power plant itself, as a piece of equipment, separately from how it is operated and supervised. The human factor was a major contributor, including insufficient training of the operators, confusing specific operating procedures, and failure to improve operations based on lessons from previous accidents. The design of the control panel may have played a part.

Fig. 27–21 The Three Mile Island Nuclear Generating Station. Unit 2, at right, began commercial operation in December 1978, and suffered a loss of coolant in March 1979. At left is Unit 1, uninvolved in the accident but nonetheless currently shut down. Note the large chimney-like cooling towers. The reactors are in the two round containment buildings in the center of this photograph.

The commission concluded that most of the radiation was contained within the plant, and that the radiation released "will have a negligible effect on the physical health of individuals." There were no immediate deaths, and the odds are less than fifty-fifty that even in the long run a single individual would die of cancer from the Three Mile Island accident.

The major harm from the accident was mental stress. This can be real enough, although much of the stress could have been alleviated by a better understanding of the pros and cons of nuclear power relative to having other kinds of power stations nearby. It is not the nicest thing to have any kind of power station next door.

Three Mile Island demonstrated the type of accident that could be potentially catastrophic. This type of accident, fortunately, has a very low probability of occurring. Another fear associated with fission reactors is that of long-term danger resulting from their normal low-level radioactive emissions. Strict controls are placed on the amount of radioactive material that can be given off by a nuclear plant. In fact, coal-burning plants give off more radioactivity than nuclear plants. (The radioactive impurities that occur naturally in coal are released when the coal is burned.)

Perhaps the most perplexing problem raised by the use of nuclear power is the disposal of radioactive waste material. When the fuel in a given sample is spent, some radioactive material is left behind. Some waste materials have very short half-lives and can be stored for a short period in safe areas until the radioactivity has decreased. However, other waste materials have very long half-lives, measured in the tens of thousands of years. These long half-lives mean the threat of dangerous levels of radioactivity for many generations to come.

For the present, wastes are stored at each nuclear plant in giant shielded underground tanks that are monitored for leaks. One generally proposed long-term solution is to encase the radioactive wastes in ceramics, glass, or some protective canisters and to bury the wastes deep underground. The burial is for secondary protection, in case the wastes should get out of the ceramics or glass, or if the canisters should leak. In both cases, the material would then be buried in places that are reasonably free of geological activity so that the protective casing is not destroyed by movements of rock. Also, the storage location would be free of flowing water, since we cannot tolerate the possibility of contaminating our water supply. It has therefore been proposed to bury the encased wastes in salt deposits, since the presence of large salt deposits shows that water has been absent in that location for tens of thousands of years. (Salt is very soluble in water, and wouldn't be there still if water had been present.)

Unfortunately, salt domes could be subject to natural stresses like earthquakes and instabilities from methane gas.

Since we must face the problem of nuclear waste disposal to deal with military wastes, the civilian waste disposal can take advantage of what we learn. The military generates about ten times more radioactive waste in the course of making bombs than civilian power plants make in generating energy. So even if we were to stop building nuclear power plants, the nuclear waste problem would still remain to be solved.

There are several other potential problems with fission power. Fuels and wastes must be transported around the country, exposing areas to the possibility of an accidental spill. There are risks in transporting all the other types of fuels as well. A problem unique to nuclear power is the fear that a sufficient amount of fissionable material can be stolen to provide another country or a terrorist group with an atomic weapon.

What about the risks of other forms of energy generation? For example, the presence of certain fossil-fuel pollutants leads to a high and predictable incidence of lung cancer and emphysema. It is also unfortunately the case that fossil-fuel power plants have potentially very serious effects on the earth's overall climate. We have already discussed the possible effects of CO_2 on the atmosphere. Further, burning coal and oil has made ordinary rain increasingly acidic. This acid rain pollutes lakes and rivers, kills fish, decreases the fertility of the soil, damages crops, and erodes buildings.

Solar energy has potential hazards as well. The accident rate in the installation of ordinary circulating-water solar energy systems—mostly ordinary plumbing—is substantial. And the idea of a set of solar collectors in space beaming down energy to earth with microwaves could lead to potential disaster if, say, the spacecraft malfunctioned and directed its beam of energy elsewhere than the intended region of reception.

Even the newly popular wood stove is bringing hundreds of deaths per year because of house fires. One might say that untrained people shouldn't use wood stoves. Another objection to wood stoves is the pollution that they release. Utilities aren't allowed to burn fuels without pollution controls, and maybe individuals shouldn't be allowed to do so either.

Oxygen must diffuse in from outside a wood flame, in contrast to a gas flame in which fuel and air are mixed before combustion. Soot and hydrocarbons form preferentially in a wood flame as a result. A quarter to a half of the particulates in the air of some towns probably arises in this way.

A potentially very hazardous way of generating electric power is hydroelectric. The number of people who live downstream from a dam is immense. A design defect or earthquake could lead to disaster on a grand scale.

So all methods of generating energy have substantial hazards. The hazards from nuclear power are different from, although not necessarily greater than, the hazards from fossil fuels. Nuclear power has been scrutinized closely. Its risks and hazards have been publicized. It could be educational to put other forms of generating energy under similar scrutiny.

27.3 *Other Alternatives*

In this section, we briefly discuss some alternatives to fossil-fuel and fission-fuel power plants.

27.3a FUSION

Nuclear energy seems like a very attractive source because a small amount of fuel yields a great amount of energy. It would be nice if these benefits could be realized without the associated disadvantages of fission reactors. This consideration brings forward the possibility of exploiting nuclear fusion, using two lighter nuclei to build a heavier one that is closer to iron than the original two nuclei. Energy is given off in the process.

The best energy yields come when we use certain light nuclei as the fuel. The most commonly discussed fuels are deuterium and lithium. One advantage is that these fuels are in plentiful supply. Although deuterium is much less abundant than normal hydrogen (only 1 part in 6600 in sea water), there is so much hydrogen in the water of the oceans that there is also a tremendous supply of deuterium. Lithium must be mined, but the deposits are sufficient to provide enough fuel to last over 100,000 years. Another advantage of these reactions is that most of the end products of the reactions are not radioactive, so waste disposal is not a problem.

Some residual radioactivity is induced in the material making up the reactor because of the neutrons that escape from the reaction, but this is thought to be a relatively minor problem.

With all these advantages, why don't we use fusion power? The problem is that it is hard to get started. In fission, the electric repulsion of the protons in the nucleus helps us because it drives the parts of the nucleus apart from each other (Fig. 27–22). However, the electric repulsion is a hindrance when we are trying to put the parts of a nucleus together. For example, to get two deuterium nuclei together, we must provide enough energy to overcome the repulsion that the protons in each of the deuterons exert on each other.

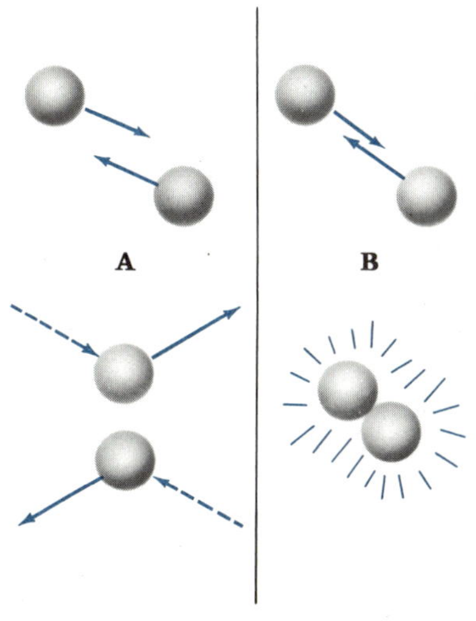

Fig. 27–22 A. At low temperatures, two protons traveling toward each other do not have enough initial speed to overcome the electric repulsion. They simply bounce off each other. B. At high temperatures the two protons can get very close together, close enough for nuclear reactions to take place. Remember, the nuclear force is a short-range force. It only acts when the two particles are very close together.

In accelerators, we shoot one particle in at a very high speed. Accelerators, however, do not accelerate very many particles, so we cannot use them in any practical way for energy generation. The other possibility is to give the particles sufficient energy to overcome the repulsion by heating the material. Remember, as we heat a material we are giving each particle, on the average, more kinetic energy.

The temperatures required to start fusion are very high—about 10 million K! At these temperatures there is also enough energy for the electrons to be completely removed from the atoms. We no longer have a gas of atoms; instead we have electrons and nuclei freely moving around. We call such a gas a plasma. One hope for nuclear fusion is to be able to control the fuel in this plasma form. But the fuel must be confined within some given region for a sufficiently long time for fusion to take place, and at 10 million K there is no material out of which we can make a container.

However, since a plasma consists of charged particles, it can be controlled somewhat by the application of magnetic fields. In principle, by appropriately shaping the magnetic field, the particles can be trapped in the same way that particles in the Van Allen belts are trapped by the earth's magnetic field, as described in Chapter 21. The *magnetic confinement* of charged particles in this way is referred to as using a magnetic bottle to hold the particles (see Fig. 21–13). Unfortunately, the process is not as easy to carry out as it

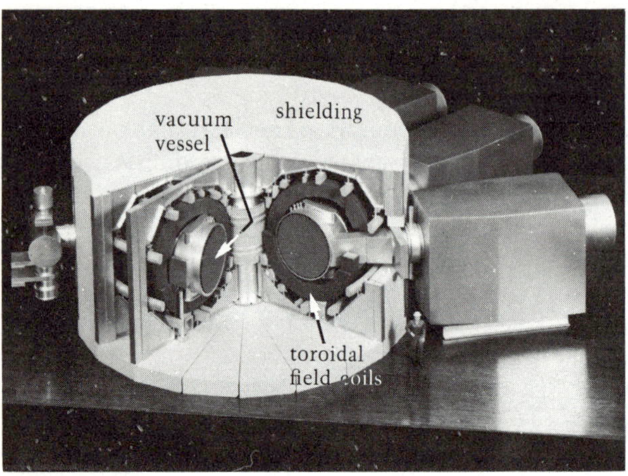

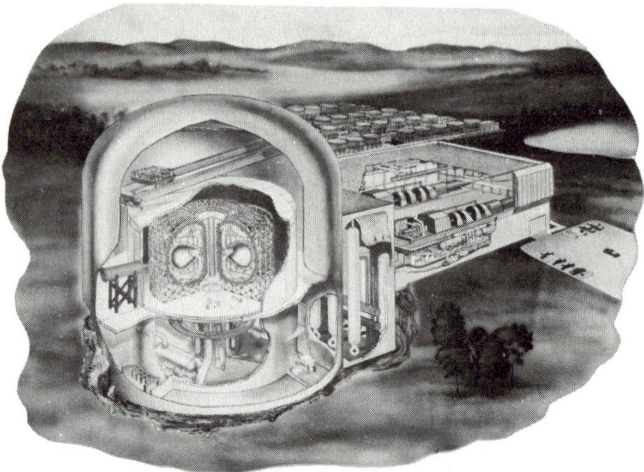

(above, left) Fig. 27–23 A model of the Tokamak fusion test reactor.

(above, right) Fig. 27–24 An artist's conception of a hypothetical fusion power plant, developed at the Princeton Plasma Physics Laboratory.

See the color essay "Fusion Research" (beginning on page 8) and Section 21.1.

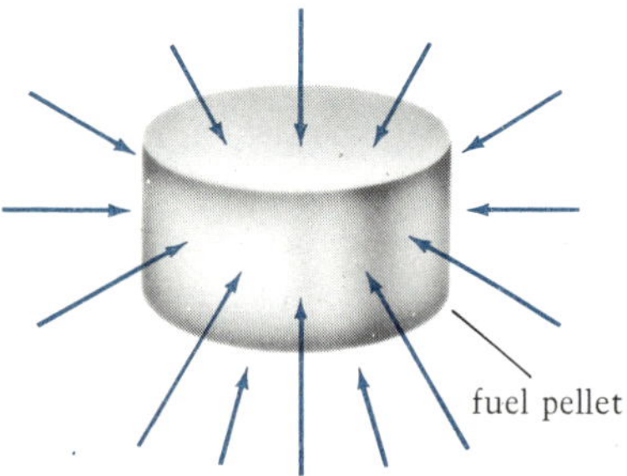

Fig. 27–25 Laser-induced fusion. The light from many powerful lasers is concentrated on all sides of a small fuel pellet. This light heats the pellet and also sends strong shock waves through the pellet, driving the protons in the fuel closer together, initiating nuclear reactions.

may sound. The confinement of plasmas at the required density and for the times necessary to get fusion reactions going and continuing has proven to be a major stumbling block.

One device for the magnetic confinement of plasmas has been developed in the Soviet Union, and is called a Tokamak. Experimental Tokamaks are now in operation in the United States and Europe, and extensive effort is being put forth to find the required containment. The improvements come slowly and in small steps. As of now, the break-even point—the point at which the Tokamak produces as much fusion energy as is required to keep it running—has not been reached. The success of this method—if indeed it is to be successful—is still many years away.

Another approach that is being tried involves the use of beams. Laser fusion, which we also discussed at the beginning of this book, involves making a small pellet of fusion material, such as solid deuterium and tritium, and shooting laser beams at it from all sides (Figs. 27–25 and 27–26). Beams of electrons or ions are also being tested. The effect is to compress the interior of the pellet in a violent manner, causing the nuclei to come close enough together for fusion to take place. The process is called *inertial confinement* because the inertia of the imploding material results in high densities at the center.

So we see that fusion provides us with a very tantalizing problem. Most scientists think that it will meet our stringent requirements for a suitable energy source. However, at this point the technical problems have not been overcome. Even optimistic projections seem to indicate that we should not count on fusion reactors prior to the twenty-first century, or maybe even the year 2020. Pessimistic projections say that we should never depend on it. There does seem to be a feeling, increasingly based on current achievements, that fusion power will eventually be available. Even so, what do we do until then?

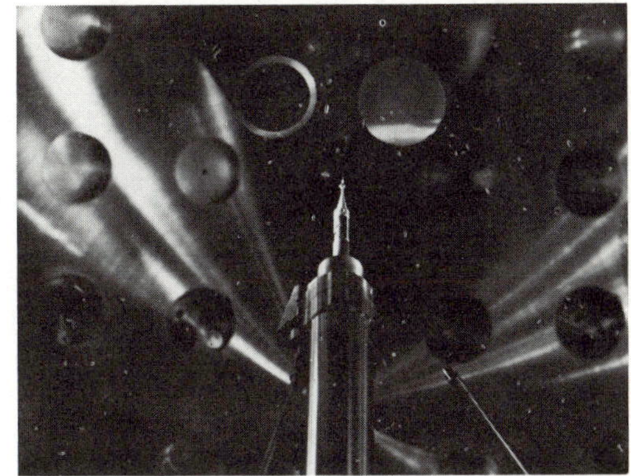

Fig. 27–26 A view inside the Shiva target chamber. A laser-fusion target, about the size of a grain of sand, is located precisely in the center of the chamber. Laser beams containing 30 trillion watts of power are focused on it.

27.3b SOLAR ENERGY

One should take account of the sun when planning the design of a house and its siting on its plot. These can be arranged, for example, so that overhangs or trees shield the interior of the house from the hot summer sun, which is high in the sky, while allowing the winter sun, which is lower in the sky, to enter.

The sun is such a dominant energy source in our everyday lives and for the earth in general that it might seem natural to be able somehow to tap the energy from the sun. The sun might take over some of the chore of heating houses and hot water, but can it reasonably be expected to supply electricity? Taking account of atmospheric effects and the day-night cycle, the average power reaching the earth is about 200 watts for every square meter of the surface. When we add this up over the whole earth, that certainly is a lot of power—26 billion kilowatts! The problem is that solar energy is very spread out. At 200 watts per square meter, it would take about 5 square kilometers to collect as much solar energy as is put out by a modern power station. And this assumes that we could convert 100 percent of the received energy into electricity. Better estimates are that the conversion efficiency could not be much greater than 10 percent, which means that we would need 50 square kilometers to equal one power station. If we were to rely on solar energy to provide all our power, by the year 2000 we would need 5000 square kilometers of collectors. This would be a square 70 km on a side. With these large areas, some people who look at the feasibility of solar energy talk about solar farms. Others want the solar cells dispersed, with some on every house. (The question of whether to distribute energy in a few centralized places or in many small places comes up with most forms of energy.)

Fig. 27–27 Skylab in orbit over the earth. Note solar cell panels projecting from the spacecraft's body.

Some solar energy schemes anticipate that the solar collectors will have the means of directly converting the solar energy to electricity. In other schemes, large fields of mirrors concentrate the solar energy to a point, where it is converted to electricity.

One way to achieve the conversion is by the use of solar cells, which directly convert the sunlight into electricity. Such solar cells do exist and have been used successfully on certain spacecraft,

Fig. 27–28 Solar cells provide electric power for a special energy exhibit at the Museum of Science and Industry in Chicago.

such as the Mariners, Vikings, and Skylab. These cells are made out of silicon that has been refined to a very high purity.

While solar cells are useful for specific applications, and are the best solution for the particular problems involved with providing power for spacecraft or other remote objects like ocean buoys or forest tower lookouts, the electricity that solar cells generate is quite expensive. Each watt now costs hundreds of dollars. Even at that rate, solar cells may still be cost effective whenever servicing is especially difficult or expensive. But the price would have to be reduced by a factor of over 100 before solar cells could become commercially feasible.

Much research by solid-state physicists and engineers is going into the development of solar cells and the costs are coming down. This holds out the hope that solar cells will provide a useful energy alternative, especially in areas that get a lot of sunlight. We still have to make sure that it doesn't take more energy to produce the high-purity silicon than we get back eventually in electricity produced.

There have also been proposals to put large plates of solar cells in orbit around the earth and then beam the power back to the surface in the form of microwaves. The advantages of this would be that the cells above the atmosphere would receive more solar radiation than we get at the ground, not subject to the nuances of weather, and could be placed at altitudes at which they spend more than half the time in the sun. In addition, we would not have to worry about having to replace solar cells because of corrosion. It has even been suggested that we could mine the material for the solar cells in space and set up space cities for the workers (Fig. 27–29). At this time, it is fair to say that all such ideas are still very speculative and, as we have seen, may carry severe risks.

Another problem with solar energy is that it is not available at times of cloudy weather or at night. We must thus develop ways of storing the energy so that it can be used later on. This turns out to be a formidable problem.

The advantages of solar energy are clear. It is clean, potentially fairly safe (depending on the way it is used), and no matter what we do, it will keep coming at us at the same rate for a long time to come. However, the difficulties in converting it to electricity are also clear. It appears that for the foreseeable future, solar-energy conversion to electricity will be only a minor source of power.

But solar energy is being used right now to heat homes. Sunlight heats a tank of water, and the hot water is circulated through the heating system whenever needed. If there is not enough heat stored in the tank, then an auxiliary heater can be used. Even when the auxiliary heater is needed, the use of solar heating greatly reduces

A

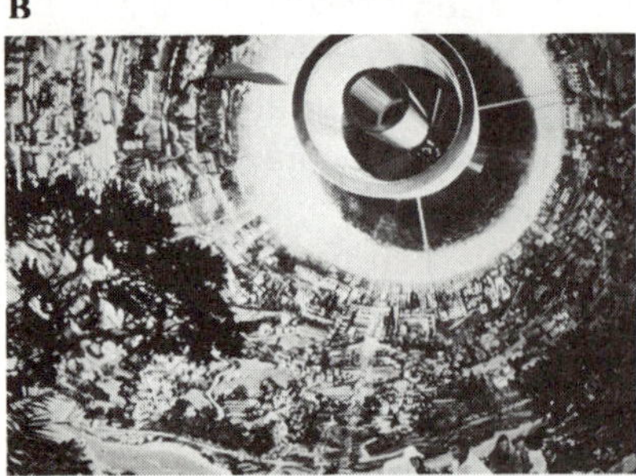

B

Fig. 27–29 An artist's conception of a solar-powered space city. A. Overall view. B. Interior view.

Fig. 27–30 This experimental windmill generates electricity on Block Island, N.Y.

the consumption of conventional fuel. These systems would obviously be most useful in locations where there was the most sunlight during seasons of cold weather.

A major way of using solar energy is to grow trees and other plants. We get the solar energy out when we burn them as fuel. The wood can be burned in individual wood stoves or in centralized plants. Wood stoves are charming for an individual, but when a whole city or town tries to use them, the wood supply becomes exhausted and the atmospheric pollution becomes intolerable. Also, wood is usually economical only if you cut it yourself.

Solar energy can also be tapped indirectly through wind power. The wind gets its energy from sunlight, blowing because of pressure differences that result from temperature differences that the sun causes between places on the earth. Like solar heating, windmills could serve as a source of auxiliary power in regions where the climate is right. Pilot projects are underway to test windmills of modern design (Fig. 27–30).

Pilot projects are also under way to extract energy from the temperature difference between layers of ocean water.

The use of solar energy, certainly in heating and perhaps even for generating electricity, will grow. Still, even the most optimistic forecasts see it meeting less than 20 percent of our energy budget by the year 2000.

Fig. 27–31 Another alternate source of power is hydroelectric, with energy extracted from falling water. Flood control is another benefit. Objections to hydroelectric power include the need to dam rivers, changing their beauty and the surrounding ecology. Also, it is quite hazardous to live below a dam.

27.3c GEOTHERMAL ENERGY

Periodically, an earthquake or the eruption of a volcano gives us an idea of the enormous amount of energy stored below the surface of the earth. The source of this energy was discussed in Chapter 12. The *average* heat flow through the surface of the earth is small, but the energy could be tapped at sites that are particularly active. Some power plants now in operation use the energy stored in underground steam wells. The steam is released by drilling into the pockets where it is trapped. Even this form of energy is not without its problems. The underground steam contains large amounts of the poisonous gas hydrogen sulfide (which gives rotten eggs their smell). Also, removing the steam can undermine the land overhead.

Another suggestion is to tap the energy stored under the earth's surface. In active regions, these can be quite hot. The supply of geothermal energy is great, but we do not as yet know how to tap that energy. Estimates are that with a concerted effort we could be producing about 10 percent of our energy needs in this way by the year 2000. Again, it appears that for some time to come this geothermal energy will serve, at best, as an auxiliary supply of power.

27.3d CONSERVATION

Many people think that an important part of the solution to the energy problem is increasing conservation. Limiting the use of electricity, curtailing driving, lowering heating levels, and improving home insulation and caulking all aid in curtailing the growth in our energy needs. "Home doctors" specially trained to advise on conservation may be worth the investment. Industrial conservation measures can save even more energy than those by individuals. In addition, proposals have been made to use electricity more efficiently. For example, motors can be made more efficient; their current designs were made when energy was relatively cheap and the copper in them relatively more expensive. Conservation measures can start immediately, so are an excellent short-term as well as long-term aid.

Computerized control of heating and cooling, and making use of heat generated by lighting and manufacturing processes can save a lot of energy in commercial buildings.

Even conservation has its hazards, however. Most people don't realize that there are radioactive elements in building materials. Brick and concrete have more of them than does wood. The radiation level in a brick house (or in granite building like New York's Grand Central Station) is much higher than it is in a frame house. Of course, all these radiation levels are below the official safety threshold.

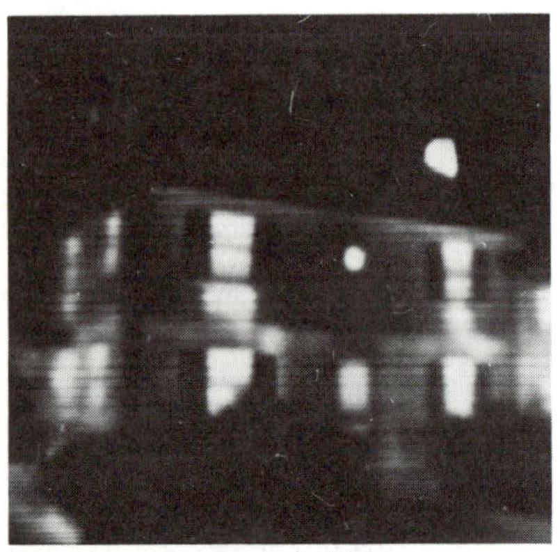

Fig. 27–32 This thermogram shows the heat escaping from a house. The brightest areas show the highest heat loss.

Building materials give off radioactive radon gas. In most houses, the radon is purged from the air periodically as the air is exchanged with the outside. But tightening the house as a conservation measure slows up the interchange of inside and outside air, allowing the radon levels to build up. Other potentially harmful gases, nonradioactive in nature, can also build up, including formaldehyde from insulation and plywood furniture, and residues from cooking gas and cleaning materials. Devices called "heat exchangers" can provide ventilation without losing all the indoor heat, and so can aid in the solution to this problem.

27.4 *The Outlook*

We can see that it isn't "radioactivity" that we are really trying to escape when we discuss alternative energy sources; radioactivity is everywhere. Even people who live at high altitudes or who fly in airplanes are exposed to more radioactivity than their counterparts at sea level. Further, even apparently benign solutions may have hidden risks. We must maximize our chance for a long, safe, healthy life by giving full consideration to hazards of all types.

We have plenty of fuel. There is certainly enough coal to last us for a long time. There is enough ^{235}U to get us through the growing stages of nuclear fission reactors, and beyond that there is a very large supply of uranium and thorium for breeder reactors. If we can control fusion, there is a virtually inexhaustible supply of deuterium and lithium. The sun also continues to shine.

The problem is that we must learn to use these fuels in a manner that does not threaten public safety or the environment, and still do this at a reasonable cost. All of our energy possibilities are at different stages of development, and it is difficult to make firm predictions on when they will be available on suitable terms. There is also a time element in all of this. Energy conservation will help, and is potentially very important. But conservation will only delay the day when our petroleum supplies run out. We will have to make some hard decisions soon.

Key Words

fossil fuels
bituminous coal
anthracite
secondary recovery techniques
shale
deep mining
strip mining
fissionable material
enrich
gas centrifuge
slow neutrons
fast neutrons
moderator
heavy water
control rods
breeder reactor
effective dose
rem
somatic effects
genetic effects
radiation sickness
meltdown
plasma
magnetic bottle
Tokamak
laser fusion
solar farms
solar cells
geothermal energy

Questions

1. What are three ways in which physics plays important roles in solutions to the "energy crisis"?
2. Why do we often find petroleum and natural gas in the same place?
3. List the principal dangers associated with the following sources of energy: (a) natural gas; (b) petroleum; (c) coal; and (d) nuclear reactions.
4. What is "enriched" uranium?
5. Why must natural uranium be "enriched" for it to be useful as a fuel in fission reactors?
6. Near the surface of the earth, the heavier molecules tend to collect closer to the surface, while the lighter molecules are more abundant

higher up in the atmosphere. How is this related to the operation of a gas centrifuge in the enrichment of uranium?

7. Why is uranium usually converted into UF_6 for enrichment?
8. What is the role of a moderator in a reactor?
9. (a) What is heavy water? (b) Why is heavy water preferable to normal water as a moderator in a reactor?
10. Why can't a reactor explode like an atomic bomb?
11. What is the role of the control rods in a reactor?
12. What features do nuclear-fission power plants and coal-burning power plants have in common?
13. How can a breeder reactor produce more fuel than it consumes?
14. Three types of "rays" have been identified in nuclear processes: alpha, beta, and gamma. Of what type of particles does each of these forms consist?
15. How does radiation from nuclear reactions heat materials nearby?
16. On the microscopic level, how can radiation alter a material?
17. Differentiate between somatic and genetic effects in exposure to radiation.
18. Why are salt deposits used for the disposal of radioactive waste?
19. What are the advantages of nuclear fusion over fission in energy generation?
20. Why does the production of energy from fission usually involve very heavy elements while the production of energy from fusion involves very light elements?
21. (a) What is a magnetic bottle? (b) How would a magnetic bottle keep a plasma confined?
22. Why is the confinement of plasmas important in developing fusion reactors?
23. If you can collect an average of 100 watts per square meter of solar energy, about how large a collector you would need to keep a normal household supplied with electricity?
24. What advances must be made before solar energy comes into widespread practical use?
25. What are the advantages and disadvantages of geothermal energy?

Part VII

Structure of the Universe

In this section we reach the frontiers of physics—the realm of the very big and the very small. On the largest scale, we want to see how all the material in the universe is put together and how it all evolves. To do this we must consider distances and time scales that are much larger than anything we come into contact with in our everyday life. Nevertheless, we will see in Chapter 29 that we can make sense out of the large-scale structure of the universe. We can understand the universe's history and can even approach the question of how the universe will evolve in the future.

On the small scale, we will continue the search for the ultimate building blocks of matter. The search is not new—the word "atom" was coined by the Greek philosopher Democritus some 2400 years ago. Democritus claimed that there had to be some smallest structure in matter, and that this would be the building block for the larger structures. In his theory, the atoms themselves had no structure; they were indivisible. It is interesting that Democritus made his prediction more on philosophical grounds than on experimental grounds. (Indeed, we'll see that many questions about the ultimate structure of matter are on the border between philosophy and physics.)

We have already seen in Chapter 23 how atoms long ago fell by the wayside as elementary particles, ever since electrons and nuclei were discovered. Even the nucleus has structure. That the nucleus contains protons and neutrons has been known for years, but we're not finished yet.

As we plunge deeper into the subatomic world, we continually find more structure and more particles that can be taken apart. We'll see that in trying to interpret the origin of all these particles, philosophy begins to creep back into the discussion. For we have to decide what constitutes a fundamental explanation and what does not, or even if a fundamental explanation is possible.

CHAPTER 28

Quarks: An Assortment of Colors and Flavors

28.1 *Particles and Forces*

The search for elementary particles is not only a search for the building blocks of matter. It is also a search for the origin of the various forces in nature. As we discussed in Sections 23.8 and 26.1b, there is a trend toward thinking of the various forces as though they arise from the exchange of virtual particles. Quantum electrodynamics (QED) is the theory that describes the *electromagnetic* interaction by the exchange of particles called virtual photons. We have already seen how Yukawa introduced the idea of virtual pions as carriers of the strong nuclear interaction. Particles have been postulated as the carriers of the other two forces as well.

The graviton is the particle that would carry the force of gravity and the W particle is the particle that would carry the weak nuclear interaction. The graviton, like the photon, should be massless, and the W, like the pion, should have some mass. Neither gravitons nor W particles have yet been discovered.

In addition to understanding the phenomena associated with the fundamental forces, theorists would like to be able to derive the properties of these forces from scratch, with some minimum number of simple, basic assumptions. An even more ambitious goal is to come up with a single general theory from which all four forces can be naturally derived. Such a theory would be called a unified field theory, and may not even exist. However, some steps in this direction have actually been made, as we'll see a little later.

28.2 *Particle Proliferation (The Particle Zoo)*

There was a time when the known fundamental particles were the neutron, the proton, the photon, and the electron. It really made sense to think of these as elementary particles. Each one served a particular purpose in the structure of matter, and it seemed that no other particles were necessary to explain that structure. In addition, there was no evidence to suggest that these particles are made up of anything smaller. Even the addition of the neutrino to this list of four particles did not change the notion that this new set of five was the basic set of particles.

However, a quick doubling of the number of fundamental particles came with the realization that for each particle there is an antiparticle. The first antiparticle was the positron (the electron with a positive charge), whose existence had been predicted by Dirac before its discovery. A particle and its antiparticle always have exactly the same mass. They have opposite electric charges and certain other properties that are opposites. If a particle and its antiparticle get close enough together, they annihilate each other, converting all of their mass into energy, according to the formula $E = mc^2$. The energy appears in the form of photons or other particles. Since the discovery of the positron, the antiparticles to many other particles have been found, and we believe that every particle has an antiparticle.

THE MICROBE, by Hilaire Belloc

The microbe is so very small
You can hardly make it out at all.
Yet many sanguine people hope
To see it through a microscope.

Its hundred tongues that lie beneath
A thousand curious rows of teeth;
Its seven-tufted tails with lots
Of lovely pink and purple spots
On each of which a pattern stands
Composed of forty separate bands;
Its eyebrows of a tender green.
All these have never yet been seen.

Yet scientists, who ought to know,
Assure us that it must be so.
Oh, let us never never doubt
What nobody is sure about.

Not even doubling the list of known particles stopped the general belief that the relatively small list of particles, now plus their antiparticles, was a complete list of truly elementary particles. However, this belief was shaken when elementary-particle physicists went on to find many more particles. The development of particle accelerators has played an important role in these discoveries.

How do we actually discover a particle in an accelerator experiment? Let's look at what happens if we use our accelerator to shoot a beam of protons at some target. Suppose that the target also consists of protons. When we set our accelerator for low energies, the two protons do not get very close together; the positive charge of one repels the positive charge of the other. With a little more energy, the two protons overcome this repulsion and get close enough for the strong force to take over. However, the strong force in this case may only cause the incoming proton to change its path. Such a change in direction is called scattering. In the scattering process, the incoming proton loses some of its energy. If it had enough energy when it arrives, then some of that energy can go into the creation of another particle. Of course, the energy available in the incoming proton must be greater than the mass of the particle that is created. (Here we are talking about mass and energy as being equivalent, $E = mc^2$.) The greater the energy of the incoming proton beam, the greater the mass of the new particles that can be created. (See Fig. 28–1.)

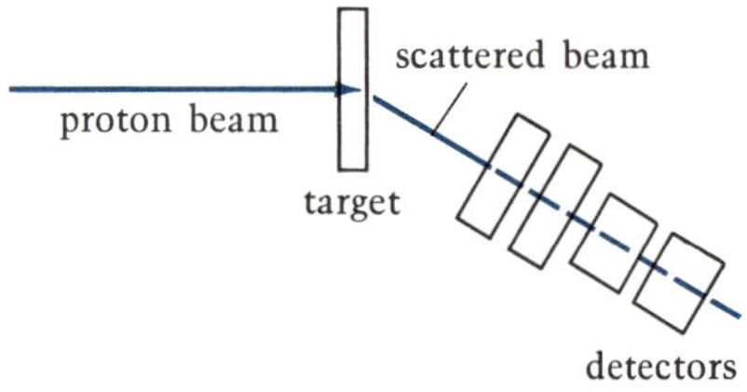

Fig. 28–1 Schematic layout of an elementary-particles high-energy experiment at an accelerator.

Once these particles are created, they come flying out of the collision region. Their motions must, of course, be consistent with the conservation of energy and momentum. Sometimes the new particles can be detected as they speed out. Various techniques are used to measure the properties of the particles. For example, they are passed through a strong magnetic field. The particles that are

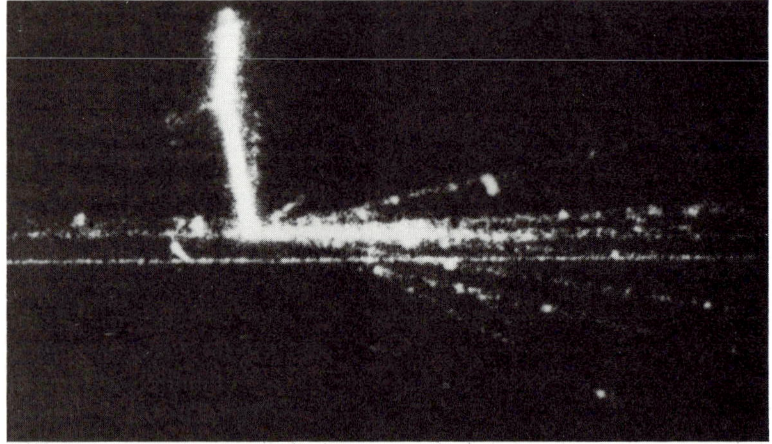

Fig. 28–2 The formation of a very short-lived particle in a spark chamber. When the proton beam from the left struck a neon nucleus, it formed other particles. One nuclear fragment shows as the bright upward jet. A spray of other particles continued to the right. A neutral particle, which is invisible, proceeded forward and slightly downward. It then broke up into two charged particles to make the lowest "v" on the right side of the picture. This neutral particle lived only about 10^{-14} seconds. It exhibits a property we shall consider later, called "charm."

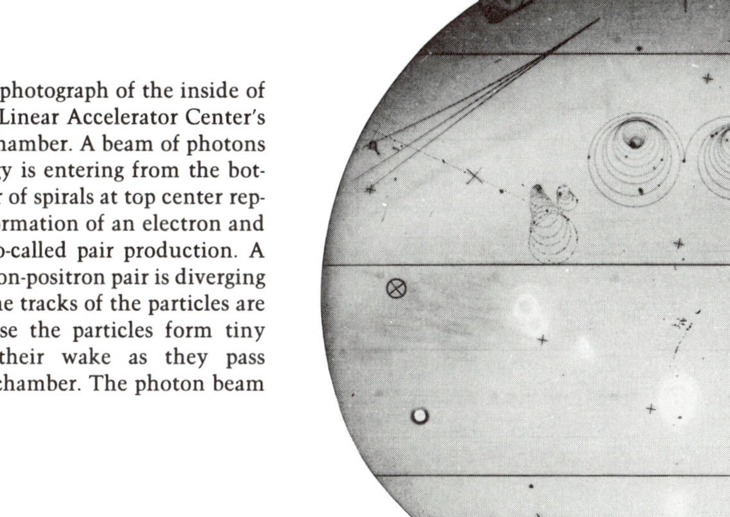

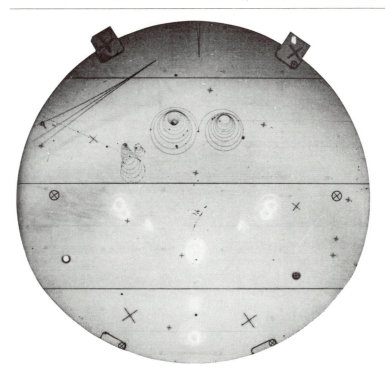

Fig. 28–3 A photograph of the inside of the Stanford Linear Accelerator Center's 1-m bubble chamber. A beam of photons of high energy is entering from the bottom. The pair of spirals at top center represents the formation of an electron and a positron, so-called pair production. A second electron-positron pair is diverging at the top. The tracks of the particles are visible because the particles form tiny bubbles in their wake as they pass through the chamber. The photon beam is invisible.

charged have their paths bent, which allows us to determine their momentum from the amount of bending. Alternatively, the paths can sometimes be followed in a bubble chamber or in spark chambers (Figs. 28–2 and 28–3). The properties of the particles can also be deduced by seeing which types of material they can penetrate. Experimental-particle physicists are like detectives trying to put together the pieces of a puzzle to see what types of particles passed through their equipment.

The situation is made harder by the fact that all the new particles that are discovered are unstable. They decay into other particles (possibly more familiar ones).

Particles that decay via the strong interaction live only about 10^{-23} seconds before decaying, an incredibly short time. Often we don't detect the particles that result from the collision because they decay too quickly. We must detect the products of the decay, and then reason backward to try to figure out what could have caused that particular combination of decay products. (Just to make things a little harder, not all the decay products can always be detected; some are invisible to our detectors.) Experimentalists must use such things as the laws of conservation of energy and momentum to try to figure out what went on in the equipment.

With all these complications, how are we ever to recognize the birth of a particle? In a typical experiment, very few of the incoming particles actually participate in reactions. One thing that the experimenter measures is the percentage of particles that actually do collide with target particles and cause a reaction of some kind. This percentage can be expressed as the probability that a given reaction will take place.

When an atomic spectroscopist shoots light at some atoms, most of the light passes right through. However, if the incoming light happens to be a frequency that corresponds to one of the differences between energy levels in these atoms, then the atoms absorb a high percentage of photons.

A similar situation occurs with elementary-particle experiments. We find that if the energy of the incoming beam is exactly right for the creation of a new particle, then the number of reactions is much larger than if we use beams with other energies. Such a large increase in the reaction rate at a particular energy is called a resonance. To say we have a resonance is to say that we have hit an energy at which the system can absorb that energy very efficiently. Finding a resonance means that a particle is being created. We can search through the tracks of the debris to find the properties of that particle, which we can call a resonance particle. All the resonance particles have higher masses than the proton, and can decay into particles of lower mass (Fig. 28–4).

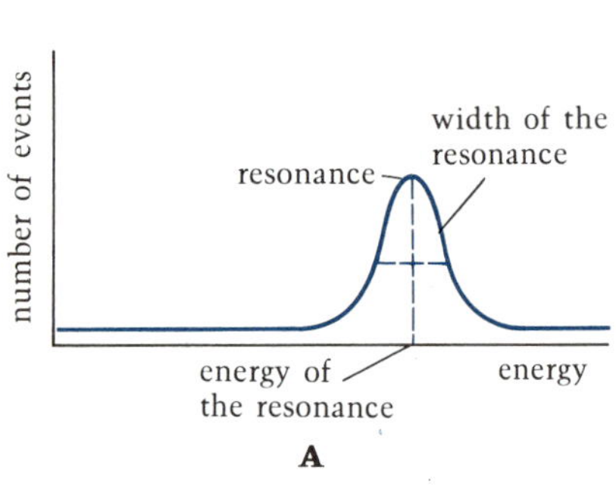

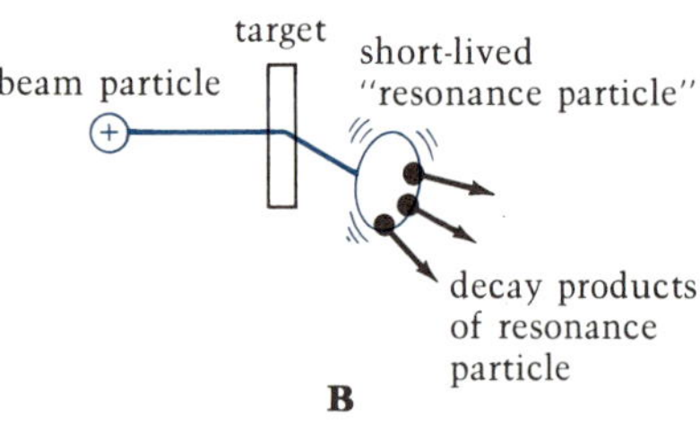

Fig. 28–4 A resonance in a scattering experiment. A. The resonance shows up as an increase in the number of absorptions at a certain energy or over a certain range in energies. B. Usually a resonance is due to the fact that a particle is created whose mass corresponds to the energy of the resonance. Often these particles are so short-lived that they are never detected directly. They decay shortly after they are formed, and the detectors only see the particles that result from the decay of the resonance particle. However, by detecting the decay products, and using the ideas of conservation of energy, momentum, and angular momentum, we can deduce the properties of the resonance particle.

The current major problem in elementary-particle physics is that well over a hundred particles have been discovered. Thee are so many that the discovery of a new "elementary" particle does not usually draw much attention any more. The proliferation of particles is fantastic. We appear to be limited only by the energies that we reach in our accelerators. Every time that new energies are reached, new particles of higher mass than those previously known are found. There doesn't seem to be any limit, which is somewhat disturbing. One can hardly think of hundreds of particles as being fundamental any more than we can think of the 92 chemical elements as being fundamental particles. As we have gone to a smaller and smaller scale, corresponding to higher and higher energies, the structure of matter seems to have gotten more complex rather than simpler. The situation led particle physicists in the 1950s to call their embarrassment of riches a particle zoo.

The first thing to do with such a collection, as in a zoo, is to try to fit the animals into classes. In the same way, particle physicists began to look for ways to classify the particles by searching for similarities and differences among the particles. A basic division is whether or not a particle can participate in the strong nuclear interaction. Particles that interact strongly are called hadrons, and most of those that don't are called leptons. (The photon is neither.)

Leptos *means "small" or "tiny" in Greek;* hadros *means "thick," "stout," or "strong" in Greek.*

The smaller category, by far, is the leptons. The electron and the neutrino are leptons. A particle called the mu meson or muon, discovered during the search for Yukawa's meson, is also a lepton. (The muon was mistakenly identified as the meson predicted by Yukawa until the pion was discovered.) We have known for some time that there are two types of neutrinos, one associated with the electron and the other with the muon. Electron, electron neutrino, muon, and muon neutrino make four fundamental leptons. Their antiparticles number four more. The feeling is widespread that the leptons are truly elementary particles. In addition to there being very few of them, experiments to probe their structure have indicated that these particles behave as though they are completely concentrated at a point, that is, they appear to have no internal structure.

A muon is like an electron except that a muon is 205 times heavier; it can be thought of as a "heavy electron." A tau is 17 times heavier than the muon. Nobody has any idea what the mass might be if another particle exists in this family.

Recently the situation with the leptons has grown a little more complex. A new member of the family, called the tau, has been discovered. It would seem that there is another neutrino to go with the tau (after all, electrons and muons have separate neutrinos). The tau's mass is much higher than the muon's, so there is no guarantee that we will not find still more leptons as we go to higher and higher energies in our accelerators.

Preliminary experimental evidence described in 1980 indicates that a given neutrino may change from one of the three neutrino types (electron, muon, tau) to another as it travels. This would mean that it is changing in mass, which would mean in turn that neutrinos have mass. (It had been thought that neutrinos had no mass.) Experiments are under way to verify the results.

Barys *means "heavy" (as in baritone);* mesos *means "in the middle" (as in* mezzo-soprano*).*

The situation with the hadrons is even more complex than that of the leptons, and has received most of the attention in elementary-particle physics. Hadrons come in two types: baryons and mesons. Baryons are particles like the proton and the neutron, and mesons are particles like the pion.

The most obvious distinction between baryons and mesons is their spin. A certain set of spins is found for the baryons; the spins of mesons have a different set of values.

However, even these basic divisions did not lead to an approximation of order within the groups. The search for underlying patterns was related to certain symmetries shown by groups of particles.

28.3 *Quarks—Order Is Restored*

By 1960 the collection of particles was formidable, but no code was known to explain why particles with certain properties exist while others don't. What made the job harder for theoreticians was that they could not even be sure that there was a code to crack. However, some order was found independently in 1961 by M. Gell-

SYMMETRY AND CONSERVATION LAWS

Look at the building shown in Fig. 28–5. What is wrong with this building? You've probably said that there is a pillar missing on the right. Why does the missing pillar constitute something wrong with the building? The problem is an aesthetic one. The building just would look more natural if the pattern of pillars were the same on the left and on the right. If the pillar were there, then we would say that the building looked symmetric. (Since the idea of *symmetry* is as much aesthetic as anything else, we shall be content here with trying to give an idea of the concept of symmetry instead of a thorough definition.)

In searching for the simplest guidelines in piecing together the laws of nature, the physicist is also guided by symmetry. Patterns are found and the equivalents of missing pillars are sought. In physics, the use of symmetries can be a great help in solving certain problems. For example, if we want to calculate the gravitational field near a perfectly spherical planet, we can cut through a lot of work by noting that the answer should be the same no matter in which direction you leave the planet. (See Fig. 28–6.)

Symmetries are important in physics because wherever there is a symmetry there is a quantity that is conserved. We have already seen conservation laws for energy, momentum, and angular momentum. When we can apply these laws, our calculations become much easier. From conservation laws applied to quantum mechanics, we can derive the fact that certain quantities are quantized and can therefore be represented by quantum numbers.

Fig. 28–5 A modified drawing of the Temple of Athena at Assos, Greece. What is wrong with this drawing?

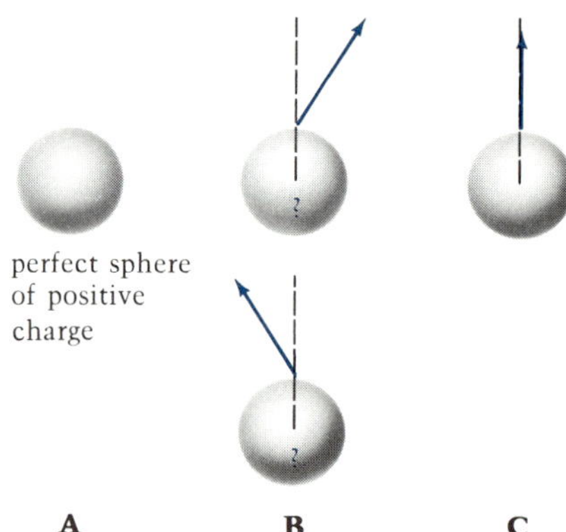

Fig. 28–6 The use of symmetry in problem solving. A. Suppose we are to find the electric field at some point on the surface of a perfect sphere. Which way will the electric field point? B. We might guess that it points off slightly to the right side. This is shown in the upper figure. However, because of the symmetry of the problem, pointing to the left looks like just as good a guess as pointing to the right. The field can't point two ways in the same place at the same time. C. The solution here and for a planet's gravitational field is that it points neither left nor right, but just straight out.

Particle physicists have sought out a variety of symmetries in describing the behavior of elementary particles. Some of these symmetries are not as obvious as the normal spatial symmetries that may come to mind in life-size architecture. The symmetries in elementary-particle physics are related to properties of the mathematical functions that describe the particles. Nevertheless, in each case in which a symmetry is present, we have an additional property that is described by a quantum number. We can completely describe a particle by giving all the quantum numbers that characterize it, including such things as electric charge, angular momentum, and the relationship of a particle to its mirror image.

If we want to see whether or not a given reaction will take place, we can check to see if any conservation laws are violated by the reaction. If none of the conservation laws is violated, then we know that the reaction is one that takes place. For example, a neutron cannot simply decay into a proton and a neutrino. Such a decay would violate conservation of charge, since the total charge before the reaction would be zero (because a neutron has no charge), and the total charge after the reaction would be $+e$ (e for the proton + no charge for the neutrino).

The relationship between conservation laws and allowed reactions is illustrated by a puzzle that bothered particle physicists in the late 1940s and early 1950s. Several particles were discovered that

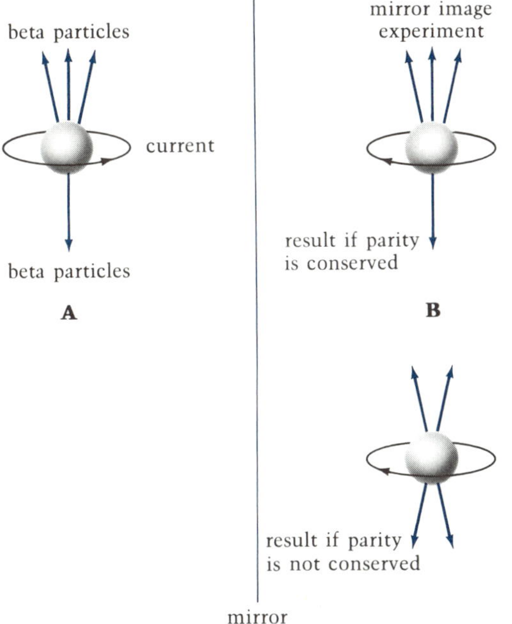

Fig. 28–7 A schematic representation of the experiment that showed that parity is not conserved. An experiment involving beta decay of cobalt-60 nuclei in a magnetic field was set up. A. Here we show the current that caused the magnetic field, rather than the magnetic field. The presence of the magnetic field caused a certain number of particles to be emitted upward and a certain number downward. B. The mirror image experiment was constructed, and the mirror image result, depicted in the upper drawing, was expected. However, it was found that the distribution of emitted particles was different from in the original experiment, as suggested in the diagram below. Thus, surprisingly enough, the mirror image of an experiment does not give the mirror-image results. As far as we know, this unusual circumstance only applies to weak interactions.

Fig. 28–8 C. N. Yang and T. D. Lee received the Nobel Prize in physics for explaining disparities in weak interaction outcomes.

did not decay as fast as they should have. These particles were therefore called strange particles. In 1952, Murray Gell-Mann at Caltech and K. Nishijima explained the reluctance of these strange particles to decay by saying that there must be some new conservation law that would be violated by such a decay reaction. Gell-Mann named this new quantity strangeness. Each particle has its own quantum numbers that correspond to strangeness. For example, the strangeness of the proton is zero. If we add up the total strangeness before a reaction and after a reaction, the two results must be the same.

Actually, not all conservation laws are perfect. There are some that are slightly violated. A famous example is that of parity—the behavior of a particle relative to its mirror image. Parity conservation says that the laws of physics should be the same when viewed in a mirror (Fig. 28–7). However, in the early 1950s there were certain decays of particles that could not be understood. Two theorists, T. D. Lee and C. N. Yang, proposed in 1956 that this puzzle could be resolved if parity was not conserved in weak interactions. An experiment done later that year showed that in beta decay, which is a weak interaction, a certain experiment and its mirror image give slightly different results. This news startled the physics community. Physicists were so used to the idea of conservation laws as absolute that the notion that one could be "imperfect" was quite alien. Lee and Yang's Nobel Prize followed quickly.

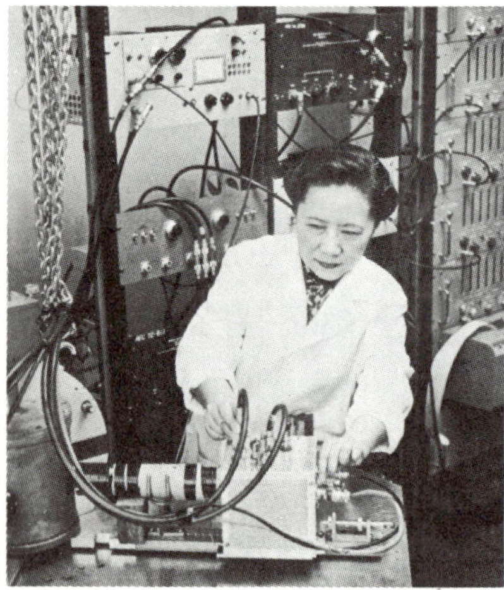

Fig. 28–9 Chien Shiung Wu, whose experiments demonstrated that parity is not always conserved. Professor Wu is shown in her laboratory at Columbia in 1958.

Mann (the author of strangeness) and Y. Ne'eman, then an Israeli army colonel. Their work was based on the application of symmetry principles, and depended on a mathematical technique known as group theory, which had been developed at the turn of the twentieth century by the Norwegian mathematician S. Lie.

In a graphical representation of their theory, Gell-Mann and Ne'eman found that the known hadrons fit very nicely into simple geometric patterns that showed certain symmetries. An inspection of the patterns showed that there were two gaps, which destroyed the symmetry in the same way that the missing pillar destroyed the symmetrical appearance of the building we discussed. Gell-Mann decided that these gaps must represent particles that exist but had not been discovered. The discovery of these particles within the next few years, with the properties predicted by the theory, was a great triumph. It is certainly important for a theory to be able to explain existing experimental evidence, but it is also important for the theory to predict the results of future experiments.

A similar theory was independently proposed by George Zweig.

Plate 47 shows combinations of quarks and antiquarks.

In 1964, Gell-Mann went a step further and pointed out that a special grouping of three particles came out of his theory. All of the properties of the other particles could be expressed in terms of the properties of these three, even though the three new particles had never been observed. Gell-Mann proposed that if these three particles exist, then all the known hadrons can be explained as combinations of the three new particles. He called these new particles quarks (from a quotation from *Finnegan's Wake* by James Joyce). We now refer to these quarks as u, d, and s (for *u*p, *d*own, and *s*ideways or *s*trange). Each quark has its own set of quantum numbers. There would also be a set of three antiquarks, with quantum numbers that are exactly the negative of those of the original three quarks.

If we put quarks together to form a particle, the charge of the particle is the sum of the charges of the quarks that make up the particle. The same would go for strangeness, or for any other quantum number that describes the particle.

Quarks do have one unusual feature—their charges are fractions of the charge on the proton. (Until the conception of quarks, all known charges were integers.) The *u* quark has a charge $+\frac{2}{3}$ that of the proton and the other two have charges $-\frac{1}{3}$ that of the proton. These fractional charges do not violate any fundamental law; they are simply unusual and surprising.

The combinations are displayed for understanding rather than for memorization.

Note that the charge works out to be $\frac{2}{3} + \frac{2}{3} - \frac{1}{3} = 1$ for the proton and $\frac{2}{3} - \frac{1}{3} - \frac{1}{3} = 0$ for the neutron, as it must.

There are also rules for how the quarks can combine. Mesons are made by combining a quark and an antiquark. (The antiquark in a meson is not necessarily the antiquark of the initial quark.) For example, the positively charged pion consists of the combination $u\bar{d}$ (where the bar over a particle represents its antiparticle). Baryons are made by combining three quarks or three antiquarks. For example, the proton is *uud* and the neutron is *udd*. Using the above rules, Gell-Mann was able to show that these quarks produced all known hadrons. Moreover, no allowed combinations of quarks resulted in particles that were not known (Fig. 28–9).

The theory seemed to explain things so well that it was hard to see how it could be wrong. Experimentalists all over the world started experiments with the hope of becoming the discoverer of the first free simple quark. They looked in accelerators, in seawater, and even in oysters, hoping to catch that telltale fraction of a charge that would unmistakably signify the presence of a quark. All these original searches failed. But the theory is still too good to give up.

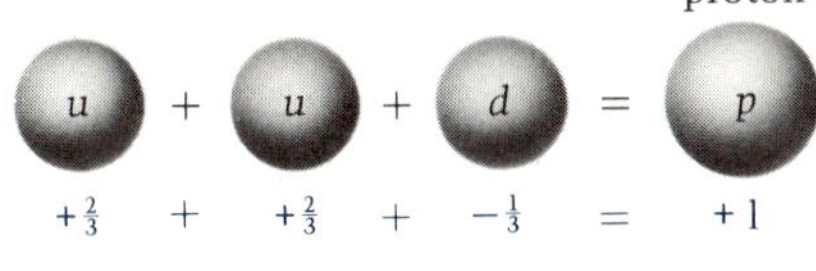

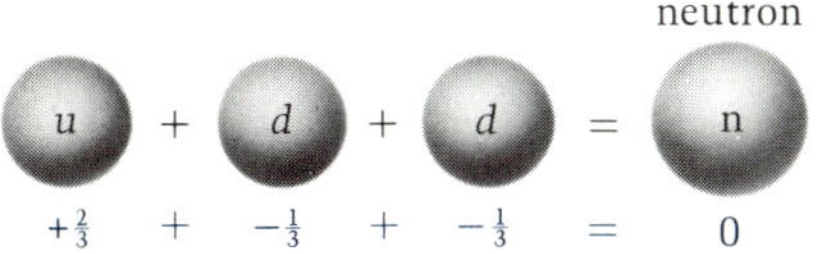

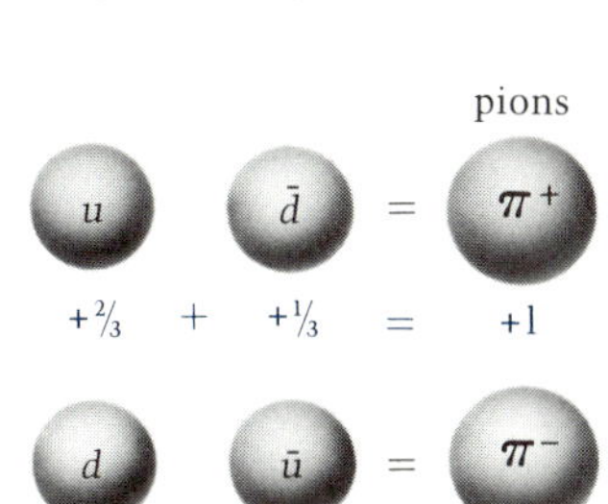

Fig. 28–10 Adding quarks together to form other particles. In this case we also show how the electrical charges (in color) add together to give the electric charge of the resulting particle.

28.4 *Color*

The failure to find free quarks is not the only problem with the theory. Another nagging problem exists from the theoretical point of view. Quarks have the same spin as the electron, and should therefore obey the Pauli exclusion principle. We have seen how this principle prevents having identical particles in the same state. However, we know that there are particles that are made up of three identical quarks, all of which are in the same state. It thus appears that quarks do not obey the exclusion principle.

One way out of this dilemma was to postulate that the three quarks were not exactly identical. This would be the case if there are really three types of *u* quarks (and three types of *d* quarks, etc.). A combination of three *u* quarks would then really consist of one of each type. If each quark can come in three forms, then there must be an additional property of these quarks that can have three different values. This property is called color. (The name "color" was chosen arbitrarily and has nothing to do with our conventional ideas of color.) The theory involving quark colors is called quantum chromodynamics. (*Chromo* is from the Greek for color.)

According to the color theory of quarks, each quark can come in three colors. The colors are arbitrarily named red, blue, and green, which is shortened to *R*, *B*, and *G*. Each of the antiquarks comes in each of three anticolors: $\bar{R}$, $\bar{B}$, and $\bar{G}$. The rules for combining quarks into particles can be translated into rules for combining colors. *Allowed particles are those with no net color.* There are two ways to get no net color. First, we can have three quarks, one of each color, as in baryons. For example, in the proton, which is *uud*, each of the three quarks must be a different color. Second, we can get no net color by combining a quark of one color with an antiquark of the appropriate anticolor. This pairing makes a meson. For example, we can combine an *R* quark with an $\bar{R}$ antiquark (Fig. 28–11).

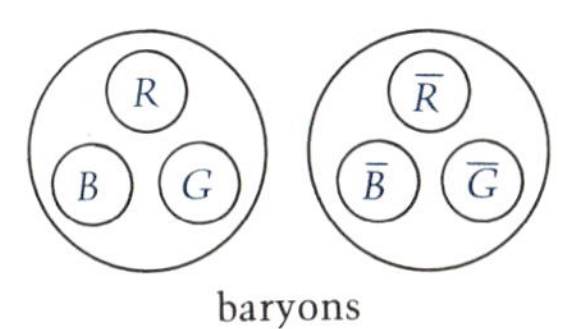

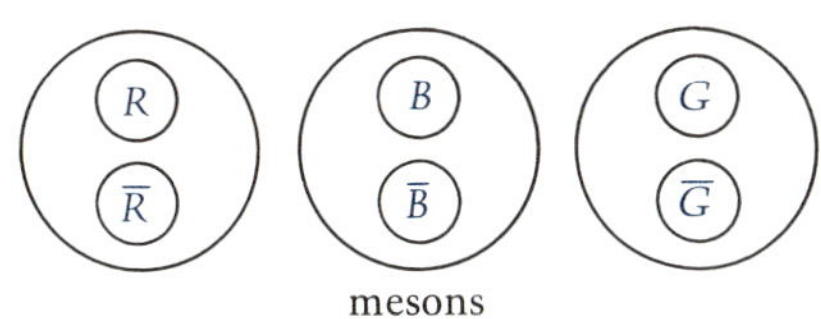

Fig. 28–11 Allowed combinations of quarks. Such combinations must be colorless. In the baryons, we get three differently colored quarks, creating no color. In the mesons, we get a quark of one color and a quark of its anticolor, creating no color.

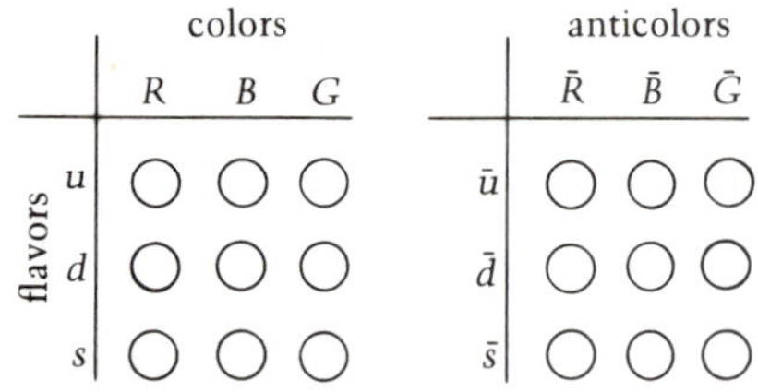

Fig. 28–12 Combinations of colors and flavors.

Plate 49 shows gluons and their effect.

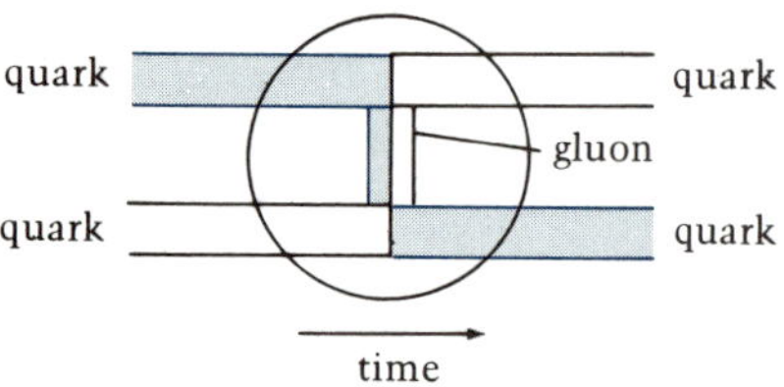

Fig. 28–13 In an interaction between two quarks, a gluon is exchanged by the two quarks. In this diagram we keep track of the two quarks as time goes on, to the right. At the beginning (on the left), the quarks are not interacting. In some time range indicated by the circle the quarks interact by having a gluon go from one to the other. For the particular gluon that is transferred in this case, the quarks exchange colors. After the interaction, the quarks continue on.

In forming these combinations, we must remember that the basic attributes of the quarks, those that determine the quantum numbers, are independent of color. For this reason, we refer to *u*, *d*, and *s* as flavors, and *R*, *G*, and *B* as colors. Each of the three quarks comes in each of the three colors. (Each antiquark comes in each anticolor.) The flavor gives the basic properties of the quark and the particles formed by the quark, and the color limits how the quarks combine into particles (Fig. 28–12).

What provides the force to keep the quarks in a given particle bound together? Since we can explain forces as the exchange of particles, the attractive force results from the exchange of virtual particles between the quarks. In the color theory, these virtual particles are called gluons. (After all, they are the glue.) There are eight different types of gluons. When a gluon travels from one quark to another, it can either leave their colors unchanged or cause the two quarks to interchange colors, depending on what type of gluon it is (see Fig. 28–13).

One interesting feature follows from some of the treatments of the color theory of quarks. The color force (the force carried by the gluons), does not change with distance between the quarks. If you pull one quark far away from another, the attractive force between them does not diminish. Because there is no reduction in the amount of work you must do to continue to separate them, to get them far enough apart to be seen individually requires an infinite amount of energy. According to this theory, quarks are permanently confined in colorless combinations, and we will never be able to detect a free quark. Theories that explain why we are not now detecting free quarks are quite new and are still really in the formative stage. (See Fig. 28–14.)

28.5 *Charm*

For a decade following Gell-Mann's presentation of the quark theory many unsolved problems remained. Developments were slow in coming. However, in late 1974 an unexpected event shook up the particles community. A new particle was discovered that was

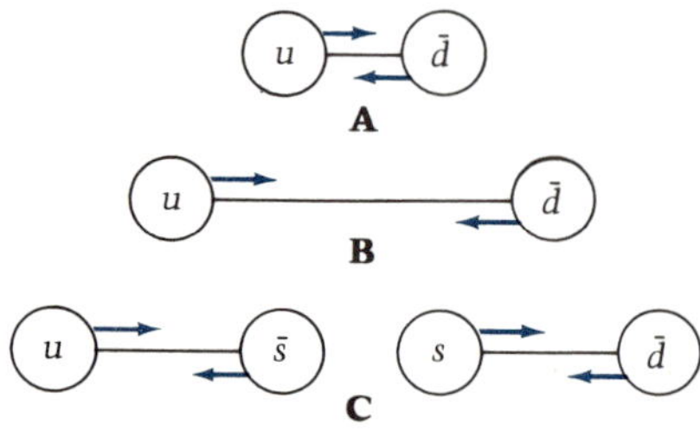

Fig. 28–14 Why we can't see a single quark. The color force between two quarks, indicated by the arrows above, does not get weaker as the quarks get farther apart. So the force in A is the same as that in B, but a lot of work must be done to move the quarks apart. So much work is done that when the quarks are far apart, as in C, the energy goes into the creation of a new quark-antiquark pair. In this case it is the s and $\bar{s}$ pair that is created. So we started with a meson, and instead of ending up with two isolated quarks, we just end up with two mesons.

A

B

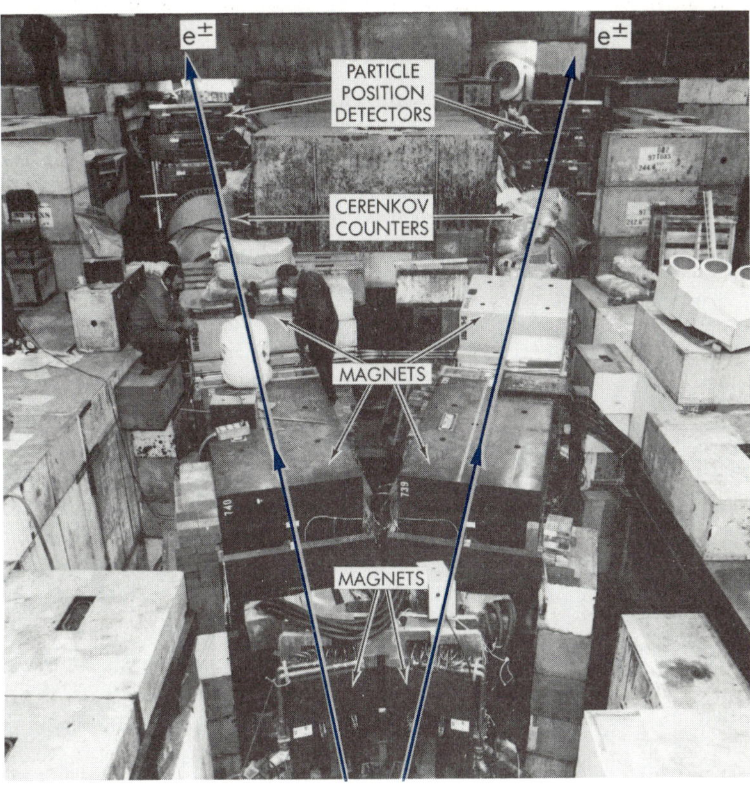

C

Fig. 28–15 A. Burton Richter, leader of the team that discovered what they called the psi particle. B. Samuel C. C. Ting, leader of the team that discovered what they called the J particle. C. The MIT/Brookhaven equipment used to discover the J particle. The blue arrows show the paths of the electron and positron as they travel 20 meters from the location of their formation. The three magnets and the counters in each arm allow the particles to be identified.

not merely another in a long list. It was a particle of very fundamental interest.

This important discovery was made almost simultaneously by two different groups, one working at Brookhaven National Laboratory on Long Island, under the direction of Samuel C. C. Ting, and the other working at Stanford, under the direction of Burton Richter. These two groups, a continent apart, were actually doing different experiments. They were studying reactions that are essentially the reverse of each other, but they came up with the same result—a new particle (Fig. 28–16).

And this new particle lives 1000 times longer than it would if it decayed via the strong interaction. The Stanford group named the

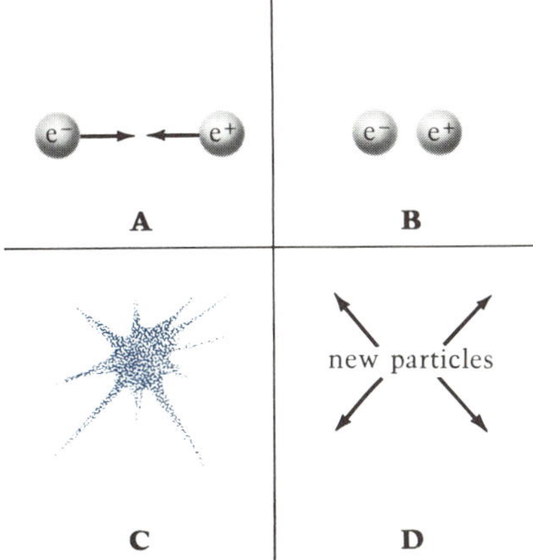

Fig. 28–16 The creation of particles in electron-positron annihilation, as in the Stanford experiment. A. The electron and positron head toward each other. B. They get very close together. C. They annihilate, leaving behind a burst of energy, photons. D. There is so much energy that it is soon converted into a variety of particles. The only requirement is that the sum total of all the new particles must have the same quantum numbers as the original electron and positron.

particle the *psi* (a Greek letter). The Brookhaven group, making the discovery at the same time, named the particle the *J*. Most people now call it the J/psi particle.

Shortly after the first discovery, the Stanford group found a second particle at a slightly higher mass, and an indication of still additional possible particles. These other particles were assumed to be excited states of the psi particle.

A number of theories were quickly proposed to explain the properties of the J/psi (which turned out to be a meson). The most commonly accepted explanation was that the J/psi consists of a new kind of quark and its own antiquark. Now, there aren't so many quarks that we can take the discovery of a new one lightly, and this discovery hit with a spectacular splash. Rumors of the discovery preceded the official announcement, and seminars on many campuses were called hastily as theorists tried to get more information and to figure out what the J/psi teaches us.

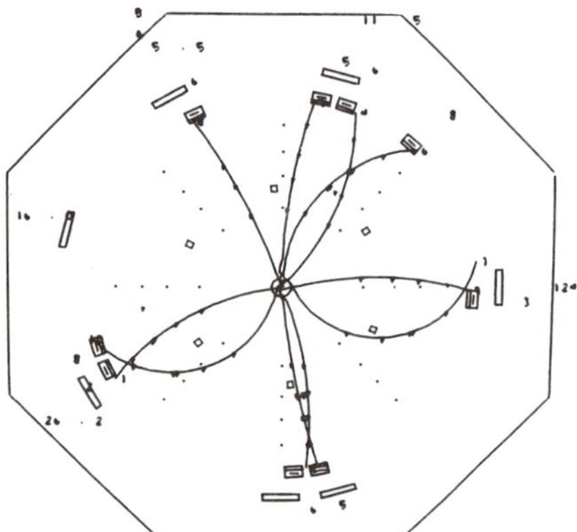

Fig. 28–17 Tracks on an oscilloscope of the SPEAR electronic detector showing the formation of various subatomic particles following the collision of a beam of electrons with a beam of positrons at the Stanford Linear Accelerator Laboratory. The first demonstration of the existence of the J/psi particle appeared on this screen.

Actually, the existence of a new quark had already been predicted a few years earlier to explain some leftover inconsistencies in the regular quark theory. Sheldon Glashow, a Harvard theoretician, gave this quark the property charm. The quark is thus designated *c*. "Charm" is a new quantum number, like strangeness. As for the other quantum numbers, charm has to be conserved in all reactions except weak ones. It was claimed that the new J/psi particle is a combination $c\bar{c}$.

One problem remained. Although the psi may have contained charmed quarks, it had no net charm, since it contained a quark and its own antiquark. The idea that a new quantum number was necessary could not be completely accepted until a particle was detected that displayed the property of charm—naked charm, it was said. It was calculated that the charmed particle should be one of the mesons: $c\bar{u}$, $c\bar{d}$, or $c\bar{s}$. The experiments to find these mesons have proven to be very difficult, but it now appears that charmed particles have been found by the Stanford group.

28.6 *Truth and Beauty*

The discovery of charm did not end the work of the particle physicists. It was only one event, however spectacular, the helped put many things in order. Both theoretical work and experimental work continued, as existing accelerators were improved and new accelerators were built.

Some theoreticians felt that there should be two more quarks to

TABLE 28–1 Quarks and Leptons

Quarks	*Leptons*	*Comments*
up and down	electron and electron neutrino	useful for making the everyday things of our world
strange and charmed	muon and muon neutrino	higher energy: as far as we know, useless for the everyday world
truth and beauty	tau and tau neutrino	still higher energy: new, and as far as we know also useless for the everyday world.

add to the previous four. Not the least of the reasons was that there now seemed to be six leptons—electron, muon, tau, and their respective neutrinos. Following the naming scheme begun with the *u* ("up") and *d* ("down") notation, the two postulated additional quarks were called *t* and *b*. Some say that these letters stand prosaically for "top" and "bottom," but many prefer to call them truth and beauty (see Table 28–1).

In 1977, an experiment was carried out by scientists from Columbia University, Fermilab, and the State University of New York at Stony Brook to look for new particles. The Fermilab accelerator was used to bombard targets of platinum or beryllium with the proton beam. In that summer, the team of scientists reported the discovery of the heaviest particle known to date, almost ten times the mass of the proton. It is also long-lived for the elementary-par-

Fig. 28–18 Part of DESY, the powerful accelerator in Hamburg, Germany.

A COLLIDING BEAM ELECTRON ACCELERATOR

The work confirming the discovery of the fifth quark was carried out at DESY (pronounced colloquially "daisy"), an acronym for Deutsches Elektronen-Synchrotron (German Electron Synchrotron). DESY, located near Hamburg, Germany, was not only already one of the world's largest facilities, but also has recently reached even higher energies. A new facility has opened there to allow beams of electrons and positrons to collide. To make such a colliding-beam facility, the beams must be dense with particles, sharply focused, and precisely aimed. In order to make the beams this way, the beams are first built up to strength and then held in "storage rings." The new colliding-beam lab, called PETRA, is thus called a storage-ring laboratory. It can provide energies of 38 billion electron volts, higher than any other. The American counterpart, PEP, will be ready in the early 1980s.

ticles world. The new particle was christened upsilon, a Greek letter (with a bottle of champagne to celebrate the event).

The upsilon particle seems to be a new quark-antiquark combination. However, whether this quark is the one called "truth" or the one called "beauty" cannot yet be determined. Since the fifth flavor of quark seems to exist, most scientists feel that the sixth probably does as well.

Fig. 28–19 The team of experimenters who discovered the upsilon particle, posing with their apparatus at Fermilab. The discovery of this particle revealed the existence of either "truth" or "beauty." The quark with this property appears together with its antiparticle in the upsilon particle.

28.7 *Fundamental Particles?*

Now things seem to be getting out of hand again. After all, with lists of quarks that seem to be growing, maybe these quarks aren't so elementary any more.

Still, we have been expecting the existence of particles within hadrons (the strongly interacting particles) since an experiment at SLAC (Stanford Linear Accelerator) in 1969 in which electrons were scattered by protons. Similarly to the way in which Rutherford had demonstrated sixty years previously that a concentration of mass exists at the center of an atom, the SLAC experiment

---

A FREE QUARK?

Although many theoreticians feel that a free, isolated quark cannot exist, experimentalists have searched for them nonetheless. The major method of search has been to look for particles of fractional charge. Since quarks would be in the minority on any charged ball, we first mean more-or-less to neutralize the charge on a ball by adding charges of the opposite sign. We can eventually hope to make the total charge almost zero—differing from zero only by the fraction that represents the quark.

In 1977, a Stanford group reported to a packed room at a meeting of the American Physical Society that they had in fact found a quark in this manner. They had floated a superconducting ball made of niobium (a rare element) on a magnetic field. The ball was $\frac{1}{4}$ mm in diameter, and contained 5×10^{19} nucleons. The number of electrons on the ball was changed one by one, and finally a charge of 0.32 of an electron charge (nearly $\frac{1}{3}$ of an electron charge) was left. Another ball also showed a similar charge, but other balls did not. These experiments have been repeated both by the same group and by other experimenters. Given the importance of the topic, most scientists feel that much more evidence is necessary before they can accept the result.

---

demonstrated that there were concentrations of mass within the proton. The concentrations were called partons at the time, and it now seems that the partons are just the quarks.

Should we be upset that we have not seen a free quark? We have already run into an example—with magnetism—where things run in pairs and never singly, so why shouldn't quarks be an example of something that always comes in twos or threes. One model, called the string model of quarks, suggests that quarks resemble the ends of pieces of string in that when you cut a string in half, each half winds up with two ends. Have you ever seen a piece of string with only one end? Alternatively, a model has been proposed in which the group of quarks that make up a particle can be thought of as being held in tiny bags; the contents of a bag make up a particle. In this quark bag model, when the contents of a bag are divided in two, we get two bags, but each is complete. The string model and the bag model are but two of the ways that have been devised to lay at rest our worry over not seeing a free quark.

Fig. 28–20 Sheldon Glashow of Harvard, who shared the 1979 Nobel Prize in physics with Steven Weinberg and Abdus Salam, awarded for their work on the unification of the electromagnetic and weak forces.

The quark theory is not the only possibility for the fundamental makeup of matter now being considered, although it now dominates. An alternative theory, advanced at about the same time, says that all of the particles in our zoo have equal importance and that none is more fundamental than any other. Thus quarks, in the sense that they are more fundamental than the hadrons, would not exist. The alternative theory is known as particle democracy and is still quite controversial.

Recent work indicates that the weak and electromagnetic interactions can be derived from a single mathematical theory. The results are mostly due to a joint theory of Steven Weinberg of Harvard and Abdus Salam of the Institute for Theoretical Physics in Trieste and Imperial College in London, and overlapping work by Sheldon Glashow of Harvard (Fig. 28–19). The Weinberg-Salam theory for what is sometimes called the electroweak interaction has had many successes in making predictions, and is growing in favor.

Progress is being made toward a theory that includes not only the electromagnetic and weak interactions but also the strong interaction. This theory, not yet fully developed, is known as grand unification. One consequence of some versions of grand unification is that the proton is not a stable particle. The period of its decay would be very long—at least 10^{30} years. (This is much longer than the age of the universe, which is about 10^{10} years). Still, if a proton lives, say, 10^{33} years, then a volume of water containing 10^{33} protons should produce one decay per year on the average. A thousand tons of water has this many protons, and experiments are under way to test the prediction.

The 1980 Nobel Prize in physics went to James Cronin of the University of Chicago and Val Fitch of Princeton for their 1964 experiment that proved that charge (C) and parity (P), the latter of which is described on page 479, are not conserved together (CP) in a certain weak interaction. Thus the sign of the charge does not always change in an interaction when parity changes to compensate for the change in charge. The Fitch-Cronin work has been incorporated into a speculative hypothesis that shows why more matter than antimatter may have survived the early universe after the big bang. Increasingly, the study of elementary particles is being related to the study of the universe as a whole. The smallest and largest scales in the universe meet.

A newer, still more complex and more controversial theory called supergravity is the most recent attempt to combine gravity with all the other forces.

You can see that whether or not the particles themselves are "fundamental," the questions being asked certainly are.

Key Words

graviton
W particle
unified field theory
scattering
bubble chamber
spark chamber
resonance
resonance particle
particle zoo
hadrons
leptons
tau
baryons
mesons
symmetry
strange
strangeness
parity
group theory
quarks
color
quantum chromodynamics
flavors
gluons
color force
J/psi particle
charm
truth
beauty
colliding beam
storage-ring laboratory
upsilon
parton
string model of quarks
quark bag model
particle democracy
Weinberg-Salam theory
electroweak interaction
grand unification
supergravity

Questions

1. Match up each of the forces with the particle that mediates (or is thought to mediate) that force:

gravitational	W particle
weak nuclear	pi meson
electromagnetic	graviton
strong nuclear	photon

2. Of the particles in the second column in Question 1, which ones are known to exist?
3. Why are accelerators used in the discovery of new particles?
4. When we create a new particle in the accelerator, where does the energy come from?
5. Describe two types of detectors that are used to keep track of particles after a collision in an accelerator.

6. How can conservation laws (energy, momentum, angular momentum) be used in determining properties of new particles, even when we don't directly observe the particles?
7. As we change the energy of our accelerator beam, how can we tell when we are at an energy at which new particles are created?
8. Classify the following particles as hadrons or leptons:
   (a) electron (b) neutrino (c) neutron (d) proton (e) pi meson (f) photon.
9. What are the reasons for believing that a tau neutrino exists?
10. Of the following three particle categories, which one includes the other two: baryons, hadrons, mesons?
11. Discuss the following statement: "The use of symmetry principles in physics is related to the philosophical idea that physical laws should be simple."
12. What is the relationship between symmetry principles and conservation laws?
13. Use the idea of symmetry to extend the following sequence of symbols:

14. Give one reason why a neutron cannot decay into a proton and a neutrino.
15. What might you conclude if you find that a reaction that should take place does not?
16. Certain reactions that cannot take place via the strong force do take place via the weak force. What does this tell you about the number of conservation laws that are honored by the strong force as opposed to the weak force?
17. In this chapter, we discussed a number of quantities that are conserved, except possibly in weak interactions. List as many of these conserved quantities as possible, and indicate which ones are not conserved in weak interactions.
18. What is the significance of the discovery of parity violation?
19. What is the experimental evidence in favor of the quark theory?
20. The u quark has a charge $+\frac{2}{3}$ (in terms of the fundamental charge); the d and s quarks have charges $-\frac{1}{3}$. Their antiquarks, $\bar{u}$, $\bar{d}$, $\bar{s}$, have charges $-\frac{2}{3}$, $+\frac{1}{3}$, and $+\frac{1}{3}$. The u, d, $\bar{u}$, and $\bar{d}$ have zero strangeness, while the s has strangeness $+1$ and the $\bar{s}$ has strangeness -1. From this list of quarks, work out all possible combinations of three quarks with zero net strangeness. Work out the net charge for each such combination.
21. For the quarks listed in Question 20, work out the allowed combinations of a quark and an antiquark (not necessarily the matching

antiquark) with zero net strangeness, and work out the net charge for these combinations.

22. In what way can we think of the color of quarks as being analogous to the spin of certain particles? (Think of the Pauli exclusion principle.)
23. What is the difference between colors and flavors for quarks?
24. What happens to the colors of two quarks when the quarks exchange a gluon?
25. Why is there a possibility that we may never see a free quark?
26. Why does the discovery of the J/psi stand out amid the discoveries of so many particles?
27. What are the reasons for believing in the existence of the fifth quark?
28. What are the reasons for believing in the existence of the sixth quark?
29. Of the six postulated quarks, how many (and which ones) would be necessary to explain everyday phenomena?
30. What are the advantages of a colliding-beam accelerator?
31. After reading this chapter, what is your definition of a fundamental particle?

CHAPTER 29

Astrophysics

When we look up at the sky at night, we see our moon, the planets, and myriad points of light. With a telescope, we see or photograph much more; for example, we can see that the planets are disks of light, we can detect glowing gas between and around some of the stars, and we can even study galaxies, groupings of billions of stars. But the objects beyond our solar system are beyond our reach. Our contact with the stars is mostly through study of their radiation, making astronomy an observational science rather than an experimental one.

Many years ago, there may have been astronomers who looked at the stars and recorded their positions without considering what the stars and the other objects in the universe were made of. Nowadays, there are almost no astronomers like that. Astronomers of today not only observe the stars but also consider theoretically what makes them shine, what holds them together, and what forces run the show that we call the universe. They are considering the physics of the universe, and are really astrophysicists.

In this chapter we'll try to set out some of the basic ideas and topics of research in modern astronomy and astrophysics, words that are now generally used interchangeably. As we go through our tour of the universe, see how many of the physical ideas that we've developed in this book come up.

29.1 The Stars

About 6000 stars are bright enough to be seen without a telescope. Each of them is a ball of gas, held together by Isaac Newton's force of gravity. But what keeps a star from collapsing? In ordinary stars, it is the pressure caused by the hot gas inside, and the gas is heated by nuclear reactions. In the sun, for example, the basic nuclear reaction is the so-called proton-proton chain, in which four protons, which are hydrogen nuclei, combine with each other one at a time to make a helium nucleus. (Two additional protons also combine, but two protons are left over at the end; Fig. 29–1.) As we saw in Chapter 26, the helium nucleus has atomic weight that is 0.7 percent lower than the sum of the atomic weights of the four protons that went into it. The difference in mass has been transformed into energy according to Einstein's formula $E = mc^2$. There was once enough hydrogen in the sun to provide about 10 billion years of life in this fashion, and about half of that hydrogen is still left.

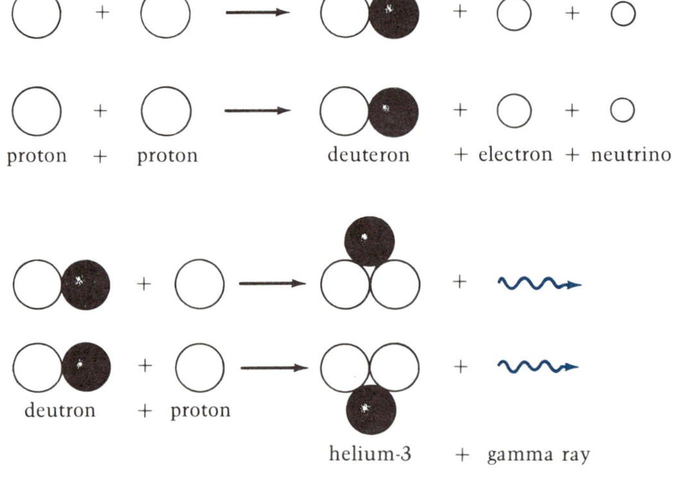

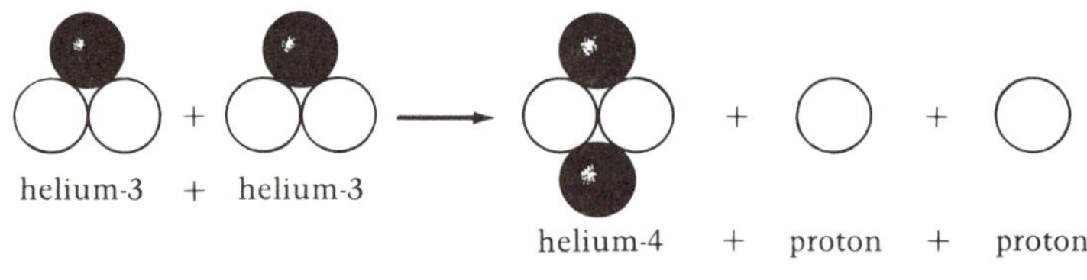

Fig. 29–1 The proton-proton chain. We believe that this chain provides most of the energy generated in the sun.

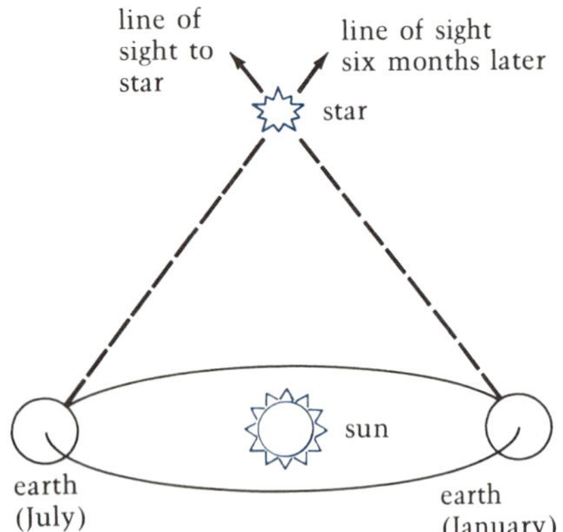

Fig. 29–2 We can measure the distances to nearby stars by triangulation. We take advantage of the motion of the earth around the sun to give us two lines of sight to the star that are slightly different. (The effect is greatly exaggerated in this figure.) Normally, the two lines of sight will make an angle of less than one second of arc with each other.

We can see by simply glancing at the sky that different stars have different brightnesses, but it is difficult to tell exactly what those

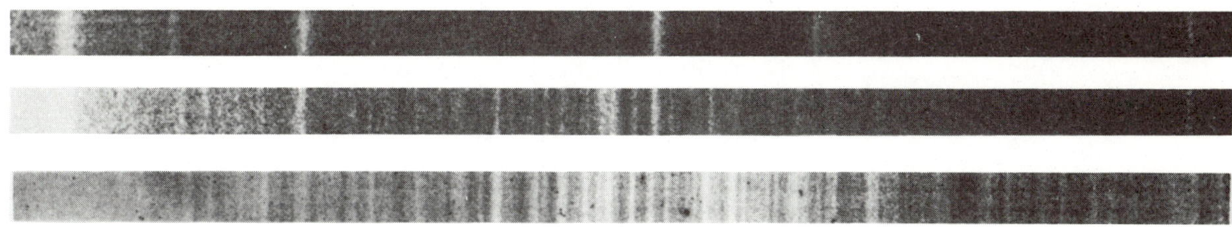

Fig. 29–3 Spectra of stars whose surfaces are at very different temperatures: 30,000 K (type B star, top); 6000 K, like the sun (type G star, middle); and less than 3500 K (type M star, bottom).

different brightnesses really mean. Some stars may appear particularly bright because they are really intrinsically bright, and others may appear particularly bright because they are merely relatively close to us. All the stars except the sun are so far away that they appear only as pinpoints of light, so we cannot compare their sizes to give us an idea of their distances.

For the closest stars, we can use the changing perspective on the sky that the earth's orbit gives us. We use this change in perspective to see how the stars change in position against the background of the distant stars as the earth moves from one side of its orbit to the other, six months later (Fig. 29–2). For farther stars, other methods have been developed to tell us distances. In any case, using the best distances that we can determine, we can correct the values we observe for the brightness of many stars—the apparent brightness—to take account of the distances and derive the intrinsic brightness—the absolute brightness.

One other quantity we can determine from the light we receive is the temperature of a star's outer layers. To do this we use the laws of spectroscopy in different ways. The *color* of a star gives us its temperature, for example, as does knowledge of just which *absorption lines* a star shows in its spectrum and how strong they are.

29.1a CLASSIFYING STARS

Using measurements of a star's intrinsic brightness and its surface temperature, two astronomers just after the turn of the century graphed one of those quantities against the other for a variety of stars. The graph that resulted, of which an example is shown in Fig. 29–4, is called a color-magnitude diagram (since the color gives us the temperature, and the "magnitude" is the scale that astronomers

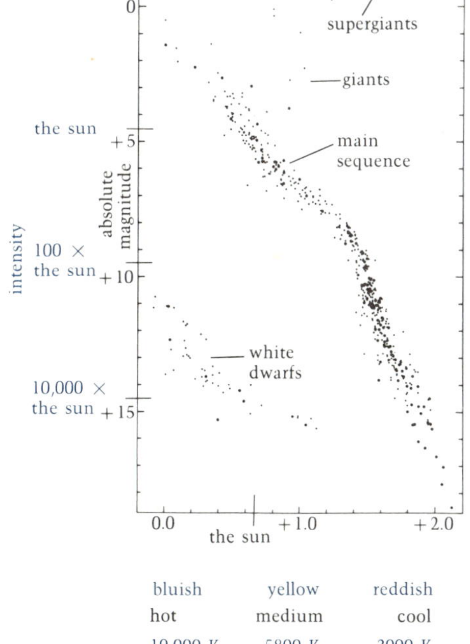

Fig. 29–4 A color-magnitude diagram of the stars nearest to us. When color or some other measure of temperature is graphed on the x-axis and magnitude or some other measure of brightness is graphed on the *y*-axis, most of the stars lie on the main sequence. Brighter stars are called giants and fainter stars are called white dwarfs. In this selection of stars nearest to us, there are no very hot stars and no supergiants. (Courtesy W. Gleise)

use to measure brightnesses). It is also called a Hertzsprung-Russell or H-R diagram, because Ejnar Hertzsprung of Sweden and Henry Norris Russell of the United States were the astronomers who first worked this out.

Even a quick glance at a color-magnitude diagram shows you that the stars don't appear randomly distributed on it. There is a diagonal band that contains most of the stars. It is called the main sequence, and stars on it are called dwarfs. The dwarfs are the normal stars, in the prime of life, undergoing nuclear fusion in a stable situation. Don't be upset that our sun is a dwarf rather than a giant.

When their hydrogen is exhausted, stars begin to collapse. The heat generated by this collapse swells the outer layers, which begin to cool. Thus these aging stars appear in the upper right section of the color-magnitude diagram. They are called giants or even supergiants because they are much larger and brighter than the dwarfs. In the next section, we will see what happens to these stars later on in life; some wind up as white dwarfs (not to be confused with the normal dwarfs we have just discussed), which are shown at the bottom left of the color-magnitude diagram. Others wind up as neutron stars or black holes, which are too faint to appear on the diagram.

From the surface temperatures and brightnesses of stars, we can use our knowledge of the processes involved to compute theoretically what must be going on inside the stars, the stellar interiors Astrophysicists use giant computers to make models of these stellar interiors, and to calculate just what nuclear fuels are being burned (a word we use colloquially to mean that nuclear reactions have taken place) in what stages of the star's life and how long the star can live.

29.1b THE NEUTRINO EXPERIMENT

It is difficult to test the validity of the theoretical calculations for stellar interiors, but twenty years ago it was pointed out that observations might be made of the neutrinos that are generated in the nuclear fusion in a star's interior. These neutrinos, once formed, would escape from the stars without being impeded by the star's outer layers; in the case of the sun, we could hope to be able to detect enough of them to verify that the specific fusion processes were taking place.

Neutrinos are so elusive that a detector of large volume has to be set up on earth, and since certain types of chlorine atoms turn into radioactive argon atoms when hit by neutrinos, a large tank of chlorine atoms would be a good detector. It proved easiest and cheapest to get these chlorine atoms in the form of a cleaning fluid, C_2C_4 (perchloroethylene), and a 400,000-liter tank of this fluid was set up deep underground in a gold mine, where no other cosmic particles could get to it.

Even in this huge tank of chlorine atoms, only about one interaction of a chlorine atom and a neutrino is expected per day, but the techniques used to measure the resulting radioactive argon are so sensitive that one per day is good enough. To everyone's surprise, the results at first were negative—no neutrinos at all were detected. By now, the experiment has run for over ten years, and a small steady rate of neutrino interactions has been detected. But it is less than half the amount predicted by the best calculations, which is worrisome. However, uncertainties in the calculations and in the measurements are large enough that the disagreement—for the moment—is not too serious.

This experiment, which would have the most important consequences should our ideas of nuclear energy in stars prove wrong, will continue, and the results are proving of great interest to those interested in the fundamental question of what holds up the stars. (The new experimental result that neutrinos may change from one type to another may provide the explanation of the discrepancy. The current experiment can detect only electron neutrinos.)

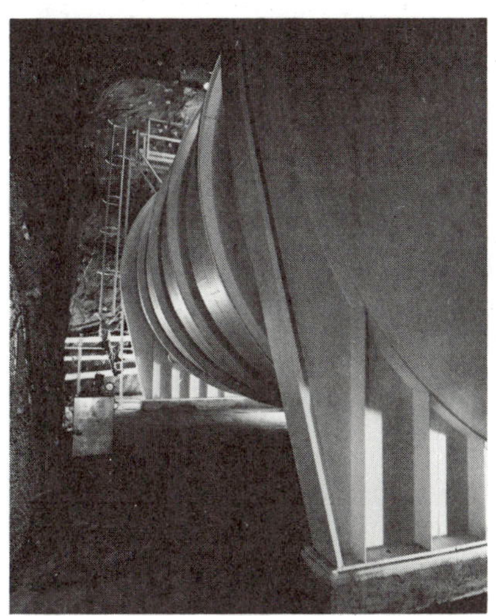

Fig. 29–5 The neutrino experiment's tank deep underground at the Homestake Gold Mine. It contains 400,000 liters of cleaning fluid that can detect one of the types of neutrinos. The region around the tank is now flooded with water to help protect the contents from particles emitted in the surrounding rock.

29.2 *Stellar Evolution*

A star is born as a region of gas, contracting under its own gravity (as described in Chapter 7). As the gas contracts, the gravitational potential energy decreases, and some of the energy heats the gas. Thus the gas becomes hot enough for nuclear reactions to begin. The star reaches the main sequence of the color-magnitude diagram.

What happens when a star begins to die? Since different things happen depending on a star's mass, let us consider three classes of stars. Using the notation of the sport of boxing, we call stars up to four times the mass of the sun lightweight stars, stars from four to eight solar masses middleweight stars, and stars more massive than eight solar masses heavyweight stars.

29.2a THE DEATH OF LIGHTWEIGHT STARS

When a lightweight star, like the sun, exhausts the hydrogen fuel at its core, it leaves the main sequence. Gravity makes the star begin to collapse, and the energy released heats up the core until it is still hotter than it had been. Hydrogen begins to burn in a shell around the helium core. The extra energy that this shell gives off swells the outer layer of the star, which grows to be bigger than the size of the earth's orbit. When the sun becomes a red giant like this, about 5 billion years from now, whatever life is on earth will be incinerated.

After a time, the heat from the continuing collapse of the core under the force of gravity causes the helium to begin to fuse into carbon. (Three helium nuclei combine to form one carbon nucleus.) Since this happens abruptly, it is called the helium flash. Eventually even the carbon core that is left heats up, and the helium begins to burn in a shell around it. The outer layers continue to expand, and eventually are blown off the star. We can see the blown-off shells as planetary nebulae (Fig. 29–6). They are called planetary nebulae because in small telescopes they appear as greenish disks, similar to the appearance of the outermost planets; there is no similarity between planetary nebulae and planets.

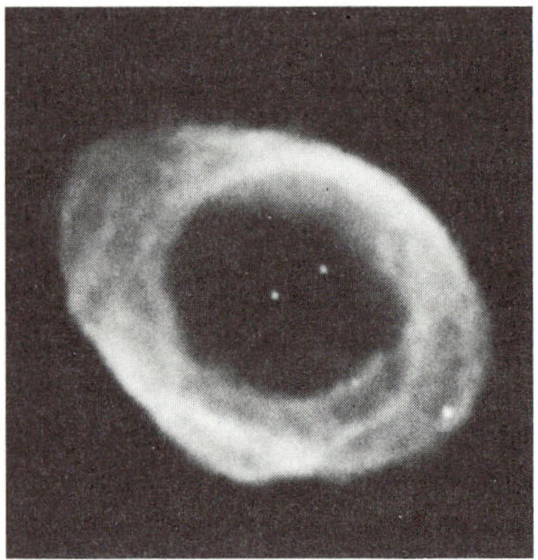

Fig. 29–6 The Ring Nebula, a planetary nebula in the constellation Lyra. The star at the center of the ring is a dying star of about the mass of the sun. It has ejected the ring.

The gas in the shell of a planetary nebula expands for 50,000 years or so, and eventually becomes transparent and disperses. Only the core of the star is left. It tries to contract because of the force of gravity, but eventually the Pauli exclusion principle prevents the electrons from coming any closer together. At this stage, the remnant of the star is called a white dwarf, and as such it will live out its days, gradually cooling into oblivion, its nuclear fires out. White dwarfs contain about as much mass as the sun in a volume the size of the earth. They are so dense that a single teaspoon contains a few thousand kilograms.

29.2b SUPERNOVAE

Middleweight and heavyweight stars have a more spectacular demise. The helium burns steadily after it is formed, and even the carbon that results can begin to fuse into heavier elements. The outer layers become much bigger even than those of red giants, and the stars become red supergiants.

Eventually the nuclei at the core of the star fuse into iron. Each fusion process gives off some energy from the mass that is transformed. But you may recall from Chapter 26 that iron is different—to fuse iron into anything heavier takes *more* energy instead of releasing energy.

A

B

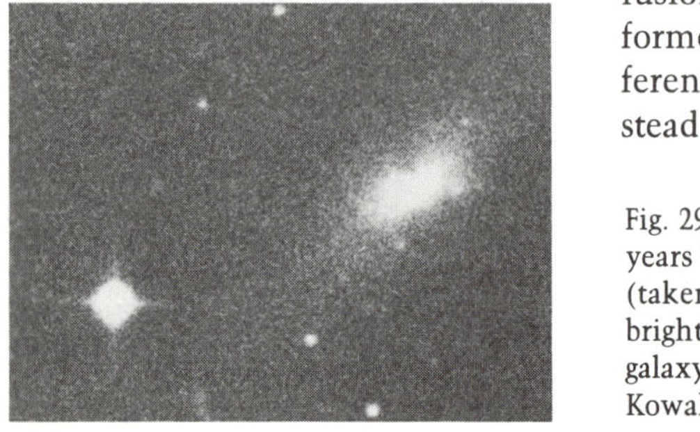

Fig. 29–7 A supernova, the bright object at lower left in B, went off in the 13 years between the two pictures. While no object is seen at that location in A (taken in 1959), the supernova in B (taken in 1972) is shining almost as brightly as the entire galaxy. The supernova is in the outer region of this galaxy, although the connection cannot be seen on the photograph. (Charles T. Kowal, Hale Observatories)

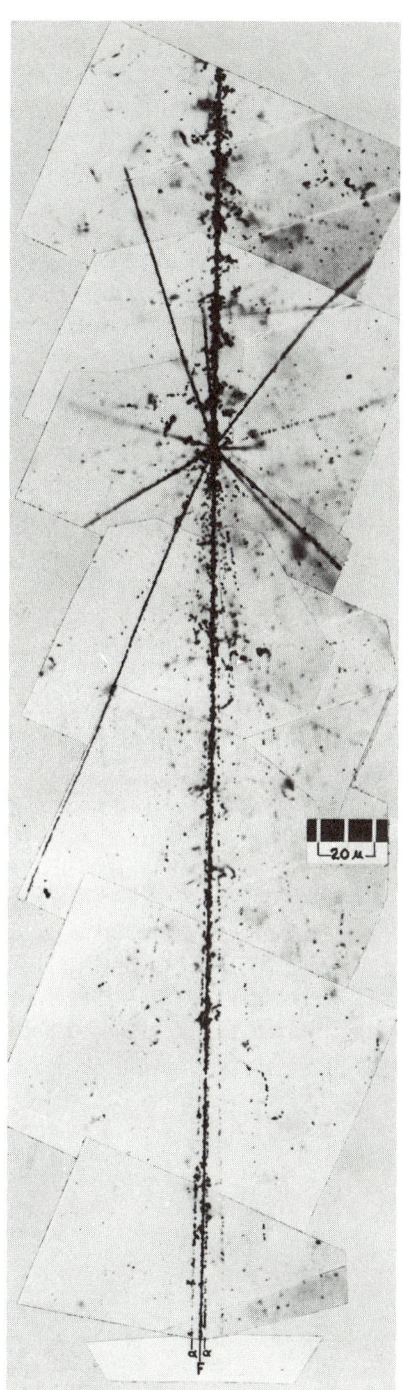

Thus the core no longer has enough energy to counteract the force of gravity, and it begins to contract again. But now it doesn't stop as easily as before. The iron nuclei may even break apart from the high-energy photons that are flying about. More energy is used up. The collapsing star goes out of control. Within seconds, a minuscule length of time for a star that has lived millions of years, the star collapses with enough force that massive numbers of neutrinos are formed. By this time the outer layers of the star are so dense that the neutrinos cannot get out and the star explodes. The star is destroyed and only the core is left behind. This core becomes a neutron star—and pulsar—as discussed in Section 6.3b.

The explosion of such a star is called a supernova, and can be as bright as the entire galaxy of 100 million stars in which the supernova is located. Astronomers now estimate that a supernova takes place in a given galaxy every twenty-five years or so. In our own galaxy, the last two supernovae we know of took place in 1572 and in 1604; we are obviously due for another.

Supernova explosions may accelerate the particles that are given off and give them very high energies. We can detect some of these particles as cosmic rays; we capture them from high-flying aircraft, balloons, or satellites. We can study their compositions, and thus find out about supernova explosions themselves.

In the explosion, the elements heavier than helium that have been "cooked" in the interiors of the stars are strewn around the universe. Thus stars that are formed relatively recently should have higher concentrations of these heavy elements than do the older stars, prediction that is confirmed by observation. All the elements in our bodies heavier than helium were probably put into space by past supernovae, so we owe our existence to explosions of stars in the distant past.

29.2c BLACK HOLES

We have seen in Chapter 9 that the presence of a large mass warps the space around it, according to the theory of general relativity.

Fig. 29–8 The transformation of a heavy cosmic ray into lighter ones. Cosmic rays produce tracks in photographic emulsions flown high in the atmosphere in balloons or above the atmosphere in satellites. In this photomicrograph (enlarged 400×), a fast nucleus of calcium leaving the dense track that comes in from above collides with a nucleus in the emulsion. A stream of nuclear breakup fragments of the original cosmic ray continue downward. In the stream are fluorine (marked F), alpha particles (marked α), and protons. The dark tracks emerging in various directions from the point of collision are the debris of the target nucleus. Which element this corresponds to is not definitely known, although it is probably silver or bromine. (Courtesy Maurice M. Shapiro, Naval Research Laboratory)

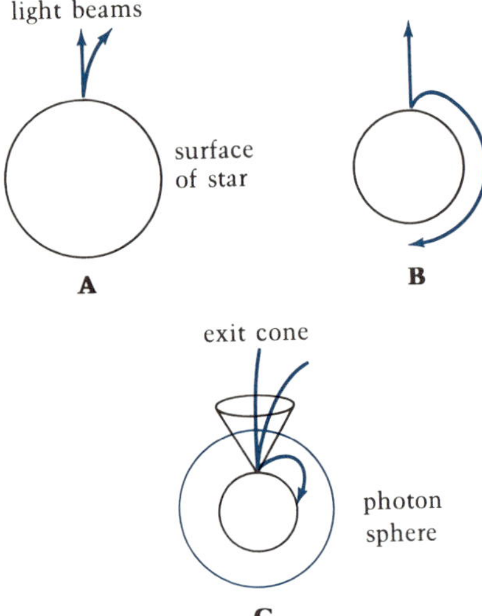

Fig. 29–9 The collapse of a star to a black hole. A. As the star gets smaller, the gravitational field gets stronger. We notice that the light that is not emitted straight up will travel in a curved path. Also, light leaving the surface will begin to have a noticeable redshift. B. The gravitational field is so strong that light is bent around into orbit around the star. C. Light can only escape if we aim it out inside a small cone called the exit cone. If we aim light right on the edge of the exit cone, it will go into orbit in the photon sphere. If we aim it completely outside the exit cone, the light will bend around and hit the surface of the star.

Even a neutron star, however, is not sufficiently dense to warp the space around it so much that light is prevented from getting out.

If the core of a supernova still has over four or five times the mass of the sun after the explosion, however, it will not stop its collapse at the stage of being a neutron star. Gravity will overwhelm the neutron degeneracy, and the star will continue to collapse for ever and ever. Nothing can stop it now.

When the star gets sufficiently small, it becomes a black hole. No light or other electromagnetic radiation can get out. This stage occurs when the star has its mass within a radius called the gravitational radius. The value of the gravitational radius is 3 km times the mass of the star in solar masses; that is, it is 9 km for a three-solar-mass star, 15 km for a five-solar-mass star, and so on.

Let us imagine that we are standing on the surface of a star as it collapses. If we try to shine a flashlight back at earth, our friends on earth will be able to receive our signal until we are at one and a half times the gravitational radius (22.5 km for a five-solar-mass star, for example). Also, the beam of light will be bent more and more as we shine it away from the vertical (Fig. 29–9A).

At this distance of one and a half times the gravitational radius, we have the photon sphere. Outside the photon sphere, our light always escapes from the vicinity of the star if we point the flashlight away from the star. When we are within the photon sphere but still outside the gravitational radius, only light that is shined sufficiently close to the vertical escapes. The directions in which it will escape mark the exit cone (Fig. 29–9C). Light emitted along the surface of the exit cone will go into orbit in the photon sphere.

As our star contracts still further, and we come closer to the gravitational radius, the exit cone becomes smaller. When we reach the gravitational radius itself, the exit cone closes entirely! Within the gravitational radius, no matter in which direction we shine our beam, it will not escape. We are within what we call the event horizon. All these properties can be predicted from a set of solutions worked out to Einstein's equations of general relativity by

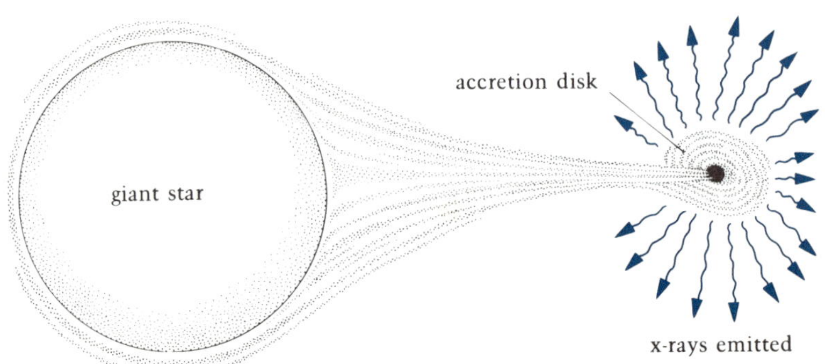

Fig. 29–10 A black hole in orbit around a giant star. The gravity of the black hole is so strong that the star is distorted and some of the atmosphere of the star is pulled away from the star toward the black hole. Some of the material falls into the black hole and other material goes into orbit around the black hole, forming an accretion disk. When the material is accelerated toward the black hole, it is heated up and gives off x-rays.

We can't see through an event horizon for a black hole, just as we can't see over an ordinary horizon on earth.

Karl Schwarzchild in 1916, immediately after Einstein's theory was published.

Although it might seem at first that we could detect a black hole when it passed between us and more distant matter, calculations show that a black hole would affect too small a region of space for this effect to be visible. We must fall back on indirect methods.

We can detect a black hole by discovering that there is an object of more than five or so solar masses that is not bright enough to be an ordinary star. Since white dwarfs and neutron stars contain less than three solar masses or so of matter, an invisible star of more than five solar masses could only be a black hole. (We say more than five solar masses so as to give ourselves a safety factor; the actual limit is the result of a theoretical calculation and is not known accurately.)

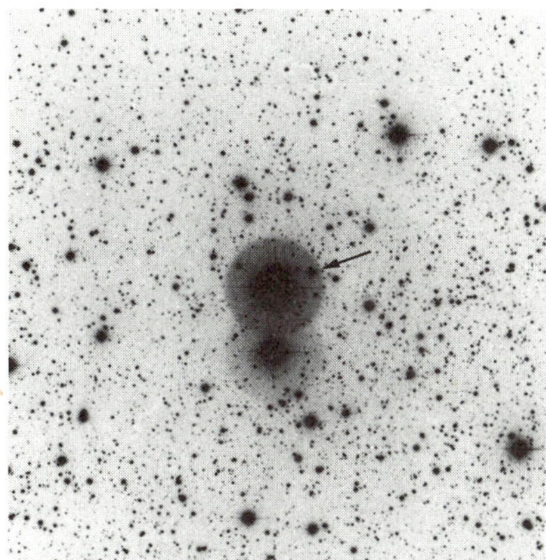

Fig. 29–11 The optical image of the region of the sky that includes the suspected black hole, Cygnus X-1. We find a blue supergiant star there, marked with an arrow on this photograph. Careful studies of this star show that it is being orbited by an invisible object of a mass so large that it must be a black hole. (Photo by Jerome Kristian, Hale Observatories)

We look in a binary system to find such an object, because we can then use the laws worked out by Kepler and Newton to calculate the masses of the objects in such a system. If one of the masses turns out to be greater than five solar masses, and is invisible but would be visible if it were an ordinary star, then we have ourselves a black hole.

We figure our where to search in the first place by looking for strong sources of x-rays in the sky. We do this because it has been calculated that even though we cannot see a black hole directly, its mass would cause particles from space nearby to spiral around it in the course of being sucked into it. These particles would heat up and radiate x-rays (Fig. 29–10).

One of the x-ray sources in particular, Cygnus X-1 (the first x-ray source to be discovered in the constellation Cygnus), meets all the above criteria. It contains about eight solar masses and is invisible, although it would be visible if it were a normal star of that mass. Most astronomers think that it is the first black hole we have detected. Several other possibilities exist, but none so convincing.

One of the major tools we have had for further study of black holes are the first two High-Energy Astronomy Observatory (HEAO) satellites, which NASA launched in 1977 and 1978. They carried sensitive telescopes that could study x-ray variations with great precision in location and good resolution in time. Since theory predicts that fluctuations in x-radiation in the region around a

Fig. 29–12 High-Energy Astronomy Observatory 2, known as the Einstein Observatory. It can make x-ray images at high spatial resolution. An Einstein Observatory image of the supernova remnant known as Cassiopeia A in Plate 73 shows a bright ring that is probably the location of the shock wave spreading outward from the explosion. The width of the image is one-sixth that of the moon in the sky. Never before have we been able to make x-ray images with such high resolution.

black hole would take place on a scale of 1 millisecond, time resolution is important. The data are still under study.

The black holes we have been discussing so far have been those that would form from collapsing stars. But it is possible for black holes to be either much larger or much smaller than star size.

The gravitational radius varies linearly with the amount of mass but the amount of mass varies with the volume, which depends on the cube of the radius. So when we go to very large radii, we are able to include quite a lot of mass even if it is quite spread out and thus at a relatively low density. This situation may be occurring in the centers of galaxies, as we shall mention in Section 29.6.

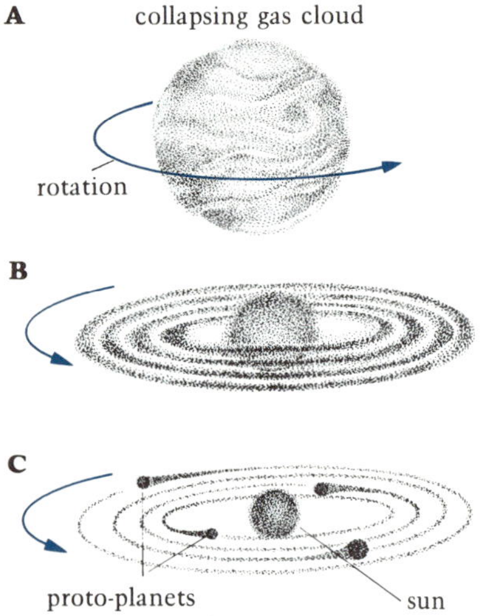

A collapsing gas cloud

rotation

B

C

proto-planets sun

Fig. 29–13 Possible steps in one scenario of the formation of the solar system. A. A rotating gas cloud begins to contract. Because of the rotation, it flattens out as it collapses. B. As the process continues, there is a bulge at the center that will eventually become the sun, and a disk of cooler material that will eventually form the planets. C. The disk begins to break up into smaller clumps, called planetesimals. These combine over eons to become proto-planets, which will eventually contract and cool as planets.

29.3 *Formation of the Solar System*

About five billion years ago, the gas and dust that filled a region of space began to collapse. Perhaps the explosion of a supernova nearby started off the collapse. In any case, the region was spinning slowly, and as the collapse continued, it began to spin more rapidly because the amount of angular momentum was conserved. The spin impeded the collapse in the direction perpendicular to the axis of spin, while nothing prevented the collapse along the axis. Thus the spinning gas and dust collapsed to a disk.

Within the disk, smaller regions of gas and dust collapsed by themselves. The largest mass, in the center, we call the proto-sun. Smaller masses, in other parts of the disk, became (we think, at the moment) small chunks of material called planetesimals which, in turn, combined to become proto-planets (see Fig. 29–13).

As the proto-sun continued to collapse, the energy from the gravitational collapse went in part to heat the gas and dust. The dust became vaporized, and still the collapse and heating went on. Eventually, the heat became great enough to overcome the electrical repulsion between nuclei, and nuclear fusion began. The sun was on the main sequence.

The proto-planets also continued to collapse, but they did not contain sufficient mass for nuclear reactions to begin. And so the planets were formed.

29.3a OTHER SOLAR SYSTEMS?

There is nothing to indicate that the above story of the origin of our solar system is unique, and most astronomers believe that other systems of planets exist about other stars. We have seen (Section 4.4) that when a system of two masses revolving around each other is left alone, its center of mass moves in a straight line. If a star in space has a planet orbiting it, and the two move across the sky, then it is the center of mass that travels in a straight line. The star, on the other hand, will appear to wobble slightly from side to side of the center of mass, and it is the star itself rather than the center of mass that we see.

Careful observations have been made over a period of over twenty years of Barnard's star, the star that appears to move most rapidly across the sky. Tentative evidence had been found that a giant planet or two are orbiting Barnard's star. However, it is currently controversial as to whether or not these planets exist. Even less conclusive evidence exists in favor of the presence of planets around other nearby stars.

Another type of study involves observation of the Doppler shifts of stars that are similar to our sun. If such a star has an invisible companion that is moving toward and away from us as the stars orbit each other, the star we can see will move away and toward us so that the center of mass can remain steady (or at least moving uniformly). Variations in the Doppler shifts from week to week were detected for one-third of the stars that were studied. These variations indicate that planets are orbiting around these stars.

A

B

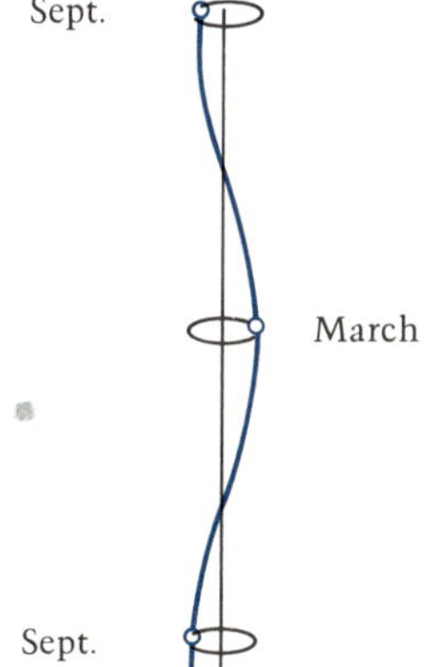

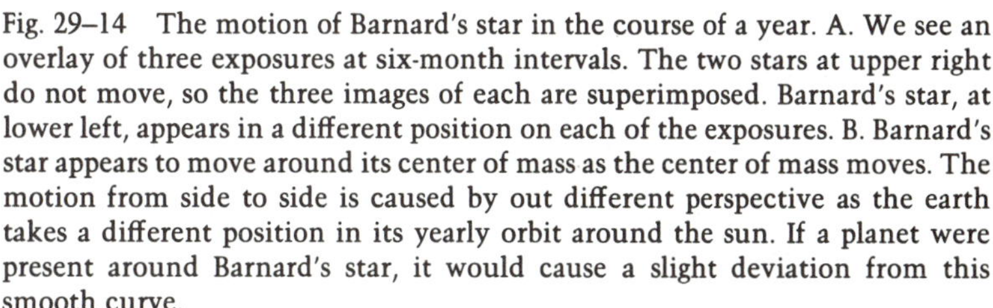

Fig. 29–14 The motion of Barnard's star in the course of a year. A. We see an overlay of three exposures at six-month intervals. The two stars at upper right do not move, so the three images of each are superimposed. Barnard's star, at lower left, appears in a different position on each of the exposures. B. Barnard's star appears to move around its center of mass as the center of mass moves. The motion from side to side is caused by out different perspective as the earth takes a different position in its yearly orbit around the sun. If a planet were present around Barnard's star, it would cause a slight deviation from this smooth curve.

29.3b LIFE IN THE UNIVERSE

Since it seems likely that other solar systems exist somewhere in the universe, it has become relatively acceptable of late to consider the possibility that life has arisen on distant planets and that we might hope to contact that extraterrestrial life. Work in the last thirty years has shown that complex molecules form fairly easily under a variety of situations, and so it may not be as hard for life to form as we had thought. Still, to say that there are complex molecules around is not to say that there is life, and we do not know just what makes things live.

Radio signals have long seemed the most obvious mode of communication and special observations of nearby stars have been undertaken to make sure we are not missing any obvious message that is being beamed to us. More complex methods of analysis of the radio signals have also been tried, and are now being furthered.

Fig. 29–15 The 305-m-diameter bowl of the radio telescope at the Arecibo Observatory, Puerto Rico, part of the National Astronomy and Ionosphere Center, which is operated by Cornell University under contract with the National Science Foundation.

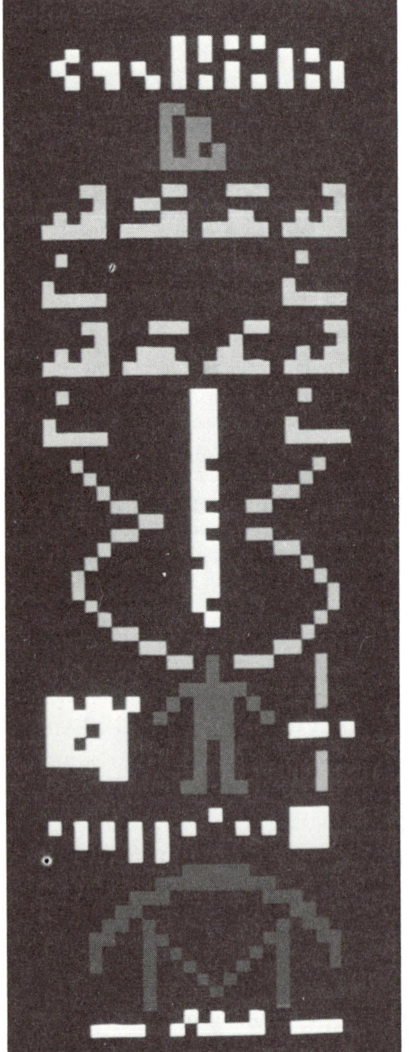

Occasionally, we even send out a signal. The giant radio telescope at Arecibo (Fig. 29–15), for example, has sent out a message (Fig. 29–16) describing life on earth and some rudiments of our knowledge.

Radio waves are not the only possible way of sending and receiving messages, although they seem to be the cheapest per bit of energy. Nevertheless, we should not reject the possibility of sending messages by neutrinos, which also travel at or near the speed of light. A project is now under way to test the feasibility of sending a neutrino message from Fermilab through the earth; remember, neutrinos are so elusive that the bulk of the earth hardly gets in the way.

29.4 *Galaxies*

In France, at the time of the American Revolution, Charles Messier was interested in discovering comets. But if he were to find a fuzzy object in the sky, he wouldn't know right off whether it was a new discovery or not. So he set out to make a list of all the nonstellar objects he could observe in the sky. Messier's list includes most of

Fig. 29–16 The message sent to contact extraterrestrial intelligence from the Arecibo Observatory. The top row shows the binary numbers 1 to 10. The figure in the second row describes atomic numbers. The signals in the third, fourth, fifth, and sixth rows show the molecules upon which life is based. The vertical wavelike structures diagram the DNA helix. The next row shows the population of the earth, the outline of a human, and his average height. Below the man is the solar system, with the earth between his feet. At bottom is the outline of the Arecibo telescope and its dimensions.

Fig. 29–17 This recent photograph of M81, the 81st object in Messier's list, clearly shows the spiral structure of this galaxy.

the interesting and unusual brighter objects that we know of, and we still refer to these objects by their position in Messier's list. We call fuzzy objects in the sky nebulae (singular, nebula), from the Latin word for cloud.

The Earl of Rosse built a giant telescope in Ireland, and in about 1850 discovered that some of Messier's objects showed a spiral form, like that of a pinwheel (Fig. 29–17). And when the new techniques of photography were applied to astronomy with large telescopes about fifty years later, many more spiral nebulae were discovered. But just how these spiral nebulae might be different from other nebulae was not known.

Whether the spiral nebulae were part of our own galaxy of stars, or whether they were, to use the term of the philosopher Kant, "island universes" of their own, was much debated in the years around 1920. But we have already mentioned that it is difficult to find the distance to astronomical objects, and the various methods that were applied to measure the size of our own galaxy and the distance to the spiral nebulae were not sufficiently reliable to allow us to decide whether or not the spiral nebulae are outside our galaxy.

The solution to the problem came in 1924, when the astronomer Edwin Hubble used the 2.5-meter (100-inch) telescope on Mt. Wilson in California, then the largest telescope in the world, to observe a certain type of variable star in some spiral nebulae. This year of star, called a Cepheid variable (pronounced "seh-fee-id") changes in brightness with a regular period, and it had earlier been realized that the duration of the period of variation was related to the intrinsic brightness of the star. Thus by measuring how long a Cepheid variable takes to go through its cycle of brightness, we know how intrinsically bright it is. Hubble detected Cepheid variables in some of the spiral nebulae, and could tell immediately how far away they were by making measurements of the period of variability. The stars, and therefore the galaxies they are in, proved to be outside our own galaxy, which is called the Milky Way Galaxy. The spiral nebulae are thus not merely clouds of gas but rather are spiral galaxies on their own.

Fig. 29–18 An edge-on view of a spiral galaxy in the constellation Coma Berenices. This edge-on view shows a dark lane of dust in the plane of the galaxy silhouetted against the galaxy's bright central region. If we could see this galaxy face on, it would look similar to the Great Galaxy in Andromeda (shown in Plate 76).

Plate 76 shows the nearest spiral galaxy to our own, the Great Galaxy in the constellation Andromeda. You can readily see that the galaxy has a central core surrounded by spiral arms. We are viewing it at an angle, but we have no choice. We can't turn the Andromeda Galaxy around, but we can look at another galaxy that happens to be oriented differently (Fig. 29–18). The fact that the core bulges above and beyond the plane of the spiral arms is obvious. We think that our own galaxy looks something like these two galaxies.

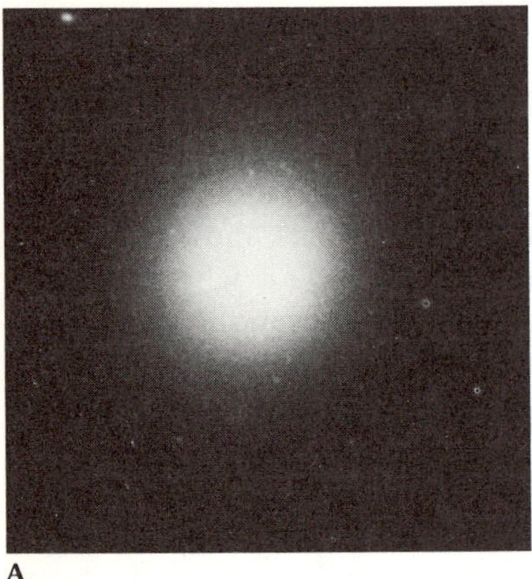

A

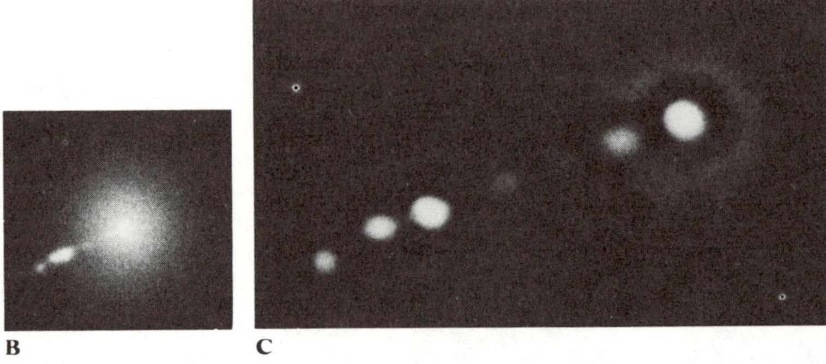

B C

Fig. 29–19 A. An elliptical galaxy, known as M87, in the constellation Virgo. Note the tiny jet of gas near the center. B. This gas jet, shown here in a shorter exposure, indicates that something active is happening. Scientists think that a giant black hole is probably present there. C. This computer enhancement of 29–19B shows that the jet is actually several discrete blobs of gas, each of which may have been ejected at a different time.

Other galaxies are elliptical in shape. A giant elliptical galaxy, M87 (the 87th object in Messier's list), is shown in Fig. 29–19. A long exposure (Fig 29–19A) shows fuzzy dots around it; these are globular clusters, clusters of a hundred thousand stars each. A shorter exposure (Fig. 29–19B) shows that a jet of gas appears to be coming out of the central region of the galaxy. Figure 29–19C shows us computer enhancement of the jet.

Recent studies of the central region of M87 have shown that the stars near the center move around as though some huge mass at the center were attracting them. Also, the number of stars appears to grow sharply at exactly the center of the galaxy. Both these observations can be interpreted to mean that there is a giant black hole, containing 100 million solar masses of material, at the center of M87. Other interpretations cannot be ruled out, but the black-hole model seems to be the most reasonable explanation known at present. (It still seems a little strange to be calling a black hole the most reasonable explanation for anything.) When Space Telescope is launched in 1983, it will be able to make observations of much higher resolution, which may be able to resolve the question.

Fig. 29–20 The 100-m radio telescope of the Max-Planck Institut für Radioastronomie, in Effelsberg, West Germany. This is the largest fully steerable antenna in the world.

M87 and other similar objects emit strongly in the radio region of the spectrum. Giant radio telescopes can be used to study them. The resolution of a radio telescope is limited by diffraction. The larger the telescope, the finer the detail we can study. The largest single antenna that can track sources continuously across the sky is the 100-meter-diameter radio telescope near Bonn, Germany (Fig. 29–20). But recently the techniques of interferometry have been used to make radio observations of much higher resolution. The idea is that if we use telescopes separated by some distance, and

Physics of the Universe

A variety of lessons for physicists and students of physics can be gained by studying astronomy. Plate 65 *(at right)* shows images of the planets made by several space probes. We see the earth rising over a lunar surface photographed by the Apollo astronauts. To the earth's left is Venus, perpetually covered with clouds. Rapidly rotating Jupiter, top left, is surrounded by clouds and has no solid surface. Comparisons of the circulation of the earth's atmosphere with those of other planets helps us understand our own planet. The circulation depends on convection, conduction, and radiation. Radioactive elements heat the earth's interior, leading to continental drift, volcanoes, and geothermal energy. Mars, the reddish planet in the middle, Venus, earth, and one of Jupiter's moons all have volcanoes. Mercury, covered with craters, appears bluish in this image. Saturn, whose ring is much more visible than those of Jupiter and Uranus, appears at top right.

Plate 66 *(below)* is a photograph of the surface of Mars taken from one of the Viking landers. Although scattering by air molecules on earth makes our own sky blue, so much pinkish dust is suspended in Mars' atmosphere that it appears pink.

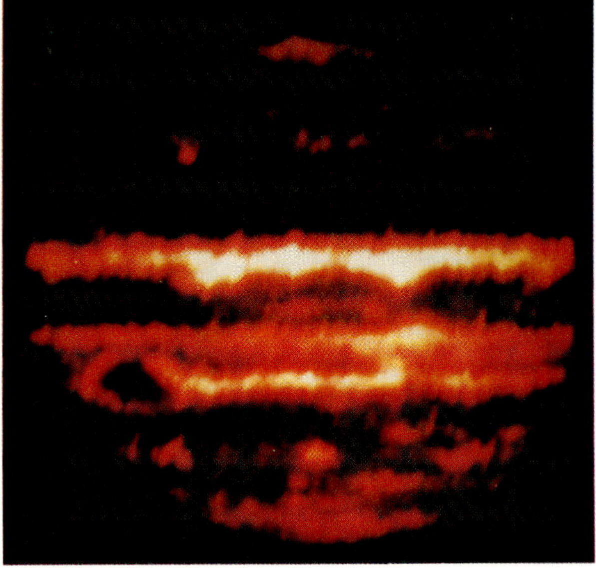

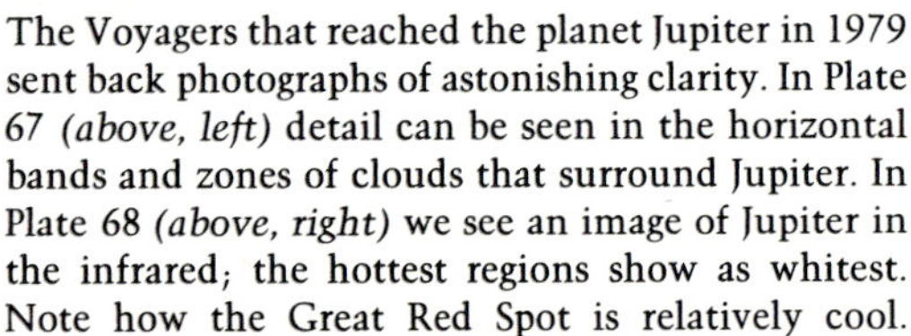

The Voyagers that reached the planet Jupiter in 1979 sent back photographs of astonishing clarity. In Plate 67 *(above, left)* detail can be seen in the horizontal bands and zones of clouds that surround Jupiter. In Plate 68 *(above, right)* we see an image of Jupiter in the infrared; the hottest regions show as whitest. Note how the Great Red Spot is relatively cool. (NASA spacecraft image; Hale Observatories infrared image; courtesy JPL)

Plate 69 *(below)* shows a close-up of the Great Red Spot, which is a rotating giant storm. Note how both the Great Red Spot and an adjacent white oval are rotating in the same direction, a sign of Coriolis forces at work. (JPL/NASA photo)

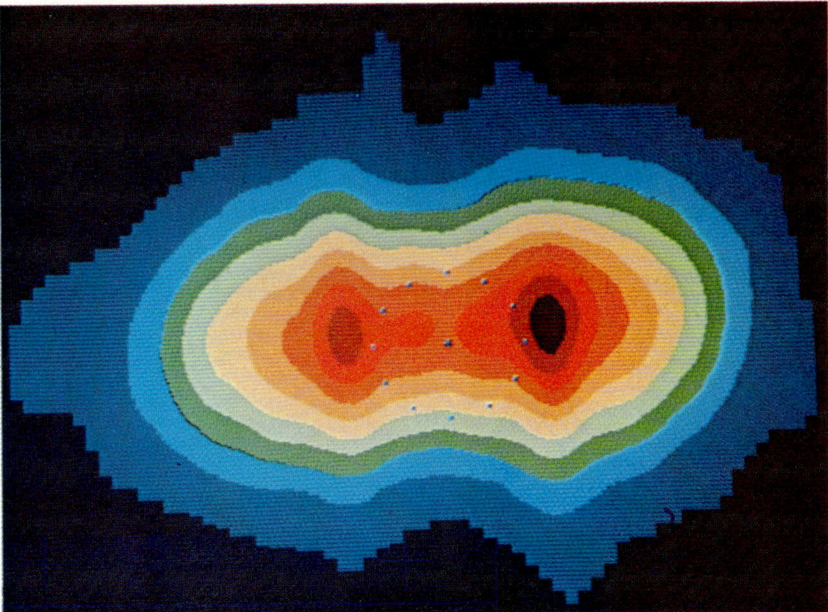

The Voyagers discovered that Jupiter's moon Io has erupting volcanoes (Plate 70, *above left*), caused by the heating of Io's interior as the moon flexes because of the gravity of Jupiter and its other moons. (JPL/NASA photo)

Radio observations of Jupiter, using an interferometer, an array of telescopes that together take advantage of the wave property of radiation known as interference, gave this view of Jupiter's radiation belts (Plate 71, *above right*). We see a cutaway view of a doughnut-shaped region surrounding Jupiter (the planet is tiny on this scale). Jupiter's magnetic field, and thus its radiation belts, are many times stronger than those of the earth. (Image by Inke de Pater, Sterrewacht Leiden, with the Westerbork array)

In Plate 72 *(at left)* we see a view of Saturn from the Voyager 1 spacecraft. Tidal forces, a gravitational effect, have given rings to Jupiter, Saturn, and Uranus. (JPL/NASA photo)

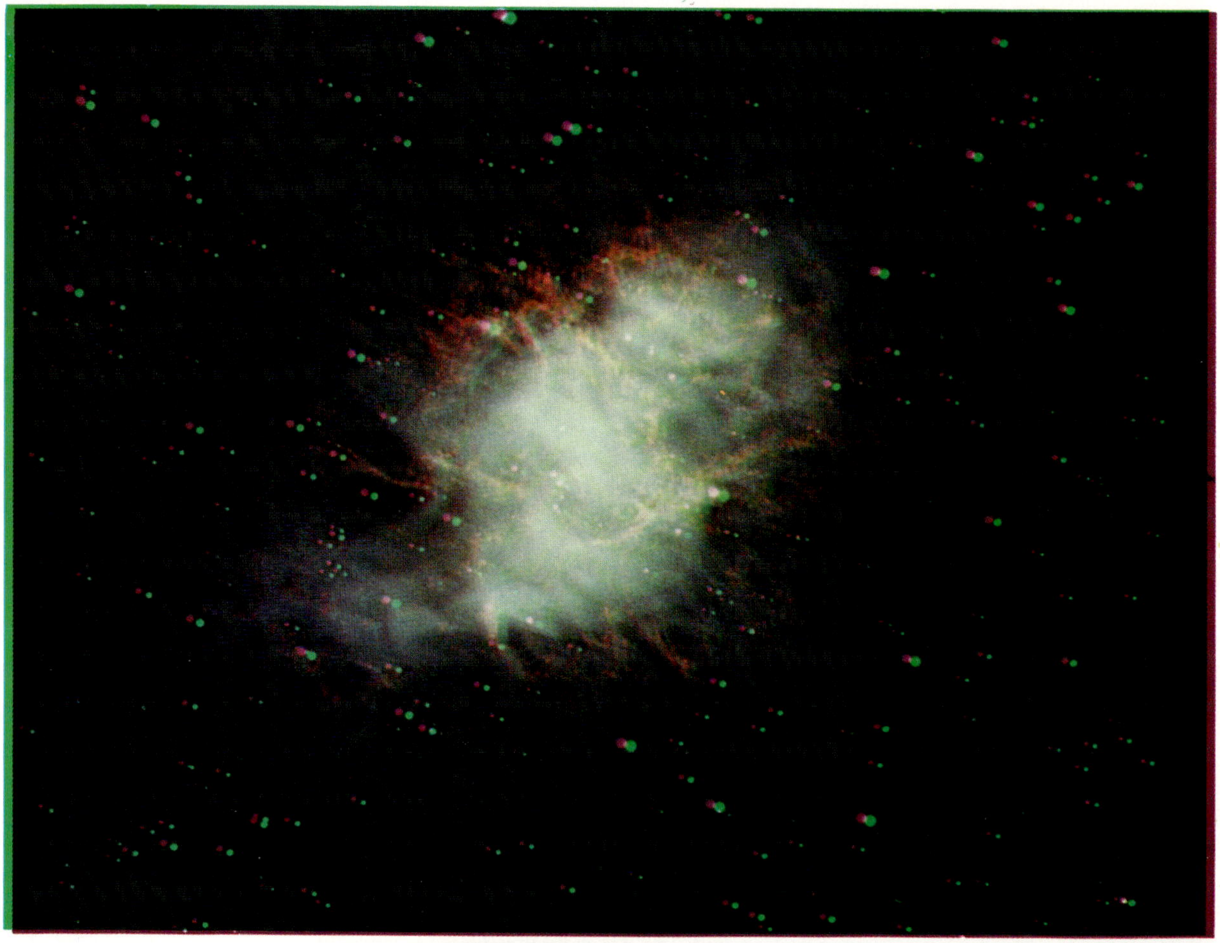

Plate 73 *(above)* is an optical image of the Crab Nebula, the remnant of a supernova that exploded in 1054 A.D. In its center is a pulsar, spinning thirty times per second to conserve its angular momentum as it contracted from a more slowly rotating state. Energy from the pulsar's slowing down makes the nebula glow. (Palomar Observatory, California Institute of Technology. © The California Institute of Technology and the Carnegie Institution of Washington; reproduced with permission. Palomar Observatory photo.)

In Plate 74 *(at right)* we see a radio image of the Cassiopeia A supernova remnant, made with the Very Large Array interferometer of the National Radio Astronomy Observatory in New Mexico. We can see a ring of heating caused by a shock wave. (Image by Philip E. Angerhofer, Richard A. Perley, Douglas Milne, and Bruce Balick).

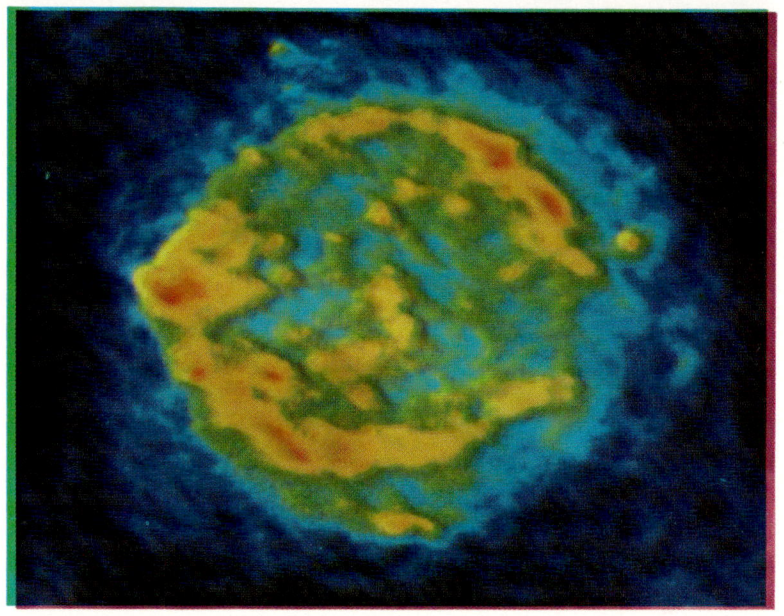

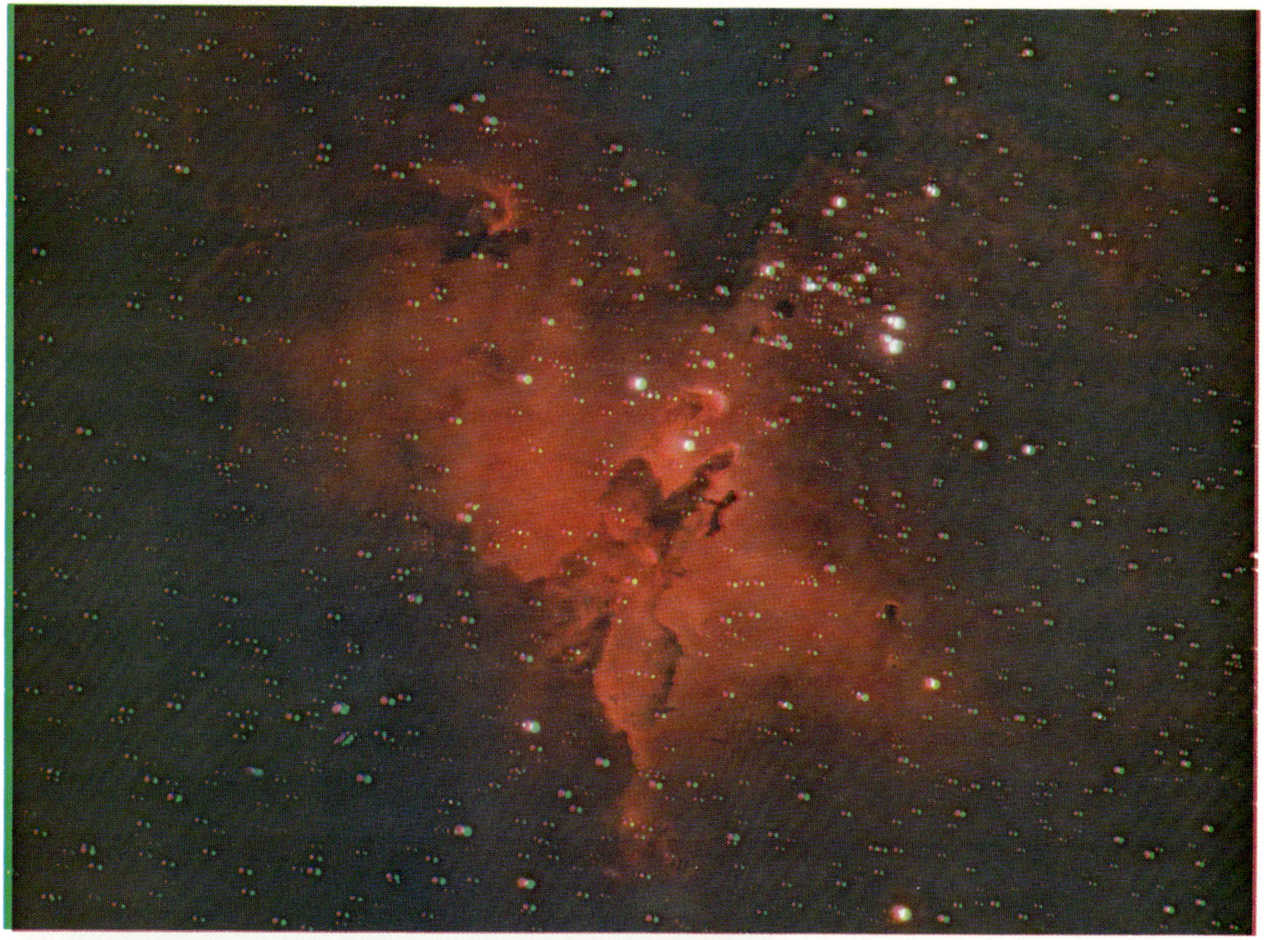

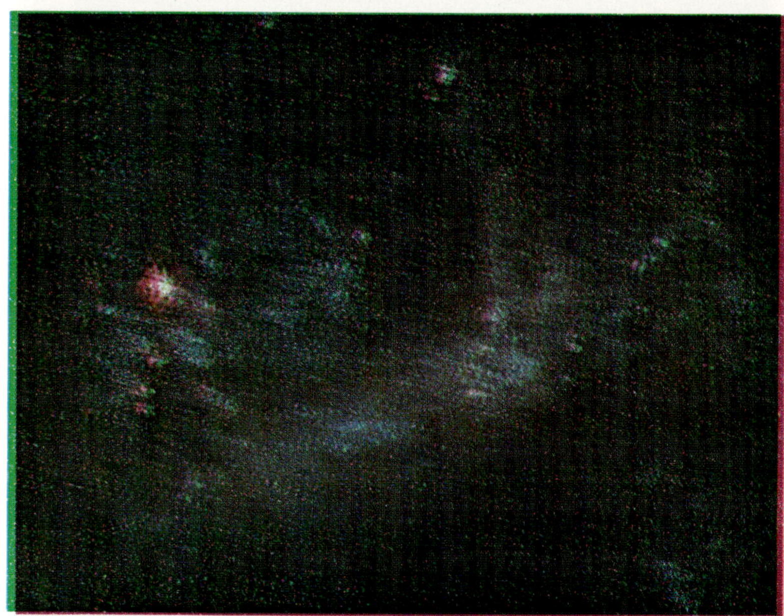

Clouds of gas in the space between the stars are known as nebulae. Plate 75 *(above)* shows the Eagle Nebula, also known as M16, in the constellation Serpens. The bright stars at top center are young and hot, and are part of a group of stars held together by gravity and known as a galactic cluster. They heat the gas so that it gives off hydrogen radiation, which is reddish. The small dark regions superimposed on the nebula are probably contracting to become stars.

In Plate 76 *(at left)* we see a small galaxy that is a satellite of our own Milky Way Galaxy. Known as the Large Magellanic Cloud, this galaxy is visible only from sites south of the United States. The hot young stars appear blue because their radiation curve peaks at relatively short wavelengths, as we saw in our discussion of radiation. Nebulae glow reddish in hydrogen light. (© The Association of Universities for Research in Astronomy, Inc. The Cerro Tololo Inter-American Observatory)

Plate 77 The Andromeda Galaxy, a relatively nearby spiral galaxy that resembles our own galaxy. It also has two satellite galaxies. Note how the hotter young stars we see in the spiral arms look bluer than the cooler older stars we see in the core. (Palomar Observatory, California Institute of Technology, © The California Institute of Technology and the Carnegie Institution of Washington; reproduced by permission. Palomar Observatory photo.)

Plate 78 *(above)* shows the Whirlpool Galaxy, photographed through ultraviolet, blue, and yellow filters to show their colors very accurately. We actually see a pair of galaxies. The spiral member contrasts strikingly in color with its companion on the end of one of its spiral arms. The blue knots in the spiral arms are clusters and associations, each containing up to fifty very massive, hot, blue stars. The yellowish light of the companion arises from a collection of cooler, much longer lived stars.

In Plate 79 *(below)* we see an edge-on view of a spiral galaxy, NGC 4631, also in the constellation Canes Venatici. Dark lanes of dust block our view of the stars in the plane of the galaxy. (Photographs by James D. Wray, McDonald Observatory, University of Texas)

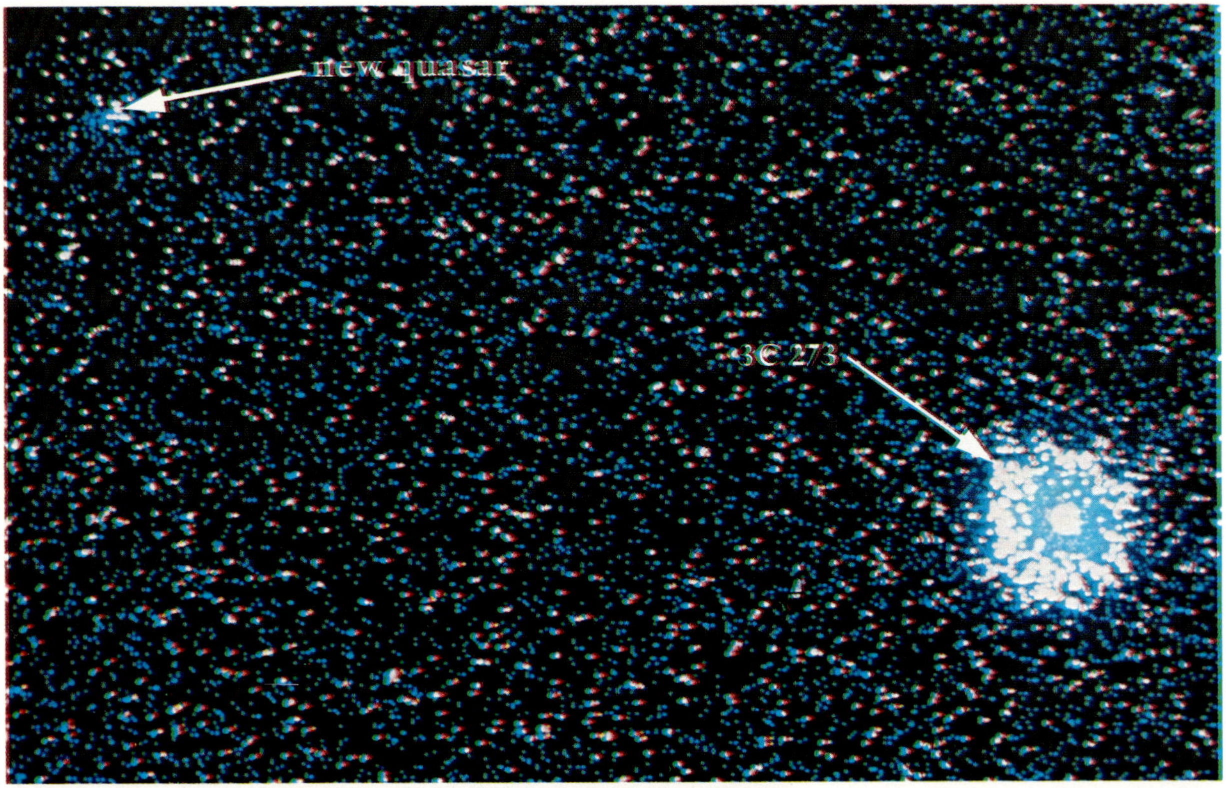

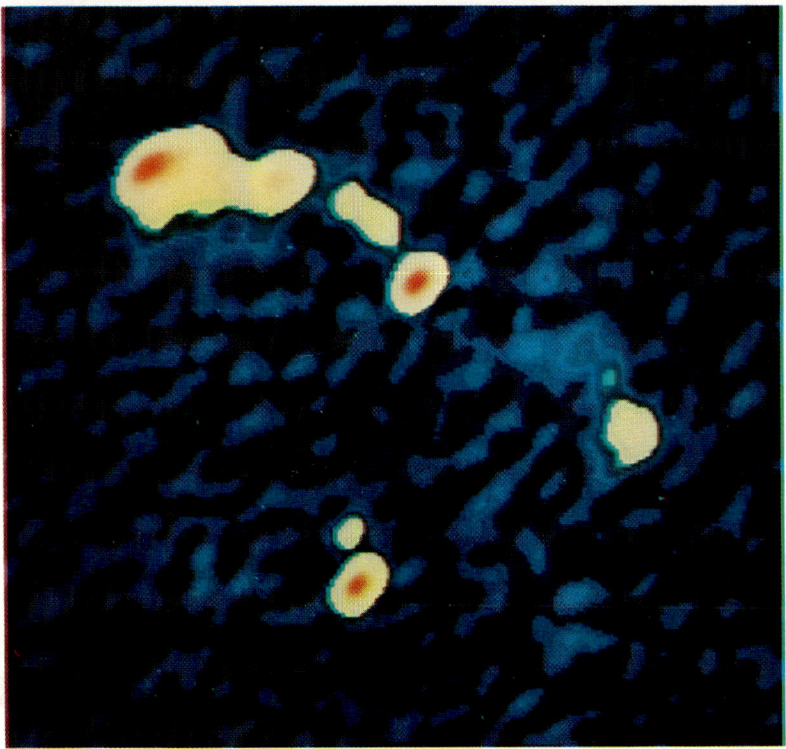

In Plate 80 *(above)* we see an x-ray view of 3C 273, the quasar that appears brightest in visible light. Studies of the spectrum of this quasar led to the realization that huge redshifts are present and that quasars are thus the farthest objects in the universe. This x-ray view from the Einstein Observatory showed that in addition to 3C 273, the brighter region of the image, a previously unknown faint quasar was present as well. The new quasar is over ten times farther away than 3C 273. (Image by Harvey Tananbaum and colleagues, Harvard-Smithsonian Center for Astrophysics)

The Very Large Array radio map in Plate 81 *(at left)* shows a source that appears as two quasars (A and B) very close together in the sky. The source is actually one quasar, split into two major images (regular ovals) by the gravitational effect of a galaxy one-third of the way from here to there. Such a bending of light and radio waves is predicted by Einstein's general theory of relativity. (Image by Bernard F. Burke, Perry E. Greenfield, and David H. Roberts, MIT)

LINES INSPIRED BY A LECTURE ON EXTRA-TERRESTRIAL LIFE

JDGM

Some time ago my late Papa
Acquired a spiral nebula.
He bought it with a guarantee
Of content and stability.
What was his undisguised chagrin
To find his purchase on the spin,
Receding from his call or beck
At several million miles per sec.,
And not, according to his friends,
A likely source of dividends.
Justly incensed at such a tort
He hauled the vendor into court,
Taking his stand on Section 3
Of Bailey 'Sale of Nebulae.'
Contra was cited Volume 4
Of Engelston's 'Galactic Law,'
That most instructive little tome
That lies uncut in every home.
'Cease' said the sage 'your quarrel base,
Lift up your eyes to Outer Space.
See where the nebulae like buns,
Encurranted with infant suns,
Shimmer in incandescent spray
Millions of miles and years away.
Think that, provided you will wait,
Your nebula is Real Estate,
Sure to provide you wealth and bliss
Beyond the dreams of avarice.
Watch as the rolling aeons pass
New worlds emerging from the gas:
Watch as the brightness slowly clots
To eligible building lots.
What matters a depleted purse
To owners of a Universe?'
My father lost the case and died:
I watch my nebula with pride
But yearly with decreasing hope
I buy a larger telescope.

compare the signals received by the two telescopes, we will get an interference pattern similar to those discussed for the double-slit experiment in Chapter 15. We can improve the resolution by moving the telescopes farther apart.

Several radio telescopes can be linked with ground lines, or the data from each can even be recorded on tape, using atomic-clock signals as time standards. Recently, satellite links have been set up. Whatever the method, when the signals from the radio telescopes are compared, an interference pattern is viewed if the object is sufficiently small in size. Whatever the interference pattern observed, it can be interpreted to map the radio sources with a resolution that corresponds to the greatest distance between radio telescopes.

In some cases, the radio telescopes have even been separated by the diameter of the earth, with one telescope in the U.S. and another in the U.S.S.R. This is Very Long Baseline Interferometry (VLBI). Such measurements have told us, for example, that the nucleus of M87 is only one-thousandth of an arc second across. (Optical measurements cannot distinguish less than one-tenth of an arc second.)

The newest system for radio interferometry is called the VLA

Fig. 29–21 Some of the 27 radiotelescopes making up the Very Large Array (VLA), operated by the National Radio Astronomy Observatory near Socorro, New Mexico. Each telescope is 26 meters in diameter, and can be moved on railroad tracks over a Y shape that is 27 km in diameter.

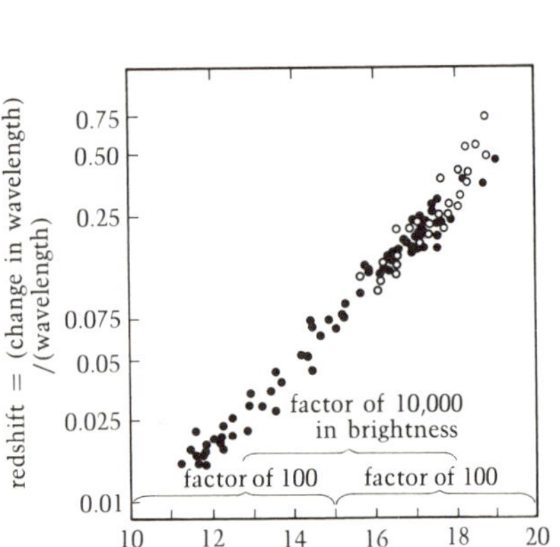

Fig. 29–22 Hubble's law, discussed in Section 15.7, is illustrated in this recent graph. The horizontal axis uses the observed brightness of a galaxy as a way to measure its distance. After all, we can calculate (with the inverse-square law) the rate at which an object would grow fainter if it were to be placed at farther and farther distances from us. A few technical corrections have also been made. The vertical axis uses the redshift as a way to measure velocity. For small velocities, the redshift is the same as the velocity divided by the speed of light. For larger velocities, a correction from the special theory of relativity must be applied.

(Very Large Array), and is shown in Fig. 29–21. This system contains 27 telescopes, each 85 feet in diameter. Located near Socorro, New Mexico, it went into full operation in 1980. The VLA can be used to map radio sources much more quickly than the continent-spanning method mentioned above, even though the ultimate resolution is not quite as great. Each method has its advantages.

29.5 *The Expansion of the Universe*

Even before it was realized that the spirals were galaxies, it had been found that their spectra showed very little redshifts. In 1929, Hubble announced that the spirals always showed redshifts (with the exceptions of the very nearest few), and that the further the galaxies were from us, the larger their redshifts. Since a redshift corresponds, via the Doppler effect, to a body that is moving away from us, Hubble's discovery means that the universe is expanding.

Hubble in fact found that the velocity of expansion is proportional to the distance of the galaxy. Hubble's law, as it is called, is written $v = Hd$, where v is the velocity, d is the distance, and H is a constant.

But does the fact that all the galaxies are moving away from us mean that we are in the center of the universe? Consideration of a raisin cake shows us that this is not the case. As we see in Fig. 29–23, as the cake rises, all the raisins move away from each other. But a pair of raisins that are twice as far away from each other as another pair move away from each other at twice the rate of the first pair. No matter which raisin we choose to stand on, all the other raisins will seem to be moving away from us at a rate that increases with their distances from us.

To establish Hubble's law, we have to measure the redshift of an object and its distance independently. We can measure the redshift for any object whose light we can receive, but we have independent measures of distance only for those objects that are not too far away. Since Hubble's law is so well established on the basis of those relatively nearby galaxies (some of which are quite far away in absolute terms), we now use Hubble's law to assign a distance to the farthest objects we observe, based on measurements of their redshifts. Thus when we are considering the outermost regions of space, measuring redshifts is as close as we can come to measuring distances.

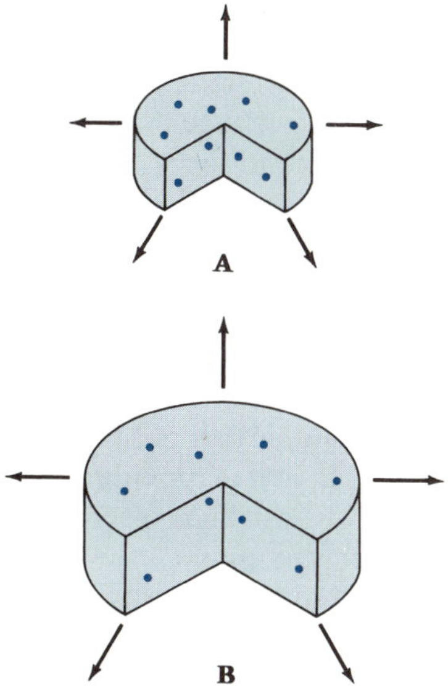

A

B

Fig. 29–23 We can think of the expansion of the universe like the expansion of a raisin cake. The raisins are like the clusters of galaxies. As the cake expands, the distance between *any two* raisins increases. The farther apart the two raisins start, the faster they will separate.

29.6 *Quasars*

Up until 1960, strong radio sources were known in the sky but their positions were not known very accurately because of the low resolution of telescopes operating at these radio wavelengths. When the first interferometers, which operated in the late 1950s, worked at specifying the positions of radio sources, one of the sources turned out to be near a faint bluish point of light that was hailed as being the first "radio star." Soon, the position of a second "radio star" became known accurately, because the moon passed in front of it. Since the position of the moon in the sky is well known, from accurate timing of when the moon first passed in front of it and when the source emerged from behind the moon, the position of the radio source was determined.

The spectra of these objects were ususual, and did not seem like those of any known objects. Thus the sources were probably not ordinary stars, and were called "quasi-stellar radio sources" ("quasi-" is a prefix drawn from the Latin for "as if"). This name was soon shortened to quasars.

But nobody knew what the quasars were, a problem that lasted three years. Then a Caltech scientist, Maarten Schmidt, realized that some of the spectral lines he had photographed in one quasar were the distinctive set of spectral lines of hydrogen (the Balmer lines) astoundingly redshifted by 16 percent of their original values. A second quasar immediately showed hydrogen lines that were redshifted by 37 percent of their original values. If the redshifts were caused by the Doppler effect, then the two quasars are receding from us at 16 percent and 37 percent of the speed of light, respectively. They were, by Hubble's law, the farthest objects then known in the universe.

Thus the quasars immediately became very important, because to study the past of the universe we want to look at objects as far out as possible. Their light, after all, has been traveling for billions of years, so we are seeing them as they were billions of years ago, when the universe was younger.

It also became clear that since the quasars were very far away, in order to appear even as bright as they do, they would have to be radiating huge amounts of energy. Also, since they were found to fluctuate in brightness within weeks, they could not be more than light-weeks across (where a light-week is the distance traveled by light in a week of time). After all, for a fluctuation to take place, all parts would have to "know" that they had to change in brightness, and that information could not travel faster than the speed of light. No explanation was known for how a quasar could generate so much energy in so small a volume. This lack of understanding is called the energy problem.

One way out of the energy problem would be to say that the quasars are closer to us than distances derived from their redshift indicate. Could the redshift be a gravitational one rather than a Doppler shift? But the spectra observed resemble that of a tenous gas rather than that of a dense object, and no model to provide suitable gravitational redshifts has ever been found. Further, recent observations of the redshift of faint gas near the quasar rule out the gravitational redshift method by showing that the redshift doesn't vary within the gas.

Most quasars appear with large redshifts; none is known very close to us. This appears to indicate (assuming we would be able to locate close quasars, which would not be very redshifted) that there were more quasars in the universe long ago. Remember, we are seeing now light from the quasars that was emitted then. The Einstein Observatory, a space telescope observing x-rays, has discovered many faint quasars very distant from us (Plate 80).

The latest ideas about quasars is that they represent explosions or other bright events in the cores of galaxies. In the last few years, it has been realized that there are certain types of galaxies that have especially bright cores relative to the arms, and that the properties of these galaxies are not very different from the properties of quasars. The arms associated with quasars may simply be too faint to be visible in most cases. The relationship between quasars and events in the cores of galaxies is growing more secure as more observations are made.

What is the solution to the energy problem? No one knows for sure. We do not have a totally satisfactory method in mind for

A

B

Fig. 29–24 These two objects look quasi-stellar in appearance but their giant redshifts show them to be very far away. They were the first quasars to be discovered. A. 3C 48. B. 3C 273. A faint jet of gas can be seen being blown off from 3C 273.

Plate 81 shows multiple images of a distant quasar apparently resulting from a gravitational lens, as predicted by the general theory of relativity.

producing so much energy in such a small volume. Most astronomers now believe that the energy comes from giant black holes in the cores of the quasars, although this is still speculative. This fits with the growing belief that black holes are present in the cores of many galaxies, our own included.

29.7 *The Big Bang*

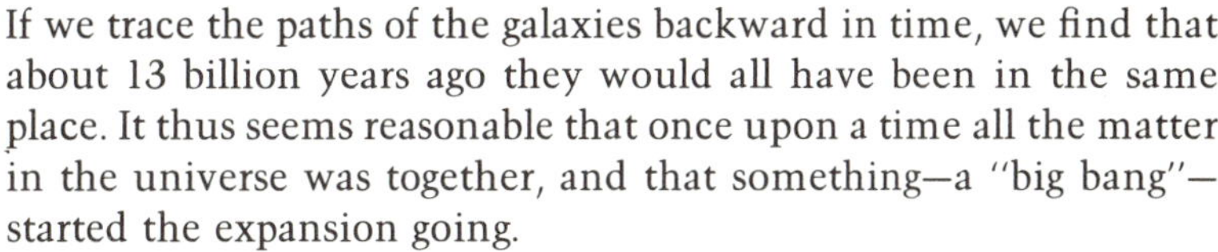

If we trace the paths of the galaxies backward in time, we find that about 13 billion years ago they would all have been in the same place. It thus seems reasonable that once upon a time all the matter in the universe was together, and that something—a "big bang"—started the expansion going.

Fig. 29–25 Robert W. Wilson (left) and Arno A. Penzias, discoverers of the cosmic background radiation, for which they received a share of the 1978 Nobel Prize in physics. They are standing in front of the antenna they used for the discovery.

Note that you mustn't think that big bang was like a bomb going off somewhere. The big bang filled all space simultaneously; there was no "center" to the explosion then and there is no center to the universe now (just as there was no time before the big bang, and there is nothing "outside" the universe).

Big-bang cosmologies is a name given to sets of mathematical solutions of Einstein's equations of general relativity and to the physical interpretation of these solutions. (*Cosmology* is the study of the universe as a whole.) Several different solutions to Einstein's equations are known, and in this section we shall discuss their general nature rather than the details of the mathematical predictions.

That a big bang occurred is accepted by almost all astrophysicists nowadays. The clinching piece of evidence was the discovery with radio telescopes of radiation that was emitted by the big bang itself. The discovery was made accidentally, but now stands alongside Hubble's law as the basis of our knowledge of the history of the universe.

The radiation was discovered in 1965, when Arno Penzias and Robert Wilson, scientists from Bell Laboratories, were investigating faint sources of static in order to improve communication techniques. They used an antenna that was pointing at the sky, and no matter how carefully they tried to account for all the sources of radio signal they received in their apparatus, there was always a slight bit left over. In the meantime, a team of Princeton scientists had calculated on theoretical grounds that such radiation should be detectable from the big bang, and were preparing to observe it. When the two groups got together, the importance of their work became clear. Actually, a similar theoretical prediction had been made seventeen years previously, but had not been known to the

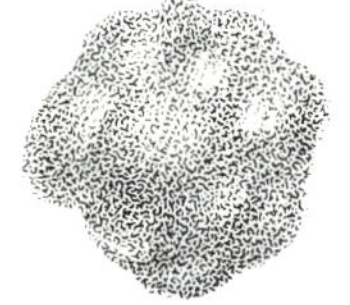

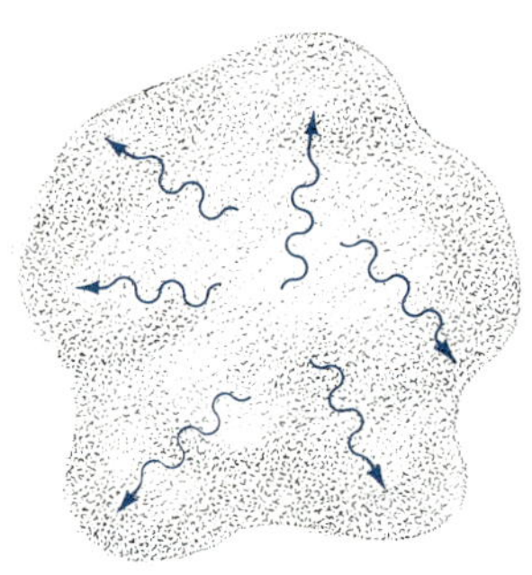

Fig. 29–26 When we talk about seeing the radiation from the big bang, that is not quite correct. For a while after the big bang the universe was opaque, so the radiation was constantly being absorbed and reemitted. When the universe cooled to about 3000 K, the electrons and protons combined together to form atoms and the universe became mostly transparent. The radiation was then left to wander around the universe on its own, and it is the cooled remnant of this radiation that we now detect.

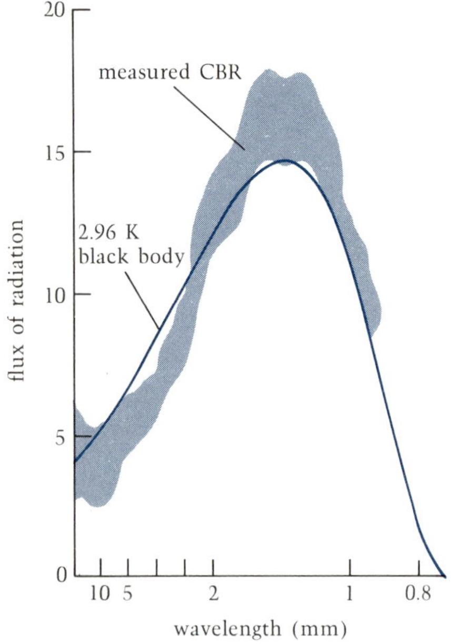

Fig. 29–27 Early measurements of cosmic background radiation (CBR) fit to a 2.96-K black-body curve quite well. However, the definitive proof awaited observations made in the mid-1970s to show that the CBR fell off at the shorter wavelengths, just like the curve.

scientists working on these projects at the time that the observational confirmation was made.

Because the universe was extremely uniform at the time of the big bang, and was in perfect equilibrium, the radiation that was emitted long ago was black-body radiation, a type that we discussed in Chapter 23. In other words, it followed Planck's law. About a million years after the big bang, when the universe had cooled to about 3000 kelvins, the electrons and protons were able to assemble themselves into atoms. Atoms are transparent to radiation at all wavelengths except those of spectral lines. The universe was therefore transparent. Black-body radiation, then at a temperature of 3000 K, was left to circulate forever through space. As the universe expanded, the radiation cooled, although it retained its black-body nature. (We can think of the individual photon wavelength expanding, shifting the entire black-body curve toward the red.) At present, after about 13 billion years of cooling, we find black-body radiation at a temperature of 3 K, not far above absolute zero. (See Fig. 29–26.)

The black-body radiation from the big bang has been measured at many different radio wavelengths (Fig. 29–27), but skeptics wanted to see not only the straight-line part (on the right side of the peak) but also the fact that the curve peaked. Unfortunately, the peak fell in the infrared and the shortest radio wavelengths, where the earth's atmosphere prevented observations from being made from the earth's surface. Finally, in 1975, observations made from balloons flying aloft to a height of 40 km measured the infrared part of the curve and confirmed the black-body nature of the radiation.

29.8 *The Future of the Universe*

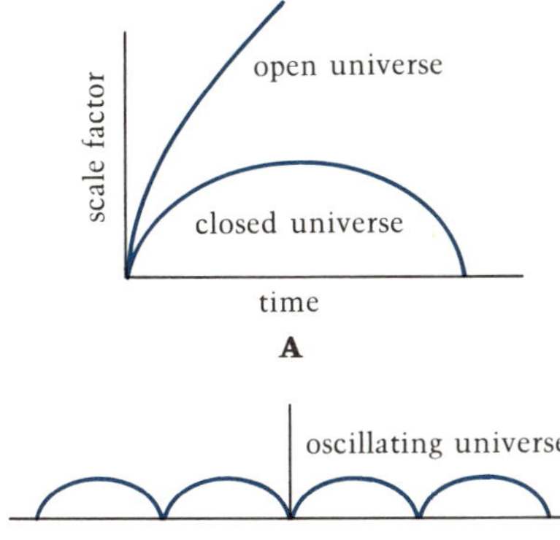

Fig. 29–28 A. An open universe will continue to get larger forever, although the rate of expansion can get smaller. In a closed universe, the expansion will stop and the universe will come back together to a point. B. An oscillating universe appears just like a closed universe to us. However, those who favor an oscillating universe (if the universe is closed) postulate that the expansion and contraction happened an infinite number of times before the current one and will happen an infinite number of times after the current one.

The end of the collapse of a closed universe could be called the big crunch.

We know that the universe is expanding now, but what will happen in the future? Although the universe has been expanding for about 13 billion years, that doesn't mean that it will keep on expanding forever.

Since mass is present in the universe, gravity is trying to slow down the expansion. We know that the expansion isn't slowing down very rapidly, or else it would show up as a major deviation from the straight-line relation between distance and redshift that we call Hubble's law. Still, a small effect could be present.

The question is, can we measure a small deviation from Hubble's law as we look to the farthest objects we can detect? If the deviation is not present or only very slight, as we see in Fig. 29–28A, then the expansion would not be slowing down enough to stop eventually. Rather, the expansion would continue throughout all time. In this case, we say that the universe is open.

If, on the other hand, the deviation from Hubble's law is sufficiently great, as we see in Fig. 29–28A, then the expansion will slow down, stop, and eventually the universe would begin to contract. In this case, we say that the universe is closed. During the period of contraction, we would detect blueshifts from distant galaxies instead of redshifts, and eventually the universe would collapse to another infinitely dense situation. Fortunately, we know that even if this is the case, the universe is expanding rapidly enough to continue its expansion for at least another 50 billion years.

If the universe is closed, and so does collapse, then two possibilities remain. We could be in the midst of the only cycle of expansion or contraction, or else we could be in the midst of only one of many such cycles. The latter case is called the oscillating universe (Fig. 29–28B). Scientists know of no way that we can distinguish between these two possible sorts of closed universe.

This method of determining whether the universe is open or closed is conceptually simple, and so astronomers for years have studied the most distant objects and tried to determine their distances so that they would test Hubble's law. The discovery of quasars gave new hope to this method, but we are only now finding ways to assess the relative distance of quasars independently of their redshifts. Mostly we use clusters of galaxies, since their properties are fairly well known.

Although the results have been inconclusive for years, recently the Hale Observatories scientists who use the 5-meter Mt. Palomar telescope to investigate this have found a definite deviation from Hubble's law on Fig. 29–22. However, this result may be telling us more about how galaxies have changed in the past billions of years.

Fig. 29–29 A cluster of galaxies in the constellation Hercules. All the objects that are not round are galaxies rather than foreground stars. We see all types of galaxies viewed from various angles.

There is another way of tackling the problem of whether the universe is open or closed. After all, the problem is analogous to predicting whether a ball we throw into the air will eventually be slowed by gravity enough to come down to earth. If we throw hard enough, the object can escape from the earth's gravity. For the universe, we know how fast the universe is expanding, so if we can somehow measure the amount of gravity, we can calculate whether it is enough to stop the expansion or not.

The most obvious way of finding the amount of gravity in the universe is to add up all the masses in the universe (planets, stars, galaxies, clusters of galaxies, quasars, interstellar gas, etc.) and see what that comes to. But this method is not definitive, because the universe includes too many invisible things. For example, we have no idea at all how many black holes there are. The amount of visible matter has only about 10 percent of the amount of gravity necessary to "close the universe," that is, to make the universe a closed one. There is some indirect evidence that enough unseen matter exists to make up the difference, but recent x-ray observations from HEAO indicate that there is not enough hot gas between the galaxies to close the universe. Now we wonder anew how much mass is in the form of neutrinos.

An important independent method of finding the amount of gravity in the universe involves the physics of the formation of the elements. Within seconds after the big bang, protons were formed. Within fifteen minutes after the big bang, some of the protons had combined to form deuterium (heavy hydrogen). If the density of the universe was very high at that time, then the deuterium quickly combined to become helium, and little deuterium was left. If, on the other hand, the density of the universe was somewhat lower, whatever deuterium was formed remained around. Since none of the processes we know in stars forms deuterium, whatever deuterium we now find in our universe presumably came from those first fifteen minutes after the big bang.

So the problem turns to finding out how much deuterium there is in the universe. The several measurements of the abundance of deuterium that have been made suggest that the density of the universe must have been relatively low soon after the big bang, much too low to close the universe. The universe must therefore be open. However, the deuterium method depends entirely on complicated theoretical calculations, and something could always be wrong with the calculations or with the values of quantities used in them. The work continues. Studies of the motions of galaxies agree with the deuterium results in indicating that the universe is open. We hope to get closer to a definite answer before 50 billion years pass, at which time the answer will become very clear.

Key Words

observational scientists
astrophysicists
proton-proton chain
apparent brightness
absolute brightness
color-magnitude diagram
main sequence
dwarfs
giants
supergiants
white dwarfs
stellar interiors
lightweight stars
middleweight stars
heavyweight stars
helium flash
planetary nebulae
red supergiants
supernova
cosmic rays
black hole
gravitational radius
photon sphere
exit cone
event horizon
proto-sun
proto-planets
nebula
Cepheid variable
elliptical galaxy
globular clusters
VLBI
VLA
Hubble's law
quasar
energy problem
big-bang cosmologies
cosmology
black-body radiation
open universe
closed universe
oscillating universe
big crunch

Questions

1. What keeps an ordinary star from collapsing?
2. How is energy being generated in a star on the main sequence?
3. What is the significance of the solar neutrino experiment and why do the results have some astrophysicists worried?
4. Why is such a large detector needed for the solar neutrino experiment?
5. After a lightweight star gives off a planetary nebula, what keeps the star from collapsing forever?
6. Why is the formation of iron in the core of a star a significant stage in the star's evolution?
7. After a star has given off most of its mass in a supernova explosion (assuming it is a middleweight star), what prevents the remainder of the star from collapsing forever?

8. If neutron degeneracy cannot keep a star from collapsing, what will?
9. What are the reasons for believing that Cygnus X-1 contains a black hole?
10. Why did the planets in our solar system become planets instead of stars?
11. What evidence is there to suggest that there may be other planetary systems?
12. (a) Why was there initially difficulty in deciding whether spiral nebulae are in our own galaxy or are far away? (b) How was this difficulty finally overcome?
13. How do we know that the universe is expanding?
14. What do we mean when we talk about the "energy problem" for quasars?
15. What observational evidence is there to favor the big-bang cosmology over the steady-state cosmology?
16. What is the difference between an open and a closed universe?
17. What techniques are used to try to decide if the universe will expand forever?

CHAPTER 30

Physics Now

We have discussed in this book a fair bit of physics, both historical and modern. We have studied certain basic physical laws, and applied some of them to understand not only machines that aid us on earth but also the very forces that run the universe. Physics has enabled us to reach the stars and beyond without leaving our home on earth.

Throughout this book we have seen that physics is a vital, exciting, and growing science. Topic after topic proved to be enlarged by recent and ongoing research, and we have every reason to believe that physics will continue to grow through the 1980s and thereafter. In fact, the basic research we are performing now may prove to be at the base of everyday life in the next century.

Physics, of course, is carried out by physicists, and physicists by and large enjoy their science. Most physicists do physics because of its fundamental appeal to their aesthetic sense of exploration of the universe on all scales. And many physicists enjoy other characteristics of their work too—the continued excitement of new discovery, for example, and the opportunity for the personal contact that takes place within a research group or on a college or university faculty or at scientific meetings.

In the following sections, we shall try to illustrate some of the everyday considerations that affect the lives of physicists who do research. Then we shall return to physics itself, and scan some of the topics of ongoing investigation.

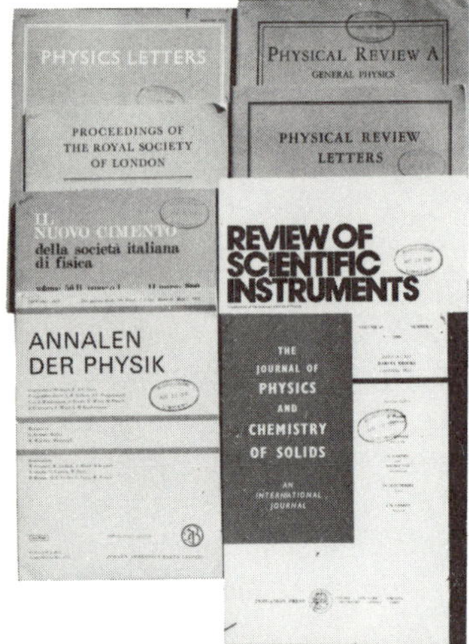

Fig. 30–1 Some of the professional journals in physics.

30.1 *Scientific Publication*

It is all very well to have a good idea, or to solve the riddle of the universe, but your ideas don't count for most people unless you somehow get them out. The history of science honors a few individuals who convinced discoverers of important ideas that their ideas should be published—both Copernicus's theory of the sun-centered universe and Newton's mathematical solutions of the laws of motion had to be pried out of the researchers by visitors.

Nowadays, there are accepted routes by which one presents the results of original research. For example, there are several specific scientific journals which one considers first for publication, because they have the widest circulation to those who would be interested in a given field of physics and/or have the highest standards. One of the prime sets of journals in the United States is the *Physical Review,* which now contains so many papers in such a wide variety of fields that it comes in six parts; many scientists subscribe to only one of these. Many of the journals are published under the auspices of the American Institute of Physics (AIP), which contains societies (such as the American Physical Society and the Optical Society of America) as members.

The standards of journals are maintained by a process of reviewers, who volunteer their time (or rather, are drafted into the work by the editors of the journals in most cases) as part of their contribution to making the scientific enterprise function. If you send in a paper to a journal, a couple of months may pass before you receive a letter in the mail from the editor of the journal that the reviewers have commented thus and so. (The reviewers remain anonymous.) About one-third of the articles to the *Physical Review,* for example, are accepted at this stage. Others are rejected entirely, in which case the editor and referees detail their reasons for this action. And many other papers are deferred, and can be resubmitted in revised form when the comments and suggestions that the reviewers have made are acted upon.

Finally, many months after it was originally submitted, the article appears. Many libraries and individuals around the world will receive their own copies, but many individuals will just notice your article in their library's copy or in a listing of titles and abstracts (a brief summary of the results) that is at the beginning of your paper. These individuals may send you form postcards asking for reprints, which your institution has ordered from the journal.

There are so many journals nowadays, how do scientists keep up? Many don't, outside their own limited field, because one could easily spend all one's waking hours reading scientific journals in physics. Not only are there many American journals but also there are many international ones, both published in English originally or else translated into English.

30.2 *Grants*

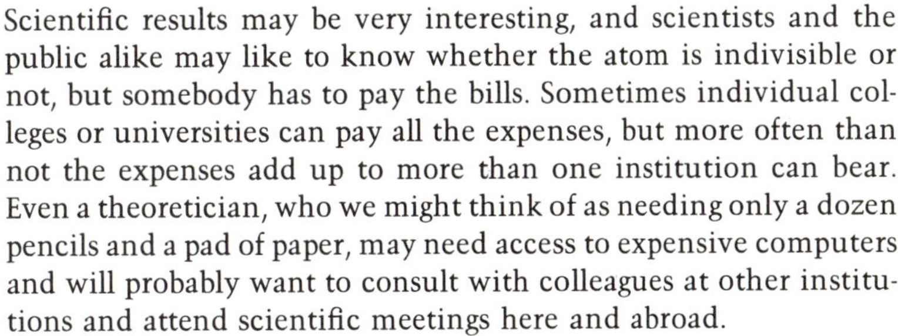

Scientific results may be very interesting, and scientists and the public alike may like to know whether the atom is indivisible or not, but somebody has to pay the bills. Sometimes individual colleges or universities can pay all the expenses, but more often than not the expenses add up to more than one institution can bear. Even a theoretician, who we might think of as needing only a dozen pencils and a pad of paper, may need access to expensive computers and will probably want to consult with colleagues at other institutions and attend scientific meetings here and abroad.

Fig. 30–2 Some of the research group of the Isabelle accelerator now under construction at the Brookhaven National Laboratory. Two beams of protons will collide with each other. One of the magnets, which will be cooled to temperatures at which it is superconducting, is in the tube at the left, and part of the refrigeration system is in the open container at the right.

As for experimentalists, much of the equipment used in physics is getting larger and larger, and more and more expensive. Thus certain national organizations have been set up to coordinate these expenditures. Fermilab, which is run by a consortium of universities, is an example.

Ultimately, the bill for most of the research is now paid in the United States by the federal government. The National Science Foundation (NSF) supports much of the basic research in general, including both support of individuals and groups at colleges and universities and of the large institutions. The Department of Energy (DOE) supports research in energy and particle accelerators as well, since it is the new cabinet department that includes the old Atomic Energy Commission (AEC). The National Aeronautics and Space Administration (NASA) supports certain projects that use space research or that apply to transportation.

Let's consider a group of physicists at a university who are involved in an ongoing research project. At some point in their work, they need money to buy equipment, to support students who work with them, to support laboratory personnel, to support the expenses of publishing their results, to allow them to consult with colleagues at other institutions, and to support their own work in the summertime (since most universities pay salaries only for the academic year). They would put together a proposal for, say, the National Science Foundation. The proposal would include a discussion of where their work fits into the field of research, and just what they propose to do and why. They would include details of how they were going to attack the problem, and just what equipment they would need. They would suggest what kind of progress they hoped to make during the period of the grant (something scientists really can't do very well but something granting agencies always want to know).

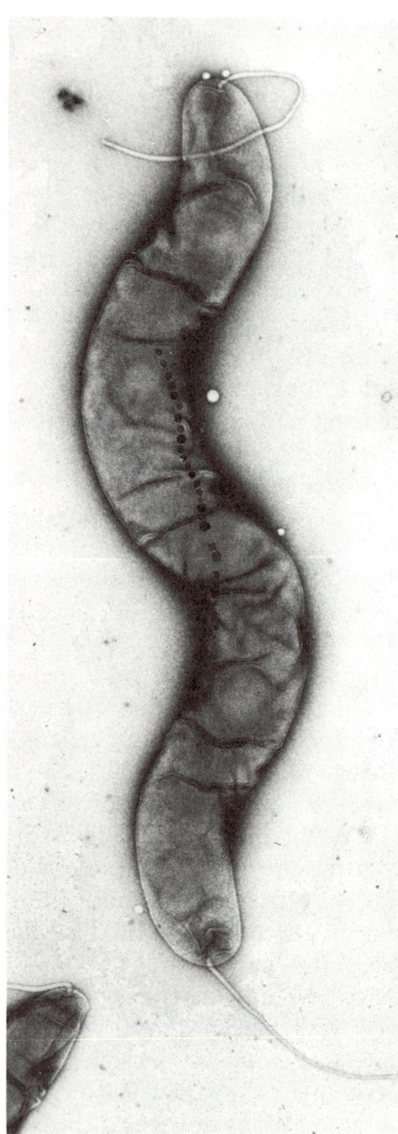

Fig. 30–3 The dark spots in this bacterium (magnified 24,000×) are a chain of little magnets, each only 50 nanometers across. The bacteria use the magnets to orient themselves with the earth's magnetic field. This may help them to find the most suitable habitat or to find food or oxygen. Bacteria in the northern hemisphere always head north, and bacteria in the southern hemisphere always head south. (Courtesy Richard Blakemore, Biomagnetic Services)

The proposal would then be sent to Washington. The proposal is actually in the name of the university, and has been approved for submission by some university administrator; the scientists are listed as investigators on the university's proposal.

In Washington, the program administrator in charge of that field at the NSF will send several copies of the proposal out for review to scientists in the field. This is the process of peer review, which is at the basis of much of American scientific funding and publication. The proposal may be read by half a dozen scientists. Their comments and ratings come back to the NSF, where the various proposals that are received by that program are ranked. Then, somehow, the always-inadequate amount of funding for the program is subdivided among the highest ranked proposals.

It may take six months to a year, but one day the president of the university will receive a letter in the mail from the NSF. If all has gone well, it will announce the award of the grant. The researchers will receive a copy of the letter. Then they can go to work officially, although practically they have usually already made much progress. By then, it is often time to write the next year's proposal, and they had better have new results and new published papers to show.

30.3 *Physics Now*

Physics is such a varied science that the results keep coming in thick and fast from every side. We have indicated future directions for research as we discussed the various topics throughout this book. Here we can merely hint at some of the directions in a quick survey.

The laws of mechanics continue to be applied in detail, now using large computers to carry out calculations to incredible accuracy. We can get a spacecraft from here to Saturn and target its position and velocity to within narrow limits.

Relativity continues to meet stern tests. Carefully controlled experiments in spacecraft to test the predictions of general relativity will be carried out in the 1980s, to add to the results from the astronomical tests that are being pursued.

Research on fluids continues to have many applications. Turbulence, for example, affects all fluid flow from blood in the human body to cooling liquids for power plants. And, similarly to the way in which all electric conductivity can be lost by certain materials at very low temperatures—superconductivity—it also turns out that all viscosity, the resistance to flow, can be lost by certain materials at very low temperatures. This new phenomenon is called superfluidity.

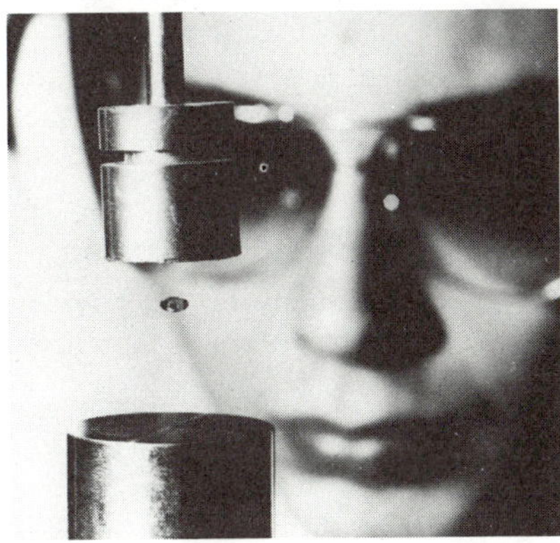

Fig. 30–4 Two ways of levitating particles against the force of gravity. A. The particle shining with a star is held aloft by laser light. (Courtesy A. Ashkin, Bell Laboratories) B. The water droplet is held aloft with sound waves in a test of a method to suspend molten materials for experiments in space.

An understanding of thermodynamics helps us design new, more energy-efficient engines that pollute the atmosphere less than present ones. Here we are limited by the second law of thermodynamics, a fundamental limitation that we must always remember.

Biophysics is one of the more exciting research fields of physics, and leads not only to basic research results but also to medical gains. Hidden damage to cells and tissue can be better detected now with new x-ray scanners, with ultrasound, or with nuclear magnetic resonance techniques, than such damage used to be. The ways that electrical impulses affect the nervous system is another topic under investigation.

"CAT" (computerized axial tomography) scanners provide x-ray images of the body. A. M. Cormack and G. N. Hounsfield won the 1979 Nobel Prize in physics for inventing the process. The techniques of reconstructing the images from the observed data turned out to be similar to those used in some kinds of radio astronomy. "PET" (positron-emission tomography) scanners use positrons (antielectrons) to map the brain.

Studies of plasma physics (where a plasma is a gas made up of charged particles) continues to be important for energy research, since we hope both to restrain fusion with a magnetic field and also to use plasma physics to provide a more efficient way of transforming the heat of fusion into electricity than we now have. The study of a plasma in a magnetic field is called magnetohydrodynamics (MHD), and also has many astrophysical applications.

Atomic and molecular spectroscopy continues to advance, as we discover more about the structure of atoms and molecules. One interesting application is to determining the content of paint and canvas or of clay or bronze to date art objects in order to validate their authenticity. Work in atomic spectroscopy led to the development of lasers, which are now an industry on their own. The development of tunable lasers is but one of the ways in which laser physics is going. Holography finds more and more uses, and will soon be all around us.

Solid-state physics now tends to include as well its extensions to the liquid state; the enlarged field is called condensed-matter physics. We want to study, for example, materials that are in an amorphous rather than in the simpler crystalline state. Coating a material with amorphous silicon, for example, is a cheaper way of making a solar energy cell than is making a crystalline silicon cell.

Cryogenics, low-temperature physics, continues to lead us to unexpected results. Bit by bit, we push upward the temperature at which we find superconductivity by finding new materials and better investigating already known ones.

Particle physics is a good field to bet on for breakthroughs, both because the situation is so muddled now that one is needed and because the completion of new accelerators under construction should certainly lead to the discovery of new particles and phenomena. The steps toward unified field theories continue. A controversial new theory called supergravity and its elaboration may well be an important step toward having such a unified theory of all the known forces.

Astrophysics is more active than ever. Black holes seem to be

Fig. 30–5 The Gossamer Penguin, a superlight airplane built by Paul MacCready of AeroVironment, Inc., under the sponsorship of the DuPont Company, is powered by a bank of about 2800 solar cells. A bicycle tows it up to takeoff speed.

appearing everywhere—as collapsed stars, in clusters of stars, at the centers of galaxies, and at the centers of quasars, for example. So relativistic astrophysics is coming into its own. Many new investigations deal with x-rays and gamma rays from space, which we can now investigate well for the first time with the new generations of high-energy astronomy observatories. Research also continues in many other fields of astrophysical research, such as space exploration of the planets. Planets, stars, galaxies, interstellar and intergalactic gas, quasars, cosmology—all have many hidden things for us to find out.

Physics can tell us about our universe and about ourselves in many ways. The work has begun, and continues. We can confidently look forward to a decade of physical discoveries as exciting as the past one.

Key Words

peer review
magnetohydrodynamics
condensed-matter physics
supergravity

Questions

1. Contrast your impressions of physics before and after reading this book.
2. Discuss one motivation for physicists to do research.
3. Why do publications play an important role in physics research?
4. Of the areas of modern physics research discussed in Section 30.3, which do you consider "pure" physics and which do you consider applied physics? Explain.

APPENDIX 1 *Physical Constants and Units*

PHYSICAL CONSTANTS

Symbol	*Physical Constant*	*Value*	*Symbol*	*Physical Constant*	*Value*
c	Speed of light	2.99792458×10^{8} m/s	e	Charge of electron	1.602×10^{-19} coulomb
G	Gravitation constant	6.672×10^{-11} m³/kg-s²	g	Acceleration of gravity	9.8062 m/s² (sea level, 45° latitude)
h	Planck's constant	6.6262×10^{-10} joule-s			
k	Boltzmann's constant	1.3806×10^{-23} joule/kelvin	$M_{\odot}$	Mass of sun	1.9891×10^{30} kg
m_p	Mass of proton	1.6726×10^{-27} kg	$M_{\oplus}$	Mass of earth	5.9742×10^{24} kg
m_n	Mass of neutron	1.6749×10^{-27} kg	1 AU	Distance of earth from sun	$1.49597870 \times 10^{11}$ m
m_e	Mass of electron	9.1096×10^{-31} kg	1 yr	1 year	3.155815×10^{7} s

CONVERSION FACTORS

1 cm = 0.3937 inch
1 m = 1.0936 yard
1 km = 0.6214 mile

1 gm = 0.0353 oz
1 inch = 2.54 cm
1 yard = 0.9144 m

1 mile = 1.6093 km
1 kg = 0.4536 kg

PREFIXES FOR USE WITH BASIC UNITS OF THE METRIC SYSTEM

Prefix	*Symbol*	*Power*		*Equivalent*
exa	E	10^{18} =	1,000,000,000,000,000,000	
peta	P	10^{15} =	1,000,000,000,000,000	
tera	T	10^{12} =	1,000,000,000,000	Trillion
giga	G	10^{9} =	1,000,000,000	Billion
mega	M	10^{6} =	1,000,000	Million
kilo	k	10^{3} =	1,000	Thousand
hecto	h	10^{2} =	100	Hundred
deca	da	10^{1} =	10	Ten
—	—	10^{0} =	1	One
deci	d	10^{-1} =	.1	Tenth
centi	c	10^{-2} =	.01	Hundredth
milli	m	10^{-3} =	.001	Thousandth
micro	μ	10^{-6} =	.000001	Millionth
nano	n	10^{-9} =	.000000001	Billionth
pico	p	10^{-12} =	.000000000001	Trillionth
femto	f	10^{-15} =	.000000000000001	
atto	a	10^{-18} =	.000000000000000001	

APPENDIX 2 *The Elements*

IA	IIA											IIIA	IVA	VA	VIA	VIIA	2 He 4.00260
								1 H 1.0079									
3 Li 6.94	4 Be 9.01218											5 B 10.81	6 C 12.011	7 N 14.0067	8 O 15.9994	9 F 18.998403	10 Ne 20.17
11 Na 22.98977	12 Mg 24.305	IIIB	IVB	VB	VIB	VIIB	VIII	VIII	VIII	IB	IIB	13 Al 26.98154	14 Si 28.0855	15 P 30.97376	16 S 32.06	17 Cl 35.453	18 Ar 39.948
19 K 39.0983	20 Ca 40.08	21 Sc 44.9559	22 Ti 47.90	23 V 50.9415	24 Cr 51.996	25 Mn 54.9380	26 Fe 55.847	27 Co 58.9332	28 Ni 58.71	29 Cu 63.546	30 Zn 65.38	31 Ga 69.735	32 Ge 72.59	33 As 74.9216	34 Se 78.96	35 Br 79.904	36 Kr 83.80
37 Rb 85.467	38 Sr 87.62	39 Y 88.9059	40 Zr 91.22	41 Nb 92.9064	42 Mo 95.94	43 Tc 98.9062	44 Ru 101.07	45 Rh 102.9055	46 Pd 106.4	47 Ag 107.868	48 Cd 112.41	49 In 114.82	50 Sn 118.69	51 Sb 121.75	52 Te 127.60	53 I 126.9045	54 Xe 131.30
55 Cs 132.9054	56 Ba 137.33	57–71 Rare Earths	72 Hf 178.49	73 Ta 180.947	74 W 183.85	75 Re 186.207	76 Os 190.2	77 Ir 192.22	78 Pt 195.09	79 Au 196.9665	80 Hg 200.59	81 Tl 204.37	82 Pb 207.2	83 Bi 208.9804	84 Po (209)	85 At (210)	86 Rn (222)
87 Fr (223)	88 Ra 226.0254	89– Acti- nides	104 (260)	105 (260)	106 (263)												

57 La 138.9055	58 Ce 140.12	59 Pr 140.9077	60 Nd 144.24	61 Pm (145)	62 Sm 150.4	63 Eu 151.96	64 Gd 157.25	65 Tb 158.9254	66 Dy 162.50	67 Ho 164.9304	68 Er 167.26	69 Tm 168.9342	70 Yb 173.04	71 Lu 174.967	Rare earths (Lanthanide series)
89 Ac (227)	90 Th 232.0381	91 Pa 231.0359	92 U 238.029	93 Np 237.0482	94 Pu (244)	95 Am (243)	96 Cm (247)	97 Bk (247)	98 Cf (251)	99 Es (254)	100 Fm (257)	101 Md (258)	102 No (259)	103 Lr (260)	Actinide series

The upper number is the *atomic number*, expressing the positive charge of the nucleus in multiples of the electron's charge e. The lower number is the *atomic mass*, weighted by isotopic abundance in the earth's surface relative to the mass of the carbon-12 isotope, which has been arbitrarily assigned a mass of 12.00000 atomic mass units (amu). The numbers in parentheses are mass numbers (the whole number nearest the value of the atomic mass, in amu) of the most stable isotope of that element.

Adapted from the *Handbook of Chemistry and Physics*, 60th ed., 1979–1980. (Particle Data Group update, April 1980.)

Atomic Number	*Element*	*Name*	*Atomic Number*	*Element*	*Name*
1	H	hydrogen	54	Xe	xenon
2	He	helium	55	Cs	caesium
3	Li	lithium	56	Ba	barium
4	Be	beryllium	57	La	lanthanum
5	B	boron	58	Ce	cerium
6	C	carbon	59	Pr	praseodymium
7	N	nitrogen	60	Nd	neodymium
8	O	oxygen	61	Pm	promethium
9	F	fluorine	62	Sm	samarium
10	Ne	neon	63	Eu	europium
11	Na	sodium	64	Gd	gadolinium
12	Mg	magnesium	65	Tb	terbium
13	Al	aluminum	66	Dy	dysprosium
14	Si	silicon	67	Ho	holmium
15	P	phosphorus	68	Er	erbium
16	S	sulfur	69	Tm	thulium
17	Cl	chlorine	70	Yb	ytterbium
18	Ar	argon	71	Lu	lutetium
19	K	potassium	72	Hf	hafnium
20	Ca	calcium	73	Ta	tantalum
21	Sc	scandium	74	W	tungsten
22	Ti	titanium	75	Re	rhenium
23	V	vanadium	76	Os	osmium
24	Cr	chromium	77	Ir	iridium
25	Mn	manganese	78	Pt	platinum
26	Fe	iron	79	Au	gold
27	Co	cobalt	80	Hg	mercury
28	Ni	nickel	81	Tl	thallium
29	Cu	copper	82	Pb	lead
30	Zn	zinc	83	Bi	bismuth
31	Ga	gallium	84	Po	polonium
32	Ge	germanium	85	At	astatine
33	As	arsenic	86	Rn	radon
34	Se	selenium	87	Fr	francium
35	Br	bromine	88	Ra	radium
36	Kr	krypton	89	Ac	actinium
37	Rb	rubidium	90	Th	thorium
38	Sr	strontium	91	Pa	protactinium
39	Y	yttrium	92	U	uranium
40	Zr	zirconium	93	Np	neptunium
41	Nb	niobium	94	Pu	plutonium
42	Mo	molybdenum	95	Am	americium
43	Tc	technetium	96	Cm	curium
44	Ru	ruthenium	97	Bk	berkelium
45	Rh	rhodium	98	Cf	californium
46	Pd	palladium	99	Es	einsteinium
47	Ag	silver	100	Fm	fermium
48	Cd	cadmium	101	Md	mendelevium
49	In	indium	102	No	nobelium
50	Sn	tin	103	Lr	lawrencium
51	Sb	antimony	104	Rf	rutherfordium
52	Te	tellurium	105	Ha	hahnium
53	I	iodine	106		(Reported 1974)

APPENDIX 3 *Nobel Prize Winners in Physics*

1901 *Wilhelm Roentgen*—Discovery of x-rays.
1902 *Hendrik A. Lorentz and Pieter Zeeman*—Research on the relationship between magnetism and radiation.
1903 *Antoine H. Becquerel*—Discovery of radioactivity.
Pierre Curie and Marie Curie—Research in radioactivity.
1904 *Lord Rayleigh*—Discovery of argon and studies of properties of certain gases.
1905 *Philipp E. A. von Lenard*—Work on cathode rays.
1906 *Joseph J. Thomson*—Studies on conduction of electricity in gases.
1907 *Albert A. Michelson*—Optical instruments and investigations using them.
1908 *Gabriel Lippmann*—Interference method for the photographic reproduction of colors.
1909 *Guglielmo Marconi and Carl Ferdinand Braun*—Contributions to wireless communications.
1910 *Johannes D. van der Waals*—Equation of state for gases and liquids.
1911 *Wilhelm Wien*—Studies of black-body radiation.
1912 *Nils G. Dalén*—Automatic regulators for use in coastal lighting.
1913 *Heike Kamerlingh Onnes*—Studies of matter at low temperatures.
1914 *Max von Laue*—Diffraction of x-rays by crystals.
1915 *William H. Bragg and William L. Bragg*—Studies of crystal structure using x-rays.
1916 No prize
1917 *Charles G. Barkla*—Studies of x-rays from various elements.
1918 *Max Planck*—Discovery of quanta of energy.
1919 *Johannes Stark*—Discovery of splitting of spectral lines in electric fields.
1920 *Charles-Édouard Guillaume*—Discovery of anomalies in nickel steel alloys.
1921 *Albert Einstein*—Explanation of the photoelectric effect.
1922 *Niels Bohr*—Investigations of the structure of atoms.
1923 *Robert A. Millikan*—Work on the elementary electric charge and on the photoelectric effect.
1924 *Karl M. G. Siegbahn*—Research in x-ray spectroscopy.
1925 *James Franck and Gustav Hertz*—Studies of collisions between electrons and atoms.
1926 *Jean B. Perrin*—Studies on the discontinuous structure of matter.
1927 *Arthur H. Compton*—Scattering of electrons and photons.
Charles T. R. Wilson—Invention of the cloud chamber to study the paths of charged particles.
1928 *Owen W. Richardson*—Work on thermionic emission.
1929 *Louis-Victor de Broglie*—Discovery of the wave nature of electrons.
1930 *Chandrasekhara Venkata Raman*—Studies of the scattering of light by atoms and molecules.
1931 No prize
1932 *Werner Heisenberg*—Creation of quantum mechanics.
1933 *Erwin Schrödinger and Paul A. M. Dirac*—Contributions to quantum mechanics.
1934 No prize
1935 *James Chadwick*—Discovery of the neutron.
1936 *Victor F. Hess*—Discovery of cosmic rays.
Carl D. Anderson—Discovery of the positron.
1937 *Clinton J. Davisson and George P. Thomson*—Discovery of diffraction of electrons by crystals.
1938 *Enrico Fermi*—New radioactive elements produced by neutron irradiation.
1939 *Ernest O. Lawrence*—Invention of the cyclotron.
1940–1942 No prize
1943 *Otto Stern*—Discovery of the magnetic moment of the proton, and contributions to molecular beams studies.
1944 *Isador I. Rabi*—Nuclear magnetic resonance.
1945 *Wolfgang Pauli*—Discovery of the exclusion principle.
1946 *Percy W. Bridgman*—Contributions to high-pressure studies.
1947 *Edward V. Appleton*—Studies of the upper atmosphere.
1948 *Patrick M. S. Blackett*—Discoveries in nuclear physics.
1949 *Hideki Yukawa*—Prediction of mesons.
1950 *Cecil F. Powell*—Photographic method of studying nuclear processes.

1951 *John D. Cockroft and Ernest T. S. Walton*—Transmutation of atomic nuclei by accelerated particles.

1952 *Felix Bloch and Edward M. Purcell*—New methods for studying nuclear magnetism.

1953 *Frits Zernike*—Invention of the phase-contrast microscope.

1954 *Max Born*—Research in quantum mechanics and statistical interpretation of the wave function.

Walther Bothe—Development of the coincidence method.

1955 *Willis E. Lamb*—Studies of the fine structure of the hydrogen spectrum.

Polykarp Kusch—Determination of the magnetic moment of the electron.

1956 *William Shockley, John Bardeen, and Walter H. Brattain*—Discovery of the transistor.

1957 *C. N. Yang and T. D. Lee*—Prediction of parity violation.

1958 *Pavel A. Čerenkov, Ilya M. Frank and Igor Tamm*—Discovery and interpretation of the Čerenkov effect.

1959 *Emilio G. Segrè and Owen Chamberlain*—Discovery of the antiproton.

1960 *Donald A. Glaser*—Invention of the bubble chamber.

1961 *Robert Hofstadter*—Studies of electron scattering by nuclei.

Rudolf L. Mössbauer—Studies of emission and absorption of gamma rays in crystals.

1962 *Lev D. Landau*—Theoretical work on condensed matter, especially liquid helium.

1963 *Eugene P. Wigner*—Contributions to nuclear and elementary particles theory, especially in the application of symmetry principles.

Maria Goeppert-Mayer and J. Hans D. Jensen—Discoveries related to nuclear shell structure.

1964 *C. H. Townes, Nikolai G. Basov, and Alexander M. Prochorov*—Discovery of the maser.

1965 *Sin-Itiro Tomonaga, Julian Schwinger, and Richard P. Feynman*—Development of quantum electrodynamics.

1966 *Alfred Kastler*—Optical methods for studying Hertzian resonance in atoms.

1967 *Hans A. Bethe*—Theory of nuclear reactions in stars.

1968 *Luis W. Alvarez*—Contributions to elementary particle physics.

1969 *Murray Gell-Mann*—Discoveries in the classification of elementary particles.

1970 *Hannes Alfvén*—Discoveries in magnetohydrodynamics.

Louis Néel—Discoveries concerning ferrimagnetism and antiferromagnetism.

1971 *Dennis Gabor*—Discovery of holography.

1972 *John Bardeen, Leon N. Cooper, and J. Robert Schrieffer*—Explanation of superconductivity.

1973 *Leo Esaki*—Discovery of tunneling in semiconductors.

Ivar Giaever—Discovery of tunneling in superconductors.

Brian D. Josephson—Prediction of properties of supercurrent through a tunnel-barrier.

1974 *Antony Hewish*—Discovery of pulsars.

Martin Ryle—Pioneering work in radio astronomy.

1975 *Aage Bohr, Ben Mottelson, and James Rainwater*—Discovery of the relationship between collective and particle motion in nuclei.

1976 *Burton Richter and Samuel C. C. Ting*—Discovery of the J/psi particle.

1977 *John N. Van Vleck, Nevill F. Mott, and Philip W. Anderson*—Studies of electrons in magnetic, noncrystalline solids.

1978 *Arno A. Penzias and Robert W. Wilson*—Discovery of the cosmic background radiation.

Peter L. Kapitza—Work in low-temperature physics.

1979 *Sheldon L. Glashow, Abdus Salam, and Steven Weinberg*—Work on the theory of the unified weak and electromagnetic interaction.

1980 *Val L. Fitch and James W. Cronin*—Experimental proof of charge-parity (CP) violation, which has applications to cosmology.

CREDITS

The authors and publisher are pleased to acknowledge the following sources of the drawings, photographs, and poetry included in this book.

Drawings—Ben Gamit, Pls. 47, 48, 49; Patricia Wynne, Figs. 1–11, 2–4, 2–10, 2–19, 2–30, 2–34, 3–4, 3–5, 3–7, 3–8, 3–11, 3–17, 3–18, 3–20, 3–26, 5–2, 5–3, 5–4, 5–7, 5–8, 5–12, 5–21, 6–5, 7–2, 8–4, 8–6, 8–9, 8–10, 8–11, 8–15, 9–1, 11–6, 11–18, 12–4, 12–12, 12–25, 13–2, 17–18, 17–25, 17–26, 17–27, 17–30, 17–33, 17–35, 18–6, 18–8, 19–19, 20–1, 20–15, 20–17, 23–22, 29–10, 29–13; all other drawings were prepared by the studio of Danmark & Michaels.

Photographs—Advanced Micro Devices, Inc., & Melgor Photographers, Fig. 25–21; AGA Corp., Pl. 38, Fig. 12–23; American Institute of Physics Niels Bohr Library (AIPNBL), Figs. 2–25, 3–2, 3–3, 15–3, 15–4, 22–12, 23–11, 23–15, 26–13; AIPNBL & the Bainbridge Collection, Fig. 23–24; AIPNBL & California Institute of Technology, Fig. 9–10A; AIPNBL & the Einstein Archives, Figs. 9–13, 9–14; AIPNBL & the Meggers Gallery of Nobel Laureates, Fig. 23–20; AIPNBL & Alan Richards, Fig. 28–8; AIPNBL & the Smithsonian Institution, Fig. 18–1; AIPNBL & the Stein Collection, Fig. 26–8; Air-Flo Co., Inc., Elkhart, Ind., Figs. 4–6, 10–34AB; Allsteel Scale Co., New York City, Fig. 3–24; Amatucci, Bill, Fig. 2–2; American Museum of Natural History, Fig. 25–1; American Petroleum Institute Photographic Film Service (APIPFS), Figs. 27–3, 27–4; APIPFS & Aleyska Pipeline Service Co., Figs. 11–4, 27–6; APIPFS & General Dynamics Corp., Fig. 27–8; APIPFS & Gulf Oil Corp., Fig. 27–10; APIPFS & Keith Hay, Fig. 27–7; APIPFS & Standard Oil Corp. of Calif., Figs. 10–15AB; Angerhoffer, Phillip E., Richard A. Perley, Douglas Milne, & Bruce Balick, Pl. 74; Argonne National Laboratory, Figs. 18–25, 27–28; Arp, Halton, & Jean Lorre, Mount Wilson & Las Campanas Observatories, The Carnegie Institution of Washington, Fig. 29–19C; Association of Universities for Research in Astronomy, Inc. The Cerro Tololo Inter-American Observatory, Pls. 75–76, Fig. 4–18; Atomic Industrial Forum, Fig. 27–16; AT&T, Figs. 17–16B, 20–9.

Backus, Peter R., & Joseph H. Taylor, University of Massachusetts–Amherst, Fig. 6–18; Baras, Irving, Miles Agalin, Kenneth R. Barasch, & Stephen A. Ostbaum, Figs. 17–40AB; Bell Laboratories, Pl. 61, Figs. 17–16A, 19–3, 20–29, 25–16, 25–20, 29–25, 30–4A; Bial, Bert, Fig. 14–14; Big Bear Solar Observatory, California Institute of Technology, Fig. 2; Biomagnetic Services, Durham, N.H., Fig. 30–3; Black Star/ Emil Schulthess, Pl. 59; Brookhaven National Laboratory, Pls. 41, 42, Figs. 11–26, 28–15BC, 29–5, 30–2; Bryant, H. Stafford, Fig. 11–11; Bundesarchiv, Fig. 23–6; Burke, Bernard F., Perry E. Greenfield, & David H. Roberts, MIT, Pl. 81.

California Institute of Technology Archives, Fig. 8–1; Calspan Corp., New York City, Fig. 3–29; Carnegie Museum of Natural History, Pittsburgh, Penna., Fig. 27–1; CERN, Fig. 8–13; Chalonge, D., & L. Divan, Fig. 23–9; Chapin Library, Williams College, Figs. 1–2, 10–22, 21–3, 21–9; Clark University Library, Fig. 7–13; Consolidated Edison Co., New York City, Fig. 27–12; Contact Press Images © 1979 & Mark Meyer, Figs. 7–19AB; Crawford Collection of the Royal Observatory, Edinburgh, Fig. 7–15; Crockett, Karen, & Ed Crockett, & the Ohio State Marching Band, Fig. 14–2.

Daytona International Speedway, Fig. 6–8; de Pater, Imke, Sterrewacht, Leiden, with Westerbork array, Pl. 71; Deutsche Bank, Pl. 50; Deutsches Elecktronen–Synchrotron (DESY), Fig. 28–18; Deutsches Museum, München, Pl. 17, Fig. 22–14; Dinsmoor, William Bell, *The Architecture of Ancient Greece* (New York: W. W. Norton & Co., 1975), Fig. 28–5.

Eames, Charles, Jagiellonian Library, Krakow, Poland, Fig. 7–6; Edgerton, Harold E., MIT, Figs. 2–23, 3–6AB, 3–9, 4–2, 4–14, 4–15, 5–14, 6–17, III–1, 13–11AB, 13–12; Eon Productions, Ltd., producer, & United Artists Corp., distributor, James Bond—007 motion picture, *Goldfinger*, Fig. 24–15; Express Newspapers Ltd., London, © 1980, Fig. 24–16; Exxon, Fig. 27–17.

Fermilab, Pls. 43–45, Figs. I–1, 28–2, 28–19; Fraley Rayburne, Fig. 27–9; Fancis Bitter National Magnet Laboratory, MIT, Figs. 21–6, 21–31; Franklin Institute, Phila., Penna., Fig. 19–2; Freelance Photographers Guild, & Richard Nowitz, Fig. 10–17; French, Harry, Figs. 5–20AB; Fundamental Photographs, New York City, Fig. 15–8.

Gingerich, Owen, Fig. 1–3; Glashow, Sheldon, Fig. 28–20; Gliese, Wilhelm, Atronomisches Rechen-Institut, Heidelberg, Fig. 29–4; Goldberg, Norman, *Popular Photography*, Pl. 60; Goodyear Tires & Jean-Valere Pignal, Fig. 2–28; Granger Collection, New York City, Figs. 10–21, 15–5, 27–11.

Haas, Ernst, Pl. 64; Hahn, Otto, *A Scientific Autobiography* (New York: Charles Scribner's Sons, 1966), Fig. 26–22; Haleakala Observatory, Institute for Astronomy, University of Hawaii, Fig. 3; Hale Papers, courtesy AIP Niels Bohr Library, by permission of the Director of the Mount Wilson & Las Campanas Observatories, and Otto Nathan, Fig. 9–10B; Harris, Sidney, in *American Scientist*, Fig. 1–1, 2–32; Harvard College Observatory, Pl. 19; Heil, Werner, & John M. Wilcox, Institute for Plasma Research, Stanford, Calif., Fig. 4; High Altitude Observatory & Southwestern at Memphis, Fig. 21–35; Horsmann, M., Ruhr University, Bochum, Germany, Pl. 36.

IBM Corp., Fig. 1–10; Institute of Theoretical Physics, Lund, Sweden, Fig. 23–27.

Jacobi, Lotte, Fig. II–1; Jaguar Rover Triumph, Inc., Fig. 2–37; Jet Records, London, & LaserMedia, Los Angeles, Pl. 63; Jones, Christopher C., Union College, Figs. 15–14, 15–19ABC.

Kammerlingh Onnes Laboratory, University of Leiden, Fig. 25–23; Karl-Schwarzschild Observatory, GDR, & S. Marx, Fig. 1–17; Kellsman Instrument Co., Figs. 10–24AB, 10–32; Roy Keppie Memorial Sloan Kettering Cancer Center, Fig. 1–7; Kowal, Charles T., California Institute of Technology, Figs. 15–27A, 29–7AB; Kramer, Peter, Department of Physics & Astronomy, Williams College, Fig. 9–16; Kristian, Jerome, Allan R. Sandage, & James A. Westphal, Mount Wilson & Las Campanas Observatories, The Carnegie Institution of Washington, & California Institute of Technology, Fig. 29–22; Kutner, Marc, Fig. 13–18.

Laboratoire de Physique Stellaire et Planètaire, France, & Lockheed Palo Alto Research Laboratory, courtesy of R. M. Bonnet, Fig. 5; Ladd, Gary, Pl. 40; Lansing, James B., Figs. 16–9AB; Leitner, Alfred, RPI, Pl. 51, Figs. 25–25AB; Lewis, E. R., et al., "Studying Neural Organization with the Scanning Electron Microscope," *Science*, 165: 1140–43, © Sept. 1969, The American Association for the Advancement of Science, Fig. 18–11; Lincoln Center for the Performing Arts & Suzanne Faulkner

Stevens, Fig. 14–16; Livingston, William, The Kitt Peak National Observatory, Ariz., Pls. 18, 20, 21, Figs. 12–16, 21–33; Los Alamos Scientific Laboratory (LASL), Pls. 1–9; LASL & Charles F. Keller, Jr., Pl. 25; Lovi, George, Vanderbilt Planetarium, Fig. 7–3; Lucasfilm Ltd. (LFL), © 1980. All rights reserved. From the motion picture: *The Empire Strikes Back*, courtesy of Lucasfilm Ltd., Fig. 24–2.

Machlis, Joseph, *The Enjoyment of Music*, by permission of W. W. Norton & Co., Inc., Fig. 13–19; Mack Trucks, Inc., Fig. 2–18; Mao, Ho-Kwang, & Peter M. Bell, Carnegie Institution of Washington, Fig. 10–5; Mariott's Great America, Fig. 6–10, Metropolitan Edison, Fig. 27–21; Michelin Tire Corp., Fig. 6–7; Miller, D. C., *The Science of Musical Sounds* (New York: Macmillan, 1922), Fig. 14–12; Miller, D. C., *Sound Waves and Their Uses* (New York: Macmillan, 1938), Fig. 14–19; Millikan, Robert A., *Electrons (Plus and Minus), Protons, Neutrons, Mesotrons, and Cosmic Rays* (Chicago: University of Chicago Press), Fig. 19–5; Milon, Dennis, Maynard, Mass., Figs. 18–19, 18–20; Minolta Corp., Fig. 17–29; Mobil Corp., Fig. 4–17; Montani, Joseph, New York City, Fig. 21–2; Moog Music, Inc., Buffalo, N.Y., Fig. 14–20; Moore, Patrick, & Irving Geis, *Earth Satellites*, with the permission of W. W. Norton & Co., Inc., copyright © 1956 by Patrick Moore & Irving Geis, Fig. 7–18; Mount Wilson & Las Campanas Observatories, The Carnegie Institution of Washington (Mount Wilson Observatory photograph), Fig. 18–21; Museum of Science, Boston, Fig. 19–27; Mystic Seaport Museum, Mystic, Conn., photo by Mary Anne Stets, Fig. 11–8.

NASA, Pls. 65, 66, Figs. 1–12, 1–13, 1–14, 2–3, 2–33, 3–27, 4–8, 6–4AB, 6–12, 6–13, 6–20, 7–17, 7–23AB, 7–28, 8–8, 12–9, 12–21, 12–22, 13–13, 18–15, 27–2, 27–27, 27–29AB, 29–12, 30–4B, 30–5; NASA–Ames Research Center, Moffett Field, Calif., Fig. 2–36; NASA–Jet Propulsion Laboratory, California Institute of Technology, Pls. 67–70, 72, Figs. 7–29, 7–30; NASA–Naval Research Laboratory/High Altitude Observatory, Pls. 26–28; NASA & Lewis House, Ernest Hildner, William Wagner, & Constance Sawyer, High Altitude Observatory/NCAR/NSF, Pl. 30; NASA–Marshall Space Flight Center, & R. M. Wilson, E. J. Reichmann, & J. E. Smith, Pl. 29; NASA–Goddard & Marshall Centers, & Bruce E. Woodgate, Einar Tandberg-Hanssen & colleagues, Pl. 31; National Astronomy & Ionosphere Center, Figs. 29–15, 29–16; National Football League (NFL) Properties & Darwin Garrison, Fig. 4–11; NFL Properties & Mike Zagaris, Fig. 3–10; National Geographic Society (NGS), © 1973, Fig. 12–10; NGS ©, & Bruce Dale, Fig. 16–3; NGS ©, & Jack Fields, Fig. 24–10B; NGS ©, & Ken Firestone, Fig. 16–1; NGS ©, & Hughes Tool Co., Fig. 24–18; Nakayama, Yasuki, Tokai University, Pl. 35, Fig. 10–37; NCAR, Boulder, Colo., Figs. 1–9, 10–26, 19–30, 20–22; NCR Corp., Fig. 24–22B; New York Times Pictures, Fig. 17–6; New York Yankees, Fig. 10–40; NFB Phototheque ONF (NFB-P-ONF) ©, photo by G. Blouin, 1949, Figs. 7–26AB; NFB-P-ONF © 1968, Harvard Project Physics, & General Radio, West Concord, Mass., Figs. 14–13AB; Nilsson, Lennart, *Behold Man*, © Albert Bonniers Forlag, Sweden (in the U.S., Little, Brown & Co., Boston; in Italy, *Questo e l'Uomo*, Edizioni Paoline, Milan), Pl. 58, Fig. 13–8; NOAA, Figs. 8–3, 12–8; Norcia, Anne, Waynesville, Ohio, Pl. 16.

Observatoire de Haute-Provence du Centre National de la Recherche Scientifique, Fig. 29–3; ONERA, France, Pls. 33, 34; Oppersdorff, Mathias, Fig. 3–22; Owens-Corning Fiberglas Corp., Fig. 12–2.

Palomar Observatory, California Institute of Technology, Pls. 73, 77, Figs. 1–8, 1–18, 6–19, 7–22B, 15–21, 15–25, 15–27AB, 18–23, 21–12, 29–6, 29–11, 29–18, 29–24, 29–29; Pasachoff, Jay M., Pls. 22, 23, 24, 39, 46, 52, 53, 54, 55, 57, Figs. 1–5, 2–1, 2–5, 2–22, 3–12, 3–23AB, 5–1, 5–5, 5–11, 5–19AB, 6–16ABC, 6–22, 7–11ABC, 10–8, 10–9, 10–10, 10–27, 10–29AB, 10–41AB, 11–1AB, 11–20, 11–22, 11–24, 11–25AB, 12–3, 12–7A, 12–11, 12–14, 12–24, 13–1, 13–4, 13–6AB, 15–2, 15–10, 15–16, 15–22AB, 15–23AB, 15–24, 16–4, 16–5AB, 16–6ABC, 16–7AB, 16–11ABC, 17–1, 17–2, 17–5, 17–17ABCDE, 17–19AB, 17–32AB, 18–3, 18–5, 18–16, 18–17, 18–18, 19–22, 19–25, 19–28, 20–7, 20–8, 20–20, 20–25, 20–26, 20–30, 20–31, 21–1, 21–4AB, 22–1, 22–2AB, 22–13, 24–21, 24–22A, 25–15, 26–17, 27–5AB, 27–30, 29–20, 29–21, 30–1; Pasachoff, Jay M., *Contemporary Astronomy* (Philadelphia: W. B. Saunders Co., 1977), Fig. 1–4, Perkin-Elmer, Danbury, Conn., Fig. 17–13; Phillips Research Laboratories & John Nijst, Pl. 62; Poskanzer, Jef A., Lawrence Berkeley Laboratory, Fig. 26–11; Power Authority of the State of New York, Figs. 11–17, 27–31; Princeton University Plasma Physics Laboratory, Pls. 14, 15, Figs. 21–14, 27–23; *PSSC Physics*, 2d ed. (Lexington, Mass.: D. C. Heath & Co., with Education Development Center, Inc. [EDC], 1965), Figs. 2–29, 5–15, 5–18, 13–5, 14–5, 14–7, 14–8, 14–9AB, 15–17, 17–11AB, 19–15ABC, 19–16ABC, 21–21ABCD; PSSC Film Series *Matter Waves*, EDC, Newton, Mass., Figs. 23–19AB; PSSC Film Series *Ripple Tank Wave Phenomena*, EDC, Figs. 13–10AB, 15–13, 17–4, 17–9.

REFAC International, Ltd., Fig. 24–11.

Sandia Laboratories, Albuquerque, N.M., Pls. 12, 13; Scharf, David, *Magnifications* (New York: Schocken Books, 1977), Fig. 18–26; *Science Digest*, photo by Gwen Akin, Fig. 13–3; Scripps Institution of Oceanography, University of California, San Diego, & Mia Tegner, Fig. 1–6; Shapiro, Maurice M., Naval Research Laboratory, Fig. 29–8; Sigelman, Jesse, New York City, Fig. 24–19; Skyviews Survey, Inc., Fig. 2–9; Smithsonian Institution, Figs. 9–12, SI#53192 Fig. 20–27; Spectra-Physics, Mountain View, Cal., frontispiece; *Sports Illustrated (SI)* & Jerry Cooke, Fig. 3–15; *SI* & Jerry Irwin, Fig. 3–25; *SI* & Mannie Millan, Fig. 5–22; *SI* & Eric Schweikhardt, Fig. 10–39; *SI* & Tony Triolo, Fig. 10–38; Sproul Observatory, Swarthmore, Penna., Fig. 29–14A; Stanford Linear Accelerator Center (SLAC), Figs. 28–3, 28–15A, 28–17; Steinberg, Saul, Fig. 6–26; Sutton, Steven E./DUOMO, Figs. 2–31, 4–16, 6–1, 6–9, 10–35.

Tananbaum, Harvey, & colleagues, Harvard-Smithsonian Center for Astrophysics, Pl. 80; Tennessee Valley Authority, Fig. 11–19; Thermography of Long Island, Fig. 27–32; TMs & © 1978 by D.C. Comics, Inc., Fig. 7–14; Toomre, Alar, & Juri Toomre, Fig. 7–22A; Triborough Bridge & Tunnel Authority, Fig. 3–21; Twentieth Century Fox Film Corp., © 1976, all rights reserved, Fig. 4–1.

Union Carbide Corp., Fig. 17–12; United States Coast Guard, Department of Transportation, Figs. 10–19, 17–21; United States Department of Energy Photographic Services (USDE), Fig. 27–24; USDE & Frank Hoffman, Fig. 27–14; United States Department of the Interior, National Parks Service, Edison National Historic Site, Figs. 20–21, 22–6; United States Department of the Navy, Figs. 10–13, 10–14; United States Environmental Protection Agency, Environmental Monitoring Systems Laboratory, Las Vegas, Nev., Fig. 11–27; University of California (UC), Lawrence Berkeley Laboratory, Fig. 23–29; UC, Lawrence Livermore Laboratory, Pls. 10, 11, Figs. 1, 27–26; UC, Lawrence Radiation Laboratory, Berkeley, Calif., Fig. 24–10A; UC, Santa Cruz, Lick Observatory, Figs. 7–7, 7–20, 7–21, 7–24, 29–17; University of Michigan Library, Dept. of Rare Books & Special Collections, Fig. 2–26B; Ursetti, Charles, Fig. I–2.

Vandiver, Kim & Harold E. Edgerton, MIT, Pl. 32.

Warder Collection, Figs. 3–1, 10–25; Western Electric, Fig. 20–28; Westinghouse Electric Corp., Fig. 24–20; Wide World Photos, Figs. 6–25, 24–1; Wilcox, John M., Institute for Plasma Research, Stanford, Calif., Fig. 4; Williams, Stanley N., Dartmouth College, from *Science*, 13 June 1980, copyright © by the American Association for the Advancement of Science, Fig. 12–7B; Williamstown Medical Associates, Fig. 19–23; Woodfin Camp & Associates, © Howard Sochurek, Pl. 37; Wray, James D., University of Texas, Austin, Pls. 78, 79, Fig. 24–17; Wu, Chien Shiung, Fig. 28–9; Wuercker, Ralph, TRW, Figs. 16–8, 16–10AB; Wyant, James C., Optical Sciences Laboratory, University of Arizona, Fig. 24–14.

Yalow, Rosalyn S., Fig. 26–20; Yarmchuk, Edward J., M. J. V. Gordon, & Richard E. Packard, University of California, Berkeley, Fig. 10–42.

Poetry—"Cosmic Gall," by John Updike, reprinted by permission of the publisher from *Telephone Poles and Other Poems* by John Updike (New York: Alfred A. Knopf, 1963); "Fission and Superstition," by HMK, reprinted by permission of the publisher from *New Statesman and Nation*, London, 14 January 1950; "Lines inspired by a lecture on extra-terrestrial life," by JDGM, reprinted by permission of the publisher from *The Observatory*, London, 65 (1943): 88; "The Microbe," by Hilaire Belloc, reprinted by permission of the publisher from *Cautionary Verses* by Hilaire Belloc (New York: Alfred A. Knopf, 1941); "Perils of Modern Living," by Harold P. Furth, reprinted by permission of the publisher from *The New Yorker*, New York, 10 November 1965.

INDEX

Note: A page number with no letter code indicates a text reference. Letter codes: m = margin note; d = diagram or photo; t = table. Pl. = color plate caption.